RICARDO MAÑÉ –
Selected Papers

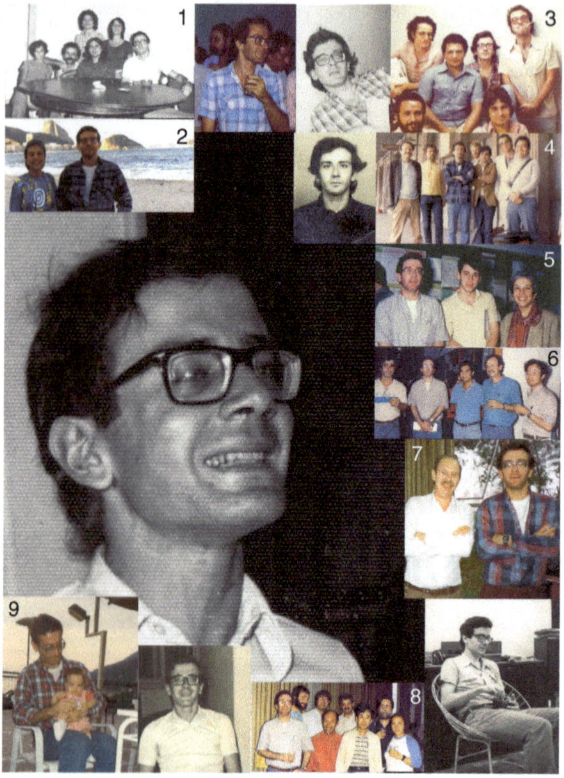

Ricardo and some of his dearest friends

1. Ada Maria de Sousa Doering, Jorge Beloqui, Artur Lopes, Claus Ivo Doering, Silvia Regina Lopes, Ricardo Mañé
2. Maria José Pacifico, Ricardo Mañé
3. Claus Ivo Doering, Ivan Tribuzi, Ricardo Mañé, Paulo Sad, Renato Tribuzi, Artur Lopes
4. Jacob Palis, Floris Takens, Ricardo Mañé, Artur Lopes, Paulo Sad, Carlos Gutierrez
5. Ricardo Mañé, Marcelo Viana, Maria José Pacifico
6. Alcides Lins Neto, Ricardo Mañé, Cesar Camacho, Jacob Palis
7. Jorge Lewowicz, Ricardo Mañé
8. Ricardo Mañé, Marco Martens, Sheldon Newhouse, John Franks, David Ruelle, Lai-Sang Young
9. Ricardo Mañé, Laura Pacifico Palis

Photos by Maria José Pacifico

RICARDO MAÑÉ – Selected Papers

Edited by
Maria José Pacifico

impa
instituto nacional de
matemática
pura e aplicada

SBM

Springer

Ricardo Mañé
IMPA
Rio de Janeiro
Brazil

Editor
Maria José Pacifico
Centro de Tecnologia
Federal University of Rio de Janeiro
Institute of Mathematics—UFRJ
Rio de Janeiro
Brazil

ISBN 978-3-319-41663-2

Library of Congress Control Number: 20122954381

Mathematics Subject Classification (2010): 28-XX, 37-XX, 58-XX, 60-XX, 70-XX, 82-XX

Printed on acid-free paper

This Springer imprint is published by Springer Nature
The registered company is Springer International Publishing AG
The registered company address is: Gewerbestrasse 11, 6330 Cham, Switzerland

Selected Works of Outstanding Brazilian Mathematicians

Mathematics in Brazil is a very young endeavor. Regular research activities and graduate studies go back to the 1950s, when the biannual Brazilian Mathematical Colloquium was launched and Brazil joined the International Mathematical Union (IMU). The Instituto de Matemática Pura e Aplicada (IMPA) was founded in 1952. Creation of the SBM, Sociedade Brasileira de Matemática, would follow, in 1969.

Half a century later, Brazil has reached Group IV in the IMU's ranking and is responsible for about 1.8 % of the world's mathematical production (according to MathSciNet). The Brazilian mathematical community is growing steadily, and has earned international reputation in several fields of research. While many great challenges remain ahead, one can certainly be proud of what has been achieved in such a short time.

This progress was largely due to a number of leading mathematicians who, in addition to their own fundamental contributions to Mathematics, created genuine "schools" around them that raised entire research fields in Brazil to the best international standards. Through this collection, the SBM and IMPA aim to highlight their scientific trajectories and to pay tribute to their pioneering efforts.

Editors
Hilário Alencar
César Camacho
André Nachbin
Marcelo Viana

Preface

Ricardo Mañé left us in March 1995, roughly a year after he prepared the text below upon request of the Third World Academy of Sciences.

"Ricardo Mañé was born in January, 1948, in Montevideo, Uruguay. He entered the University of the Republic of Uruguay in 1967. In 1969 he became an assistant lecturer in the department of mathematics, where, under the guidance of J. Lewowicz he found his vocation for mathematics and specially for dynamical systems. Through him he got in contact in 1971 with J. Palis, who was then starting a lively research group in this field at the Institute for Pure and Applied Mathematics (IMPA) in Rio de Janeiro. He encouraged Mañé to pursue his studies at IMPA, which he did, obtaining his PHD in 1973, having Palis as thesis advisor. His thesis was connected with the vast research that had taken place in the previous decade but was still being developed around the concept of hyperbolic dynamics, introduced by S. Smale around 1963. One of the main aspects of this research was the use of hyperbolicity to grant the persistence of specially interesting features of a dynamical system under small perturbations. Uppermost in this line of research was structural stability, roughly speaking the property of a dynamical system having an orbit structure that remains topologically unchanged when the system is slightly perturbed.

In 1968, Palis and Smale, in the light of their own joint work on structural stability of Morse-Smale systems, and that of Anosov on the class of systems that nowadays carries his name, conjectured that a combination of hyperbolicity of the recurrente orbits plus certain transversally assumptions on the transient trajectories were necessary and sufficient conditions for structural stability: This became a fundamental question in dynamics known as the Stability Conjecture. The sufficiency for Morse-Smale systems in low dimensions was provided by Palis (1968), in all dimensions by Palis and Smale (1968). In particular, Palis and Smale provided the sufficiency for gradient dynamical systems. In full generality, the sufficiency was proved in papers of Robbin (1971), de Melo (1972) and Robinson (1973). The necessity remained open until much later. Its conceptual nature and part of its technical difficulties were present in Mañé's thesis, where he solved a similar but simpler problem, proving that the hyperbolicity assumption, shown by several authors, among them Hirsch, Pugh and Shub, to be sufficient for persistence of an invariant manifold of a dynamical systems, was also necessary. After many partial results, notably those of Franks, Pliss, Liao and Sannami, the completion of stability conjecture for diffeomorphisms was done by Mañé in 1986.

Based on that, Palis extended the result to the limit set. The long path to this theorem put Mañé in contact with many other challenging phenomena of which several are still topics of active research nowadays, like persistently transitive dynamical systems, the creation of attracting periodic orbits by small perturbations, the generic theory of Lyapunov exponents. This research inspired Mañé to write several works on ergodic theory (mainly on entropy and Lyapunov exponents) of differentiable dynamical systems, geodesic flows and rational maps.

In 1982, he published the book Introdução à Teoria Ergódica, that in 1987 was translated into English and published by Springer-Verlag in the Ergebnisse der Mathematik series. Two essays were to follow: Global variational methods in conservative dynamics, Notes of the Brazilian Mathematical Colloquium, 1991, and The dynamics of inner functions, with C. Doering, Ensaios Matemáticos-Brazilian Mathematical Society, 1991. Since he became an assistant professor at IMPA, in 1978, all the academic career of Mañé was developed at this admirable Institute, where he has been thesis adviser of eleven doctorate students C. Doering, A. Araujo, R. Ruggiero, M. Craizer, G. Contreras, M. Paternain, M. Carvalho, A. Rovella, J. Delgado, R. Iturriaga and H. Enrich. The topics of their theses included persistently transitive flows, Lorenz like attractors, existence of hyperbolic attractors for diffeomorphisms of surfaces, the dynamics of inner functions and the variational ergodic theory of Lagrangian ows. Mañé was invited to speak twice in the section of Ordinary Differential Equations and Dynamical Systems of the International Congress of Mathematics, in 1983 at Warsaw and in 1994 at Zurich. He was also speaker at the Colloquium organized by the Societé Mathématique de France, celebrating R. Thom's 65 anniversary. In 1994, he became a member of the Brazilian Academy of Sciences and was awarded the Third World Academy of Sciences Prize for Mathematics."

Note by Maria José Pacifico

Mañé 's proof of the stability conjecture for diffeomorphisms (1986) was a masterpiece, and in Mañé own words, the focusing point of his contribution to dynamical systems.

Remarkably, the techniques developed by Mañé in the course of his scientific career are still being used to establish other important results in the theory, especially in C^1 generic dynamics. Mañé's Ergodic Closing Lemma is a fundamental and important advance and was a key element in several of his subsequent results. His research also dealt with another open problem analogous to the stability conjecture for rational maps, which explores the question as to whether the hyperbolicity of the Julia set is necessary for its stability (the sufficiency was proved in 1968 by Guckenheimer). A combination of the joint work of Mañé, P. Sad and D. Sullivan, with his proof of the instability of Herman rings (1985) and Sullivan's important papers on the role of quasi-conformal mappings in dynamics, prove that when the Julia set has measure zero, stability implies hyperbolicity. The Mañé-Sad-Sullivan Theorem has since become a fundamental tool in the theory of rational maps. Mañé also worked with collaborators on the ergodic theory of rational maps and made important contributions in thermodynamic formalism, mainly concerning the question of differentiability under varying parameters. In his later years, Mañé worked on the variational ergodic theory of conservative dynamical systems. Drawing on the Mather theory of minimizing measures of Lagrangian systems, he introduced a natural concept of generic Lagrangians and arrived at a fundamental result: for generic Lagrangians the minimizing Mather measure is unique. The results of this work inspired several other important papers on weak KAM theory and Hamilton-Jacobi theory by A. Fathi, G. Contreras, R. Iturriaga, P. Bernard, D. Gomes, D. Massart, M. Carneiro, G. Paternain and R. Ruggiero. The Mañé conjecture, which claims that for a generic Lagrangian the minimizing probability has support on a periodic trajectory, was recently solved in dimension two by G. Contreras.

Mañé's only foray into one-dimensional dynamics produced a famous result, remarkable for its generality; it states that a compact set forward invariant under a C^2 interval map and not containing critical points, stable or indifferent periodic orbits, is hyperbolic. In particular, the techniques in this paper inspired the proof, by E. Pujals and M. Sambarino (2009), that in the complement of the closure of the hyperbolic surface diffeomorphisms, those that exhibit a homoclinic tangency are C^1 dense. This result implies the proof for surface diffeomorphisms, of a conjecture by Palis, establishing that Every C^r diffeomorphism of a compact manifold can be C^r approximated by one which is hyperbolic or by one exhibiting a heterodimensional cycle or a homoclinic tangency. Recently (2015), S. Crovisier and E. Pujals proved this conjecture in the C^1-topology for a weaker form of hyperbolicity (essential hyperbolicity), and Mañé-like techniques for the proof of the Stability Conjecture can be found in the proof.

A result announced, without a proof, in a note by Mañé to accompany the Proceedings of the International Congress of Mathematicians held in Warsaw in 1983 also deserves mention. There, Mañé argued that for any compact surface there is a residual set of C^1 area-preserving diffeomorphisms that are either Anosov or have zero Lyapunov exponents. The complete proof of this result was later provided by J. Bochi (2002), sparking a series of impressive results on Lyapunov exponents by J. Bochi and M. Viana (2003, 2004 and 2005).

Another example is his insightful concept in volume preserving dynamics, now known as Mañé's critical value (1997), which plays a fundamental role in understanding dynamical as well as symplectic aspects of energy levels for Euler Lagrange flows. Essentially, this critical value has proved to be a "watershed" in the dynamical behavior of Euler-Lagrange flows in energy levels and has been extensively used to describe changes in the dynamics of these flows when the threshold value is surpassed: one example is the recent work of G.P. Paternain, L.

Macarini (2015) and collaborators on the symplectic topology of Mañé's critical value and the existence of infinitely many periodic orbits of exact magnetic flows on surfaces for nearly every energy level below Mañé's critical value.

Rio de Janeiro, Brazil
April 2016

Maria José Pacifico

Acknowledgments

I would like to thank Marcelo Viana for inviting me to edit this volume. This gave me the opportunity to review and read many of the articles by Ricardo Mañé, who besides being a brilliant mathematician was one of my dearest and best friends. I would like also to express my profound gratitude to Isaia Nisoli, for his valued help throughout the process of selecting the papers for this volume.

Contents

TRANSACTIONS OF THE
AMERICAN MATHEMATICAL SOCIETY
Volume 246, December 1978

PERSISTENT MANIFOLDS ARE NORMALLY HYPERBOLIC

BY

RICARDO MAÑÉ

ABSTRACT. Let M be a smooth manifold, f: $M \hookleftarrow$ a C^1 diffeomorphism and $V \subset M$ a C^1 compact submanifold without boundary invariant under f (i.e. $f(V) = V$). We say that V is a persistent manifold for f if there exists a compact neighborhood U of V such that $\bigcap_{n \in \mathbf{Z}} f^n(U) = V$, and for all diffeomorphisms g: $M \hookleftarrow$ near to f in the C^1 topology the set $V_g = \bigcap_{n \in \mathbf{Z}} g^n(U)$ is a C^1 submanifold without boundary C^1 near to V. Several authors studied sufficient conditions for persistence of invariant manifolds. Hirsch, Pugh and Shub proved that normally hyperbolic manifolds are persistent, where normally hyperbolic means that there exist a Tf-invariant splitting $TM/V = N^sV \oplus N^uV \oplus TV$ and constants $K > 0$, $0 < \lambda < 1$ such that:

$$\|(Tf)^n/N^s_xV\| \leq K\lambda^n, \qquad \|(Tf)^{-n}/N^u_xV\| \leq K\lambda^n,$$
$$\|(Tf)^n/N^s_xV\| \cdot \|(Tf)^{-n}/T_{f^n(x)}V\| \leq K\lambda^n$$

for all $n > 0$, $x \in V$. In this paper we prove the converse result, namely that persistent manifolds are normally hyperbolic.

Let M be a manifold, $V \subset M$ a submanifold and f: $M \hookleftarrow$ a diffeomorphism satisfying $f(V) \subset V$. Several authors [1]–[9] have considered under which conditions for every diffeomorphism g nearby f there exists a submanifold $V_g \subset M$ nearby V satisfying $g(V_g) \subset V_g$. This kind of question arises frequently in ordinary, functional [18] and partial [16] differential equations as well as in the stability theory of group actions [19], bifurcations [3] and the construction of diffeomorphisms exhibiting certain persistent properties [9].

Answers to this problem are usually given in terms of exponential rates of the iterates of the derivative of f at V. In this paper we shall consider the converse problem, proving that the persistence of V (see definition below) is equivalent to these conditions.

To give the precise statement of the results let us denote by $\mathrm{Diff}^1(M)$ the space of C^1 diffeomorphisms of M endowed with the topology of C^1 convergence on compact subsets.

DEFINITION. V is a compact invariant manifold of $f \in \mathrm{Diff}^1(M)$ if it is a compact boundaryless C^1 submanifold of M satisfying $f(V) = V$. We say that V is persistent if there exist a neighborhood U of V in M and a

Received by the editors September 19, 1977.
AMS (MOS) subject classifications (1970). Primary 58F15; Secondary 58F10.

Originally published in: *Trans. Amer. Math. Soc. 246 (1978), 261-283*,

neighborhood $\mathfrak{U}$ of f in $\mathrm{Diff}^1(M)$ such that:

(a) For all $g \in \mathfrak{U}$ the set $V_g = \bigcap_{n \in \mathbf{Z}} g^n(U)$ is a C^1 submanifold of M and $V_g = V$.

(b) V_g is C^1 near to V if g is C^1 near to f.

Finally let us say that V is normally hyperbolic if there exist constants $C > 0$, $0 < \lambda < 1$ and a continuous (Tf)-invariant splitting $TM/V = N^sV \oplus N^uV \oplus TV$ satisfying

$$\|(Tf)^n/N_x^sV\| \leqslant C\lambda^n, \quad \|(Tf)^{-n}/N_x^uV\| \leqslant C\lambda^n,$$

$$\|(Tf)^n/N_x^sV\| \cdot \|(Tf)^{-n}/T_{f^n(x)}V\| \leqslant C\lambda^n,$$

$$\|(Tf)^{-n}/N_x^uV\| \cdot \|(Tf)^n/T_{f^{-n}(x)}V\| \leqslant C\lambda^n$$

for all $x \in \mathbf{Z}^+$.

In [9] it is proved that normal hyperbolicity implies persistence. Here we shall prove the converse, thus obtaining

THEOREM A. *A compact invariant manifold of a diffeomorphism is persistent if and only if it is normally hyperbolic.*

This theorem follows from a stronger version given in §2 that states that a Lipschitz persistent compact invariant manifold (same definition as above requiring V_g Lipschitz near to V instead of C^1 near) is normally hyperbolic.

Recalling that a compact subset Λ of a diffeomorphism $f \in \mathrm{Diff}^1(M)$ is said to be isolated if it has a neighborhood U such that $\bigcap_{n \in \mathbf{Z}} f^n(U) = \Lambda$, the Lipschitz version of Theorem A suggests the following conjecture:

CONJECTURE. There exists a residual subset $\mathfrak{B} \subset \mathrm{Diff}^1(M)$ such that every isolated compact invariant manifold of a diffeomorphism in $\mathfrak{B}$ is normally hyperbolic.

Hyperbolicity hypotheses (i.e. hypotheses involving exponential rates like normal hyperbolicity) are widely used in perturbation theory. However, the generalized feeling that they are good hypotheses in the sense that they are equivalent to the stability property they seek to grant, has, in terms of theorems, an incomplete justification. These problems when settled in $\mathrm{Diff}^r(M)$ with $r \geqslant 1$ are even harder for several reasons, among them the lack of a C^r closing lemma [14]. This is, for instance, the main stumbling block to prove a C^r version of Theorem A. Such a version would state a C^r persistent invariant compact manifold (obvious definition) is r-normally hyperbolic in the sense defined in [9].

Finally I wish to thank Jacob Palis for his advice during the preparation of my thesis [10] (of which Theorem A is part) and to C. Pugh for several useful talks on Theorem A.

1. Domination and invariant subbundles. Let K be a compact metric space. Denote by $\mathfrak{L}(K)$ the set of continuous finite-dimensional vector bundles F on

K endowed with a Riemann structure, i.e. a continuous map B: $F \times F \to \mathbf{R}$ such that $B/(F_x \times F_x)$ is an inner product on F_x for all $x \in K$. If $F \in \mathcal{E}(K)$ let $\mathcal{E}(F)$ be the set of continuous vector bundle isomorphisms of F. If $\Phi \in \mathcal{E}(F)$ we say that a subbundle $E \subset F$ is Φ-invariant if $\Phi(E) = E$. In the next proposition we give a sufficient condition for the existence of an invariant complement of E, i.e. a subbundle $G \subset F$ such that G is Φ-invariant and $\Phi(G) = G$. For the statement of the result we need the following definitions:

Definition 1.1. $F_1, F_2 \in \mathcal{E}(K)$ and $\Phi_i \in \mathcal{E}(F_i)$, $i = 1, 2$. We say that Φ_1 dominates Φ_2, denoted $\Phi_1 > \Phi_2$, if there exist constants $C > 0$, $0 < \lambda < 1$ such that

$$\|\Phi_1^n/F_{1,x}\| \cdot \|\Phi_2^{-n}/F_{2,\Phi^n(x)}\| \leqslant C\lambda^n \tag{1}$$

for all $x \in K$, $n \geqslant 0$. Moreover, if $\Phi_1 > \Phi_2$ and (1) is satisfied we say that Φ_1 (C, λ)-dominates Φ_2.

Definition 1.2. Let E be a finite-dimensional Hilbert space, $E_1 \subset E$ a subspace and $E_1^\perp$ its orthogonal complement. If $S \subset E$ is a subspace such that $\dim S = \dim E_1$ and $S \cap E_1^\perp = \{0\}$, we define the angle $\alpha(E_1, S)$ by $\alpha(E_1, S) = \|L\|$, where L: $E_1 \to E_1^\perp$ is a linear map satisfying $S = \operatorname{graph}(L)$.

Proposition 1.1. *Let $F \in \mathcal{E}(K)$, $\Phi \in \mathcal{E}(F)$, $E \subset F$ a continuous Φ-invariant subbundle, $\tilde{F} = F/E$ the quotient bundle and $\tilde{\Phi} \in \mathcal{E}(\tilde{F})$ be the vector bundle isomorphism induced by Φ.*

If $\tilde{F}$ and E have Φ-invariant splittings $\tilde{F} = \tilde{F}_1 \oplus \cdots \oplus \tilde{F}_k$, $E = E_1 \oplus \cdots \oplus E_l$ such that for all $1 \leqslant i \leqslant k$, $1 \leqslant j \leqslant l$ either $\tilde{\Phi}/\tilde{F}_i$ $(\ , \lambda)$-dominates Φ/E_j or Φ/E_j (C, λ)-dominates $\tilde{\Phi}/\tilde{F}_i$, then there exists a Φ-invariant continuous subbundle $\hat{F} \subset F$ satisfying $\hat{F} \oplus E = F$ and such that for all $x \in K$,

$$\alpha(E_x^\perp, \hat{F}_x) \leqslant \bar{C}d^2/(1 - \lambda),$$

where

$$d = \sup\left\{\|\Phi/F_x\|, \|\Phi^{-1}/F_x\| \,|\, x \in K\right\}$$

and $\bar{C}$ depends only on C and

$$\delta = \sup\Big\{\alpha\big((\tilde{F}_1 \oplus \cdots \oplus \tilde{F}_j)_x^\perp, (\tilde{F}_{j+1} \oplus \cdots \oplus \tilde{F}_k)\big),$$

$$\alpha\big((E_1 \oplus \cdots \oplus E_1)_x^\perp, (E_{i+1} \oplus \cdots \oplus E_l)_x\big) |$$

$$1 \leqslant j < k, 1 \leqslant i < l, x \in K\Big\}.$$

Remark. It follows from this proposition that if π: $\hat{F} \to \tilde{F}$ is the canonical projection and $\hat{F}_j = \pi^{-1}\tilde{F}_j$, $j = 1, \ldots, k$, we have a Φ-invariant continuous splitting $F = E \oplus \hat{F}_1 \oplus \cdots \oplus \hat{F}_j$, and if $\tilde{\Phi}/\tilde{F}_i > \Phi/E_j$ $(\Phi/E_j > \Phi/\tilde{F}_i)$ then $\Phi/\tilde{F}_i > \Phi/E_j$ $(\Phi/E_j > \Phi/\hat{F}_i)$.

PROOF. Let G be the continuous vector bundle on K whose fiber on $x \in K$ is the space of linear transformations $L\colon E_x^\perp \to E_x$. Let $\Gamma^0(G)$ be the space of continuous sections of G endowed with the norm $\|\eta\| = \{\|\eta(x)\| \mid x \in K\}$. Define $\hat{\Phi}\colon \Gamma^0(G) \hookleftarrow$ by

$$\hat{\Phi}(\eta)(x) = \Phi \circ \eta(\Phi^{-1}(x)) \circ (\hat{\pi}\Phi/E^\perp_{\Phi^{-1}(x)})^{-1}$$

where $\hat{\pi}\colon F \to E^\perp$ is the orthogonal projection. Let $\tilde{\pi}\colon F \to \tilde{F}$ be the canonical projection. Define

$$\hat{F}_j = E^\perp \cap \tilde{\pi}^{-1}(\tilde{F}_j), \qquad 1 \leqslant j \leqslant k.$$

Let Γ^s (resp. Γ^u) be the space of sections $\eta \in \Gamma^0(G)$ such that $\eta(x)\hat{F}_{j,x} \subset \bigoplus\{E_{i,x} | \Phi/E_i > \tilde{\Phi}/\tilde{F}_j\}$ ($\eta(x)\hat{F}_{j,x} \subset \bigoplus\{E_{i,x} | \tilde{\Phi}/\tilde{F}_j > \Phi/E_i\}$) for all $x \in K$. It is easy to see that

$$\|\hat{\Phi}^n/\Gamma^s\| \leqslant \hat{C}\lambda^n, \qquad (1)$$

$$\|\hat{\Phi}^{-n}/\Gamma^u\| \leqslant \hat{C}\lambda^n \qquad (2)$$

for all $n \in \mathbf{Z}^+$. If we find $\eta_0 \in \Gamma^0(G)$ satisfying

$$\eta_0(x) = \hat{\Phi}(\eta_0)(x) + \pi\Phi \circ (\hat{\pi}\Phi/E^\perp_{\Phi^{-1}(x)})^{-1}, \qquad x \in K,$$

where $\pi\colon F \to E$ is the orthogonal projection, then the subbundle $\hat{F} \subset F$ with fibers $\hat{F}_x = \text{graph}(\eta_0(x))$ is continuous and Φ-invariant. To solve (3) observe that (1) and (2) imply (see [9] for details) that $I - \hat{\Phi}$ has an inverse that satisfies

$$\|(I - \hat{\Phi})^{-1}\| \leqslant \frac{\overline{C}}{1 - \lambda}.$$

Let $\xi \in \Gamma^0(G)$ be defined by

$$\xi(x) = \pi\Phi \circ (\hat{\pi}\Phi/E^\perp_{\Phi^{-1}(x)})^{-1}, \qquad x \in K.$$

Then $\eta_0 = (I - \hat{\Phi})^{-1}\xi$ satisfies (3) and

$$\|\eta_0\| \leqslant (C/(1 - \lambda))\|\xi\| \leqslant (C/(1 - \lambda))d^2$$

where $d = \sup\{\|\Phi/F_x\|, \|\Phi^{-1}/F_x\| \mid x \in K\}$.

PROPOSITION 1.2. *Let $F \in \mathfrak{L}(K)$, $\Phi \in \mathfrak{L}(F)$ and let $K_0 \subset K$ be a Φ-invariant compact subset. Suppose that F/K_0 has a continuous Φ-invariant splitting $F/K_0 = E_0 \oplus F_0$ such that $\Phi/E_0 > \Phi/F_0$. Then there exist a compact neighborhood U of K_0 and a continuous subbundle $E \subset F/\Lambda^s$, where $\Lambda^s = \bigcap_{n>0}\Phi^{-n}(U)$, satisfying $\Phi(E) \subset E$, $E/K_0 = E_0$.*

PROOF. Take a compact neighborhood U_0 of K_0 and let $\Lambda_0^s = \bigcap_{n\in\mathbf{Z}^+}\Phi^{-n}(U_0)$. If U_0 is small enough there exists a continuous splitting $F/\Lambda_0^s = \hat{E} \oplus \hat{F}$ such that $\hat{E}/K_0 = E_0$, $\hat{F}/K_0 = F_0$. Let $\pi_1\colon F/\Lambda_0^s \to \hat{E}$, $\pi_2\colon F/\Lambda_0^s \to \hat{F}$ be the projections associated with this splitting. Let G be the continuous vector bundle on Λ_0^s whose fiber on x is the space of linear maps

$L: \hat{E}_x \to \hat{F}_x$. If $U \subset U_0$ is a compact neighborhood of Λ_0^s, let $\Gamma^0(U)$ be the space of continuous sections η of $G/\Lambda^s(U)$, such that $\eta(x) = 0$ if $x \in K_0$, where $\Lambda^s(U) = \bigcap_{n \in \mathbf{Z}^+} \Phi^{-n}(U)$, endowed with the norm $\|\eta\| = \sup\{\|\eta(x)\| \mid x \in \Lambda^s(U)\}$. Define $\Phi_1: \hat{E} \hookleftarrow$ by $\Phi_1 = \pi_1 \Phi/\hat{E}$, $\Phi_2: \hat{F} \hookleftarrow$ by $\Phi_2 = \pi_2 \Phi/\hat{F}$ and $\hat{\Phi}: \Gamma^0(U) \hookleftarrow$ as $\hat{\Phi}(\eta)(x) = \Phi_2^{-1} \circ \eta(\Phi(x)) \circ \Phi_1$.

If U is small enough $\hat{\Phi}$ is well defined.

LEMMA 1.1. *There exist constants $C > 0$, $0 < \lambda < 1$ and a neighborhood $U_1 \subset U_0$ of K_0 such that for all compact neighborhoods $U \subset U_1$ of K_0 the linear map $\hat{\Phi}: \Gamma^0(U) \hookleftarrow$ satisfies $\|\hat{\Phi}^n\| \leqslant C\lambda^n$ for all $n \in \mathbf{Z}^+$.*

PROOF. There exists $n_0 \in \mathbf{Z}^+$ such that

$$\|\Phi^{n_0}/E_{0,x}\| \cdot \|\Phi^{-n_0}/F_{0,\Phi^{n_0}(x)}\| \leqslant \tfrac{1}{3}$$

for all $x \in K$. Then, there exists a neighborhood U_1 of K_0 such that

$$\|\Phi_1^{n_0}/\hat{E}_x\| \cdot \|\Phi_2^{-n_0}/\hat{F}_{\Phi^{n_0}(x)}\| \leqslant \tfrac{1}{2}$$

for all $x \in U_1$. Let $U \subset U_1$ be a compact neighborhood of K_0. If $x \in \Lambda^s(U)$, $\Phi^n(x) \in U$ for all $x \in \mathbf{Z}^+$. Therefore

$$\|\Phi^{kn_0}/\hat{E}_x\| \cdot \|\Phi_2^{-kn_0}/F_{\Phi^{kn_0}(x)}\| \leqslant (1/2)^k$$

for all $k \in \mathbf{Z}^+$. Then, for some constant $C > 0$, $0 < \lambda < 1$ we have

$$\|\Phi_1^n/\hat{E}_x\| \cdot \|\Phi_2^{-n}/\hat{F}_{\Phi^n(x)}\| \leqslant C\lambda^n$$

for all $x \in \Lambda^s(U)$, $n \in \mathbf{Z}^+$, and this implies

$$\|\hat{\Phi}^n\| \leqslant C\lambda^n$$

for all $n \in \mathbf{Z}^+$.

If we find $\eta_0 \in \Gamma^0(U)$ such that $\Phi(\text{graph}(\eta_0(x))) = \text{graph}(\eta_0(\Phi(x)))$ for all $x \in \Lambda^s(U)$, the lemma is proved, defining a subbundle E of $F/\Lambda^s(U)$ by $E_x = \text{graph}(\eta_0(x))$. Let $\Gamma_\varepsilon^0(U) = \{\eta \in \Gamma^0 \mid \|\eta\| \leqslant \varepsilon\}$. It is easy to see that there exist $\varepsilon_2 > \varepsilon_1 > 0$ such that for all $\delta > 0$ we can find a compact neighborhood $U_\delta \subset U$ of K_0 and a map $P: \Gamma_{\varepsilon_1}^0(U_\delta) \to \Gamma_{\varepsilon_2}^0(U_\delta)$ satisfying:

(a) $\Phi^{-1}(\text{graph}(\eta(x))) = \text{graph}(P(\eta)(\Phi^{-1}(x)))$ for all $x \in \Phi(\Lambda^s(U_\delta))$.

(b) $\|P(0)\| \leqslant \delta$.

(c) $\|(P - \hat{\Phi})(\eta_1) - (P - \hat{\Phi})(\eta_2)\| \leqslant \delta\|\eta_1 - \eta_2\|$ for all $\eta_1, \eta_2 \in \Gamma_{\varepsilon_1}^0(U_\delta)$.

Take $\delta \leqslant (1 - \lambda)/2C(1 + \varepsilon_1)$. From Lemma 1.1 it follows [4] that $I - \hat{\Phi}$ has an inverse satisfying $\|(I - \hat{\Phi})^{-1}\| \leqslant C/(1 - \lambda)$. Moreover, if $\eta \in \Gamma_{\varepsilon_1}$ and $Q = P - \hat{\Phi}$,

$$\begin{aligned}
&\|(I - \Phi)^{-1}Q(\eta)\| \\
&\quad \leqslant \|(I - \Phi)^{-1}Q(\eta) - (I - \Phi)^{-1}Q(0)\| + \|(I - \Phi)^{-1}Q(0)\| \\
&\quad \leqslant C\delta\varepsilon_1/(1 - \lambda) + C\delta/(1 - \lambda) \leqslant \tfrac{1}{2}\varepsilon_2;
\end{aligned}$$

hence, $(I - \Phi)^{-1}Q(\Gamma_{\varepsilon_1}^0(U_\delta)) \subset \Gamma_{\varepsilon_2}^0(U_\delta)$, and if $\eta_1, \eta_2 \in \Gamma_{\varepsilon_1}^0(U_\delta)$ we have

$$\|(I-\Phi)^{-1}Q(\eta_1)-(I-\Phi)^{-1}Q(\eta_2)\| \leqslant \frac{C\delta}{1-\lambda}\|\eta_1-\eta_2\| \leqslant \tfrac{1}{2}\|\eta_1-\eta_2\|.$$

By the fixed point theorem for contractions there exists $\eta_0 \in \Gamma^0_{\varepsilon_1}(U_\delta)$ such that $(I-\hat{\Phi})^{-1}Q(\eta_0)=\eta_0$. Therefore $P(\eta_0)=\eta_0$ and, by (a), the proposition is proved.

PROPOSITION 1.3. *Let $F \in \mathcal{E}(K)$, $\Phi \in \mathcal{E}(F)$ and $K \subset K_1$ be a Φ-invariant dense subset. Suppose that for all $x \in K_1$ there exists a splitting $F_x = E_x \oplus G_x$ satisfying*:

(a) $\Phi(E_x)=E_{\Phi(x)}$, $\Phi(G_x)=G_{\Phi(x)}$.

(b) *There exist $C>0$, $0<\lambda<1$ such that $\|\Phi^n/E_x\|\cdot\|\Phi^{-n}/G_{\Phi^n(x)}\| \leqslant C\lambda^n$ for all $x \in \mathbf{Z}^+$, $x \in K_1$.*

(c) $\dim E_x$ *is constant.*

Then the families of subspaces $\{E_x \mid x \in K\}$, $\{G_x \mid x\in K\}$ extend to Φ-invariant continuous subbundles E and G of F such that $F = E \oplus G$ and $\Phi/E > \Phi/G$.

PROOF. Let $\Sigma \subset K - K_1$ be a subset such that each Φ-orbit contained in $K-K_1$ intersects Σ at exactly one point. For each $x \in \Sigma$ chose a sequence $\{x_n \mid n \in \mathbf{Z}^+\} \subset K_1$ such that $x_n \to x$ when $n \to +\infty$ and the sequences $\{G_{x_n} \mid n \in \mathbf{Z}^+\}$, $\{E_{x_n} \mid n \in \mathbf{Z}^+\}$ are convergent. For $x \in \Sigma$ define

$$G_x = \lim_{n\to+\infty} G_{x_n}, \qquad E_x = \lim_{n\to+\infty} E_{x_n},$$

and for each $x \in K - K_1$, choose $x_k = \Phi^k(x) \in \Sigma$ and define

$$G_x = \Phi^{-k}(G_{x_k}), \qquad E_x = \Phi^{-k}(E_{x_k}).$$

Let P be the projective bundle associated to F and $d(\cdot,\cdot)$ a metric on P inducing its topology. If A, B are subsets of P define

$$d(A,B)=\sup\{d(a,b) \mid a \in A, b \in B\}.$$

Suppose that $\dim E_x = m$, $\dim G_x = n$ for all $x \in K$. Let $S \in \mathcal{E}(K)$ be the vector bundle whose fiber on x is the set of subspaces (L_1, L_2) of F_x such that $\dim L_1 = m$, $\dim L_2 = n$. For $x \in K$ let H_x be the set of $(L_1,L_2) \in S_x$ such that there exists a sequence $\{x_n \mid n \in \mathbf{Z}^+\} \subset K$ satisfying $x_n \to x$, $G_{x_n} \to L_2$, $E_{x_n} \to L_1$ when $n \to +\infty$. Proving that for all $x \in K$, $H_x = \{(E_x, G_x)\}$ we are done. Observe that $\bigcup_{x\in K}\tilde{H}_x$, where

$$\tilde{H}_x = \{\theta \in P_x \mid \theta \subset L_1 \cup L_2, (L_1,L_2) \in H_x\}$$

is a closed subset of P. Moreover, if $x \in K$, and $(L_1,L_2)\in H_x$ we have

$$\|\Phi^n/L_1\|\cdot\|\Phi^{-n}/\Phi^n(L_2)\| = C\lambda^n \tag{1}$$

for all $n \in \mathbf{Z}^+$. In particular, $L_1 \cap L_2 = \{0\}$; hence $F_x = L_1 \oplus L_2$. Let $x \in K$, $(L_1,L_2) \in H_x$ with $(L_1,L_2) \neq (E_x,G_x)$; let us say $L_1 \neq E_x$. Then there exists $\theta \in P_x \cap L_1$ such that $\theta \cap E_x = \{0\}$. Since $(E_x, G_x) \in H_x$, (1) implies that $d(\Phi^n(\theta), G_{\Phi^n(x)}) \to 0$ when $n \to +\infty$. Let y be an ω-limit point of

x. Take a sequence $\{n_k \mid k \in \mathbf{Z}^+\}$ such that $\lim_{k\to+\infty} n_k = +\infty$ and the sequences $\Phi^{n_k}(L_2)$, $\Phi^{n_k}(G_x)$, $\Phi^{n_k}(E_x)$ converge to subspaces L_1', L_2', S_1', S_2' of F_x. Then $(L_1', L_2') \in H_y$, $(S_1', S_2') \in H_y$ and $L_1' \cap S_2' \neq \{0\}$. Since $L_1' \oplus L_2' = F_y$ we must have $L_2' \not\subset S_1'$. Take $\theta \in P_y$ such that $\theta \subset L_2'$ and $\theta \cap S_1' \neq \{0\}$. By (1) $d(\Phi^{-n}(\theta), \Phi^{-n}(S_1')) \to 0$ if $n \to +\infty$. Take an α-limit point z of y and a sequence $\{n_k \mid k \in \mathbf{Z}^+\} \subset \mathbf{Z}^+$ such that $\lim_{k\to+\infty} n_k = +\infty$ and $\Phi^{-n_k}(L_1')$, $\Phi^{-n_k}(L_2')$, $\Phi^{-n_k}(S_1')$, $\Phi^{-n_k}(S_2')$ converge to subspaces L_1'', L_2'', S_1'', S_2''. Then $(L_1'', L_2'') \in H_z$, $(S_1'', S_2'') \in H_z$, $L_2'' \cap S_1'' \neq \{0\}$; take $0 \neq v_1 \in L_1'' \cap S_2''$, $0 \neq v_2 \in L_2'' \cap S_1''$. Since $v_1 \in L_1''$, $v_2 \in L_2''$, (1) implies

$$\|\Phi^n(v_1)\| / \|\Phi^n(v_2)\| \leqslant K\lambda^n(\|v_1\| / \|v_2\|) \tag{2}$$

for all $n \in \mathbf{Z}^+$. But $v_1 \in S_2$, $v_2 \in S_1$. Hence (1) implies

$$\|\Phi^n(v_2)\| / \|\Phi^n(v_1)\| \leqslant K\lambda^n(\|v_2\| / \|v_1\|),$$

clearly contradicting (2).

2. Proof of Theorem A. Let M be a C^∞ boundaryless manifold and $V \subset M$ a C^1 compact boundaryless submanifold. Assume that M is a submanifold of $\mathbf{R}^n$. Let NV be a C^1 subbundle of TM/V satisfying $TV \oplus NV = TM/V$. If η is a section of NV define the Lipschitz constant of η by

$$\operatorname{Lip}(\eta) = \sup\{\|\eta(x) - \eta(y)\| / \|x - y\| \mid x, y \in V, x \neq y\}.$$

We say that η is a Lipschitz section if $\operatorname{Lip}(\eta) < +\infty$. Let $\Gamma_{\mathcal{L}}(NV)$ be the space of Lipschitz sections of NV endowed with the norm

$$\|\eta\|_{\mathcal{L}} = \sup\{\|\eta(x)\| \mid x \in V\} + \operatorname{Lip}(\eta).$$

Let $\operatorname{Diff}^1(M)$ be the space of C^1 diffeomorphisms with the topology of the C^1 convergence on compact subsets.

DEFINITION 2.1. Let $f \in \operatorname{Diff}^1(M)$. We say that V is a Lipschitz persistent invariant manifold of f if there exists a neighborhood U of V such that for all $\delta > 0$ there exists a neighborhood $\mathcal{U}_\delta$ of f such that if $g \in \mathcal{U}_\delta$ there exists $\eta \in \Gamma_{\mathcal{L}}(NV)$ with $\|\eta\|_{\mathcal{L}} \leqslant \delta$ satisfying $V_g = \operatorname{graph}(\eta)$, where $\operatorname{graph}(\eta) = \{\exp_x(\eta(x)) | x \in V\}$, $V_g = \bigcap_{n\in\mathbf{Z}^+} g^n(U)$.

Observe that this definition implies $V_f = V$, hence $f(V) = V$. Moreover, the Lipschitz persistence is independent of the bundle NV.

In this section we shall prove the following proposition, which clearly implies Theorem A.

PROPOSITION 2.1. *If V is a Lipschitz persistent invariant manifold of f then V is normally hyperbolic.*

To prove this proposition we shall start showing some properties of the action of Tf on the Grassmannian bundle of dim V-dimensional subspaces of the tangent space of M at points of V. More precisely, let $S_x(M)$ be the set of dim V-dimensional subspaces of T_xM. For $x \in V$, $S \in S_x(M)$ define $\alpha(S)$

$= \infty$ if $S \cap N_x V \neq \{0\}$ and $\alpha(S) = \|L\|$ if $S \cap N_x V = \{0\}$ and $L\colon T_x V \to N_x V$ is the linear map satisfying graph$(L) = S$. Let $\mathcal{O}$ be the set of diffeomorphisms of M that leave V invariant.

LEMMA 2.1. *For all $\delta > 0$ there exist $\varepsilon = \varepsilon(\delta)$ and a neighborhood $\mathcal{U} = \mathcal{U}(\delta)$ of f such that if $g \in \mathcal{U}$, $x \in V$, $S \in S_x(V)$, $m \in \mathbf{Z}^+$ and $\alpha(S) \leqslant \varepsilon$, $\alpha((Tg)^m S) \leqslant \varepsilon$, then $\alpha((Tg)^j S) \leqslant \delta$ for all $0 \leqslant j \leqslant m$.*

PROOF. Suppose the lemma false. Then there exists $\delta > 0$ such that for all neighborhoods $\mathcal{U}$ of f and $\varepsilon > 0$ we can find $y_1 \in V$, $S \in S_{y_1}(V)$, $g \in \mathcal{U} \cap \mathcal{O}$ and $0 < n < m$ such that $\alpha(S) < \varepsilon$, $\alpha((Tg)^m S) < \varepsilon$ and $\alpha((T_g)^n S) > \delta$. We can assume without loss of generality that there exists a neighborhood W_1 of y_1 such that $W_1 \cap g^j(W_1) = \varnothing$ for all $0 < j < m$. Let $W_2 = g^m(W_1)$. We can suppose that there exist C^∞ diffeomorphisms $\varphi_i\colon W_i \to \{x \in \mathbf{R}^{l_1} \times R^{l_2} \mid \|x\| \leqslant 1\}$, $i = 1, 2$, satisfying $\varphi_i(y_i) = 0$ and $\varphi_i(W_i \cap V) = \{x \in \mathbf{R}^{l_1} \mid \|x\| \leqslant 1\}$. Take $L\colon \mathbf{R}^{l_1} \to \mathbf{R}^{l_2}$, $P\colon \mathbf{R}^{l_1} \to \mathbf{R}^{l_2}$ linear maps satisfying $(T\varphi_1)S = \text{graph}(L)$, $(T\varphi_2)^m S = \text{graph}(P)$. We can assume that $T\varphi_i / T_{y_i} M$ is an isometry for $i = 1, 2$. Hence $\|P\| < \varepsilon$, $\|L\| < \varepsilon$. Let E_1 be the kernel of L and $E_2 = E_1^\perp$ its orthogonal complement. Take a C^∞ function $\psi\colon \mathbf{R} \to \mathbf{R}$ satisfying $\psi(0) = 1$, $\psi(t) = 0$ for $|t| > 1$, $|\psi'(t)| \leqslant 2$, and $|\psi(t)| \leqslant 1$ for all t.

Define: $B_i = \{x \in E_i \mid \|x\| \leqslant \frac{1}{2}\}$, $i = 1, 2$, $B_3 = \{x \in R^{l_2} \mid \|x\| \leqslant \frac{1}{2}\}$; $B = B_1 \times B_2 \times B_3$. Let $F\colon B \to \mathbf{R}^{l_2}$ defined by

$$F(x_1, x_2, x_3) = \psi(\|x_1\|^2/\lambda)\psi(\|x_2\|^2/\lambda)\psi(\|x_3\|^2/\lambda^2)Lx_2$$

where λ is a positive number. Denoting by F_1' the partial derivative of F with respect to the ith variable, it is easy to see that

$$\sup_B \|F_i'\| \to 0, \qquad i = 1, 3, \tag{1}$$

$$\sup_B \|F_2'\| \leqslant 2\varepsilon, \tag{2}$$

$$\sup_B \|F\| \to 0 \tag{3}$$

when $\lambda^{-1} \to 0$. Now define $\varphi = \varphi_2 \circ g^m \circ \varphi_1^{-1}$, $E_1' = (T\varphi)E_1$, $E_2' = (T\varphi)E_2$, $B_i' = \{x \in E_1| \|x\| \leqslant 1\}$, $i = 1, 2$, $B' = B_1' \times B_2'$. Take $\tilde{H}\colon B' \to \mathbf{R}^{l_2}$ such that $\varphi(\text{graph}(F/B_1 \times B_2)) = \text{graph}(\tilde{H})$. For small values, by (1)–(3), $\tilde{H}$ is well defined and satisfies

$$\sup_{B'} \|\tilde{H}_1'\| \to 0, \tag{4}$$

$$\sup_{B'} \|\tilde{H}_2'\| \leqslant \varepsilon, \tag{5}$$

$$\sup_{B'} \|\tilde{H}\| \to 0 \tag{6}$$

when $\lambda \to 0$. Take $H\colon B' \times B_3 \to \mathbf{R}^{l_2}$ defined as $H(x_1, x_2, x_3) = \psi(4x_3)\tilde{H}(x_1, x_2)$. If F', H' denote the derivatives of F and H we can rewrite

(4)–(6) as

$$\sup_{B} \|F'\| \leqslant 2\varepsilon, \tag{7}$$

$$\sup_{B' \times B_3} \|H'\| \leqslant 2\varepsilon, \tag{8}$$

$$F'(0)/B_1 \times B_2 = L, \qquad H'(0)/B' = P.$$

Take $k \in \mathrm{Diff}^1(M)$ defined by $k(x) = x$ if $x \notin W_1 \cup W_2$, $k(x) = (\varphi_1^{-1} \circ (F + I) \circ \varphi_1)(x)$ if $x \in W_1$, $k(x) = (\varphi_2^{-1} \circ (I - H) \circ \varphi_2)(x)$ for $x \in W_2$. Follows from (7) and (8) that k is C^1 near to the identity if ε and λ are small. Hence, taking g near enough to f and ε and λ sufficiently small, it follows that $\bar{g} = k \circ g$ belongs to the neighborhood $\mathfrak{U}_\delta$ given by the definition of persistence. But observe that

$$V_{\bar{g}} = \left(V - \bigcup_{0}^{m} \bar{g}^j(W_1 \cap V)\right) \cup \left(\bigcup_{0}^{m} \bar{g}^j(D)\right)$$

where $D = \varphi_1^{-1}(\mathrm{graph}(F))$. Then

$$\alpha\left(T_{\bar{g}^n(y_1)} V_{\bar{g}}\right) = \alpha\left((Tg)^n S\right) > \delta,$$

thus contradicting the definition of $\mathfrak{U}_\delta$.

Now let $\tilde{N}V$ be the quotient bundle $(TM/V)/TV$ and $L(V)$ the vector bundle on V whose fibers $L_x(V)$ are the spaces of linear transformations L: $T_x V \to \tilde{N}_x V$. For $g \in \mathcal{O}$ define the vector bundle isomorphism Φ_g: $L(V) \hookleftarrow$ by $\Phi_g(L) = \tilde{N}g \circ L \circ (Tg)^{-1}$ where $\tilde{N}g$: $\tilde{N}V \hookleftarrow$ is the vector bundle isomorphism induced by Tg. Let $\Phi = \Phi_f$.

LEMMA 2.2. *Φ is a quasi-Anosov vector bundle isomorphism* (*i.e.* **[13]** *for all* $0 \neq L \in L(V)$ *the set* $\{\|\Phi^n(L)\| \mid n \in \mathbf{Z}\}$ *is unbounded*).

PROOF. Suppose that Φ is not quasi-Anosov. Then there exist $\bar{x} \in V$ and $0 \neq L \in L_{\bar{x}}(V)$ such that $\{\|\Phi^n(L)\| \mid n \in \mathbf{Z}\}$ is bounded, let us say by $K > 0$. Take a linear map L_0: $T_{\bar{x}}V \to N_{\bar{x}}V$ such that $\pi L_0 = L$, where π: $TM/V \to \tilde{N}V$ is the canonical projection. Let $S = \mathrm{graph}(L_0)$ and $\delta = \|L_0\| = \alpha(S)$. Let $\mathfrak{U}(\delta/2)$ and $\varepsilon(\delta/2)$ be given by Lemma 2.1. As in **[20**, Lemma 1.1] we can find $1 > \varepsilon_0 > 0$ such that given any finite subset $\Sigma \subset V$ and a family of linear maps L_x: $T_x M \to T_{f(x)}M$, $x \in \Sigma$, such that $L_x(T_x V) = T_{f(x)}V$ and $\|L_x - Tf/T_x M\| \leqslant \varepsilon_0$ for all $x \in \Sigma$, then there exists $g \in \mathfrak{U}(\delta/2)$ satisfying $g(V) = V$, $g(x) = f(x)$ and $Tg/T_x M = L_x$ for all $x \in \Sigma$. Choose $n \in \mathbf{Z}^+$ satisfying

$$\left(1 - \frac{\varepsilon_0}{3}\right)^n \leqslant \frac{\varepsilon_0 K^{-1}\varepsilon(\delta/2)}{12 \dim V} \tag{9}$$

and (x, L') nearby $(\bar{x}, L)$ and $g \in \mathcal{O} \cap \mathfrak{U}(\delta/3)$, such that $g^j(x) \neq g^i(x)$ for all $-n \leqslant j < i \leqslant n$, $\|\Phi_g^j(L')\| \leqslant K$ for all $|j| \leqslant n$ and $\|L_0'\| > \delta/2$ where

$L'_0\colon T_xV \to N_xV$ is a linear map satisfying $L' = \pi L'_0$. Let $S' = \operatorname{graph}(L'_0)$ and $\Sigma = \{ g^j(x) \mid |j| \leqslant n\}$. By the definition of ε_0 there exists a map $g\colon [0, \varepsilon_0/3] \to \mathcal{O} \cap \mathcal{U}(\delta/2)$ satisfying:

(1) $g_0 = g$.

(2) $g_\lambda(g^j(x)) = g^{j+1}(x)$ for all $|j| \leqslant n$.

(3)

$$\tilde{N}g_\lambda / \tilde{N}_{g^j(x)}V = \left(1 - \frac{\varepsilon_0}{3}\right)\tilde{N}g \qquad \text{for all } 0 \leqslant j \leqslant n,$$

$$(Ng_\lambda)^{-1}/N_{g^j(x)}V = \left(1 - \frac{\varepsilon_0}{3}\right)(Ng)^{-2} \quad \text{for all } -n \leqslant j \leqslant 0.$$

(4)

$$(Tg_\lambda)^n/T_xV = (1+\lambda)(Tg)^n, \quad (Tg_\lambda)^n/T_xV = (1+\lambda)(Tg)^n,$$

$$(Tg_\lambda)^{-n}/T_xV = (1+\lambda)(Tg)^{-n}.$$

We claim that for some $\lambda_0 \in [0, \varepsilon_0/3]$ we have $\alpha((Tg_{\lambda_0})^{-n}S') \leqslant \varepsilon(\delta/2)$ and $\alpha((Tg_{\lambda_0})^nS') \leqslant \varepsilon(\delta/2)$. Since $\alpha(S') > \delta/2$ this contradicts $g_{\lambda_0} \in \mathcal{U}(\delta/2) \cap \mathcal{O}$ and the lemma is proved. To find λ_0 we shall use the following property: Let $L_\lambda^+\colon T_{g^n(x)}V \to N_{g^n(x)}V$, $L_\lambda^-\colon T_{g^n(x)}V \to N_{g^{-n}(x)}V$ be linear maps such that $\operatorname{graph}(L_\lambda^+) = (Tg_\lambda)^nS'$, $\operatorname{graph}(L_\lambda^-) = (Tg_\lambda)^{-n}S'$ and let $\{v_1^+, \dots, v_l^+\}$, $\{v_1^-, \dots, v_\lambda^-\}$ be orthonormal bases of $T_{g^n(x)}V$ and $T_{g^{-n}(x)}V$. Then if $\bar{\lambda} \in [0, \varepsilon_0/3]$, $1 \leqslant j \leqslant l$ and $\|L_{\bar\lambda}^+ v_j^+\| \geqslant \varepsilon(\delta/2)$ (resp. $\|L_{\bar\lambda}^- v_j^-\| \geqslant \varepsilon(\delta/2)$), it follows that $\|L_\lambda^+ v_j^+\| \leqslant \varepsilon(\delta/2)$ ($\|L_\lambda^- v_j^-\| \leqslant \varepsilon(\delta/2)$) for all $\bar\lambda + \varepsilon_0/6 \dim V \leqslant \lambda \leqslant \varepsilon_0/3$. With this property it is easy to find the desired λ_0 as follows: Suppose that for some j we have $\|L_0^+ v_j^+\| \leqslant \varepsilon(\delta/2)$. Then $\|L_{\lambda_1}^+ v_j^+\| \leqslant \varepsilon(\delta/2)$ where $\lambda_1 = \varepsilon_0/6 \dim V$. Now suppose that $\|L_{\lambda_1}^+ v_i^+\| \geqslant \varepsilon(\delta/2)$. Then $\|L_{2\lambda}^+ v_i^+\| \leqslant \varepsilon(\delta/2)$ and $\|L_{2\lambda_1}^+ v_j^+\| \leqslant \varepsilon(\delta/2)$. At the end of this method we find $\lambda_0 = m\lambda_1$, with $0 < m < 2\dim V$ (so $\lambda_0 \in [0, \varepsilon_0/3]$). It remains to prove the property above. For this define the linear maps $A_\lambda\colon N_xV \to N_{f^n(x)}V$, $P_\lambda\colon N_xV \to T_{f^n(x)}V$, $Q_\lambda\colon T_xV \to T_{f^n(x)}V$ such that

$$(Tg_\lambda)^n/T_xM = \begin{pmatrix} A_\lambda & P_\lambda \\ 0 & Q_\lambda \end{pmatrix}.$$

Then, if $w_j = Q_0^{-1}v_j^+$,

$$\|L_\lambda^+ v_j^+\| = \frac{\|A_\lambda L'_0 w_j\|}{\|Q_0(1+\lambda)w_j + P_\lambda L'_0 w_j\|} = \frac{\|(1-\varepsilon_0/3)^n A_0 L'_0 w_j\|}{\|Q_0(1+\lambda)w_j + P_0 L'_0 w_j\|}.$$

Suppose $\|L_{\bar\lambda}^+ v_j^+\| \geqslant \varepsilon(\delta/2)$. Then, if $\lambda > \bar\lambda + \varepsilon_0/6 \dim V$,

$$\|Q_0(1+\lambda)w_j + P_0L'_0w_j\| \geqslant (\lambda - \bar\lambda)\|Q_0 w_j\| - \|Q_0(1+\bar\lambda)w_j + P_0L'_0w_j\|;$$

hence,

$$\|L_\lambda^+ v_j^+\| \leqslant \frac{1}{\dfrac{\lambda-\bar{\lambda}}{(1-\varepsilon_0/3)^n}\cdot\dfrac{\|Q_0 w_j\|}{\|A_0 L_0' w_j\|}-\dfrac{1}{\|L_\lambda^+ v_j\|}}$$

$$\leqslant \frac{1}{\dfrac{\lambda-\bar{\lambda}}{(1-\varepsilon_0/3)^n}K^{-1}-\left(\varepsilon(\delta/2)\right)^{-1}}. \tag{10}$$

By (9),

$$\frac{\lambda-\bar{\lambda}}{(1-\varepsilon_0/3)^n}K^{-1} \geqslant \frac{\varepsilon_0 K^{-1}}{6\dim V(1-\varepsilon_0/3)^n} \geqslant 2\left(\varepsilon(\delta/2)\right)^{-1};$$

hence (10) implies

$$\|L_\lambda^+ v_j^+\| \leqslant \varepsilon(\delta/2).$$

In the next lemma we show that $TM/\Omega(f/V)$ (where $\Omega(f/V)$ denotes the set of nonwandering points of f/V) has a splitting satisfying the domination conditions required by the definition of normal hyperbolicity.

LEMMA 2.3. *There exists a continuous Tf-invariant splitting $TM/\Omega(f/V) = TV/\Omega(f/V) \oplus N^s \oplus N^u$ such that $Tf/N^s > Tf/TV > Tf/N^u$.*

PROOF. If $g \in \mathcal{O}$ define $E_x^s(g)$ (resp. $E_x^u(g)$) as the set of $L \in L_x(V)$ such that $\{\|\Phi_g^n(L)\| \,|n \geqslant 0\}$ ($\{\|\Phi_g^{-n}(L)\| \,|n \geqslant 0\}$) is bounded,

$$\tilde{N}_x^s(g) = \cup\{L(T_x V) | L \in E_x^s(g)\},$$

$$N_x^u(g) = \cup\{L(T_x V) | L \in E_x^u(g)\}$$

and $\mathrm{Per}(g)$ as the set of periodic points of g/V. In the proof of this lemma we shall use the following property whose proof will be given later:

LEMMA 2.4. *There exist a neighborhood $\mathcal{U}$ of f and a dense subset $\mathcal{S}$ of $\mathcal{U} \cap \mathcal{O}$ such that for all $g \in \mathcal{S}$ and $x \in \mathrm{Per}(g)$ we have $\tilde{N}_x^s(g) \cap \tilde{N}_x^u(g) = \{0\}$.*

Suppose we show that

$$\tilde{N}_x^s(f) \oplus \tilde{N}_x^u(f) = \tilde{N}_x V \tag{11}$$

for all $x \in \Omega(f/V)$ and that there exist $C > 0$, $0 < \lambda < 1$ satisfying

$$\|(\tilde{N}f)^n/\tilde{N}_x^s(f)\| \cdot \|(Tf)^{-n}/T_{f^n(x)}V\| \leqslant C\lambda^n, \tag{12}$$

$$\|(\tilde{N}f)^{-n}/\tilde{N}_x^u(f)\| \cdot \|(Tf)^n/T_{f^{-n}(x)}V\| \leqslant C\lambda^n \tag{13}$$

for all $x \in \Omega(f/V)$, $n \geqslant 0$. Then by Proposition 1.1 the maps $x \to \tilde{N}_x^s(f)$, $x \to \tilde{N}_x^u(f)$ define continuous $\tilde{N}f$-invariant subbundles $\tilde{N}^s$, $\tilde{N}^u$ of $\tilde{N}V/\Omega(f/V)$ such that $\tilde{N}f/N^s > Tf/(TV/\Omega(f/V)) > \tilde{N}f/N^u$. Hence, by Lemma 1.1, there exists a continuous Tf-invariant splitting satisfying our statement. So if we prove (11)–(13) we are done. By [**13**, Proposition 1.1 and

Lemma 1.2] there exist a neighborhood $\mathcal{U}$ of f, $C > 0$ and $0 < \lambda < 1$ such that Φ_g is quasi-Anosov and

$$\|\Phi_g^n / E_x^s(g)\| \leqslant K\lambda^n, \tag{14}$$

$$\|\Phi_g^{-n} / E_x^u(g)\| \leqslant K\lambda^n \tag{15}$$

for all $g \in \mathcal{U} \cap \mathcal{O}$, $x \in V$ and $n \geqslant 0$. Moreover, since Φ_g is quasi-Anosov, it is easy to see that $E_x^s(g) \oplus E_x^u(g) = L_x(V)$, hence $\tilde{N}_x^s(g) + \tilde{N}_x^u(g) = \tilde{N}_x V$ for all $g \in \mathcal{U} \cap \mathcal{O}$, $x \in \mathrm{Per}(g)$. Let $G_x^s(g)$, $G_x^u(g)$ be the spaces of linear maps from $T_x V$ into $\tilde{N}_x^s(g)$ and $\tilde{N}_x^u(g)$, respectively. We claim that $g \in \mathcal{U} \cap \mathcal{S}$ and $x \in \mathrm{Per}(g)$ imply

$$G_x^s(g) = E_x^s(g), \tag{16}$$

$$G_x^u(g) = E_x^u(g). \tag{17}$$

Suppose, for instance, that (16) is false. Then $\dim G_x^s(g) > \dim E_x^s(g)$. Therefore,

$$\begin{aligned} \dim T_x V \cdot \dim \tilde{N}_x V = \dim L_x(V) &= \dim E_x^s(g) + \dim E_x^u(g) \\ &< \dim G_x^s(g) + \dim G_x^u(g) \\ &= \dim T_x V(\dim \tilde{N}_x^s(g)) + \dim \tilde{N}_x^u(g) \\ &= \dim T_x V \cdot \dim \tilde{N}_x V. \end{aligned}$$

Now consider $x \in \Omega(f/V)$. There exist sequences $\{g_n | n \in \mathbf{Z}^+\} \subset \mathcal{S}$, $\{x_n | n \in \mathbf{Z}^+\} \subset V$ such that $x_n \in P(g_n)$, $g_n \to g$ and $x_n \to x$ when $n \to +\infty$ [**14**]. We can assume that there exist subspaces E^+, E^- of $L_x(V)$ such that

$$\lim E_{x_n}^s(g_n) = E^+, \tag{18}$$

$$\lim E_{x_n}^u(g_n) = E^-. \tag{19}$$

Since $E_{x_n}^s(g_n) \oplus E_{x_n}^u(g_n) = L_x(V)$ we obtain

$$\dim E^+ + \dim E^- = \dim L_x(V) \tag{20}$$

and (14), (15) imply $\|\Phi^n / E^+\| \leqslant K\lambda^n$, $\|\Phi^{-n} / E^-\| \leqslant K\lambda^n$ for all $n \geqslant 0$. Since Φ is quasi-Anosov this implies $E^+ \cap E^- = \{0\}$, which together with (20) gives $L_x(V) = E^+ \oplus E^-$, hence $E^+ = E_x^s(f)$, $E^- = E_x^u(f)$; then

$$E_x^s(f) \oplus E_x^u(f) = L_x(V). \tag{21}$$

Moreover, (16)–(19) and the elementary properties

$$G_x^s(f) = \lim G_{x_n}^s(g_n), \quad G_x^u(f) = \lim G_{x_n}^u(g_n)$$

prove that

$$G_x^s(f) = E_x^s(f), \tag{22}$$

$$G_x^u(f) = E_x^u(f). \tag{23}$$

By (21) $\tilde{N}_x^s(f) + \tilde{N}_x^u(f) = \tilde{N}_x V$ and by (21)–(23), $\tilde{N}_x^s(f) \cap \tilde{N}_x^u(f) = \{0\}$. This proves (11). (12) and (13) follow from (14), (15), (22) and (23).

PROOF OF LEMMA 2.4. Let $\mathfrak{U}$ be a neighborhood of f such that Φ_g is quasi-Anosov for all $g \in \mathcal{O} \cap \mathfrak{U}$. Let $\mathcal{S}$ be the set of diffeomorphisms $g \in \mathfrak{U} \cap \mathcal{O}$ satisfying the following properties:

(I) $\tilde{N}_g$ is a C^1 vector bundle isomorphism.

(II) $\operatorname{Per}(g)$ is dense in $\Omega(g/V)$.

(III) Every periodic point of g/V is hyperbolic.

(IV) If $x \in \operatorname{Per}(g)$ and m is its period, all the eigenvalues of $(Tf)^m/T_xM$ are simple and two of them have equal modulus if and only if they are conjugates.

Now observe that for all $g \in \mathcal{S}$, the set of points $x \in \operatorname{Per}(g)$ such that $\tilde{N}_x^s(g) \cap \tilde{N}_x^u(g) = \{0\}$ is closed in $\operatorname{Per}(g)$. To see this suppose that $\{x_n | n \in \mathbf{Z}^+\}$ is a sequence contained in $\operatorname{Per}(g)$, $g \in \mathcal{S}$, and $x_n \to x \in \operatorname{Per}(g)$ when $n \to +\infty$. If $\tilde{N}_{x_n}^s(g_n) \cap \tilde{N}_{x_n}^u(g_n) = \{0\}$, we deduce, as in the proof of Lemma 2.3, that $G_{x_n}^s(g) = E_{x_n}^s(g)$, $G_{x_n}^u(g) = E_{x_n}^u(g)$ and

$$\lim E_{x_n}^s(g_n) = E_x^s(g), \quad \lim E_{x_n}^u(g_n) = E_x^u(g).$$

Since, clearly, $\lim G_{x_n}^s(g) = G_x^s(g)$ and $\lim G_{x_n}^u(g) = G_x^u(g)$, it follows that $E_x^s(g) = G_x^s(g)$, $E_x^u(g) = G_x^u(g)$. If $\tilde{N}_x^s(g) \cap \tilde{N}_x^u(g) \neq \{0\}$, these relations imply $E_x^s(g) \cap E_x^u(g) \neq \{0\}$, contradicting the fact that Φ_g is quasi-Anosov. Suppose then that for some $g_0 \in \mathcal{S}$ there exists $a \in \operatorname{Per}(g_0)$ satisfying $\tilde{N}_a^s(g) \cap \tilde{N}_a^u(g) \neq \{0\}$. By the previous remark this implies that there exists an open subset W of $\operatorname{Per}(g_0)$ such that $\tilde{N}_x^s(g_0) \cap \tilde{N}_x^u(g_0) \neq \{0\}$ for every $x \in W$. To exhibit a contradiction between this fact and our definition of $\mathcal{S}$ we need the following definitions and Lemmas 2.5–2.7.

DEFINITION 2.2. Let $g \in \mathcal{S}$, $x \in \operatorname{Per}(g)$. We say that

$$\tilde{N}_x V = \tilde{N}_1 \oplus \cdots \oplus \tilde{N}_l, \tag{24}$$

$$T_x V = T_1 \oplus \cdots \oplus T_k \tag{25}$$

are the canonical splittings at x if they satisfy the following conditions:

(I) $(Tg)^m T_i = T_i$, $(\tilde{N}g)^m \tilde{N}_j = \tilde{N}_j$ for all $1 \leqslant i \leqslant k$, $1 \leqslant j \leqslant l$, where m is the period of x.

(II) $(Tg)^m/T_{i_1} > (Tg)^m/T_{i_2}$, $(\tilde{N}g)^m/\tilde{N}_{j_1} > (\tilde{N}g)^m/\tilde{N}_{j_2}$ for all $1 \leqslant i_1 < i_2 \leqslant k$, $1 \leqslant j_1 < j_2 \leqslant l$.

(III) For all $1 \leqslant i \leqslant k$, $1 \leqslant j \leqslant l$, $(\tilde{N}g)^m/\tilde{N}_j > (Tg)^m/T_i$ or $(Tg)^m/T_i > (\tilde{N}g)^m/\tilde{N}_j$.

(IV) For all $1 \leqslant i_1 < i_2 \leqslant k$ there exists $1 \leqslant j \leqslant l$ such that $(Tg)^m/T_{i_1} > (\tilde{N}g)^m/\tilde{N}_j > (Tg)^m/T_{i_2}$.

LEMMA 2.5. *There exists $\gamma > 0$ such that if $g \in \mathcal{S}$, $x \in \operatorname{Per}(g)$ and (24), (25) are the canonical splittings at x, then*

$$\alpha\big((T_1 \oplus \cdots \oplus T_i)^\perp, (T_{i+1} \oplus \cdots \oplus T_k)\big) \leqslant \gamma,$$

$$\alpha\big((\tilde{N}_1 \oplus \cdots \oplus \tilde{N}_j)^\perp, (\tilde{N}_{j+1} \oplus \cdots \oplus \tilde{N}_l)\big) \leqslant \gamma$$

for all $1 \leqslant i < l$, $1 \leqslant j < l$.

PROOF. We shall prove only the first inequality. The second one is proved applying the same method. From (14) and (15) it follows that there exists $k > 0$ such that

$$\|L_1 - L_2\| \geqslant k \tag{26}$$

for all $g \in \mathfrak{U}$, $x \in V$, $L_1 \in E_x^s(g)$, $L_2 \in E_x^u(g)$, $\|L_1\| = \|L_2\| = 1$. Suppose that, contradicting the lemma, we can find for $\varepsilon = k/3$ a diffeomorphism $g \in \delta$ and $x \in \operatorname{Per}(g)$ with canonical splittings (24), (25) such that $\alpha(T_-^{\perp}, T_+) \leqslant \varepsilon$, where $T_+ = T_{l+1} \oplus \cdots \oplus T_k$, $T_- = T_1 \oplus \cdots \oplus T_i$. Suppose, for instance, that $\dim T_-^{\perp} \leqslant \dim T_+$. Let $T' \subset T_+$, $T'' \subset T_+$ be subspaces such that $\dim T' = \dim T_-$, $T'' = (T' \oplus T_-)^{\perp}$. Take an isometry G: $T_x V \hookleftarrow$ such that $\|G - I\| \leqslant 2\varepsilon$, $G(T'') = T''$, $G(T') = T_-$. Let $1 \leqslant j \leqslant l$ satisfy $(Tg)^m/T_- > (\tilde{N}g)^m \tilde{N}_j > (Tg)^m/T_+$ and L: $T_x V \to \tilde{N}_j$ be a linear map with $\|L\| = 1$, $L/T_+ = 0$, $LG/T' \neq 0$. Then $L \in E_x^u(g)$, $LG \in E_x^s(g)$ and $\|L - G\| \leqslant \|L\| \cdot \|I - G\| \leqslant 2\varepsilon = \frac{2}{3}k$, contradicting (26).

LEMMA 2.6. *There exist $C' > 0$, $0 < \lambda < 1$ such that if $g \in \mathbb{S}$, $x \in \operatorname{Per}(g)$ and* (24), (25) *are the canonical splittings at x, then for all $1 \leqslant j \leqslant l$, $1 \leqslant i \leqslant l$ we have*

$$\|(\tilde{N}g)^n/\tilde{N}_j\| \cdot \|(Tg)^{-n}/(Tg)^n T_i\| \leqslant C'\lambda^n$$

for all $n \in \mathbf{Z}^+$ or

$$\|(\tilde{N}g)^{-n}/(\tilde{N}g)^n \tilde{N}_j\| \cdot \|(Tg)^n/T_i\| \leqslant C'\lambda^n$$

for all $n \in \mathbf{Z}^+$.

PROOF. Suppose that $(\tilde{N}g)^m/\tilde{N}_j < (Tg)^m/T_i$ for all $i \leqslant i_0$ and $(Tg)^m/T_{i'} < (\tilde{N}g)^m N_j$ if $i' > i_0$. We claim that

$$\|(\tilde{N}g)^n/\tilde{N}_j\| \cdot \|(Tg)^{-n}/(Tg)^n T_-\| \leqslant C'\lambda^n, \tag{27'}$$

$$\|(\tilde{N}g)^{-n}/(\tilde{N}g)^n N_j\| \cdot \|(Tg)^n/T_+\| \leqslant C'\lambda^n \tag{27''}$$

for all $n \in \mathbf{Z}^+$, where $T^+ = T_1 \oplus \cdots \oplus T_{i_0}$, $T^- = T_{i_0+1} \oplus \cdots \oplus T_k$, where $C' > 0$ is a constant independent of $g \in \mathbb{S}$, $x \in \operatorname{Per}(g)$ and $n \in \mathbf{Z}^+$ and λ satisfies (14) and (15). If we prove this claim we are done. Let us prove (27′) ((27″) follows in a similar way). Take $v \in \tilde{N}_j$, $w \in T'$ with $\|w\| = 1$. Let $\mathfrak{B}$ be an orthonormal basis of T_- containing w. Define a linear map L: $T^- \to \tilde{N}_j$ by $Lu = v$ if $u \in \mathfrak{B}$ and $Lu = 0$ if $u \in T_+$. There exists a constant $\gamma' > 0$ depending only on γ such that $\|L\| \leqslant \gamma' \|L/T^-\|$. Hence $\|L\| \leqslant \gamma' \|v\|$. Clearly $L \in E_x^s(g)$. Therefore, by (14),

$$\|(\tilde{N}g)^n v\| / \|(\tilde{N}g)^n w\| \leqslant \|\Phi_g^n(L)\| \leqslant C\lambda^n \|L\| \leqslant C\gamma'\lambda^n \|v\|. \tag{28}$$

Since

$$\|(\tilde{N}g)^n/\tilde{N}_j\|\cdot\|(Tg)^{-n}/(Tg)^nT^-\|$$

$$= \sup\left\{\frac{\|(\tilde{N}g)^n v\|}{\|v\|\cdot\|(Tg)^n w\|}\,\middle|\, v\in\tilde{N}_j, w\in T^-, \|w\|=1\right\},$$

putting $C' = C\gamma'$, (27') follows from (28).

Finally we want to prove that if $g \in \mathfrak{S}$ and x is a periodic point of g/V with period m there exists a subspace $N_x \subset T_xM$ that is a $(Tg)^m$-invariant complement of T_xV, and we shall give an upper bound for $\alpha((T_xV)^\perp, N_x)$. For this let us say that if $g \in \mathfrak{S}$ and $x \in \mathrm{Per}(g)$, the multindex $\sigma = (n_1, \dots, n_k, m_1, \dots, m_l)$ is the signature of x if the canonical splittings at x are (24) and (25) and satisfy $\dim \tilde{N}_i = n_i$, $\dim T_j = m_j$ for all $1 \leqslant i \leqslant k$, $1 \leqslant j \leqslant l$. Let $\Lambda_\sigma(g)$ be the set of periodic points of g/V with signature σ and let $\overline{\Lambda}_\sigma(g)$ be its closure. By Proposition 1.2 and Lemma 2.6 there exist splittings $\tilde{N}V/\overline{\Lambda}_\sigma(g) = \tilde{N}_1 \oplus \cdots \oplus \tilde{N}_l$, $TV/\overline{\Lambda}_\sigma(g) = T_1 \oplus \cdots \oplus T_k$ satisfying

$$\tilde{N}g/\tilde{N}_{i'} > \tilde{N}g/\tilde{N}_{i''}, \qquad Tg/T_{j'} > Tg/T_{j''}$$

for all $1 \leqslant i' \leqslant i'' \leqslant l$, $1 \leqslant j' < j'' \leqslant k$ and $\tilde{N}g/\tilde{N}_i > Tg/T_j$ or $Tg/T_j > \tilde{N}g/\tilde{N}_i$ for all $1 \leqslant i \leqslant l$, $1 \leqslant j \leqslant k$. By Proposition 1.1 there exist continuous Tf-invariant subbundles $N_1, \dots, N_l$ of $TM/\overline{\Lambda}_\sigma(g)$ such that

$$T_1 \oplus \cdots \oplus T_k \oplus N_1 \oplus \cdots \oplus N_l = TM/\overline{\Lambda}_\sigma(g).$$

By Lemma 2.6 and Proposition 1.1 there exists $\gamma_1 > 0$ depending on γ (given by Lemma 2.5), C' and λ (given by 2.6) such that

$$\alpha\big((T_xV)^\perp, (N_1 \oplus \cdots \oplus N_l)_x\big) \leqslant \gamma_1$$

for all $x \in \overline{\Lambda}_\sigma(g)$. Resuming we have proved the following property:

LEMMA 2.7. *There exists $\gamma_1 > 0$ such that for all $g \in \mathfrak{S}$ and every signature σ, there exists a continuous Tf-invariant subbundle N of $TM/\overline{\Lambda}_\sigma(g)$ such that*

$$TV/\overline{\Lambda}_\sigma(g) \oplus N = TM/\overline{\Lambda}_\sigma(g), \quad \alpha\big((T_xV)^\perp, N_x\big) \leqslant \gamma_1$$

for all $x \in \overline{\Lambda}_\sigma(g)$.

Let us return now to the situation we were considering before Lemma 2.5: we have $g_0 \in \mathfrak{S}$ near to f such that $\mathrm{Per}(g_0)$ contains a subset W, open in the relative topology in $\mathrm{Per}(g_0)$ such that $\tilde{N}_x^s(g_0) \cap \tilde{N}_x^u(g_0) \neq \{0\}$ for all $x \in W$. Since the set of possible signatures is finite, for some σ the set $W \cap \Lambda_\sigma(g)$ contains an open subset W_0 of $\mathrm{Per}(g_0)$. Let γ_1 be the constant given in Lemma 2.7. Take $\beta > 0$ and $\rho > 0$ satisfying the following property: If $x \in V$, $S \subset T_xM$ is a suspace satisfying $S \oplus T_xV = T_xM$, $\alpha((T_xV)^\perp, S) \leqslant 2\gamma_1$, $v \in S$, $y \in V$, $w \in N_yV$ and $\|v\| \leqslant \rho$, $\|w\| \leqslant \rho$ and $\exp(v) = \exp(w)$:

$$\|x - y\| \leqslant \beta\|v\|, \qquad \|v\| \leqslant \beta\|w\|.$$

Let $\delta = \min(\rho, \frac{2}{3}\beta)$. Since g_0 is near to f we can suppose that $g_0 \in \mathfrak{U}_{\delta/2}$

(see Definition 2.1). Using g_0 we shall construct $\bar{g} \in \mathfrak{U}_{2\delta/3}$ such that $V_{\bar{g}} = \{\exp h(p)/p \in V\}$ where h is a section of NV with $\mathrm{Lip}(h) \geqslant \delta$, contradicting $\bar{g} \in \mathfrak{U}_{2\delta/3}$. For this take the subbundle N of $TM/\overline{\Lambda}_\sigma(g_0)$ given by Lemma 2.7. There exist a neighborhood B of $\overline{\Lambda}_\sigma(g)$ in M, a C^∞ subbundle N' of TM/B satisfying $N'/\overline{\Lambda}_\sigma(g_0) \oplus TV/\overline{\Lambda}_\sigma(g_0) = TM/\overline{\Lambda}_\sigma(g_0)$, $\alpha((T_xV)^\perp, N'_x) \leqslant 2\gamma_2$ for all $x \in \overline{\Lambda}_\sigma(g_0)$ and a diffeomorphism $g \in \mathfrak{U}_{3\delta/5}$ such that $g \in \mathbb{S}$, $g/V = g_0$, $N'/\overline{\Lambda}_\sigma(g_0)$ is Tg-invariant, $\tilde{N}g = \tilde{N}g_0$ and $g(\exp v) = \exp((Tg)v)$ for all $v \in N'/\overline{\Lambda}_\sigma(g_0)$ small enough, say with $\|v\| \leqslant r$. Denote $\Lambda = \overline{\Lambda}_\sigma(g_0)$. We start the construction of $\bar{g}$ proving the following property:

LEMMA 2.8. *For all* $0 \neq v \in N'/\Lambda$ *the set* $\{\|(Tg)^n v\| \mid n \in \mathbf{Z}\}$ *is unbounded.*

PROOF. Suppose that $0 \neq v \in N'/\Lambda$ and $\{\|(Tg)^n v\| \mid n \in \mathbf{Z}\}$ is bounded, say by K. Take $w = \lambda v$ such that $K\lambda \leqslant r/2$. Then $d(g^n(\exp w), V) = d(\exp(Tg)^n w, V) \leqslant K\lambda$ for all $n \in \mathbf{Z}^+$, hence $g^n(\exp w) \in \bigcap_{n \in \mathbf{Z}} g^n(U)$ if λ is small enough (where U is given by Definition 2.1). Therefore $V_g \supset \{w\} \cup V$, contradicting Definition 2.1.

By the results in **[13]** or **[21]** this lemma implies that there exists a continuous Tf-invariant splitting $N'/\Lambda = N^s \oplus N^u$ and constants $C > 0$, $0 < \sigma < 1$ such that

$$\|(Tg)^n/N_x^s\| \leqslant C\sigma^n, \qquad \|(Tg)^{-n}/N_x^u\| \leqslant C\sigma^n$$

for all $n \in \mathbf{Z}^+$. Take $x_0 \in W_0$. The condition $\tilde{N}_{x_0}^s(g) \cap \tilde{N}_{x_0}^u(g) \neq \{0\}$ implies that there exists an eigenvalue λ of $(\tilde{N}g)^m/\tilde{N}_{x_0}V$ (m being the period of x_0) and eigenvalues λ_1, λ_2 of $(Tg)^m/T_xV$ such that $|\lambda_1| < |\lambda| < |\lambda_2|$. Otherwise $E_x^s(g)$ (resp. $E_x^u(g)$) would be the space of linear maps from T_xV in the subspace $\tilde{N}^+$ ($\tilde{N}^-$) of $\tilde{N}_xV$ spanned by the invariant subspaces associated to eigenvalues with modulus smaller (greater) than the modulus of every eigenvalue of $(Tg)^m/T_xV$; hence $\tilde{N}_{x_0}^s(g) = \tilde{N}^+$, $\tilde{N}_{x_0}^u(g) = \tilde{N}^-$. Since every period of g is hyperbolic (because $g \in \mathbb{S}$) there exist two possible cases: $|\lambda_1| < |\lambda| < |\lambda_2|$ and $1 < |\lambda| < |\lambda_2|$. Consider the case $|\lambda_1| < |\lambda| < 1$. The other case is handled with the same method. Let $N_0 \subset N'_{x_0}$ be the subspace associated to the eigenvalue λ, and $T_0 \subset T_{x_0}V$ be the subspace associated with λ_1. There exists an embedded disc $D \subset V$ containing x_0 and satisfying $T_{x_0}D = T_0$, $g(D) \subset D$. We have two possible situations: either every neighborhood of x_0 in D contains a wandering point of g_0/V or it does not. Suppose that we have the second situation. Then $\Lambda \cap D$ is a neighborhood of x_0 in D. Take $y \in \Lambda \cap D$. By **[9]** there exists a continuous subbundle N'_0 of N'/D such that $(Tg)^m N'_0 \subset N'_0$ and $N'_{0,x_0} = N_0$. The properties of D imply $\lim\|(Tg)^n/N'_{0,x_0}\| = 0$ for all $x \in D$. Hence $N'_0 \subset N^s/D$. Take a C^∞ section ξ of TM such that ξ/B is a section of N', the support of ξ is contained in a neighborhood $U_0 \subset \Lambda$ of a point $y \in D$ satisfying $x_0 \notin U_0$, $\xi(y) \neq 0$, $\bigcup_1^k g^j(U_0) \cap U_0 = \varnothing$, where k satisfies $\sigma^{k+1} r \leqslant 2\|\xi(y)\|$. Define a section η of $N'V$ by

$$\eta(x) = \sum_{0}^{\infty} (Tg)^n \xi(g^{-n}(x)).$$

For all $x \in \Lambda$ we have

$$\|\eta(x)\| \leqslant \sum_{0}^{\infty} \|(Tg)^n \xi(g^{-n}(x))\| \leqslant C \sum_{0}^{\infty} \sigma^n \|((Tg)^{-n}(x))\|.$$

Suppose that $\|\xi(p)\| \leqslant r(1-\sigma)/2C$ for all $p \in V$. Then

$$\|\eta(x)\| \leqslant r/2 \tag{29}$$

for all $x \in V$. The condition $g^j(U_0) \cap U_0 = \varnothing$ for all $1 \leqslant j \leqslant k$ implies

$$\begin{aligned} \|\eta(y)\| &\geqslant \|\xi(y)\| - \sum_{k+1}^{\infty} \|(Tg)^n \xi(g^{-n}(y))\| \\ &\geqslant \|\xi(y)\| - C\sigma^{k+1}/(1-\sigma) \cdot r(1-\sigma)/2C > 0 \end{aligned} \tag{30}$$

and it follows from the definition of η that

$$(Tg)\eta(x) + \xi(g(x)) = \eta(g(x)) \tag{31}$$

for all $x \in \Lambda$. Given $\mu \in R$ take $\bar{g}$ such that

$$\bar{g} \in \mathfrak{U}_\delta, \tag{32}$$

$$\bar{g}(\exp v) = \exp(\mu\xi(g(x)) + (Tg)v) \tag{33}$$

for all $x \in \Lambda$, $v \in N_x'$ with $\|v\| \leqslant \mu r/2$. If $|\mu|$ is small there exists $\bar{g}$ satisfying (32) and (33). By (33) and (31) we have

$$\bar{g}(\exp \mu\eta(x)) = \exp(\mu\xi(g(x)) + (Tg)\eta(x)) = \exp \mu\eta(g(x)).$$

Hence,

$$\bar{g}^n(\exp \mu\eta(x)) = \exp \mu\eta(g^n(x)) \tag{34}$$

for all $n \in \mathbf{Z}$, $x \in \Lambda$. This and (29) imply that the distance between $\bar{g}^n(\exp \mu\eta(x))$ and V is $\leqslant \mu r/2$ for all $n \in \mathbf{Z}$. Therefore, by Definition 2.1, this proves that $\exp \mu\eta(x) \in V_{\bar{g}}$ for all $x \in \Lambda$. Now take $|\lambda_1| < c_1 < c < |\lambda|$ and $K > 0$ satisfying $\|(Tg)^n v\| \geqslant Kc^n\|v\|$ for all $n \in \mathbf{Z}^+$, $v \in N_{0,y}'$ and $\|g^n(y) - g^n(x_0)\| \leqslant Kc_1^n$ for all $n \in \mathbf{Z}^+$. By the condition $g^n(y) \notin U_0$ for all $n \geqslant 1$ and (30),

$$\eta(g^n(y)) = (Tg)^n \eta(y).$$

Hence

$$\frac{\|\eta(g^n(y))\|}{\|g^n(y) - g^n(x_0)\|} = \frac{\|(Tg)^n \eta(y)\|}{\|g^n(y) - g^n(x_0)\|} \geqslant \left(\frac{c}{c_1}\right)^n \|\eta(y)\|.$$

Now let $h \in \Gamma_{\mathcal{C}}(NV)$ such that $V_{\bar{g}} = \{\exp h(p)/p \in V\}$ and $\|h\|_{\mathcal{C}} \leqslant \delta$. We claim that $\exp \mu\eta(y) \in V_{\bar{g}}$. This follows from (34), which gives $\bar{g}^n(\exp \mu\eta(y)) = \exp \mu\eta(g^n(y))$ for all $n \in \mathbf{Z}$. Hence if μ is small enough, $\bar{g}^n(\exp \mu\eta(y)) \in U$ (U given by Definition 2.1); thus $\exp \mu\eta(y) \in V_{\bar{g}}$. Take $z_n \in V$ such that

$\exp h(z_n) = \bar{g}^n(\exp \mu\eta(y)) = \exp \mu\eta(g^n(y))$. Observing that the g-orbit and $\bar{g}$-orbit of x_0 are the same we obtain $h(g^n(x_0)) = 0$ for all $n \in \mathbf{Z}$. Therefore,

$$\operatorname{Lip}(\eta) \geqslant \frac{\|h(z_n) - h(g^n(x_0))\|}{\|z_n - g^n(x_0)\|} = \frac{\|h(z_n)\|}{\|z_n - g_n(x_0)\|}$$

$$\geqslant \frac{1}{\|z_n - g^n(y)\|/\|h(z_n)\| + \|g^n(y) - g^n(x_0)\|/\|h(z_n)\|}$$

$$\geqslant \frac{1}{\beta(1 + \|g^n(y) - g^n(x_0)\|/\|h(z_n)\|)} \geqslant \frac{1}{\beta\big(1 + (c_1/c)^n/\mu\|\eta(y)\|\big)}.$$

Since by (30) $\|\eta(y)\| > 0$, taking $n \to +\infty$ we obtain $\operatorname{Lip}(\eta) > 1/2\beta = \delta$, contradicting (32). This completes the proof when x_0 is in the closure of $\operatorname{Per}(g_0) - \{x_0\}$.

It remains to consider the case when $\Omega(g/V) \cap D$ is not a neighborhood of x_0 in D. This case is considered in the next lemma.

LEMMA 2.8. *There exists a neighborhood* $\mathcal{V}$ *of* f *such that if* $g \in \mathcal{V} \cap \mathcal{S}$, $x \in \operatorname{Per}(g)$ *and* m *is the period of* x, *then for every eigenvalue* λ_1 *of* $(Tg)^m/T_xV$ *and every eigenvalue* λ *of* $(\tilde{N}g)^m/N_xV$ *if we have* $|\lambda_1| < |\lambda| < 1$ $(|\lambda_1| > |\lambda| > 1)$, *then every embedded disc* $D \subset V$ *such that* $g^m(D) \subset D$ $(g^{-m}(D) \subset D)$ *and* T_xD *is the subspace of* T_xV *associated to* λ_1, *is contained in* $\Omega(g/V)$.

PROOF. If the lemma is false we can find $g \in \mathcal{U}_\delta$, where $\delta = 1/2\beta$ (β and ρ satisfying the same properties as in the previous proof), and $x \in \operatorname{Per}(g)$ with an eigenvalue λ_1 of $(Tg)^m/T_xV$ (m the period of x) and an eigenvalue λ of $(\tilde{N}g)^m/\tilde{N}_xV$ such that, for instance, $|\lambda_1| < |\lambda| < 1$ and there exists an embedded disc $D \subset V$ satisfying $g^m(D)$, T_xD is the subspace associated to the eigenvalue λ_1 and $D - \Omega(g/V) \neq \varnothing$. Take $y \in D - \Omega(g/V)$, a neighborhood U_0 of y satisfying $g^n(U_0 \cap V) \cap U_0 \neq \varnothing$ for all $n \in \mathbf{Z}$ and a disc $D_0 \subset M$ containing y such that $\lim_{n\to+\infty}\|g^n(z) - g^n(y)\| = 0$ for all $z \in D_0$ and there exist $k > 0$, $|\lambda_1| < c < |\lambda|$ satisfying

$$\|g^n(z) - g^n(y)\| \geqslant Kc^n\|z - y\| \tag{35}$$

for all $n \in \mathbf{Z}$, $z \in D_0$. Moreover, if $|\lambda_1| < c_1 < c$ there exist $K_1 > 0$, such that

$$\|g^n(z) - g^n(x)\| \leqslant K_1c_1^n\|z - x\| \tag{36}$$

for all $z \in D$, $n \in \mathbf{Z}^+$. Choose $\bar{g} \in \mathcal{U}_\delta$ with $\bar{g}(g^{-1}(y)) = \bar{y} \in D_0 - V, \bar{y} \neq y$ and $g(p) = \bar{g}(p)$ for all $p \notin g^{-1}(U_0)$.

The condition $g^n(U_0) \cap U_0 = \varnothing$ for all $n \in \mathbf{Z}$ implies that $\{g^n(\bar{y}) \mid n \in \mathbf{Z}^+\} \cup \{g^{-n}(g^{-1}(y)) \mid n \in \mathbf{Z}^+\}$ is a $\bar{g}$-orbit. Hence $\bar{y} \in V_{\bar{g}}$. Let $h \in \Gamma_{\mathcal{E}}(NV)$ satisfy $V_{\bar{g}} = \{\exp h(p) \mid p \in V\}$. Let $z_n \in V$ be such that $h(z_n) = g^n(\bar{y})$. Using (35), (36) as in the previous proof, we conclude $\|h\|_{\mathcal{E}} > 1/2\beta$, contradicting $g \in \mathcal{U}_\delta$.

PROOF OF PROPOSITION 2.1. Define N_x^sV (N_x^uV) as the set of $v \in T_xM$ such that $\lim\|(Tf)^n V\| \cdot \|(Tf)^{-n}/T_{f^n(x)}V\| = 0$ $(\lim\|(Tf)^{-n}v\| \cdot \|(Tf)^n/T_{f^{-n}(x)}V\| = 0)$. Observe that when $x \in \Omega(f/V)$, $N_x^s V$, $N_x^u V$ are the fibers at x of the subbundles N^s, N^u of $TM/\Omega(f/V)$ given by Lemma 2.3. We want to prove that the maps $x \to N_x^s V$, $x \to N_x^u V$ define continuous subbundles of TM/V satisfying all the conditions required by the definition of normal hyperbolicity. Clearly $(Tf)N_x^s V = N_{f(x)}^s V$, $(Tf)N_x^u V = N_{f(x)}^u V$ for all $x \in V$. Let us prove that for all $x \in V$, $N_x^s V \cap N_x^u V = \{0\}$. Let π: $TM/V \to \tilde{N}V$ be the canonical projection. Let $v \in N_x^s V \cap N_x^u V$ and $\tilde{v} = \pi v$. If $0 \neq v$ we can take a linear map $0 \neq L$: $T_x V \to \{\lambda v \mid \lambda \in \mathbf{R}\}$. Then $L \in E_x^s(f) \cap E_x^u(f)$, contradicting Lemma 2.3. Now let us show that $T_x M = T_x V \oplus N_x^s V \oplus N_x^u V$ for all $x \in V$. We start with the following proposition:

Lemma 2.9. *Let* π_0: $TM/V \to TV$, π: $TM/V \to NV$ *be the projections associated to the splitting* $TM/V = TV \oplus NV$. *There exists* $\delta > 0$ *such that if* $x \in V$, $v \in T_x M$ *and* $\|\pi(Tf)^n v\| \leqslant \delta\|\pi_0(Tf)^n v\|$ *for all* $n \in \mathbf{Z}^+$ *then* $v \in N_x^s V \oplus T_x V$.

Proof. By Proposition 1.3 there exist a neighborhood U of $\Omega(f/V)$ and a continuous subbundle $\hat{N}^s$ of TV/Λ^s, where $\Lambda^s = \bigcap_{n \geqslant 0} f^{-n}(U)$, satisfying

$$(Tf)\hat{N}^s \subset \hat{N}^s, \tag{37}$$

$$\hat{N}^s/\Omega(f/V) = N^s. \tag{38}$$

From these relations it follows that $\hat{N}_x^s \subset N_x^s V$ for all $x \in V$ (in fact it is possible to prove $\hat{N}_x^s = N_x^s V$, but we shall not need this property). Moreover, we can assume that there exists a continuous subbundle $\hat{N}^u$ of TM/Λ^s such that $\hat{N}^u/\Omega(f/V) = N^u$ and $\hat{N}^u \oplus \hat{N}^s \oplus TV/\Lambda^s = TM/\Lambda^s$. Let π^s: $TM/\Lambda^s \to \hat{N}^s$, π^u: $TM/\Lambda^s \to \hat{N}^u$, π^c: $TM/\Lambda^s \to TV/\Lambda^s$ be the projections associated to this splitting. Define $\pi^{cs} = \pi^c + \pi^s$ and for $x \in \Lambda^s$, $\varepsilon > 0$ define the cones

$$S_\varepsilon(x) = \{v \in T_x M / \|\pi^{cs} v\| \leqslant \varepsilon\|\pi^u v\|\}.$$

There exist $m \in \mathbf{Z}^+$, $\varepsilon > 0$ satisfying

$$(Tf)S_\varepsilon(x) \subset S_{\varepsilon/2}(f(x)),$$

$$\|(Tf)^m v\|/\|(Tf)^m w\| \geqslant 2\|v\|/\|w\|$$

for all $x \in \Omega(f/V)$, $v \in S_\varepsilon(x)$, $0 \neq w \in N_x^s \oplus TV$. Let $U_0 \subset U$ be a neighborhood of $\Omega(f/V)$ such that if $\Lambda_0^s = \bigcap_{n > 0} f^{-n}(U_0)$ then these relations remain true for all $x \in \Lambda_0^s$, $v \in S_\varepsilon(x)$, $0 \neq w \in N_x^s \oplus T_x V$. From (37) and (38) it follows that there exist $C > 0$, $0 < \sigma < 1$ satisfying

$$\|(Tf)^n v\|/\|(Tf)^n w\| \geqslant C(1/\sigma)^n\|v\|/\|w\| \tag{39}$$

for all $x \in \Lambda_0^s$, $v \in S_\varepsilon(x)$, $0 \neq w \in \hat{N}_x^s \oplus T_x V$. Take $\delta > 0$ such that $v \notin S_\varepsilon(x)$ for all $x \in V$ and $v \in T_x M$ such that $\|\pi v\| \leqslant \delta\|\pi_0 v\|$. Now suppose that $y \in V$, $v \in T_y M$ satisfy $\|\pi(Tf)^n v\| \leqslant \delta\|\pi_0(Tf)^n v\|$ for all $n \in \mathbf{Z}^+$. We want to prove $v \in N_y^s V \oplus T_y V$. Suppose $v \notin N_y^s V \oplus T_y V$. Hence if $f^{n_1}(y) = x \in \Lambda_0^s$ we have $(Tf)^{n_1} v \notin N_x^s V \oplus T_x V$. Write $(Tf)^{n_1} v = v' + v''$ where $v' \in \hat{N}_x^u$, $v'' \in N_x^s \oplus T_x V$, $v' \neq 0$. By (39) we have

$$\|(Tf)^n v'\| / \|(Tf)^n v''\| \geqslant C(1/\sigma)^n \|v'\| / \|v''\| \tag{40}$$

for all $n \in \mathbf{Z}^+$. Then

$$\|\pi^{cs}(Tf)^{n+n_1} v\| = \|\pi^{cs}(Tf)^n (v' + v'')\| \leqslant \|\pi^{cs}(Tf)^n v'\| + \|(Tf)^n v''\|.$$

Since $v' \in \hat{N}_x^u \subset S_\varepsilon$, we obtain

$$\|\pi^{cs}(Tf)^n v'\| \leqslant \varepsilon \|\pi^u (Tf)^n v'\| = \varepsilon \|\pi^u (Tf)^n (v' + v'')\| = \varepsilon \|\pi^u (Tf)^{n+n_1} v\|;$$

hence,

$$\|\pi^{cs}(Tf)^{n+n_1} v\| \leqslant \left[\varepsilon + \frac{\|(Tf)^n v''\|}{\|\pi^u (Tf)^{n+n_1} v\|} \right] \|\pi^u (Tf)^{n+n_1} v\|.$$

This inequality, together with (40), gives

$$\|\pi^{cs}(Tf)^{n+n_1} v\| \leqslant \left(\varepsilon + \frac{\sigma^n \|v''\|}{C \|v'\|} \right) \|\pi^u (Tf)^{n+n_1} v\|.$$

Hence $(Tf)^{n+n_1} v \in S_{2\varepsilon}(x)$ for large values of n, thus contradicting the definition of δ.

Using this lemma we shall prove

$$N_x^s V + N_x^u V + T_x V = T_x M$$

for all $x \in V$. Together with the property $N_x^s V \cap N_x^u V = \{0\}$, this relation proves

$$N_x^s V \oplus N_x^u V \oplus T_x V = T_x M$$

for all $x \in V$. Suppose that for some x we have $N_x^s V + N_x^u V + T_x V \neq T_x M$. By Lemma 2.4, $x \notin \Omega(f/V)$. Let $\delta > 0$ be the number given by Lemma 2.9 and $\mathfrak{U}_\delta$ the neighborhood of f given by Definition 2.1. Let W be a neighborhood of x such that $f^n(W) \cap W = \varnothing$ for all $n \in \mathbf{Z}$ and let $g \in \mathfrak{U}_\delta$ be a diffeomorphism satisfying $g(y) = f(y)$ for all y in a neighborhood of $V - W$, $g(x) = f(x)$, $g(f^{-1}(x)) = x$ and

$$\dim\big((Tg)(N_{f^{-1}(x)}^u V \oplus T_{f^{-1}(x)} V) \cap (N_x^s V \oplus T_x V)\big) < \dim V; \tag{41}$$

let $h \in T_{\mathfrak{L}}(NV)$ such that $V_g = \{\exp h(p) | p \in V\}$. Since $h \in \Gamma_{\mathfrak{L}}$, given any basis $\{v_1, \ldots, v_l\}$ of $T_x V$, there exist $\{w_1, \ldots, w_l\} \subset N_x V$ such that

$$\liminf_{\lambda \to 0} \left\| \frac{h(\exp \lambda v_i) - h(x)}{\lambda} - w_i \right\| = 0$$

for $1 \leqslant i \leqslant l$, where exp: $TV \to V$ is the exponential mapping of the Riemann manifold V. The condition $\|h\|_{\mathfrak{L}} \leqslant \delta$ implies $\|w_i\| / \|v_i\| \leqslant \delta$ for all $1 \leqslant i \leqslant l$. For all $n \in \mathbf{Z}^+$, by the invariance of V_g, we have

$$\liminf_{\lambda\to 0}\left\|\frac{h\big(\exp\lambda\pi_0(Tg)^n(v_i+w_i)\big)-h\big(g^n(x)\big)}{\lambda}-\pi(Tf)^n(v_i+w_i)\right\|=0.$$

Since $\|h\|_{\mathcal{C}} \leqslant \delta$ we obtain

$$\|\pi(Tg)^n(v_i+w_i)\| \leqslant \delta\|\pi_0(Tg)^n(v_i+w_i)\|.$$

But g satisfies $(Tg)^n/(T_{f(x)}M) = (Tf)^n$ and $(Tg)^{-n}/T_xM = (Tf)^{-n}/T_xM$ for all $n \in \mathbf{Z}^+$. Therefore, by Lemma 2.9,

$$\{(Tg)(v_i+w_i) \mid i = 1,\ldots,l\} \subset N^s_{f(x)}V \oplus T_{f(x)}V,$$

$$\{v_i+w_i \mid i = 1,\ldots,l\} \subset N^u_xV \oplus T_xV.$$

Let S be the space spanned by $\{v_i+w_i \mid i = 1,\ldots,l\}$. Then

$$\dim((Tg)(N^u_xV \oplus T_xV) \cap (N^s_xV \oplus T_xV)) \geqslant \dim S = \dim V,$$

contradicting (41).

Now we know that $N^s_xV \oplus N^u_xV \oplus T_xV = T_xM$ for all $x \in V$. Our next step is to show that the maps $x \to N^s_xV$, $x \to N^u_xV$ define continuous subbundles of TM/V. For this define, as before, G^s_x (G^u_x) as the set of linear maps L: $T_xV \to \pi(N^s_xV)$ (L: $T_xV \to \pi(N^u_xV)$). It is clear that

$$G^s_x \oplus G^u_x = L_x(V)$$

for all $x \in V$ and

$$G^s_x = E^s_x(f), \tag{42'}$$

$$G^u_x = E^u_x(f). \tag{42''}$$

Then Φ: $L_x(V) \hookleftarrow$ (defined as in the proof of Lemma 2.3) is a hyperbolic vector bundle isomorphism and the maps $x \to E^s_x(f)$, $x \to E^u_x(f)$ define continuous subbundles of $L_x(V)$ [**13**, Proposition 1.1]. From this and (42′), (42″) it follows that the maps $x \to \pi(N^s_xV)$, $x \to \pi(N^u_xV)$ define a continuous $\tilde{N}f$-invariant splitting $\tilde{N}V = \tilde{N}^sV \oplus \tilde{N}^uV$ satisfying

$$\tilde{N}f/\tilde{N}^sV > Tf/TV > \tilde{N}f/N^uV.$$

From Proposition 1.2 it follows that there exist continuous subbundles N^sV, N^uV invariant under Tf and satisfying $\pi(N^sV) = \tilde{N}^sV$, $\pi(N^uV) = \tilde{N}^uV$ and

$$Tf/N^sV > Tf/TV > Tf/N^sV. \tag{43}$$

From this it follows easily that the fibers at $x \in V$ of N^sV and N^uV are the spaces N^s_xV, N^u_xV defined before.

It remains to prove that there exist constants $C > 0$, $0 < \lambda < 1$ satisfying

$$\|(Tf)^n/N^s_xV\| \leqslant C\lambda^n, \tag{44}$$

$$\|(Tf)^{-n}/N^u_xV\| \leqslant C\lambda^n \tag{45}$$

for all $n \in \mathbf{Z}^+$, $x \in V$. Let us prove (45). If we show that $\{\|(Tf)^nv\| \mid n \in \mathbf{Z}^+\}$ is unbounded for all $0 \neq v \in N^u_xV$, $x \in \Omega(f/V)$ it follows [**13**, Proposition 1.1] that there exist $C_1 > 0$, $0 < \lambda_1 < 1$ satisfying $\|(Tf)^{-n}/N_x^uV\| \leqslant C_1\lambda_1^n$ for all $x \in \Omega(f/V)$, $n \in \mathbf{Z}^+$. Then it is easy to see that $\{\|(Tf)^n v\| \mid n \in \mathbf{Z}^+\}$ is unbounded for all $0 \neq v \in N^uV$. Applying again [**13**, Proposition 1.1], there exist $C > 0$, $0 < \lambda < 1$ satisfying (45). Hence let us suppose that for some $x \in \Omega(f/V)$ there exists $0 \neq v \in N_x^uV$ such that for some $K > 0$, $\|(Tf)^n v\| \leqslant K\|v\|$ for all $n \in \mathbf{Z}^+$. It is not difficult to show that there exist $C_1 > 0$, $0 < \lambda_1 < 1$, and $r > 0$ satisfying

$$\|f^n(x_1) - f^n(x_2)\|/\|(Tf)^n w\| \leqslant C_1\lambda_1^n\|x_1 - x_2\|/\|w\|$$

for all $x_i \in V$, $i = 1, 2$, $0 \neq w \in N_{x_1}^uV$ and $n \in \mathbf{Z}^+$ such that $\|x_1 - x_2\| < \delta(n)$, where $\delta(n)$ is so small that $\|f^j(x) - f^j(x_2)\| \leqslant r$ for all $0 \leqslant j \leqslant n$, if x, x_2 satisfy $\|x_1 - x_2\| < \delta(n)$. Then if $y \in V$ and $\|y - x\| \leqslant \delta(n)$, we conclude that

$$\|f^n(y) - f^n(x)\| \leqslant C_1\lambda_1^n \frac{\|y - x\|}{\|v\|} \cdot \|(Tf)^n v\| \leqslant KC_1\lambda_1^n\|y - x\|.$$

Take $\delta(1) > r_1 > 0$ satisfying $KC_1\lambda_1^n r_1 \leqslant \delta(1)$ for all $n \in \mathbf{Z}^+$. Defining $\bar{K} = KC_1$, we obtain

$$\|f^n(y) - f^n(x)\| \leqslant \bar{K}\lambda_1^n\|y - x\|$$

for all $n \in \mathbf{Z}^+$ and $y \in V$ such that $\|y - x\| \leqslant r_1$. Since x is nonwandering this implies that x is periodic and there exists an eigenvalue λ of $(Tf)^m/N_x^uV$ (m the period of x) with $|\lambda| \leqslant 1$. Since $Tf/N^uV > Tf/TV$ it follows that $(Tf)^m/N_x^uV > (Tf)^m/T_xV$ and this proves that $|\mu| < |\lambda|$ for every eigenvalue $|\mu|$ of $(Tf)^m/T_xV$, and now it is easy to approximate f by $g \in \mathfrak{S}$ such that $g(y) = f(y)$ for every y in the orbit of x and such that $(\tilde{N}g)^m/\tilde{N}_xV$ has an eigenvalue λ' with $|\lambda'| < 1$ such that $|\mu| < |\lambda'|$ for all eigenvalues of $(Tg)^m/T_xV$, thus contradicting Lemma 2.9.

References

1. J. Hadamard, *Sur l'itération et les solutions asymptotiques des équations différentielles*, Bull. Soc. Math. France **29** (1901), 224–228.

2. O. Perron, *Über Stabilität und asymptotischen Verhalten der Integrals von Differentialgleichnungssystemen*, Math. Z. **29** (1928), 129–160.

3. D. Ruelle and F. Takens, *On the nature of turbulence*, Comm. Math. Phys. **20** (1971), 167–192.

4. J. Hale, *Integral manifolds of perturbed differential systems*, Ann. of Math. (2) **73** (1961), 496–531.

5. S. Diliberto, *Perturbation theorems for periodic surfaces*. I, II, Rend. Circ. Mat. Palermo **9** (1960), 265–299; **10** (1961), 111–161.

6. R. Sacker, *A perturbation theorem for invariant Riemannian manifolds*, Academic Press, New York, 1967, pp. 43–54.

7. I. Kupka, *Stabilité des variétés invariantes d'un champ de vecteurs pour les petites perturbations*, C. R. Acad. Sci. Paris **258** (1964), 4197–4200.

8. N. Fenichel, *Persistence and smoothness of invariant manifolds of flows*, Indiana Univ. Math. J. **21** (1971–1972), 193–226.

9. M. Hirsch, C. Pugh and M. Shub, *Invariant manifolds* (to appear).

10. R. Mañé, *Variedades persitentes*, Thesis, Inst. Mat. Pura Apl., Rio de Janeiro, 1973.

11. ______, *Persistent manifolds are normally hyperbolic*, Bull. Amer. Math. Soc. **80** (1974), 90–91.

12. A. Gottlieb, *Converses to the Ω-stability and invariant lamination theorems*, Trans. Amer. Math. Soc. **202** (1975), 369–383.

13. R. Mañé, *Quasi-Anosov diffeomorphisms and hyperbolic manifolds*, Trans. Amer. Math. Soc. **229** (1977), 351–370.

14. C. Pugh, *An improved closing lemma and a general density theorem*, Amer. J. Math. **89** (1967), 1010–1021.

15. J. Franks, *Necessary conditions for stability of diffeomorphisms*, Trans. Amer. Math. Soc. **158** (1971), 301–308.

16. R. Mañé, *Reduction of semilinear parabolic equations to finite dimensional C^1-flows* (Proc. III ELAM), Lecture Notes in Math., vol. 597, Springer-Verlag, Berlin and New York, 1977, pp. 361–379.

17. J. Selgrade, *Isolated invariant sets for flows on vector bundles*, Trans. Amer. Math. Soc. **203** (1975), 359–390.

18. J. Kurzweil, *Invariant manifolds for flows* (Proc. Sympos., Universidad Puerto Rico), Academic Press, New York, 1967, pp. 431–469.

19. C. Pugh and M. Shub, *Axiom* A *actions*, Invent. Math. **29** (1975), 7–38.

Instituto de Matemática Pura e Aplicada, Luiz de Camões, 68 CEP 20000, Rio de Janeiro, Brazil

Inventiones math. 46, 147–152 (1978)

Inventiones mathematicae

Invariant Sets of Anosovs Diffeomorphisms

Ricardo Mañé

Instituto de Matemática pura e Aplicada, Rua Luiz de Camões, 68, Rio de Janeiro-RJ, Brasil

Let M be a compact smooth boundaryless manifold. We say that a diffeomorphism $f: M \hookleftarrow$ is Anosov if there exists a Riemann structure on M and a continuous Tf-invariant splitting $TM = E^s \oplus E^u$ such that $\|(Tf)/E^s_x\| < 1$, $\|(Tf)^{-1}/E^u_x\| < 1$ for all $x \in M$. The simplest examples of Anosov diffeomorphisms are the toral hyperbolic automorphisms i.e. Anosov automorphisms of the group $T^n = \mathbb{R}^n/\mathbb{Z}^n$. They are induced by linear automorphisms $L: \mathbb{R}^n \hookleftarrow$ with determinant one, integer entries, and no eigenvalues of absolute value one.

In this paper we shall consider the problem of describing the compact invariant sets of Anosov diffeomorphisms. A good example of this kind of question is the following: Can an Anosov diffeomorphism have an invariant set homeomorphic to S^2? Replacing S^2 by S^1 the answer is negative because S^1 doesn't admit expansive homeomorphisms. In the case of S^2 this argument fails because it is unknown whether S^2 admits expansive homeomorphisms. In the case that f is a hyperbolic toral automorphism this question was answered by Hirsch in [5] where, he proved that a compact invariant topological surface of a hyperbolic toral automorphism can't be homeomorphic to a sphere, Klein bottle or projective space, but it remains unknown whether there can exist (even for hyperbolic toral automorphisms of T^4) a compact invariant topological surface with genus >1. In the same paper Hirsch asked whether the restriction of a diffeomorphism to a hyperbolic manifold is Anosov. The answer is negative as shown in [3] but the following stronger version of the problem remains open:

Conjecture ([10] Problem 10). If $f: M \hookleftarrow$ is an Anosov diffeomorphism and $V \subset M$ is a C^1 compact invariant boundaryless manifold then f/V is Anosov.

This conjecture was proved in some cases by Hirsch [5] and by Franks [4] when M is a torus, f is an automorphism and V is class C^r with $r > \dim M$. Here we shall prove it assuming only that $M = T^n$.

Theorem A. *If $f: T^n \hookleftarrow$ is an Anosov diffeomorphism and $V \subset M$ a boundaryless compact invariant C^1 manifold then f/V is Anosov and each connected component of V is homeomorphic to a torus.*

0020-9910/78/0046/0147/$01.20

The main tool for the proof of A will be the next theorem. For its statement recall that a compact invariant set Λ of a diffeomorphism f is isolated if it has a neighborhood U such that $\bigcap_n f^n(U)=\Lambda$.

Theorem B. *Every compact connected isolated invariant set of Λ a toral hyperbolic automorphism has the form $\Lambda=x+G$ where x is a fixed point and G an invariant compact subgroup.*

Most of the problems considered here were originated in a question of Smale [5] asking if a hyperbolic toral automorphism can have a compact 1-dimensional invariant set. Franks [2] proved that is impossible when the set contains a C^2 arc. But recently, Bowen and Hancock obtained examples of one dimensional invariant sets for hyperbolic automorphisms of T^3. Hancock showed that given an Anosov diffeomorphism on T^3, there exist arcs $\gamma\colon[0,1]\to T^3$ such that $\bigcup_{n\in\mathbb{Z}} f^n(\gamma[0,1])$ is not dense. Then he considers the set $\Sigma=\overline{\bigcup_{n\in\mathbb{Z}} f^n(\gamma[0,1])}$. It is invariant and then so is its interior. But an open invariant set of f must be dense (because f is topologically transitive). So the interior of Σ is empty and this implies $\dim\Sigma<3$ [7]. By [5] Theorem 9 $\dim\Sigma\neq 2$ and since Σ contains $\gamma([0,1])$ we have $\dim\Sigma\geqq 1$. Then $\dim\Sigma=1$. Hancock's construction gives also arcs satisfying the previous condition and containing some fixed point x_0. Then it is easy to see that the set $\bigcup f^n(\gamma[0,1])$ is connected (because $x_0\in f^n(\gamma[0,1])$ for all $n\in\mathbb{Z}$), and then Σ is alo connected. This example shows that Theorem A is false without the isolation hypothesis.

Proof of Theorem B. Let $f\colon T^n \hookleftarrow$ be a hyperbolic toral automorphism and $\Lambda\subset T^n$ a compact connected isolated invariant set of f. For $x\in\Lambda$ and $\varepsilon>0$ define

$$W^s_\varepsilon(x)=\{y\in\Lambda \mid d(f^n(y),f^n(x))\leqq\varepsilon \quad \text{for all } n\geqq 0\},$$

$$W^u_\varepsilon(x)=\{y\in\Lambda \mid d(f^{-n}(y),f^{-n}(x))\leqq\varepsilon \quad \text{for all } n\geqq 0\}.$$

Suppose that $x,y\in\Lambda$ and $W^s_\varepsilon(x)\cap W^u_\varepsilon(y)\neq\emptyset$. Then the orbit of points in $W^s_\varepsilon(x)\cap W^u_\varepsilon(y)\neq\emptyset$ remain at distance $\leqq\varepsilon$ of Λ. Since Λ is isolated this implies (taking ε small enough) that $W^s_\varepsilon(x)\cap W^u_\varepsilon(y)\subset\Lambda$ for all $x,y\in\Lambda$. Hence f/Λ has canonical coordinates and a hyperbolic metric in the sense defined in [1]. Therefore Λ contains periodic points because they are dense in the nonwandering set of f/Λ [1]. Let $x\in\Lambda$ be one of them. Take m such that $f^m(x)=x$. Then $\Lambda-x$ is invariant under f^m. If we prove that $\Lambda-x$ is a toral subgroup it will follow that Λ is a torus and f/Λ an Anosov diffeomorphism of Λ. Then f will have a fixed point $x_0\in\Lambda$ and we can write

$$\Lambda=x_0+(\Lambda-x_0)=x_0+(\Lambda-x)+(x-x_0)=x_0+(\Lambda-x).$$

This argument reduces the proof of Theorem A to the case when the identity e of T^n belongs to Λ.

Let us say that a continuous arc $\gamma\colon[0,1]\to T^n$ is δ-adapted to Λ if there exist $0=t_0<t_1<\dots<t_m=1$ such that $\gamma(t_j)\in\Lambda$ for all $0\leqq j\leqq m$ and $d(\gamma(t),\gamma(t_j))<\delta$ for all $t_j\leqq t\leqq t_{j+1}$, $0\leqq j<m$. Let Γ_δ be the subgroup of $\pi_1(T^n,e)=\mathbb{Z}^n$ spanned by arcs $\gamma\colon[0,1]\to T^n$ δ-adapted to Λ and satisfying $\gamma(0)=\gamma(1)=e$. Let N_δ be the subspace

spanned by Γ_δ. Now observe that $\Gamma_{\delta_1} \subset \Gamma_{\delta_2}$ if $\delta_1 < \delta_2$ and that given $\delta > 0$ there exists $\delta_1 > 0$ such that $L(\Gamma_{\delta'}) \subset \Gamma_\delta$ if $0 < \delta' < \delta_1$, where L denotes the lifting of f to $\mathbb{R}^n$. Let $N_0 = \bigcap_{\delta > 0} N_\delta$. Then $L(N_0) = N_0$. Let $\Gamma_0 = N_0 \cap \mathbb{Z}^n$.

Lemma 1. *Let $\rho: (\mathbb{R}^n/\Gamma_0, e) \to (\mathbb{R}^n/\mathbb{Z}^n, e)$ the canonical covering map and let $\tilde{f}$: $(\mathbb{R}^n/\Gamma_0, e) \hookleftarrow$ the lifting of $f: (\mathbb{R}^n/\mathbb{Z}^n, e)$. Then there exists a continuous map $\varphi: (\Lambda, e) \to (\mathbb{R}^n/\Gamma_0, e)$ such that the following diagram conmutes:*

$$\begin{array}{ccccc}
 & & (\mathbb{R}^n/\Gamma_0, e) & \xrightarrow{\tilde{f}} & (\mathbb{R}^n/\Gamma_0, e) \\
 & \varphi \nearrow & \downarrow p & & \downarrow p \\
(\Lambda, e) & \xrightarrow[i]{} & (\mathbb{R}^n/\mathbb{Z}^n, e) & \xrightarrow[f]{} & (\mathbb{R}^n/\mathbb{Z}^n, e)
\end{array}$$

Proof. Since $N_{\delta_1} \subset N_{\delta_2}$ if $\delta_1 \leqq \delta_2$ it follows that for some δ_0 we have $N_\delta = N_0$ for all $0 < \delta \leqq \delta_0$. Then $\Gamma_\delta \subset \mathbb{Z}^n \cap N_\delta = \mathbb{Z}^n \cap N_0 = \Gamma_0$ if $0 < \delta \leqq \delta_0$. To define $\varphi(x)$ take a continuous arc $\gamma: [0,1] \to T^n$ δ-adapted to Λ with $0 < \delta < \delta_0$, $\gamma(0) = e$, $\gamma(1) = x$. Take $\tilde{\gamma}: [0,1] \to \mathbb{R}^n/\Gamma_0$ such that $\rho\,\tilde{\gamma} = \gamma$ and $\tilde{\gamma}(0) = e$. Since $\Gamma_\delta \subset \Gamma_0$ it follows that $\tilde{\gamma}(1)$ is independent of γ. Then define $\varphi(x) = \tilde{\gamma}(1)$.

Lemma 2. $\varphi(\Lambda) \subset N_0/\Gamma_0$.

Proof. Let $\pi: \mathbb{R}^n/\Gamma_0 \to \mathbb{R}^n/N_0$ the canonical homomorphism. Let $\hat{f}: \mathbb{R}^n/N_0 \supset$ be the automorphism satisfying $\hat{f}\pi = \pi\tilde{f}$. Then $\hat{f}$ is hyperbolic. Let $\Lambda_0 = \pi\,\varphi(\Lambda)$. Λ_0 is compact and $\hat{f}(\Lambda_0) = \hat{f}\pi\,\varphi(\Lambda) = \pi\tilde{f}\,\varphi(\Lambda) = \pi\,\varphi(\Lambda) = \Lambda_0$. But the only compact invariant set of a hyperbolic automorphism of a vector space is the identity. Hence $\pi\,\varphi(\Lambda) = \{e\}$ and this implies $\varphi(\Lambda) \subset N_0/\Gamma_0$.

Since $\Lambda = p\varphi(\Lambda)$ to prove that Λ is a subgroup it is sufficient to show that $\varphi(\Lambda) = N_0/\Gamma_0$. Let $W^s = \{x \in N_0/\Gamma_0 |\ \lim_{n \to +\infty} \tilde{f}^n(x) = e\}$. W is dense in N_0/Γ_0 because $\tilde{f}/(N_0/\Gamma_0)$ is a hyperbolic toral automorphism. Then to show $\varphi(\Lambda) = N_0/\Gamma_0$ we have only to prove that $\varphi(\Lambda) \supset W^s$. Let

$$E^s = \{v \in N_0 |\ \lim_{n \to +\infty} L^n v = 0\}, \qquad E^u = \{v \in N_0 |\ \lim_{n \to +\infty} L^{-n} v = 0\}.$$

Then if $q: N_0 \to N_0/\Gamma_0$ is the canonical homomorphism we have $q(E^s) = W^s$. Hence the property $\varphi(\Lambda) \supset W^s$ is implied by the property $q^{-1}\,\varphi(\Lambda) \supset E^s$. This relation follows easily form the next Lemma:

Lemma 3. *If π^s: $N_0 \to E^s$ is the projection associated with the splitting $N_0 = E^s \oplus E^u$, then $\pi^s(\Gamma_\delta)$ is dense in E^s for all $0 < \delta < \delta_0$, and if δ is small enough $\pi^s(\Gamma_\delta) \subset q^{-1}\,\varphi(\Lambda)$.*

Proof. If $0 < \delta \leqq \delta_0$ Γ_δ is uniform in N_0 and $\Gamma_\delta \cap E^u \subset \Gamma_0 \cap E^u = \{0\}$. Let $q: N_0 \to N_0/\Gamma_0$, $P_1: N_0 \to N_0/\Gamma_\delta$, $P_2: N_0/\Gamma_\delta \to N_0/\Gamma_0$ be the canonical homomorphisms. $\overline{P_1(E^u)}$ is a closed subgroup of N_0/Γ_δ such that $P_2(\overline{P_1(E^u)}) = \overline{q(E^u)} = \overline{W^u}$ $= N_0/\Gamma_0$. Since P_2 is a local diffeomorphism and $\overline{P_1(E^u)}$ is a submanifold it follows that $\dim \overline{P_1(E^u)} = \dim N_0/\Gamma_0 = \dim N_0/\Gamma_\delta$. Hence $\overline{P_1(E^u)} = N_0/\Gamma_\delta$. Then given $x \in E^s$ and $\varepsilon > 0$ we can take $y \in P_1(x+E^u)$ such that $d(y,e) < \varepsilon$ because $P_1(x+E^u)$ is dense in N_0/Γ_δ). Then there exist $z \in x+E^u$ and $w \in \Gamma_\delta$ such that $\|z-w\| < \varepsilon$. Therefore $\|x - \pi^s w\| = \|\pi^s z - \pi^s w\| \leqq \|\pi^s\| \varepsilon$. To prove the second part of the Lemma take $\varepsilon > 0$ and define that two points $x, y \in q^{-1}(\varphi(\Lambda))$ are n-related if we can find $\{x = x_0, x_1, \dots, x_n = y\}$ such that

$$\|\pi^s(x_{j+1} - x_j)\| \leqq \varepsilon, \tag{1}$$

$$\|\pi^u(x_{j+1} - x_j)\| \leqq \varepsilon \tag{2}$$

for all $0 \leqq j < n$, where $\pi^u = I - \pi^s$. Take $\varepsilon > 0$ such that if $x, y \in q^{-1}\varphi(\Lambda)$ are 1-related then the point $(x+E) \cap (y+E^u)$ belongs to $q^{-1}(\varphi(\Lambda))$. Such ε exists because Λ is isolated and then if ε is small enough the point $z = pq((x+E^s) \cap (y+E^u))$ satisfies that

$$d(f^n(pq(x)), f^n(z)) \quad \text{and} \quad d(f^{-n}(pq(y)), f^{-n}(z))$$

remain small for all $n \geqq 0$, therefore z must belong to Λ and $(x+E^s) \cap (y+E^u) \in q^{-1}\varphi(\Lambda)$. We claim that if $x, y \in q^{-1}(\varphi(\Lambda))$ and x and y are n-related for some n, then $(x+E^s) \cap (y+E^u) \in q^{-1}(\varphi(\Lambda))$. The claim implies that $\pi^s(\Gamma_\delta) \subset q^{-1}(\varphi(\Lambda))$ because if δ is small enough every point $x \in \Gamma_\delta$ is n-related with the identity and then $\pi^s(x) = (x+E^u) \cap E^s \in q^{-1}(\varphi(\Lambda))$. The claim will be proved by induction on n. For $n=1$ it is true by the way ε was chosen. Suppose the property proved for $n = m-1$ and let $x, y \in q^{-1}(\varphi(\Lambda))$ be points m-related by the chain $\{x = x_0, x_1, \dots, x_m = y\}$ satisfying (1), (2). Define $\bar{x}_j = (x_j + E^s) \cap (x_{j+1} + E^u)$ for $0 \leqq m-1$. Then $\bar{x}_j \in q^{-1}(\varphi(\Lambda))$ for all $0 \leqq j \leqq m-1$ and $\bar{x}_0$, $\bar{x}_{m-1}$ are $(m-1)$-related by the chain $\{\bar{x}_0, \bar{x}_1, \dots, \bar{x}_{m-1}\}$ because:

$$\|\pi^s(\bar{x}_{j+1} - \bar{x}_j)\| = \|\pi^s(x_{j+1} - x_j)\| \leqq \varepsilon$$

$$\|\pi^u(\bar{x}_{j+1} - \bar{x}_j)\| = \|\pi^u(x_{j+2} - x_j)\| \leqq \varepsilon$$

for all $0 \leqq j < m-1$. By the induction hypothesis

$$(\bar{x}_0 + E^s) \cap (\bar{x}_{m-1} + E^u) \in q^{-1}(\varphi(\Lambda)).$$

But $\bar{x}_0 + E^s + x + E^s$ and $\bar{x}_{m+1} + E^u = x_m + E^u$. Then

$$(x+E^s) \cap (y+E^u) \in q^{-1}(\varphi(\Lambda)).$$

Proof of Theorem A. Let $f\colon T^n \hookleftarrow$ be an Anosov diffeomorphism and V a C^1 compact invariant boundaryless manifold. It is not difficult to see that is sufficient to prove Theorem A assuming that V is connected. By Franks-Manning Theorem [8] there exists a hyperbolic automorphism $f_1\colon T^n \hookleftarrow$ and a homeomorphism $h\colon T^n \hookleftarrow$ such that $f_1 h = hf$. By [9] f/V satisfies Axiom A and the no cycles condition. Let Σ be an attractor of f/V. By [1] we can write $\Sigma = \Sigma_1 \cup \dots \cup \Sigma_n$ where each Σ_i is compact, $f(\Sigma_j) = \Sigma_{j+1}$ for $0 \leqq j < m$ $f(\Sigma_m) = \Sigma_1$, and the unstable manifold of each $x \in \Sigma_1$ is dense in Σ_1. Since the unstable manifold of x is contained in Σ_1 (because Σ is an attractor) it follows that Σ_1 is connected. Then Σ_1 is a connected attractor for f^m. If $\alpha(\cdot,\cdot)$ is a metric and T^n define the local stable and unstable manifolds of $x \in T^n$ as

$$W_\varepsilon^s(x)=\{y\in T^n/d(f_0^k(x), f_0^k(y))\leqq\varepsilon, \forall k\geqq 0\},$$

$$W_\varepsilon^u(x)=\{y\in T^n/d(f_0^{-k}(y))\leqq\varepsilon, \forall k\geqq 0\},$$

where $f_0=f^m$. We claim that Σ has local product structure i.e. that there exists $\varepsilon>0$ such that $W_\varepsilon^s(x)\cap W_\varepsilon^u(y)\subset\Sigma_1$ for all $x,y\in\Sigma_1$. To prove this, chose $\varepsilon_0>0$ such that $V\cap W_{\varepsilon 0}^u(y)\subset\Sigma_1$ for all $y\in\Sigma_1$. Take $r>0$ such that $x,y\in\Sigma_1$ and $d(x,y)\leqq r$ imply that $(W_{\varepsilon 0}^s(x)\cap V)\cap(W_{\varepsilon 0}^u(x)\cap V)$ contains a unique point. Finally chose $0<\varepsilon<\varepsilon_0$ by the property $W_\varepsilon^s(x)\cap W_\varepsilon^u(y)\neq\emptyset$ implies $d(x,y)\leqq r$. Then if $x,y\in\Sigma_1$ and $W_\varepsilon^s(x)\cap W_\varepsilon^u(y)\neq\emptyset$ it follows that $d(x,y)\leqq r_1$ and then $W_{\varepsilon 0}^s(x)\cap W_{\varepsilon 0}^u(y)\cap V$ contains a unique point. Then

$$W_\varepsilon^s(x)\cap W_\varepsilon^u(y)=W_{\varepsilon 0}^s(x)\cap W_{\varepsilon 0}^u(y)\cap V\subset W_{\varepsilon 0}^s(x)\cap\Sigma_1.$$

By [6] Σ_1 is isolated, hence $h(\Sigma_1)$ is isolated, connected, invariant under f_1^m and then by Theorem B it has the form $h(\Sigma_1)=x+G$ where G is a toral subgroup under f_1^m and x a fixed point of f_1^m. Then we can assume that $h(\Sigma_1)$ is a toral subgroup invariant under f_1^m. Therefore to prove that f/V is Anosov and V homeomorphic to a torus is sufficient to show that $V=\Sigma_1$. Let

$$W^s(x)=\{y\in V|\lim_{n\to+\infty} d(f_0^n(x), f_0^n(y))=0\},\qquad W^s(\Sigma_1)=\bigcup\{W^s(x)|x\in\Sigma_1\}.$$

Then $W^s(\Sigma_1)$ is an open set and by elementary properties of Axiom A and cycles diffeomorphism the relation $V=\Sigma_1$ is equivalent to $W^s(\Sigma_1)=V$. Suppose that $x\in V\setminus W^s(\Sigma_1)$. Take a continuous are γ: $[0,1]\to V$ such that $\gamma(0)=h^{-1}(e)$, $\gamma(1)=x$. Let $0<t_0<1$ such that $\gamma(t_0)\notin W^s(\Sigma_1)$ and $\gamma(t)\in W^s(\Sigma_1)$ if $0\leqq t<t_1$. Let $\tilde\gamma=h\gamma$ and L: $\mathbb{R}^n\hookleftarrow$ be a hyperbolic isomorphism such that if p: $\mathbb{R}^n\to\mathbb{R}^n/\mathbb{Z}^n=T^n$ the canonical homomorphism then $f_1^m p=pL$. Let

$$E^s=\{v\in\mathbb{R}^n|\lim_{n\to+\infty} L^n v=0\}\qquad E^u=\{v\in\mathbb{R}^n|\lim_{n\to+\infty} L^{-n}v=0\}.$$

Then if N_0 is a subspace satisfying $pN_0=h(\Sigma_1)$ it is easy to see that $p^{-1}(h(W^s(\Sigma_1))\subset E^s+N_0+\mathbb{Z}^n$. But the arc connected component of $N_0+E^s+\mathbb{Z}^n$ is N_0+E^s. Then if $\hat\gamma$: $[0,1]$ is a continuous arc such that $\hat\gamma(0)=0$ and $p\hat\gamma=\tilde\gamma$ it follows that

$$\hat\gamma([0,t_0))\subset p^{-1}(\tilde\gamma([0,t_0)))\subset p^{-1}(h(W^s(\Sigma_1)))\subset E^s+N_0+\mathbb{Z}^n.$$

Then $\hat\gamma(t)\in E^s+N_0$ for all $0\leqq t<t_0$. Hence

$$\hat\gamma(t_0)\in E^s+N_0,\qquad \tilde\gamma(t_0)\in p(E^s+N_0)\cap h(V)\subset h(W^s(\Sigma_1))$$

and then $\gamma(t_0)\in W^s(\Sigma_1)$.

References

1. Bowen, R.: Periodic points and measures for Axiom A diffeomorphisms. Trans. Amer. Math. Soc. **154**, 377–397 (1971)
2. Franks, J.: Invariant set of hyperbolic toral automorphisms. Amer. J. of Math. (to appear)

3. Franks, J., Robinson, C.: A quasi-Anosov diffeomorphism that is not Anosov. Trans. Amer. Meth. Soc. **223**, (1976)
4. Hancock, S.G.: Orbits of paths under hyperbolic toral automorphisms (to appear)
5. Hirsch, M.: On invariant subsets of hyperbolic sets. Essays on topology and related topics, pp. 126–135. Berlin-Heidelberg-New York: Springer 1970
6. Hirsch, M., Palis, J., Pugh, G., Shub, M.: Neighborhoods of hyperbolic sets. Invent. Math. **9**, 121–134 (1970)
7. Hurewicz, W., Wallman, H.: Dimension Theory. Princeton: University Press 1948
8. Manning, A.: There are no new Anosov diffeomorphisms on Tori. Amer. J. of Math. **96**, 422–429 (1974)
9. Mañé, R.: Quasi-Anosov diffeomorphisms and hyperbolic manifolds. Trans. Amer. Math. Soc. **229**, 351–370 (1977)
10. Palis, J., Pugh, C.: Fifty problems on dynamical systems. Dynamical Systems, Warwick 1974. Berlin-Heidelberg-New York: Springer Lecture Notes on Mathematics Vol. 468 pp. 345–353, 1975

Received September 13, 1977

Annals of Mathematics, 116 (1982), 503–540

An ergodic closing lemma

By Ricardo Mañé

Introduction

A classical subject of the theory of dynamical systems has been whether interesting features of a dynamical system persist under perturbations and appear also in every nearby system. Especially interesting from this viewpoint are those dynamical systems that are identical, up to a continuous change of coordinates of the phase space, to every nearby system. Such systems will have the remarkable property that every topological property of their orbit structures will remain unchanged by perturbations. Let us make precise this property for the case of discrete dynamical systems, i.e., diffeomorphisms of a compact smooth boundaryless manifold M. Let Diff(M) be the space of C^1 diffeomorphisms of M endowed with the C^1 topology. We say that f is *structurally stable* if it has a neighborhood $\mathcal{U}$ such that for every $g \in \mathcal{U}$ there exists a homeomorphism h: $M \hookleftarrow$ such that $g = hfh^{-1}$. Sufficient conditions for the structural stability of a diffeomorphism are known.

Whether these conditions are also necessary is still an open problem known as the stability conjecture. One of the aims of this paper is to solve this conjecture on two dimensional manifolds and to prove a result about the general n-dimensional case that seems a promising step for the complete solution of the conjecture.

Let us briefly recall the main definitions required to state the sufficient conditions for structural stability. The key concept is that of hyperbolicity. A set $\Lambda \subset M$ is a *hyperbolic set* [17] of $f \in$ Diff(M) if it is compact, f-invariant (i.e., $f(\Lambda) = \Lambda$) and there exists a continuous splitting $TM/\Lambda = E^s \oplus E^u$, invariant under the derivative of f, such that there exist $K > 0$, $0 < \lambda < 1$, satisfying

$$\|(D_x f^n) \mid E_x^s\| \leq K\lambda^n$$

and

$$\|(D_x f^{-n}) \mid E_x^u\| \leq K\lambda^n$$

003-486X/82/0116-3/0503/038$03 80/1

for all $x \in \Lambda$, $n \geq 0$. Recall also the definition of *nonwandering point*: $x \in M$ is a nonwandering point of $f \in \mathrm{Diff}(M)$ if for every neighborhood U of x there exists $n \neq 0$ such that $f^n(U) \cap U \neq \varnothing$. The set of nonwandering points of f is denoted $\Omega(f)$. Always contained in $\Omega(f)$ is the set $\mathrm{Per}(f)$ of periodic points. We say that $f \in \mathrm{Diff}(M)$ satisfies Axiom A [17] if $\Omega(f)$ is hyperbolic and $\overline{\mathrm{Per}(f)} = \Omega(f)$. If f satisfies Axiom A, it has the remarkable property that for all $x \in M$ the sets $W^s(x), W^u(x)$ defined by

$$W^s(x) = \left\{y \mid \lim_{n \to +\infty} d(f^n(x), f^n(y)) = 0\right\}$$

and

$$W^u(x) = \left\{y \mid \lim_{n \to +\infty} d(f^{-n}(x), f^{-n}(y)) = 0\right\},$$

are injectively immersed Euclidean spaces [17]. For this reason these sets are called the stable and unstable manifolds of x. An Axiom A diffeomorphism f is said to satisfy the *strong transversality condition* if for every x its stable and unstable manifolds intersect transversally at x, i.e., if $T_xW^s(x) + T_xW^u(x) = T_xM$. It was proved by Robbin [13] and Robinson [14] that Axiom A and the strong transversality condition imply structural stability. Closely related to structural stability is the concept of Ω-stability: a diffeomorphism f is *Ω-stable* if for every g near f there exists a homeomorphism h: $\Omega(f) \to \Omega(g)$ such that $g \mid \Omega(g) = hfh^{-1} \mid \Omega(g)$. Sufficient conditions for Ω-stability were found by Smale [19]. For their statement let us recall the definition of the no cycles condition. If f is an Axiom A diffeomorphism, $\Omega(f)$ has a unique decomposition $\Omega(f) = \Omega_1 \cup \cdots \cup \Omega_l$ in disjoint, compact and transitive sets (transitive meaning that each Ω_i contains a point whose orbit is dense in Ω_i). These sets are called basic sets. The stable (resp. unstable) manifold of a basic set is defined as the union of the stable (resp. unstable) manifolds of its points. We say that f satisfies the *no cycles condition* if for every family $\Omega_{i_1}, \ldots, \Omega_{i_n}$ of basic sets such that the stable manifold of Ω_{i_j} has nonempty intersection with the unstable manifold of $\Omega_{i_{j+1}}$ for all $1 \leq j < n$, the stable manifold of Ω_{i_n} does not intersect the unstable manifold of Ω_{i_1}. Smale's Ω-stability theorem states that a diffeomorphism satisfying Axiom A and the no cycles condition is Ω-stable.

The problems of whether structural stability implies Axiom A and the strong transversality condition or Ω-stability implies Axiom A and the no cycles condition can be unified in the following conjecture due to Palis and Smale:

Conjecture I (Stability Conjecture). Ω-stable diffeomorphisms satisfy Axiom A.

This reduction follows from the facts that an Ω-stable diffeomorphism that satisfies Axiom A also satisfies the no cycles condition (Palis [8]) and a structurally stable diffeomorphism that satisfies Axiom A also satisfies the strong transversality condition (Robinson [15]; observe also that structural stability clearly implies Ω-stability.

To study this conjecture we shall let $\mathcal{F}(M)$ denote the set of diffeomorphisms $f \in \text{Diff}(M)$ that have a neighborhood $\mathcal{U}$ such that for all $g \in \mathcal{U}$ every periodic point x of g is hyperbolic; i.e., the eigenvalues of $D_x g^m$: $T_x M \hookleftarrow$, where m is the period of x, have modulus $\neq 1$. It is known (Franks [2]) that Ω-stable diffeomorphisms belong to $\mathcal{F}(M)$. Moreover if $f \in \mathcal{F}(M)$ then $\Omega(f) = \overline{\text{Per}(f)}$ ([4], Lemma 3.1). Thus the stability conjecture is contained in the following one:

Conjecture II. If $f \in \mathcal{F}(M)$ then $\Omega(f)$ is a hyperbolic set.

Moreover, if this conjecture is true, $\mathcal{F}(M)$ is characterized as the set of diffeomorphisms satisfying Axiom A and the no cycles condition. To prove this, observe that all the periodic points of an Axiom A diffeomorphism are hyperbolic and that the set of diffeomorphisms satisfying Axiom A and the no cycles condition is open [19]. Thus it is contained in $\mathcal{F}(M)$. Conversely, if $f \in \mathcal{F}(M)$ then f satisfies Axiom A (assuming that the conjecture is true and using the fact mentioned above that $f \in \mathcal{F}(M)$ implies $\Omega(f) = \overline{\text{Per}(f)}$). It remains to show that f satisfies the no cycles condition. Suppose it does not. Then, if we use the method in [8], a diffeomorphism g arbitrarily near to f can be constructed satisfying $\text{Per}(g) - \Omega(f) \neq \varnothing$ and $g \mid \Omega(f) = f \mid \Omega(f)$. Suppose that $x \in \text{Per}(g) - \Omega(f)$ and let N be the g-period of x. Join g to f by a small continuous arc g_t, $0 \leq t \leq 1$, $g_0 = f$, $g_1 = g$. Since $f \in \mathcal{F}(M)$, we can suppose that all the periodic points of every g_t are hyperbolic. Define $\nu(t) = \#\{x \mid g_t^N(x) = x\}$. By well known properties of hyperbolic periodic points, the function ν is continuous; being integer valued, it is constant. But the properties of g imply that $\nu(1) > \nu(0)$. Hence f satisfies the no cycles condition.

The natural approach to Conjecture II is by contradiction: to show that if $\Omega(f)$ is not hyperbolic then non-hyperbolic periodic orbits can be created by perturbations. The main result about creating periodic orbits by perturbations is Pugh's Closing Lemma which states that if $x \in \Omega(f)$ there exists g arbitrarily near to f such that $x \in \text{Per}(g)$. Unfortunately no information is obtained about the relative positions of the f-orbit and g-orbit of x. More precisely, consider the following question: Suppose x is a recurrent point of $f \in \text{Diff}(M)$, i.e., $\liminf_{n \to +\infty} d(f^n(x), x) = 0$; is it true that for all $\varepsilon > 0$ there exist g arbitrarily near to f and $y \in \text{Per}(g)$ such that $d(f^j(x), g^j(y)) \leq \varepsilon$ for all $0 \leq j \leq m$, where m is the minimum g-period of y? This question is one of the several challenging open problems concerning possible strengthenings of Pugh's Closing Lemma. Several partial results about the stability conjecture have been obtained using the hypothesis $f \in \mathcal{F}(M)$ in order to show that f has properties that amount to some sort of "weak hyperbolicity" which grant that f satisfies the above property (see for instance [6] or [9]); then these properties are used to exhibit a contradiction between $f \in \mathcal{F}(M)$ and the lack of hyperbolicity of $\Omega(f)$. The novelty in the approach we shall follow here is that we can prove that the question above holds for every diffeomorphism, at least for a total probability set of points. Recall that, given $f \in \mathrm{Diff}(M)$, a set $\Lambda \subset M$ is said to have *total probability* if $\mu(\Lambda) = 1$ for every f-invariant probability measure μ on the Borel sets of M. To state precisely this property, define $B_\varepsilon(f, x)$ as the set of points $y \in M$ such that $d(f^n(x), y) \leq \varepsilon$ for some $n \in \mathbf{Z}$ and define $\Sigma(f)$ as the set of points $x \in M$ such that for every neighborhood $\mathcal{U}$ of f and every $\varepsilon > 0$ there exist $g \in \mathcal{U}$ and $y \in M$ such that $y \in \mathrm{Per}(g)$, $g = f$ on $M - B_\varepsilon(f, x)$ and $d(f^j(x), g^j(y)) \leq \varepsilon$ for all $0 \leq j \leq m$, where m is the g-period of y.

Theorem A. *$\Sigma(f)$ is a total probability set.*

Using this theorem we shall prove the following partial answer to Conjecture II. If $f \in \mathrm{Diff}(M)$ let $\Lambda_j(f)$ denote the closure of the set of hyperbolic periodic points of f whose stable manifold has dimension j; clearly $\Lambda_j(f) \cap \Lambda_i(f) = \varnothing$ for every $0 \leq i < j \leq \dim M$ whenever f satisfies Axiom A.

Theorem B. *A diffeomorphism f satisfies Axiom A and the no cycles condition if and only if $f \in \mathcal{F}(M)$ and $\Lambda_j(f) \cap \Lambda_i(f) = \varnothing$ for every $0 \leq i < j \leq \dim M$.*

In [10] Pliss proved that $f \in \mathcal{F}(M)$ implies that the sets of sinks and sources (i.e., periodic points whose stable manifolds have respectively dimensions $\dim M$ and 0) are finite. In particular $\Lambda_{\dim M}(f)$ and $\Lambda_0(f)$ coincide with the sets of sinks and sources respectively. Therefore $\Lambda_j(f) \cap \Lambda_i(f) = \varnothing$ when $i \neq j$ and $j = 0$ or $\dim M$. If $\dim M = 2$ this proves that $\Lambda_j(f) \cap \Lambda_i(f) = \varnothing$ for all $0 \leq i < j \leq \dim M$; hence Conjectures I and II are true for $\dim M = 2$ and we have proved:

Corollary I. *If $\dim M = 2$, f satisfies Axiom A and the no cycles condition if and only if $f \in \mathcal{F}(M)$.*

Also in the case of two dimensional manifolds we have the following genericity theorem:

Corollary II. *If $\dim M = 2$ there exists a residual subset $\mathcal{O} \subset \mathrm{Diff}(M)$ such that every $f \in \mathcal{O}$ satisfies one of the following properties:*

I) *f has infinitely many sinks,*
II) *f has infinitely many sources or*
III) *f satisfies Axiom A and the no cycles condition.*

Let us see how Corollary II follows from Corollary I. If $f \in \mathrm{Diff}(M)$ let $\Lambda(f)$ denote the set of hyperbolic periodic points of f that are sinks or sources and $\overline{\Lambda}(f)$ its closure. It is clear that the set-valued function $f \mapsto \overline{\Lambda}(f)$ is lower semicontinuous; i.e., if U is an open set and for some f, $\overline{\Lambda}(f) \cap U \neq \varnothing$ then $\overline{\Lambda}(g) \cap U \neq \varnothing$ if g is near to f. By well known properties about set-valued functions, there exists a residual subset $\mathcal{O} \subset \mathrm{Diff}(M)$ such that every f in $\mathcal{O}$ is a point of semicontinuity of $\overline{\Lambda}$; i.e., if $K \subset M$ is compact and $\overline{\Lambda}(f) \cap K = \varnothing$ then $\overline{\Lambda}(g) \cap K = \varnothing$ for all g near to f. We claim that every $f \in \mathcal{O}$ satisfies one of the three properties of Corollary II. Suppose that $f \in \mathcal{O}$ and does not satisfy (I) and (II); then $\Lambda(f)$ is finite. To show that f satisfies (III) it is sufficient, by Corollary I, to prove that $f \in \mathcal{F}(M)$. For each $x \in \Lambda(f)$ take an open neighborhood U_x of x and a neighborhood $\mathcal{U}_x$ of f such that for all $g \in \mathcal{U}_x$ the set $U_x \cap \mathrm{Per}(g)$ contains a unique point and that point is in $\Lambda(g)$. Since $\Lambda(f)$ is finite, $\mathcal{U}_0 = \cap\{\mathcal{U}_x \mid x \in \Lambda(f)\}$ is a neighborhood of f and if we denote $U = U\{U_x \mid x \in \Lambda(f)\}$ then $\mathrm{Per}(g) \cap U \subset \Lambda(g)$ if $g \in \mathcal{U}_0$. If we use the fact that the function $\overline{\Lambda}$ is upper semicontinuous at f, there exists an open neighborhood $\mathcal{U}_1$ of f such that $\overline{\Lambda}(g) \cap U^c = \varnothing$ if $g \in \mathcal{U}_1$. Then $\Lambda(g) \subset U$ for all $g \in \mathcal{U}_0 \cap \mathcal{U}_1$. Now suppose that $f \notin \mathcal{F}(M)$. Then we can find $g \in \mathcal{U}_0 \cap \mathcal{U}_1$ having a non-hyperbolic periodic point x. Since $g \in \mathcal{U}_0$, the point x cannot belong to U because $\mathrm{Per}(g) \cap U \subset \Lambda(g)$ and $x \notin \Lambda(g)$. But the non-hyperbolicity of x implies that if m is its period all the eigenvalues of $D_x g^m$: $T_x M \hookleftarrow$ have modulus ≥ 1 or ≤ 1 (here the hypothesis $\dim M = 2$ is used again). Then we can find $\tilde{g}$ arbitrarily near to g such that $\tilde{g}(g^n(x)) = g^{n+1}(x)$ for all n and with all the eigenvalues of $D_x \tilde{g}^m$ simultaneously > 1 or < 1. Then $x \in \Lambda(\tilde{g})$. But $\mathcal{U}_1$ is open and then if $\tilde{g}$ is near enough to g we have $\tilde{g} \in \mathcal{U}_1$. This implies $\Lambda(\tilde{g}) \cap U^c = \varnothing$, contradicting $x \in \Lambda(\tilde{g}) \cap U^c$.

In [18] Smale asked whether the set of diffeomorphisms of S^2 having sinks is dense in $\mathrm{Diff}(S^2)$. Plykin ([11]) showed that it is not. However he proved that Axiom A diffeomorphisms of S^2 always have sinks or sources. Together with Corollary II this result shows that the set of diffeomorphisms of S^2 having sinks or sources is dense (and obviously open) in $\mathrm{Diff}(S^2)$.

All the properties and problems we considered here have corresponding versions for flows. However in the case of flows, Conjecture II and Corollaries I and II are false. This follows from an example due to Guckenheimer [3] which shows that there exists an open set $\mathcal{U}$ of flows on S^3 having all its periodic orbits hyperbolic but not satisfying Axiom A, neither having sinks nor sources. However Conjecture II and Corollary I hold for flows on three dimensional manifolds when every point is nonwandering [1]. Moreover, in Guckenheimer's example a fundamental role is played by singularities that produce a behaviour whose corresponding version in the discrete case would be to have a diffeomorphism

$f \in \mathfrak{F}(M)$ with $\Lambda_j(f) \cap \Lambda_i(f) \neq \emptyset$ for some $i \neq j$. Therefore Theorem B is still a candidate to be true for flows. However in the methods that we shall use for diffeomorphisms, there is a crucial step that is the fact that $f \in \mathfrak{F}(M)$ implies $\Omega(f) = \overline{\mathrm{Per}(f)}$. We don't know whether this property is true for flows. Therefore the extension of Theorem B for flows seems to require nontrivial improvements in the techniques used for diffeomorphisms.

All our results depend heavily on the fact that we are working in the C^1 topology. In the C^2 case the knowledge about the stability conjecture is very poor. For instance it is not known whether a C^2 structurally stable diffeomorphism must have periodic orbits. An extension to the C^2 topology of the Closing Lemma would settle this question but there are other equally basic problems that would not be directly solved by a C^2 Closing Lemma. For instance, can a C^2 structurally stable diffeomorphism f have a sequence of periodic points x_n with $D_{x_n}^{m_n} f\colon T_{x_n}M \hookleftarrow$ (where m_n is the period of x_n) having an eigenvalue with modulus λ_n and $\lambda_n \to 1$? The answer seems to be no because in this case, by small perturbations, a non-hyperbolic periodic point could be created, and that would contradict the C^2 stability (Robinson [15]). However certain surprising obstructions appear when we try to create the non-hyperbolic periodic point. When stability is replaced by the stronger concept of absolute stability then complete characterization is possible independently of the topology [7].

Theorem A will be proved in Section I. A fundamental tool for its proof will be a lemma (Lemma I.3) essentially contained in Pugh's proof of the Closing Lemma. Theorem B will be proved in Section II. As we remarked above, the only part of its statement that has to be proved is that $\Omega(f)$ is a hyperbolic set. This will be done by first constructing (using only the hypothesis $f \in \mathfrak{F}(M)$) a continuous splitting of $TM \mid \Omega(f)$ exhibiting certain special properties. These properties together with the hypothesis $\Lambda_j(f) \cap \Lambda_i(f) = \emptyset$ for $0 \leq i < j \leq \dim M$ will be used to prove, with the help of Theorem A, that the splitting is actually hyperbolic. An important part of the construction of this splitting can be found in the works of Pliss [9], [10] and in [6], but here we shall present a different self-contained version.

I wish to thank Clark Robinson and Sheldon Newhouse for several important remarks and corrections, Charles Pugh for very useful comments on the proof of the Closing Lemma, and Claus Doering for thoroughly revising the first draft of this work.

I. Proof of Theorem A

If $f \in \mathrm{Diff}(M)$ let $\mathfrak{M}(f)$ be the set of all f-invariant probability measures defined on the Borel subsets of M. $B_\varepsilon(f, x)$ and $\Sigma(f)$ have the same meaning as in the introduction; it is easy to see that $\Sigma(f)$ is f-invariant. If $\mathcal{U}$ is a neighborhood of f and $\varepsilon > 0$, let $\Sigma(\mathcal{U}, \varepsilon)$ be the set of points $x \in M$ such that there exist $g \in \mathcal{U}$, $y \in M$ and $m \in \mathbf{Z}^+$ satisfying $g^m(y) = y$, $g = f$ on $M - B_\varepsilon(f, x)$ and $d(f^n(x), g^n(y)) \leq \varepsilon$ for all $0 \leq n \leq m$. Observe that if $\mathcal{U}_n$ and $\varepsilon_n > 0$, $n \geq 0$, are a basis of neighborhoods of f and a sequence converging to 0 then:

$$\Sigma(f) = \bigcap_{n \geq 0} \Sigma(\mathcal{U}_n, \varepsilon_n). \tag{1}$$

It is easy to see that $\Sigma(\mathcal{U}, \varepsilon)$ is always a Borel set. Property (1) together with the well known property that states that if a set has total measure for every ergodic measure in $\mathcal{M}(f)$ then it has total probability, show that the proof of Theorem A is reduced to proving the following proposition:

PROPOSITION I.1. *For every $f \in \mathrm{Diff}(M)$, every neighborhood $\mathcal{U}$ of f and every $\varepsilon > 0$ we have*

$$\mu(\Sigma(\mathcal{U}, \varepsilon)) = 1$$

for every ergodic $\mu \in \mathcal{M}(f)$.

In the proof of this proposition a key role will be played by the next lemma which is formally stronger than the Closing Lemma [12] but whose proof is essentially contained in the proof of the Closing Lemma.

LEMMA I.2. *Given $f \in \mathrm{Diff}(M)$, $p \in M$, $\varepsilon > 0$ and a neighborhood $\mathcal{U}$ of f, there exist $r > 0$, $\rho > 1$ such that if $x \in B_{\bar{r}}(p)$ with $0 < \bar{r} \leq r$ and $f^m(x) \in B_{\bar{r}}(p)$ for some $m > 0$ then there exist $0 \leq m_1 < m_2 \leq m$ and $g \in \mathcal{U}$ such that $f^{m_1}(x) \in \overline{B_{\rho\bar{r}}(p)}$, $f^{m_2}(x) \in \overline{B_{\rho\bar{r}}(p)}$, $g^{m_2 - m_1}(f^{m_2}(x)) = f^{m_2}(x)$, $g(w) = f(w)$ for $w \notin B_\varepsilon(f, p)$ and $d(g^j(f^{m_2}(x)), f^j(f^{m_1}(x))) \leq \varepsilon$ for $0 \leq j \leq m_2 - m_1$.*

To prove this property we shall use the following lemma that summarizes the perturbation theory developed by Pugh in [12]. To the reader not acquainted with the proof of the Closing Lemma we suggest either to admit Lemma I.2 or, before trying to go through its proof, to read Robinson's exposition in [16] of the basic ideas in the proof of the Closing Lemma.

LEMMA I.3. *Let $f \in \mathrm{Diff}(M)$ and $p \in M$. Suppose that a compact neighborhood of p is identified by the exponential map with the ball $B_R = \{x \in T_pM \mid \|x\| \leq R\}$. Then there exists a splitting $T_pM = E_1 \oplus \cdots \oplus E_l$ such that given a neighborhood $\mathcal{U}$ of f and constants $C > 1$, $2 > \delta > 1$ there exist $N > 0$, $0 < r_1 <$*

$r_0 < R$ *with* r_0 *arbitrarily small and* $\lambda_i > 0$, $i = 1,\ldots,l$, *satisfying the following properties*:

I) *If* $\pi_i \colon T_pM \to E_i$, $i = 1,\ldots,l$, *denote the projections associated with the splitting* $T_pM = E_1 \oplus \cdots \oplus E_l$, *then*

$$\left\{x \mid \sup_i \lambda_i \|\pi_i(x-u)\| \le 2r_1\right\} \subset B_{r_0} \text{ if } \|u\| \le r_1.$$

II) *If* $\|y\| \le r_1$, $\|z\| \le r_1$, *and* $1 \le \mu_i \le C$, $i = 1,\ldots,l$, *there exists* $g \in \mathcal{U}$ *such that*

$$g^N(y) = f^N(z) \tag{2}$$

and

$$g(w) = f(w) \tag{3}$$

when

$$w \notin \bigcup_{j=0}^{N-1} f^j\big(\tilde{B}_{\delta r}(q)\big),$$

where $q = 2^{-1}(y+z)$, $r = 2^{-1}\|y-z\|$, $\tilde{B}_{\delta r}(q) = \{x \mid \sup_i \mu_i \lambda_i \|\pi_i(x-q)\| \le \delta r\}$.

Observe that if besides f, p and $\mathcal{U}$ we also have some $\varepsilon > 0$, then take r_0 so small that

$$\sup_{0 \le j \le N} \operatorname{diam} f^j\big(B_{r_0}\big) \le \varepsilon, \tag{4}$$

then, since $\tilde{B}_{\delta r}(q) \subset B_{r_0}$ (by property I), we obtain from (3) that:

$$g(w) = f(w) \tag{5}$$

when $w \notin B_\varepsilon(f, p)$. Moreover if p is not f-periodic we may choose r_0 so small that

$$f^{j_1}\big(B_{r_0}\big) \cap f^{j_2}\big(B_{r_0}\big) = \varnothing \tag{6}$$

for all $0 \le j_1 < j_2 \le N$; now if for some m we have

$$y = f^m(z) \tag{7}$$

and

$$f^j(z) \notin \tilde{B}_{\delta r}(q) \tag{8}$$

for all $0 < j < m$ then (2), (3), (5), (7) and (8) imply that $m \ge N$ and

$$g^m(y) = y \tag{9}$$

and

$$d\big(g^j(y), f^j(z)\big) \le \varepsilon \tag{10}$$

for all $0 \le j \le m$.

For the proof of I.2 we shall also need the following version of the Fundamental Lemma of [12].

LEMMA I.4. *Given $m \in \mathbf{Z}^+$ there exist $C = C(m) > 1$, $A = A(m) > 1$ and $2 > \delta = \delta(m) > 1$ such that if $E = E_1 \oplus \cdots \oplus E_m$ is a vector space and $|\cdot|_i$ a norm on E_i generated by an inner product then for any set $x_0, x_1, \ldots, x_n$ of distinct points in E there exist $0 \le j_1 < j_2 \le n$ and $1 \le \mu_i \le C$, $i = 1, \ldots, m$ such that by definition of a norm $\|\cdot\|_1$ on E by $\|v\|_1 = \sup_i \mu_i \|\pi_i v\|$, where π_i: $E \to E_i$ are the projections associated with the splitting of E, the following properties hold*:

$$\|x_{j_i} - x_0\|_1 \le A\|x_n - x_0\|_1, \qquad i = 1,2 \tag{11}$$

and

$$\|x_j - p\|_1 > \delta \tfrac{1}{2}\|x_{j_1} - x_{j_2}\|_1, \qquad j_1 < j < j_2, \tag{12}$$

where $p = 2^{-1}(x_{j_1} + x_{j_2})$.

Proof. Take $1 < \delta_0 < \sqrt{5}/2$ and choose $C > 1$ and $1 < \delta < \delta_0$ such that $C^2 > 4m(\delta_0^2 - 1)^{-1}$ and $2\delta^2 < \delta_0^2 + 1$. It follows that $\delta^2\left(1 + \dfrac{m-1}{C^2}\right) \le \delta_0^2$ and that

$$\delta\left(\sum_i \max\{\beta_i^2, C^{-2}\}\right)^{1/2} \le \delta_0\left(\sum_i \beta_i^2\right)^{1/2} \tag{13}$$

for all $0 \le \beta_i \le 1$, $i = 1, \ldots, m$, with $\beta_1 = 1$. Define

$$A = 4Cm^{1/2}.$$

Let $|\cdot|$ be the norm on E given by $|v| = \left(\Sigma_i |\pi_i v|_i^2\right)^{1/2}$ and if $\mu = (\mu_1, \ldots, \mu_m)$, with $C \ge \mu_i \ge 1$ for all $1 \le i \le m$ define the norm $\|\cdot\|_\mu$ by $\|v\|_\mu = \sup_i \mu_i |\pi_i v|_i$. For all $v \in E$ and μ we have:

$$C^{-1}\|v\|_\mu \le |v| \le m^{1/2}\|v\|_\mu.$$

We claim that there exist j_1, j_2 $(0 \le j_1 < j_2 \le n)$ such that

$$|x_{j_i} - x_0| \le 4|x_0 - x_n|, \qquad i = 1,2$$

and

$$|x_j - p| > \delta_0 2^{-1}|x_{j_1} - x_{j_2}|, \qquad j_1 < j < j_2. \tag{14}$$

Suppose that the claim is true. Then for any choice of μ we have

$$\|x_{j_i} - x_0\|_\mu \le C|x_{j_i} - x_0| \le 4Cm^{1/2}\|x_n - x_0\|_\mu = A\|x_n - x_0\|_\mu$$

for $i = 1,2$. Then to complete the proof of the lemma we have only to find μ such that $\|x - p\|_\mu \le \delta 2^{-1}\|x_{j_1} - x_{j_2}\|_\mu$ implies $|x - p| \le \delta_0 2^{-1}|x_{j_1} - x_{j_2}|$

because then (14) will imply (12). Suppose, to simplify the notation, that $|\pi_1(x_{j_1} - x_{j_2})|_1 = \sup_i |\pi_i(x_{j_1} - x_{j_2})|_i$. Define μ_i by C if $|\pi_i(x_{j_1} - x_{j_2})|_i = 0$ and by

$$\mu_i = \min\left\{C, \left|\pi_1(x_{j_1} - x_{j_2})\right|_1 \cdot \left|\pi_i(x_{j_2} - x_{j_2})\right|_i^{-1}\right\}$$

otherwise. Then

$$|x - p| = \left(\sum_i |\pi_i(x-p)|_i^2\right)^{1/2} \le \left(\sum_i \mu_i^{-2}\right)^{1/2} \|x - p\|_\mu$$

$$\le \left(\sum_i \mu_i^{-2}\right)^{1/2} \delta 2^{-1} \|x_{j_1} - x_{j_2}\|_\mu = \left(\sum_i \mu_i^{-2}\right)^{1/2} \delta 2^{-1} \left|\pi_1(x_{j_1} - x_{j_2})\right|_1.$$

But (13) implies:

$$\left(\sum_i \mu_i^{-2}\right)^{1/2} \delta < \delta_0 \left[\sum_i \frac{\left|\pi_i(x_{j_1} - x_{j_2})\right|_i^2}{\left|\pi_1(x_{j_1} - x_{j_2})\right|_1^2}\right]^{1/2}.$$

Hence:

$$|x - p| \le \delta_0 2^{-1} \left(\sum_i \left|\pi_i(x_{j_1} - x_{j_2})\right|_i^2\right)^{1/2} = \delta_0 2^{-1} |x_{j_1} - x_{j_2}|.$$

It remains to prove the claim. Take by induction a set

$$0 = i_0 \le i_1 \le \cdots \le i_s < l_s \le l_{s-1} \le \cdots \le l_0 = n$$

such that

(15a) $$x_{i_{k+1}} = x_{i_k}$$

or

(15b) $$x_{l_{k+1}} = x_{l_k}$$

for all $0 \le k < s$ and, also for all $0 \le k < s$,

(15c) $$|x_{i_{k+1}} - x_{l_{k+1}}| \le 2^{-1}(1 + \delta_0^2)^{1/2} |x_{i_k} - x_{l_k}|.$$

Suppose that we have taken a maximal set of indexes satisfying these properties. Then if $p = 2^{-1}(x_{i_s} + x_{l_s})$, we have

$$|x_j - p| > \delta_0 2^{-1} |x_{i_s} - x_{l_s}|$$

for all $i_s < j < l_s$, because if $|x_j - p| \le \delta_0 2^{-1} |x_{i_s} - x_{l_s}|$ for some $i_s < j < l_s$, using the parallelogram identity we obtain:

$$|x_j - x_{i_s}|^2 + |x_j - x_{l_s}|^2 = 2|x_j - p|^2 + \tfrac{1}{2}|x_{i_s} - x_{l_s}|^2$$
$$\le \tfrac{1}{2}(\delta_0^2 + 1)|x_{i_s} - x_{l_s}|^2$$

and then either for i_s or l_s (suppose that for l_s) we have:

$$|x_j - x_{l_s}|^2 \le \tfrac{1}{4}(1+\delta_0^2)\,|x_{i_s} - x_{l_s}|^2.$$

Then with $i_{s+1} = j$, $l_{s+1} = l_s$, conditions (15a), (15b), (15c) are satisfied, contradicting the maximality hypothesis of the set of indexes $0 \le i_0 \le \cdots \le i_s < l_s \le \cdots \le l_0 = n$. Moreover (15c) implies that

$$|x_{i_k} - x_{l_k}| \le \left(\tfrac{1}{2}(1+\delta_0^2)^{1/2}\right)^k |x_0 - x_n|\,.$$

Using this property together with (15a) and (15b), we obtain

$$\begin{aligned}|x_{i_s} - x_0| &\le |x_0 - x_n| \sum_0^\infty \left(\tfrac{1}{2}(1+\delta_0^2)^{1/2}\right)^k\\ &\le |x_0 - x_n| \sum_0^\infty \left(\tfrac{3}{4}\right)^k = 4\,|x_0 - x_n|\,.\end{aligned}$$

In a similar way the same inequality is proved with x_{l_s} instead of x_{i_s}.

Now let us prove I.2. Given f, $\mathcal{U}$, $\varepsilon > 0$ and p as in I.2 and R as in I.3, take the splitting $T_pM = E_1 \oplus \cdots \oplus E_l$ given by Lemma I.3 and $C = C(l)$, $A = A(l)$, $\delta = \delta(l)$ given by I.4. Let $N > 0$, $0 < r_1 < r_0 < R$ and $\lambda_i > 0$, $i = 1,\ldots,l$ be given by I.3 for $\mathcal{U}$ and these values of C and δ. If p is periodic there is nothing to prove; therefore suppose that p is not periodic and also that r_0 is so small that (4) and (6) are also satisfied. Let $S = [1, C] \times \overset{l}{\cdots} \times [1, C]$. If $\mu \in S$ define a norm $\|\cdot\|_\mu$ on T_pM by:

$$\|v\|_\mu = \sup_i \mu_i \lambda_i \|\pi_i v\|$$

and take $0 < k < K$, $0 < b < B$ such that

$$kd(x,y) \le \|x - y\| \le Kd(x,y)$$

for all $x, y \in B_R$, where $d(\cdot,\cdot)$ denotes the metric of M, and

$$b\|x\|_\mu \le \|x\| \le B\|x\|_\mu$$

for all $x \in T_pM$, $\mu \in S$. Take $r > 0$ and $\rho > 1$ satisfying:

$$(2ABb^{-1} + 1)Kr \le r_1, \tag{16}$$

$$(2ABb^{-1}(1 + B^\delta b^{-1}) + 1)Kr \le r_0 \tag{17}$$

and

$$\rho = (2ABb^{-1} + 1)Kk^{-1}. \tag{18}$$

Suppose now that $x \in M$, $m > 0$ and $0 < \bar{r} \le r$ satisfy $x \in B_{\bar{r}}(p)$ and $f^m(x) \in$

$B_{\bar r}(p)$. Let $0 = k_1 < \cdots < k_n = m$ be the integers in $[0, m]$ such that $f^{k_i}(x) \in B_{r_0}$. If $f^{k_i}(x) = f^{k_j}(x)$ for some $i \neq j$ then we are done; therefore, we may suppose the $f^{k_i}(x)$ distinct. Applying I.4 to the points $x_0 = x$, $x_j = f^{k_j}(x)$, $1 \leq j \leq n$, the splitting $T_pM = E_1 \oplus \cdots \oplus E_l$, and the norms $|v|_i = \lambda_i \|v\|$ on E_i, we find $0 \leq m_1 = k_{j_1} < k_{j_2} = m_2 \leq m$ and $\mu \in S$ such that defining $\|\cdot\|_1 = \|\cdot\|_\mu$ we have:

$$\|f^{m_i}(x) - x\|_1 \leq A\|f^m(x) - x\|_1, \qquad i = 1, 2$$

and

$$f^{k_j}(x) \notin \tilde{B}_{\delta r_2}(q) \tag{19}$$

for all $j_1 < j < j_2$, where $q = 2^{-1}(f^{m_1}(x) + f^{m_2}(x))$, $r_2 = 2^{-1}\|f^{m_1}(x) - f^{m_2}(x)\|_1$ and $\tilde{B}_{\delta r_2}(q) = \{v \mid \sup_i \mu_i \lambda_i \|\pi_i(v - q)\| \leq \delta r_2\}$. Then, for $i = 1, 2$,

$$\begin{aligned} \|f^{m_i}(x) - x\| &\leq B\|f^{m_i}(x) - x\|_1 \leq AB\|f^m(x) - x\|_1 \\ &\leq ABb^{-1}(\|f^m(x)\| + \|x\|) \leq 2ABb^{-1}K\bar r. \end{aligned} \tag{20}$$

We claim that

$$f^j(x) \notin \tilde{B}_{\delta r_2}(q)$$

for all $m_1 < j < m_2$. If $\|f^j(x) - q\|_1 \leq \delta r_2$, by (20) and (17) we have:

$$\begin{aligned} \|f^j(x)\| &\leq \|f^j(x) - q\| + \|q - x\| + \|x\| \\ &\leq B\|f^j(x) - q\|_1 + \tfrac{1}{2}\|f^{m_1}(x) - x\| + \tfrac{1}{2}\|f^{m_2}(x) - x\| + \|x\| \\ &\leq B\delta\tfrac{1}{2}\|f^{m_1}(x) - f^{m_2}(x)\|_1 + 2ABb^{-1}K\bar r + Kd(x, p) \\ &\leq B\delta\tfrac{1}{2}b^{-1}(\|f^{m_1}(x) - x\| + \|f^{m_2}(x) - x\|) + 2ABb^{-1}K\bar r + K\bar r \\ &\leq B\delta b^{-1}2ABb^{-1}K\bar r + 2ABb^{-1}K\bar r + K\bar r \\ &= \big(2ABb^{-1}(1 + B\delta b^{-1}) + 1\big)K\bar r \leq r_0. \end{aligned}$$

Hence $f^j(x) \in B_{r_0}$ and $j = k_i$ for some $j_1 < i < j_2$, contradicting (19). Moreover (20) and (16) imply, for $i = 1, 2$,

$$\begin{aligned} \|f^{m_i}(x)\| &\leq 2ABb^{-1}K\bar r + \|x\| \leq 2ABb^{-1}K\bar r + Kd(x, p) \\ &\leq (2ABb^{-1} + 1)K\bar r \leq r_1. \end{aligned}$$

Then we can apply Lemma I.3 to $y = f^{m_2}(x)$, $z = f^{m_1}(x)$ and we obtain $g \in \mathfrak{U}$ satisfying (2), (3) and (8). But since we took r_0 so small that (4) and (6) hold, then, as we observed after Lemma I.3, the diffeomorphism g satisfies $g(w) = f(w)$ for $w \notin B_\varepsilon(f, p)$,

$$g^{m_2 - m_1}(f^{m_2}(x)) = f^{m_2}(x) \quad \text{and} \quad d\big(g^j(f^{m_2}(x)), f^j(f^{m_1}(x))\big) \leq \varepsilon$$

for $0 \leq j \leq m_2 - m_1$. Finally by (20) and (18), for $i = 1, 2$,

$$d(f^{m_i}(x), p) \leq k^{-1} \| f^{m_i}(x) \| \leq (2ABb^{-1} + 1)Kk^{-1}\bar{r} = \rho\bar{r},$$

and the proof of I.2 is complete.

Define $\Sigma(\mu, \varepsilon, r, \rho, m)$, where $r > 0$, $\rho > 1$ and $m \in \mathbf{Z}^+$ as the set of points $x \in M$ such that if $y \in B_{r_1}(x)$ for some $0 < r_1 \leq r$ and $f^m(y) \in B_{r_1}(x)$ there exist $0 \leq m_1 < m_2 \leq m$, $g \in \mathcal{U}$ and $z \in M$ such that $g = f$ on $M - B_\varepsilon(f, x)$,

$$g^{m_2 - m_1}(z) = z, \, d(g^n(z), f^n(f^{m_1}(y))) \leq \varepsilon$$

for all $0 \leq n \leq m_2 - m_1$ and $f^{m_1}(y) \in \overline{B_{\rho r_1}(x)}$. (Observe that in this case $f^{m_1}(y) \in \Sigma(\mathcal{U}, \varepsilon)$.) It is easy to see that $\Sigma(\mathcal{U}, \varepsilon, r, \rho, m)$ is closed. Then the set $\Sigma(\mathcal{U}, \varepsilon, r, \rho)$ defined by $\Sigma(\mathcal{U}, \varepsilon, r, \rho) = \bigcup_{m \geq 1} \Sigma(\mathcal{U}, \varepsilon, r, \rho, m)$ is a Borel Set. Lemma I.2 implies that if $r_n > 0$, $\rho_n > 1$ are monotone sequences converging to 0 and $+\infty$ respectively then:

$$M = \bigcup_{n \geq 0} \bigcup_{m \geq 0} \Sigma(\mathcal{U}, \varepsilon, r_n, \rho_m) \tag{21}$$

for every neighborhood $\mathcal{U}$ of f and every $\varepsilon > 0$.

The proof of Proposition I.1 also requires a measure theoretical property about certain partitions of the torus $T^s = S^1 \times \overset{s}{\cdots} \times S^1$. First let us introduce some definitions. We say that a set $A \subset T^s$ is a cube if it can be written as $A = I_1 \times \cdots \times I_s$ where the sets I_i are intervals in S^1 (containing both, none or one of its boundary points) with equal lengths. If p_i is the middle point of I_i we say that the point $(p_1, \ldots, p_s)$ is the center of the cube A. The length of the intervals I_i is called the side of the cube. For each $k \in \mathbf{Z}^+$ let $\mathcal{P}_1^{(k)} \leq \mathcal{P}_2^{(k)} \leq \cdots$ be a sequence of partitions of T^s each one containing a finite set of disjoint cubes whose union is T^s. Suppose also that the side of the atoms of $\mathcal{P}_j^{(k)}$ is $2\pi/k^j$. For every atom Q of a partition $\mathcal{P}_j^{(k)}$ we can associate cubes $\hat{Q}$ and $\tilde{Q}$ having the same center as Q and sides $2\pi/k^{j-1}$ and $6\pi/k^{j-1}$ respectively. If $x \in T^s$, denote by $\mathcal{P}_j^{(k)}(x)$ the atom of $\mathcal{P}_j^{(k)}$ containing x. We shall suppose M isometrically embedded in T^s.

Lemma I.5. *For every probability measure μ on the Borel sets of T^s, every $0 < \delta$ and every odd integer k the following inequalities hold for any $j \geq 1$:*

$$\mu\left(\left\{x \mid \mu\left(\mathcal{P}_j^{(k)}(x)\right) \geq \delta\mu\left(\hat{\mathcal{P}}_j^{(k)}(x)\right)\right\}\right) \geq 1 - \delta K^s$$

and

$$\mu\left(\left\{x \mid \mu\left(\mathcal{P}_j^{(k)}(x)\right) \geq \delta\mu\left(\tilde{\mathcal{P}}_j^{(k)}(x)\right)\right\}\right) \geq 1 - \delta 3^s k^s.$$

Proof. Let $\{x_1,\dots,x_l\}$ be a set of points, one in each atom of $\mathcal{P}_l^{(k)}$ (hence $l = k^l$). Let

$$S = \left\{1 \le i \le l \mid \mu\left(\mathcal{P}_l^{(k)}(x_i)\right) < \delta\mu\left(\hat{\mathcal{P}}_l^{(k)}(x_i)\right)\right\}.$$

Then:

$$\begin{aligned}\mu\left(\left\{x \mid \mu\left(\mathcal{P}_l^{(k)}(x)\right) < \delta\mu\left(\hat{\mathcal{P}}_l^{(k)}(x)\right)\right\}\right) &= \sum_{i\in S} \mu\left(\mathcal{P}_l^{(k)}(x_i)\right) < \delta \sum_{i\in S} \mu\left(\hat{\mathcal{P}}_l^{(k)}(x_i)\right)\\ &\le \delta \sum_{i=1}^{l} \mu\left(\hat{\mathcal{P}}_l^{(k)}(x_i)\right)\\ &= \delta \sum_{i=1}^{l} k^s \mu\left(\mathcal{P}_l^{(k)}(x_i)\right) = \delta k^s.\end{aligned}$$

We used the fact that, since k is odd, the sets $\hat{\mathcal{P}}_l^{(k)}(x_i)$, $i = 1,\dots,l$, cover each atom of $\mathcal{P}_l^{(k)}$ exactly k^s times. This proves the first inequality. The second one follows by use of the same method.

Now we shall prove I.1. Let $f \in \mathrm{Diff}(M)$, $\varepsilon > 0$, a neighborhood $\mathcal{U}$ of f and an ergodic $\mu \in \mathcal{M}(f)$ be given. Extend μ to a measure on T^s defining $\mu(A) = \mu(A \cap M)$ for every Borel set A of T^s. Take monotone sequences $r_n > 0$, $\rho_n > 1$ converging to 0 and $+\infty$ respectively. For each couple of integers $n > 0$, $m > 0$ we can find $k = k(n, m)$ and $j(n, m)$ such that if $j \ge j(n, m)$ and $x \in T^s$ there exists $0 < r \le r_n$ satisfying

$$\mathcal{P}_j^{(k)}(x) \subset B_r(x) \tag{22}$$

and

$$\hat{\mathcal{P}}_l^{(k)}(x) \supset \overline{B_{\rho_m r}(x)}\,. \tag{23}$$

Now the notation $B_t(z)$ denotes the open ball in T^s with radius t and center z. We shall always choose $k(n, m)$ odd.

LEMMA I.6. *If* $x \in \Sigma(\mathcal{U}, \varepsilon, r_n, \rho_m)$, $j \ge j(n, m)$, $k = k(n, m)$ *and* $\mu(\mathcal{P}_l^{(k)}(x)) \ge \delta\mu(\hat{\mathcal{P}}_l^{(k)}(x))$ *we have:*

$$\mu\left(\hat{\mathcal{P}}_j^{(k)}(x) \cap \Sigma(\mathcal{U}, \varepsilon)\right) \ge \delta\mu\left(\hat{\mathcal{P}}_j^{(k)}(x)\right).$$

Proof. Take $0 < r \le r_n$ satisfying properties (22) and (23). Since μ is ergodic there exists $y \in M$ such that:

$$\lim_{l\to+\infty} \frac{1}{l}\#\left\{1 \le i \le l \mid f^i(y) \in \Sigma(\mathcal{U}, \varepsilon) \cap \hat{\mathcal{P}}_j^{(k)}(x)\right\} = \mu\left(\Sigma(\mathcal{U}, \varepsilon) \cap \hat{\mathcal{P}}_l^{(k)}(x)\right)$$

and

$$\lim_{l\to+\infty}\frac{1}{l}\#\{1\le i\le l\,|\,f^i(y)\in\mathcal{P}_j^{(k)}(x)\}=\mu(\mathcal{P}_j^{(k)}(x)).$$

By properties (22) and (23) and the definition of $\Sigma(\mathcal{U},\varepsilon,r_n,\rho_m)$ between any pair of integers $0\le i_1<i_2$ such that $f^{i_1}(y)$ and $f^{i_2}(y)$ belong to $\mathcal{P}_j^{(k)}(x)$ (and then to $B_r(x)$), there exists i_3, $i_1\le i_3<i_2$, such that $f^{i_3}(y)\in\overline{B_{\rho_m r}(x)}\cap\Sigma(\mathcal{U},\varepsilon)$ (and then to $\hat{\mathcal{P}}_j^{(k)}(x)\cap\Sigma(\mathcal{U},\varepsilon)$). Hence:

$$\#\{1\le i\le l\,|\,f^i(y)\in\hat{\mathcal{P}}_j^{(k)}(x)\cap\Sigma(\mathcal{U},\varepsilon)\}$$
$$\ge\#\{1\le i\le l\,|\,f^i(y)\in\mathcal{P}_j^{(k)}(x)\}-1.$$

Therefore

$$\begin{aligned}\mu(\hat{\mathcal{P}}_j^{(k)}(x)\cap\Sigma(\mathcal{U},\varepsilon))&=\lim_{l\to+\infty}\frac{1}{l}\{1\le i\le l\,|\,f^i(y)\in\Sigma(\mathcal{U},\varepsilon)\cap\hat{\mathcal{P}}_j^{(k)}(x)\}\\&\ge\lim_{l\to+\infty}\frac{1}{l}\{1\le i\le l\,|\,f^i(y)\in\mathcal{P}_j^{(k)}(x)\}\\&=\mu(\mathcal{P}_j^{(k)}(x))\ge\delta\mu(\hat{\mathcal{P}}_j^{(k)}(x)).\end{aligned}$$

Now define $\Lambda_\delta^0(n,m)$, for $\delta>0$, as the set of points $x\in T^s$ such that for $k=k(n,m)$ the inequalities

$$\mu(\mathcal{P}_j^{(k)}(x))\ge\delta\mu(\hat{\mathcal{P}}_j^{(k)}(x))$$

and

$$\mu(\mathcal{P}_j^{(k)}(x))\ge\delta\mu(\tilde{\mathcal{P}}_j^{(k)}(x))$$

hold for an infinite sequence $\nu(x)$ of values of j. By Lemma I.5 we have

$$\mu(\Lambda_\delta^0(n,m))\ge 1-\delta(k^s+3^sk^s). \tag{24}$$

Define $\Lambda_\delta(n,m)=\Sigma(\mathcal{U},\varepsilon,r_n,\rho_m)\cap\Lambda_\delta^0(n,m)$. By (24)

$$\bigcup_{r=1}^{\infty}\Lambda_{1/r}(n,m)=\Sigma(\mathcal{U},\varepsilon,r_n,\rho_m)\mathrm{mod}(0). \tag{25}$$

Therefore if we prove that

$$\mu(\Sigma(\mathcal{U},\varepsilon)^c\cap\Lambda_\delta(n,m))=0 \tag{26}$$

for all $0<\delta<1$ it will follow from (25) and (26) that

$$\Sigma(\mathcal{U},\varepsilon)\supset\Sigma(\mathcal{U},\varepsilon,r_n,\rho_m)\mathrm{mod}(0).$$

Since this property will hold for every $n>0$, $m>0$, using (21) we have $\mu(\Sigma(\mathcal{U},\varepsilon))=\mu(M)=1$ and the proof of I.1 will be completed.

The proof of (26) requires a property loosely resembling Vitali's Covering Lemma:

LEMMA I.7. *Given a neighborhood U of $\Lambda_\delta(n,m) \cap \Sigma(\mathcal{U},\varepsilon)^c$ there exist sequences $x_i \in \Lambda_\delta(n,m) \cap \Sigma(\mathcal{U},\varepsilon)^c$, $j_i \in \nu(x_i)$, $i = 1,2,3,\ldots$, such that*

I) *The sets $\hat{\mathcal{P}}_{j_i}^{(k)}(x_i)$, $i = 1,2,3,\ldots$ are disjoint and contained in U.*

II) $\mu(\Sigma(\mathcal{U},\varepsilon)^c \cap \Lambda_\delta(n,m) - \bigcup_i \hat{\mathcal{P}}_{j_i}^{(k)}(x_i)) = 0.$

Proof. Using standard measure theoretical arguments a translation $\tau\colon T^s \hookleftarrow$ can be found such that

$$\mu\Big(\tau\Big(\cup\,\big\{\partial\hat{A}\,|\,A \in \mathcal{P}_i^{(l)}, l \geq 1, i \geq 1\big\}\Big)\Big) = 0,$$

where $\partial\hat{A}$ denotes the boundary of $\hat{A}$. Then we can suppose, without loss of generality, that:

$$\mu\Big(\cup\,\big\{\partial\hat{A}\,|\,A \in \mathcal{P}_i^{(l)}, l \geq 1, i \geq 1\big\}\Big) = 0. \tag{27}$$

Denote by $\mathcal{F}$ the family of sets $\mathcal{P}_j^{(k)}(x)$ with $x \in \Lambda_\delta(n,m) \cap \Sigma(\mathcal{U},\varepsilon)^c$ and $j \in \nu(x)$. Take a sequence $A_i \in \mathcal{F}$, $i \geq 1$, satisfying:

a) $\hat{A}_i \subset U$ for all $1 \leq i$;

b) $\mu(\hat{A}_i \cap \hat{A}_l) = 0$ for all $1 \leq l < i$;

c) For all $1 < i$:

$$\operatorname{diam} A_i = \max\big\{\operatorname{diam} A\,|\,A \in \mathcal{F}, \hat{A} \subset U, \mu\big(\hat{A} \cap \hat{A}_l\big) = 0 \text{ for all } 1 \leq l < i\big\}.$$

It is easy to see that these properties imply:

$$\lim_{j \to +\infty} \operatorname{diam} A_j = 0 \tag{28}$$

and

$$\sum_i \mu(A_i) = \mu\Big(\bigcup_i A_i\Big) \leq 1. \tag{29}$$

We claim that for any $N \geq 1$

$$\Lambda_\delta(n,m) \cap \Sigma(\mathcal{U},\varepsilon)^c - \bigcup_{i=1}^{N} \overline{\hat{A}} \subset \bigcup_{i>N} \tilde{A}_i. \tag{30}$$

Suppose that $x \in \Lambda_\delta(n,m) \cap \Sigma(\mathcal{U},\varepsilon)^c - \bigcup_{i=1}^N \overline{\hat{A}}_i$. Then there exists $A \in \mathcal{F}$ with $x \in A$ and

$$\hat{A} \cap \bigcup_{i=1}^{N} \overline{\hat{A}}_i = \varnothing.$$

Take $N_1 > N$ such that

$$\hat{A} \cap \hat{A}_i = \varnothing$$

for $1 \leq i < N_1$ and

$$\hat{A} \cap \hat{A}_{N_1} \neq \varnothing .$$

The existence of N_1 follows from condition (c) and property (28). Also from (c), it follows that

$$\operatorname{diam} \hat{A} \leq \operatorname{diam} \hat{A}_{N_1}.$$

Then $\hat{A} \subset \tilde{A}_{N.}$ and

$$x \in A \subset \tilde{A}_{N_1} \subset \bigcup_{i>N} \tilde{A}_i ,$$

which completes the proof of (30). By (27) and (30)

$$\mu\left(\left(\Lambda_\delta(n, m) \cap \Sigma(\mathfrak{U}, \varepsilon)^c \right) - \bigcup_{i=1}^{N} \hat{A}_i \right)$$

$$= \mu\left(\left(\Lambda_\delta(n, m) \cap \Sigma(\mathfrak{U}, \varepsilon)^c \right) - \bigcup_{i=1}^{N} \overline{\hat{A}}_i \right)$$

$$\leq \mu\left(\bigcup_{i>N} \tilde{A}_i \right) \leq \sum_{i>N} \mu(\tilde{A}_i) \leq \delta^{-1} \sum_{i>N} \mu(A_i).$$

By (29) the lemma is proved.

It follows from Lemmas I.6 and I.7 that

$$\mu(U) \geq \sum_i \mu\left(\hat{\mathcal{P}}^{(k)}_{j_i}(x_i) \right) \geq \frac{1}{1-\delta} \sum_i \mu\left(\hat{\mathcal{P}}^{(k)}_{j_i}(x_i) \cap \Sigma(\mathfrak{U}, \varepsilon)^c \right)$$

$$= \frac{1}{1-\delta} \mu\left(\bigcup_i \hat{\mathcal{P}}^{(k)}_{j_i}(x_i) \cap \Sigma(\mathfrak{U}, \varepsilon)^c \right)$$

$$\geq \frac{1}{1-\delta} \mu\left(\Lambda_\delta(n, m) \cap \Sigma(\mathfrak{U}, \varepsilon)^c \right).$$

But if $\mu(\Lambda_\delta(n, m) \cap \Sigma(\mathfrak{U}, \varepsilon)^c) > 0$ we can take U satisfying

$$\mu(U) < \frac{1}{1-\delta} \mu\left(\Lambda_\delta(n, m) \cap \Sigma(\mathfrak{U}, \varepsilon)^c \right),$$

contradicting the last inequality. Hence $\mu(\Lambda_\delta(n, m) \cap \Sigma(\mathfrak{U}, \varepsilon)^c) = 0$.

II. Proof of Theorem B

We shall deduce Theorem B from Theorem A and the next proposition, in whose statement we shall use the following notation: if x is a periodic point of a diffeomorphism $f\colon M \hookleftarrow$ then $E^s_x(f)$ (resp. $E^u_x(f)$) denotes the set of vectors $v \in T_x M$ such that $\sup_{n \geq 0} \|(D_x f^n) v\| < \infty$ (resp. $\sup_{n \geq 0} \| D_x f^{-n}) v \| < \infty$). Moreover $\mathfrak{F}(M)$ and $\Lambda_j(f)$ will have the same meaning as in the introduction.

PROPOSITION II.1. *If $f \in \mathfrak{F}(M)$ there exists a neighborhood $\mathfrak{U}$ of f and constants $K > 0$, $m \in \mathbf{Z}^+$ and $0 < \lambda < 1$ such that:*

a) *If $g \in \mathfrak{U}$ and $x \in \mathrm{Per}(g)$ has minimum period $n \geq m$ then*

$$\prod_{i=0}^{k-1} \|(D_{g^{mi}(x)}g^m)/E^s_{g^{mi}(x)}(g)\| \leq K\lambda^k$$

and

$$\prod_{i=0}^{k-1} \|(D_{g^{-mi}(x)}g^{-m})/E^u_{g^{-mi}(x)}(g)\| \leq K\lambda^k,$$

where $k = [n/m]$.

b) *For all $0 < j < \dim M$ there exists a continuous splitting $TM/\Lambda_j(f) = G^s \oplus G^u$ such that*

$$\|(D_x f^m)/G^s_x\| \cdot \|(D_{f^m(x)} f^{-m})/G^u_{f^m(x)}\| \leq \lambda$$

for all $x \in \Lambda_j(f)$ and $G^s_x = E^s_x(f)$, $G^u_x = E^u_x(f)$ when $x \in \mathrm{Per}(f)$ and $\dim E^s_x(f) = j$.

c) *For all $x \in \mathrm{Per}(f)$*

$$\limsup_{n \to +\infty} \frac{1}{n} \sum_{i=0}^{n-1} \log\|(D_{f^{mi}(x)} f^m)/E^s_{f^{mi}(x)}(f)\| < 0$$

and

$$\limsup_{n \to +\infty} \frac{1}{n} \sum_{i=0}^{n-1} \log\|(D_{f^{-mi}(x)} f^{-m})/E^u_{f^{-mi}(x)}\| < 0.$$

Before proving Proposition II.1 let us show how it implies Theorem B. If $f \in \mathfrak{F}(M)$ then $\Omega(f) = \overline{\mathrm{Per}(f)}$ ([4], Lemma 3.1) and $\Lambda_0 f$ and $\Lambda_{\dim M}(f)$ are finite sets of sources and sinks [10]. Hence to show that f satisfies Axiom A, it is sufficient to show that $\Lambda_j(f)$ is a hyperbolic set for each $0 < j < \dim M$. Fix any $0 < j < \dim M$ and any compact neighborhood U_j of $\Lambda_j(f)$. Since $\Lambda_j(f) \cap \Lambda_i(f) = \varnothing$ for each $i \neq j$, we may take U_j such that $U_j \cap \Lambda_i(f) = \varnothing$ for $i \neq j$. First we claim that if $\mathfrak{U}_0 \subseteq \mathfrak{F}(M)$ is the connected component of $\mathfrak{F}(M)$ containing f and $g \in \mathfrak{U}_0$ is such that $g = f$ on $M - U_j$ then $\Lambda_i(f) = \Lambda_i(g)$ for all $i \neq j$. Suppose the property false. Then there exists $g \in \mathfrak{U}_0$ such that $g = f$ on $M - U_j$ and $\Lambda_{i_0}(f) \neq \Lambda_{i_0}(g)$ for some $i_0 \neq j$; it follows that $\Lambda_{i_0}(f) \subsetneq \Lambda_{i_0}(g)$ and there exists $p \in \mathrm{Per}(g) - \Lambda_{i_0}(f)$ with $\dim E^s_x(g) = i_0 \neq j$. Suppose that $g^n(x) = x$. Define ν: $\mathfrak{U}_0 \to \mathbf{Z}$ by

$$\nu(g) = \#\{y \in M \mid g^n(y) = y, \dim E^s_y(g) = i_0\}.$$

Since $\mathfrak{U}_0 \subset \mathfrak{F}(M)$, the function ν is continuous and since $\mathfrak{U}_0$ is connected it follows that it is constant. But the properties of g imply $\nu(g) > \nu(f)$ and this contradiction completes the proof of the claim.

Now we are ready to prove that the splitting $TM/\Lambda_i(f) = G^s \oplus G^u$ given by Proposition II.1 is a hyperbolic splitting. By compactness of $\Lambda_i(f)$ it is easy to see that if

$$\text{(1)} \qquad \liminf_{n\to+\infty} \|(D_x f^n)/G_x^s\| = 0$$

and

$$\liminf_{n\to+\infty} \|(D_x f^{-n})/G_x^u\| = 0$$

for all $x \in \Lambda_i(f)$, then the splitting $TM/\Lambda_i(f) = G^s \oplus G^u$ is hyperbolic. Hence everything is reduced to proving these properties. In fact we shall prove only the first one because the second follows if we apply the same methods to f^{-1} instead of f.

If (1) does not hold for every $x \in \Lambda_i(f)$, we can find $x \in \Lambda_i(f)$ and a sequence $j_n \to +\infty$ such that

$$\lim_{n\to+\infty} \frac{1}{j_n} \log\|(D_x f^{mj_n}) \mid G_x^s\| \geq 0,$$

where m is the integer given by II.1. Without loss of generality we can suppose that the sequence j_n is such that there exists an f^m-invariant probability measure μ on $\Lambda_i(f)$ such that

$$\int_{\Lambda_i(f)} \varphi \, d\mu = \lim_{n\to+\infty} \frac{1}{j_n} \sum_{i=0}^{j_n-1} \varphi(f^{mi}(x))$$

for every continuous φ: $\Lambda_i(f) \to \mathbf{R}$. Applying this property to φ defined by $\varphi(p) = \log\|(D_p f^m)/G_p^s\|$ we obtain:

$$\text{(2)} \qquad \int_{\Lambda_i(f)} \varphi \, d\mu = \lim_{n\to+\infty} \frac{1}{j_n} \sum_{i=0}^{j_n-1} \log\|(D_{f^{mi}(x)} f^m)/G^s_{f^{mi}(x)}\|$$

$$\geq \lim_{n\to+\infty} \frac{1}{j_n} \log\|(D_x f^{mj_n})/G_x^s\| \geq 0.$$

On the other hand, by Birkhoff's Theorem:

$$\text{(3)} \qquad \int_{\Lambda_i(f)} \varphi \, d\mu = \int_{\Lambda_i(f)} \lim_{n\to+\infty} \frac{1}{n} \sum_{i=0}^{n-1} \log\|(D_{f^{mi}(p)} f^m)/G^s_{f^{mi}(p)}\| \, d\mu(p).$$

By Theorem A, $\Lambda_i(f) \cap \Sigma(f)$ is an f-invariant total probability subset of $\Lambda_i(f)$.

Hence if we define an f-invariant probability measure ν on $\Lambda_j(f)$ by

$$\nu = \frac{1}{m}\sum_{l=0}^{m-1} f_*^l(\mu)$$

we obtain

$$0 = \nu(\Lambda_j(f) - \Lambda_j(f) \cap \Sigma(f)) = \frac{1}{m}\sum_{l=0}^{m-1} \mu(f^l(\Lambda_j(f) - \Lambda_j(f) \cap \Sigma(f)))$$

$$= \frac{1}{m}\sum_{l=0}^{m-1} \mu(\Lambda_j(f) - \Lambda_j(f) \cap \Sigma(f)) = \mu(\Lambda_j(f) - \Lambda_j(f) \cap \Sigma(f)).$$

This, together with (2) and (3) implies:

$$0 \leq \int_{\Sigma(f)\cap\Lambda_j(f)} \lim_{n\to+\infty} \frac{1}{n}\sum_{i=0}^{n-1} \log\|(D_{f^{mi}(p)}f^m)/G^s_{f^{mi}(p)}\| \, d\mu(p).$$

Hence there exists $p \in \Sigma(f) \cap \Lambda_j(f)$ such that:

(4) $$\lim_{n\to+\infty} \frac{1}{n}\sum_{i=0}^{n-1} \log\|(D_{f^{mi}(p)}f^m)/G^s_{f^{mi}(p)}\| \geq 0.$$

Now take $\lambda < \lambda_0 < 1$ and $n_0 > 0$ such that:

(5) $$\frac{1}{n}\sum_{i=0}^{n-1} \log\|(D_{f^{mi}(p)}f^m)/G^s_{f^{mi}(p)}\| \geq \log\lambda_0$$

when $n \geq n_0$. Observe that the point p is not periodic because if it were, it should be hyperbolic with $E^s_p(f) = G^s_p$ and then (4) would contradict part (c) of Proposition II.1.

Since $p \in \Sigma(f) \cap \Lambda_j(f)$ and $\Lambda_j(f) \cap \Lambda_i(f) = \varnothing$ for $i \neq j$ we can find g arbitrarily near f and $\bar{p} \in \mathrm{Per}(g)$ such that $g = f$ on $M - U_j$, $\bar{p} \in U_j$ and the distance between $f^l(p)$ and $g^l(\bar{p})$ is small for all $0 \leq l \leq n$, where n denotes the minimum g-period of $\bar{p}$. Since p is not f-periodic, the period n goes to ∞ when g approaches f. Hence we shall assume that:

(6) $$n \geq m,$$

(7) $$k \geq n_0,$$

(8) $$K\lambda^k < \lambda_0^k,$$

and

(9) $$\left(\frac{\lambda}{\lambda_0}\right)^k C^m \leq 1/2,$$

where $k = [n/m]$, K and λ are given by II.1 and $C = \sup_{x\in M}\|(D_x f^{-1})\|$. These properties imply that

$$\|(D_{f^n(p)}f^{-n})/G^u_{f^n(p)}\| \leq 1/2, \tag{10}$$

since part b) of II.1 and inequality (5) imply

$$\begin{aligned}
&\|(D_{f^n(p)}f^{-n})/G^u_{f^n(p)}\| \\
&\quad \leq \left(\prod_{l=0}^{k-1} \|(D_{f^{n-ml}(p)}f^{-m})/G^u_{f^{n-ml}(p)}\|\right) \cdot \|(D_{f^{n-mk}(p)}f^{-(n-km)}/G^u_{f^{n-mk}}\| \\
&\quad \leq \lambda^k C^m \lambda_0^{-k} \leq 1/2.
\end{aligned}$$

The next step of the proof requires a lemma due to Franks ([2] Lemma 1.1). The version that we shall use here is slightly more general but the proof in [2] can be easily extended to include this statement.

LEMMA II.2. *If $f \in \mathrm{Diff}(M)$ for any neighborhood $\mathcal{U}$ of f there exist $\varepsilon > 0$ and a neighborhood $\mathcal{U}_0 \subset \mathcal{U}$ of f such that given $g \in \mathcal{U}_0$, a finite set $\{x_1, \ldots, x_N\} \subset M$, a neighborhood U of $\{x_1, \ldots, x_N\}$ and linear maps $L_l\colon T_{x_l}M \to T_{g(x_l)}M$ such that $\|L_l - (D_{x_l}g)\| \leq \varepsilon$ for all $1 \leq l \leq N$, then there exists $\bar{g} \in \mathcal{U}$ such that $\bar{g}(x) = g(x)$ if $x \in \{x_1, x_2, \ldots, x_N\} \cup (M - U)$ and $D_{x_l}\bar{g} = L_l$ for all $1 \leq l \leq N$.*

Using this lemma we can perturb g to obtain a diffeomorphism that coincides with g on $M - U$ and on the g-orbit of $\bar{p}$ but such that its derivative at points $g^l(\bar{p})$, $1 \leq l \leq n$, is essentially a translation of $D_{f^l(p)}f$. More precisely it follows from II.2 that there exists $\bar{g} \in \mathcal{U}$ (where $\mathcal{U}$ is given by II.1) such that $\bar{g} = g = f$ on $M - U_i$ (with $U_i \supseteq \Lambda_i(f)$ given above), $g^l(\bar{p}) = \bar{g}^l(\bar{p})$ for all $0 \leq l \leq n$ and

$$D_{g^l(\bar{p})}\bar{g} = A_{l+1}^{-1}(D_{f^l(p)}f)A_l \quad \text{for } 0 \leq l \leq n-1;$$

here $A_l\colon T_{g^l(\bar{p})}M \to T_{f^l(p)}M$ are isomorphisms such that A_l^{-1} restricted to $G^s_{f^l(p)}$ and to $G^u_{f^l(p)}$ are isometries, for each $0 \leq l \leq n$, and such that

$$A_0^{-1}G^u_p = A_n^{-1}(G^u_{f^n(p)})$$

and

$$A_0^{-1}G^s_p = A_n^{-1}(G^s_{f^n(p)}).$$

These properties imply that

$$T_{\bar{p}}M = A_0^{-1}G^u_p \oplus A_0^{-1}G^s_p, \tag{11}$$

$$(D_{\bar{p}}\bar{g}^n)A_0^{-1}G^u_p = A_0^{-1}G^u_p \tag{12}$$

and

$$(13)\qquad (D_{\bar p}\bar g^n)A_0^{-1}G_p^s = A_0^{-1}G_p^s.$$

Moreover by (10):

$$\begin{aligned}\|(D_{\bar p}\bar g^{-n})/A_0^{-1}G_p^u\| &= \|(D_{\bar g^n(\bar p)}\bar g^{-n})/A_n^{-1}(G^u_{f^n(p)})\|\\ &= \|(D_{f^n(p)}f^{-n})/G^u_{f^n(p)}\| \le 1/2.\end{aligned}$$

But then $A_0^{-1}G_p^u \subset E^u_{\bar p}(\bar g)$. By the claim, $\Lambda_i(\bar g) = \Lambda_i(f)$ for $i \neq j$ and therefore $\bar p \in \Lambda_j(\bar g)$. Hence

$$\dim A_0^{-1}G_p^u = \dim G_p^u = (\dim M) - j = \dim E^u_{\bar p}(\bar g)$$

and then $A_0^{-1}G_p^u = E^u_{\bar p}(\bar g)$. Since $\bar p$ is a hyperbolic periodic point, properties (11), (12) and (13) imply that $A_0^{-1}G_p^s = E^s_{\bar p}(\bar g)$. But then for any $i \in \mathbf{Z}^+$,

$$\begin{aligned}E^s_{\bar g^{mi}(\bar p)}(\bar g) &= (D_{\bar p}\bar g^{mi})E^s_{\bar p}(\bar g)\\ &= (D_{\bar p}\bar g^{mi})A_0^{-1}G_p^s\\ &= A_{mi}^{-1}(D_pf^{mi})G_p^s = A_{mi}^{-1}G^s_{f^{mi}(p)}\end{aligned}$$

and this implies

$$\|(D_{\bar g^{mi}(\bar p)}\bar g^m)/E^s_{\bar g^{mi}(\bar p)}(\bar g)\| = \|(D_{f^{mi}(p)}f^m)/G^s_{f^{mi}(p)}\|.$$

Hence, by (5) and (7):

$$\prod_{i=0}^{k-1}\|(D_{\bar g^{mi}(\bar p)}\bar g^m)/E^s_{\bar g^{mi}(\bar p)}(\bar g)\| = \prod_{i=0}^{k-1}\|(D_{f^{mi}(p)}f^m)/G^s_{f^{mi}(p)}\| \ge \lambda_0^k.$$

But by II.1 the first product is $< K\lambda^k$ thus contradicting property (8).

Now we shall prove Proposition II.1. The main tool for its proof is the following result about families of periodic sequences of linear maps. Let GL(N) be the group of linear isomorphisms of $\mathbf{R}^N$. If $\xi\colon \mathbf{Z} \to \mathrm{GL}(N)$ is a sequence of isomorphisms of $\mathbf{R}^N$, denote $E_j^s(\xi)$ (resp. $E_j^u(\xi)$) the space of vectors $v \in \mathbf{R}^N$ such that

$$\sup\left\{\left\|\left(\prod_{i=0}^{n}\xi_{j+i}\right)v\right\| \,\Big|\, n \ge 0\right\} < \infty$$

$$\left(\text{resp. } \sup\left\{\left\|\left(\prod_{i=0}^{n}(\xi_{j-1-i})^{-1}\right)v\right\| \,\Big|\, n \ge 0\right\} < \infty\right).$$

We say that the sequence ξ is hyperbolic if $E_j^s(\xi) \oplus E_j^u(\xi) \equiv \mathbf{R}^N$ for all $j \in \mathbf{Z}$. It is equivalent to require this condition for only some value of j. When the sequence is periodic, i.e., if there exists $n_0 \geq 1$ such that $\xi_{j+n_0} = \xi_j$ for all j, then the hyperbolicity of the sequence is equivalent to the hyperbolicity of the linear map $\prod_{j=0}^{n_0-1} \xi_j$. We say that the sequence ξ is contracting if $E_j^s = \mathbf{R}^N$ for every $j \in \mathbf{Z}$. Let $\{\xi^{(\alpha)} \mid \alpha \in \mathcal{C}\}$ be a family of periodic sequences of linear maps. We say that it is hyperbolic (resp. contracting) if every sequence of the family is hyperbolic (resp. contracting) and $\sup\{\|\xi_n^{(\alpha)}\| \mid \alpha \in \mathcal{C},\ n \in \mathbf{Z}\} < \infty$. If $\{\xi^{(\alpha)} \mid \alpha \in \mathcal{C}\}$, $\{\eta^{(\alpha)} \mid \alpha \in \mathcal{C}\}$ are families of periodic sequences of linear maps of $\mathbf{R}^N$, we define

$$d(\xi, \eta) = \sup\{\|\xi_n^{(\alpha)} - \eta_n^{(\alpha)}\| \mid \alpha \in \mathcal{C}, n \in \mathbf{Z}\}$$

and we say that the two families are periodically equivalent if for every $\alpha \in \mathcal{C}$ the minimum periods of $\xi^{(\alpha)}$ and of $\eta^{(\alpha)}$ coincide. Finally we say that a hyperbolic family $\{\xi^{(\alpha)} \mid \alpha \in \mathcal{C}\}$ is uniformly hyperbolic if there exists $\varepsilon > 0$ such that any periodically equivalent family $\{\eta^{(\alpha)} \mid \alpha \in \mathcal{C}\}$ satisfying $d(\xi, \eta) \leq \varepsilon$ is also hyperbolic (resp. contracting). The key property of uniformly hyperbolic families is the following:

LEMMA II.3. *If $\{\xi^{(\alpha)} \mid \alpha \in \mathcal{C}\}$ is a uniformly hyperbolic family of periodic sequences of isomorphisms of $\mathbf{R}^N$, then there exist constants $K > 0$, $m \in \mathbf{Z}^+$ and $0 < \lambda < 1$ such that:*

a) *If $\alpha \in \mathcal{C}$ and $\xi^{(\alpha)}$ has minimum period $n \geq m$, then*

$$\prod_{j=0}^{k-1} \left\| \left(\prod_{i=0}^{m-1} \xi_{mj+i}^{(\alpha)} \right) \Big/ E_{mj}^s(\xi^{(\alpha)}) \right\| \leq K\lambda^k$$

and

$$\prod_{j=0}^{k-1} \left\| \left(\prod_{i=0}^{m-1} \xi_{mj+i}^{(\alpha)} \right)^{-1} \Big| E_{m(j+1)}^u(\xi^{(\alpha)}) \right\| \leq K\lambda^k,$$

where $k = [n/m]$.

b) *For all $\alpha \in \mathcal{C}$, $j \in \mathbf{Z}$:*

$$\left\| \prod_{i=0}^{m-1} \xi_{j+i}^{(\alpha)} / E_j^s(\xi^{(\alpha)}) \right\| \cdot \left\| \left(\prod_{i=0}^{m-1} \xi_{j+i}^{(\alpha)} \right)^{-1} \Big/ E_{j+m}^u(\xi^{(\alpha)}) \right\| \leq \lambda.$$

c) *For all $\alpha \in \mathcal{C}$*

$$\limsup_{n \to +\infty} \frac{1}{n} \sum_{j=0}^{n-1} \log \left\| \left(\prod_{i=0}^{m-1} \xi_{mj+i}^{(\alpha)} \right) \Big/ E_{mj}^s(\xi^{(\alpha)}) \right\| < 0$$

and

$$\limsup_{n \to +\infty} \frac{1}{n} \sum_{j=0}^{n-1} \log \left\| \left(\prod_{i=0}^{m-1} \xi_{mj+i}^{(\alpha)} \right)^{-1} \Big/ E_{m(j+1)}^u(\xi^{(\alpha)}) \right\| < 0.$$

To prove II.1 from this lemma we begin by taking the neighborhood $\mathfrak{U}_0$ of f and the constant $\varepsilon > 0$ given by Lemma II.2 when applied to f and $\mathfrak{U} = \mathfrak{F}(M)$. We claim that the family of periodic sequences $\{\xi^{(x,g)} \mid g \in \mathfrak{U}_0, x \in \mathrm{Per}(g)\}$, where $\xi_j^{(x,g)}$ is the matrix of $D_{g^j(x)}g$ written with respect to orthonormal basis of $T_{g^j(x)}M$ and $T_{g^{j+1}(x)}M$, is uniformly hyperbolic. If it was not, we could find a family $\{\eta^{(x,g)} \mid g \in \mathfrak{U}_0, x \in \mathrm{Per}(g)\}$ such that $d(\xi, \eta) \leq \varepsilon$ and such that for some $g \in \mathfrak{U}_0$, $x \in \mathrm{Per}(g)$, the sequences $\eta^{(x,g)}$ and $\xi^{(x,g)}$ have the same minimum period and $\eta^{(x,g)}$ is not hyperbolic. Then by II.2 we could find $\bar{g} \in \mathfrak{U} = \mathfrak{F}(M)$ such that $\bar{g}(g^j(x)) = g^{j+1}(x)$ for all j and $D_{g^j(x)}\bar{g} = \eta_j^{(x,g)}$. Hence x would be a non-hyperbolic periodic point of $\bar{g}$, contradicting $\bar{g} \in \mathfrak{F}(M)$.

Now applying II.3 to the family $\{\xi^{(x,g)} \mid g \in \mathfrak{U}_0, x \in \mathrm{Per}(g)\}$ we obtain property (c) of II.1 as a corollary of (c) of II.3 and property (a) of II.1 as a corollary of (a) of II.3. To prove (b) of II.1 requires a little more work. To define the splitting $G^s \oplus G^u = TM/\Lambda_j(f)$, we define $G_x^s = E_x^s(f)$, $G_x^u = E_x^u(f)$ if $x \in \mathrm{Per}(f)$ and $\dim E_x^s(f) = j$. To extend the definition to points of $\Lambda_j(f)$ that do not satisfy these properties, we fix, for each orbit $\{f^n(x) \mid n \in \mathbf{Z}\}$ in $\Lambda_j(f)$ such that x is not a periodic point with $\dim E_x^s(f) = j$, a sequence $x_n \in \mathrm{Per}(f)$ with $\dim E_{x_n}^s(f) = j$ such that $\lim x_n = x$ (and then $\lim f^k(x_n) = f^k(x)$ for all k) and such that for all k the subspaces $E_{f^k(x_n)}^s(f)$, $E_{f^k(x_n)}^u(f)$ converge to subspaces of $T_{f^k(x)}M$ which we define as $G_{f^k(x)}^s$, $G_{f^k(x)}^u$. Now we have attached to every $x \in \Lambda_j(f)$ two subspaces G_x^s, G_x^u with the following properties:

$$\dim G_x^s + \dim G_x^u = \dim M, \tag{14}$$

and

$$(D_x f) G_x^\sigma = G_{f(x)}^\sigma, \qquad \sigma = s, u \tag{15}$$

for all $x \in \Lambda_j(f)$ and, from property (b) of II.3,

$$\|(D_x f^m)/G_x^s\| \cdot \|(D_{f^m(x)} f^{-m})/G_{f^m(x)}^u\| \leq \lambda \tag{16}$$

if $x \in \mathrm{Per}(f)$ and $\dim E_x^s(f) = j$. Since these points are dense in $\Lambda_j(f)$ it follows that (16) holds for every $x \in \Lambda_j(f)$. From (16) it follows that $G_x^s \cap G_x^u = \{0\}$ which, together with (14), implies

$$G_x^s \oplus G_x^u = T_x M \tag{17}$$

for all $x \in \Lambda_j(f)$. By Proposition 1.3 of [5] properties (15)–(17) imply that the subspaces G_x^s, G_x^u depend continuously on x. By (16) these subspaces satisfy the inequality in II.1 (b) and by definition $E_x^s(f) = G_x^s$, $E_x^u(f) = G_x^u$ when $x \in \mathrm{Per}(f)$ and $\dim E_x^s(f) = j$, thus completing the proof of II.1.

For the proof of II.3 we shall need certain properties of uniformly contracting families of periodic sequences.

LEMMA II.4. *If* $\{\xi^{(\alpha)} \mid \alpha \in \mathcal{Q}\}$ *is a uniformly contracting family of periodic sequences of isomorphisms of* $\mathbf{R}^N$, *there exist* $\varepsilon > 0$ *and* $K > 0$ *such that if* $\{\eta^{(\alpha)} \mid \alpha \in \mathcal{Q}\}$ *is a family of periodic sequences such that* $d(\xi, \eta) \leq \varepsilon$ *then*

$$\left\| \prod_{j=0}^{n-1} \eta_j^{(\alpha)} \right\| \leq K$$

for all $\alpha \in \mathcal{Q}$, *where n denotes the minimum period of the sequence* $\eta^{(\alpha)}$.

Proof. If the lemma is false, for every $\varepsilon > 0$ and $K > 0$ we can find a family $\{\eta^{(\alpha)} \mid \alpha \in \mathcal{Q}\}$ with $d(\xi, \eta) \leq \varepsilon$ such that for some α_0 the matrix $\prod_{j=0}^{n-1} \eta_j^{(\alpha_0)}$ has an entry b with $|b| \geq K$, where n denotes the minimum period of the sequence $\eta^{(\alpha)}$. Take an $N \times N$ matrix A such that all its entries are 0 except one whose value we shall denote δ. If this nonzero entry is correctly placed we have

$$\operatorname{tr}\left(\prod_{j=0}^{n-1} \eta_j^{(\alpha_0)} \right)(I + A) = \operatorname{tr}\left(\prod_{j=0}^{n-1} \eta_j^{(\alpha_0)} \right) + b\delta.$$

If ε is small enough, $\{\eta^{(\alpha)} \mid \alpha \in \mathcal{Q}\}$ is a contracting family. Hence

$$\left| \operatorname{tr}\left(\prod_{j=0}^{n-1} \eta_j^{(\alpha_0)} \right) \right| < N$$

because all its eigenvalues have modulus < 1. Take $\delta = 2N/K$. Then:

$$\left| \operatorname{tr}\left(\prod_{j=0}^{n-1} \eta_j^{(\alpha_0)} \right)(I + A) \right| \geq |b\delta| - N \geq N. \tag{18}$$

Now define a periodic sequence $\tilde{\eta}^{(\alpha_0)}\colon \mathbf{Z} \to \mathrm{GL}(N)$ with period n such that $\tilde{\eta}_j^{(\alpha_0)} = \eta_j^{(\alpha_0)}$ for $0 \leq j \leq n-2$, $\tilde{\eta}_{n-1}^{(\alpha_0)} = \eta_{n-1}^{(\alpha_0)}(I + A)$. Let $C = \sup\{\|\xi_j^{(\alpha)}\| \mid \alpha \in \mathcal{Q},\ j \in \mathbf{Z}\}$. Then $\sup\{\|\eta_j^{(\alpha)}\| \mid \alpha \in \mathcal{Q},\ j \in \mathbf{Z}\} \leq C + \varepsilon$. Hence if ε is small, say $\varepsilon < C$:

(19)

$$\begin{aligned}\sup\left\{\|\tilde{\eta}_j^{(\alpha_0)} - \eta_j^{(\alpha_0)}\| \mid j \in \mathbf{Z}\right\} &= \|\tilde{\eta}_{n-1}^{(\alpha_0)} - \eta_{n-1}^{(\alpha_0)}\| \\ &= \|\eta_{n-1}^{(\alpha_0)}\|\|A\| \leq (C + \varepsilon)\delta = (C + \varepsilon)\frac{2N}{K} \leq \frac{4NC}{K}.\end{aligned}$$

By (18) the sequence $\tilde{\eta}^{(\alpha_0)}$ is not contracting and by (19) the family $\{\tilde{\eta}^{(\alpha)} \mid \alpha \in \mathcal{Q}\}$ obtained by replacing $\eta^{(\alpha_0)}$ with $\tilde{\eta}^{(\alpha_0)}$ satisfies $d(\tilde{\eta}, \eta) \leq 4NC/K$ and is not contracting. Hence $d(\tilde{\eta}, \xi) \leq \varepsilon + 4NC/K$. Since ε is arbitrarily small and K arbitrarily large it follows that $\{\xi^{(\alpha)} \mid \alpha \in \mathcal{Q}\}$ is not uniformly contracting.

LEMMA II.5. *Let* $\{\xi^{(\alpha)} \mid \alpha \in \mathcal{Q}\}$ *be a uniformly contracting family of periodic sequences of isomorphisms of* $\mathbf{R}^N$. *Then there exist* $K_0 > 0$, $0 < \lambda < 1$ *and* $m_0 \in \mathbf{Z}^+$ *such that if* $\alpha \in \mathcal{Q}$ *and* $\xi^{(\alpha)}$ *has minimum period* $n \geq m_0$, *then*

$$\prod_{j=0}^{k-1} \left\| \prod_{i=0}^{m_0-1} \xi^{(\alpha)}_{i+m_0 j} \right\| \leq K_0 \lambda^k$$

where $k = [n/m_0]$.

Proof. First let us prove the following lemma.

LEMMA II.6. *If* $A \in \mathrm{GL}(N)$, $v \in \mathbf{R}^N$ *and* $\delta > 0$, *there exists* $P \in \mathrm{GL}(N)$ *with* $\| P - I \| \leq N\delta$ *such that*

$$\| APv \| \geq \delta \| A \| \| v \|.$$

Proof. $A = TS$ where T is an isometry and S a positive symmetric matrix which without loss of generality we can suppose to be diagonal: $S = \mathrm{diag}(\lambda_1, \dots, \lambda_N)$ with $|\lambda_1| = \max\{|\lambda_j| \mid 1 \leq j \leq N\} = \| A \|$. Suppose that $v = (v_1, \dots, v_N)$ and that $|v_{j_0}| = \sup\{|v_j| \mid 1 \leq j \leq N\}$. Then $|v_{j_0}| \geq N^{-1}(\Sigma v_j^2)^{1/2} = N^{-1} \| v \|$. Let Q be a matrix with all its entries 0 except one which is $\sigma\delta N$ where σ is $+1$ or -1 according to the sign of v_1. If this entry is correctly placed we have $(I + Q)v = (v_1 + \delta N v_{j_0}, v_2, \dots, v_N)$. Then

$$\begin{aligned} \| A(I + Q)v \| = \| S(I + Q)v \| &\geq |\lambda_1(v_1 + \sigma\delta N v_{j_0})| \\ &\geq \delta N |\lambda_1| |v_{j_0}| \geq \delta \| A \| \| v \|. \end{aligned}$$

Hence, taking $P = I + Q$, the lemma is proved.

To prove II.5 take $\varepsilon > 0$ and K given by Lemma II.4. Let

$$C = \inf\{ \| Tv \| \mid \| v \| = 1, \| T - \xi^{(\alpha)} \| \leq \varepsilon \text{ for some } \alpha \in \mathcal{Q} \}$$

and

$$C_0 = \sup\{ \| T \| \mid \| T - \xi^{(\alpha)} \| \leq \varepsilon \text{ for some } \alpha \in \mathcal{Q} \}.$$

Let $\varepsilon_0 = \varepsilon / 2C_0$ and m_0 be such that $(\varepsilon_0/N)(1 + \varepsilon_0)^{m_0} > 1$. Define $\lambda = ((\varepsilon_0/N)(1 + \varepsilon_0)^{m_0})^{-1}$ and put $K_0 = K/C^{m_0}$. We shall prove that these constants, K_0, λ and m_0, satisfy all the required properties. Let $\alpha \in \mathcal{Q}$ be such that $\xi^{(\alpha)}$ has minimum period $n \geq m_0$. Take any $v \neq 0$ in $\mathbf{R}^N$ and let $\eta = (1 + \varepsilon_0)\xi^{(\alpha)}$.

By II.6 we can take matrices $P_0, \dots, P_{k-1}$, where $k = [n/m_0]$ such that $\| P_j - I \| \leq \varepsilon_0$ for all $0 \leq j \leq k - 1$ and:

$$\left\| \left(\prod_{i=0}^{m_0-1} \eta_{i+m_0 j} \right) P_j v_j \right\| \geq \frac{\varepsilon_0}{N} \left\| \prod_{i=0}^{m_0-1} \eta_{i+m_0 j} \right\| \cdot \| v_j \|,$$

where $v_0 = v$ and, for $j \geq 1$,

$$v_j = \left(\prod_{l=0}^{j-1}\left(\prod_{i=0}^{m_0-1} \eta_{i+m_0 l}\right) P_l\right) v_0.$$

Now take the sequence $\tilde{\eta}$ defined by the conditions of having period n and

$$\tilde{\eta}_{m_0 j} = \eta_{m_0 j} P_j$$

for $0 \leq j \leq k-1$ and $\tilde{\eta}_i = \eta_i$ when $0 \leq i \leq n-1$ is not a multiple of m_0. Then $\|\tilde{\eta}_i - \eta_i\| \leq \varepsilon_0 C_0$ and $\|\eta_i - \xi_i^{(\alpha)}\| \leq \varepsilon_0 C_0$ for all i. Hence $\|\tilde{\eta}_i - \xi_i^{(\alpha)}\| \leq 2\varepsilon_0 C_0 = \varepsilon$ for all i. Therefore by II.4,

$$\left\|\left(\prod_{j=0}^{n-1} \tilde{\eta}_j\right) v\right\| \leq K \|v\|.$$

But by the definition of the matrices P_j:

$$\left\|\left(\prod_{j=0}^{n-1} \tilde{\eta}_j\right) v\right\| \geq C^{m_0} \left\|\left(\prod_{j=0}^{km_0-1} \tilde{\eta}_j\right) v\right\|$$

$$\geq C^{m_0} \prod_{j=0}^{k-1}\left(\left\|\prod_{i=0}^{m_0-1} \eta_{i+m_0 j}\right\| \frac{\varepsilon_0}{N}\right) \|v\|.$$

Hence:

$$K_0 = \frac{K}{C^{m_0}} \geq \prod_{j=0}^{k-1}\left(\left\|\prod_{i=0}^{m_0-1} \eta_{i+m_0 j}\right\| \frac{\varepsilon_0}{N}\right)$$

$$= \prod_{j=0}^{k-1}\left(\left\|\prod_{i=0}^{m_0-1} \xi^{(\alpha)}_{i+m_0 j}\right\| (1+\varepsilon_0)^{m_0} \frac{\varepsilon_0}{N}\right) = \lambda^{-k} \prod_{j=0}^{k-1}\left\|\prod_{i=0}^{m_0-1} \xi^{(\alpha)}_{i+m_0 j}\right\|.$$

Lemma II.7. *Let $\{\xi^{(\alpha)} \mid \alpha \in \mathcal{Q}\}$ be a uniformly contracting family of periodic sequences of isomorphisms of $\mathbf{R}^N$. There exist $K > 0$, $0 < \lambda < 1$, $m \in \mathbf{Z}^+$ such that*

a) *If $\alpha \in \mathcal{Q}$ and $\xi^{(\alpha)}$ has minimum period $n \geq m$, then*

$$\prod_{j=0}^{k-1}\left\|\prod_{i=0}^{m-1} \xi^{(\alpha)}_{i+mj}\right\| \leq K\lambda^k,$$

where $k = [n/m]$.

b) *For all $\alpha \in \mathcal{Q}$*

$$\limsup_{n\to+\infty} \frac{1}{n} \sum_{j=0}^{n-1} \log\left\|\prod_{i=0}^{m-1} \xi^{(\alpha)}_{i+mj}\right\| < 0.$$

Proof. If for some $m \in \mathbf{Z}^+$ we can find $K > 0$ and $0 < \lambda < 1$ satisfying (a) then for any multiple of m there exist $K > 0$ and $0 < \lambda < 1$ that also satisfy the property. The same is true for property (b): if the property holds for some value of m then it holds also for any multiple of m. Therefore if we find some value $m = m_1$ satisfying (b), any common multiple of m_1 and the constant m_0 given by II.5 will satisfy both (a) and (b). Therefore the proof of II.7 is reduced to proving part (b). Let m_0 be given by II.5 and let $\mathcal{Q}_0$ be the set of $\alpha \in \mathcal{Q}$ such that property (b) holds for $m = m_0$. It is easy to see that $\mathcal{Q}_0$ contains every α such that $\xi^{(\alpha)}$ has large minimum period. We claim that there exists $m_1 \in \mathbf{Z}^+$ such that given $\alpha \in \mathcal{Q} - \mathcal{Q}_0$ and $j \in \mathbf{Z}$ there exists $0 < p = p(\alpha, j) \leq m_1$ such that $\left\|\prod_{i=0}^{p-1}\xi_{j+i}^{(\alpha)}\right\| \leq 1/2$. If the claim is false, for every $n \in \mathbf{Z}^+$ there exist $\alpha_n \in \mathcal{Q} - \mathcal{Q}_0$ and $j_n \in \mathbf{Z}$ such that $\left\|\prod_{i=0}^{p-1}\xi_{j_n+i}^{(\alpha_n)}\right\| > 1/2$ for all $0 < p \leq n$. Since the minimum periods of the sequences $\xi^{(\alpha_n)}$ are bounded (because $\alpha_n \in \mathcal{Q} - \mathcal{Q}_0$), we can suppose that they are constant, say $\bar{n}$. Since the set $\{\xi^{(\alpha)} \mid \alpha \in \mathcal{Q}\}$ is bounded there exists a sequence of positive integers $n_1, n_2, n_3, \ldots$ such that setting $j_{n_l} = k_l$, for every $i \geq 0$ the sequence $\xi_{k_l+i}^{(\alpha_{n_l})}$ in l converges to some matrix ξ_i. Then the sequence ξ_i has period $\bar{n}$ and

$$\left\|\prod_{i=0}^{n-1} \xi_i\right\| \geq 1/2 \tag{20}$$

for all $n > 0$. Moreover

$$\sup_{i \geq 0} \|\xi_i - \xi_{k_l+i}^{(\alpha_{n_l})}\| = \sup_{0 \leq i \leq n} \|\xi_i - \xi_{k_l+i}^{(\alpha_{n_l})}\|.$$

The second term goes to zero when $l \to +\infty$. Hence the fact that $\{\xi^{(\alpha)} \mid \alpha \in \mathcal{Q}\}$ is uniformly contracting implies that ξ_i is contracting, contradicting (20) and completing the proof of the claim. Now let $C = \sup\{\|\xi_i^{(\alpha)}\| \mid \alpha \in \mathcal{Q}, i \in \mathbf{Z}\}$ and take $\bar{m} \in \mathbf{Z}^+$ such that

$$(1/2)^{[\bar{m}/m_1]} C^{m_1} \leq 1/2.$$

Then for all $\alpha \in \mathcal{Q} - \mathcal{Q}_0$ and $j \in \mathbf{Z}$ we have

$$\left\|\prod_{i=0}^{\bar{m}-1} \xi_{i+j}^{(\alpha)}\right\| \leq 1/2, \tag{21}$$

because by the claim we can find $0 = p_0 < p_1 < \cdots < p_L < p_{L+1} = \bar{m}$ such that for each $0 \leq l \leq L$

$$p_{l+1} - p_l \leq m_1$$

and

$$\left\|\prod_{i=0}^{a_l} \xi_{p_l+j+i}^{(\alpha)}\right\| \leq 1/2,$$

where $a_l = p_{l+1} - p_l - 1$. Then $L \geq [\bar{m}/m_1]$ and:

$$\left\| \prod_{i=0}^{\bar{m}-1} \xi_{i+j}^{(\alpha)} \right\| \leq C^{\bar{m}-p_L} \prod_{l=0}^{L} \left\| \prod_{i=0}^{a_l} \xi_{p_l+j+i}^{(\alpha)} \right\| \leq C^{m_1}(1/2)^L$$

$$\leq C^{m_1}(1/2)^{[\bar{m}/m_1]} \leq 1/2.$$

This completes the proof of (21) and from (21) it is clear that (b) holds for $m = \bar{m}$ and $\alpha \in \mathcal{Q} - \mathcal{Q}_0$. Then choosing as m a common multiple of $\bar{m}$ and m_0, property (b) is satisfied for every $\alpha \in \mathcal{Q}$.

LEMMA II.8. *If $\{\xi^{(\alpha)} \mid \alpha \in \mathcal{Q}\}$ is a uniformly hyperbolic family of periodic sequences of isomorphisms of $\mathbf{R}^N$, there exist $\varepsilon > 0$, $K > 0$, $0 < \lambda < 1$ such that if $\{\eta^{(\alpha)} \mid \alpha \in \mathcal{Q}\}$ is a periodically equivalent family with $d(\eta, \xi) \leq \varepsilon$, then $\{\eta^{(\alpha)} \mid \alpha \in \mathcal{Q}\}$ is hyperbolic and satisfies*

$$\left\| \prod_{j=0}^{n-1} \eta_j^{(\alpha)} / E_0^s(\eta^{(\alpha)}) \right\| \leq K\lambda^n \tag{22}$$

and

$$\left\| \left(\prod_{j=0}^{n-1} \tilde{\eta}_j^{(\alpha)} \right)^{-1} \right\| \leq K\lambda^n \tag{23}$$

for all $\alpha \in \mathcal{Q}$, where n is the minimum period of $\eta^{(\alpha)}$ and $\tilde{\eta}_j^{(\alpha)}$: $\mathbf{R}^N/E_j^s(\eta^{(\alpha)}) \to \mathbf{R}^N \mid E_{j+1}^s(\eta^{(\alpha)})$ is the map induced by $\eta_j^{(\alpha)}$.

Proof. For each $\alpha \in \mathcal{Q}$ take orthonormal bases of $\mathbf{R}^N/E_j^s(\xi^{(\alpha)})$, $j \in \mathbf{Z}$, varying periodically with j with the same minimum period of $\xi^{(\alpha)}$. Let $\tilde{\xi}_j^{(\alpha)}$ be the matrix with respect to these bases of the map from $\mathbf{R}^N/E_j^s(\xi^{(\alpha)})$ into $\mathbf{R}^N/E_{j+1}^s(\xi^{(\alpha)})$ induced by $\xi_j^{(\alpha)}$. Now take $\varepsilon > 0$ and for each $0 < m < N$ define a family $\{\varphi^{(\beta)} \mid \beta \in \mathcal{B}\}$ containing every periodic sequence φ: $\mathbf{Z} \to \mathrm{GL}(N-m)$ such that for some $\alpha \in \mathcal{Q}$ the minimum periods of $\xi^{(\alpha)}$ and φ are the same and $\sup_j \|\varphi_j - (\tilde{\xi}_j^{(\alpha)})^{-1}\| \leq \varepsilon$. We claim that if $\varepsilon > 0$ is sufficiently small, $\{\varphi^{(\beta)} \mid \beta \in \mathcal{B}\}$ is a uniformly contracting family. If it is not we can find a sequence ψ: $\mathbf{Z} \to \mathrm{GL}(N-m)$ such that for some β the periods of ψ and $\varphi^{(\beta)}$ coincide, $\sup_j \|\varphi_j^{(\beta)} - \psi_j\|$ is arbitrarily small and $\prod_{j=0}^{n-1} \psi_j$ has an eigenvalue with modulus 1. Suppose that for $\alpha \in \mathcal{Q}$ the periods of $\tilde{\xi}^{(\alpha)}$ and $\varphi^{(\beta)}$ coincide and $\sup_j \|(\tilde{\xi}_j^{(\alpha)})^{-1} - \varphi_j^{(\beta)}\| \leq \varepsilon$. Then it is easy to find a sequence ξ: $\mathbf{Z} \to \mathrm{GL}(N)$ with the same minimum period of $\xi^{(\alpha)}$ such that $\xi_j^{(\alpha)}/E_j^s(\xi^{(\alpha)}) = \xi/E_j^s(\xi^{(\alpha)})$ for every $j \in \mathbf{Z}$ and the matrix of the map from $\mathbf{R}^N/E_j^s(\xi^{(\alpha)})$ into $\mathbf{R}^N/E_{j+1}^s(\xi^{(\alpha)})$ induced by ξ_j is ψ_j^{-1}. Then the map $\prod_{j=0}^{n-1}\xi_j$ has an eigenvalue with modulus 1. If $\varepsilon > 0$ is small enough, this contradicts the uniform hyperbolicity of the family $\{\xi^{(\alpha)} \mid \alpha \in \mathcal{Q}\}$ and proves the claim. Then if $0<\lambda<1$ and $\lambda^{-1}-1$ is small enough, the family $\{\lambda^{-1}\varphi^{(\beta)} \mid \beta \in \mathfrak{B}\}$ is also uniformly contracting and then by II.4 there exists $K>0$ such that for all $\beta \in \mathfrak{B}$

$$\left\| \prod_{i=0}^{n-1} \lambda^{-1}\varphi_i^{(\beta)} \right\| \leq K,$$

where n is the minimum period of $\varphi^{(\beta)}$; it follows that

$$\left\| \prod_{i=0}^{n-1} \varphi_i^{(\beta)} \right\| \leq K\lambda^n.$$

By the definition of the family $\{\varphi^{(\beta)} \mid \beta \in \mathfrak{B}\}$, this inequality proves (23). Property (22) is proved by similar methods.

For the next lemma we shall need the concept of angle between subspaces of a Euclidean space. If E_1, E_2 are subspaces of $\mathbf{R}^N$ such that $E_1 \oplus E_2 = \mathbf{R}^N$ we define the *angle* $\alpha(E_1, E_2)$ as $\alpha(E_1, E_2) = \|L\|^{-1}$, where $L\colon E_1^{\perp} \to E_1$ is such that $E_2 = \{v + Lv \mid v \in E_1^{\perp}\}$; in particular, $\alpha(E_1, E_1^{\perp}) = +\infty$.

LEMMA II.9. *If $\{\xi^{(\alpha)} \mid \alpha \in \mathcal{Q}\}$ is a uniformly hyperbolic family of periodic sequences of isomorphisms of $\mathbf{R}^N$, there exist $\varepsilon>0$, $\gamma>0$ and $n_0 \in \mathbf{Z}^+$ such that if $\{\eta^{(\alpha)} \mid \alpha \in \mathcal{Q}\}$ is a periodically equivalent family with $d(\eta, \xi) \leq \varepsilon$ then $\{\eta^{(\alpha)} \mid \alpha \in \mathcal{Q}\}$ is hyperbolic and*

$$\alpha\big(E_0^s(\eta^{(\alpha)}), E_0^u(\eta^{(\alpha)})\big) > \gamma$$

for every $\alpha \in \mathcal{Q}$ such that the minimum period of $\eta^{(\alpha)}$ is larger than n_0.

Proof. If the property is false there exist hyperbolic periodic sequences $\eta\colon \mathbf{Z} \to \mathrm{GL}(N)$ with period n arbitrarily large, $\alpha(E_0^s(\eta), E_0^u(\eta))$ arbitrarily small and such that for some $\alpha \in \mathcal{Q}$ the periods of $\xi^{(\alpha)}$ and η coincide and $\sup_i \|\eta_i - \xi_i^{(\alpha)}\|$ is arbitrarily small. Suppose that with respect to the splitting $R^N = E_0^s(\eta)^{\perp} \oplus E_0^s(\eta)$, the matrix of $\prod_{j=0}^{n-1} \eta_j$ is

$$\begin{pmatrix} A & 0 \\ P & B \end{pmatrix}.$$

By II.8 there exist $K>0$ and $0<\lambda<1$ such that

(24) $$\|A^{-1}\| \leq K\lambda^n$$

and

(25) $$\|B\| \leq K\lambda^n.$$

Let $L\colon E_0^s(\eta)^{\perp} \to E_0^s(\eta)$ be such that $\{v + Lv \mid v \in E_0^s(\eta)^{\perp}\} = E_0^u(\eta)$. From $(\prod_{j=0}^{n-1} \eta_j) E_0^u(\eta) = E_0^u(\eta)$, it follows that:

$$LA = P + BL.$$

Hence

$$L = PA^{-1} + BLA^{-1}$$

and by (24), (25),

$$\|L\| \le \|PA^{-1}\| + K^2\lambda^{2n}\|L\|.$$

If n is large enough, $K^2\lambda^{2n} \le \frac{1}{2}$; since

$$\tfrac{1}{2}\|PA^{-1}\|^{-1} \le \|L\|^{-1} = \alpha(E_0^s(\eta), E_0^u(\eta)),$$

the norm $\|PA^{-1}\|^{-1}$ is arbitrarily small because the angle $\alpha(E_0^s(\eta), E_0^u(\eta))$ is arbitrarily small. Now define a sequence ξ: $\mathbf{Z} \to \mathrm{GL}(N)$ with minimum period n, $\xi_j = \eta_j$ for $0 < j \le n-1$ and

$$\xi_0 = \eta_0 \begin{pmatrix} I & C \\ 0 & I \end{pmatrix}.$$

Then

$$\prod_{j=0}^{n-1} \xi_j = \begin{pmatrix} A & 0 \\ P & B \end{pmatrix}\begin{pmatrix} I & C \\ 0 & I \end{pmatrix} = \begin{pmatrix} A & AC \\ P & PC+B \end{pmatrix}.$$

If we can find C with $\|C\|$ arbitrarily small such that the last matrix has eigenvalue 1 we shall obtain a contradiction with the uniform hyperbolicity of $\{\xi^{(\alpha)} \mid \alpha \in \mathcal{Q}\}$. Therefore let us consider the system:

$$\begin{cases} Ax + ACy = x \\ Px + (PC+B)y = y. \end{cases}$$

Then $x = (I-A)^{-1}ACy = -(I-A^{-1})^{-1}Cy$ and

$$(I-B)^{-1}P\big(I - (I-A^{-1})^{-1}\big)Cy = y.$$

But $I - (I-A^{-1})^{-1} = -A^{-1}(I-A^{-1})^{-1}$ Then

$$-(I-B)^{-1}PA^{-1}(I-A^{-1})^{-1}Cy = y. \tag{26}$$

Now let v be such that $\|v\| = \|PA^{-1}\|^{-1}$ and $\|PA^{-1}v\| = 1$. Let $y = -(I-B)^{-1}PA^{-1}v$. By (25) we can suppose $\|I-B\| \le 2$. Hence $\|y\|^{-1} \le 2$. Now take w such that $(I-A^{-1})^{-1}w = v$. By (24) $(I-A^{-1})^{-1}$ is near to the identity and we can suppose that $\|w\| \le 2\|v\|$. Take C such that

$$Cy = w$$

and

$$\|C\| = \frac{\|w\|}{\|y\|}.$$

Then

$$\|C\| \le 4\|v\| = 4\|PA^{-1}\|^{-1},$$

which is arbitrarily small. For these C and y (26) is satisfied and the proof is complete.

Now we are ready to prove Lemma II.3. From II.7 it follows that (a) and (c) of II.3 hold for E^s and some values of K, λ and m, say K_1, λ_1 and m_1; similarly for E^u. If we show that there exist $0 < \lambda_2 < 1$, $m_2 \in \mathbf{Z}$ such that (b) holds for $\lambda = \lambda_2$, $m = m_2$, then it is easy to see that by taking as m a common multiple of m_1 and m_2 and as λ the maximum between λ_1 and λ_2, we can find a suitable value of K such that (a), (b) and (c) are satisfied. Therefore the proof of II.3 is reduced to proving only part (b). But it is clear that to prove (b) it is enough to show the existence of $m \in \mathbf{Z}^+$ such that for every $\alpha \in \mathcal{Q}$ and $i \in \mathbf{Z}$ there exists $0 < m_1 \leq m$ such that

$$\left\| \prod_{j=0}^{m_1-1} \xi_{i+j}^{(\alpha)}/E_i^s(\xi^{(\alpha)}) \right\| \cdot \left\| \left(\prod_{j=0}^{m_1-1} \xi_{i+j}^{(\alpha)} \mid E_i^u(\xi^{(\alpha)}) \right)^{-1} \right\| \leq 1/2.$$

If such an m does not exist we can find sequences $\alpha_n \in \mathcal{Q}$ and $i_n \in \mathbf{Z}$ satisfying

$$(27) \qquad \left\| \prod_{j=0}^{m-1} \xi_{i_n+j}^{(\alpha_n)}/E_{i_n}^s(\xi^{(\alpha_n)}) \right\| \cdot \left\| \left(\prod_{j=0}^{m-1} \xi_{i_n+j}^{(\alpha_n)}/E_{i_n}^u(\xi^{(\alpha_n)}) \right)^{-1} \right\| \geq 1/2$$

for every $0 < m \leq n$. Suppose first that the minimum periods of the sequences $\xi^{(\alpha_n)}$ are bounded. Then we can assume that they are all the same, say n_0. Therefore we can suppose that $0 \leq i_n < n_0$ for all n. Hence we can also assume that the sequence i_n is constant, say i_0. Since the set $\{\xi_j^{(\alpha)} \mid \alpha \in \mathcal{Q},\ j \in \mathbf{Z}\}$ is bounded we can suppose that for every j the sequence $\xi_{i_0+j}^{(\alpha_n)}$ converges to some linear map ξ_{i_0+j} and that $E_{i_0+j}^s(\xi^{(\alpha_n)})$ and $E_{i_0+j}^u(\xi^{(\alpha_n)})$ converge to subspaces $E_{i_0+j}^+$, $E_{i_0+j}^-$ satisfying $E_{i_0+j+n_0}^+ = E_{i_0+j}^+$, $E_{i_0+j+n_0}^- = E_{i_0+j}^-$ for every j and, from (27),

$$(28) \qquad \left\| \prod_{j=0}^{n-1} \xi_{j+i_0}/E_{i_0}^+ \right\| \left\| \left(\prod_{j=0}^{n-1} \xi_{j+i_0}/E_{i_0}^- \right)^{-1} \right\| \geq 1/2$$

for every $n > 0$. Moreover the uniform hyperbolicity of the family $\{\xi^{(\alpha)} \mid \alpha \in \mathcal{Q}\}$ implies the hyperbolicity of the sequence ξ. Then $\prod_{j=0}^{n_0-1} \xi_{j+i_0}$ is a hyperbolic matrix limit of the sequence of hyperbolic matrices $\prod_{j=0}^{n_0-1} \xi_{j+i_0}^{(\alpha_n)}$. Hence $E_{i_0}^+ = \lim E_{i_0}^s(\xi^{(\alpha_n)})$, $E_{i_0}^- = \lim E_{i_0}^u(\xi^{(\alpha_n)})$ are stable and unstable spaces of the hyperbolic matrix $\prod_{j=0}^{n_0-1} \xi_{j+i_0}$. But this contradicts (28).

Now suppose that the minimum periods of the sequences $\xi^{(\alpha_n)}$ are unbounded. To exhibit a contradiction in this case we shall need the following elementary lemma:

LEMMA II.10. *Let* $\mathbf{R}^N = E_1 \oplus E_2$. *Then*

a) *For all* $v \in E_2$, $u \in E_1$

$$\|v - u\| \geq \frac{\alpha(E_1, E_2)}{1 + \alpha(E_1, E_2)} \|v\|.$$

b) *If* $T\colon \mathbf{R}^N \hookleftarrow$ *is a linear map*:

$$\|T\| \leq \frac{1 + \alpha(E_1, E_2)}{\alpha(E_1, E_2)} (\|T/E_1\| + \|T/E_2\|).$$

c) *If* $A\colon E_2 \to E_1$ *is a linear map and* $G = \{v + Av \mid v \in E_2\}$, *there exist linear maps* $T_i\colon \mathbf{R}^N \hookleftarrow$, $i = 1, 2$, *satisfying*:

c1) $$\|T_i\| \leq \frac{1 + \alpha(E_1, E_2)}{\alpha(E_1, E_2)} \|A\| \quad \textit{and} \quad T_i \mid E_1 = 0 \qquad \textit{for } i = 1, 2.$$

c2) $$(T_1 + I)G = E_2 \quad \textit{and} \quad (T_2 + I)E_2 = G.$$

Remark: If $E_2 = E_1^{\perp}$ then $\alpha = \alpha(E_1, E_2) = +\infty$ and $1 + \alpha/\alpha = 1 = \alpha/1 + \alpha$.

Proof. a) Let $L\colon E_1^{\perp} \to E_1$ be such that $E_2 = \{w + Lw \mid w \in E_1^{\perp}\}$. Then if $w \in E_1^{\perp}$ and $w + Lw = v$:

$$\|v - u\| = \|w + Lw - u\| \geq \|w\|.$$

But $\|v\| \leq \|w\| + \|L\|\|w\| \leq \|w\|(1 + \|L\|)$. Hence: $\|w\| \geq (1/1 + \|L\|)\|v\|$,

$$\|v - u\| \geq \frac{1}{1 + \|L\|}\|v\| = \frac{\alpha(E_1, E_2)}{1 + \alpha(E_1, E_2)}\|v\|.$$

b) If $v \in \mathbf{R}^N$, write $v = v_1 + v_2$ with $v_1 \in E_1$, $v_2 \in E_1^{\perp}$. Then if L is as in the proof of (a):

$$\begin{aligned}\|Tv\| &= \|T(v_1 + v_2)\| = \|T(v_2 + Lv_2) + T(v_1 - Lv_2)\| \\ &\leq \|T/E_2\|(1 + \|L\|)\|v_2\| + \|T/E_1\|(\|v_1\| + \|L\| \cdot \|v_2\|).\end{aligned}$$

Since $\|v_1\| \leq \|v\|$, $\|v_2\| \leq \|v\|$ and $\|L\| = \alpha(E_1, E_2)^{-1}$, the inequality is proved.

c) Let $\pi\colon \mathbf{R}^N \to E_2$ be the projection along E_1 and define $T_1 = -A\pi$ and $T_2 = A\pi$. Then $T_i \mid E_1 = 0$, $T_i \mid E_2 = \pm A$ and c1) follows by (b). Given $v \in E_2$ we have $(I + T_1)(v + Av) = v + Av - A\pi v - A\pi Av = v$ and $(I + T_2)v = v + A\pi v = v + Av$. Lemma II.10 is proved.

Now let $\varepsilon_1' > 0$, $K_1' > 0$ and $0 < \lambda < 1$ be the constants given by Lemma II.8 and $\varepsilon'' > 0$, $\gamma > 0$ and $\bar{n} \in \mathbf{Z}^+$ be the constants given by Lemma II.9. Let $\varepsilon = \min\{\varepsilon', \varepsilon''\}$, $K = K' + 1$ and $C = \sup\{\|\xi_i^{(\alpha)}\| \mid \alpha \in \mathcal{C}, i \in \mathbf{Z}\}$.

Without loss of generality we can suppose that the family $\{\xi^{(\alpha)} \mid \alpha \in \mathcal{C}\}$ has the following property: for every $\alpha \in \mathcal{C}$ and $j \in \mathbf{Z}$ the sequence ξ defined by $\xi_n = \xi_{j+n}^{(\alpha)}$ for all $n \in \mathbf{Z}$ is also in the family. This is true because if the family does not satisfy this property we can replace it by an extended family $\{\xi^{(\alpha, j)} \mid \alpha \in \mathcal{C}, j \in \mathbf{Z}\}$ defined by $\xi_n^{(\alpha, j)} = \xi_{j+n}^{(\alpha)}$ and this new family is also uniformly hyperbolic. Therefore we assume that the family $\{\xi^{(\alpha)} \mid \alpha \in \mathcal{C}\}$ has the property mentioned above and then, by II.8 and II.9, for all $\alpha \in \mathcal{C}$ and $i \in \mathbf{Z}$

$$\left\| \prod_{j=0}^{n-1} \xi_{j+i}^{(\alpha)} \mid E_i^s(\xi^{(\alpha)}) \right\| \leq K\lambda^n \tag{29}$$

and

$$\left\| \left(\prod_{j=0}^{n-1} \xi_{j+i}^{(\alpha)} \mid E_i^u(\xi^{(\alpha)}) \right)^{-1} \right\| \leq K\lambda^n, \tag{30}$$

where n is the minimum period of $\xi^{(\alpha)}$; moreover if η: $\mathbf{Z} \to \mathrm{GL}(N)$ is any periodic sequence with the same period n of $\xi^{(\alpha)}$ and such that

$$\sup\left\{ \|\eta_j - \xi_j^{(\alpha)}\| \mid j \in \mathbf{Z} \right\} \leq \varepsilon \tag{31}$$

then η is hyperbolic and if $n \geq \bar{n}$ then for all $j \in \mathbf{Z}$,

$$\alpha\left(E_j^s(\eta), E_j^u(\eta)\right) > \gamma. \tag{32}$$

Now take $\varepsilon_0 > 0$ satisfying $(2\varepsilon_0 + \varepsilon_0^2)C \leq \varepsilon$ and $\varepsilon_1 > 0$, $m \in \mathbf{Z}^+$ such that $(1 + \varepsilon_1)\lambda < 1$, $m \geq \bar{n}$,

$$\varepsilon_1 \leq \frac{\gamma}{1 + \gamma} \varepsilon_0, \tag{33}$$

$$(1 + \varepsilon_1)^m \geq 4 + \frac{2}{\gamma} \tag{34}$$

and

$$K^2((1 + \varepsilon_1)\lambda)^m < 1. \tag{35}$$

Since the minimum periods of the sequences $\xi^{(\alpha_n)}$ are unbounded we can choose $l > m$ such that the minimum period of $\xi^{(\alpha_l)}$ is $n_0 > m$. Put $\alpha = \alpha_l$ and $i = i_l$. By (27):

$$\left\| \prod_{j=0}^{m-1} \xi_{i+j}^{(\alpha)} / E_i^s(\xi^{(\alpha)}) \right\| \cdot \left\| \left(\prod_{j=0}^{m-1} \xi_{i+j}^{(\alpha)} / E_i^u(\xi^{(\alpha)}) \right)^{-1} \right\| \geq 1/2.$$

Therefore there exist $v \in E_i^u(\xi^{(\alpha)})$ and $w \in E_i^s(\xi^{(\alpha)})$ with $\|v\| = \|w\| = 1$ and such that

$$\frac{1}{2}\left\| \prod_{j=0}^{m} \xi_{i+j}^{(\alpha)} v \right\| \leq \left\| \prod_{j=0}^{m} \xi_{i+j}^{(\alpha)} w \right\|. \tag{36}$$

Take a linear map $L: E_i^u(\xi^{(\alpha)}) \to E_i^s(\xi^{(\alpha)})$ satisfying

$$Lv = \varepsilon_1 w$$

and

$$\|L\| = \varepsilon_1.$$

Define $\hat{L}: E_{i+n_0}^u(\xi^{(\alpha)}) \to E_{i+n_0}^s(\xi^{(\alpha)})$ by

$$\hat{L} = (1+\varepsilon_1)^{n_0}\left(\prod_{j=0}^{n_0-1} \xi_{i+j}^{(\alpha)}\right) \cdot L \cdot \left(\prod_{j=0}^{n_0-1} \xi_{i+j}^{(\alpha)} \,|\, E_i^u(\xi^{(\alpha)})\right)^{-1}.$$

By (29), (30) and (35),

$$\|\hat{L}\| \leq (1+\varepsilon_1)^{n_0}(K\lambda^{n_0})^2\|L\| = K^2\big((1+\varepsilon_1)\lambda^2\big)^{n_0}\|L\|$$
$$\leq K^2\big((1+\varepsilon_1)\lambda\big)^m\|L\| < \varepsilon_1.$$

Now define, for $0 \leq t \leq 1$,

$$G_t = \left\{u + tLu \,|\, u \in E_i^u(\xi^{(\alpha)})\right\}$$

and

$$\hat{G}_t = \left\{u + t\hat{L}u \,|\, u \in E_{i+n_0}^u(\xi^{(\alpha)})\right\}$$

and take linear maps P_t, S_t of $\mathbf{R}^N$, $0 \leq t \leq 1$, satisfying

$$P_t \,|\, E_i^s(\xi^{(\alpha)}) = 0,$$
$$(I + P_t)\cdot E_i^u(\xi^{(\alpha)}) = G_t,$$
$$S_t \,|\, E_{i+n_0}^s(\xi^{(\alpha)}) = 0$$

and

$$(I + S_t)\hat{G}_t = E_{i+n_0}^u(\xi^{(\alpha)}) \qquad \text{for all } 0 \leq t \leq 1.$$

By Lemma II.10 c), (32) and (33), it is possible to find these maps satisfying:

$$\|P_t\| \leq \frac{1+\gamma}{\gamma}\|tL\| \leq \frac{1+\gamma}{\gamma}\varepsilon_1 \leq \varepsilon_0 \tag{37}$$

and

$$\|S_t\| \leq \frac{1+\gamma}{\gamma}\|t\hat{L}\| \leq \frac{1+\gamma}{\gamma}\varepsilon_1 \leq \varepsilon_0.$$

Finally take linear maps T_j of $\mathbf{R}^N$, $j \in \mathbf{Z}$, satisfying

$$T_j \mid E_j^s(\xi^{(\alpha)}) = \varepsilon_1 I$$

and

$$T_j \mid E_j^u(\xi^{(\alpha)}) = 0.$$

The norm of T_j can be estimated with Lemma II.10 b), (32) and (33):

$$\|T_j\| \leq \frac{1 + \alpha\big(E_j^s(\xi^{(\alpha)}), E_j^u(\xi^{(\alpha)})\big)}{\alpha\big(E_j^s(\xi^{(\alpha)}), E_j^u(\xi^{(\alpha)})\big)}\varepsilon_1 \leq \frac{1+\gamma}{\gamma}\varepsilon_1 \leq \varepsilon_0. \tag{38}$$

With these linear maps we shall construct a perturbation of the sequence $\xi^{(\alpha)}$ such that the angle between its stable and unstable subspaces at $i + m$ is very small, thus contradicting (32). Define $\eta^{(t)}$: $\mathbf{Z} \to \mathrm{GL}(N)$, $0 \leq t \leq 1$, with minimum period n_0 and

$$\eta_i^{(t)} = (I + T_i)\xi_i^{(\alpha)}(I + P_t),$$

$$\eta_{i+j}^{(t)} = (I + T_{j+i})\xi_{i+j}^{(\alpha)}$$

for $1 \leq j < n_0 - 1$ and

$$\eta_{i+n_0-1}^{(t)} = (I + S_t)T_{i+n_0-1}\xi_{i+n_0-1}^{(\alpha)}.$$

Then by (37) and (38),

$$\|\eta_i^{(t)} - \xi_i^{(\alpha)}\| \leq (\|T_i\| + \|P_t\| + \|T_i\|\|P_t\|)\|\xi_i^{(\alpha)}\|$$
$$\leq (2\varepsilon_0 + \varepsilon_0^2)C \leq \varepsilon.$$

In a similar way it is proved that

$$\|\eta_{i+n_0-1}^{(t)} - \xi_{i+n_0-1}^{(\alpha)}\| \leq \varepsilon$$

and, if $0 < j < n_0 - 1$,

$$\|\eta_{i+j}^{(t)} - \xi_{i+j}^{(\alpha)}\| \leq \|T_{i+j}\|\|\xi_{i+j}^{(\alpha)}\| \leq \varepsilon_0 C \leq \varepsilon.$$

Thus, for each $0 \leq t \leq 1$, $\eta^{(t)}$ satisfies (31) and is a hyperbolic sequence. Let

$$\beta = \alpha\big(E_{i+m}^s(\eta^{(1)}), E_{i+m}^u(\eta^{(1)})\big);$$

by (32) it follows that

$$\beta > \gamma. \tag{39}$$

To show that this is a contradiction, let us first prove that for any $0 \leq t \leq 1$,

$$E_i^u(\eta^{(t)}) = E_i^u(\xi^{(\alpha)}), \tag{40}$$

and for any $0 \le t \le 1$ and $j \in \mathbf{Z}$,

$$E_j^s(\eta^{(t)}) = E_j^s(\xi^{(\alpha)}). \tag{41}$$

Let $0 \le t \le 1$ be given; for every $0 \le s \le 1$ the sequence $s\eta^{(t)} + (1-s)\xi^{(\alpha)}$ is also hyperbolic by (31) and it follows that $\dim E_j^s(\eta^{(t)}) = \dim E_j^s(\xi^{(\alpha)})$ for any $j \in \mathbf{Z}$. Hence it is enough to prove (40) and (41) replacing $=$ by $\supset$.

From the definition of $\eta^{(t)}$ it follows that for all $j \in \mathbf{Z}$, $\eta_j^{(t)} \mid E_j^s(\xi^{(\alpha)}) = (1+\varepsilon_1)\xi_j^{(\alpha)} \mid E_j^s(\xi^{(\alpha)})$. Hence by (29) and (35),

$$\left\| \prod_{k=0}^{n_0-1} \eta_{j+k}^{(t)} \mid E_j^s(\xi^{(\alpha)}) \right\| \le (1+\varepsilon_1)^{n_0} K\lambda^{n_0} < 1,$$

which implies $E_j^s(\xi^{(\alpha)}) \subseteq E_j^s(\eta^{(t)})$. To prove (40) observe that

$$\left(\prod_{j=0}^{n_0-1} \eta_{i+j}^{(t)} \right) \cdot E_i^u(\xi^{(\alpha)}) = E_{i+n_0}^u(\xi^{(\alpha)}) = E_i^u(\xi^{(\alpha)}),$$

and since the spectral radius of

$$\left(\prod_{j=0}^{n_0-1} \eta_{i+j}^{(t)} \right)^{-1} \mid E_i^u(\xi^{(\alpha)})$$

depends continuously on t, is < 1 for $t = 0$ and $\neq 1$ for every t, it follows that it is < 1 for all t; therefore $E_i^u(\xi^{(\alpha)}) \subseteq E_i^u(\eta^{(t)})$ for all $0 \le t \le 1$.

Finally let us estimate the angle β between $E_{i+m}^s(\eta^{(1)})$ and $E_{i+m}^u(\eta^{(1)})$. Since $v \in E_i^u(\xi^{(\alpha)})$ and $w \in E_i^s(\xi^{(\alpha)})$, (40) and (41) imply

$$u_1 = \left(\prod_{j=0}^{m-1} \eta_{i+j}^{(1)} \right) v \in E_{i+m}^u(\eta^{(1)})$$

and

$$u_2 = \left(\prod_{j=0}^{m-1} \eta_{i+j}^{(1)} \right) w \in E_{i+m}^s(\eta^{(1)}).$$

But by the way $\eta^{(t)}$ was defined, we have that

$$u_1 = \left(\prod_{j=0}^{m-1} \xi_{i+j}^{(\alpha)} \right) v + (1+\varepsilon_1)^m \left(\prod_{j=0}^{m-1} \xi_{i+j}^{(\alpha)} \right) w$$

and

$$u_2 = (1+\varepsilon_1)^m \left(\prod_{j=0}^{m-1} \xi_{i+j}^{(\alpha)} \right) w.$$

Hence, using Lemma II.10 a), we have

$$\left\| \left(\prod_{j=0}^{m-1} \xi_{j+1}^{(\alpha)} \right) v \right\| = \| u_1 - u_2 \| \geq \frac{\beta}{1+\beta} \| u_1 \|$$

$$\geq \frac{\beta}{1+\beta} \left| (1+\varepsilon_1)^m \left\| \left(\prod_{j=0}^{m-1} \xi_{j+1}^{(\alpha)} \right) w \right\| - \left\| \left(\prod_{j=0}^{m-1} \xi_{j+1}^{(\alpha)} \right) v \right\| \right|,$$

and using (36) we obtain

$$\frac{1+\beta}{\beta} \geq \frac{(1+\varepsilon_1)^m}{2} - 1.$$

Thus, by (34),

$$\beta \leq \frac{2}{(1+\varepsilon_1)^m - 4} \leq \gamma,$$

contradicting (39).

INSTITUTO DE MATEMÁTICA PURA E APLICADA, RIO DE JANEIRO, BRASIL

REFERENCES

[1] C. DOERING, Campos de Vetores sem pontos errantes em variedades tridimensionais compactas, Thesis, IMPA, 1979.
[2] J. FRANKS, Necessary conditions for stability of diffeomorphisms, Trans. A. M S. **158** (1971), 301–308.
[3] J. GUCKENHEIMER, A strange, strange attractor, The Hopf bifurcation and its applications, Applied Mathematical Series **19** (1976), Springer-Verlag
[4] R. MAÑÉ, Contributions to the stability conjecture, Topology **17** (1978), 383–396.
[5] ———, Persistent manifolds are normally hyperbolic, Trans A. M. S. **246** (1978), 261–283
[6] ———, Expansive diffeomorphisms, Proc. Warwick Symposium on Dynamical Systems, Lecture Notes in Math. **468** (1975), Springer-Verlag.
[7] ——— Characterization of AS diffeomorphisms, Proc. ELAM III, Lecture Notes in Math. **597** (1977) Springer-Verlag
[8] J. PALIS, A note on Ω-stability, *Global Analysis*, Proc Symp. Pure Math. XIV (1970) AMS
[9] V. A. PLISS, Analysis of the necessity of the conditions of Smale and Robbin for structural stability of periodic systems of differential equations, Diff. Uravnenija **8** (1972), 553–555.
[10] ———, On a conjecture due to Smale, Diff. Uravnenija, **8** (1972), 268–282.
[11] R. V. PLYKIN, Sources and sinks of *A*-diffeomorphisms, Math. USSR Sbornik **23** (1974).
[12] C. PUGH, The closing lemma, Amer J. Math **89** (1967), 956–1009.
[13] J. ROBBIN, A structural stability theorem, Ann. of Math **94** (1971), 447–493.
[14] C. ROBINSON, Structural stability of C^1 diffeomorphisms, J Diff. Eq **22** (1976), 28–73.
[15] ———, C' structural stability implies Kupka-Smale, *Dynamical Systems*, (1975) Academic Press.
[16] ———, Introduction to the Closing Lemma, *The Structure of Attractors in Dynamical Systems*, Lecture Notes in Math. **668** (1978), Springer-Verlag
[17] S. SMALE, Differentiable dynamical systems, Bull A. M S. **73** (1967), 747–817.
[18] ———, Problems of Present Day Mathematics (XVII), Mathematical developments arising from Hilbert problems, Proc. Symp. Pure Math XXVIII (1976) AMS.
[19] ———, The Ω-stability theorem, *Global Analysis*, Proc. Symp. Pure Math. XIV (1970) AMS

(Received June 19, 1981)

Invent. math. 69, 375–392 (1982)

Inventiones mathematicae
© Springer-Verlag 1982

On the Entropy of the Geodesic Flow in Manifolds Without Conjugate Points

A. Freire and R. Mañé *

Institute de Matematica, Pura e Aplicada, Dona Castorina 110, Jardim Botanico, 22460 Rio de Janeiro, RJ, Brasilien

Introduction

Let M be a compact Riemannian manifold and $\tilde{M}$ its universal covering. In [7] Manning introduced the volume growth rate λ of M defined by:

$$\lambda = \lim_{r \to +\infty} \frac{1}{r} \log \operatorname{vol} B_r(p)$$

where $p \in \tilde{M}$ and $B_r(p)$ denotes the ball with center p and radius r. He proved that this limit exists and is independent of p. In fact the relation between the growth of the function $\operatorname{vol} B_r(p)$, the curvature of M and its fundamental group had already been considered by several authors (Milnor [10], Margulis [9], Dinaburg [2]). The constant λ and $\pi_1(M)$ can be related as follows: Let N be a fundamental domain of $\tilde{M}$ and let $G = \{g \in \pi_1(M) \mid gN \cap N \neq \emptyset\}$. Milnor proved in [10] that G is a generator of $\pi_1(M)$. Define the growth rate α of $\pi_1(M)$ with respect to the generator G as $\lim_{k \to +\infty} k^{-1} \log N(k)$ where $N(k)$ is the number of elements of $\pi_1(M)$ which can be written in the form $\prod_{i=1}^{k} g_i$, $g_i \in G$, $1 \leqq i \leqq k$. This limit always exists and depends on G. However, the fact that α is zero or positive is independent of the generator [10]. In the second case we say that $\pi_1(M)$ has exponential growth. Then, if d_1 and d_2 denote the minimum and maximum distances between the sets N and $\bigcup \{gN \mid gN \cap N = \emptyset\}$, the following inequality holds:

$$\frac{\alpha}{d_2} \leqq \lambda \leqq \frac{\alpha}{d_1}.$$

These inequalities were essentially proved by Dinaburg [2] and Milnor [10] respectively. In particular $\lambda > 0$ if and only if $\pi_1(M)$ has exponential growth. In [7] Manning related the dynamics of the geodesic flow $\varphi\colon R \times TM_1 \to TM_1$ to λ in the following way: If $h_{\text{top}}(\varphi)$ denotes the topological entropy of φ:

$$h_{\text{top}}(\varphi) \geqq \lambda \tag{1}$$

* *Present address:* c/o Professor S. Newhouse, Department of Mathematics, University of North Carolina, Chapel Hill, NC 27514, USA

0020-9910/82/0069/0375/$03.60

and

$$h_{\rm top}(\varphi)=\lambda \tag{2}$$

if the manifold has non positive sectional curvature. In fact his proof holds without any change for the more general case of manifolds without focal points and, with some alterations, it can be extended to Riemannian manifolds whose geodesic flow is Anosov. It is then natural to ask whether (2) holds for manifolds without conjugate points. This class of manifolds obviously contains the manifolds without focal points and, by a result of Klingenberg [6], manifolds whose geodesic flow is Anosov. The main objective of this work is to give a positive answer to this question under some mild regularity assumptions on the Riemannian structure:

Theorem I. $h_{\rm top}(\varphi)=\lambda$ *if M has no conjugate points and its Riemannian metric is Hölder C^3.*

Manning's proof of (2) uses the hypothesis on the curvature of M only to insure that if γ_i: $[a,b]\to\tilde{M}$, $i=1, 2$ are geodesic arcs, then:

$$d(\gamma_1(t),\gamma_2(t))\leqq d(\gamma_1(a),\gamma_2(a))+d(\gamma_1(b),\gamma_2(b)). \tag{3}$$

This property is also true for manifolds without focal points (see [5] for a proof). When the geodesic flow is Anosov it is not difficult to prove, using the results of Eberlein [3] on the behaviour of Jacobi vector fields for these manifolds, that (3) holds if we multiply its second member by a certain uniersal constant independent of the geodesic arcs. This modified version of (3) is enough to prove (2) applying the same method used by Manning. As far as we know no simple modification of (3) holds for manifolds without conjugate points, but it can be expected that the essential geometric meaning of (3), namely that geodesic arcs do not spread apart more than their endpoints, is in some weaker form true. Our proof of Theorem I follows a different method. We shall use Przytycki's inequality for the topological entropy [14]. Let N be a compact boundaryless manifold and let φ: $R\times N\to N$ be a Hölder C^1 flow. Przytycki's inequality states that:

$$h_{\rm top}(\varphi)\leqq \lim_{t\to+\infty}\frac{1}{t}\log\int_N \|(D_x\varphi_t)^\wedge\|\,d\lambda_0(x)$$

where

$$(D_x\varphi_t)^\wedge\colon (T_xN)^\wedge\to(T_{\varphi_t(x)}N)^\wedge$$

denotes the linear map induced by $D_x\varphi_t$ on the exterior algebra of the tangent space T_xN and λ_0 is the Lebesgue measure on N. If $N=TM_1$ and φ is the geodesic flow we shall prove that:

$$\lim_{t\to+\infty}\frac{1}{t}\log\int_{TM_1}\|(D_x\varphi_t)^\wedge\|\,d\lambda_0(x)\leqq\lambda$$

which together with (1) implies:

$$h_{\rm top}(\varphi)=\lim_{t\to+\infty}\frac{1}{t}\log\int_N\|(D_x\varphi_t)^\wedge\|\,d\lambda_0(x)=\lambda.$$

The limit in Przytycki's inequality had already been considered by Shub and Sacksteder [16] in connection with the Entropy Conjecture and has interesting geometrical connotations.

We now consider the problem of finding estimates for the metric entropies $h_\mu(\varphi)$ where μ is a φ-invariant probability on TM_1. Let us recall the main results of Oseledec theory [11] which are fundamental for this kind of problems. Consider a C^1 flow φ: $\mathbb{R}\times N\to N$, where N is a compact boundaryless manifold. It follows from Oseledec's results that there exists a total probability Borel set $\Lambda\subset N$ (i.e. such that $\mu(\Lambda^c)=0$ for any φ-invariant probability μ on N) such that at each $x\in\Lambda$ there exists a unique splitting $T_xN=E^u(x)\oplus E^c(x)\oplus E^s(x)$ satisfying:

$$\lim_{t\to\pm\infty}\frac{1}{t}\log\|(D_x\varphi_t)v\|=0 \qquad v\in E^c(x),$$

$$\lim_{t\to\pm\infty}\frac{1}{t}\log\|(D_x\varphi_t)v\|<0 \qquad v\in E^s(x),$$

$$\lim_{t\to\pm\infty}\frac{1}{t}\log\|(D_x\varphi_t)v\|>0 \qquad v\in E^u(x).$$

Moreover the limit:

$$\chi(x)=\lim_{t\to\pm\infty}\frac{1}{t}\log|\det(D_x\varphi_t)/S| \tag{4}$$

exists and is independent of S, for any $x\in\Lambda$ and any subspace $E^u(x)\subset S\subset E^u(x)\oplus E^c(x)$. The importance of $\chi(x)$ is that if μ is a φ-invariant probability measure on N (Ruelle [15]):

$$h_\mu(\varphi)\leqq\int_N\chi\,d\mu \tag{5}$$

and if φ is Hölder C^1 and μ is absolutely continuous with respect to Lebesgue measure (Pesin [12]):

$$h_\mu(\varphi)=\int_N\chi\,d\mu \tag{6}$$

We shall try to estimate the metric entropies of the geodesic flow using these formulas. If $(p,v)\in TM_1$, denote by $E(p,v)$ the subspace of T_pM orthogonal to v and by $K(p,v)$: $E(p,v)\hookleftarrow$ the linear map defined by $K(p,v)\,w=R_p(v,w)\,v$ where R_p is the curvature tensor of M at p. The Ricatti equation of M is:

$$\dot{U}+U^2+K=0. \tag{7}$$

A solution of the Ricatti equation is a function that to each $(p,v)\in TM_1$ assigns a self adjoint linear map $U(p,v)$: $E(p,v)\hookleftarrow$ such that if τ_t: $T_pM\to T_{\gamma(t)}M$ denotes the parallel transport along the geodesic with initial condition (p,v) then the limit

$$\dot{U}(p,v)=\lim_{t\to 0}\frac{1}{t}(\tau_t^{-1}\circ U(\varphi_t(p,v))\circ\tau_t-U(p,v))$$

exists for all $(p,v)\in TM_1$ and satisfies:

$$\dot{U}(p,v)+U^2(p,v)+K(p,v)=0.$$

The results of Green in [4] can be reformulated as saying that if M has no conjugate points (7) has measurable solutions and any solution is bounded. There is a close relationship between solutions of (7) and invariant subbundles of the tangent flow $T\varphi\colon R\times T(TM_1)\to T(TM_1)$, as we shall explain in Sect. I.

The function χ and the solutions of the Ricatti equation are connected by the following result:

Theorem II. *If M has no conjugate points there exists a measurable solution U^+ of the Ricatti equation such that the equality:*

$$\chi(x)=\lim_{T\to+\infty}\frac{1}{T}\int_0^T \operatorname{tr}(U^+\circ\varphi_t(x))\,dt$$

holds for a total probability set of points x in TM_1.

Then, by (5) and the ergodic theorem:

$$h_\mu(\varphi)\leqq\int_{TM_1}\operatorname{tr}U^+\,d\mu \tag{8}$$

and if the Riemannian metric is Hölder C^3 and λ_0 denotes the Lebesgue measure of TM_1, by (6):

$$h_{\lambda_0}(\varphi)=\int_{TM_1}\operatorname{tr}U^+\,d\lambda_0. \tag{9}$$

Using (8) and (9) some interesting estimates for the metric entropies of φ can be obtained. Consider the space of bounded measurable functions that to each $(p,v)\in TM_1$ associate a self adjoint linear map $A(p,v)$ of $E(p,v)$ and identify two such functions which coincide μ-a.e. In this space we can define an inner product:

$$\langle U,V\rangle=\int_{TM_1}\operatorname{tr}UV\,d\mu.$$

Then Cauchy-Schwarz's inequality implies:

$$\int_{TM_1}\operatorname{tr}UV\leqq\Big(\int_{TM_1}\operatorname{tr}U^2\,d\mu\Big)^{1/2}\Big(\int_{TM}\operatorname{tr}V^2\,d\mu\Big)^{1/2}$$

and equality holds if and only if U is a scalar multiple of V. Then (8) implies

$$\begin{aligned}h_\mu(\varphi)&\leqq\Big(\int_{TM_1}\operatorname{tr}U^{+2}\,d\mu\Big)^{1/2}\Big(\int_{TM_1}\operatorname{tr}I^2\,d\mu\Big)^{1/2}\\&=(n-1)^{1/2}\Big(\int_{TM_1}\operatorname{tr}U^{+2}\,d\mu\Big)^{1/2}\end{aligned}$$

where $n=\dim M$. The φ-invariance of μ implies

$$\int_{TM_1}\operatorname{tr}\dot{U}^+\,d\mu=0.$$

Then, integration of (7) with respect to μ yields

$$\int_{TM_1} \operatorname{tr} U^{+2}\, d\mu = -\int_{TM_1} \operatorname{tr} K\, d\mu.$$

But $(n-1)^{-1} \operatorname{tr} K(\theta)$ is the Ricci tensor $\operatorname{Ric}(\theta)$. Hence, we have proved the following Corollary:

Corollary II.1. *If M has no conjugate points:*

$$h_\mu(\varphi) \leqq (n-1)\Big(-\int_{TM_1} \operatorname{Ric}\, d\mu\Big)^{1/2}$$

for any φ-invariant probability μ on TM_1. If the equality holds, the sectional curvatures at points of the support of μ are constant.

When μ is the Lebesgue measure λ_0 on TM_1 and λ_1 is the Lebesgue measure on M (normalized in order to have $\lambda_0(TM_1) = \lambda_1(M) = 1$) we have the following result:

Corollary II.2. *If M has no conjugate points and $S(p)$ denotes the scalar curvature at $p \in M$:*

$$h_{\lambda_0}(\varphi) \leqq (n-1)\Big(-\int_{TM_1} S\, d\lambda_1\Big)^{1/2}$$

and the equality holds if and only if M has constant sectional curvatures.

To prove this corollary denote by λ_2 the Lebesgue measure on the $(n-1)$-dimensional sphere of R^n and recall the equality:

$$\int_{S^{n-1}} \langle Ax, x\rangle\, d\lambda_2(x) = \frac{1}{n} \lambda_2(S^{n-1}) \operatorname{tr} A. \qquad (10)$$

Moreover, if $(p, v) \in TM_1$ and $\{x_1, \dots, x_{n-1}\}$ is an orthonormal basis of $E(p, v)$:

$$\operatorname{Ric}(p, v) = \frac{1}{n-1} \sum_{i=1}^{n-1} \langle R_p(v, x_i) v, x_i\rangle.$$

Then if $A\colon T_pM \hookleftarrow$ is defined by

$$Av = \frac{1}{n-1} \sum_{i=1}^{n-1} R_p(x_i, v) x_i$$

we have:

$$\operatorname{Ric}(p, v) = \langle Av, v\rangle$$

and then by (10) and setting $x_n = v$:

$$\int_{S^{n-1}} \operatorname{Ric}(p, v)\, d\lambda_2(v) = \frac{1}{n} \lambda_2(S^{n-1}) \frac{1}{n-1} \sum_{j=1}^{n} \sum_{i=1}^{n-1} \langle R_p(x_j, x_i) x_i, x_j\rangle$$
$$= \lambda_2(S^{n-1})\, S(p),$$

Then:

$$\int_{TM_1} \mathrm{Ric}\, d\lambda_0 = \frac{1}{\lambda_2(S^{n-1})} \int_M \Big(\int_{S^{n-1}} \mathrm{Ric}(p,v)\, d\lambda_2(v)\Big)\, d\lambda_1(p) = \int_M S\, d\lambda_1 .$$

The proof is now completed using Corollary II.1.

Another easy corollary of Theorem II is the following lower bound for $h_{\lambda_0}(\varphi)$ when the curvature of M is negative:

Corollary II.3. *If M is Hölder C^3 and all its sectional curvatures are negative:*

$$h_{\lambda_0}(\varphi) \geqq (n-1)^{1/2} \int_{TM_1} (-\mathrm{Ric})^{1/2}\, d\lambda_0 .$$

Moreover the equality holds if and only if the sectional curvature is constant.

This inequality was proved in the 2-dimensional case by Manning [8] and its proof is based on essentially the same method. P. Sarnak also obtained weaker forms of Corollaries II.2 and II.3 for manifolds of negative curvature [17]. We shall use the following stronger version of Theorem II: when M has negative curvature then there exists a measurable solution U of the Ricatti equation satisfying the statement of the theorem and such that there exists $c>0$ such that $\langle U(\theta)x, x\rangle \geqq c\,\|x\|^2$ for every $\theta \in TM_1$, $x \in E(\theta)$. In Sect. II, after proving Theorem II we shall show that the solution of the Ricatti equation constructed in its proof satisfies this extra condition if M has negative curvature. In particular it follows that the function $\det U^+$ is bounded away from zero and then $\log \det U^+$ is integrable. Moreover

$$\begin{aligned} \frac{d}{dt}(\log \det U^+(\varphi_t(\theta)))|_{t=0} &= \mathrm{tr}\, \dot{U}(\theta)\, U^{-1}(\theta) \\ &= -\mathrm{tr}\, U^+(\theta) - \mathrm{tr}\, K(\theta)\, U^{+^{-1}}(\theta). \end{aligned}$$

Integrating with respect to λ_0 we obtain, by the φ-invariance of λ_0:

$$\int_{TM_1} \mathrm{tr}\, U^+\, d\lambda_0 = -\int_{TM_1} \mathrm{tr}\, K U^{+^{-1}}\, d\lambda_0 .$$

Now observe that if A and B are positive self adjoint maps of a Hilbert space:

$$\mathrm{tr}\, AB \leqq (\mathrm{tr}\, A^2)^{1/2} (\mathrm{tr}\, B^2)^{1/2} \leqq \mathrm{tr}\, A\, \mathrm{tr}\, B. \tag{11}$$

In particular $\mathrm{tr}\, AB = \mathrm{tr}\, A\, \mathrm{tr}\, B$ implies that A is a scalar multiple of B. From this property and Cauchy-Schwarz's inequality we obtain:

$$\begin{aligned} \int_{TM_1} (-\mathrm{tr}\, K)^{1/2}\, d\lambda_0 &= \int_{TM_1} (-\mathrm{tr}\, K U^{+^{-1}} U^+)^{1/2}\, d\lambda_0 \\ &\leqq \int_{TM_1} (\mathrm{tr}(-K U^{+^{-1}}))^{1/2} (\mathrm{tr}\, U^+)^{1/2}\, d\lambda_0 \\ &\leqq \Big(\int_{TM_1} \mathrm{tr}(-K U^{+^{-1}})\, d\lambda_0 \Big)^{1/2} \Big(\int_{TM_1} \mathrm{tr}\, U^+\, d\lambda_0 \Big)^{1/2} \\ &= \int_{TM_1} \mathrm{tr}\, U^+\, d\lambda_0 = h_{\lambda_0}(\varphi). \end{aligned} \tag{12}$$

Replacing $\operatorname{tr} K$ by $(n-1)$ Ric the inequality of the corollary is proved. If the equality holds we must have:

$$\operatorname{tr}(-K U^{+^{-1}} U^{+}) = \operatorname{tr}(-K U^{+^{-1}}) \operatorname{tr} U^{+} \quad \text{a.e.}$$

Then there exists a measurable function λ: $TM_1 \to R$ such that $KU^{+^{-1}} = \lambda U^{+}$ a.e. But if the first and last terms of (12) coincide we must also have that $\operatorname{tr}(-KU^{+^{-1}})$ is a constant scalar multiple of $\operatorname{tr} U^{+}$. Since $-KU^{+^{-1}} = -\lambda U^{+}$ this implies that λ is constant curvature. The converse property is well known.

We didn't find any significant lower estimates for metric entropies but we conjecture that for manifolds without conjugate points either $h_{\lambda_0}(\varphi) > 0$ or M is flat. Replacing $h_{\lambda_0}(\varphi)$ by $h_{\text{top}}(\varphi)$ we obtain a weaker conjecture that is by Theorem I equivalent to the question of whether the fundamental group of a non flat manifold without conjugate points has exponential growth. For non flat manifolds without focal points it is known that $h_{\lambda_0}(\varphi) > 0$ (Pesin [13]). In particular $\lambda = h_{\text{top}}(\varphi) \geqq h_{\lambda_0}(\varphi) > 0$ and the fundamental group has exponential growth. This property had already been proved by Avez [1].

I. Proof of Theorem I

Suppose that γ: $\mathbb{R} \to M$ is a geodesic arc parametrized by arc length and that for each $t \in \mathbb{R}$ we give a linear map $V(t)$: $E(\gamma(0), \dot\gamma(0)) \to E(\gamma(t), \dot\gamma(t))$. We define the derivative of V as the linear map:

$$\left(\frac{DV}{Dt}\right)(t): E(\gamma(0), \dot\gamma(0)) \to E(\gamma(t), \dot\gamma(t))$$

defined by:

$$\left(\frac{DV}{Dt}\right)(t)\, w = \frac{D(V(t)\, w)}{Dt}.$$

For each $\theta \in TM_1$ let $Y_\theta(t)$: $E(\theta) \to E(\varphi_t(\theta))$, $t \in \mathbb{R}$, be the solution of the Jacobi equation:

$$\left(\frac{D^2 Y_\theta}{Dt^2}\right)(t) = -K(\varphi_t(\theta))\, Y_\theta(t) \tag{1}$$

with initial condition:

$$Y_\theta(0) = 0,$$

$$\left(\frac{DY_\theta}{Dt}\right)(0) = I.$$

Then for all $\theta \in TM_1$ and $v \in E(\theta)$, the Jacobi perpendicular vector field J along the geodesic determined by θ with initial conditions $J(0) = 0$, $(DJ/Dt)(0) = v$ is given by $J(t) = Y_\theta(t) v$. It is well known that for all $p \in M$ the derivative of the exponential map $\exp_p$: $T_p M \to M$ is given by the expression:

$$(D_{tv} \exp_p)\, w = \frac{1}{t}\, Y_{(p,v)}(t)\, w \tag{2}$$

where $t>0$, $v\in T_pM$ has $\|v\|=1$ and $w\in E(p,v)$. Then:

$$\det(D_{tv}\exp_p)=\frac{1}{t^{n-1}}\det Y_{(p,v)}(t). \tag{3}$$

Moreover, if $S_p=\{w\in T_pM \mid \|w\|=1\}$, it follows from (3) that

$$\operatorname{vol} B_r(p)=\int_{S_r}\left(\int_0^r |\det Y_{(p,v)}(t)\,|\,dt\right)dv. \tag{4}$$

If $\theta\in TM_1$ and $t\neq 0$ define the linear map $U_\theta(t)$: $E(\varphi_t(\theta))\hookleftarrow$ by:

$$U_\theta(t)=\left(\frac{DY_\theta}{Dt}\right)(t)\,Y_\theta(t)^{-1}.$$

Since M has no conjugate points $Y_\theta(t)$ is an isomorphism for all $\theta\in TM_1$ and $t\neq 0$. Hence $Y_\theta(t)^{-1}$ exists and $U_\theta(t)$ is well defined. Moreover $U_\theta(t)$ is self adjoint. A proof of this elementary fact can be found in Green [4] or directly checked by the reader, observing that the derivatives of the functions of t $\langle U_\theta^{-1}(t)v,w\rangle$ and $\langle v,U_\theta(t)^{-1}w\rangle$ coincide and these functions converge to 0 if $t\to 0$. Green also proved in [4] that there exists $A>0$ such that

$$\|U_\theta(t)\|\leqq A \tag{5}$$

for all $\theta\in TM_1$ and $t\geqq 1$. To prove Theorem I we shall need the following lemmas. The first one is a slightly improved version of Manning's proof of the existence of the growth rate λ.

Lemma I.1. *For every $\varepsilon>0$ there exists $C_\varepsilon>0$ such that:*

$$\operatorname{vol} B_r(p)\leqq C_\varepsilon\exp(\lambda+\varepsilon)r$$

for every $r>0$ and $p\in\tilde M$.

Proof. Choose a fundamental domain N of $\tilde M$. Let $a=\operatorname{diam}(\tilde N)$. Then $B_r(q)\subset B_{r+a}(p)$ for every p and q in N and all $r>0$. Therefore $\operatorname{vol} B_r(q)\leqq \operatorname{vol} B_{r+a}(q)$. Fix $p\in N$ and choose $r_0>0$ such that $\operatorname{vol} B_r(p)\leqq\exp(\lambda+\varepsilon)r$ if $r\geqq r_0$. Then if $q\in N$:

$$\operatorname{vol} B_r(q)\leqq\operatorname{vol} B_{r+a}(p)\leqq\exp(\lambda+\varepsilon)a\cdot\exp(\lambda+\varepsilon)r. \tag{6}$$

Since for every $q\in\tilde M$ there exists $q'\in N$ with $\operatorname{vol} B_r(q)=\operatorname{vol} B_r(q')$ for all $r>0$, the inequality (6) holds for all $q\in\tilde M$ and every $r\geqq r_0$. Putting $C_\varepsilon=\max\{\exp(\lambda+\varepsilon)a, \max\{\exp(-(\lambda+\varepsilon)r)\operatorname{vol} B_r(q)\mid 0<r\leqq r_0,\ q\in\tilde M\}\}$ the lemma is proved.

The second lemma is much more delicate and we shall give its proof only after completing the demonstration of Theorem I:

Lemma I.2. *There exists $C>0$ such that:*

$$\|(D_\theta\varphi_t)^\wedge\|\leqq C\,|\det Y_\theta(t)|$$

for every $\theta\in TM_1$ and t with $|t|\geqq 1$.

Then, by Przytycki's inequality and Lemma I.2:

$$h(\varphi) \leqq \limsup_{t\to+\infty} \frac{1}{t} \log \int_{TM_1} \|(D_x\varphi_t)^\wedge\| \, d\lambda_0(x)$$

$$\leqq \limsup_{t\to+\infty} \frac{1}{t} \log \int_{TM_1} |\det Y_x(t)| \, d\lambda_0(x).$$

Since $\det Y_x(t) \neq 0$ for all $x \in TM_1$ and $t>0$ it follows that $\det Y_x(t)$ has constant sign. Then the definition of $U_x(t)$ implies:

$$\frac{d}{dt}|\det Y_x(t)| = \frac{d}{dt}(\det Y_x(t)) = \operatorname{tr} U_x(t) \det Y_x(t)$$

$$= \operatorname{tr} U_x(t) \, |\det Y_x(t)|$$

and

$$h(\varphi) \leqq \limsup_{t\to+\infty} \frac{1}{t} \log \int_{TM_1} \Big(|\det Y_x(1)|$$

$$+ \int_1^t \operatorname{tr} U_x(s) \, |\det Y_x(s)| \, ds\Big) \, d\lambda_0(x).$$

By (5):

$$h(\varphi) \leqq \limsup_{t\to+\infty} \frac{1}{t} \log \int_{TM_1} (n-1) A \int_1^t |\det Y_x(s)| \, ds) \, d\lambda_0(x)$$

$$= \limsup_{t\to+\infty} \frac{1}{t} \log \int_{TM_1} \Big(\int_1^t |\det Y_x(s)| \, ds\Big) \, d\lambda_0(x)$$

$$= \limsup_{t\to+\infty} \frac{1}{t} \log \int_M \int_{S_p} \int_1^t |\det Y_{(p,v)}(s)| \, ds \, dv \, d\lambda_1(p).$$

By (4):

$$h(\varphi) \leqq \limsup_{t\to+\infty} \frac{1}{t} \log \int_M (\operatorname{vol} B_t(p) - \operatorname{vol} B_1(p)) \, d\lambda_1(p)$$

$$= \limsup_{t\to+\infty} \frac{1}{t} \log \int_M \operatorname{vol} B_t(p) \, d\lambda_1(p).$$

Given $\varepsilon>0$ and taken $C_\varepsilon>0$ with the property of Lemma I.1 we obtain:

$$h(\varphi) \leqq \limsup_{t\to+\infty} \frac{1}{t} \log \int_M C_\varepsilon \exp(\lambda+\varepsilon) t \, d\lambda_1 = \lambda+\varepsilon.$$

Since ε is arbitrarily small the theorem is proved.

Now we shall prove I.2. For its proof, as well as for the proof of Theorem II that we shall give in the next section, the crucial step is the following lemma:

Lemma I.3. *There exists $B>0$ such that:*

$$\|Y_\theta(t) v\| \geqq B \|v\|$$

for all $\theta \in TM_1$, $|t| \geqq 1$ and $v \in E(\theta)$.

Proof. Define $K_0 = \sup\{\|K(\theta)\| \mid \theta \in TM_1\}$ and $A' = 3^{-1}K_0 + A + 1$, where A is given by (5). Fix a C^∞ function ψ: $R \hookleftarrow$ such that $\psi(0) = 1$ and $\psi(t) = 0$ for $|t| \geqq 1$. Define:

$$C = \int_{-1}^{1} (|\psi''\psi| + K_0\psi^2)\, dt$$

and choose B satisfying

$$0 < B < (A'C)^{-\frac{1}{2}}. \tag{7}$$

Suppose that $\|Y_\theta(T)v\| < B\|v\|$ for some $\theta \in TM_1$, $v \in E_\theta$ and $|T| > 1$. Assume that $T > 1$ (the other case follows applying the same method). Let γ: $R \to M$ be the geodesic with initial condition θ and J the perpendicular Jacobi vector field with initial conditions $J(0) = 0$, $(DJ/Dt)(0) = v$. Then $J(t) = Y_\theta(t)v$ and $\|J(T)\| < B\|v\|$. On the space Ω_T of continuous piecewise C^2 perpendicular vector fields along the arc $\gamma/[-1, T+1]$ that vanish at the endpoints define the index form:

$$I_T(V, W) = -\int_{-1}^{T+1} \left\langle \frac{D^2 V}{Dt^2}(t) + K(\varphi_t(\theta))\, V, W \right\rangle - \sum_i \left\langle \Delta_i \left(\frac{DV}{Dt}\right), W(t_i) \right\rangle$$

where $\Delta_i(DV/Dt)$ denotes the jump of the function DV/Dt at the discontinuity t_i. Following Klingenberg [6] define an element $\tilde{J} \in \Omega_T$ by $\tilde{J}(t) = 0$ for $-1 \leqq t \leqq 0$, $\tilde{J}(t) = J(t)$ for $0 \leqq t \leqq T$ and $\tilde{J}(t) = (T+1-t)\,\tau_T^t J(T)$ for $T \leqq t \leqq T+1$, where τ_T^t: $T_{\gamma(T)}M \to T_{\gamma(t)}M$ denotes the parallel transport along γ. Then:

$$I_T(\tilde{J}, \tilde{J}) = -\int_T^{T+1} (T+1-t)^2 \langle K(\varphi_t(\theta))\, \tau_T^t J(T) \rangle\, dt$$
$$+ \left\langle \left(\frac{DJ}{Dt}\right)(T), J(T) \right\rangle + \|J(T)\|^2.$$

But $(DJ/Dt)(T) = U_\theta(T)J(T)$. Then, since $T \geqq 1$, (5) implies

$$\left| \left\langle \left(\frac{DJ}{dt}\right)(T), J(T) \right\rangle \right| \leqq A\|J(T)\|^2 \leqq AB^2.$$

Then:

$$I_T(\tilde{J}, \tilde{J}) \leqq \tfrac{1}{3}K_0\|J(T)\|^2 + (A+1)B^2 \leqq A'B^2.$$

Define $Z \in \Omega_T$ by $Z(t) = \psi(t)\,\tau_0^t((DJ/Dt)(0))$ for $|t| \leqq 1$ and $Z(t) = 0$ otherwise. It is easy to verify that:

$$I_T(\tilde{J}, Z) = -\|(DJ/Dt)(0)\|^2 = -1,$$

$$I_T(Z, Z) = -\int_{-1}^{1} (\psi''(t)\psi(t) + \psi^2(t) \langle K(\varphi_t(\theta))\, \tau_0^t(DJ/Dt)(0),$$
$$\tau_0^t(DJ/Dt)(0)\rangle)\, dt \leqq \left(\int_{-1}^{1} |\psi''\psi|\, dt + K_0 \int_{-1}^{1} \psi^2\, dt\right) = C,$$

Then, if $\lambda \in \mathbb{R}$:

$$I_T(Z-\lambda\tilde{J}, Z-\lambda\tilde{J}) \leqq C+2\lambda+\lambda^2 A' B^2.$$

Since $4-4A'CB^2>0$ (because B was chosen satisfying (7)) this polynomial in λ has two real roots. Therefore $I_T(Z-\lambda\tilde{J},\ Z-\lambda\tilde{J})<0$ for some value of λ, thus proving the existence of conjugate points in the arc $\gamma/[-1, T+1]$.

Now we are ready to prove I.2. Consider the vector bundle F on TM_1 whose fiber F_0 over 0 is the subspace of $T_\theta(TM_1)$ orthogonal to the vector field associated to the geodesic flow. It is well known that F is invariant under the derivative $D\varphi_t$ of the flow φ_t. Consider also the vector bundle G on TM_1 whose fiber G_θ over θ is $E(\theta)\times E(\theta)$. If $t\in\mathbb{R}$ define the vector bundle isomorphism ψ_t: $G\hookleftarrow$ as follows: if $(v,w)\in G_\theta$ let J be the Jacobi vector field along the geodesic determined by θ, with initial conditions $J(0)=v$, $(DJ/Dt)(0)=w$. Set $\psi_t(v,w)=(J(t),(DJ/Dt)(t))$. The transformations ψ_t define a flow of vector bundle isomorphisms covering φ_t. In [3] Eberlein proved that there exists a vector bundle isomorphism H: $G\to F$, covering the identity and isometric in each fiber, such that

$$H\circ\psi_t=((D\varphi_t)/F)\circ H$$

for all $t\in R$. Therefore:

$$\|(\psi_t/G_\theta)^\wedge\| = \|((D_\theta\varphi_t)/F)^\wedge\|$$

and since:

$$\|((D_0\varphi_t)/F)^\wedge\| = \|(D_\theta\varphi_t)^\wedge\|$$

we obtain:

$$\|(\psi_t/G_\theta)^\wedge\| = \|(D_\theta\varphi_t)^\wedge\|.$$

Hence our problem is now to find $C>0$ satisfying

$$\|(\psi_t/G_\theta)^\wedge\| \leqq C\,|\det Y_\theta(t)|$$

for all $|t|\geqq 1$, $\theta\in TM_1$. From now on we shall consider the case $t\geqq 1$. The case $t\leqq -1$ follows applying the same methods. Write $\operatorname{graph} U_\theta(t) =\{(u, U_\theta(t)u)\in G_{\varphi_t(0)}\,|\,u\in E(\varphi_t(\theta))\}$. The definitions of ψ_t and $U_\theta(t)$ imply:

$$\psi_t(\{0\}\times E(\theta))=\operatorname{graph} U_\theta(t) \tag{8}$$

for all $t\in\mathbb{R}$. In particular:

$$\psi_{-t}(\{0\}\times E(\varphi_t(\theta)))=\operatorname{graph} U_{\varphi_t(\theta)}(-t). \tag{9}$$

Consider the splittings:

$$G_\theta=(\{0\}\times E(\theta))\oplus\operatorname{graph} U_{\varphi_t(\theta)}(-t), \tag{10}$$

$$G_{\varphi_t(\theta)}=(\{0\}\times E(\varphi_t(\theta)))\oplus\operatorname{graph} U_\theta(t). \tag{11}$$

Observe that by (8) and (9) ψ_t/G_θ sends the factors in the first decomposition onto those of the second. Moreover if $u\in E(\varphi_t(\theta))$:

$$\psi_{-t}(0,u)=(Y_{\varphi_t(\theta)}(-t)\,u,\ U_{\varphi_t(\theta)}(-t)\,Y_{\varphi_t(\theta)}(-t)\,u).$$

Then, if $t \geqq 1$:

$$\|\psi_{-t}(0, u)\| \geqq \| Y_{\varphi_t(\theta)}(-t) u\| \geqq B \|u\|.$$

This inequality, together with (9), implies:

$$\|(\psi_t/\text{graph}\, U_{\varphi_t(\theta)}(-t))\| \leqq B^{-1}. \tag{12}$$

We can always assume $B<1$. Then (12) and the observation above about how ψ_t transforms the splitting (10) in the splitting (11), implies:

$$\|(\psi_t/G_\theta)^\wedge\| \leqq B^{-(n-1)} \|(\psi_t/\{0\} \times E(\theta))^\wedge\|. \tag{13}$$

Moreover, if π_t: $G_{\varphi_t(\theta)} \to E(\varphi_t(\theta))$ is the canonical projection onto the first factor, we have:

$$\pi_t \circ (\psi_t/(\{0\} \times E(\theta)) = Y_\theta(t)$$

which means:

$$\psi_t/(\{0\} \times E(\theta)) = (\pi_t/\text{graph}\, U_\theta(t))^{-1} \circ Y_\theta(t). \tag{14}$$

But (5) implies that if $t \geqq 1$:

$$\|(\pi_t/\text{graph}\, U_\theta(t))^{-1}\| = (1 + \|U_\theta(t)\|^2)^{1/2} \leqq (1+A^2)^{1/2}. \tag{15}$$

Hence (13), (14) and (15) imply:

$$\|(\psi_t/G_\theta)^\wedge\| \leqq B^{-(n-1)}(1+A^2)^{1/2} \| Y_\theta(t)^\wedge\|.$$

Now the problem has been reduced to finding $A'>0$ satisfying

$$\| Y_\theta(t)^\wedge\| \leqq A' |\det Y_\theta(t)|$$

for every $\theta \in TM_1$ and $t \geqq 1$. Observe that $\| Y_\theta(t)^\wedge\|$ is the maximum of $|\det Y_\theta(t)/S|$ where S varies in the set of all the subspaces of $E(\theta)$. Choose S such that $\| Y_\theta(t)^\wedge\| = |\det Y_\theta(t)/S|$. With respect to the splittings $E(\theta) = S \oplus S^\perp$, $E(\varphi_t(\theta)) = Y_\theta(t) S \oplus (Y_\theta(t) S)^\perp$, $Y_\theta(t)$ can be written in the form

$$Y_\theta(t) = \begin{pmatrix} Y_1 & Y_3 \\ 0 & Y_2 \end{pmatrix}.$$

Then:

$$|\det Y_1| = |\det Y_\theta(t)/S| = \| Y_\theta(t)^\wedge\|$$

and

$$|\det Y_\theta(t)| = |\det Y_1|\, |\det Y_2| = |\det Y_2|\, \| Y_\theta(t)^\wedge\|.$$

From Lemma I.3 it follows that:

$$\| Y_1 v + Y_3 w\|^2 + \| Y_2 w\|^2 \geqq B^2(\|v\|^2 + \|w\|^2)$$

for all $v \in S$, $w \in S^\perp$. Take $v = -Y_1^{-1} Y_3 w$. Then $Y_1 v + Y_3 w = 0$ and:

$$\| Y_2 w\|^2 \geqq B^2(\|v\|^2 + \|w\|^2) \geqq B^2 \|w\|^2.$$

Then the image by Y_2 of the unit ball of $S^\perp$ contains a ball of radius B. This means that:

$$|\det Y_2| \geqq B^k$$

where $k = \dim S$. Hence:

$$\| Y_\theta(t)^\wedge \| = |\det Y_2|^{-1} |\det Y_\theta(t)| \leqq B^{-k} |\det Y_\theta(t)|.$$

II. Proof of Theorem II

First recall the construction due to Green [4] of solutions of the Ricatti equation of M (Eq. (7) of the Introduction). If $\theta \in TM_1$ and $s \in \mathbb{R}$ let $Y_{\theta,s}(t)$ be the solution of the Jacobi equation (Eq. (1) of Sect. I) such that $Y_{\theta,s}(s)=0$, $Y_{\theta,s}(0)=I$. In [4] Green proves that $\lim_{s\to-\infty} Y_{\theta,s}(t)$ exists for all $\theta \in TM_1$ and $t \in \mathbb{R}$ (see also Eberlein [3], Sect. 2). He also shows that defining:

$$Y_\theta^+(t) = \lim_{s\to-\infty} Y_{\theta,s}(t) \tag{1}$$

we obtain a solution of Jacobi equation such that $\det Y_\theta^+(t) \neq 0$. Moreover it is proved in [4] and [3] that $(DY_\theta^+/Dt)(t) = \lim_{s\to-\infty} (DY_{\theta,s}/Dt)(t)$. Then, if we define

$$U_s(\theta) = \frac{DY_{\theta,s}}{Dt}(0), \tag{2}$$

$$U^+(\theta) = \frac{DY_\theta^+}{Dt}(0) \tag{3}$$

we obtain

$$U^+(\theta) = \lim_{s\to-\infty} U_s(\theta). \tag{4}$$

Moreover it is easy to check that

$$Y_{\varphi_h(\theta),s-h}(t) = Y_{\theta,s}(t+h)\, Y_{\theta,s}^{-1}(h).$$

Then

$$Y_{\varphi_h(\theta)}^+(t) = Y_\theta^+(t+h)\, Y_\theta^{+-1}(h)$$

and

$$U^+(\varphi_h(\theta)) = \frac{DY_\theta^+}{Dt}(h)\, Y_\theta^{+-1}(h)$$

for all $h \in \mathbb{R}$. From this it follows that U^+ is a solution of the Ricatti equation of M. From (4) follows that it is measurable (because $U_s(\theta)$ is continuous on θ) and from Lemma 2 of [4] (or [3] Sect. 6) it is bounded. We shall prove that this solution satisfies Theorem II.

Given $\theta \in TM_1$ and $s \in \mathbb{R}$ define

$$V_{\theta,s} = \operatorname{graph} U_s(\theta) = \{(x, U_s(\theta)x) \mid x \in E(\theta)\}$$
$$V_\theta = \operatorname{graph} U^+(\theta).$$

From (4) it follows that:

$$V_\theta = \lim_{s\to -\infty} V_{\theta,s}.$$

We claim that:

$$\psi_{-s}(\{0\}\times E(\varphi_s(\theta))) = V_{\theta,s}. \tag{5}$$

To prove this take $v\in E(\varphi_s(\theta))$ and let $J(t)$ be a Jacobi perpendicular vector field along the geodesic $\gamma(t)$ with initial condition $(\gamma(0), \dot\gamma(0)) = \varphi_s(\theta)$ such that $J(0)=0$, $(DJ/Dt)(0)=v$. The by the definition of ψ:

$$\psi_{-s}(0, v) = (J(-s), (DJ/Dt)(-s)).$$

Consider the Jacobi vector field $\hat J(t) = J(t-s)$ along the geodesic $\hat\gamma(t)=\gamma(t-s)$. Then $\hat J(s)=0$ and $\hat J(0)=J(-s)$. Hence:

$$\hat J(t) = Y_{\theta,s}(t)\,J(-s)$$

or

$$J(t) = Y_{\theta,s}(t+s)\,J(-s).$$

Then

$$\left(\frac{DJ}{Dt}\right)(-s) = \frac{DY_{\theta,s}}{Dt}(0)\,J(-s) = U_s(\theta)\,J(-s)$$

which means by the definition of $V_{\theta,s}$ that:

$$\psi_{-s}(\{0\}\times E(\varphi_s(\theta))) \subset V_{\theta,s}.$$

The equality now follows from the fact that both members have the same dimension.

By Oseledec's theory there exists a total probability Borel set $\Lambda\subset M$ such that for every $x\in\Lambda$ there is a unique splitting $G_x = E^s(x)\oplus E^c(x)\oplus E^u(x)$ such that:

$$\lim_{t\to\pm\infty}\frac{1}{t}\log\|\psi_t v\| = 0 \qquad v\in E^c(x),$$

$$\lim_{t\to\pm\infty}\frac{1}{t}\log\|\psi_t v\| > 0 \qquad v\in E^u(x),$$

$$\lim_{t\to\pm\infty}\frac{1}{t}\log\|\psi_t v\| < 0 \qquad v\in E^s(x)$$

and for every subspace $E^u(x)\subset S\subset E^u(x)\oplus E^c(x)$ the limit

$$\chi(x) = \lim_{t\to+\infty}\frac{1}{t}\log(\det(\psi_t/S)) \tag{6}$$

exists and is independent of S.

We claim that for any φ-invariant probability μ on M the following properties hold for μ-a.e. $\theta\in\Lambda$:

$$E^u(\theta)\subset V_\theta\subset E^u(\theta)\oplus E^c(\theta), \tag{7}$$

$$\chi(\theta) = \lim_{T\to+\infty}\frac{1}{T}\int_0^T \operatorname{tr} U^+(\varphi_t(\theta))\,dt. \tag{8}$$

From (8) follows Theorem II because it is easy to prove, using the isomorphism $H: G \to F$ defined in Sect. I, that the function χ associated to the linear flow $\psi: \mathbb{R} \times G \to G$ coincides with that associated to the linear flow $D\varphi: \mathbb{R} \times T(TM_1) \to TM_1$.

To prove the second inclusion in (7) observe that if $v \in E(\theta)$:

$$\psi_t(0, v) = \left(Y_\theta(t)v, \left(\frac{DY_\theta}{Dt}\right)(t)v\right),$$

Then Lemma I.3 implies that $(0, v) \notin E^s(\theta)$ if $v \neq 0$. Hence

$$E^s(\theta) \cap (\{0\} \times E(\theta)) = \{0\}.$$

Therefore for every $\theta \in \Lambda$ there exist a subspace $T(\theta) \subset E^u(\theta) \oplus E^c(\theta)$ and a linear map $L(\theta)$: $T(\theta) \to E^s(\theta)$ such that

$$\{0\} \times E(\theta) = \operatorname{graph} L(\theta).$$

Given $\varepsilon > 0$ take a compact subset $K \subset \Lambda$ with $\mu(K) \geqq 1 - \varepsilon$ and satisfying:

a) $\sup\{\|L(\theta)\| \,|\, \theta \in K\} < +\infty$.

b) There exists $T > 0$ and $0 < \xi < 1 < \tilde{\xi} < \xi^{-1}$ such that:

$$\|\psi_t / E^s(\theta)\| \leqq \xi^t,$$
$$\|\psi_t^{-1} / (E^u(\theta) \oplus E^c(\theta))\| \leqq \tilde{\xi}^t$$

for every $\theta \in K$, $t \geqq T$.

Then, if $\theta \in \Lambda$:

$$\begin{aligned} V_{\theta,s} &= \psi_s(\{0\} \times E(\varphi_{-s}(\theta))) = \psi_s(\operatorname{graph} L(\varphi_{-s}(\theta))) \\ &= \operatorname{graph}(\psi_s \circ L(\varphi_{-s}(\theta)) \circ (\psi_{-s} / L(\theta))). \end{aligned}$$

If $\theta \in K$ take a sequence $s_n \to -\infty$ such that $\varphi_{s_n}(\theta) \in K$. Then by property (a):

$$\|\psi_{s_n} \circ L(\varphi_{-s_n}(\theta)) \circ (\psi_{-s_n} / L(\theta))\| \leqq \xi^{s_n} \tilde{\xi}^{s_n} \sup_{\theta \in K} \|L(\theta)\|.$$

But $\xi^{s_n} \tilde{\xi}^{s_n} \to 0$ if $n \to +\infty$. Then:

$$V_\theta = \lim_{n \to +\infty} V_{\theta, s_n} = T(\theta) \subset E^u(\theta) \oplus E^c(\theta)$$

therefore the second inclusion in (7) is satisfied by any $\theta \in K$. Since $\mu(K) \geqq 1 - \varepsilon$ and ε is arbitrary this inclusion holds for a.e. $\theta \in \Lambda$.

To prove the first inclusion of (7) consider the splitting:

$$G_\theta = V_\theta + V_\theta^\perp$$

for $\theta \in \Lambda$. Since V_θ is ψ_t-invariant, with respect to this splitting ψ_t can be written in block form:

$$\psi_t = \left(\begin{array}{c|c} A_t(\theta) & C_t(\theta) \\ \hline 0 & B_t(\theta) \end{array}\right).$$

Now recall that if $J_1(t)$, $J_2(t)$ are perpendicular Jacobi vector fields of M it is known that:

$$\left\langle J_1(t), \frac{DJ_2}{Dt}(t)\right\rangle - \left\langle J_2(t), \frac{DJ_1}{Dt}(t)\right\rangle$$

is constant in t. This means that the vector bundle isomprphism $J: G \hookleftarrow$ covering the identity and acting on the fibers G_θ as:

$$J(v, w) = (-w, v)$$

satisfies:

$$\langle J\psi_t u, \psi_t v\rangle = \langle Ju, v\rangle$$

for all $u, v \in G_\theta$, $t \in \mathbb{R}$, $\theta \in M$. This means that

$$\psi_t^* J \psi_t = J$$

for all t, or:

$$A_t(\theta) = B_t^{*-1}(\theta)$$

for all $\theta \in \Lambda$, $t \in \mathbb{R}$. Since $V_\theta \subset E^u(\theta) \oplus E^c(\theta)$:

$$\lim_{t \to \pm\infty} \frac{1}{t} \log \|B_t^{*-1}(\theta) w\| = \lim_{t \to \pm\infty} \frac{1}{t} \log \|A_t(\theta) w\| \geqq 0$$

for all $w \in V_\theta^\perp$. Then it is easy to see for all $w \in V_\theta^\perp$:

$$\lim_{t \to \pm\infty} \frac{1}{t} \log \|B_t^{-1}(\theta) w\| \geqq 0. \tag{9}$$

Now suppose that $E^u(\theta) \not\subset V_\theta$. Take $0 \neq v \in E^u(\theta)$, $v \notin V_\theta$. Then $v = w + z$, $0 \neq w \in V_\theta^\perp$, $z \in V_\theta$. Moreover:

$$\psi_{-t}(v) = B_t^{-1}(\theta)\, w + u_t$$

for some

$$u_t \in V_{\varphi_{-t}}(\theta).$$

Then:

$$\|\psi_{-t}(v)\| \geqq \|B_t^{-1}(\theta) w\|$$

and by (9):

$$\lim_{t \to +\infty} \frac{1}{t} \log \|\psi_{-t}(v)\| \geqq 0$$

contradicting $v \in E^u(\theta)$.

Finally we shall prove (8). Let $\pi_\theta \colon V_\theta \to E(\theta)$ be the restriction to $V_\theta \subset E(\theta) \times E(\theta)$ of the canonical projection onto the first factor. Then:

$$\pi_\theta^{-1}(v) = (v, U_\theta^+(v)).$$

Therefore:

$$\sup_{\theta \in \Lambda} \|\pi_\theta^{-1}\| < +\infty. \tag{10}$$

Moreover it is easy to check that:

$$\psi_t/V_\theta = \pi^{-1}_{\varphi_t(\theta)} \circ Y^+_\theta(t) \circ \pi_\theta \tag{11}$$

for all $\theta \in \Lambda$, $t \in \mathbb{R}$. From (10, (11) and Lemma I.3 we conclude that:

$$\lim_{t\to+\infty} \frac{1}{t} \log\det(\psi_t/V_\theta) = \lim_{t\to+\infty} \frac{1}{t} \log\det Y^+_\theta(t)$$
$$= \lim_{t\to+\infty} \frac{1}{t} \int_0^t \operatorname{tr} Y^+_\theta(s) \frac{DY^+_\theta}{Dt}(s)\,ds = \lim_{t\to+\infty} \frac{1}{t} \int_0^t \operatorname{tr} U^+(\varphi_\theta(s))\,ds.$$

This equality, together with (6) and (7) implies (8) for μ-a.e. $\theta \in V$.

Remark I. When M has negative sectional curvatures there exists $c>0$ such that $\langle U^+(\theta)v, v\rangle \geqq c\|v\|^2$ for all $\theta \in TM_1$, $v \in E(\theta)$. To prove this set $A = \sup\{\|U^+(\theta)\| \mid \theta \in TM_1\}$ and take $\delta>0$ such that for any perpendicular Jacobi vector field J with $\|J(0)\|=1$ and $\|(DJ/Dt)(0)\| \leqq 2A$ then $\|J(t)\| \geqq \frac{1}{2}$ for $-\delta \leqq t \leqq 0$. Then, given $\theta \in TM_1$, it follows that

$$\|Y^+_\theta(t)v\| \geqq \tfrac{1}{2}\|v\|$$

for all $\theta \in TM_1$, $v \in E(\theta)$, $-\delta \leqq t \leqq 0$. Then (1) and (1) imply that given $\theta \in TM_1$ there exists $C>0$ such that if $-s \geqq C$:

$$\|Y_{\theta,s}(t)v\| \geqq \tfrac{1}{2}\|v\|$$

for every $-\delta \leqq t \leqq 0$ and $v \in E(\theta)$. Fix any $v \in E(\theta)$ and define:

$$u(t) = \langle Y_{\theta,s}(t)v, Y_{\theta,s}(t)v\rangle.$$

Then:

$$u(s) = \dot u(s) = 0,$$
$$\dot u(0) = 2\langle U_s(\theta)v, v\rangle.$$

Hence:

$$\langle U_s(\theta)v, v\rangle = \tfrac{1}{2} \int_s^0 \ddot u(r)\,dr.$$

Choose $k>0$ satisfying $\langle K(x)w, w\rangle \leqq -k\|w\|^2$ for all $x \in TM_1$ and $w \in E(x)$. It is easy to verify that:

$$\ddot u(r) \geqq k u(r).$$

Then, if $-s \geqq C$:

$$\langle U_s(\theta)v, v\rangle \geqq \tfrac{1}{2} k \int_s^0 u(r)\,dr = \tfrac{1}{2} k \int_s^0 \langle Y_{\theta,s}(r)v, Y_{\theta,s}(r)v\rangle\,dr$$
$$\geqq \tfrac{1}{2} k \int_{-\delta}^0 \langle Y_{\theta,s}(r)v, Y_{\theta,s}(r)v\rangle\,dr \geqq \tfrac{1}{8}\delta k\|v\|^2.$$

Therefore $\langle U^+(\theta)v, v\rangle \geqq \lim_{s\to-\infty} \langle U_s(\theta)v, v\rangle \geqq 8^{-1}\delta k\|v\|^2$.

References

1. Avez, A.: Variétés riemanniennes sans points focaux. C.R. Acad. Sci. Paris, Ser A-B **270**, A188–A191 (1970)
2. Dinaburg, E.I.: On the relations among various entropy characteristics of dynamical systems. Math. USSR Izv. **5**, 337–388 (1971)
3. Eberlein, P.: When is a geodesic flow of Anosov type? I. J. Differential Geometry **8**, 437–463 (1973)
4. Green, L.A.: A theorem of E. Hopf. Michigan Math. J. **5**, 31–34 (1958)
5. Katok, A.: Entropy and closed geodesics. Ergodic Theory and Dynamical Systems in press (1982)
6. Klingenberg, W.: Riemannian manifolds with geodesic flows of Anosov type. Ann. Math. **99**, 1–13 (1974)
7. Manning, A.: Topological entropy for geodesic flows. Ann. Math. **110**, 567–573 (1979)
8. Manning, A.: Curvature bounds for the entropy of the geodesic flow on a surface. J. of the London Math. Soc. **24**, 351–358 (1981)
9. Margulis, G.A.: Applications of ergodic theory to the investigation of manifolds of negative curvature. Funct. Anal. Appl. **3**, 335–336 (1969)
10. Milnor, J.: A note on curvature and fundamental group. J. Differential Geometry **2**, 1–17 (1968)
11. Oseledec, V.I.: A multiplicative ergodic theorem. Tras. Moscow Math. Soc. **19**, 197–231 (1968)
12. Pesin, Ya.: Lyapounov characteristic exponents and smooth ergotid theory. Usp. Mat. Nauk **32-4**, 55–111 (1977)
13. Pesin, Ya.: Equations for the entropy of a geodesic flow on a compact Riemannian manifold without conjugate points. Notes in Mathematics **24**, 796–805 (1978)
14. Przytycki, F.: An Upper Estimation for Topological Entropy of Diffeomorphisms, Inventiones math. **59**, 205–213 (1980)
15. Ruelle, D.: An inequality for the entropy of differentiable maps. Bol. Soc. Bras. Mat. **9**, 83–87 (1978)
16. Sacksteder, R., Shub, M.: Entropy of a differentiable map. Advances in Mathematics **28**, 181–185 (1978)
17. Sarnak, P.: Entropy estimates for geodesic flows, Thesis, Stanford University

Oblatum 24-II-1982

An invariant measure for rational maps

Alexandre Freire, Artur Lopes and Ricardo Mañé

Introduction

Let $\overline{\mathbb{C}} = \mathbb{C} \cup \{\infty\}$ be the Riemann sphere and $f : \overline{\mathbb{C}} \circlearrowleft$ be an analytic endomorphism of degree $d \geq 2$. Then f can be written as a rational function $f(z) = P(z)/Q(z)$ where P and Q are relatively prime polynomials and either P or Q has degree d. Set $f^n = f^n \circ \overset{n}{\cdots} \circ f$. *The purpose of this paper is to construct an f-invariant probability that describes the asymptotic random distribution of the roots of the equation* $f^n(z) = a$, *when* $n \to +\infty$. More precisely, denote $z_i^{(n)}(a)$, $i = 1, \ldots, d^n$, the roots of the equation $f^n(z) = a$ (counted with algebraic multiplicity), and define a probability $\mu_n(a)$ by:

$$\mu_n(a) = \frac{1}{d^n} \sum_{i=0}^{d^n} \delta_{z_i^{(n)}(a)}$$

Let $\mathcal{M}$ be the space of probabilities on the Borel σ-algebra of $\overline{\mathbb{C}}$ endowed with the weak topology, i.e., the unique metrizable topology on $\mathcal{M}$ such that a sequence $\{\mu_n \mid n \geq 1\} \subset \mathcal{M}$ converges to $\mu \in \mathcal{M}$ if and only if:

$$\lim_{n \to +\infty} \int_{\overline{\mathbb{C}}} \phi d\mu_n = \int_{\overline{\mathbb{C}}} \phi d\mu$$

for every continuous $\phi : \overline{\mathbb{C}} \to \mathbb{R}$. We shall prove that for *every* $a \in \overline{\mathbb{C}}$ *(with the possible exception of two values that can be explicitly characterized) the sequence* $\mu_n(a)$ *converges to an f-invariant probability* $\mu_f \in \mathcal{M}$, *independent of a, that exhibits certain interesting ergodic properties.* To give the full statement of our theorem, we have to recall first the definition of the Julia set $J(f)$ of f. $J(f)$ is the set of points $z \in \overline{\mathbb{C}}$ such that for every neighborhood U of z, the family $\{f^n/U \mid n \geq 0\}$ is not normal. It is easy to check that $J(f)$ is compact and satisfies $f^{-1}(J(f)) = J(f)$. Moreover, $J(f)$ is the closure of the set of sources of f, i.e., points z such that $f^n(z) = z$ and $|(f^n)'(z)| > 1$ for some $n \geq 1$ (Julia [4], Fatou [3]). The definition of $J(f)$ easily implies that every $z \notin J(f)$ has a neighborhood U where the family of iterates $f^n/U \to \overline{\mathbb{C}}$ is equicontinuous. In other words, in the complement of $J(f)$ the dynamics of f is extremely stable (in the sense of Lyapounov). On the other hand, every neighborhood of a point in

Recebido em 29/4/83.

Originally published in: *Bol. Soc. Brasil. Mat. 14 (1983), no. 1, 45–62,* the Julia set is expanded under forward interation. In fact, if U is a neighborhood of $z \in J(f)$, the non-normality of $\{f^n/U \mid n \geq 1\}$ implies (by Montel's characterization of normal families) that for large values of n, $\bigcup_{m \geq n} f^m(U)$ covers the whole sphere except for at most two points. With some more work (see Brolin [1]), it can be proved that there exists a set $\mathrm{Exc}(f) \subset \overline{\mathbb{C}}$, whose elements are called exceptional points, containing at most two points, and such that for every neighborhood U of a point in $J(f)$, there exists $N > 0$ such that $f^n(U) = \mathrm{Exc}(f)^c$ for every $n \geq N$. Moreover, the points of $\mathrm{Exc}(f)$ can be described as follows. If $\mathrm{Exc}(f)$ contains only one point p, then it must satisfy $f^{-1}(p) = \{p\}$. Then if $L : \overline{\mathbb{C}} \circlearrowleft$ is a Möbius transformation such that $L(p) = \infty$, it is easy to check that $(LfL^{-1})^{-1}(\infty) = \{\infty\}$, and this implies that LfL^{-1} is a polynomial. If $\mathrm{Exc}(f)$ contains two points p and q, they must satisfy $f^{-1}(\{p, q\}) = \{p, q\}$. Taking a Möbius transformations $L : \overline{\mathbb{C}} \circlearrowleft$ such that $L(p) = \infty$, $L(q) = 0$, it follows that $(L^{-1}fL)^{-1}(\{0, \infty\}) = \{0, \infty\}$. This property implies that $(L^{-1}fL)^{-1}(z) = \alpha z^{\pm d}$, for some $\alpha \in \overline{\mathbb{C}}$.

Theorem. *There exists an f-invariant probability μ_f satisfying the following properties:*

a) $\lim_{n \to +\infty} \mu_n(a) = \mu_f$ *for every* $a \notin \mathrm{Exc}(f)$. *Moreover, this convergence is uniform when a varies in a compact subset of* $\mathrm{Exc}(f)^c$.

b) *The support of* μ_f *if* $J(f)$.

c) f *is a K-system with respect to* μ_f.

d) $u_f(f(A)) = d\mu_f(A)$ *for every Borel set* $A \subset \overline{\mathbb{C}}$ *such that* f/A *is injective. Conversely,* μ_f *is the unique f-invariant probability satisfying this property.*

e) $h_{u_f}(f) = \log d$.

Since the definition of K-system used sometimes in Ergodic Theory applies only to invertible transformations, that is not our case, it is perhaps useful for the reader to explain property (c). We shall that if $\mathcal{A}$ is the Borel σ-algebra of $\overline{\mathbb{C}}$, then $\bigcap_{n \geq 0} f^{-n}(\mathcal{A})$ contains only sets of measure 0 or 1. It is natural to include these transformations (as is done by several authors) in the class of K-systems. As in the invertible case (and for the same reasons), it implies the mixing property but is much stronger.

In [1], Brolin proved the existence of μ_f satisfying (*a*), (*b*) and a weaker form of (*c*) (with mixing instead of K-system) for the case of polynomial mappings. His methods, based in Potential Theory, do not extend to general rational maps. On the other hand, when f is a polynomial, these methods give a remarkable identification of μ_f, namely, that μ_f

is the equilibrium distribution (in the sense of Potential Theory) associated to $J(f)$. This means, roughly speaking, that μ_f describes the way a unit positive electric charge would be distributed in $J(f)$ under equilibrium conditions (for the formal definition, see Brolin [1]). Unfortunately, this beautiful characterization of μ_f doesn't extend to general rational maps. For instance, take

$$f(z) = \prod_{i=1}^{m} \left(\frac{z - a_i}{1 - \bar{a}_i z} \right)$$

where $|a_i| < 1$ for all i and not all are zero. It is easy to check that $J(f)$ is the unit circle and that $f/J(f)$ is an expanding endomorphism, i.e., $\lim_{n \to +\infty} |(f^n)'(z)v| = +\infty$ for every $z \in J(f)$ and every $0 \neq v$ tangent to $J(f)$ at z. The equilibrium distribution in the unit circle is the Lebesgue measure λ. We shall show that μ_f is singular with respect to λ. By the theory of expanding endomorphisms of manifolds, there exists an f-invariant ergodic probability ν on $J(f)$ that is equivalent to λ and such that $d\nu/d\lambda$ is a strictly positive continuous function. Since both ν and μ_f are ergodic, they are either singular or equal. Suppose that $\nu = \mu_f$ and set $H = d\mu_f/d\lambda$. From property (*d*) it follows that:

$$m = |f'(z)| \frac{H(f(z))}{H(z)}$$

for a.e. z. Hince H is continuous, this property holds for every z. Therefore

$$|f'(z)| = m\,,$$

for every fixed point z of f. It is not difficult to show that there exist values of $a_1, \ldots, a_m$ such that this condition is not satisfied. Then, for these values the probabilities μ_f and ν cannot coincide. Hence, they are singular and μ_f is singular with respect to the Lebesgue measure of the unit circle.

I. **Proof of the Theorem**

The proof of the Theorem will be based in the following definitions and lemma. We say that a set $\gamma \subset \overline{\mathbb{C}}$ is an arc if it is homeomorphic to the interval $[0, 1]$. A set $U \subset \overline{\mathbb{C}}$ is a topological disk if it is homeomorphic to the disk $D = \{z \mid |z| < 1\}$.

Definition I. We say that a set $U \subset \overline{\mathbb{C}}$ is (N, ε)-adapted if for all $n \geq N$ there exist topological disks $S_i^{(n)}$, $i = 1, \ldots, \ell_n$ and intergers $1 \leq k_i^{(n)}$ such that:

a) $f^n/S_i^{(n)}$ is a $k_i^{(n)}$-to-1 map onto $\dot{U}$.
b) $\text{diam}(S_i^{(n)}) \leq \varepsilon$
c) $\sum_{i=1}^{\ell_n} k_i^{(n)} \geq (1-\varepsilon)d^n$.
d) $\lim_{n\to+\infty} (\sup_i \text{diam}(S_i^{(n)})) = 0$.

Definition II. We say that two points $\dot{z}_i \in \overline{\mathbb{C}}$, $i = 1, 2$, are (N, ε)-related if for all $n \geq N$ the roots $z_i^{(n)}(z_1)$, $i = 1, \ldots, d^n$ of the equation $f^n(z) = z_1$ and the roots $z_i^{(n)}(z_2)$, $i = 1, \ldots, d^n$ of the equation $f^n(z) = z_2$ can be indexed in such a way that:

$$d(z_i^{(n)}(z_1), z_i^{(n)}(z_2)) \leq \varepsilon$$

for all $1 \leq i \leq t_n$, where t_n satisfies

$$t_n \geq (1-\varepsilon)d^n.$$

An important property that links these two definitions is that if z_i belongs to an (N, ε)-adapted set U_i, $i = 1, 2$, and $U_1 \cap U_2 \neq \emptyset$, then z_1 and z_2 are $(N, 2\varepsilon)$-related. The proof is immediate and we leave it to the reader.

Fundamental Lemma. *Given $\varepsilon > 0$, $z \notin Exc(f)$ and an arc γ containing z and such that $\gamma - \{z\}$ doesn't contain critical values of f^n for all $n \geq 1$, there exists an (N, ε)-adapted set $U \supset \{\gamma\}$ for some $N \geq 1$.*

The proof of this Lemma will be given in the next section. Now let us show some of its corollaries.

Corollary I. *Given a compact set $K \subset Exc(f)^c$ and $\varepsilon > 0$, there exists $N = N(K, \varepsilon) > 0$ such that any couple of points in K is (N, ε)-related.*

Proof. First, we shall prove that if z_1 and z_2 are in K, there exist $N > 0$ and an open set $V \supset \{z_1, z_2\}$ such that any couple of points in V is $(N, \varepsilon/2)$-related. For this, take arcs $\gamma_i \supset \{z_i\}$, $i = 1, 2$ satisfying the hypothesis of the Fundamental Lemma and $\gamma_1 \cap \gamma_2 \neq \emptyset$. Then there exist $(N_i, \varepsilon/2)$-adapted sets $U_i \supset \gamma_i$, $i = 1, 2$. Taking $N = \max(N_1, N_2\}$, the remark after Definition II concludes the proof of the property. Now define $\widetilde{N} : K \times K \to \mathbb{Z}^+$ by the following property: $N(z_1, z_2)$ is the minimum $N > 0$ such that there exist neighborhoods V_i of z_i, $i = 1, 2$, such that every point in V_1 is $(N(z_1, z_2), \varepsilon)$-related to every point in V_2. The previous property shows that N is well defined. Moreover, it is obviously upper semicontinuous. Then it is bounded. Let $N > 0$ be an upper bound. Clearly, N satisfies the required property.

Corollary II. *If $K \subset Exc(f)^c$ is compact, for all $\varepsilon_1 > 0$ and every continuous function $\phi : \overline{\mathbb{C}} \to \mathbb{R}$, there exists $N = N(K, \varepsilon_1, \phi) > 0$ such that:*

$$\left| \int_{\overline{\mathbb{C}}} \phi d\mu_n(z_1) - \int_{\overline{\mathbb{C}}} \phi d\mu_n(z_2) \right| \leq \varepsilon_1$$

for every z_1 and z_2 in K and $n \geq N$.

Proof. Take $\varepsilon > 0$ such that:

$$\varepsilon \sup_z |\phi(z)| \leq \varepsilon_1/2$$

and $|\phi(z') - \phi(z'')| \leq \varepsilon_1/2$ if $d(z', z'') \leq \varepsilon$. Take $N = N(K, \varepsilon)$ given by Corollary I. If $n \geq N$, by the definition of (N, ε)-related points, given z_1 and z_2 in K, we can arrange the roots $z_i^{(n)}(z_1)$, $i = 1, \ldots, d^n$ of $f^n(z) = z_1$, and the roots $z_i^{(n)}(z_2)$, $i = 1, \ldots, d^n$ of $f^n(z) = z_2$, in such a way that $d(z_i^{(n)}(z_1), z_i^{(n)}(z_2)) \leq \varepsilon$ for $i = 1, \ldots, s_n$, where s_n satisfies $s_n \geq (1 - \varepsilon)d^n$. Then:

$$\left| \int_{\overline{\mathbb{C}}} \phi d\mu_n(z_1) - \int_{\overline{\mathbb{C}}} \phi d\mu_n(z_2) \right| \leq \frac{1}{d^n} \sum_{i=1}^{d^n} |\phi(z_i^{(n)}(z_1)) - \phi(z_i^{(n)}(z_2))| \leq$$

$$\leq \frac{1}{d^n} \sum_{i=1}^{s_n} |\phi(z_i^{(n)}(z_1)) - \phi(z_i^{(n)}(z_2)| + \frac{d^n - s_n}{d^n} \sup_z |\phi(z)| \leq$$

$$\leq \frac{1}{d^n} \frac{\varepsilon_1}{2} s_n + \varepsilon \sup_z |\phi(z)| \leq \frac{\varepsilon_1}{2} + \varepsilon \sup_z |\phi(z)| \leq \frac{\varepsilon_1}{2} + \frac{\varepsilon_1}{2} = \varepsilon_1.$$

Corollary III. *Given a compact set $K \subset Exc(f)^c$, $\varepsilon_1 > 0$ and a continuous function $\phi : \overline{\mathbb{C}} \to \mathbb{R}$, there exists $N > 0$ such that:*

$$\left| \int_{\overline{\mathbb{C}}} \phi d\mu_n(z) - \int_{\overline{\mathbb{C}}} \phi d\mu_m(z) \right| \leq \varepsilon_1$$

for every $z \in K$ and $m \geq n \geq N$.

Proof. Set $\widetilde{K} = \overline{\bigcup_{n \geq 0} f^{-n}(K)}$ and take $N = N(\widetilde{K}, \phi, \varepsilon_1)$ given by Corollary II. Using the notation of the introduction, we can write for all $m > n$ and $z \in \overline{\mathbb{C}}$:

$$\int_{\overline{\mathbb{C}}} \phi(z) d\mu_m(z) = \frac{1}{d^k} \sum_{i=1}^{d^k} \int_{\overline{\mathbb{C}}} \phi_n d\mu_n(z_i^{(k)}(z))$$

where $k = m - n$. Then, if $z \in K$,

$$\left| \int_{\overline{\mathbb{C}}} \phi d\mu_n(z) - \int_{\overline{\mathbb{C}}} \phi d\mu_m(z) \right| \leq \frac{1}{d^k} \sum_{i=1}^{d^k} \left| \int_{\overline{\mathbb{C}}} \phi d\mu_n(z) - \int_{\overline{\mathbb{C}}} \phi d\mu_n(z_i^{(k)}(z)) \right|.$$

But $z_i^{(k)} \in \widetilde{K}$ for all $n \geq 0$. Hence, if $n \geq N$, the last term above is bounded by $d^{-k}(d^k \varepsilon_1) = \varepsilon_1$ by Corollary II.

Now we are ready to prove the theorem. By Corollary III, if $z \in \text{Exc}(f)^c$, the sequence $\{\mu_n(z) \mid n \geq 0\}$ converges, in the topology of $\mathcal{M}$, to a probability $\mu_f(z)$. Moreover, also by Corollary III, this convergence is uniform on compact sets of $\text{Exc}(f)^c$. By Corollary II, the probability $\mu_f(z)$ is independent of z. Denote it μ_f. Moreover, μ_f is invariant because for every continuous $\phi : \overline{\mathbb{C}} \to \mathbb{R}$, taking any $a \in \text{Exc}(f)^c$

$$\int_{\overline{\mathbb{C}}} (\phi \circ f) d\mu_f = \lim_{n \to +\infty} \frac{1}{d^n} \sum_{i=1}^{d^n} \phi(f(z_i^{(n)}(a)) =$$

$$= \lim_{n \to +\infty} \frac{1}{d^{n-1}} \sum_{i=1}^{d^{n-1}} \phi(z_i^{(n-1)}(a)) = \int_{\overline{\mathbb{C}}} \phi d\mu_f.$$

This completes the proof of the existence of an f-invariant probability μ_f satisfying (a). Now let us prove property (b). Take $a \in J(f)$. Then the support of $\mu_n(a)$ is contained in $J(f)$ for all $n \geq 1$. Then the support of μ_f is contained in $J(f)$. Conversely, if $p \in J(f)$, we shall prove that it belongs to the support of μ_f. Take a point q in the support of μ_f. Since p and q are not exceptional points (because they belong to $J(f)$), there exist a sequence $p_n \to p$ and a sequence of integers $m_n \to +\infty$ such that $f^{m_n}(p_n) = q$. Since the support is closed, it is sufficient to show that p_n belongs to the support of μ_f for all n. Given any neighborhood V of p_n, we take a neighborhood U of p_n such that $U \subset V$, $\mu_f(\partial U) = 0$ and also $\mu_f(\partial(f^{m_n}(U))) = 0$. Moreover, $f^{m_n}(U)$ is an open set containing q. Since the boundaries of U and $f^{m_n}(U)$ have measure zero, we can write:

(1) $$\mu_f(U) = \lim_{n \to +\infty} \mu_n(a)(U) = \lim_{j \to +\infty} \frac{1}{d^j} \# \{z \in U \mid f^j(z) = a\}$$

(2) $$\mu_f(f^{m_n}(U)) = \lim_{j \to +\infty} \mu_j(a)(f^{m_n}(U)) = \lim_{j \to +\infty} \frac{1}{d^j} \# \{z \in f^{m_n}(U) \mid f^j(z) = a\}$$

where a is any point in $\text{Exc}(f)^c$ that is not a critical value of any f^j (in order to grant that every root of $f^n(z) = a$ is simple). Moreover:

(3) $$\# \{z \in U \mid f^j(z) = a\} \geq \# \{z \in f^{m_n}(U) \mid f^{j-m_n}(z) = a\}.$$

From (1), (2) and (3) it follows that $\mu_f(U) \geq d^{m_n} \mu_f(f^{m_n}(U))$. But $\mu_f(f^{m_n}(U)) > 0$ because $f^{m_n}(U)$ is a neighborhood of q that is in the support of μ_f. Hence, $\mu_f(U) > 0$ and then $\mu_f(V) \geq \mu_f(U) > 0$.

To prove (c), we shall show that if $\mathcal{A}$ denotes the Borel σ-algebra of $J(f)$, then $\bigcap_{n \geq 0} f^{-n}(\mathcal{A})$ contains only sets of measure 0 or 1. Suppose that $S \in \bigcap_{n\geq 0} f^{-n}(\mathcal{A})$ and $\mu_f(S)>0$. We have to prove that $\mu_f(S)=1$. Take $0<\varepsilon<\mu_f(S)/2$. Take $N>0$, $m>0$ such that there exists a family $\mathcal{P}=\{P_1,\ldots,P_r\}$ of disjoint $(N,\varepsilon/2)$-adapted sets such that:

$$\mu_f((\bigcup_i P_i)^c)=0.$$

The existence of this partition is based on the fact that a topological disk contained in an $(N,\varepsilon/2)$-adapted set is also an $(N,\varepsilon/2)$-adapetd set. Then to every point z in $J(f)$, by the Fundamental Lemma and the previous remark, we can associate an integer $N(z)$ and a disk $B(z)$ centered at z, with $\mu_f(\partial B(z))=0$, that is $(N(z),\varepsilon/2)$-adapted. Take $z_i \in J(f)$, $i=1,\ldots,t$ such that $J(f)\subset \bigcup_i B(z_i)$ and set $N=\max_i N(z_i)$. Finally, take as sets P_i, all the intersections of sets $B(z_i)$ or $B^c(z_i)$ that have measure $\neq 0$. Denote $P_{i,j}^{(n)}$, $k_{i,j}^{(n)}$, $j=1,\ldots,\ell(n,i)$ the sets and integers associated to P_i by Definition I. We shall need two lemmas:

Lemma I. *Let $V\subset\overline{\mathbb{C}}$ be an open set with $\mu_f(\partial V)=0$ such that there exist open sets $V_1,\ldots,V_d$ such that for all $1\leq i\leq d$, f/V_i is a homeomorphism of V_i onto V. Then*

$$d\mu_f(f^{-1}(A)\cap V_i)=\mu_f(A)$$

for every Borel set $A\subset V$ and all $1\leq i\leq d$.

Proof. First we shall prove the lemma when A is open and $\mu_f(\partial A)=0$. By the hypothesis $\mu_f(\partial A)=0$, we can calculate $\mu_f(A)$ by:

$$\text{(1)}\qquad \mu_f(A)=\lim_{n\to+\infty}\mu_n(a)(A)=\lim_{n\to+\infty}\frac{1}{d^n}\#\{z\in A\mid f^n(z)=a\}$$

where a is any point that is not a critical value of f^n for all $n\geq 1$. If $\mu_f(\partial A)=0$, then $\mu_f(\partial(f^{-1}(A)\cap V_i))\leq\mu_f(f^{-1}(\partial A))=\mu_f(\partial A)=0$. Hence,

$$\text{(2)}\qquad \mu_f(f^{-1}(A)\cap V_i)=\lim_{n\to+\infty}\frac{1}{d^n}\#\{z\in f^{-1}(A)\cap V_i\mid f^n(z)=a\}.$$

By the relation $f(f^{-1}(A)\cap V_i)=A$ and the injectivity of f/V_i,

$$\text{(3)}\qquad \#\{z\in f^{-1}(A)\cap V_i\mid f^n(z)=a\}=\frac{1}{d}\#\{z\in A\mid f^{n-1}(z)=a\}.$$

Then, (1), (2) and (3) prove the lemma in this case. If $A\subset V$ is any Borel set, we can write it as $A=(\bigcap_{n\geq 0}A_n)\cup A$ where $A_1\supset A_2\supset\ldots$ is a sequence of open sets with $\mu(\partial A_i)=0$ for all i and $\mu_f(A_\infty)=0$. Then:

$$\mu_f(f^{-1}(A)\cap V_i) = \lim_{n\to+\infty} \mu_f(f^{-1}(A_n)\cap V_i)) - \mu_f(f^{-1}(A_\infty)\cap V_i) =$$
$$= \lim_{n\to+\infty} \mu_f(f^{-1}(A_n)\cap V_i) = \frac{1}{d}\lim_{n\to+\infty} \mu_f(A_n) = \frac{1}{d}\mu_f(A).$$

Corollary. *$\mu_f(f(A)) \le d\mu_f(A)$ for every Borel set A.*

Proof. Take a family $V_1, V_2, \ldots$ of disjoint topological disks satisfying:

a) f/V_i is a homeomorphism onto $f(V_i)$.
b) $\mu_f((\bigcup_i V_i)^c) = 0$ and $\mu_f(\partial f(V_i)) = 0$ for all i.
c) $f(V_i)$ doesn't contain critical values of f, for all i. Then,

$$\mu_f(f(A)) = \mu_f(f((\bigcup_i (A\cap V_i))\cup(\bigcup_i A\cap\partial V_i))) \le$$
$$\le \sum_i \mu_f(f(A\cap V_i)) + \sum_i \mu_f(f(A\cap\partial V_i)) \le \sum_i \mu_f(f(A\cap V_i)) +$$
$$+ \sum_i \mu_f(f(A)\cap\partial f(V_i)) = \sum_i \mu_f(f(A\cap V_i)).$$

But by (a) and (c), we can apply Lemma I to obtain:

$$\mu_f(f(A\cap V_i)) = d\mu_f(A\cap V_i).$$

Hence, by (b):

$$\mu_f(f(A)) \le d\sum_i \mu_f(A\cap V_i) = d\mu_f(A\cap(\bigcup V_i)) = d\mu_f(A).$$

Lemma II. *If A is a Borel set contained in a set $P^{(n)}_{i,j}$, $n \ge N$:*

$$\mu_f(A) = k^{(n)}_{i,j} d^{-n} \mu_f(f^n(A)).$$

Proof. Take a topological disk $V \subset P_i$ with $\mu_f(\partial V) = 0$, $\mu_f(P_i\setminus V) = 0$ and not containing critical values of f^n. To obtain V, it is enough to join the critical values of f^n in P_i to its boundary with disjoint arcs having zero measure and the defining V as the complement in P_i of theses curves. Define $W = f^{-n}(V)\cap P^{(n)}_{i,j}$. Since there are no critical values of f in V, there exist $k^{(n)}_{i,j}$ branches of f^{-n} on V, i.e., analytic functions $\phi_i : V \to W$, $i = 1,\ldots,k^{(n)}_{i,j}$, such that $\phi_i(f(z)) = z$ for all $z\in W$ and $\phi_i(w) \ne \phi_j(w)$ for all $w\in V$, $1 \le i < j \le k^{(n)}_{i,j}$. It is clear that:

(1) $$\phi_i(V)\cap\phi_j(V) = \varnothing,\ 1 \le i < j \le k^{(n)}_{i,j}.$$

(2) $$f^n(\phi_i(V)) = V,\ i = 1,\ldots,k^{(n)}_{i,j}.$$

(3) $$f^n/\phi_i(V) \text{ is injective, } i = 1,\ldots,k^{(n)}_{i,j}.$$

(4) $$\bigcup_i \phi_i(V) = W.$$

By properties (1)-(4), we can apply Lemma I to obtain:

(5) $$\mu_f(A \cap \phi_i(V)) = d^{-n}\mu_f(f^n(A) \cap V).$$

By (1), (4) and (5):

$$\mu_f(A \cap W) = \sum_i \mu_f(A \cap \mu_i(V)) = k_{i,j}^{(n)} d^{-n}\mu_f(f^n(A) \cap V) = k_{i,j}^{(n)} d^{-n}\mu_f(f(A)).$$

But $\mu_f(A \cap W) = \mu_f(A)$ because:

$$\mu_f(A) - \mu_f(A \cap W) \leq \mu_f(P_{i,j}^{(n)} \backslash W) \leq \mu_f(f^{-n}(P_i \backslash V)) = \mu_f(P_i \backslash V) = 0.$$

Now set $U_n = \bigcup_{i,j} P_{i,j}^{(n)}$. By Lemma II:

$$\mu_f(U_n) = \sum_{i,j} \mu_f(P_{i,j}^{(n)}) = d^{-n} \sum_{i,j} k_{i,j}^{(n)} \mu_f(P_i) =$$

$$= d^{-n} \sum_i \left(\sum_j d_{i,j}^{(n)} \right) \mu_f(P_i) \geq d^{-n}\left(1 - \frac{\varepsilon}{2}\right) d^n \sum_i \mu_f(P_i) = 1 - \frac{\varepsilon}{2}.$$

For each $n \geq N$, the sets $P_{i,j}^{(n)} \cap J(f)$, $i = 1, \dots, 4$, $j = 1, \dots, \ell(n, i)$ are a partition of $U \cap J(f)$. This partition can be extended to a partition P_n of $J(f)$ in such a way that:

$$\lim_{n \to +\infty} (\sup_{\mathscr{P} \in \mathscr{P}_n} \operatorname{diam}(P)) = 0.$$

This property and standard derivation theorems imply that if $\mathscr{P}_n(x)$ denotes the atom of $\mathscr{P}_n$ containing x, then the sequence of functions $F_n : J(f) \to \mathbb{R}$ defined by:

$$F_n(x) = \lim_{n \to +\infty} \frac{\mu_f(S \cap \mathscr{P}_n(x))}{\mu_f(\mathscr{P}_n(x))}$$

converges in measure to the characteristic function f_S of S. From this property, we shall prove the following claim: if we take Borel sets A_n, $n \geq 1$, such that $f^{-n}(A_n) = S$, there exists an atom P_i and a sequence $n_j \to +\infty$ such that $\mu_f(P_i \backslash A_{n_j}) \to 0$ when $j \to +\infty$. This implies that $\mu_f(S) = 1$ because there exists $n > 0$ such that $f^n(P_i) \supset J(f)$, and by the corollary of Lemma I:

$$\lim_{j \to \infty} \mu_f(f^n(P_i \cap A_{n_j})) = \lim_{j \to +\infty} \mu_f((f^n(P_i) - f^n(P_i \backslash A_{n_j})) =$$

$$= 1 - \lim_{j \to +\infty} \mu_f(f^n P_i \backslash A_{n_j})) \geq 1 - d^n \lim_{j \to +\infty} \mu_f(P_i \backslash A_{n_j}) = 1.$$

But:

$$f^n(P_i \cap A_{n_j}) \subset f^n(P_i) \cap f^n(A_{n_j}) = f^n(A_{n_j}) = A_{n_j + n}.$$

Then $\lim_{j\to+\infty} \mu_f(A_{n_j+n}) = 1$ and:

$$\mu_f(S) = \mu_f(f^{-(n_j+n)}(A_{n_j+n})) = \mu_f(A_{n_j+n}).$$

Hence, $\mu_f(S) = 1$. To prove the claim observe that the convergence in measure of the sequence F_n to f_S implies that for every $k > 0$ there exists n_k such that the set of points x satisfying:

$$\left| \frac{\mu_f(S \cap \mathscr{P}_{n_k}(x))}{\mu_f(\mathscr{P}_{n_k}(x))} - f_S(x) \right| \leq \frac{1}{k} \tag{6}$$

has measure $\geq 1 - (\varepsilon/2)$. Then this set intersects $S \cap U_{n_k}$ because

$$\mu_f(S \cap U_{n_k}) = \mu_f((S^c \cup U^c_{n_k})^c) \geq \left((1 - 2\varepsilon) - \frac{\varepsilon}{2}\right) = \frac{3}{2}.$$

Let x_k be a point in the intersection. Since it belongs to U_{n_k}, we have $\mathscr{P}_{n_k}(x_k) = P^{(n_k)}_{i,j}$ for some i and j. By (6):

$$\frac{\mu_f(S \cap P^{(n_k)}_{i,j})}{\mu_f(P^{(n_k)}_{i,j})} \geq 1 - \frac{1}{k}. \tag{7}$$

By Lemma II:

$$d^{-n_k} k^{(n_k)}_{i,j}\, \mu_f(f^{n_k}(S \cap P^{(n_k)}_{i,j})) = \mu_f(S \cap P^{(n_k)}_{i,j}) \tag{8}$$

$$d^{-n_k} k^{(n_k)}_{i,j}\, \mu_f(f^{n_k}(P^{(n_k)}_{i,j})) = \mu_f(P^{(n_k)}_{i,j}). \tag{9}$$

But $f^{n_k}(P^{(n_k)}_{i,j}) = P_i$. Hence, (7), (8) and (9) imply:

$$1 - \frac{1}{k} \leq \frac{\mu_f(S \cap P^{(n_k)}_{i,j})}{\mu_f(P^{(n_k)}_{i,j})} = \frac{\mu_f(f^{n_k}(S \cap P^{(n_k)}_{i,j}))}{\mu_f(P_i)} \leq$$

$$\leq \frac{\mu_f(f^{n_k}(S) \cap f^{n_k}(P^{(n_k)}_{i,j}))}{\mu_f(P_i)} =$$

$$= \frac{\mu_f(A_{n_k} \cap P_i)}{\mu_f(P_i)}.$$

Hence:

$$\mu_f(P_i \backslash A_{n_k}) = \mu_f(P_i) - \mu_f(P_i \cap A_{n_k}) \leq$$

$$\leq \frac{1}{k} \mu_f(P_i) \leq \frac{1}{k} \sup_i \mu_f(P_i)$$

thus completing the proof of the claim.

To prove (d) take a family $\{U_1, \ldots, U_m\}$ of topological disks not containing critical values of f and such that:

$$(1) \qquad \mu_f\Big(\big(\bigcup_{i=1}^{m} U_i\big)^c\Big) = 0.$$

Then for every $1 \le i \le m$, there exist d branches $g_j^{(i)} = U_i \to \overline{\mathbb{C}}$, $j = 1, \ldots, d$ of $f^{-1} \backslash U_i$. Set $U_j^{(i)} = g_j^{(i)}(U_i)$. From (1) it follows that:

$$(2) \qquad \mu_f\left(\left(\bigcup_{i=1}^{m} \bigcup_{j=1} U_j^{(i)}\right)^c\right) = 0.$$

Suppose $A \subset \overline{\mathbb{C}}$ is a Borel set such that f/A is injective. It follows from the injectivity of f/A that the sets $f(A \cap U_j^{(i)})$, $1 \le i \le m$, $1 \le j \le d$, are disjoint. This property together with (1), (2) and Lemma I yield

$$\mu_f(f(A)) = \mu_f\big(\bigcup_i (f(A) \cap U_i)\big) = \mu_f\big(\bigcup_{i,j} f(A \cap U_j^{(i)})\big) =$$

$$= \sum_{i,j} \mu_f(f(A \cap U_j^{(i)})) = d \sum_{i,j} \mu_f(A \cap U_j^{(i)}) =$$

$$= d\,\mu_f\big(A \cap (\bigcup_{i,j} U_j^{(i)})\big) = d\,\mu_f(A).$$

To prove that μ_f is the unique f-invariant probability satisfying (d), consider another f-invariant probability μ satisfying (d). We shall prove that $\mu \ll \mu_f$. Then the ergodicity of μ_f implies $\mu = \mu_f$. To show $\mu \ll \mu_f$, we have to find for every $\varepsilon > 0$, a $\delta > 0$ such that if $K \subset J(f)$ is a compact set whith $\mu_f(K) \le \delta$, then $\mu(K) \le \varepsilon$. Given $\varepsilon' > 0$ take $N > 0$ such that there exists a family $\mathcal{P} = \{P_1, \ldots, P_r\}$ of $(N, \varepsilon/2)$-adapted sets such that

$$\mu_f\big((\bigcup_i P_i)^c\big) = 0$$

$$\mu\big((\bigcup_i P_i)^c\big) = 0.$$

This family is constructed as the family used in the proof of (c). Take $\delta > 0$ satisfying:

$$(1) \qquad \delta\,\frac{\mu(P_i)}{\mu_f(P_i)} \le \frac{\varepsilon}{4}$$

for every $1 \le i \le r$. As in the proof of (c), denote $P_{i,j}^{(n)}$ and $k_{i,j}^{(n)}$ $j = 1, \ldots, \ell(n, i)$ the sets and integers associated to P_i by Definition I. By Lemma II, we have:

$$(2) \qquad \mu_f(P_{i,j}^{(n)}) = k_{i,j}^{(n)}\, d^{-n} \mu_f(P_i).$$

Since in the proof of Lemma II the only property of μ_f used is precisely (d), we can apply Lemma II to μ instead of μ_f. Hence:

(3) $$\mu(P_{i,j}^{(n)}) = k_{i,j}^{(n)} d^{-n}\mu(P_i).$$

From (3) and part (c) of Definition I, we obtain:

$$\mu(\bigcup_j P_{i,j}^{(n)}) = d^{-n}\mu(P_i)\sum_j k_{i,j}^{(n)} \geq d^{-n}\mu(P_i)\left(1-\frac{\varepsilon}{2}\right)d^n =$$
$$= \left(1-\frac{\varepsilon}{2}\right)\mu(P_i).$$

The same argument, replacing (3) by (2), shows that:

$$\mu_f(\bigcup_j P_{i,j}^{(n)}) \geq \left(1-\frac{\varepsilon}{2}\right)\mu_f(P_i).$$

Then

(4) $$\mu(\bigcup_{i,j} P_{i,j}^{(n)}) \geq 1-\frac{\varepsilon}{2},$$

(5) $$\mu_f(\bigcup_{i,j} P_{i,j}^{(n)}) \geq 1-\frac{\varepsilon}{2}.$$

Suppose that $K \subset J(f)$ is a compact set with $\mu_f(K) \leq \delta$. We want to show that $\mu(K) \leq \varepsilon$. By (5):

$$\mu(K \cap (\bigcup_{i,j} P_{i,j}^{(n)})^c) \leq \frac{\varepsilon}{2}.$$

It remains to prove:

(6) $$\mu(K \cap (\bigcup_{i,j} P_{i,j}^{(n)})) \leq \frac{\varepsilon}{2}.$$

Set

$$\mathscr{S}_n = \{(i,j) \mid P_{i,j}^{(n)} \cap K \neq \emptyset\}$$
$$K_n = \bigcup \{P_{i,j}^{(n)} \mid (i,j) \in \mathscr{S}_n\}$$

Since the diameters of the atoms of $P_{i,j}^{(n)}$ converge to zero uniformely in (i,j) (by part (d) of Definition I) and by the compactness of K, it follows that

$$\lim_{n\to+\infty} \mu_f(K_n) = \mu_f(k),$$

(7) $$\lim_{n\to+\infty} \mu(K_n) = \mu(K).$$

Then $\mu_f(K_n) \leq 2\delta$ if n is large. It follows that:

$$2\delta \geq \mu_f(K_n) \geq \sum_{(i,j)\in \mathscr{S}_n} \mu_f(P_{i,j}^{(n)}).$$

By (2) and (3):

$$2\delta \geq \sum_{(1,j)\in \mathscr{S}_n} \mu_f(P_{i,j}^{(n)}) = \sum_{(i,j)\in \mathscr{S}_n} \frac{\mu_f(P_i)}{\mu(P_i)} \mu(P_{i,j}^{(n)}).$$

Hence:

$$2\delta \sup_i \frac{\mu(P_i)}{\mu_f(P_i)} \geq \sum_{(i,j)\in \mathscr{S}_n} \mu(P_{i,j}^{(n)}) = \mu(K_n).$$

By (1):

$$\frac{\varepsilon}{2} \geq 2\delta \sup_i \frac{\mu(P_i)}{\mu_f(P_i)} \geq \mu(K_n).$$

Then, by (7), $\mu(K) \leq \varepsilon/2$.

To prove (c), take a family $\mathscr{P} = \{P_1, \ldots, P_r\}$ of disjoint topological disks such that f/P_i is injective for all $1 \leq i \leq r$ and $\mu_f((\bigcup_i P_i)^c) = 0$.

Denote $\mathscr{P}_n = \bigvee_{j=0}^{n} f^n(\mathscr{P})$ and let $\mathscr{P}_n(x)$ be the atom of $\mathscr{P}_n$ containing x. Observe that $f^n/\mathscr{P}_n(x)$ is injective for all $n \geq 1$. In fact, this property holds for $n = 0$ and if it is true for $n = m$ then

$$f^{m+1}/\mathscr{P}_{m+1}(x) = (f/\mathscr{P}_{m+1}(x)) \circ (f^m/\mathscr{P}_m(f(x))).$$

But $f/\mathscr{P}_{m+1}(x)$ is injective because $\mathscr{P}_{m+1}(x) \subset \mathscr{P}(x)$ and $f^m/\mathscr{P}_m(f(x))$ is injective by the induction hypothesis. Now we shall prove that:

$$\text{(10)} \qquad \lim_{n\to+\alpha} \sup\left(-\frac{1}{n}\log \mu_f(\mathscr{P}_n(x))\right) \geq \log d$$

for μ_f – a.e. x. We have: $f^n(\mathscr{P}_n(x)) \subset \mathscr{P}(f^n(x))$. Then

$$\mu_f(f^n(\mathscr{P}_n(x)) \leq \mu_f(\mathscr{P}(f^n(x))).$$

By (d) and by the injective of $f^n/\mathscr{P}_n(x)$:

$$\mu_f(\mathscr{P}_n(x)) = d^{-n}\mu_f(f^n(\mathscr{P}_n(x))) \leq d^{-n}\mu_f(\mathscr{P}(f^n(x))).$$

Then:

$$-\frac{1}{n}\log \mu_f(\mathscr{P}_n(x)) \geq \log d - \frac{1}{n}\log \mu_f(\mathscr{P}(f^n(x))).$$

From this (10) follows. But (10) implies that

$$h_{\mu_f}(f) \geq \log d.$$

On the other hand, Gromov proved in [3] that $h_{top}(f) = \log d$. Then $h_{\mu_f}(f) \leq \log d$.

II. Proof of the Fundamental Lemma

If $w \in \overline{\mathbb{C}}$ and $n \in \mathbb{Z}^+$, denote $\widehat{m}_n(w)$ the multiplicity of w as root of the equation $f^n(z) = f^n(w)$. Set $m_n(z) = \max\{\widehat{m}_n(w) \mid w \in f^{-n}(z)\}$. We shall need the following lemma:

Lemma. *For every $z \notin Exc(f)$ there exist $N_1 > 0$ and $1 < d_0 < d$ such that $m_n(z) \leq d_0^n$ for all $n \geq N_1$.*

Proof. Define $\mathscr{B}(z)$ as the set of functions $\theta : \mathbb{Z}^+ \to \overline{\mathbb{C}}$ such that $\theta(0) = z$ and $f(\theta(j+1)) = \theta(j)$ for all $j \geq 0$. Define $\mathscr{B}_0(z)$ as the set of $\theta \in \mathscr{B}(z)$ such that $\theta(j)$ is a critical value of f only for a finite set of values of j. Then it is easy to see that $\mathscr{B}(z) \neq \mathscr{B}_0(z)$ if and only if z belongs to the orbit of a periodic critical point, and that in this case, there is only one element $\gamma \in \mathscr{B}(z) \backslash \mathscr{B}_0(z)$ that is periodic i.e., for some t, satisfies $\gamma(t + j) = \gamma(j)$ for all $j \geq 0$. Now define

$$\alpha_n(z) = \{\theta(n) \mid \theta \in \mathscr{B}(z)\},$$
$$\alpha_n^0(z) = \{\theta(n) \mid \theta \in \mathscr{B}_0(z)\}.$$

From the fact that f has only finitely many critical points it follows that there exists $N_2 > 0$ such that $\theta(n)$ is not a critical point of f for all $n \geq N_2$, $\theta \in \mathscr{B}_0(z)$. Hence

$$\widehat{m}_1(w) = 1$$

for all $w \in \alpha_n^0(z)$ and $n \geq N_2$. Then, if $w \in \alpha_n^0(z)$ with $n \geq N_2$, we obtain

$$\widehat{m}_n(w) = \prod_{j=0}^{n-1} \widehat{m}_1(f^j(w)) = \prod_{j=0}^{N_2-1} \widehat{m}_1(f^j(w)) \lesssim d^{N_2}.$$

If $\mathscr{B}_0(z) = \mathscr{B}(z)$, this concludes the proof because

$$m_n(z) = \max\{\widehat{m}_n(w) \mid w \in \alpha_n(z)\} = \max\{\widehat{m}_n(w) \mid w \in \alpha_n^0(z)\} \leq d^{N_2}$$

for all $n \geq N_2$. If $\mathscr{B}_0(z) \neq \mathscr{B}(z)$ and if γ is the unique element in $\mathscr{B}(z) \backslash \mathscr{B}_0(z)$:

$$m_n(z) = \max\{\widehat{m}_n(w) \mid w \in \alpha_n(z)\} \leq \max\{d^{N_2}, \widehat{m}_n(\gamma(n))\}.$$

Therefore the problem is reduced to show that there exist $G > 0$ and $1 < d_1 < d$ such that $\widehat{m}_n(\gamma(n)) \leq G d_1^n$, for large values of n. But if $n = kt + r$, $k \in \mathbb{Z}^+$, $r \in \mathbb{Z}^+$, $0 \leq r < t$, we have, using the periodicity of γ, that

$$\hat{m}_n(\gamma(n)) = \prod_{j=0}^{n-1} \hat{m}_1(\gamma(n-j)) = \prod_{j=0}^{r} \hat{m}_1(\gamma(n-j)) \left(\prod_{j=0}^{t-1} \hat{m}_1(\gamma(j)) \right)^k.$$

Set:

$$d_1 = \left(\prod_{j=0}^{t-1} \hat{m}_1(\gamma(j)) \right)^{\frac{1}{t}}.$$

Then:

$$\widehat{m}_n(\gamma(n)) \leq d^r d_1^{tk} \leq d^t d_1^n.$$

for every $n \geq t$. This reduces our problem to prove that $d_1 < d$. But since every factor in the definition of d_1 is $\leq d$, it is sufficient to show that we cannot have $\hat{m}_1(\gamma(j)) = d$ for all $0 \leq j \leq t$. But $\hat{m}_1(\gamma(j)) = d$ for all $0 \leq j \leq t$ implies that $f^{-1}(\pi) = \pi$, where π is the periodic orbic $\pi = \{\gamma(0), \ldots, \gamma(t-1)\}$. Then $f^{-1}(\pi^c) \subset \pi^c$. Therefore $(\bigcup_{n \geq 0} f^n(U)) \cap \pi = \varnothing$ for every subset U of π^c. But it is clear that $\pi \cap J(f) = \varnothing$. Hence, we can take U as an open set intersecting $J(f)$. Then, as we explained in the introduction, $\bigcup_{n \geq 0} f^n(U) \supset \mathrm{Exc}(f)^c$. This, together with $(\bigcup_{n \geq 0} f^n(U)) \cap \pi = \varnothing$ yields $\pi \subset \mathrm{Exc}(f)$, contradicting the assumption $z \notin \mathrm{Exc}(f)$.

Now let us prove the Fundamental Lemma. Take N_0 so large that the sequence nd^{-n}, $n \geq 1$, is decreasing for $n \geq N_0$ and $m_{N_0}(z) N_0 d^{-N_0} \leq \varepsilon/2$. Such N_0 exists by the previous Lemma. Then

$$(*) \qquad m_{N_0}(z) n d^{-n} \leq \frac{\varepsilon}{2}.$$

for every $n \geq N_0$. Assume that N_0 is large enough to satisfy

$$(**) \qquad 4 m_{N_0}(z) d^{-N_0} \frac{d^2}{1-d} \leq \frac{\varepsilon}{2}.$$

Set $m = m_{N_0}(z)$. Since the only critical value of f^{N_0} contained in γ can be z, it follows that the connected components $\gamma_1, \ldots, \gamma_r$ of $f^{-N_0}(\gamma)$ are either arcs or a union of arcs with a unique point of intersection. Therefore, each γ_i is simply connected. We can then take a topological disk $U_0 \supset \gamma$, so thin that there exist disjoint topological disks $V_i \supset \gamma_i$, $i = 1, \ldots, r$ such that

$$f^{N_0}(V_i) = U_0$$

and $f^{N_0}/V_i : V_i \to U_0$ is a k_i-to-1 map for all $1 \le i \le r$. Now set:

$$\varepsilon_{N_0} = 0$$

$$\varepsilon_{n+1} = \varepsilon_n + 4md^{-nd^2}$$

for $n \ge N_0$. Observe that, by (**):

$$\varepsilon_n \le \varepsilon_{N_0} + 4md^2 \sum_{j=N_0}^{\infty} d^{-j} = 4md^{-N_0} \frac{d^3}{d-1} \le \frac{\varepsilon}{2}.$$

We claim that for every $n \ge N_0$, $f^{-n}(U_0)$ contains a union of disjoint topological disks $W_i^{(n)}$, $i = 1, \ldots, \tilde{\ell}_n$, such that, for all i, $f^n(W_i^{(n)}) = U_0$ and $f^n/W_i^{(n)} : W_i^{(n)} \to U_0$ is a $k_i^{(n)}$-to-1 map, where $1 \le k_i^{(n)} \le m$, $1 \le i \le \tilde{\ell}_n$ are integers satisfying

$$\sum_{i=1}^{\tilde{\ell}_n} k_i^{(n)} \ge \left(1 - \frac{1}{2}\varepsilon_n\right) d^n.$$

Clearly, the property is true for $n = N_0$ just taking $W_i^{(N_0)} = V_i$, $i = 1, \ldots, r$, $k_i^{(N_0)} = k_i$ because

$$\sum_{i=1}^{r} k_i^{(N_0)} = \sum_{i=1}^{r} k_i = d^n.$$

The proof of the claim will now be completed by induction. Suppose constructed $W_i^{(n)}$, $k_i^{(n)}$, $1 \le i \le \tilde{\ell}_n$. Let H be the set of integers t between 1 and $\tilde{\ell}_n$ such that $W_t^{(n)}$ doesn't contain critical values of f. For every $t \in H$ there exist disjoint topological disks $D_j^{(t)}$, $1 \le j \le d$, such that f maps $D_j^{(t)}$ homeomorphically onto $W_t^{(n)}$. Define as $W_i^{(n+1)}$, $i = 1, \ldots, \tilde{\ell}_{n+1}$, the sets $D_j^{(t)}$, $t \in H$, $1 \le j \le d$, and, if $W_i^{(n+1)} = D_j^{(t)}$, set $k_i^{(n+1)} = k_t^{(n)}$. Then:

$$\sum_{i=1}^{\tilde{\ell}_{n+1}} k_i^{(n+1)} = d \sum_{i \in H} k_i^{(n)} = d \sum_{i=1}^{\tilde{\ell}_n} k_i^{(n)} - d \sum_{i \notin H} k_i^{(n)} \ge$$

$$\ge d \sum_{i=1}^{\tilde{\ell}_n} k_i^{(n)} - d(\tilde{\ell}_n - \# H)m \ge \left(1 - \frac{1}{2}\varepsilon_n\right) d^{n+1} - d(\tilde{\ell}_n - \# H)m.$$

But $\tilde{\ell}_n - \# H$ is bounded by the number of critical values of f, that is, $2d - 2$. Hence:

$$\sum_{i=1}^{\tilde{\ell}_{n+1}} k_i^{(n+1)} \ge \left(1 - \frac{\varepsilon_n}{2}\right) d^{n+1} - 2d^2 m =$$

$$= \left(1 - \frac{1}{2}(\varepsilon_n + 4d^2 md^{-(n+1)})\right) d^{n+1} =$$

$$= \left(1 - \frac{1}{2}\varepsilon_{n+1}\right) d^{n+1}.$$

This completes the proof of the claim. The next step is to restrict, for each n, the family $W_i^{(n)}$, $1 \le i \le \widetilde{\ell}_n$, to those values of i that satisfy:

$$\lambda(W_i^{(n)}) \le \frac{1}{n},$$

where λ denotes the Lebesgue measure.

Suppose that those values of i are $1, \ldots, \ell_n$. Then:

$$\sum_{i=1}^{\ell_n} k_i^{(n)} \ge \sum_{i=1}^{\widetilde{\ell}_n} k_i^{(n)} - (\ell_n - \widetilde{\ell}_n)m.$$

To bound $(\ell_n - \widetilde{\ell}_n)$, observe that

$$1 \ge \lambda(\bigcup_{i > \ell_n} W_i^{(n)}) \ge (\widetilde{\ell}_n - \ell_n)\frac{1}{n}.$$

Then:

$$(1) \qquad \sum_{i=1}^{\ell_n} k_i^{(n)} \ge \left(1 - \frac{1}{2}\varepsilon\right)d^n - nm = \left(1 - \frac{1}{2}(\varepsilon + 2mnd^{-n})\right)d^n.$$

By (*), the factor of d^n is bounded by $(1 - \varepsilon)$. Finally, to complete the proof of the Lemma, we shall prove that for any topological disk U whose closure is contained in U_0, we have:

$$(2) \qquad \lim_{n \to +\infty} (\sup_i \operatorname{diam}(f^{-n}(U) \cap W_i^{(n)})) = 0.$$

If this property is true, the lemma is proved taking U containing γ and with closure contained in U_0. Then we define:

$$S_i^{(n)} = f^{-n}(U) \cap W_i^{(n)}.$$

By (2), there exists $N \ge N_0$ such that property (b) of Definition I is satisfied. Property (d) also follows from (2). Finally, $f^n/S_i^{(n)}$ is a $k_i^{(n)}$-to-1 map onto U, and from (1) and (*) it follows that the integers $k_i^{(n)}$ satisfy property (c). It remains to prove (2). By the way the sets $W_i^{(n)}$ were constructed, we know that $f^{n-N_0}/W_i^{(n)}$ is a conformal representation onto some V_j. Let $\phi_i^{(n)} : V_j \to W_i^{(n)}$ be its inverse. Set $D_r = \{z \mid |z| < r\}$. Let $\alpha_j : D_1 \to V_j$ be a conformal representation. Define $\psi_i^{(n)} : D_1 \to W_i^{(n)}$ as $\psi_i^{(n)} = \phi_i^{(n)}\alpha_j$. Instead of (2), we shall prove

$$(3) \qquad \lim_{n \to +\infty} (\sup_i \operatorname{diam} \psi_i^{(n)}(D_r)) = 0$$

for all $0 < r < 1$. This implies (2) because $\psi_i^{(n)}(D_r) \supset f^{-(n-N_0)}(U) \cap W_i^{(n)}$ for all $n \ge N$, if r is near enough to 1. To prove (3), recall that the Distortion Theorem for univalent functions states that for all $0 < r < 1$

there exists $K(r)$ such that every univalent function $\phi : D_1 \to \mathbb{C}$ satisfies $|\phi'(a)/\phi'(b)| \leq K(r)$ for all a and b in D_r. In particular, if $\lambda(\cdot)$ denotes Lebesgue measure, $\lambda(\phi(D_r)) \geq K(r)^{-1}|\phi'(a)|^2\lambda(D_r)$, for all $a \in D_r$. In our case:

$$\frac{1}{n} \geq \lambda(W_i^{(n)}) \geq \lambda(\psi_i^{(n)}(D_r)) \geq K(r)^{-1}\lambda(D_r)|(\psi_i^{(n)})'(z)|^2$$

for all $0 < r < 1$, $z \in D_r$. Then:

$$\lim_{n \to +\infty} \left(\sup_{i,\, z \in D_r} |(\psi_i^{(n)})'(z)| \right) = 0$$

for all $0 < r < 1$, and this obviously implies (3).

References

[1] H. Brolin, *Invariant sets under iteration of rational functions*, Arkiv für Mathematik, Band G, nr. 6 (1966).

[2] M. Fatou, *Sur les fonctions satisfaisant certes equations fonctionelles*, Bull. de la Soc. Math. de France 47-48 (1919-1920).

[3] M. Gromov, *On the entropy of holomorphic maps*, Preprint I.H.E.S.

[4] G. Julia, *Sur l'iteration des fonctions rationelles*, Journal de Mathematique Pure et Appliquée, 8 (1918).

Artur Lopes
Instituto de Matemática
Universidade Federal de Rio Grande do Sul
Porto Alegre, Brazil

Alexandre Freire
Instituto de Matemática Pura e Aplicada
Rio de Janeiro, Brazil

Ricardo Mañé
Instituto de Matemática Pura e Aplicada
Rio de Janeiro, Brazil

On the uniqueness of the maximizing measure for rational maps

Ricardo Mañé

I. Introduction

If f is a continuous transformation of a compact metric space X, the Dinaburg-Goodman-Goodwyn theorem states that

$$h_{top}(f) = sup\{h_\mu(f) \mid \mu \in \mathcal{M}(f)\}$$

where $h_{top}(f)$ is the topological entropy of f, $\mathcal{M}(f)$ is the space of f-invariant probabilities on the Borel σ-algebra of X and $h_\mu(f)$ denotes the μ-entropy of f. A measure $\mu \in \mathcal{M}(f)$ such that $h_\mu(f) = h_{top}(f)$ is called a *maximizing measure*. It is not difficult to exhibit examples of continuous maps without maximizing measures (see [2] p. 148 for instance). More delicate is the construction of C^r diffeomorphisms of compact manifolds $(1 \leq r < \infty)$ not having maximizing measures (Misiurevicz [6]) and it is still unknown whether there exist similar examples in the C^∞ class. Even more particular are the continuous transformations having a *unique* maximizing measure. Several interesting classes of transformations with this property are known, like transitive finite type subshifts, basic sets of Axiom A diffeomorphism and automorphisms of compact groups. The purpose of this paper is to prove that this property holds for analytic endomorphisms of the Riemann sphere.

Theorem. *Analytic endomorphisms of the Riemann Sphere* $\overline{\mathbb{C}} = \mathbb{C} \cup \{\infty\}$ *with degree* ≥ 2 *have a unique maximizing measure.*

Analytic endomorphisms of the Riemann sphere are given by rational functions $f(z) = P(z)/Q(z)$, where the polynomials P and Q have no common roots. Its degree (that coincides with its topological degree) is defined as the maximum of the degrees of P and Q. According to the results of Gromov [4] and Misiurevicz-Przyticki [7], its topological entropy is given by the logarithm of the degree.

Recebido em 29/4/83.

Originally published in: *Bol. Soc. Brasil. Mat. 14 (1983), no. 1, 27–43,*

Given an analytic endomorphism $f:\overline{\mathbb{C}}\to\overline{\mathbb{C}}$ with degree $d\geq 2$, an f-invariant measure μ_f with maximum entropy has been constructed by Freire, Lopes and Mañé [5] (and, according with the theorem above, is the unique one with this property). Let us briefly explain the construction of μ_f. Given $a\in\overline{\mathbb{C}}$ denote by $z_1^{(n)}(a),\ldots,z_{d^n}^{(n)}(a)$ the roots of the equation $f^n(z)-a=0$. Define a probability $\mu^{(n)}(a)$ by

$$\mu^{(n)}(a)=\frac{1}{d^n}\sum_{i=1}^{d^n}\delta_{z_i^{(n)}(a)}.$$

In [5], it is proved that for every $a\in\overline{\mathbb{C}}$ (with two possible exceptions easy to describe), the sequence of probabilities $\{\mu^{(n)}(a)\mid n\geq 1\}$ converges in the weak topology to an f-invariant measure μ_f, independent of a. Moreover it is proved that, with respect to μ_f, f is a K-system and that μ_f is the *unique f-invariant probability satisfying*

$$\mu_f(f(A))=d\mu_f(A) \tag{1}$$

for every Borel set where f/A is injective. This characterization will play a key role in the proof of the Theorem above, whose proof we now proceed partially outline.

Suppose that μ is an f-invariant probability. Assume that f is absolutely continuous with respect to μ, i.e. that $\mu(f(A))=0$ if $\mu(A)=0$. Then it is not difficult to find a μ-integrable function $J_\mu:\overline{\mathbb{C}}\to[0,+\infty)$ such that:

$$\mu(f(A))=\int_A J_\mu d\mu \tag{2}$$

for every Borel set $A\subset\overline{\mathbb{C}}$ where f/A is injective. Using a convenient partition of $\overline{\mathbb{C}}$, it is easy to prove that:

$$\int_{\overline{\mathbb{C}}} J_\mu d\mu = d. \tag{3}$$

Much more delicate is the proof of the formula for its entropy:

$$\int_{\overline{\mathbb{C}}}\log J_\mu d\mu = h_\mu(f). \tag{4}$$

Then, if μ is a maximizing measure:

$$\log d = h_\mu(f)=\int_{\overline{\mathbb{C}}}\log J_\mu d\mu\leq\log\int_{\overline{\mathbb{C}}}J_\mu d\mu=\log d. \tag{5}$$

Therefore:

(6) $$\log \int_{\overline{\mathbb{C}}} J_\mu d\mu = \int_{\overline{\mathbb{C}}} \log J_\mu d\mu = \log d.$$

But this implies that $J_\mu = d$ μ-a.e.. Hence, from (2) and the absolute continuity of f, it follows that μ satisfies property (1) and we conclude that $\mu = \mu_f$. However, to prove the theorem, we have to introduce some changes in this method in order to make it work also when f is not absolutely continuous with respect to μ. Without the absolute continuity hypotheses, (2) cannot be expected to hold. What can be actually proved is that there exists a Borel set $A_\mu \subset \overline{\mathbb{C}}$, with $\mu(A_\mu^c) = 0$ and such that (2) holds if f/A is injective and $A \subset A_\mu$. Equality (3) must then be corrected. It still holds replacing $=$ by $\leq$ and that is enough to get (6) using (5). Then we can reach the conclusion $J_\mu = d$ μ-a.e. With some minor technical work, this property still implies that μ has property (1), but is slightly less immediate than in the absolutely continuous case. Let us remark that (4) will be in fact proved only when μ is ergodic and $h_\mu(f) > 0$. This is not an obstruction in our scheme because when a transformation has more than one maximizing measure, it has more than one maximizing ergodic measure ([2], Prop. 13.3).

The core (and most difficult part) of the proof is the entropy formula. As the rest of the proof of the Theorem, it becomes much simpler to prove in the absolutely continuous case.

There is a criteria, due to Bowen [1] and based on the method used by Parry to prove the uniqueness of the maximizing measure for finite type subshifts, from which this property can be deduced for finite type subshifts, basic sets of Axiom A diffeomorphisms and group automorphisms. We weren't able to deduce our theorem from this criteria. It is not clear at all whether the measure μ_f satisfies the very strong uniform estimates that are required by that criteria.

II. Proof of the Theorem

Let $f : \overline{\mathbb{C}} \circlearrowleft$ be an analytic endomorphism of degree $d \geq 2$. To formalize the sketch of the proof given in the introduction, the first step will be to prove existence of J_μ and A_μ.

Lemma II.1. *For every $\mu \in \mathcal{M}(f)$ without atoms, there exist a μ-integrable function $J_\mu : \overline{\mathbb{C}} \to [1, +\infty)$ and a Borel set A_μ with $\mu(A_\mu) = 1$ such that*

(*) $$\mu(f(A)) = \int_A J_\mu d\mu$$

for every Borel set $A \subset A_\mu$ such that f/A is injective.

Proof. Using that μ has no atoms, it is easy to construct a family of disjoint topological disks (i.e., sets homeomorphic to a disk) not containing critical values of f and satisfying $\mu\left(\left(\bigcup_{i=1}^{m} U_i\right)^c\right)=0$. Let $g_j^{(i)} : U_i \to \overline{\mathbb{C}}, j=1,\ldots,d$, $i=1,\ldots,m$, be the branches of f^{-1}/U_i. Then if $A \subset U_i$ is a Borel set, we have

$$\mu(g_j^{(i)}(A)) \le \mu(A) \tag{1}$$

for every j because

$$\mu(g_j^{(i)}(A)) \le \mu(f^{-1}(A)) = \mu(A).$$

In particular, every $g_j^{(i)}$ transforms measure zero sets in measure zero sets. Then, by Radon-Nykodim theorem, there exist μ-integrable functions $H_j^{(i)} : U_i \to [0, +\infty)$ such that

$$\mu(g_j^{(i)}(A)) = \int_A H_j^{(i)} d\mu \tag{2}$$

for every $1 \le i \le m$, $1 \le j \le d$ and every Borel set $A \subset U_i$. Define

$$A_\mu = f^{-1}(\bigcup_i U_i) \cup \{g_j^{(i)}(x) \mid 1 \le i \le i \le m,\ 1 \le j \le d,$$

$$x \in U_i,\ H_j^{(i)}(x) \neq 0\}.$$

From (2) it follows easily that $\mu(A_\mu^c) = 0$ or $\mu(A_\mu) = 1$. Moreover, (1) and (2) imply:

$$\int_A H_j^{(i)} d\mu \le \mu(A)$$

for every Borel set $A \subset U_i$. Hence, $H_j^{(i)}(x) \le 1$ for a.e. $x \in U_i$. Without loss of generality, we can then assume that

$$H_j^{(i)}(x) \le 1 \tag{3}$$

for all $1 \le i \le m$, $1 \le j \le d$, $x \in U_i$. Define $J_\mu : \overline{\mathbb{C}} \to [1, +\infty)$ as

$$J_\mu(x) = (H_j^{(i)}(f(x)))^{-1} \tag{4}$$

if $x \in A_\mu \cap g_j^{(i)}(U_i)$ and

$$J_\mu(x) = 1$$

if $x \notin A_\mu$. The definition of A_μ grants that $H_j^{(i)}(f(x)) \neq 0$ in (4). Moreover, (3) implies that $J_\mu(x) \ge 1$. To verify (*) is sufficient to consider the case when A is contained in a set $g_j^{(i)}(U_i) \cap A_\mu$. To obtain from this the general case observe that if $A \subset A_\mu$ is a Borel set and f/A is injective, then the sets

$$A_j^{(i)} = g_j^{(i)}(f(A) \cap U_i)$$

satisfy

$$\bigcup_{i,j} A_j^{(i)} = A \cap f^{-1}\left(\bigcup_i U_i\right) \tag{5}$$

$$\bigcup_{i,j} f(A_j^{(i)}) = f(A) \cap \left(\bigcup_i U_i\right) \tag{6}$$

and

$$f(A_j^{(i)}) \cap f(A_\ell^{(k)}) = \emptyset \tag{7}$$

if $(i, j) \neq (k, \ell)$. The first two equalities are trivial. To prove the third suppose that the intersection is $\neq \emptyset$. Then $i = k$ because $f(A_j^{(i)}) \subset U_i$ and $f(A_\ell^{(k)}) \cap U_k$. Take y in the intersection. We can write $y = f(x_1) = f(x_2)$ with $x_1 \in A_j^{(i)}$, $x_2 \in A_\ell^{(i)}$. Since f/A is injective, it follows that $x_1 = x_2$. Hence, $A_j^{(i)} \cap A_\ell^{(i)} \neq \emptyset$. This means $g_j^{(i)}(U_i) \cap g_\ell^{(i)}(U_i) \neq \emptyset$ that implies $j = \ell$ completing the proof of (7). From (5), (6), (7), it follows that:

$$\mu(f(A)) = \sum_{i,j} \mu(f(A_j^{(i)})) = \sum_{i,j} \int_{A_j^{(i)}} J_\mu d\mu = \int_{\bigcup_{i,j} A_j^{(i)}} J_\mu d\mu = \int_A J_\mu d\mu.$$

This completes the proof of the reduction step.

Suppose now that $A \subset g_j^{(i)}(U_i) \cap A_\mu$. Set

$$S_n = \{x \in f(A) \mid H_j^{(i)}(x) \geq 1/n\}.$$

The property $A \subset A_\mu$ implies that

$$\bigcup_{n \geq 1} S_n = f(A).$$

Moreover, the definition of $H_j^{(i)}$ implies:

$$\mu(S_n) = \int_{S_n} d\mu = \int_{S_n} H_j^{(i)}(x)\,(H_j^{(i)}(x))^{-1} d\mu(x) =$$

$$= \int_{S_n} H_j^{(i)}(J_\mu \circ g_j^{(i)}) d\mu = \int_{g_j^{(i)}(S_n)} J_\mu d\mu.$$

Hence:

$$\mu(f(A)) = \lim_{n \to +\infty} \mu(S_n) = \lim_{n \to +\infty} \int_{g_j^{(i)}(S_n)} J_\mu d\mu.$$

By the monotone convergence theorem, it follows that J_μ is integrable on

$$\bigcup_{n \geq 1} g_j^{(i)}(S_n) = g_j^{(i)}\left(\bigcup_{n \geq 1} S_n\right) = g_j^{(i)}(f(A)) = A$$

and

$$\int_A J_\mu d\mu = \lim_{n\to+\infty} \int_{g_j^{(i)}(S_n)} J_\mu d\mu = \mu(f(A)).$$

$A_{\mu_2} = \overline{\mathbb{C}}$. Applying the Lemma to μ_1, we obtain J_{μ_1} and A_{μ_1}. Then it is easy to see that $J_\mu = \lambda J_{\mu_1} + (1-\lambda)J_{\mu_2}$ and $A_\mu = A_{\mu_1}$ satisfy the required properties.

Corollary.
$$\int_{\overline{\mathbb{C}}} J_\mu d\mu \le d.$$

Proof.

$$\int_{\overline{\mathbb{C}}} J_\mu d\mu = \sum_{i,j} \int_{g_j^{(i)}(U_i)} J_\mu d\mu = \sum_{i,j} \int_{g_j^{(i)}(U_i)\cap A_\mu} J_\mu d\mu =$$
$$= \sum_{i,j} \mu(f(g_j^{(i)}(U_i) \cap A_\mu)) \le \sum_{i,j} \mu(f(g_j^{(i)}(U_i))) =$$
$$= \sum_i (\sum_j \mu(f(g_j^{(i)}(U_i)))) = d \sum_i \mu(U_i).$$

Lemma II.2. *If $\mu \in \mathcal{M}(f)$ is ergodic and $h_\mu(f) > 0$, then:*

$$h_\mu(f) = \int_{\overline{\mathbb{C}}} \log J_\mu d\mu.$$

Proof. Endow $\overline{\mathbb{C}}$ with the standard Riemann structure. Then, if $\|(f^n)'(z)\|$ denotes the norm with respect to this structure of the linear maps $(f^n)'(z) : T_z\overline{\mathbb{C}} \to T_{f^n(z)}\overline{\mathbb{C}}$, we have

$$\|(f^n)'(z)\| = \prod_{j=0}^{n-1} \|f'(f^j(z))\|.$$

Denote by $d(\cdot\,,\cdot\,)$ the metric on $\overline{\mathbb{C}}$ associated to this Riemann structure. We shall need several lemmas to complete the proof of II.2. The first one concerns the integrability of $\log\|f'\|$ and the Lyapounov exponents of f.

Lemma II.3. *For every ergodic $\mu \in \mathcal{M}(f)$ with $h_\mu(f) > 0$, the function $z \to \log\|f'(z)\|$ is μ-integrable and*

$$(*) \qquad \int_{\overline{\mathbb{C}}} \log\|f'(z)\|\, d\mu(z) \ge \frac{1}{2} h_\mu(f).$$

Moreover,

$$(**)\qquad \lim_{n\to+\infty} \frac{1}{n} \log \|(f^n)'(x)\| = \int_{\overline{\mathbb{C}}} \log \|f'(z)\|\, d\mu(z)$$

for $\mu-a.e.\ x\in\overline{\mathbb{C}}$.

Proof. The function $\log\|f'\|$ is obviously measurable and upper bounded. Then, since μ is ergodic, either $\log\|f'\|$ is not μ-integrable and then

$$(1)\qquad \lim_{n\to+\infty} \frac{1}{n}\log\|(f^n)'(x)\| = \lim_{n\to+\infty} \frac{1}{n}\sum_{j=0}^{n-1} \log\|f'(f^j(x))\| = -\infty$$

for μ-a.e $x\in\overline{\mathbb{C}}$, or $\log\|f'\|$ is μ-integrable and:

$$(2)\qquad \lim_{n\to+\infty} \frac{1}{n}\log\|(f^n)'(x)\| = \int_{\overline{\mathbb{C}}} \log\|f'\|\, d\mu$$

for μ-a.e. $x\in\overline{\mathbb{C}}$. When (1) holds, the Lyapounov exponent of f is $-\infty$. By Ruelle inequality [9], this implies $h_\mu(f)=0$, contradicting the hypothesis. Then $\log\|f'\|$ is μ-integrable and (2) holds. Clearly, (2) and (**) state the same property. Moveover, (2) implies that the Lyapounov exponent (with multiplicity 2) of f is

$$\chi = \int_{\overline{\mathbb{C}}} \log\|f'\|\, d_\mu.$$

Using again Ruelle's inequality, we get $h_\mu(f)\le \max\{2\chi, 0\}$. From $h_\mu(f)>0$, we obtain $\chi>0$ and then

$$h_\mu(f)\le 2\chi = 2\int_{\overline{\mathbb{C}}} \log\|f'\|\, d_\mu.$$

Corollary. *If z_0 is a critical point of f, the function $z\to\log d(z,z_0)$ is μ-integrable for every ergodic $\mu\in\mathcal{M}(f)$ with $h_\mu(f)>0$.*

Proof. The μ-integrability of $z\to\log d(z,z_0)$ follows from the fact that it is bounded below by a constant times $\log\|f'(z)\|$.

Lemma II.3 and its Corollary will be used to prove the following result that is the key step in the proof of II.2.

Lemma II.4. *If $\mu\in\mathcal{M}(f)$ is ergodic and $h_\mu(f)>0$, there exists a partition $\mathcal{P}$ of $\overline{\mathbb{C}}$ and a sequence of partitions $\mathcal{R}_0\le\mathcal{R}_1\le\dots$ satisfying the following properties:*

a) $\bigvee_{n=0}^{\infty} f^{-n}(\mathscr{P})$ *is the Borel* σ*-algebra of* $\bar{\mathbb{C}}$.

b) *If* $\mathscr{P}_n = \bigvee_{j=0}^{n} f^{-j}(\mathscr{P})$ *and we denote by* $\mathscr{P}_n(x)$, $\mathscr{R}_n(x)$ *the atoms of* $\mathscr{P}_n$ *and* $\mathscr{R}_n$ *containing x, then, for every* $n \geq 0$ *and* μ*-a.e. x, f maps* $\mathscr{R}_n(x)$ *bijectively onto almost all* $\mathscr{P}_n(f(x))$ *and* $\mathscr{P}_n(x) = \mathscr{R}_{n-1}(x) \cap \mathscr{P}(x)$.

c) $\bigvee_{n=0}^{\infty} \mathscr{R}_n$ *is the Borel* σ*-algebra.*

d) *The* μ*-entropy of* $\mathscr{P}$ *and* $\mathscr{R}_0$ *is finite.*

Before proving this lemma, we shall show how it is used to complete the proof of II.2. By properties (a) and (d) of II.4, in order to prove II.2 it is sufficient to show that the sequence of functions:

$$\phi_n(x) = -\frac{1}{n}\log \mu(\mathscr{P}_n(x))$$

converge in $\mathscr{L}^1$ to the μ-integral of $\log J_\mu$. Set:

$$F_n(x) = \log \frac{\mu(\mathscr{P}_{n-1}(f(x)))}{\mu(\mathscr{P}_n(x))}.$$

Then:

$$\phi_n(x) = -\frac{1}{n}\sum_{j=0}^{n-1} \log \frac{\mu(\mathscr{P}_{n-j}(f^j(x)))}{\mu(\mathscr{P}_{n-j-1}(f^{j+1}(x)))} - \frac{1}{n}\log \mu(\mathscr{P}(f^n(x))) =$$

$$= \frac{1}{n}\sum_{j=0}^{n-1} F_{n-j}(f^j(x)) - \frac{1}{n}\log \mu(\mathscr{P}(f^n(x))).$$

But since the μ-entropy of $\mathscr{P}$ is finite, the function $x \to |\log \mu(\mathscr{P}(x))|$ is μ-integrable and its integral is the μ-entropy of $\mathscr{P}$. The f-invariance of μ implies that the same is true for the functions $x \to |\log \mu(\mathscr{P}(f^n(x)))|$. Then the last term converges to zero in $\mathscr{L}^1$. Our problem is now reduced to show that the sequence of functions:

$$\frac{1}{n}\sum_{j=0}^{n-1} F_{n-j} \circ f^j$$

belong to $\mathscr{L}^1$ and converges in $\mathscr{L}^1$ to the integral of $\log J_\mu$. In order to insure this, it is enough to prove that the functions F_n are in $\mathscr{L}^1$ and converge in $\mathscr{L}^1$ to $\log J_\mu$, because if this is true, we write

$$\frac{1}{n}\sum_{j=0}^{n-1} F_{n-j} \circ f^j = \frac{1}{n}\sum_{j=0}^{n-1} (F_{n-j} - \log J_\mu) \circ f^j + \frac{1}{n}\sum_{j=0}^{n-1} (\log J_\mu) f^j$$

and by the ergodicity of μ the last term converges in $\mathscr{L}^1$ to the integral of $\log J_\mu$ and the first term satisfies:

$$\left\| \frac{1}{n} \sum_{j=0}^{n-1} (F_{n-j} - \log J_\mu) f^j \right\|_1 \leq \frac{1}{n} \sum_{j=0}^{n-1} \| F_{n-j} - \log J_\mu \|_1 =$$

$$= \frac{1}{n} \sum_{j=1}^{n} \| F_j - \log J_\mu \|_1$$

that converges to zero if $F_n \to \log J$ in $\mathscr{L}^1$. To prove this convergence write:

$$F_n(x) = \log \frac{\mu(\mathscr{P}_{n-1}(f(x)))}{\mu(\mathscr{P}_n(x))} = \log \frac{\mu(\mathscr{P}_{n-1}(f(x)))}{\mu(\mathscr{R}_{n-1}(x) \cap \mathscr{P}(x))} =$$

$$= -\log \frac{\mu(\mathscr{R}_{n-1}(x) \cap \mathscr{P}(x))}{\mu(\mathscr{R}_{n-1}(x))} + \log \frac{\mu(\mathscr{P}_{n-1}(f(x)))}{\mu(\mathscr{R}_{n-1}(x))} =$$

(1)
$$= -\log \frac{\mu(\mathscr{R}_{n-1}(x) \cap \mathscr{P}(x))}{\mu(\mathscr{R}_{n-1}(x))} - \log \frac{\mu(f(A_\mu \cap \mathscr{R}_{n-1}(x)))}{\mu(\mathscr{P}_{n-1}(f(x)))} +$$

$$+ \log \frac{\mu(f(A_\mu \cap \mathscr{R}_{n-1}(x)))}{\mu(\mathscr{R}_{n-1}(x))}.$$

Observe that:

$$\int_{\bar{\mathbb{C}}} \left| \log \frac{\mu(\mathscr{R}_{n-1}(x) \cap \mathscr{P}(x))}{\mu(\mathscr{R}_{n-1}(x))} \right| d\mu(x) = -\int_{\bar{\mathbb{C}}} \log \frac{\mu(\mathscr{R}_{n-1}(x) \cap \mathscr{P}(x))}{\mu(\mathscr{R}_{n-1}(x))} d\mu(x)$$

$$= H_\mu(\mathscr{P}/\mathscr{R}_{n-1}).$$

But by II.4 (c) we know that $H(\mathscr{P}/\mathscr{R}_{n-1}) \to 0$ if $n \to +\infty$. This shows that the first term in (1) converges to zero in $\mathscr{L}^1$. The same technique can be used to show the convergence to zero in $\mathscr{L}^1$ of the second term. However, this case is slightly more delicate. Take an atom $P \in \mathscr{P}$ and $R \in \mathscr{R}_0$ such that f maps R bijectively onto P. Consider the partition $\mathscr{M}$ of P with atoms $f(A_\mu \cap R)$ and $f(A_\mu^c \cap R)$. Consider also the partitions $\mathscr{M}_n$ of P whose atoms are those atoms of $\mathscr{P}_n$ contained in P. If $\mathscr{B}$ denotes the Borel σ-algebra of P, it follows from II.4 (a) that $\bigvee_0^\infty \mathscr{M}_n = \mathscr{B}$ and obviously $\mathscr{M}_0 \leq \mathscr{M}_1 \leq \ldots$. Define a probability μ_0 on $\mathscr{B}$ by $\mu_0(S) = \mu(S)/\mu(P)$. It is easy to check that:

$$\int_P \left| \log \frac{\mu(\mathscr{M}_n(x) \cap \mathscr{M}(x))}{\mu(\mathscr{M}_n(x))} \right| d\mu(x) = \mu(P) H_{\mu_0}(\mathscr{M}/\mathscr{M}_n).$$

The map $f/R : R \to P$ satisfies $\mu((f/R)^{-1}(S)) \le \mu(S)$ for every Borel set S contained in P. Hence:

$$\int_R \left|\log \frac{\mu(\mathcal{M}_n(f(x)) \cap \mathcal{M}(f(x)))}{\mu(\mathcal{M}_n(f(x)))}\right| d\mu(x) \le \int_P \left|\log \frac{\mu(\mathcal{M}_n(x) \cap \mathcal{M}(x))}{\mu(\mathcal{M}_n(x))}\right| d\mu(x)$$

$$= \mu(P) H_{\mu_0}(\mathcal{M}/\mathcal{M}_n).$$

Now observe that if $x \in R \cap A_\mu$, we have:

$$\mathcal{M}_n(f(x)) \cap f(A_\mu \cap R) = \mathcal{P}_n(f(x)) \cap f(A_\mu \cap R) =$$

$$= f(\mathcal{R}_n(x)) \cap f(A_\mu \cap R) = f(A_\mu \cap R \cap \mathcal{R}_n(x)) =$$

$$= f(A_\mu \cap \mathcal{R}_n(x)).$$

Then, using that $\mu(A_\mu \cap R) = \mu(R)$, we obtain

$$\int_R \left|\log \frac{\mu(f(A_\mu \cap \mathcal{R}_n(x)))}{\mu(\mathcal{P}_n(f(x)))}\right| d\mu(x) = \int_R \left|\log \frac{\mu(\mathcal{M}_n(f(x)) \cap \mathcal{M}(f(x)))}{\mu(\mathcal{M}_n(f(x)))}\right| d\mu(x) \le$$

$$\le \mu(P) H_{\mu_0}(\mathcal{M}/\mathcal{M}_n).$$

Since $f^{-1}(P)$ contains d atoms of $\mathcal{R}$, we conclude that:

$$\text{(2)} \qquad \int_{f^{-1}(P)} \left|\log \frac{\mu(f(A_\mu \cap \mathcal{R}_n(x)))}{\mu(\mathcal{P}_n(f(x)))}\right| d\mu(x) \le d\mu(P) H_{\mu_0}(\mathcal{M}/\mathcal{M}_n).$$

Given $\varepsilon > 0$ take a set S that is a finite union of atoms of $\mathcal{P}$ (say k atoms) and such that $\mu(S^c) \le \varepsilon/(2d \log 2)$. Since $H_{\mu_0}(\mathcal{M}/\mathcal{M}_n) \to 0$ for every atom P, it follows from (2) that we can choose N such that if $n \ge N$ then

$$\int_{f^{-1}(P)} \left|\log \frac{\mu(f(A_\mu \cap \mathcal{R}_n(x)))}{\mu(\mathcal{P}_n(f(x)))}\right| d\mu(x) \le \frac{\varepsilon}{k}$$

for every atom $P \in \mathcal{P}$ contained in S. Then

$$\text{(3)} \qquad \int_{f^{-1}(S)} \left|\log \frac{\mu(f(A_\mu \cap \mathcal{R}_n(x)))}{\mu(\mathcal{P}_n(f(x)))}\right| d\mu(x) \le \varepsilon$$

if $n \ge N$. Moreover, since the partition $\mathcal{M}$ contains only two atoms, (2) implies that

$$\int_{f^{-1}(P)} \left|\log \frac{\mu(f(A_\mu \cap \mathcal{R}_n(x)))}{\mu(\mathcal{P}_n(f(x)))}\right| d\mu(x) \le d\mu(P) \log 2.$$

Then

$$(4)\qquad \int_{f^{-1}(S^c)} \left|\log \frac{\mu(f(A_\mu \cap \mathcal{R}_n(x)))}{\mu(\mathcal{P}_n(f(x)))}\right| d\mu(x) \leq d \log 2\ \mu(S^c) \leq \frac{\varepsilon}{2}.$$

From (3) and (4), it follows that the second term in (1) converges to zero in $\mathcal{L}^1$ as we wished to show. To complete the proof of II.2, it remains to show that the third term in (1) converges to $\log J_\mu$ in $\mathcal{L}^1$. But

$$\mu(\mathcal{R}_n(x)) = \mu(\mathcal{R}_n(x) \cap A_\mu) \leq \mu(f^{-1}(f(A_\mu \cap \mathcal{R}_n(x)))) = \\ = \mu(f(A_\mu \cap \mathcal{R}_n(x))).$$

Then

$$(5)\qquad \frac{\mu(f(A_\mu \cap \mathcal{R}_n(x)))}{\mu(\mathcal{R}_n(x))} \geq 1.$$

Moreover,

$$(6)\qquad \frac{\mu(f(A_\mu \cap \mathcal{R}_n(x)))}{\mu(\mathcal{R}_n(x))} = \frac{1}{\mu(\mathcal{R}_n(x))} \int_{A_\mu \cap \mathcal{R}_n(x)} J_\mu d\mu = \frac{1}{\mu(\mathcal{R}_n(x))} \int_{\mathcal{R}_n(x)} J_\mu d\mu.$$

Then (6), together with the fact that $\bigvee_0^\infty \mathcal{R}_n$ is the Borel σ-algebra, implies that

$$(7)\qquad \frac{\mu(f(A_\mu \cap \mathcal{R}_n(x)))}{\mu(\mathcal{R}_n(x))} \to J_\mu(x)$$

in $\mathcal{L}^1$ (also for a.e. x, but this will not be used). But $J_\mu(x) \geq 1$ for a.e. x and the same is true for the functions $x \to \mu(f(A_\mu \cap \mathcal{R}_n(x)))/\mu(\mathcal{R}_n(x))$ by (5). Hence, (7) implies that

$$\log \frac{\mu(f(A_\mu \cap \mathcal{R}_n(x)))}{\mu(\mathcal{R}_n(x))} \to \log J_\mu(x)$$

in $\mathcal{L}^1$.

Now we shall prove Lemma II.4. We shall need the following property:

Lemma II.5. *For all* $0 < k < 1$ *there exists a continuous function* $\rho : \overline{\mathbb{C}} \to [0, +\infty)$ *satisfying the following properties:*

a) There exist constants $G > 0$, $\alpha > 0$ *such that:*

$$\rho(x) \geq G \prod_{i=1}^m d(x, x_i)^\alpha,$$

where $x_1, \ldots, x_m$ *are critical points of* f.

b) *If x is not a critical point of f and $d(y_1, x) < \rho(x)$, $d(y_2, x) < \rho(x)$, then*

$$d(f(y_1), f(y_2)) \geq k \, \| f'(x) \| \, d(y_1, y_2).$$

Proof. If x is not a critical point, let $\rho_1(x)$ be the maximum positive number such that f is injective when restricted to the disk $\{y \mid d(y, x) < \rho(x)\}$. Set $\rho_1(x) = 0$ when x is a critical point. It is easy to see that the function $\rho_1 : \overline{\mathbb{C}} \to \mathbb{R}$ is continuous, bounded away from zero on compact sets not containing critical points, and, using that, nearby a critical point, f acts as $z \to z^{\ell}$ with $2d - 1 \geq \ell \geq 2$, that in a neighborhood of a critical point, x_i is bounded below by $G_i d(x, x_i)^{\alpha_i}$, where G_i and α_i are positive. Therefore, there exist $A > 0$, $\alpha > 0$ such that:

$$\rho_1(x) \geq A \prod_{i=1}^{m} d(x, x_i)^{\alpha} \tag{1}$$

for all $x \in \overline{\mathbb{C}}$. We claim that there exists a continuous function $\rho_0 : \overline{\mathbb{C}} \to [0, +\infty)$ satisfying (a), upper bounded by ρ_1 and such that

$$\| f'(y) \| \geq k \, \| f'(x) \| \tag{2}$$

if $d(y, x) \leq \rho_0(x)$. To construct ρ_0, we take disks $D_1, \ldots, D_m$ centered at the critical points $x_1, \ldots, x_m$ such that f'/D_i can be written as

$$f'(z) = (z - x_i)^{d_i} g_i(z),$$

where $g_i : D_i \to \mathbb{C}$ has no zeroes. Then there exists $r > 0$ such that if $1 \leq i \leq m$, $x \in D_i$, $y \in D_i$ and $d(x, y) < r$ then:

$$|g_i(y)| > \sqrt{k} \, |g_i(x)|. \tag{3}$$

Now take the continuous functions $\rho^{(i)} : D_i \to [0, +\infty)$ defined by:

$$\rho^{(i)}(x) = \min \{r, (1 - k^{1/2d_i}) \, |x - x_i|\}. \tag{4}$$

Let $\widehat{D}_1, \ldots, \widehat{D}_m$ be disks centered at $x_1, \ldots, x_m$ whose closures are contained in $D_1, \ldots, D_m$ respectively. Set $r_0 = \min_i d(\widehat{D}_i, D_i^c)$ and choose $\delta > 0$ such that if $x \in (\bigcup_i \widehat{D}_i)^c$ and $d(y, x) \leq \delta$ then $\| f'(y) \| \geq k \, \| f'(x) \|$. If $x \in \widehat{D}_i$ and $d(y, x) < \min \{r_0, \rho^{(i)}(x)\}$ it follows that $y \in D_i$ (because $d(y, x) < r_0$) and $d(y, x) < \rho^{(i)}(x) < r$. Hence (3) implies:

$$\frac{\| f'(y) \|}{\| f'(x) \|} = \left(\frac{|y - x_i|}{|x - x_i|} \right)^{d_i} \cdot \frac{|g_i(y)|}{|g_i(x)|} \geq \left(\frac{|y - x_i|}{|x - x_i|} \right)^{d_i} \sqrt{k}.$$

If we now use (4), we obtain:

$$\text{(5)}\qquad \frac{\| f'(y) \|}{\| f'(x) \|} \geq \left(1 - \frac{|y-x|}{|x-x_i|}\right)^{d_i} \sqrt{k} \geq \left(1 - \frac{\rho^{(i)}(x)}{|x-x_i|}\right)^{d_i} \sqrt{k} \geq$$
$$\geq (1-(1-k^{1/2d_i}))^{d_i} \sqrt{k} = k.$$

Now take a continuous function $\rho_0 : \bar{\mathbb{C}} \to [0, +\infty)$ such that:

$$\rho_0(x) = \min \{\rho_1(x), \delta\} \text{ if } x \notin \bigcup_i D_i,$$

$$\rho_0(x) = \min \{\rho_1(x),\ r_0, \rho^{(i)}(x)\}$$

if $x \in U_i$, $1 \leq i \leq m$ where U_i is neighborhood of x_i with closure contained in $\hat{D}_i$ and

$$0 < \rho_0(x) \leq \min \{\rho_1(x),\ r_0,\ \rho^{(i)}(x)\}$$

if $x \in \hat{D}_i - U_i$, $1 \leq i \leq m$. By (5) and by the way we choose δ it follows that ρ_0 satisfies the required properties. Now define ρ by

$$\rho(x) = \frac{1}{16} \rho_0(x)$$

Clearly ρ satisfies (a) because ρ_0 does. To complete the proof of II.5, we have to show that it also satisfies (b). Suppose that x is not a critical point. Since $\rho_0(x) < \rho_1(x)$, we know that $f/\{y \mid d(y,x) < \rho_0(x)\}$ is injective. By Koebe's theorem [3] we have:

$$\text{(6)}\qquad f(\{y \mid d(y,x) < \rho_0(x)\}) \supset \{y \mid d(y, f(x)) < \frac{1}{4} \rho_0(x) \| f'(x) \| \}.$$

Let g be the branch of $f^{-1}/\{y \mid d(y, f(x)) < \left(\frac{1}{4}\right) \rho_0(x) \| f'(x) \| \}$ that maps $f(x)$ in x. Applying Koebe's theorem to g, we obtain:

$$g(\{y \mid d(y, f(x)) < \frac{1}{4} \rho_0(x) \| f'(x) \| \}) \supset \{y \mid d(y,x) < \frac{1}{16} \rho_0(x)\} =$$
$$= \{y \mid d(y,x) < \rho(x)\}.$$

Then:

$$\text{(7)}\qquad f(\{y \mid d(y,x) < \rho(x)\}) \subset \{y \mid d(y, f(x)) < \frac{1}{4} \rho_0(x) \| f'(x) \| \}.$$

Therefore, if $d(y_1, x) < \rho(x)$ and $d(y_2, x) < \rho(x)$, we obtain from (7) that

$$d(f(y_i), f(x)) < \frac{1}{4}\rho_0(x) \| f'(x) \|, \; i = 1,2.$$

Hence, all the segment joining $f(y_1)$ and $f(y_2)$ is contained in the disk $\{y \mid d(y, x) \leq \left(\frac{1}{4}\right)\rho_0(x) \| f'(x) \| \}$. Then:

$$d(y_1, y_2) = d(g(f(y_1)), g(f(y_2))) \leq$$

$$\leq d(f(y_1), f(y_2)) \sup \{ \| g'(z) \| \mid d(z, f(x)) < \frac{1}{4}\rho_0(x) \| f'(x) \| \}$$

But $d(z, f(x)) < \left(\frac{1}{4}\right)\rho_0(x) \| f'(x) \|$ implies by (6) that:

$$d(g(z), x) \leq \rho_0(x).$$

Hence

$$\| g'(z) \| = \| (f'(g(z)))^{-1} \| \leq k^{-1} \| f'(x) \|^{-1}$$

and then

$$d(y_1, y_2) \leq k^{-1} \| f'(x) \|^{-1} d(f(y_1), f(y_2)),$$

as we wished to prove.

Continuing with the proof of II.4, we take a partition $\mathscr{P}$ with finite μ-entropy and satisfying:

(8) $$\operatorname{diam} \mathscr{P}(x) < \rho(x).$$

Such a partition exists because the function $x \to \log \rho(x)$ is integrable by II. 3. Therefore we can apply Lemma 2.3 of [8]. For each $P \in \mathscr{P}$, we choose a point $a \in P$ and then (8) implies that P is contained in the disk $D(P) = \{x \mid d(x, a) < \rho(x)\}$. Take a topological disk $\hat{D}(P) \subset D(P)$ with $\mu(\hat{D}(P)) = \mu(D(P))$ not containing critical values of f. Then there exist branches $g_i^{(P)} : \hat{D}(P) \to \overline{\mathbb{C}}$ of $f^{-1}/\hat{D}(P)$, $i = 1, \ldots, d$. Let $\mathscr{R}_n$ be the partition with atoms of the form $g_i^{(P)}(A)$, where P is an atom of $\mathscr{P}$, A an atom of $\mathscr{P}_n$ contained in P and $1 \leq i \leq d$. From this definition, it is clear that $\mathscr{P}$ and $\mathscr{R}_0 \leq \mathscr{R}_1 \leq \ldots$ satisfy condition (b) of II.4. Condition (d) holds by the way we choose $\mathscr{P}$. To prove conditions (a), it is sufficient to show that:

(9) $$\lim_{n \to +\infty} \operatorname{diam} \mathscr{P}_n(x) = 0$$

for μ-a.e x. Once this is proved, by the way $\mathscr{R}_n$ was constructed it follows that $\lim_{n\to+\infty} \operatorname{diam} \mathscr{R}_n(x) = 0$ for μ-a.e. x. In order to show (9), observe that if $y_i \in \mathscr{P}_n(x)$, $i=1,2$, then $f^j(y_i) \in \mathscr{P}(f^j(x))$, $i=1,2$, for all $0 \le j \le n$. Hence, since $\operatorname{diam} \mathscr{P}(f^j(x)) \le \rho(f^j(x))$, we have:

$$d(f^j(y_i),\ f^j(y_2)) \ge k \,\|\, f'(f^{j-1}(x)) \,\|\, d(f^{j-1}(y_1),\ f^{j-1}(y_2))$$

for all $1 \le j \le n$. Then:

$$d(f^n(y_1),\ f^n(y_2)) \ge k \,\|\, (f^n)'(x) \,\|\, d(y_1, y_2).$$

But $d(f^n(y_1),\ f^n(y_2)) \le \operatorname{diam} \mathscr{P}_n(f^n(x))$. The last term is bounded, say by $G > 0$, for every x and $n \ge 0$. Hence,

$$d(y_1, y_2) \le \frac{G}{k} \,\|\, (f^n)'(x) \,\|^{-1}$$

and then

$$\text{(10)} \qquad \operatorname{diam} \mathscr{P}_n(x) \le \frac{G}{k} \,\|\, (f^n)'(x) \,\|^{-1}$$

By (*) and (**) of II.3, (10) implies (9) completing the proof of II.4.

Now we shall prove the theorem. As we mentioned in the introduction, the f-invariant measure μ_f constructed in [5] satisfies $h_{\mu_f}(f) = \log d$ and is characterized by the property

$$\mu_f(f(A)) = d_{\mu_f}(A),$$

for all Borel set A such that f/A is injective. If by contradiction we assume there exists another f-invariant measure $v \ne \mu_f$ with $h_v(f) = \log d$ then it is well known that there exists an f-invariant *ergodic* measure $\mu_f \ne \mu$ with $h_\mu(f) = \log d$ (this follows for instance from [2] Prop. 13.3). Associated to μ, we have the function J_μ and the set A_μ given by Lemma II.1. Then applying II.2, the Corollary of II.1 and the convexity of the logarithm we get

$$\log d = h_\mu(f) = \int_{\bar{\mathbb{C}}} \log J_\mu d_\mu \le \log \int_{\bar{\mathbb{C}}} J_\mu d_\mu \le \log d.$$

Hence:

$$\int_{\bar{\mathbb{C}}} \log J_\mu d_\mu = \log \int_{\bar{\mathbb{C}}} J_\mu d_\mu = \log d.$$

Then $J_\mu = d$ μ*-a.e.* Now let us show that $\mu(f(A_\mu^c)) = 0$. Take a family of disjoint connected and simply connected open sets $\{U_1, \ldots, U_\ell\}$ such that they do not contain critical values and

$$\mu((\bigcup_i U_i)^c) = 0. \tag{11}$$

Then for each $1 \le i \le k$, we can find open sets $D_1^{(i)}, \ldots, D_d^{(i)}$ such that $f/D_j^{(i)}$ is a homeomorphism onto U_i and $\bigcup_i D_j^{(i)} = f^{-1}(U_i)$. It follows from (11) that

$$\mu(\bigcup_{i,j} D_j^{(i)}) = 1. \tag{12}$$

Hence:

$$\mu(f(A_\mu \cap D_j^{(i)})) = \int_{A_\mu \cap D_j^{(i)}} J_\mu d\mu =$$
$$= d\mu(A_\mu \cap D_j^{(i)}) = d\mu(D_j^{(i)}).$$

Therefore,

$$\sum_j \mu(f(A_\mu \cap D_j^{(i)})) = d\mu \sum_j \mu(D_j^{(i)}) = d\mu(f^{-1}(U_i)) = d\mu(U_i). \tag{13}$$

Since:

$$\mu(f(A_\mu \cap D_j^{(i)})) \le \mu(f(D_j^{(i)})) = \mu(U_i) \tag{14}$$

for all i, j, we obtain from (13) and (14):

$$\mu(f(A_\mu \cap D_j^{(i)})) = \mu(U_i)$$

for all i, j. Then:

$$\begin{aligned} 0 = \mu(U_i) - \mu(f(A_\mu \cap D_j^{(i)})) &= \mu(f(D_j^{(i)})) - \mu(f(A_\mu \cap D_j^{(i)})) = \\ &= \mu(f(A_\mu^c \cap D_j^{(i)})). \end{aligned} \tag{15}$$

Since this holds for all i and j, it follows from (15) and (12) that $\mu(f(A_\mu^c)) = 0$. Now consider a Borel set A where f/A is injective. We have:

$$f(A) = f(A \cap A_\mu) \cup f(A \cap A_\mu^c)$$

and then:

$$\mu(f(A)) = \mu(f(A \cap A_\mu)) + \mu(f(A \cap A_\mu^c)) =$$
$$= \mu(f(A \cap A_\mu)) = \int_{A \cap A_\mu} J_\mu d\mu = \int_A J_\mu d\mu = d\mu(A).$$

Therefore $\mu = \mu_f$.

References

[1] R. Bowen, *Some systems with unique equilibrium states*, Math. Syst. Theory 8 (1974), 193-202.

[2] M. Denker, C. Grillemberg, K. Sigmund, *Ergodic Theory in Compact Spaces*, Lecture Notes in Mathematics 527, Springer-Verlag.

[3] E. Hille, *Analytic Functions Theory*.

[4] M. Gromov, *On the entropy of holomorphic maps*, Preprint, I.H.E.S.

[5] A. Lopes, A. Freire, R. Mañé, *An invariant measure for rational functions* (to appear).

[6] M. Misiurevicz, *Diffeomorphisms without any invariant measures with maximum entropy*, Bull. Acad. Pol. Sci. 21 (1973), 903-910.

[7] M. Misiurevicz, F. Przyticki, *Topological entropy and degree of smooth mappings*, Bull. Acad. Pol. Sci 25 (1977), 573-574.

[8] R. Mañé, *A proof of Pesin's formula*, Ergodic Theory and Dynamical Systems, 1 (1981), 95-103.

[9] D. Ruelle, *An equality for the entropy of differentiable maps*, Bol. Soc. Bras. Mat 9 (1978), 83-87.

Ricardo Mañé
I.M.P.A., Rio de Janeiro

Ann. scient. Éc Norm. Sup.,
4e série, t. 16, 1983, p. 193 à 217.

ON THE DYNAMICS OF RATIONAL MAPS

BY R. MAÑÉ, P. SAD AND D. SULLIVAN

I. – Introduction

1. It is a remarkable fact that each analytic [1] endomorphism f of the Riemann sphere $\mathbb{C}$ exhibits highly non-trivial dynamical phenomena. In this paper we first describe classical and recent results which give the basic dynamical picture of these mappings. Then we make use of this picture to construct topological conjugacies or partial topological conjugacies between certain nearby endomorphisms in analytic families of endomorphisms.

The partial or global conjugacies we will construct depend analytically on the parameters of the family and these satisfy the interesting geometric property of *quasi-conformality*. One corollary will be – *there is an open dense set* $\mathbb{C}$ *of degree* d *polynomial mappings of* $\mathbb{C}$ *such that all mappings in each connected component of* C *are conjugate by quasi-conformal homeomorphisms.*

Another corollary will be – *an open dense set of polynomial mappings satisfies the Axiom* A *expanding property iff there is an open dense set of polynomial mappings where the Julia set has Lebesgue measure zero.* A precise statement of results comes later in this introduction.

2. Our construction of conjugacies depends on a simple but at first surprising proposition concerning analytic perturbations of the inclusion of an arbitrary subset A of the sphere.

λ-LEMMA. – *Let* A *be a subset of* $\overline{\mathbb{C}}$, D *the open unit disk of* $\overline{\mathbb{C}}$ *and* $i_\lambda : A \to \overline{\mathbb{C}}$ *a family of injections depending analytically on* $\lambda \in D$ (*i.e. the function* $\lambda \to i_\lambda(z)$ *is analytic for all* $z \in A$). *Suppose that* i_0 *is the inclusion map* $A \hookrightarrow \overline{\mathbb{C}}$. *Then every* i_λ *has a quasi-conformal extension* $i_\lambda : \overline{A} \to \overline{\mathbb{C}}$ *which is a topological embedding depending analytically on* $\lambda \in D$ *and so that the map* $D \times A \ni (\lambda, z) \to i_\lambda(z) \in \overline{\mathbb{C}}$ *is continuous.*

Using the λ-Lemma and the general picture of the dynamics of endomorphisms of $\overline{\mathbb{C}}$ the topological conjugacies between endomorphisms are built in two steps. First, one builds analytically varying partial conjugacies on some dynamically easy subset A of the sphere

---

[1] Without mention to the contrary analytic here means complex analytic.

Originally published in: *Ann. Sci. Ecole Norm. Sup. (4) 16 (1983), no. 2, 193–217,*

(like the countable set of expanding periodic points union the basins of attracting periodic cycles). Second, one saturates by the mapping and applies the λ-Lemma to extend this partial conjugacy to the closure of A.

It is easy to see that from the above version of the λ-Lemma it can be deduced the more general statement obtained by replacing the unit disk in $\mathbb{C}$ by the unit ball in $\mathbb{C}^n$.

After this paper was written, Sullivan and Thurston [14] proved an extension Lemma that states that given a compact set $A \subset \mathbb{C}$ and a family of injections i_λ depending analytically on λ and satisfying the hypothesis of the λ-Lemma, then there exist a family of quasi-conformal homeomorphisms $i_\lambda: \overline{\mathbb{C}} \circlearrowleft$ extending $i_\lambda: A \to \overline{\mathbb{C}}$ and depending analytically on λ for λ in a disk $D_0 \subset D$, whose radius is a universal constant independent of the original family. Combining this Lemma with the techniques developed in this paper Sullivan obtained in [13] a substantial improvement of the stability Theorem presented here (Theorem D).

3. We will work with analytic families $f: W \times \overline{\mathbb{C}} \to \overline{\mathbb{C}}$ of endomorphisms where W is a connected complex manifold and $f = f(w, z)$ is analytic in two variables. For the global family of all endomorphisms of degree d, $\mathrm{End}_d(\mathbb{C})$, W is an open connected subset of the complex projective space $\mathbb{C}P^{2d+1}$, with the inherited topology equivalent to the C^0 topology on $\mathrm{End}_d(\overline{\mathbb{C}})$. The three-dimensional Moebius group $\{A : z \to (az+b)/(cz+d)\}$ acts by conjugation $f \to A.f.A^{-1}$ on the global family showing the space of analytically inequivalent endomorphisms has dimension $2d-2$. The number $2d-2$ coincides with the number of critical points $\{c \mid f'(c)=0\}$ of any endomorphism f of degree d. Critical points have topological dynamical meaning (f is not locally injective there) and the reader will observe in the course of the discussion below a relationship between the structure of the orbits of critical points and the number of essential analytic parameters for perturbations of f.

Similarly, the analytic family of all polynomial mappings of degree d is parameterized by an open connected dense subset of $\mathbb{C}P^{d+1}$ in which the similarity group $\{z \to az+b\}$ acts by conjugation. The quotient analytic space of analytically inequivalent polynomial mappings has dimension $d-1$ which coincides with the number of critical points of a polynomial mapping of degree d.

4. The Julia set and the stable regions. — Now we describe the dynamical picture of an analytic endomorphism f of $\overline{\mathbb{C}}$. Say that a point $x \in \overline{\mathbb{C}}$ is *stable* for f if on some neighborhood of x the family of interates $f, f^2, f^3, \ldots$, is an equicontinuous family of mappings of a neighborhood into the sphere. Note that when x is not stable i. e. *unstable*, for any neighborhood the union of images of iterates must cover $\overline{\mathbb{C}}$ except two points at most. Fatou [3] and Julia [6] showed *the set of unstable points* $J(f)$ (now called the Julia set) coincides with the *closure of the expanding periodic points* (sources).

The open set of stable points consists of countably many connected components, *the stable regions of* f, which are transformed among themselves by f. In Sullivan [9] it is shown that under the iteration of f *each stable region is eventually cyclic*. The cycles of stable regions are classified into five types (Sullivan [9]). The first two types *attractive basins* and *parabolic basins* have fundamental domains for the equivalence relation $x \sim y$ iff $f^n x = f^m y$ for some n, $m \geqq 0$. The third type, *superattractive basins* do not, but they are foliated by the closures of the classes of the equivalence relation, $x \approx y$ iff $f^n x = f^n y$ for some $n \geqq 0$. The last two types are rotation domains, *Siegel disks* or *Herman rings*, which are foliated by the closures of forward orbits.

(i) An *attractive basin* D arises from an attractive periodic cycle γ with non zero derivative of modulus less than one, $\gamma = \{z, f(z), \ldots, f^{n-1}(z)\}$, $f^n(z) = z$, $0 < |(f^n)(z)| < 1$, and D consists of the components of $W_s(\gamma) = \bigcup_{x \in \gamma} \{y \mid \lim_{n \to +\infty} \text{distance}\,(f^n y, f^n x) = 0\}$ containing points of γ. Fatou [3] showed that such a D must contain a critical point of f. Thus there are no more than $2d-2$ attractive basins for an endomorphism of degree d.

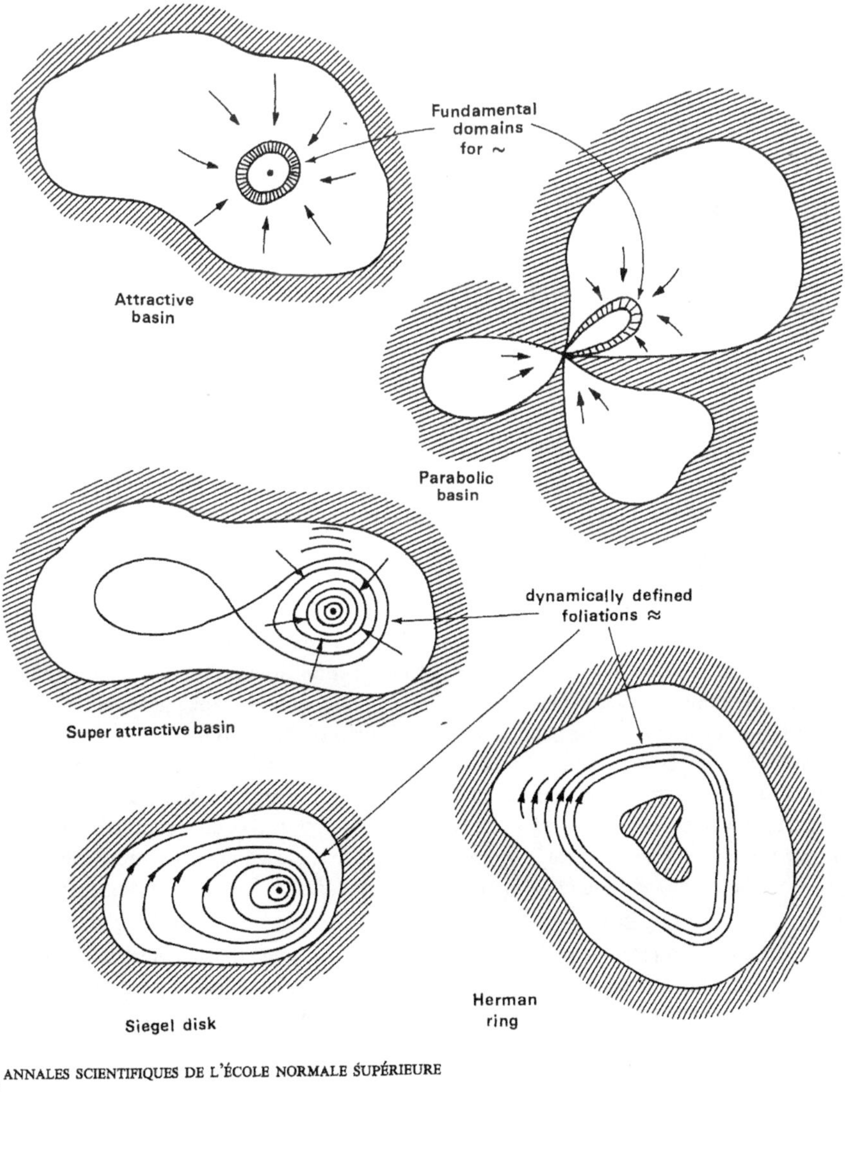

If we remove from D the inverse orbit of γ, $\{\bigcup_{n\geq 0} f^{-n}\gamma\}$, the set of $\sim$ equivalence classes ($x\sim y$ iff $f^n x = f^m y$) defines a torus with branch points corresponding to the critical points of f. This follows easily from the local model of f near γ, where near a fixed point of a power of f we have $z \to \lambda z$, $0<|\lambda|<1$.

(ii) A *parabolic basin* D arises from a non-hyperbolic periodic cycle γ with derivative a root of unity, $\gamma = \{z, f(z), \ldots, f^{n-1}(z)\}$, $f^n(z)=z$, $((f^n)'(z))^m = 1$, γ is contained in the frontier of D, and each compact in D converges to γ under forward iteration of f (Fatou [3]). The local picture of the dynamics consists of parabolic sectors arranged around the fixed point of a power of f which in local coordinates is $z \to z + z^l + \ldots$ and topologically equivalent to $z \to z + z^l$ (Fatou [3], Camacho [2]).

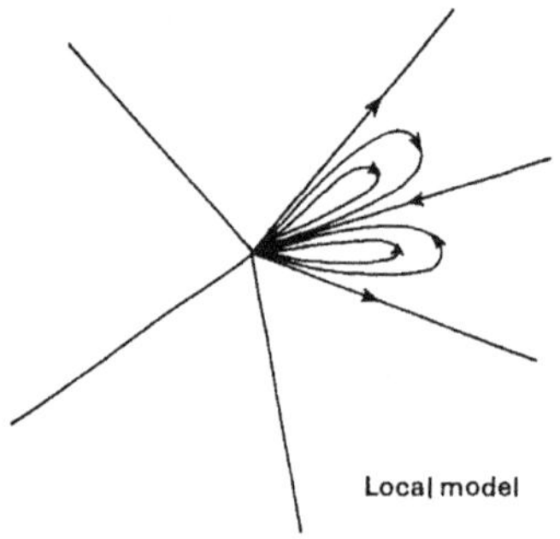

Local model

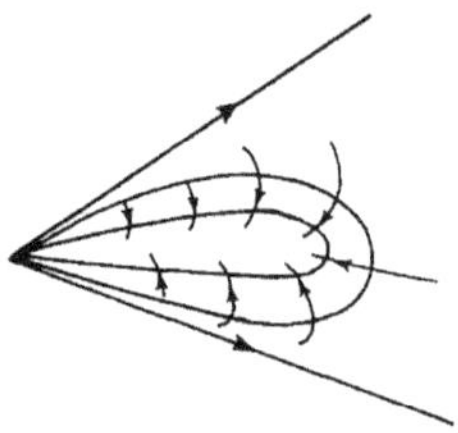

Fundamental domain

The local model produces a fundamental domain for the global dynamics on D because all orbits in D tend to γ. Looking at the local picture then shows the quotient of D by the $x\sim y$ equivalent classes is a union of twice punctured sphere with branched points coming from the critical points of f lying in D (there must be at least one critical point in D, Fatou [3]).

(iii) *A superattractive basin* D is defined just like an attractive basin but now the derivative of the power of f having a fixed point is zero. Now points arbitrarily near the attracting cycle are identified by f and there is no true fundamental domain for the $\sim$ equivalence classes. The more precise relation $x \approx y$ iff $f^n x = f^n y$ for some $n \geq 0$ defines a foliation with singularities of $D' = D -$ inverse orbit of γ by the closures of the $\approx$ equivalence classes. The leaves are 1-manifolds which are not necessarily compact and which have singularities at the inverse orbit of other critical points in D. The local analytic "linearization" near a superattractive fixed point or, more precisely, its analytical equivalence to $z \to z^m$ for some $m>0$ shows the leaves near γ are nearly concentric closed curves around the points of γ. The rest of the foliation of D' is obtained by applying f^{-1} to this concentric foliation near γ.

(iv) A *Siegel disk* is a stable region which is cyclic and on which the appropriate power of f is analytically conjugate to a rotation of the standard unit disk. Siegel [1942] proved these occur near a non-hyperbolic periodic point if the argument u of its derivative satisfies the following diophantine condition: there exists $c>0$ and $\nu \geqq 2$ such that $|u-(p/q)| \geqq c/q^{\nu}$ for every relatively prime integers p and q.

Fatou and Julia showed that if such regions existed their frontiers were contained in the union of the ω-limit sets of critical points.

Siegel disks around the origin may occur already in the family $z \to \lambda z+z^2$, $|\lambda|=1$. However, they do not occur when u is sufficiently Liouville because then there are periodic points tending to zero in this case (an easy calculation).

(v) A *Herman ring* is a stable region similar to a Siegel disk. Now we have a periodic cycle of annuli and a power of f which restricted to any of these annuli is analytically equivalent to an irrational rotation of the standard annulus. Again the frontier is contained in the ω-limit sets of critical points (Fatou [3]). Such regions were found by M. Herman for the map:

$$z \to \frac{e^{i\theta}}{z} \cdot \left(\frac{z-a}{1-\bar{a}z}\right)^2,$$

for appropriate θ and a. Herman uses Arnold's theorem about real analytic conjugations of real analytic diffeomorphisms of the circle to rigid rotations when the rotation number is like a Siegel number. Note that both Siegel disks and Herman rings are foliated by the closures of orbits and the leaves are closed real analytic curves.

5. MORE DYNAMICAL PROPERTIES. — (i) One knows there are only finitely many cyclic stable regions described in 4 (Sullivan [9]). But it is a *problem* to find the sharp upper bound for the number of cycles in terms of the degree. Is it $2d-2$?

(ii) Also for polynomials one knows *each bounded stable region is simply connected* (apply the maximum principle to $f, f^2, \ldots$). *Thus polynomials do not have Herman rings.*

(iii) An amusing corollary of the classification of stable regions in 4 is the following — *if all critical points of f are eventually periodic but none are periodic then the Julia set of f is all of $\overline{\mathbb{C}}$* (because each type of cyclic region besides the superattractive basin requires a critical point with an infinite forward orbit). Examples of this type are $z \to ((z-2)/z)^2$ and the quotient of some higher degree endomorphism of a one-dimensional torus by the equivalence relation $x \sim -x$. See for instance the example due to Lattes [4].

(iv) Fatou and Julia showed that f on $J(f)$ is topologically transitive. In fact, for any z in $J(f)$ the inverse orbit $\bigcup_{n \geqq 0} f^{-n}(z)$ is dense in $J(f)$. If no critical points tend to $J(f)$ or touch it, Fatou showed some power of f is expanding on $J(f)$. He surmised *the dynamical structure was continuous in the coefficients for such examples* (now called Axiom A or hyperbolic systems, see below) *and guessed that this property should be true except for special values of the parameters.*

Even when $J(f)$ is contaminated by critical points one may think of $J(f)$ as the *hyperbolic* part (2) of the dynamics. The Siegel disks and Herman rings are in the *elliptic* part of the dynamics. The attractive basins and the parabolic basins are the *properly discontinuous* part of the dynamics.

(2) The words "hyperbolic" and "elliptic" are meant to suggest chaotic and rigid structure respectively in the dynamics.

6. Now we state our theorems about partial conjugacies between members of analytic families $f: W \times \overline{\mathbb{C}} \to \overline{\mathbb{C}}$ of endomorphisms. First we shall introduce the concept of *persistently non-hyperbolic periodic point.* We shall give first the definition in a particular case where it is easier to understand. If z_0 is a periodic point of f_{w_0}, say $(f_{w_0}^n(z_0)=z_0$, and if $(f_{w_0}^n)'(z_0)\neq 1$, then when w moves in a small neighborhood W_0 of w_0 we can find $z(w)$, depending analytically on w and such that $z(w_0)=z_0$ and $(f_w^n)(z(w))=z(w)$ for every w in the neighborhood. We say that z_0 is a persistently non-hyperbolic periodic point of f_{w_0} if $|(f_w^n)'(z(w))|=1$ for every w in a neighborhood of w_0 i.e. if we cannot destroy the non hyperbolicity of z_0 by moving the parameter. Observe that the condition $|(f_w^n)'(z(w))|=1$ for every w nearby w_0 implies that in fact $(f_w^n)'(z(w))$ is constant in that neighborhood. Unfortunately this definition is not sufficient because we shall need to handle the case $(f_{w_0}^n)'(z_0)=1$, when it is not always possible to find $z(w)$ as before. To state the general definition we first introduce the analytic sets:

$$M_n=\{(w, z)\in W\times\overline{\mathbb{C}}\,|\,f_w^n(z)=z, f_w^j(z)\neq z,\ 0\leqq j<n\}.$$

Define the projection $P_n: M_n\to W$ as $P_n(w, z)=w$ and the eigenvalue function $\lambda_n: M_n\to\mathbb{C}$ by:

$$\lambda_n(w, z)=(f_w^n)'(z).$$

If P_n is injective when restricted to a neighborhood of a point $(w_0, z_0)\in M_n$, then there exists a neighborhood W_0 of w_0 and an analytic function $\varphi: W_0\to\overline{\mathbb{C}}$, with $\varphi(w_0)=z_0$, such that its graph $\{(w, \varphi(w))\,|\,w\in W_0\}$ is a neighborhood of (w_0, z_0) in M_n. We say that a periodic point z_0 of f_{w_0} is persistently non-hyperbolic if it is non-hyperbolic and:

(i) There exists a neighborhood W_0 of w_0 and an analytic function $\varphi: W_0\to\overline{\mathbb{C}}$ such that $\varphi(w_0)=z_0$ and its graph $\{(w, \varphi(w))\,|\,w\in W_0\}$ is a neighborhood of (w_0, z_0) in M_n;

(ii) λ_n is constant in a neighborhood of (w_0, z_0) in M_n.

By the analicity of f we can reformulate (i) in a weaker form: P_n is injective on a neighborhood of (z_0, w_0). In fact, since f is analytic and the function $f_{w_0}^n(z)-z$ is not identically zero, we can find neighborhoods W_0 of w_0 and U_0 of z_0 such that on $W_0\times U_0$ we can factorize $f_w^n(z)-z$ (assuming $z_0=0$) as:

$$f_w^n(z)-z=(z^k+\sum_{j=0}^{k-1} a_j(w)\,z^j)\,g(w, z),$$

where the coefficients a_j are analytic functions of w and g is analytic and $\neq 0$ in $W_0\times U_0$. If $P_n/((W_0\times U_0)\cap M_n)$ is injective then for every $w\in W_0$ there exists a unique $\varphi(w)\in U_0$ such that $(w, \varphi(w))\in M_n$. Then $\varphi(w)$ is the unique element of U_0 such that $f_w^n(\varphi(w))-\varphi(w)=0$. By Rouche's Theorem, if W_0 is small enough, the order of $\varphi(w)$ as zero of $z\to f_w^n(z)-z$ must be the same of $z_0=0$ as root of $z\to f_{w_0}^n(z)-z$, that is k. But $g\neq 0$ in a neighborhood of (w_0, z_0). Hence $\varphi(w)$ is a root of order k of $z^k+\sum_{j=0}^{k-1} a_j(w)\,z^j$. Hence this polynomial is $(z-\varphi(w))^k$. This means that $\varphi(w)=ka_{k-1}(w)$ and then φ is analytic.

Moreover observe that λ_n is analytic on the analytic set M_n [since it is the restriction of $(w, z) \to (f_w^n)'(z)$]. Then, if λ_n is constant in a neighborhood of a point, it is constant on the whole connected component of M_n containing that point. Since every connected component of M_n projects onto W, it follows that if for some w_0, f_{w_0} has a persistently non-hyperbolic periodic point, then f_w has a non-hyperbolic periodic point for every $w \in W$. Therefore, if for some $w_1 \in W$, all the periodic points of f_{w_1} are hyperbolic, then, for all $w \in W$, f_w has no persistently non-hyperbolic periodic points.

Now define the set $H(f) \subset W$ as the set of values $w_0 \in W$ that have a neighborhood W_0 such that for every $w \in W$ every periodic point of f_w is either hyperbolic or persistently non-hyperbolic. Clearly this set is open. It is also dense:

THEOREM A. — *For any analytic family* $f: W \times \overline{\mathbb{C}} \to \overline{\mathbb{C}}$, $H(f)$ *is an open dense subset of* W.

Thus we have an open dense set of parameters W for which the nature of each periodic point of f_w stays constant. We can show that in this open dense set of parameters the dynamical structure of f_w in its Julia set $J(f)$ remains topologically unchanged.

DEFINITIONS. — Two endomorphisms f, g in End $\overline{\mathbb{C}}$ are J-equivalent if there exists a homeomorphism h: $J(f) \to J(g)$ such that $hf = gh$. Given an analytic family $f: W \times \overline{\mathbb{C}} \to \overline{\mathbb{C}}$ we say that $w_0 \in W$ is *J-stable* if w_0 has a neighborhood W_0 such that f_w is J-equivalent to f_{w_0} for all w in W_0 and $J(f_w)$ depends continuously on $w \in W_0$ in the sense of the Hausdorff distance between two closed sets. Say that a map φ of a subset $X \subset \overline{\mathbb{C}}$ into the sphere is *quasi-conformal* if it is a topological embedding and:

$$\sup_{x \in X} \limsup_{t \to 0} \frac{\sup_{y \in S_t(x)} d(\varphi(y), \varphi(x))}{\inf_{Y \in S_t(x)} d(\varphi(y), \varphi(x))} \leqq \infty,$$

where $S_t(x) = \{ y \in X \mid d(y, x) = t \}$.

We ignore whether a quasi-conformal map $\varphi: X \to \overline{\mathbb{C}}$ can be extended to a quasi-conformal map of a neighborhood of X.

THEOREM B. — *For every analytic family of endomorphisms* $f: W \times \overline{\mathbb{C}} \to \overline{\mathbb{C}}$, $H(f) \subset W$ *coincides with the set of J-stable values of the parameter. Moreover, if* w_0 *belongs to* $H(f)$ *there exists a neighborhood* W_0 *in* W *of* w_0 *and a continuouu conjugacy function* $h: W_0 \times J(f_{w_0}) \to \overline{\mathbb{C}}$ *so that for all* w *in* W_0;

(i) h_w *is a conjugacy between* f_{w_0} *on* $J(f_{w_0})$ *and* f_w *on* $J(f_w)$ *and* h_{w_0} *is the identity;*

(ii) *for each* z, $h_w(z)$ *is analytic in* w;

(iii) *For each* w, h_w *is quasi-conformal;*

(iv) *the set of J-stable points coincide with the interior of the set of parameters where the Julia set moves continuously.*

7. Now we will enlarge our topological conjugacies beyond the Julia set using the structure described in 4. We obtain almost the natural expected result. Whenever Siegel disks or Herman rings are present there is, however, a glueing problem near their frontiers. On this problem, see remark after the statement of Theorem D.

Now we discuss the orbit structure of critical points: we say $a \approx b$ if either $a=b$ or a and b lie in the same leaf of the dynamically defined foliations of superattractive basins, Siegel disks, or Herman rings discussed in 4. We define a subset $C(f) \subset W$ for the analytic family $f: W \times \overline{\mathbb{C}} \to \overline{\mathbb{C}}$ by $w_0 \in C(f)$ iff there is a neighborhood $W_0 \subset W$ of w_0 and analytic functions $c_1, \ldots, c_k$ on W_0 so that:

(i) $c_1(w), \ldots, c_k(w)$ are all the critical points of f_w;

(ii) either $f_w^m(c_j(w)) \approx f_w^n(c_i(w))$ for some $m \geq 0$, $n \geq 0$ and all $w \in W_0$, or $f_w^m(c_j(w)) \not\approx f_w^n(c_i(w))$ for all $w \in W_0$, $m \geq 0$, $n \geq 0$.

Remark. — Note that for w in a component of $H(f)$ the critical points in $J(f)$ satisfy (ii) by Theorem B.

THEOREM C. — *For any analytic family $f: W \times \overline{\mathbb{C}} \to \overline{\mathbb{C}}$, $C(f)$ is an open dense subset of $H(f)$ (which is open and dense in W).*

Now we will state global or almost global stability results for endomorphisms of $C(f)$. Given an endomorphism f let $\Lambda(f)$ denote the collection of completely invariant compact sets Λ [i. e. $f(\Lambda)=\Lambda=f^{-1}(\Lambda)$] which intersect each (open) Siegel disk or (open) Herman ring in a compact set. For instance, the closure of all stable regions which are eventually attractive, superattractive or parabolic basins belongs to $\Lambda(f)$.

THEOREM D. — *For any analytic family $f: W \times \overline{\mathbb{C}} \to \overline{\mathbb{C}}$; if w_0 belongs to the open dense subset $C(f) \subset W$ then for any choice of $\Lambda \in \Lambda(f_{w_0})$ there is a neighborhood W_0 of w_0 and a continuous mapping $h: W_0 \times \Lambda \to \overline{\mathbb{C}}$ satisfying:*

(i) *h_w is a conjugacy between f_{w_0} on Λ and f_w on $h_w(\Lambda)$; and $h_w(\Lambda)$ belongs to $\Lambda(f_w)$;*

(ii) *for each z, $h_w(z)$ is analytic in w;*

(iii) *For each w, h_w is quasi-conformal.*

Conversely, if for any choice $\Lambda \in \Lambda(f_{w_0})$ there exists a neighborhood W_0 of w_0 and a continuous mapping $h: W_0 \times \Lambda \to \overline{\mathbb{C}}$ satisfying (i) *then $w_0 \in C(f)$.*

Observe that when f_{w_0} has no Siegel disks or Herman domains then $\overline{\mathbb{C}} \in \Lambda(f_{w_0})$, and this is the best choice to which Theorem D applies. It always happens for the family of polynomials of a given degree. As we explained in the introduction, in [13] Sullivan, combined the extension Lemma in [14] with the results above to obtain an improvement of Theorem D that states that it holds globally i. e. if $w \in C(f)$ then we can find $h: W_0 \times \overline{\mathbb{C}} \to \overline{\mathbb{C}}$ satisfying all the properties in Theorem D. Therefore, the set of values $w_0 \in W$ such that f_{w_0} is stable in the family (i. e. topologically equivalent to any f_w with w near to w_0) is open and dense.

8. Say that an endomorphism f is expanding on, the Julia set if for each z in $J(f)$ there is an n so that $|(f^n)'(z)|>1$. An easy argument shows the expanding property is equivalent to the Axiom A property: there exist $c>1$ and $N>0$ such that $|(f^k)'(z)|>c^k$, for all $k>N$ and $z \in J(f)$.

The classification of 4 shows immediately the *Axiom A expanding property implies all critical points are contained in attractive or superattractive basins and these are the only periodic stable regions.* The converse as remarked above follows from Fatou [3].

Note that the Axiom A expanding property implies membership in $H(f)$ for any family f. Thus we have an intrinsic property, Axiom A, which implies a property, membership in $H(f)$, determined by perturbations. For example we don't know if membership in $H(f)$ for the global family f implies the same is true for the iterates. However, it is obvious the Axiom A expanding property passes to iterates.

A second favorable feature of this property is that there is a powerful theory [Anosov, Sinai, Smale, Bowen, ...] for treating the dynamics of these hyperbolic systems. Markov partitions and symbolic dynamics can be used to describe f on $J(f)$ (Jacobson [5]).

Also in the conformal case the Axiom A expanding property for f implies $J(f)$ is a quasi-self similar fractal. It is not hard to show $J(f)$ has finite positive Hausdorff measure in its dimension which is strictly less than 2 (Sullivan [10], see also Bowen [1] and Ruelle [7]).

For all these reasons it is important to be able to verify the Axiom A expanding property. It would be important *to know whether or not the Axiom A expanding property is true for an open dense set of endomorphisms* (in reasonable families). The openess is known (see Jacobson [5]). But the density has defied verification.

Say that an *endomorphism f has a invariant line field on* $J(f)$ if there is a completely invariant subset $A \subset J(f)$ of positive Lebesgue measure, and a measurable family of tangent lines defined a. e. in A invariant by the tangent action of f.

Say that an analytic family of endomorphisms $f: W \times \overline{\mathbb{C}} \to \overline{\mathbb{C}}$ is *reduced* if different members cannot be analytically conjugate. Define $r(w)$ for $w \in W$ as the number of equivalence classes of not eventually periodic critical points in $J(f)^c$ under the equivalence relation $c_i \approxeq c_j$ iff $f^n_w(c_i) \approx f^m_w(c_j)$ for some $n \geqq 0$, $m \geqq 0$ (where $\approx$ is defined before the statement of Theorem C) plus the number of critical points in $J(f_w)$.

THEOREM E. — *Suppose* $f: W \times \overline{\mathbb{C}} \to \overline{\mathbb{C}}$ *is a reduced analytic family and* w *is a point of* $C(f)$ *where* $r(w) \leqq$ *dimension* W. *Then* f_w *satisfies the Axiom* A *expanding property iff*:

(i) *all periodic points of* f_w *are hyperbolic*;

(ii) f_w *has no Herman rings*;

(iii) f_w *has no invariant line fields on the Julia set.*

COROLLARY. — *The Axiom* A *expanding property is true for an open dense set of polynomial mappings of degree d iff the Lebesgue measure of the Julia set is zero for an open dense set of polynomial mappings of degree d.*

II. — Proof of the λ-Lemma and Theorem A and B

The proof of the λ-Lemma is based in the following: any analytic map of the unit λ-disk into the triply punctured sphere $\overline{\mathbb{C}} - \{0, 1, \infty\}$ is distance non increasing for the complete Poincaré metrics on the unit disk and punctured sphere (Schwarz's Lemma). Choose three points from A and renormalize i so their images by i_λ are constantly 0, 1, and ∞.

For any three other distinct finite points x, y, z of A, consider the functions $x(\lambda) = i_\lambda(x)$, $y(\lambda) = i_\lambda(y)$, $z(\lambda) = i_\lambda(z)$, $\lambda \to (y(\lambda) - x(\lambda))/y(\lambda)$. Let $0 < R < 1$ and $0 < m < M$ be given. These functions avoid 0, 1 and ∞. Applying the above to the second function we see that $|y(0)|<M$ implies that $y(\lambda)$ is not too large for $|\lambda|\leqq R<1$ (i.e. bounded by a function of M). Applying the above to the fourth function we see that if $0<m\leqq|y(0)|\leqq M<\infty$ and $|x(0)-y(0)|$ is small then $|(x(0)-v(0))/y(0)|$ is small, which implies $|(x(\lambda)-v(\lambda))/y(\lambda)|$ is small if $|\lambda|\leqq R<1$, which implies $|x(\lambda)-y(\lambda)|$ is small by the first remark.

Thus each i_λ is uniformly continuous on $A\cap\{z \mid m\leqq|z|\leqq M\}$.

Such annuli cover the sphere (permuting the roles of 0, 1, ∞) so i_λ has a continuous extension $i_\lambda:\overline{A}\to\overline{\mathbb{C}}$. Since 0 and any other particular λ_0 play symmetric roles in the hypothesis, the φ_λ have continuous inverses. For each z in $\overline{A}$, $i_\lambda(z)$ is analytic in λ because it is a uniform limit of analytic functions on each disk $|\lambda|\leqq R<1$.

To prove $i_\lambda(x)$ is quasiconformal apply the non-increasing property to the function $g(\lambda)=(x(\lambda)-y(\lambda))/(x(\lambda)-z(\lambda))$ when $|x(0)-y(0)|=|x(0)-z(0)|$ and conclude that $|g(\lambda)|$ is bounded for $|\lambda|\leqq R$. To prove the continuity of $i:D\times\overline{A}\to\overline{\mathbb{C}}$ is sufficient to show that the family of functions $\lambda\to i_\lambda(x)$, $x\in\overline{A}$, is equicontinuous. Again this is a consequence of the non-increasing property that grants $|i_{\lambda_1}(x)-i_{\lambda_2}(x)|=|x(\lambda_1)-x(\lambda_2)|$ is small uniformely in x if $|\lambda_1-\lambda_2|$ is small.

To prove Theorem A take $w_0\in W$ and any open neighborhood W_0 of w_0. We shall show that W_0 contains points of $H(f)$. If $w\in W_0$ denote $\alpha(w)$ the number of attractive and superattractive periodic orbits of f_w, and $\beta(w)$ the number of non-hyperbolic periodic orbits. By Fatou [3], $\alpha(w)+\beta(w)\leqq 4(d-1)$. Choose $w_1\in W_0$ such that $\alpha(w_1)=\max\{\alpha(w)\mid w\in W_0\}$. Then, since attractive and superattractive periodic orbits are persistent, there exists an open neighborhood $W_1\subset W_0$ of w_1 such that α/W_1 is constant. Choose $w_2\in W_1$ such that $\beta(w_2)=\max\{\beta(w)\mid w\in W_1\}$. Denote $p_1,\ldots,p_l$ the non-hyperbolic periodic orbits of f_{w_2}. We claim there exists a neighborhood $W_2\subset W_1$ of w_2 and neighborhoods U_i of p_i, $1\leqq i\leqq l$, such that if n_i is determined by $(w_2,p_i)\in M_{n_i}$, then $P_n/((W_2\times U_i)\cap M_{n_i})$ is injective and:

$$\bigcup_{i=1}^{l}(\{w\}\times\overline{\mathbb{C}})\cap((W_2\times U_i)\cap M_{n_i}),$$

is the set of non-hyperbolic periodic points of f_w, for all $w\in W_2$. Since the attractive and superattractive periodic points of f_w move analytically with w and their number is constant in W_1, it follows that we can find neighborhoods U_i of p_i and a neighborhood $W_2\subset W_1$ of w_2 such that for every $w\in W_2$ and $1\leqq i\leqq l$, U_i doesn't contain attractive or superattractive periodic points of f_w. Suppose W_2 and the U_i's $i=1,\ldots,l$ so small that $M_{n_i}\cap(W_2\times U_i)$ is connected for every $1\leqq i\leqq l$. The absence of attractive or superattractive periodic points in U_i for all $w\in W_2$ means that the analytic function λ_{n_i} on the analytic set $M_{n_i}\cap(W_2\times U_i)$ satisfies $|\lambda_{n_i}(w,z)|\geqq 1$. Hence $\lambda_{n_i}^{-1}$ is also an analytic function on this analytic set and is bounded by 1. But $|\lambda_{n_i}^{-1}(w_2,p_i)|=1$. Hence it attains its maximum at (w_2,p_i). Therefore it is constant in the connected analytic set $(W_2\times U_i)\cap M_{n_i}$. Then, for every $w\in W_2$, the points of $(\{w\}\times\overline{\mathbb{C}})\cap((W_2\times U_i)\cap M_{n_i})$ are non-hyperbolic periodic points of f_w. Since β attains a maximum at $w=w_2$ it is easy to see that the number of non-hyperbolic periodic points of f_w also attains a maximum at $w=w_2$. Then:

$$\sum_{i=1}^{l} \#(\{w\}\times\overline{\mathbb{C}})\cap((W_2\times U_i)\cap M_{n_i})\leqq\beta(w_2)=l,$$

for every $w\in W_2$. But $\#(\{w\}\times\overline{\mathbb{C}})\cap((W_2\times U_i)\cap M_{n_i})\geqq 1$. Hence it must be equal to one for all $w\in W_2$. This means that $P_n/((W_2\times U_i)\cap M_{n_i})$ is injective for all $1\leqq i\leqq l$. The previous arguments also show that points in the set:

$$S(w)=\bigcup_{i=1}^{l}(\{w\}\times\overline{\mathbb{C}})\cap((W_2\times U_i)\cap M_{n_i}),$$

are non-hyperbolic and that $\# S(w)=l$. But since as we observed above the number of non-hyperbolic periodic points of f_w attains a maximum l at $w=w_2$, it follows that number is bounded by l. Hence $S(w)$ is exactly the set of *all* the non-hyperbolic periodic points of f_w and this completes the Proof of the Claim. Now we claim that $w_2\in H(f)$. If it doesn't, there exists $w_3\in W_2$ such that f_{w_3} has a non-hyperbolic periodic point p that is not persistently non-hyperbolic. But by the Claim, p must have the form $p=(\{w_3\}\times\overline{\mathbb{C}})\cap((W_2\times U_i)\cap M_{n_i})$ for some $1\leqq i\leqq l$. Then P_{n_i} is injective in a neighborhood of (w_2, p) in M_{n_i} [in fact, the neighborhood $(W_2\cap U_i)\cap M_{n_i}$ satisfies this property]. Hence (w_2, p) satisfies condition (i) of the Definition of persistently non-hyperbolic periodic point (in the equivalent formulation that we gave after the Definition). But we proved that λ_{n_i} is constant in $(W_2\times U_i)\cap M_{n_i}$. Then it also satisfies part (ii) of the Definition and p is persistently non-hyperbolic.

To prove Theorem B suppose w_0 belongs to $H(f)\subset W$ for the analytic family f: $W\times\overline{\mathbb{C}}\to\overline{\mathbb{C}}$. Let W_0 containing w_0 be a simply connected neighborhood of w_0 in $H(f)$. We claim each expanding periodic point x_n of f_{w_0} defines an analytic function x_n : $W_0\to\overline{\mathbb{C}}$ such that $x_n(w)$ is a periodic point of f_w of the same period of x_n. The implict function theorem tells us we can analytically continue a transversal fixed point of f^n_w uniquely for some neighborhood of parameters. Thus by following the periodic point determined by $x_n(w_0)$ we locally define $x_n(w)$ on an open set. At a frontier point w_1 of such an open set, the limit of $x_n(w)$ is still a periodic point which is hyperbolic if it has period n because $W_0\subset H(f)$.

Actually, the period can not drop neither when $x=\lim_{w\to w_1} x_n(w)$ has period $k<n$ $(f^k_{w_1})'(x)$ not a root of unit (if $m\neq k$ is given there exists a neighborhood U of x so that no point in $U-\{x\}$ has period m) nor when $(f^k_w)'(x)$ is a root of unit [use the local model for the dynamics and the fact that $W_0\subset H(f)$].

Thus $x_n(w_1)$ is a hyperbolic periodic point of order n which may be analytically continued on a neighborhood of w_1. This definition agrees with the previous one by the uniqueness. So we can define $x_n(w)$ on all of W_0 which is simply connected.

Since hyperbolic periodic points cannot collide, as we have already remarked, we may apply the λ-Lemma to the set A of expanding periodic points. We obtain $h(w, z)$: $W_0\times J(f_{w_0})\to\overline{\mathbb{C}}$, analytic in w, quasi-conformal in z.

Since the role of w_0 and any particular w_1 in W_1 may be reversed it is clear the $x_n(w_1)$ must be all of the expanding periodic points of f_{w_1}. Thus for each w, $h(w, z)$ defines a homeomorphism between Julia sets. By definition $h(w, z)$ is a conjugacy between sets of expanding periodic points. By continuity $h(w, z)$ is a conjugacy between Julia sets.

This proves the first part of Theorem B.

To prove the converse property, let w_0 be a J-stable value of W. Since the number of periodic cycles of f_w in $J(f_w)$ stays constant for w in a small neighborhood W_0 of w_0, we see that all periodic cycles of f_{w_0} with derivative equal to one have an analytic continuation through W_0. Now observe that the function $W_0 \ni w \to J(f_w)$ is continuous with respect to the Hausdorff metric. We shall show that $w \in H(f)$. Obviously this shows that J-stable values of the parameter belong to $H(f)$. If $w_0 \notin H(f)$, there exists $w_1 \in W_0$ such that for some $n \geqq 0$, $f^n_{w_1}$ has a non hyperbolic fixed point z_1 that is not persistently non hyperbolic. This implies that we can find w_2 near to w_1 such that $f^n_{w_2}$ has a fixed point z_2 (near to z_1) such that $(f^n_{w_2})'(z_2)$ is a Siegel number and z_2 is not a persistently non hyperbolic periodic point. Then $z_2 \notin J(f_{w_2})$.

But since z_2 is not persistently non hyperbolic, we can find w_3 arbitrarily near to w_2 and fixed points z_3 of $f^n_{w_3}$, arbitarily near to z_2, such that $|(f^n_{w_3})'(z_3)| \geqq 1$. Hence $z_3 \in J(f_{w_3})$. We have thus proved that there exist arbitrarily small perturbations of $w_2 \in W_0$ that make the Julia set reach points (like z_3) that are bounded away from $J(f_{w_2})$ [because z_3 is arbitrarily near to z_2, and $z_2 \notin J(f_{w_2})$]. This concludes the proof of Theorem B.

III. – Proof of Theorem C

First we shall show that $C(f)$ is contained in $H(f)$. We shall use the following standard Lemma:

LEMMA III.1. – *If $W_0 \subset W$ is an open simply connected set and $\varphi : W_0 \to \overline{\mathbb{C}}$ is an analytic function such that for every $w \in W_0$ the point $\varphi(w)$ doesn't belong to any forward f_w-orbit of a critical point of f_w, then there exist analytic functions $\varphi_{n,i} = W_0 \to \overline{\mathbb{C}}$, $n \geqq 0$, $i \leqq i \leqq d^n$, satisfying $f^n_w(\varphi_{n,i}(w)) = \varphi(w)$ and $\varphi_{n,i}(w) \neq \varphi_{n,j}(w)$ for all $n \geqq 0$, $1 \leqq i \leqq d^n$, $w \in W_0$.*

The λ-Lemma now yields the following property:

LEMMA III.2. – *Suppose that W_0 and φ satisfy the hypothesis of Lemma III.1. Moreover suppose that either $\varphi(w)$ is not f_w-periodic for any $w \in W_0$ or that for some $N \geqq 0$ we have $f^N_w(\varphi(w)) = \varphi(w)$ for all $w \in W_0$. Then $W_0 \subset H(f)$.*

Proof. – We shall prove the Lemma only in the case when $\varphi(w)$ is not f_w-periodic for any $w \in W_0$. The other case reduces to this just by replacing φ by $\varphi_{1,i}$, where i is chosen taking any $w_0 \in W_0$ and i such that $f_{w_0}(\varphi_{1,i}(w_0)) \neq f^{N-1}_{w_0}(\varphi(w_0))$. Then the same relation holds for all $w \in W_0$ because by III.1 preimages of $\varphi(w_0)$ don't collide. Therefore $\varphi_{1,i}(w)$ is not f_w-periodic for any $w \in W_0$. Now fix some $w_0 \in W_0$ and set $\Lambda = \bigcup_{n \geqq 0} f^{-n}_{w_0}(\varphi(w_0))$. Define h : $W_0 \times \Lambda \to \overline{\mathbb{C}}$ by:

$$h_w(z) = \varphi_{n,i}(w),$$

if $z=\varphi_{n,i}(w_0)$. Observe that since $\varphi(w_0)$ is not f_{w_0}-periodic, and $\varphi_{n,j}(w_0)\neq\varphi_{n,i}(w_0)$ for all $n>0$, $1\leqq i<j\leqq d^n$, the n and i satisfying $z=\varphi_{n,i}(w_0)$ are unique. This shows that $h_w(z)$ is well defined and depends analytically in w. Moreover, every h_w is injective. In fact if $z_1\neq z_2$ belong to Λ, then:

$z_1=\varphi_{n_1,i_1}(w_0)$, $z_2=\varphi_{n_2,i_2}(w_0)$, where either $n_1\neq n_2$ or $n_1=n_2$ and $i_1\neq i_2$. If $n_1\neq n_2$, the equality $h_w(z_1)=h_w(z_2)$ implies:

$$\varphi_{n_1,i_1}(w)=\varphi_{n_2,i_2}(w)$$

and then:

$$f_w^{n_2}(\varphi_{n_2,i_2}(w))=\varphi(w)=f_w^{n_1}(\varphi_{n_1,i_1}(w))=f_w^{n_1}(\varphi_{n_2,i_2}(w)).$$

Suppose $n_2\geqq n_1$. Then:

$$f_w^{n_2-n_1}(\varphi(w))=f_w^{n_2-n_1}(f_w^{n_1}(\varphi_{n_2,i_2}(w)))=f_w^{n_2}(\varphi_{n_2,i_2}(w))=\varphi(w),$$

thus contradicting that $\varphi(w)$ is not f_w-periodic. If $n_1=n_2$ then:

$$h_w(z_1)=\varphi_{n_1,i_1}(w)\neq\varphi_{n_1,i_2}(w)=h_w(z_2).$$

Now we can apply the λ-Lemma to extend every map $h_w:\Lambda\to h_w(\Lambda)$ to a homeomorphism $h_w:\overline{\Lambda}\to\overline{h_w(\Lambda)}$ that obviously satisfies:

$$(\star)\qquad f_w(h_w(z))=h_w(f_w(f_{w_0}(z)),$$

for every $z\in f_{w_0}^{-1}(\Lambda)$. But $\overline{\Lambda}$ contains $J(f_{w_0})$ because it contains the full backward orbit of a point. Therefore, if we show that $h_w(J(f_{w_0}))=J(f_w)$ we shall be done. If $z\in J(f_{w_0})$, it is the limit of a sequence z_n of f_{w_0}-periodic points different from z. Then $h_w(z_n)$ is a sequence of f_w-periodic points different from $h_w(z)$ and converging to $h_w(z)$. Hence $h_w(z)\in J(f_w)$ thus proving $h_w(J(f_{w_0}))\subset J(f_w)$. To prove $J(f_w)\subset h_w(J(f_{w_0}))$ we just observe that interchanging the roles of w and w_0 we obtain h_w^{-1} instead of h_w. Then $h_w^{-1}(J(f_w))\subset J(f_{w_0})$ and $J(f_w)=h_w(h_w^{-1}(J(f_w)))\subset h_w(J(f_{w_0}))$ thus concluding the Proof of Lemma III.2.

To prove $C(f)\subset H(f)$ we start observing than any endomorphism $g:\overline{\mathbb{C}}\hookleftarrow$ not satisfying Axiom A, has a critical point z_0 such that $\bigcup_{n\geqq 0} g^{-n}(z_0)$ doesn't contain critical points. In fact, if such a critical point doesn't exist, it is easy to see that every critical point must be periodic. Therefore g satisfies Axiom A. Now suppose that $w_0\in C(f)$. Then either f_{w_0} satisfies Axiom A (and then $w_0\in H(f)$) or it doesn't, in which case we can take z_0 with the above property. The definition of $C(f)$ grants the existence of a neighborhood W_0 of w_0 and an analytic function $\psi:W_0\to\overline{\mathbb{C}}$ such that $\psi(w_0)=z_0$ and $(f_w)'(\psi(w))=0$ for all $w\in W_0$. Also from the definition of $C(f)$ we know that $\psi(w)$ doesn't belong to the forward f_w-orbit of any critical point because $z_0=\psi(w_0)$ is not in the f_{w_0}-forward orbit of any critical points. Applying Lemma III.2 to $\psi:W_0\to\overline{\mathbb{C}}$ it follows that $W_0\subset H(f)$.

The openness of $C(f)$ is trivial. To prove its density we shall produce a dense set $S \subset H(f)$ whose elements can be aproximated by elements in $C(f)$. First define the critical set:

$$C_s = \{(w, z) \mid (f_w)'(z) = 0\}.$$

If $\pi : W \times \overline{\mathbb{C}} \to W$ is the projection, let S_0 be the set of regular values of π/C_s. Then S_0 is open and dense in W. Moreover given any $w_0 \in S_0$ there exist a neighborhood $W_0 \subset S_0$ of w_0 and analytic functions $c_i : W_0 \to \overline{\mathbb{C}}$, $1 \leqq i \leqq k$, such that $\{c_1(w), \ldots, c_k(w)\}$ is the set of critical points for all $w \in W_0$. Define $S(i, j, n, m)$ as the set of values $w_0 \in S_0 \cap H(f)$ such that either $f^n_{w_0}(c_i(w_0)) \approx f^m_{w_0}(c_j(w_0))$ or $f^n_w(c_i(w)) \not\approx f^m_w(c_j(w))$ holds for every w in a neighborhood of w_0. Now we set:

$$S(i, j) = \bigcap_{n, m} S(i, j, n, m)$$

We claim that each $S(i, j, n, m)$ is open and dense and that if $w_0 \in S(i, j)$, it has a neighborhood W_0 such that for all $n \geqq 0$, $m > 0$ either $f^n_w(c_i(w)) \approx f^m_w(c_j(w))$ or $f^n_w(c_i(w)) \not\approx f^m_w(c_j(w))$ holds for every $w \in W_0$. It follows from the claim that the residual (thus dense) set $S = \bigcap_{i, j} S(i, j, n, m)$ is contained in $C(f)$. The Proof of the Claim requires a series of Lemmas that describe how cyclic domains vary with the parameter.

LEMMA III.3. — *Let z_0 be an attractive periodic point of f_{w_0} with period m. Then there exist neighborhoods V_0 and W_0 of z_0 and w_0 and analytic functions $h : W_0 \times V_0 \to \mathbb{C}$, $\varphi : W_0 \to V_0$ such that:*

(a) $f^m_w(V_0) \subset V_0$ for all $w \in W_0$;

(b) $f^m_w(\varphi(w)) = \varphi(w)$ for all $w \in W_0$;

(c) $h_w(\varphi(w)) = 0$, $w \in W_0$;

(d) $(f^m_w)'(\varphi(w))\, h_w(z) = h_w(f^m_w(z))$ for all $w \in W_0$, $z \in V_0$.

LEMMA III.4. — *Suppose that for some i and $m > 0$, $f^m_w(c_i(w)) = c_i(w)$ for every w in a neighborhood of some $w_0 \in S_0$. Then there exist neighborhoods V_0 and W_0 of z_0 and w_0 respectively and an analytic function $h : W_0 \times V_0 \to \mathbb{C}$ such that:*

(a) $f^m_w(V_0) \subset V_0$ for all $w \in W_0$;

(b) $h(c_i(w)) = 0$, $w \in W_0$;

(c) $h_w(z)^n = h_w(f^m_w(z))$ for all $w \in W_0$, $z \in V_0$,

where n is the multiplicity of $c_i(w_0)$ as a critical point of $f^m_{w_0}$.

LEMMA III.5. — *Suppose that $w_0 \in H(f)$ and z_0 is a parabolic periodic point of f_{w_0} of period m. Then there exist neighborhoods V_0 and W_0 of z_0 and w_0 respectively and a continuous map $h : W_0 \times V_0 \to \mathbb{C}$ such that:*

(a) $h_w(.) = h(w, .)$ is continuous and injective for all $w \in W_0$;

(b) for all $z \in V_0$, the map $w \to h_w(z)$ is analytic;

(c) $f_w h_w(z) = h_w f_{w_0}(z)$ for every $w \in W_0$ and $z \in V_0 \cap f^{-1}_{w_0}(V_0)$.

LEMMA III.6. — *Suppose that $w_0 \in \mathrm{H}(f)$ and* A *is a Herman ring or Siegel disk satisfying* $f^m(\mathrm{A}) = \mathrm{A}$. *Let* $\mathrm{U} \subset \mathrm{A}$ *be an open* f^m*-invariant (i.e.* $f^m(\mathrm{U}) = \mathrm{U}$) *disk or annulus whose closure is contained in* A *and whose boundary consists in one or two* f^m*-invariant analytic curves. Then there exist a neighborhood* W_0 *of* w_0, *an open disk or ring* $\mathrm{T} \subset \mathbb{C}$ *and an analytic function* $h : \mathrm{W}_0 \times \mathrm{T} \to \mathrm{C}$ *such that every* h_w *is a conformal representation satisfying:*

$$h_w(e^{i\theta} z) = f_w(h_w(z)),$$

for all $z \in \mathrm{U}$, *where* θ *is the rotation number of* $f^m_{w_0}/\mathrm{A}$, *and* $h_0(\mathrm{T}) = \mathrm{U}$.

Lemma III.3 is an analytically parametrized version of Poincaré's linearization Theorem, and its Proof follows immediately from the usual technique used to prove this Theorem. Lemma III.4 is in a similar situation with respect to the analytic linearization Theorem of super attractive periodic points. The Proofs of Lemmas III.5 and III.6 will be presented after completing the Proof of the Claim. First observe the Lemmas above imply easily that each $\mathrm{S}(i, j, n, m)$ is open and dense. Suppose that $w_0 \in \mathrm{S}(i, j)$. The Proof of the Claim will be divided according to the following cases:

(I) There exist $n \geqq 0$, $m > 0$, such that $f^n_{w_0}(c_i(w_0)) \approxeq f^m_w(c_j(w_0))$. Suppose that n_0 is the minimum positive integer such that the equality above is satisfied for some $m > 0$. Then take as m_0 the minimum integer such that the equality $f^{n_0}_{w_0}(c_i(w_0)) \approxeq f^{m_0}_{w_0}(c_j(w_0))$ holds. Since $w_0 \in \mathrm{S}(i, j, n_0, m_0)$ there exists a neighborhood W_0 such that $f^{n_0}_w(c_i(w)) \approxeq f^{m_0}_w(c_j(w))$ for all $w \in \mathrm{W}_0$. Restricting W_0 if necessary, we can grant that $f^n_w(c_i(w)) \not\approxeq f^m_w(c_j(w))$ for all $w \in \mathrm{W}_0$ and $0 \leqq n \leqq n_0$, $0 \leqq m < m_0$. From these facts it follows that W_0 satisfies the property required by the Claim.

(II) $f^n_{w_0}(c_i(w)) \not\approxeq f^m_w(c_j(w))$ for every $n \geqq 0$, $m > 0$. This case is subdivided in two situations:

(II *a*) $c_i(w_0)$ or $c_j(w_0)$ (perhaps both) belong to $\mathrm{J}(f_{w_0})$. Then it suffices to takes as W_0 the connected component of $\mathrm{H}(f)$ that contains w_0. Then the topological equivalence between the Julia sets of f_{w_0} and f_w, for any $w \in \mathrm{W}_0$, given by Theorem B, plus the fact that the number (counted with multiplicity) of critical points in $\mathrm{J}(f_w)$ is constant on connected components of $\mathrm{H}(f)$ (again a Corollary of Theorem B), implies the Claim.

(II *b*) $c_i(w_0) \notin \mathrm{J}(f_{w_0})$, $c_j(w_0) \notin \mathrm{J}(f_{w_0})$ and they eventually belong to orbits of different cyclic domains of $\mathrm{J}(f_{w_0})^c$. Then Lemmas III.1, 2, 3 or 4 imply that $f^n_w(c_i(w)) \not\approxeq f^m_w(c_j(w))$ for every $n \geqq 0$, $m \geqq 0$ and all w in a neighborhood of w_0.

(II *c*) $c_i(w_0) \notin \mathrm{J}(f_{w_0})$, $c_j(w_0) \notin \mathrm{J}(f_{w_0})$ and they eventually belong to the orbit of the same cyclic domain. Taking as W_0 the neighborhood given by Lemma I.1, 2, 3 or 4 (according to which type belongs the cyclic domain that eventually contain the orbits of $c_i(w_0)$ and $c_j(W_0)$) the parametrized linearization shows that $f^n_w(c_i(w)) \not\approxeq f^m_w(c_j(w))$ for every $w \in \mathrm{W}_0$.

Let us prove Lemma III.5. Suppose that $f_w(z_0) = z_0$ to simplify the discussion. Set $\lambda = (f_{w_0})'(z_0)$. If $\lambda^q = 1$ and $\lambda^j \neq 1$ for $0 < j < q$, we can write, after an analytic change of coordinates in a neighborhood of z_0 (*see* [2] for details):

$$f_w(z) = \lambda z + \sum_{j=1}^{\infty} a_j(w) z^{q+j},$$

where the series converge in a fixed disk $B_r(z_0)$ for every $w \in W_0$ (taking W_0 sufficiently small) and the functions a_j are analytic. Observe that we are here using that z_0 is a persistently
where the series converge in a fixed disk $B_r(z_0)$ for every $w \in W_0$ (taking W_0 sufficiently small) and the functions a_j are analytitic. Observe that we are here using that z_0 is a persistently
non hyperbolic fixed point of f_{w_0} because $w_0 \in H(f)$. This grants that the fixed point z_0 has an analytic continuation as fixed point $z(w)$ of f_w and that $f'_w(z'(w))=\lambda$ for every w near w_0. We claim that if $a_j(w_0)=0$ for $1 \leqq j<m$ and $a_m(w_0) \neq 0$ then $a_j(w)=0$ for all $1 \leqq j<m$ and every w near to w_0. To prove this we shall again use the hypothesis $w_0 \in H(f)$. Write:

$$f^q_w(z)=z+\sum_{j=1}^{\infty} \tilde{a}_j(w)\, z^{q+j}$$

and observe that $\tilde{a}_j(w_0)=0$ for $1 \leqq j<m$ and $\tilde{a}_m(w_0) \neq 0$. Then:

$$f^q_{w_0}(z)-z=\sum_{j=1}^{\infty} \tilde{a}_j(w_0)\, z^{q+j}=z^{q+m}\left(\sum_{j=m}^{\infty} \tilde{a}_j(w_0)\, z^{j-m}\right).$$

If V_0 is a neighborhood of 0 where the second factor in the last term is $\neq 0$, then $z=0$ is the unique fixed point of f_{w_0} in V_0. The J-stability of f_{w_0} implies that f_w has a unique fixed point in V_0 for w near to w_0. This implies that $a_j(w)=0$ for every $1 \leqq j<m$ and every w in a neighborhood of w_0. Hence $a_j(w)=0$ for every w in the same neighborhood and $1 \leqq j<m$ completing the Proof of the Claim. Now with a linear change of coordinates we can write:

$$f_w(z)=\lambda z+\sum_{j=m+1}^{\infty} a_j(w)\, z^{q+j}.$$

By [2], f_w is equivalent to the map $z \to \lambda z+z^{q+m}$ in a neighborhood of 0 and the homeomorphism h_w that conjugates both maps can be chosen depending analytically in w.

Now let us prove Lemma III.6. To simplify the notation suppose that $m=1$. Let $U_1 \subset U$ be an invariant annulus whose boundary has two analytic closed curves. We shall need the following Lemma, to be proved later:

LEMMA III.7. — *There exists a neighborhood* W_0 *of* w_0 *such that for every* $w \in W_0$, $\overline{U}_1$ *doesn't contain eventually periodic points of* f_w *or points of* $J(f_w)$.

Now take some point $z_1 \in U_1$ and if $C=\{e^{in\theta} \mid n \geqq 0\}$ define $h_w : C \to \overline{C}$, $w \in W_0$, by $h_w(e^{in\theta})=f^n_w(z_1)$. This family of maps depends analytically on the parameter $w \in W_0$. To apply the λ-Lemma to this family we have only to check that every h_w is injective. If h_w is not injective, there exist n and m such that $f^n_w(z_1)=f^m_w(z_1)$. Then z_1 is an eventually periodic point of f_w, contradicting Lemma III.5. Now, applying the λ-Lemma we obtain a conjugacy h_w between the rotation $z \to e^{i\theta} z$ in the circle $\overline{C}$ and the restriction of f_w to the Jordan f_w-invariant curve $h_w(\overline{C})$. By Lemma III.5 this Jordan f_w-invariant curve must belong to some fixed connected component of $J(f_w)^c$. Clearly it must be either a Herman ring or a Siegel's disk. But $h_0(\overline{C})$ is a Jordan f_{w_0}-invariant curve and has points of $J(f_{w_0})$ in its interior if and only if A is a Herman ring. Moreover $h_w(\overline{C})$ contains points of $J(f_w)$ in its interior if and only if $h_0(\overline{C})$ does. Therefore $h_w(\overline{C})$ belongs to a Siegel disk or Herman ring if and only if $h_0(\overline{C})$ belongs to the same type of component. Moreover it is an analytic curve and since it belongs to a Herman ring or Siegel disk, the restriction of f_w to $h_w(\overline{C})$ is real analytically equivalent to a rotation of the circle. Say that the circle is $\overline{C}$ and let $F: h_w(\overline{C}) \to \overline{C}$ be this real analytic conjugacy. Then $F^{-1} h_w$ is a conjugacy between two rotations. Therefore the rotations must coincide. In particular h_w is real analytic. Now let $B_w = \{ z \mid r_1(w) < |z| < r_2(w) \}$ be the maximal ring where an analytic extension $H_w : B_w \to \overline{\mathbb{C}}$ of $h_w : \overline{C} \to \overline{\mathbb{C}}$ exists. From the identity principle it is easy to conclude that $f_w h_w(z) = h_w(e^{i\theta} z)$ for every $z \in B_w$. We claim that for values of w near to w_0 the relation $h_w(B_w) \supset U$ holds. This will clearly prove Lemma III.6. In fact we can show that $h_w(B_w)$ is the Herman ring $A(w)$ containing $h_w(\overline{C})$. This follows from the fact that if $\overline{h_w(B_w)} \subset A(w)$ then one of the boundaries of $h_w(B_w)$ must be an f_w-invariant real analytic curve or reduce to a point. In the first case, the extension property of conformal representations would show that B_w is not maximal. We leave to the reader to verify that the second only arises when $r_1(w) = 0$ and then h_w extends to an analytic map of $B_w \cup \{0\}$.

To prove Lemma III.7, take W_0 so small that $J(f_w) \cap \overline{U} = \emptyset$ for all $w \in W_0$. This is granted by Theorem B. It remains to show that there are no eventually periodic points of f_w in U_1, for w in W_0.

Take a neighborhood V of the periodic orbits of f_w not contained in $J(f_{w_0})$ or in A if A is a Siegel disk. Take V so small that $V \cap A = \emptyset$. Since $w_0 \in H(f)$, all the f_w-periodic orbits not in $J(f_w)$ are contained in V for all $w \in W_0$, if W_0 is chosen small enough. Now suppose that for some $w \in W_0$ and $z \in U_1$ there exists $N > 0$ such that $f_w^N(z)$ is f_w-periodic. Since $J(f_w) \cap U = \emptyset$, $f_w^N(z) \notin J(f_w)$. Hence $f_w^N(z) \in V$.

Moreover if $h_w: J(f_{w_0}) \to J(f_w)$ is the homeomorphism given by Theorem B, the fact that $\overline{V}$ and $\overline{U}_1$ belong to different connected components of the complement of $J(f_{w_0})$, implies that they are also in different connected components of $h_w(J(f_{w_0})) = J(f_w)$ for all $w \in W_0$, if W_0 is small enough.

But we can take W_0 such that $f_w(U_1) \cap U_1 \neq \emptyset$ all $w \in W_0$. Then $f_w^{n+1}(U_1) \cap f_w^n(U_1) \neq \emptyset$ for all n, thus implying that $\bigcup_{n \geqq 0} f_w^n(U_1)$ is connected. On the other hand the property $f_w^n(z) \in V$ implies that $\bigcup_{n \geqq 0} f_w^n(U_1)) \cap V \neq \emptyset$ then U and V belong to the same connected component of the complement of $J(f_w)$.

Lastly, suppose A is Siegel disk with fixed point z_0; let $z_0(w)$ be its local analytic continuation. If U_1 still belongs to the Siegel disk corresponding to $z_0(w)$, clearly $\bigcup_{n \geqq 0} f_w^n(U) \not\ni z_0(w)$ for w close to w_0.

IV. – Proof of Theorem D

Suppose that $w_0 \in C(f)$ and that Λ is a compact totally invariant set of f_{w_0} i. e. $f_{w_0}^{-1}(\Lambda)=\Lambda$ such that its intersection with any Siegel disk or Herman ring is compact. We shall construct a set $\Lambda_0 \subset \Lambda$, a simply connected open neighborhood W_0 of w_0 and a map $h: W_0 \times \Lambda_0 \to \overline{\mathbb{C}}$ with the following properties:

(a) $f_{w_0}(\Lambda_0)=\Lambda_0$;

(b) $\Lambda=\overline{\bigcup_{n\geq 0} f_{w_0}^{-n}(\Lambda_0)}$;

(c) The map $w \to h_w(z)$ is analytic for every $z \in \Lambda_0$, $h_{w_0}(z)=z$ and h_w is injective for all $w \in W_0$;

(d) $h_w(z)$ belongs to the orbit of a critical point of f_w if and only if z belongs to the orbit of a critical point of f_{w_0}.

(e) $f_w(h_w(z))=h_w(f_{w_0}(z))$ for every $w \in W_0$, $z \in \Lambda_0$.

Let us show how to prove the partial stability follows from the existence of h, W_0 and Λ_0. Let $\hat{\Lambda}_0$ be Λ_0 minus the intersection of forward f_{w_0}-orbits of critical points with Λ_0. Set $\hat{\Lambda}=\bigcup_{n\geq 0} f_{w_0}^{-n}(\hat{\Lambda}_0)$. Then $\hat{\Lambda}$ is dense in Λ. Define $h_w: \hat{\Lambda} \to \overline{\mathbb{C}}$ as follows: if $z \in \hat{\Lambda}$ choose n such that $f_{w_0}^n(z) \in \hat{\Lambda}_0$ and let $\varphi: W_0 \to \overline{\mathbb{C}}$ be an analytic function such that $\varphi(w_0)=z$ and:

$$f_w^n(\varphi(w))=h_w(f_{w_0}^n(z)),$$

for all $w \in W_0$. Then define $h_w(z)=\varphi(w)$. The existence of φ in a neighborhood od w_0 is granted by the implicit function theorem and the fact that $(f_{w_0}^n)'(z)\neq 0$ (because orbits of critical points don't intersect $\hat{\Lambda}_0$). Now let $W_1 \subset W_0$ be the maximal simply connected open set to which φ can be extended. Suppose that $W_1 \neq W_0$. Then there exists a sequence $w_n \in W_1$ converging to some $w \in W_0 \setminus W_1$. Suppose that $\varphi(w_n)$ converges to z_∞. Then $f_{w_\infty}^n(z_\infty)=h_{w_\infty}(f_{w_0}^n(z))$. We claim that $(f_{w_\infty}^n)'(z)\neq 0$. If not, $h_{w_\infty}(f_{w_0}^n(z))$ belongs to the orbit of a critical point of f_{w_∞}. By (e) this can happen only if $f_{w_0}^n(z)$ has same property with respect to f_{w_0}. But this is impossible because $f_{w_0}^n(z) \in \hat{\Lambda}_0$. Then $(f_{w_\infty}^n)'(z_\infty)\neq 0$ and the implicit function theorem gives an analytic function $\hat{\varphi}: W_\infty \to \overline{\mathbb{C}}$, where W_∞ is a neighborhood of w_∞, satisfying $\hat{\varphi}(w_\infty)=z_\infty$ and $(f_w^n)\,\hat{\varphi}(w))=h_w(f_{w_0}^n(z))$ for all $w \in W$. Moreover, the uniqueness property of the implicit function theorem grants $\hat{\varphi}(w)=\varphi(w)$ for all $w \in W_\infty \cap W_1$. Then this contradicts the maximality of W_1 thus proving that $W_1=W_0$.

Now that we have well defined maps $h_w: \hat{\Lambda} \to \overline{\mathbb{C}}$, that depend analytically in w, let us show that every h_w is injective. Suppose that $z_1 \neq z_2$ and $h_w(z_1)=h_w(z_2)$. Take $n \geq 0$ such that $f^n(z_i) \in \hat{\Lambda}_0$, $i=1, 2$. Then:

$$h_w(f_{w_0}^n(z_1))=f_w^n(h_w(z_1))=f_w^n(h_w(z_2))=h_w(f_{w_0}^n(z_2)).$$

But h_w is injective on $\hat{\Lambda}_0$. Then $f_{w_0}^n(z_1)=f_{w_0}^n(z_2)$. Take the minimum $n_0>0$ such that $f_{w_0}^{n_0}(z_1)=f_{w_0}^{n_0}(z_2)$. On the other hand:

$$h_w(f_{w_0}^{n_0-1}(z_1))\neq h_w(f_{w_0}^{n_0-1}(z_2))$$

for values of w near to w_0. Then we can assume that there exists a sequence $w_n \to w$ such that:

$$h_{w_n}(f_{w_0}^{n_0-1}(z_1))\neq h_{w_n}(f_{w_0}^{n_0-1}(z_2)).$$

But:

$$(\star) \qquad f_{w_n}(h_{w_n}(f_{w_0}^{n_0-1}(z_1)))=h_{w_n}(f_{w_0}^{n_0}(z_1)) =h_{w_n}(f_{w_0}^{n_0}(z_2))=f_{w_n}(h_{w_n}(f_{w_0}^{n_0-1}(z_2)))$$

Hence $h_{w_n}(f_{w_0}^{n_0-1}(z_1))$ and $h_{w_n}(f_{w_0}^{n_0-1}(z_2))$ are different, converge to the same point:

$$h_w(f_{w_0}^{n_0-1}(z_1))=f_w^{n_0-1}(h_w(z_1))=f_w^{n_0-1}(h_w(z_2))=h_w(f_{w_0}^{n_0-1}((z_2))),$$

and its images under f_{w_n} are by $(\star)$ the same. This means that:

$h_w(f_{w_0}^{n_0-1}(z_1))=h_w(f_{w_0}^{n_0-1}(z_2))$ is a critical point of f_w. But:

$$f^{n-(n_0-1)}(h_w(f_{w_0}^{n_0-1}(z_1)))=h_w(f_{w_0}^n(z_1)).$$

By (*d*) this means that $f_{w_0}^{n_0}(z_1)$ belongs to the orbit of a critical point, contradicting the definition of $\hat{\Lambda}_0$.

To finish the Proof of Theorem D we have only to show the existence of Λ_0, W_0 and $h: W_0\times\Lambda_0\to\overline{\mathbb{C}}$. To simplify the exposition suppose that every non hyperbolic periodic point or Herman ring is fixed. Without loss of generality we can suppose that Λ contains the attractive regions of attractive and superattractive fixed points. Let $S_1,\ldots, S_m$ be the Siegel disks of f_{w_0} and $H_1,\ldots, H_r$ its Herman rings. Let $z_1,\ldots, z_s$ be the attracting cycles and superattracting fixed points of f_{w_0} and let $V_1,\ldots, V_s$ be disjoint neighborhoods of $z_1,\ldots, z_s$ obtained by the application of Lemmas III.3 and III.4 to these attracting and superattracting fixed points. Finally, if $z_{s+1},\ldots, z_{s+l}$ are the parabolic periodic points, we take open sets $\Lambda_j\subset V_{s+j}$, $j=1,\ldots, l$ where V_{s+j} is obtained applying III.5 to these points, such that $f_{w_0}(\Lambda_j)\subset\Lambda_j$ and $\bigcup_{n=0} f^{-n}(\Lambda_j)$ is the union of all the parabolic domains with z_{s+j} in its boundary. The existence of these sets follows from the discussion in the introduction. First we shall show that if we take a neighborhood W_0 of w_0 contained in all the neighborhoods of w_0 given be Lemmas III.3 to III.6, and setting:

$$\Lambda_0=(\bigcup_i S_i\cap\Lambda)\cup(\bigcup_i H_i\cap\Lambda)\cup(\bigcup_i V_i)\cup(\bigcup_i \Lambda_i),$$

then, it is possible to define the map $h: W_0\times\Lambda_0\to\overline{\mathbb{C}}$ with the desired properties [that Λ_0 satisfies (*a*) and (*b*) is obvious]. If we forget about condition (*d*), the construction of h_w can be easily completed as follows. Suppose we want to define h_w on S_1. Then look at the preimage of z under the map $\tilde{h}_{w_0}$ given by Lemma III.4 and compose it with $\tilde{h}_w$

i. e. $h_w(z)=\hat{h}_w \hat{h}_{w_0}^{-1}(z)$. This definition satisfies (*c*) and (*e*) (on S_1) but (*d*) may fail to be true. To satisfy also (*d*) we define h_w as the composition $\hat{h}_w g_w \hat{h}_{w_0}^{-1}$, where g_w is a homeomorphism of the domain D of the map $\hat{h}_{w_0}$ that maps the $\hat{h}_{w_0}$-preimage of the set of points where f_{w_0}-forward orbits of critical points of f_{w_0} first hit S_1 onto the $\hat{h}_w$-preimage of the set of points where f_w-forward orbits of critical points of f_w first hit $\hat{h}_w(T)$. Require also that g_w depends analytically in w and that commutes with the rotation $z \to ze^{i\theta}$, where θ is the rotation number of f_w/S_1. Then $h_w=h_w g_w h_{w_0}^{-1}$ satisfies also property (*d*). To formalize the construction of the rearrangement map g_w we need the following Lemma:

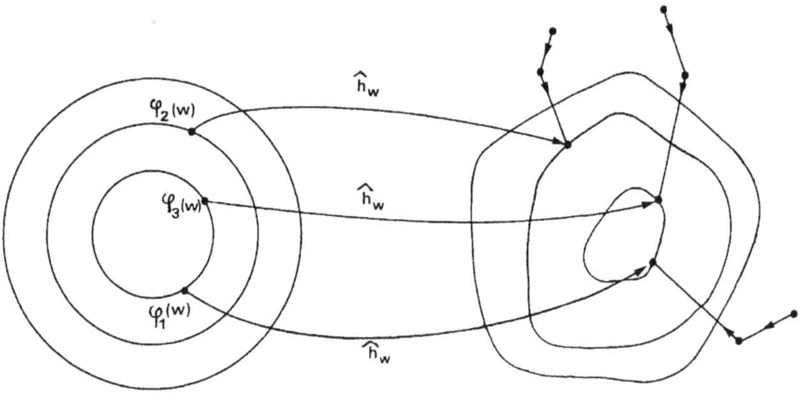

LEMMA IV.1. – *Let* D *be the unit disk,* U *a complete manifold, and* $\varphi_i: U \to D$, $1 \leq i \leq k$ *analytic functions without zeroes such that for all* $1 \leq i < j \leq k$ *either* $|\varphi_i(w)| \neq |\varphi_j(w)|$ *for all* $w \in U$ *or there exists* $\alpha \in \mathbb{R}$ *such that* $\varphi_i(w)=e^{i\alpha}\varphi_j(w)$ *for all* $w \in U$. *Then given* $w_0 \in U$ *there exists* $g: U \times D \to D$ *such that:*

(*a*) *for all* $w \in U$, $g_w(.)=g(w, .)$ *is a homeomorphism of* D *and* $g_{w_0}=Identity$;

(*b*) *for every* $z \in D$ *the map* $w \to g_w(z)$ *is analytic;*

(*c*) $g_w(\varphi_i(w_0))=\varphi_i(w)$ *for all* $w \in U$, $1 \leq i \leq k$;

(*d*) $g_w(e^{i\theta}z)=e^{i\theta}g_w(z)$ *for all* $w \in U$, $z \in D$, $\theta \in \mathbb{R}$.

Proof. – Arrange the indexes of the family $\{\varphi_i\}$ in such a way that:

$$|\varphi_1(w_0)| < |\varphi_2(w_0)| < \ldots < |\varphi_n(w_0)|$$

and for all $n < j \leq k$, there exists $1 \leq i(j) \leq n$ satisfying:

$$|\varphi_j(w_0)| = |\varphi_{i(j)}(w_0)|.$$

Then

$$(\star) \qquad 0 < |\varphi_1(w)| < |\varphi_2(w)| < \ldots < |\varphi_n(w)| < 1$$

and

$$|\varphi_j(w)| = |\varphi_{i(j)}(w)|,$$

for every $w \in U$, $n<j \leqq k$. Write the functions φ_i in polar coordinates as:

$$\varphi_i(w)=(r_i(w), \theta_i(w)).$$

Let F_w: [0, 1] be the unique monotone continuous function such that $\log F_w$ is piecewise linear and $\log F_w(r_i(w_0))=\log r_i(w)$ for $1 \leqq i \leqq n$, $F_w(0)=0$, $F_w(1)=1$.

The existence of F_w follows from ($\star$).

Let G_w: $[0, 1] \rightarrow \mathbb{R}$ be the continuous piecewise linear function satisfying:

$$G_w(1)=0,$$

$$G_w(0)=0,$$

$$G_w(r_i(w_0))=\theta_i(w)-\theta_i(w_0).$$

Now define (in polar coordinates) $g_w(r, \theta)=(F_w(r), G_w(r)+\theta)$. Clearly every g_w is a homeomorphisms satisfying (*a*), (*c*) and (*d*). To prove (*b*) observe that in each annulus $\{(r, \theta) \mid r_i(w_0)<r<r_{i+1}(w_0)\}$ we can write:

$$\log F_w(r)=[r_{i+1}(w_0)-r_i(w_0)]^{-1}[(r-r_i(w_0))\log r_{i+1}(w)-(r-r_{i+1}(w_0))\log r_i(w)]$$

and:

$$\log G_w(r)=[r_{i+1}(w_0)-r_i(w_0)]^{-1}[(r-r_i(w_0))\,\theta_{i+1}(w)-(r-r_{i+1}(w_0))\,\theta_i(w)].$$

Clearly, for each fixed r, these are harmonic conjugate functions. Hence (*b*) is satisfied in all these annulus. A similar computation shows that it is also satisfied in the disk $r<r_i(w_0)$ and in the annulus $r_n(w_0)<r<1$, so the Lemma is proved.

Now take analytic functions $\varphi_i: W_0 \rightarrow D$, $i=1, \ldots, k$ such that $h_w(\varphi_i(w))$, $1 \leqq i \leqq k$, are the points where the f_w-forward orbits of critical points of f_w first hit S_1. If we show that these functions satisfy the hypothesis of Lemma IV.1 we are done. Restrict W_0 to grant that if $|\varphi_i(w_0)| \neq |\varphi_j(w_0)|$ for some $1 \leqq i<j \leqq n$, then $|\varphi_i(w)| \neq |\varphi_j(w_0)|$ for all $w \in W_0$. Now suppose that $|\varphi_i(w_0)|=|\varphi_j(w_0)|$ for some $1 \leqq i<j \leqq n$. Since $w_0 \in C(f)$ we must have $|\varphi_i(w)|=|\varphi_j(w)|$ for every $w \in W_0$. Hence φ_i/φ_j has constant modulus. Therefore it is constant i. e. $\varphi_i(w)=e^{i\alpha}\varphi_j(w)$ for all $w \in W_0$.

Now we shall show how to construct h_w on a set of type Λ_i. To simplify the notation set $i=1$ and let V a neighborhood of z_1 is defined a map $h: W_0 \times V \rightarrow \overline{\mathbb{C}}$ such that $\hat{h}_w$ is a local conjugacy between f_{w_0} and f_w. This map is given by Lemma III.5. We can take Λ_1

such that $\Lambda_1/f_{w_0}(\Lambda_1)$ is a union of sets whose boundaries are closed arcs: as in the Figure (*see* the discusion in the Introduction). Then every f_{w_0}-orbit intersects $U(w_0)=\Lambda_1 \setminus f_{w_0}(\Lambda_1)$ in at most one point, and, by the existence of a local conjugacy, every f_w-orbit intersects $U(w)=\hat{h}_w(U(w_0))$ in at most one point. Take the images by $\hat{h}_w^{-1}$ of the intersections of the forward f_w-orbits of critical points of f_w with $U(w)$.

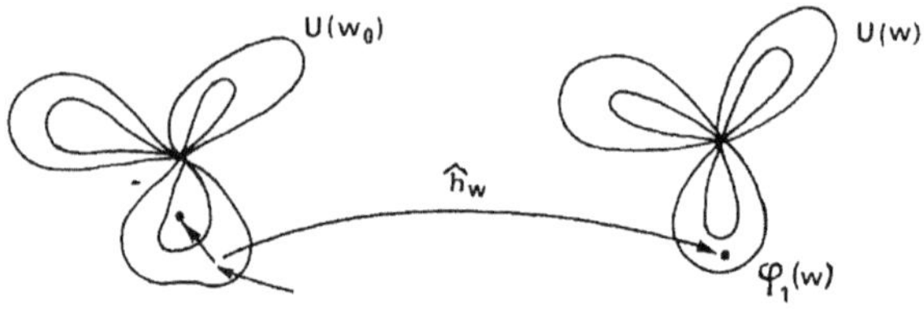

Describe these images by analytic functions $\varphi_1(w), \ldots, \varphi_m(w)$, $w \in W_0$. Restricting W_0 if necessary it is easy to construct a map $g_w: U(w_0) \circlearrowleft$, that is a homeomorphism for all $w \in W_0$, is the identity in the boundary of $U(w_0)$, depends analytically in w, is the identity for $w=w_0$ and maps $\varphi_i(w_0)$ in $\varphi_i(w)$, $1 \leqq i \leqq m$. Extend $g_w: U(w_0) \circlearrowleft$ to $g_n: \Lambda_1 \circlearrowleft$ using f_{w_0} i.e.:

$$g_w(f_{w_0}^n(z)) = f_{w_0}^n(g_w(z)) \quad \text{if} \quad n \geqq 0, \quad w \in W_0, \quad z \in U(w_0).$$

Now define $h_w: \Lambda_i \to \overline{\mathbb{C}}$ by $h_w = \hat{h}_w g_w \hat{h}_w^{-1}$. Clearly h_w satisfies conditions (*a*)-(*d*). As in the case of Siegel disks and Herman rings the role of the rearrangement map g_w is to make h_w satisfy condition (*d*).

To construct h_w on sets of type V_i, associated to an attractive fixed point, the construction is exactly the same replacing Lemma III.5 by Lemma III.3. Let $\hat{h}: W_0 \times V_i \to \overline{\mathbb{C}}$ be the parameterized linearization given by III.3. Set $\lambda = (f_{w_0})'(z_i)$. By composing $\hat{h}_w$ with a convenient map we can suppose that $\hat{h}_w$ is a conjugacy between $f_w: V_i \circlearrowleft$ and $D \ni z \to \lambda z \in D$, where D is a disk. Set $U = D \setminus \lambda D$. This set is a ring and we have analytic functions $\varphi_1(w), \ldots, \varphi_m(w)$ indicating the images by $\hat{h}_w$ of the points where forward f_w-orbits of critical points of f_w hit $V_i \setminus f_w(V_i)$. Take $g_w: U \circlearrowleft$ as before and extend it to $g_w: D \circlearrowleft$ by: $g_w(\lambda^n z) = \lambda^n g_w(z)$. Finally define: $h_w = \hat{h}_w^{-1} g_w h_{w_0}$.

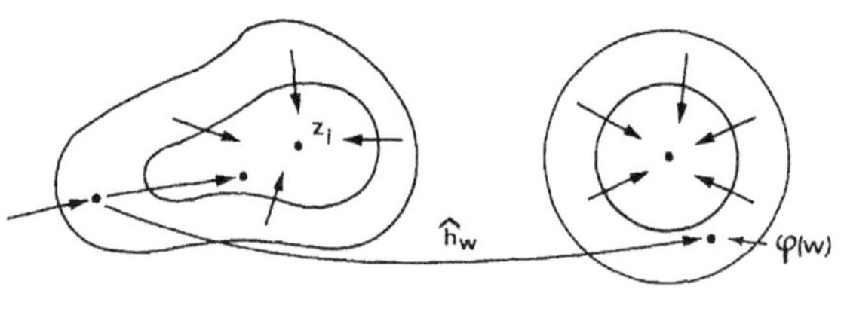

This map satisfies the required conditions. Once more observe that $\hat{h}_w^{-1} \hat{h}_{w_0}$ would satisfy (*a*), (*b*), (*c*) and (*e*), and that the role of g_w is to make it satisfy also (*d*). It remains to consider the case of super attracttive fixed points.

By Lemma III.4 there is a map $\hat{h}: W_0 \times V_i \to D$, where D is a disk centered at 0, such that each h_w conjugates $f_w: V_i \circlearrowleft$ with $D: z \to z^n \in D$. Set $U = D \setminus \{z^n | z \in D\}$. Let $\varphi_1(w), \ldots, \varphi_m(w)$ be analytic functions describing the images by h_w of the points of f_w intersect U. Let $g_w: U \circlearrowleft$ be a family of homeomorphisms of U, that are the identity in the boundary of U, depending analytically in w, and satisfying $g_w(\varphi_i(w_0)) = \varphi_i(w)$. Require also that g_w acts as a linear map in circles $z = \text{Const}$. Such family is given by Lemma IV.1. Now extend g_w to D by setting $g_w(z) = g_w(z_0)^{n^m}$ if $z \in D$, $z_0 \in U$ and $z_0^{n^m} = z$. The fact that g_w acts as a linear map on circles grants that this definition is independent of the root z_0 of z used. Now define $h_w = \hat{h}_w^{-1} g_w \hat{h}_{w_0}$.

To prove the converse property is sufficient to observe that a map h_w satisfying (i) maps the set of critical points of f_{w_0} onto the set of critical points of f_w, preserving their orders, and maps the foliations of superattractive regions, Herman rings and Siegel disks of f_{w_0} onto the corresponding foliations for f_w. Those properties immediately yield that $w_0 \in C(f)$.

V. – Proof of Theorem E

We start proving the following property:

LEMMA V.1. – *Suppose that f and g are analytic endomorphisms of $\overline{\mathbb{C}}$ such that there exists a quasi-conformal homeomorphism $h: \overline{\mathbb{C}} \circlearrowleft$ satisfying $gh = hf$ and analytic in the complement of $J(f)$. Then, if f has no invariant line fields in $J(f)$, h is analytic.*

Proof. – Associate to a. e. $z \in \overline{\mathbb{C}}$ the ellipse $C(z) = h(z)^{-1} C_0(h(z))$ where $C_0(h(z))$ is the unit circle of $T_{h(z)} \overline{\mathbb{C}}$. The relation $gh = hf$ implies:

$$(\star) \qquad f'(z) C(z) = C(f(z)),$$

for a. e. $z \in \overline{\mathbb{C}}$. If $J(f)$ has measure zero, the Lemma is proved because h is in this case 1-quasi-conformal, hence conformal. Then we can suppose that $J(f)$ has non zero measure. Let Σ be the set of points of $z \in J(f)$ where $C(z)$ is defined and is not a circle. Then $(\star)$ shows that $f^{-1}(\Sigma) = \Sigma$. If $z \in \Sigma$ define $E(z)$ as the one dimensional subspace of $T_z \overline{\mathbb{C}}$ containing the major axis of $C(z)$. Again, $(\star)$ proves that $f'(z) E(z) = E(f(z))$ for $z \in \Sigma$. Since f has no invariant line fields in $J(f)$, it follows that the measure of Σ is zero. This means that $C(z)$ is a circle for a. e. $z \in J(f)$. Outside $J(f)$, $C(z)$ is always a circle by the analyticity of h in the complement of $J(f)$. Therefore $C(z)$ is a circle for a. e. $z \in \overline{\mathbb{C}}$. This means that h is 1-quasi-conformal, therefore conformal as we wanted to prove.

Now suppose that $w_0 \in C(f)$ satisfies the hypothesis of Theorem E. First of all observe that the hypothesis $w_0 \in C(f)$ plus the non existence of Herman rings and non hyperbolic periodic points imply that $J(f_{w_0}) \neq \overline{\mathbb{C}}$ because by Theorem D, every f_w with w near w_0, is topologically equivalent to f_{w_0} *via* a quasi-conformal map. By the previous Lemma this means that these maps are analytic, thus implying that every f_w, with w near w_0, is analytically equivalent to f_{w_0} contradicting the hypothesis that the family is reduced. Then

$J(f_{w_0})=\overline{\mathbb{C}}$, and since there are no Herman rings or Siegel disks, there exist attracting cycles and superattracting cycles whose attractive sets cover the complement of $J(f_{w_0})$. Recall $x \approxeq y$ if $f_w^n(x) \approx f_w^m(y)$ for some $n \geqq 0$, $m \geqq 0$. Let $r_1(w)$ be the number of $(\approxeq)$-equivalence classes of critical points not contained in $J(f_w)$ and not eventually periodic. The next step of the proof is the construction of an analytic function $\Psi: W_1 \to M$ of constant rank, where $W_1 \subset W$ is open and M is a complex manifold with dimension $r_1(w_0)$, such that for every value $c \in \Psi(W_1)$ and every connected component S of $\Psi^{-1}(c)$, all the endomorphisms f_w with $w \in S$ are analytically conjugate. If such a function exists, then the fact that the family of endomorphisms that we are considering is reduced implies that every S must consist of at most one point. Hence $\dim W \leqq r_1(w_0)$. But on the other hand $\dim W \geqq r(w_0)$ by hypothesis. Then $r_1(w_0)=r(w_0)$. This means that there are no critical points in $J(f_{w_0})$, hence, as observed in the introduction, f_{w_0} satisfies Axiom A.

Therefore the proof of Theorem E is now reduced to construct the function Ψ. In fact we shall construct this function satisfying the *a priori* weaker property that all the endomorphisms f_w with w in a set S as above, are topologically equivalent *via* a conjugacy that is quasi-conformal and analytic outside $J(f_w)$. But then Lemma V.1 shows that the conjugacy is in fact analytic. To simplify the construction of Ψ, we shall suppose that every attractive periodic point of f_{w_0} is fixed. Then the same property holds for every w in W_0, if W_0 is small enough. Let $\varphi_i: W_0 \to \overline{\mathbb{C}}$, $i=1,\ldots,k$ be analytic functions describing the position of the attractive and superattractive fixed points $\varphi_1(w),\ldots,\varphi_k(w)$ of f_w, $w \in W_0$. Suppose that $\varphi_i(w_0)$ is superattractive for $i=1,\ldots,k_1$ and that $(f_{w_0})'(\varphi_i(w_0)) \neq 0$ for $k_1 < i \leqq k$. Define $\Psi_i: W_0 \to \overline{\mathbb{C}}$, $k_1 < i \leqq k$, by $\Psi_i(w)=(f_w)'(\varphi_i(w))$. Let $V_1,\ldots,V_k$ be neighborhoods of $\varphi_1(w_0),\ldots,\varphi_k(w_0)$ given by Lemma III.4. Assume that W_0 is so small that the maps $\hat{h}_w$ given by this Lemma are defined for $w \in W_0$. If $1 \leqq i \leqq k_1$, let n_1 be the number of $(\approxeq)$-equivalence classes of critical points of f_{w_0} whose orbits intersect V_i and don't coincide eventually with $\varphi_i(w_0)$. For each $1 \leqq i \leqq k_1$ take an analytic function $\hat{\Psi}_i: W_0 \to \mathbb{C}^{n_i}$ such that each coordinate is the image under $\hat{h}_w$ of a point in the forward f_w-orbit of a critical point of f_w and moreover different coordinates correspond to different $(\approxeq)$-equivalence classes of critical points. For $k_1 < i \leqq k$, define n_i as before and let M_i be the quotient of $(\mathbb{C}-\{0\})^{n_i}$ by the action of $\mathbb{C}-\{0\}$ given by $(z_1,\ldots,z_{n_i})=(\lambda z_1,\ldots,\lambda z_{n_i})$. Define $\hat{\Psi}_i: W_0 \to M_i$ as an analytic function that to every $w \in W_0$ associates the element of M_i determined by the images under $\hat{h}_w$ of the forward f_w-orbits of critical points. Finally, define:

$$\Psi: \quad W_0 \to \mathbb{C}^{n_1} \times \ldots \times \mathbb{C}^{n_{k_1}} \times M_{k_1+1} \times \ldots \times M_k \times \mathbb{C} \times \ldots \times \mathbb{C},$$

by $\Psi(w)=(\hat{\Psi}_1(w),\ldots,\hat{\Psi}_k(w),\Psi_{n_k+1}(w),\ldots,\Psi_k(w))$. The manifold at right has dimension:

$$n_1+\ldots+n_{k_1}+(n_{k_1+1}-1)+\ldots+(n_k-1)+(k-k_1)=r_1(w_0),$$

as we wished. Now restrict Ψ to $W_1 \subset W_0$ where it has maximal rank. Let S be a connected component of $\Psi^{-1}(c)$ where $c \in \Psi(W_1)$. Fix some $w_1 \in S$. Let Λ be the union of the attractive sets of the attractive fixed points of f_w minus the orbits of the eventually fixed points. In a neighborhood W_2 of w_1 there exist, for every $w \in W_2$, an analytic conjugacy

$h_w: \Lambda \to h_w(\Lambda)$ between f_{w_0}/Λ and $f_w/h_w(\Lambda)$, depending analytically in w, constructed by the same method used in Section IV to prove Theorem D i. e., starting with local conjugacies nearby the attractive fixed points that map f_{w_0}-orbits of critical points of f_{w_0} onto f_w-orbits of critical points of f_w, and then pulling back these conjugacies by f_{w_0} and f_w, in order to fill Λ. The new fact we have now is that w varies in S. Then the property $\Psi(w)=c=\Psi(w_0)$ makes it possible to take these local conjugacies to be analytic because now that there are only attractive and superattractive fixed points to consider and the rearrangement functions g_w can be taken as linear maps depending in w. Then $h_w: \Lambda \to h_w(\Lambda)$ is analytic. The extension of h_w to $\overline{\mathbb{C}}$ (granted as before by the λ-Lemma) is (also by the λ-Lemma) quasi-conformal and analytic in Λ. Since Λ is $\overline{\mathbb{C}}-\mathrm{J}(f_{w_0})$ minus a discrete set of points, it follows that h_w is analytic in $\Lambda-\mathrm{J}(f_{w_0})$ as we wished.

REFERENCES

[1] R. Bowen, *Hausdorff Dimension of Quasi Circles* (*Publications Mathematiques I.H.E S.*, Vol 50, 1979, pp. 11-25)

[2] C. Camacho, *On the Local Structure of Conformal Mappings and Holomorphic Vector Fields in* C^2 (*Astérisque*, Vol. 59-60, 1978, pp. 83-94).

[3] M Fatou, *Sur les équations fonctionnelles* (*Bull. Soc Math. Fr.*, Vol 47, 1919, pp. 161-271, Vol. 48, 1920, pp. 33-94 and pp. 208-314).

[4] S. Lattes, *Sur l'itération des substitutions rationnelles et les fonctions de Poincaré* (*C.R. Acad. Sc.*, Paris, T. 166, 1918, pp. 26-28).

[5] M. V. Jacobson, *On the Question of the Classification of Rational Maps of Riemann Sphere* (*Uspehi Mat. Nauk.*, Vol. 28, 1973, pp. 247-248).

[6] G. Julia, *Sur l'itération des fonctions rationnelles* (*Journal de Math. Pure et Appliquée*, Vol. 8, 1918, pp. 47-245).

[7] D. Ruelle, *Analytic Repellors, Ergodic Theory and Dynamical Systems* (to appear).

[8] C. L. Siegel, *Iteration of Analytic Functions* (*Annals of Mathematics*, Vol. 43, 1942, pp. 607-612).

[9] D. Sullivan, *Quasi Conformal Homeomorphisms and Dynamics II* (to appear).

[10] D. Sullivan, *Conformal Dynamical Systems* [*Proc. of the International Symposium on Dynamical Systems*, Rio de Janeiro, 1981 (to appear)].

[11] D. Sullivan, *Density at Infinity* (*Publications Mathématiques I.H.E.S.*, Vol. 50, 1979, pp. 419-450).

[12] D. Sullivan, *Discrete Conformal Groups and Measurable Dynamics* (*Bull. of the A.M.S.*, Vol. 6, 1982)

[13] D. Sullivan, *Quasi Conformal Homeomorphisms and Dynamics III* (to appear).

[14] D. Sullivan and W. Thurston, *Extending Holomorphic Motions* (to appear).

(Manuscrit reçu le 27 mai 1982,
révisé le 22 mars 1983).

R. Mañé, P. Sad and D. Sullivan,
I.M.P A.,
Estrada Dona Castorina 110,
22460 Rio de Janeiro, RJ
Brasil.

LYAPOUNOV EXPONENTS AND STABLE MANIFOLDS FOR COMPACT TRANSFORMATIONS

by

Ricardo Mañé

Instituto de Matemática Pura e Aplicada (IMPA)
Rio de Janeiro, Brazil

Introduction

The aim of this work is the extension of the Lyapounov exponents and stable manifolds theories developed by Oseledec [4], Pesin [5] and Ruelle [6], [7], to compact non linear maps of infinite dimensional Banach spaces. The interest of such extension is its possibility of application to improve the understanding of the dynamics associated to parabolic semilinear equations or retarded functional differential equations, that are described by non linear maps of open sets in Banach spaces with the remarkable property of having compact derivative at every point of its domain. Such feature will be the main hypothesis in all the theorems we shall prove. Unfortunately, for the best and more precise results, hypothesis like the injectivity of these derivatives, density of its ranges and certain injectivity hypothesis of the transformation itself are needed. These kind of hypothesis, quite reasonable for transformations arising from semilinear parabolic equations, are not always satisfied, or at least its validity is difficult to decide, in the case of retarted functional differential equations. The reader will find the basic information about how these equations generate compact transformations in Hale [1] (for retarded functional differential equations) and Henry [2] (for semilinear parabolic equations).

Originally published in: *Geometric dynamics (Rio de Janeiro, 1981), 522–577, Lecture Notes in Math. 1007*,
Reprint by Springer International Publishing 2017

In this section we shall only give a brief description of the main results. Precise complete statements will be given in the next section.

Let E be a Banach space, $U \subset E$ an open set and $f: U \to E$ a C^1 map with compact derivatives. Let $K \subset E$ be a compact set such that $f(K) \subset K$. We want to study the asymptotic behaviour of the linear maps $D_x f^n$ for $x \in K$ and $n \geq 0$. Following Oseledec we introduce the concept of Lyapounov exponents: A Lyapounov exponent of f at x is a number of the form:

$$\limsup_{n\to+\infty} \frac{1}{n} \log\|(D_x f^n)v\| \tag{1}$$

where $0 \neq v \in E$. We shall agree in setting $\log 0 = -\infty$. Now observe that when x is a fixed point there are at most a countable number of Lyapounov exponents of f at x and that they are the limits (and not only the superior limits) of expressions like (1). All these properties follow from the spectral theorem for compact linear transformations (applied to $(D_x f)$) and the Lyapounov exponents at x are precisely the logaritms of the absolute values of the numbers in the spectrum of $D_x f$. Such nice property cannot be expected to hold for any $x \in K$ (as we shal see in a while). The point is that it does hold for a set of points $x \in K$ that is large enough from the probabilistic (or ergodic) view point. The exact statement of this properties require the introduction of the concept of total probability set: We say that a set $K_o \subset K$ has total probability if every f-invariant probability μ on the Borel sets of K (i.e. such that $\mu(f^{-1}(A)) = \mu(A)$ for every Borel set $A \subset K$) satisfies $\mu(K_o) = 1$. The dynamical meaning of this property is the following: If K_o has total probability then:

$$\limsup_{n\to+\infty} \frac{1}{n} \#\{0 \leq i \leq n \mid d(f^i(x), K_o) > \epsilon\} = 0$$

for every $x \in K$ and $\epsilon > 0$. In other words if every point of K

spends most of the time nearby K_o. It is well know that the set of recurrent points of K (i.e. points $x \in K$ that satisfy $\liminf_{n\to+\infty} d(f^n(x),x) = 0$) has total probability. Periodic points are contained in every total probability set and Birkhoff's ergodic theorem states that there exists a total probability set such that for any x in it, the orbital average:

$$\lim_{n\to+\infty} \frac{1}{n} \sum_{j=0}^{n} \varphi(f^n(x))$$

exists for any continuous $\varphi: K \to R$.

The aim of this work is to prove that there exists a total probability set $K_o \subset K$ such that for all $x \in K_o$ the limit

$$\lim_{n\to+\infty} \frac{1}{n} \log\|(D_x f^n)v\|$$

exists for every $v \neq 0$ (it can be $-\infty$) and can take only a countable or finite set of values $\lambda_1(x) > \lambda_2(x) > \ldots$ in $[-\infty,+\infty)$. Moreover there exist subspaces

$$E = G_1(x) \supset G_2(x) \supset \ldots$$

such that $v \neq 0$ satisfies $\lim_{n\to+\infty} n^{-1} \log\|(D_x f^n)v\| = \lambda_j(x)$ if and only if $v \in G_j(x)\setminus G_{j+1}(x)$. When f/K and $D_x f$ are <u>injective</u> for all $x \in K$ this result can be considerably improved as follows: There exist subspaces $E_1(x), E_2(x), \ldots, F_1(x) \subset F_2(x) \subset \ldots$ such that:

$$E = E_1(x) \oplus \ldots \oplus E_j(x) \oplus F_j(x)$$

for all j and

$$\lim_{n\to\pm\infty} \frac{1}{n} \log\|(D_x f^n)v\| = \lambda_j(x)$$

for all j and $0 \neq v \in E_j(x)$. Formally the expression $(D_x f^n)v$ doesn't make sense for $n < 0$ because compact linear maps are never invertible, but we shall agree that to write $(D_x f^n)v$ with $n < 0$ means that there exists w satisfying $(D_{f^n(x)} f^{-n})w = v$ and we define $(D_x f^n)v = w$. The injectivity of $D_x f$ for all x grants the uniqueness of such w.

Once having a Lyapounov exponents theory is natural to develop a stable manifold theory. If $x \in K$ and $\lambda \geq 0$ define $W^s_\lambda(x)$ as the set of points $y \in \bigcap_{n\geq 0} f^{-n}(K)$ (i.e. such that $f^n(y)$ is well defined for all $n \geq 0$) satisfying

$$\limsup_{n\to+\infty} \frac{1}{n} \log\|f^n(y) - f^n(x)\| < -\lambda$$

when f/K is a homeomorphism and $D_x f$ is injective and has dense range for all $x \in K$ we shall prove that if f is $C^{k,\gamma}$ (i.e. class C^k with k-th derivative Hölder continuous with Hölder constant γ) then for a total probability set $K_1 \subset K_o$ of points $x \in K$ the set $W^s_\lambda(x)$ is a $C^{k,\gamma}$ finite codimensional immersed submanifold of E for all $\lambda \geq 0$. This means that $W^s_\lambda(x)$ is an increasing union of $C^{k,\gamma}$ submanifolds of E with the same finite codimension. Moreover it will be proved that if $\lambda' > \lambda'' \geq 0$ then $W^s_{\lambda'}(x) = W^s_{\lambda''}(x)$ if and only if $(-\lambda', \lambda'']$ doesn't contain any Lyapounov exponent of f at x. The codimension of $W^s_\lambda(x)$ can vary with λ and x but we shall prove that for each λ is uniformely bounded on $x \in K_1$.

Similar results hold for the unstable manifolds $W^u_\lambda(x)$ defined, when f is injective and $\lambda \geq 0$, as the set of points $y \in \bigcap_{n\geq 0} f^n(U)$ such that:

$$\limsup_{n\to+\infty} \frac{1}{n} \log\|f^{-n}(y) - f^{-n}(x)\| < -\lambda.$$

If $D_x f$ is injective for all $x \in U$ and f is $C^{k,\gamma}$ then there exists a total probability set $K_2 \subset K_o$ such that if $x \in K_2$ then $W^u_\lambda(x)$ is a finite dimensional $C^{k,\gamma}$ immerse submanifold for all $\lambda \geq 0$ and given $\lambda' > \lambda'' \geq 0$, the equality $W^u_{\lambda'}(x) = W^u_{\lambda''}(x)$ holds if and only if $[-\lambda', -\lambda')$ doesn't contain Lyapounov exponents of f at x.

The methods we shall use to prove the results on Lyapounov exponents are radically different of those used by Oseledec in [3] or Ruelle in [6] and [7]. The stable manifold theory follows more conventional techniques. Essentially we shall use a suitable adaptation of the implicit function method of constructing invariant manifolds (or Perron method). In the works of Pesin [4] and Ruelle [6] the graph transform (or Hadamard) method is used.

I - Basic definitions and statement of the theorems

Let K be a topological space. A Banach vector bundle on K is a triad $(F,\pi,\|\cdot\|)$ where F is a topological space, $\pi: F \to K$ a continuous map (called the canonical projection), $\|\cdot\|: F \to \mathbb{R}$ a continuous function and for every $x \in K$ the set $F_x = \pi^{-1}(x)$ is endowed with a vector space structure such that $\|\cdot\|/F_x$ is a Banach norm on F_x and the topology induced by this norm coincides with the relative topology of F_x. Moreover we require that for all $x \in K$ there exist a neighbourhood U and a homeomorphism $\varphi: F_x \times U \to \pi^{-1}(U)$ such that for all $p \in U$ the map $\varphi(\cdot,p): F_x \to \pi^{-1}(U)$ is an isomorphism between the Banach spaces F_x and $\pi^{-1}(p)$. The set F_x is called the fiber of x. As usual we shall shorten all this just saying that F is a Banach vector bundle on K, the maps π and $\|\cdot\|$ assumed implicit in the definition.
If F_i, $i = 1,2$ are Banach bundles on K_i, $i = 1,2$, we say that a continuous map $L: F_1 \to F_2$ is a homomorphism convering a continuous map $\varphi: K_1 \to K_2$ if $\pi_2 L = \varphi\pi_1$ (where π_i, $i = 1,2$, are the canonical projection associated to F_i, $i = 1,2$) and $L/F_{1,x}$ is bounded linear map of $F_{1,x}$ into $F_{2,\varphi(x)}$ for all $x \in K_1$.
Finally we shall say that a subset $F \subset F_o$ is a subbundle if $(F_o,\|\cdot\|/F_o,\pi/F_o)$ is a Banach vector bundle on K. If K_o is a subset of K we shall denote F/K_o the Banach vector bundle isomorphism given by $(\pi^{-1}(K_o),\ \pi/\pi^{-1}(K_o),\ \|\cdot\|/\pi^{-1}(K_o))$.

In what follows we shall consider a Banach vector bundle F on a compact metric space K and a homomorphism $L: F \circlearrowleft$ covering a homeomorphism $\varphi: K \circlearrowleft$ such that L/F_x is injective for all $x \in K$. If $n > 0$ and $v \in F$ we define $L^{-n}v = w$ if $L^n v = w$. Obviously w doesn't necessarily exist but we shall adopt the convention that the expression $L^{-n}v$ carries the claim that $L^{-n}v$ exists. The injectivity of L/F_x for all x grants the uniqueness of $L^{-n}v$.

Now we are ready to introduce the central definition of this paper:

Definition I.1. We say that $x \in K$ is a regular point of L if satisfies one of the following hypothesis:

I) $\lim_{n\to+\infty} \frac{1}{n} \log\|L^n/F_x\| = -\infty$

II) There exists $k \in \mathbb{Z}^+$ and a splitting

$$F_x = E_1(x) \oplus \ldots \oplus E_k(x) \oplus F_\infty(x)$$

and numbers $\lambda_1(x) > \ldots > \lambda_k(x)$ such that:

a) $\dim E_j(x) < +\infty$ for all $1 \le j \le k$

b) $\lim_{n\to+\infty} \frac{1}{n} \log\|L^n v\| = \lambda_j(x)$ for all $0 \ne v \in E_j(x)$, $1 \le j \le k$.

c) $\lim_{n\to+\infty} \frac{1}{n} \log\|L^n/F_\infty(x)\| = -\infty$.

d) If $0 \ne v \in F_\infty(x)$ and $L^{-n}v$ exists for all $n \ge 0$ then:

$$\lim_{n\to+\infty} \frac{1}{n} \log\|L^{-n}v\| = +\infty.$$

III) There exist subspaces $E_j(x)$, $F_j(x)$, $j = 1,2,\ldots$ and numbers $\lambda_1(x) > \lambda_2(x) > \ldots$ such that:

a) $\dim E_j(x) < +\infty$ for all j

b) $\lim_{j\to+\infty} \lambda_j(x) = -\infty$

c) $E_1(x) \oplus \ldots \oplus E_j(x) \oplus F_j(x) = F_x$ for all j

d) For all j and $0 \neq v \in E_j(x)$:

$$\lim_{n\to\pm\infty} \frac{1}{n} \|L^n v\| = \lambda_j(x)$$

e) For all j:

$$\lim_{n\to+\infty} \frac{1}{n} \log\|L^n/F_j(x)\| = \lambda_{j+1}(x)$$

f) For all j and $0 \neq v \in F_j(x)$ such that $L^{-n}v$ exists for all $n \geq 0$:

$$\liminf_{n\to+\infty} \frac{1}{n} \log\|L^{-n}v\| \geq -\lambda_{j+1}(x).$$

Let us observe some elementary useful properties of regular points that can be deduced directly from the definition. Denote Σ_i, $i = 1,2,3$ the sets of regular points satisfying (I), (II) or (III) respectively. Then these sets are disjoint and $\varphi(\Sigma_i) = \Sigma_i$, $i = 1,2,3$. Moreover if $x \in \Sigma_2 \cup \Sigma_3$ it is easy to see that the subspaces $E_j(x)$, $F_j(x)$ and the numbers $\lambda_1(x) > \ldots$ are unique and satisfy:

$$L(E_j(x)) = E_j(\varphi(x))$$

$$L^{-1}(F_j(x)) = F_j(\varphi^{-1}(x))$$

$$L^{-1}(F_\infty(x)) = F_\infty(\varphi^{-1}(x))$$

$$\lambda_j(\varphi(x)) = \lambda_j(x).$$

If $x \in \Sigma_3$ and we <u>define</u> $F_\infty(x) = \bigcap_j F_j(x)$ (that can be $\{0\}$) then the third equality also holds for x in Σ_3. If x is a regular point we say that the Lyapounov exponents of L at x are the numbers $\lambda_j(x)$ if $x \in \Sigma_2 \cup \Sigma_3$, plus $-\infty$ if either $x \in \Sigma_1 \cup \Sigma_2$ or $x \in \Sigma_3$ and $F_\infty(x) \neq \{0\}$. In other words the Lyapounov exponents of L at a regular point $x \in K$ are the possible values of

$$\limsup_{n\to+\infty} \frac{1}{n} \log\|L^n v\|$$

for $0 \neq v \in F_x$. Denote $\Sigma(L) = \Sigma_1 \cup \Sigma_2 \cup \Sigma_3$.

<u>Theorem A. If</u> L/F_x <u>is compact for all</u> $x \in K$ <u>then</u> $\Sigma(L)$ <u>is a total probability Borel set</u>.

The next theorem examines the dependence of the Lyapounov exponents and the spaces $E_j(x)$, $F_j(x)$ and $F_\infty(x)$ on $x \in \Sigma(L)$. Let $G^+(n)$, $G^-(n)$ denote the Grassmannian bundles of n-dimensional and closed n-codimensional subspaces of the fibers of F i.e. $G^+(n)$ (resp. $G^-(n)$) is the set of n-dimensional subspaces (resp. closed n-codimensional subspaces) of the fibers of F endowed with the unique metrizable topology such that a sequence $E_n \subset F_{x_n}$ converges to $E \subset F_x$ if and only if $x_n \to x$ and there exists a neighborhood U of x, a canonical chart $\varphi: F_x \times U \to \pi^{-1}(U)$ and a splitting $F_x = E \oplus G$ such that for large values of n the subspace $\varphi^{-1}(E_n)$ is the graph of a linear map $T_n: E \to G$ and $\lim_{n\to+\infty} \|T_n\| = 0$. Let $G_2(n_1,\ldots,n_k)$ be the set of points in Σ_2 with k finite Lyapounov exponents and $\dim E_j(x) = n_j$ for $1 \le j \le k$. Let $G_3(n_1,\ldots,n_k)$ be the set of points in Σ_3 with $\dim E_j(x) = n_j$ for $1 \le j \le k$. When X and Y are topological spaces we say that a map $f: X \to Y$ is <u>strongly measurable</u> if there exist Borel sets $X_n \subset X$, $n = 1,2,3,\ldots$ such that $\bigcup_n X_n = X$ and f/X_n is continuous for very n. Obviously strongly measurable maps are measurable but the converse property is false. (The reader can easily verify that the function $f: [0,1] \circlearrowleft$ defined as $f(t) = 0$ when t is rational and $f(t) = 1$ if t is irrational is not strongly measurable).

<u>Theorem B. For every set of positive integers</u> $n_1,\ldots,n_k$ <u>the sets</u> $G_i(n_1,\ldots,n_k)$ <u>are Borel sets and the functions</u>:

$$G_i(n_1,\ldots,n_k) \ni x \mapsto E_j(x) \in G^+(n_j) \qquad i = 2,3, \qquad j = 1,\ldots,k$$

$$G_2(n_1,\ldots,n_k) \ni x \mapsto F_\infty(x) \in G^-(\sum_j n_j)$$

$$F_3(n_1,\ldots,n_k) \ni x \mapsto F_k(x) \in G^-(\sum_j n_j)$$

<u>are strongly measurable.</u> <u>The functions</u>

$$G_i(n_1,\dots,n_k) \ni x \mapsto \lambda_j(x) \qquad j = 1,\dots,k$$

<u>are measurable.</u>

As a corollary to this theorem follows that $\Sigma(L)$ is a Borel set. This is proved observing that Σ_1 is Borel (because is the preimage of $-\infty$ under the measurable function $K \ni x \mapsto \lim_{n\to+\infty} n^{-1} \log\|L^n/F\|$) and

$$\Sigma_2 = \bigcup G_2(n_1,\dots,n_k)$$

$$\Sigma_3 = \bigcup G_3(n_1,\dots,n_k)$$

where the unions are taken over all k-uples of positive integers and $k \in \mathbb{Z}^+$. Denote $\Sigma_2^{(k)}$ the set of points in Σ_2 with exactly k finite Lyapounov exponents. Since the dependence of the subspaces $E_j(x)$, $F_j(x)$ on x is not continuous, the "angles" between these subspaces can be very small.
To study this problem we consider for a point $x \in \Sigma_2^{(k)} \cup \Sigma_3$ the projection $\pi_k(x)\colon F_x \to F_k(x)$ (if $x \in \Sigma_3$) or $\pi_k(x)\colon F_x \to F_\infty(x)$ (if $x \in \Sigma_3$) associated to the splitting $F_x = E_1(x) \oplus\dots\oplus E_k(x) \oplus \oplus F_k(x)$ (if $x \in \Sigma_3$) or to the splitting $F_x = E_1(x) \oplus\dots\oplus E_k(x) \oplus \oplus F_\infty(x)$ (if $x \in \Sigma_2^{(k)}$).

<u>Theorem C.</u> <u>The equality:</u>

$$\lim_{n\to\pm\infty} \frac{1}{n} \log\|\pi_k(\varphi^n(x)\| = 0$$

<u>holds for a total probability set in</u> $\bar{\Sigma}_2^{(k)} \cup \bar{\Sigma}_3$.

Now we shall consider the case of a vector bundle homomorphism $L\colon F \circlearrowleft$ covering a continuous map $\varphi\colon K \circlearrowleft$ <u>without any assumption on the injectivity of</u> φ <u>or the maps</u> L/F_x. <u>We still assume that</u> L/F_x <u>is compact for every</u> $x \in K$. We say that $x \in K$ is <u>almost regular</u> if there exist a countable or finite set $\lambda_1(x)>\lambda_2(x)>\dots\subset[-\infty,+\infty)$

with $\inf \lambda_j(x) = -\infty$ and a family of subspaces $F_x = F_1(x) \supset F_2(x) \supset \ldots$ with finite condimensions and such that

$$\lim_{n\to+\infty} \frac{1}{n} \log\|L^n v\| \leq \lambda_j(x)$$

for every j and $0 \neq v \in F_j(x)$ and

$$\lim_{n\to+\infty} \frac{1}{n} \log\|L^n v\| = \lambda_j(x)$$

if $0 \neq v \in F_j(x)\setminus F_{j+1}(x)$. It is easy to see that the numbers $\lambda_1(x) > \lambda_2(x) > \ldots$ and the subspaces $F_1(x) \supset F_2(x) \supset \ldots$ are unique. The numbers $\lambda_1(x) > \ldots$ are called the Lyapounov exponents of L at x and we shall agree in including $-\infty$ as a Lyapounov exponent when $\bigcap_{j\geq 1} F_j(x) \neq \{0\}$. As before it follows that the Lyapounov exponents of L at a point $x \in \Lambda$ are possible values of (1). Denote $\Lambda(L)$ the set of almost regular points of L.

Theorem D. $\Lambda(L)$ is a total probability set.

The following result shows that the codimension of the subspace $F_j(x)$ are bounded by a constant determined by $\lambda_j(x)$.

Theorem E. For all $\lambda \geq 0$ there exists $N(\lambda)$ such that if $x \in \Lambda(L)$ and $\lambda_j(x) \geq -\lambda$ then codim $F_j(x) \leq N(\lambda)$.

Finally let us give the precise statements of the theorems on stable and unstable manifolds that we sketched in the introduction. Let E be a Banach space $U \subset E$ an open set and $f\colon U \to E$ a C^1 map. If $x \in \bigcap_{n\geq 0} f^{-n}(U)$ and $\lambda \geq 0$ we define:

$$W^s_\lambda(s) = \{y \in \bigcap_{n\geq 0} f^{-n}(U) \mid \limsup_{n\to+\infty} \frac{1}{n} \log\|f^n(y)-f^n(x)\| < -\lambda\}.$$

If f is injective and $x \in \bigcap_{n\geq 0} f^n(U)$ we define:

$$W^u_\lambda(x) = \{y \in \bigcap_{n\geq 0} f^n(U) \mid \limsup_{n\to+\infty} \frac{1}{n} \log\|f^{-n}(y)-f^{-n}(x)\| < -\lambda\}.$$

Before stating the results let us briefly recall the definitions of $C^{r,\gamma}$ maps and submanifolds. If E_1, E_2 are Banach spaces and $V \subset E_1$ is an open set we say that $f\colon V \to E_2$ is a $C^{r,\gamma}$ map (or a Hölder C^r map with Hölder constant γ) if it is C^r and there exists $A > 0$ such that $\|(D_x^r f) - (D_x^r f)\| \le A\|x-y\|^{\gamma}$ for all x, y in V. We say that a subset $M \subset E_1$ is a $C^{r,\gamma}$ submanifold if it is a C^1 submanifold and every point $x \in M$ has a neighbourhood W such that $W \cap M$ is the image of $\{v \in T_xM \mid \|v\| < 1\}$ by a $C^{r,\gamma}$ homeomorphism. We say that $N \subset E_1$ <u>is an immerse</u> $C^{r,\gamma}$ <u>submanifold if it is the union of a monotone sequence</u> $N_1 \subset N_2 \subset \dots$ <u>of</u> $C^{r,\gamma}$ <u>submanifolds such that</u> N_j <u>is open in</u> N_{j+1} <u>for all</u> $j \ge 1$. In this case we say that codim $N = n$ (resp. dim $N = n$) if codim $N_j = n$ (resp. dim $N_j = n$) for all $j \ge 1$.

<u>Theorem F</u>. Let $K \subset U$ be a compact set such that $f(K) = K$. Then if f is $C^{r,\gamma}$ and D_xf <u>is injective and compact for all</u> $x \in K$ <u>and has dense range for all</u> $x \in U$ <u>there exists a total probability set of points</u> $\Gamma \subset K$ <u>contained in the set of regular points of</u> $Df\colon E\times K \supset$ (where $Df\colon E\times K \supset$ is the homomorphism of the Banach vector bundle $E\times K$ defined by $(x,v) \to (f(x), (D_xf)v))$ <u>such that the following properties hold for every</u> $x \in \Gamma$:

a) For every $\lambda \ge 0$ $W_\lambda^s(x)$ is a finite codimensional immerse $C^{r,\gamma}$ submanifold.

b) For all $\lambda' > \lambda'' \ge 0$, $W_{\lambda'}^s(x) = W_{\lambda''}^s(x)$ if and only if $(-\lambda', -\lambda'']$ doesn't contain any Lyapounov exponent of Df at x.

c) If $\lambda_1(x) > \lambda_2(x) > \dots$ are the Lyapounov exponents of Df at x and $F_j(x)$, $j = 1,2,\dots$ the subspaces given in the definition of regular points, then $T_xW_\lambda^s(x) = F_{j_0}(x)$ where $j_0 = \min\{j/\lambda_j(x) < -\lambda\}$.

Finally we state the version for unstable manifolds of Theorem F.

Theorem F. If f is $C^{r,\gamma}$ and $K \subset U$ is a compact set such that $f(K) = K$, then if f is injective and $D_x f$ is compact for all $x \in U$, there exists a total probability set Γ^u contained in the set of regular points of $Df: E\times K \circlearrowleft$ such that the following properties hold for every $x \in \Gamma^u$:

a) For every $\lambda \geq 0$, $W^u_\lambda(x)$ is a finite dimensional immerse $C^{r,\gamma}$ submanifold.

b) For all $\lambda' > \lambda'' \geq 0$, $W^u_{\lambda'}(x) = W^u_{\lambda''}(x)$ if and only if $(-\lambda',\lambda'']$ doesn't contain any Lyapounov exponent of Df at x.

c) If $\lambda_1(x) > \lambda_2(x) > \ldots$ are the Lyapounov exponents at x and $E_1(x), E_2(x), \ldots$ are the subspaces given by the definition of regular points, then

$$TW^u_\lambda(x) = \bigoplus_{\lambda_j(x)>\lambda} E_j(x).$$

II - Some technical lemmas on compact homomorphisms.

Let K and H be compact metric spaces, F and G Banach vector bundles on K and H respectively and $L: F \to G$ a homomorphism covering a continuous map $\varphi: K \to G$. Suppose that L/F_x is compact for every $x \in K$.

Lemma II.1. $L(S)$ is relatively compact if $S \subset F$ is bounded.

Proof. If $v_n \in F_{x_n}$, $n \in \mathbb{Z}^+$ is a sequence in S we can assume that $x_n \to x \in K$. Since all our arguments will involve only fibers at points of the convergent sequences x_n and $\varphi(x_n)$ we can suppose that F and G are trivial bundles, say $F = F_o \times K$, $G = G_o \times H$. Then, since $L_x = L/F_o \times \{x\}$ is compact we can assume that the sequence $L_x v_n$ converges to some $v \in G_o$. Therefore:

$$\|L_{x_n} v_n - v\| \leq \|L_{x_n} v_n - L_x v_n\| + \|L_x v_n - v\| \leq$$

$$\leq \|L_{x_n} - L_x\| \sup_n \|v_n\| + \|L_x v_n - v\|.$$

Since $L_{x_n} \to L_x$ and $L_x v_n \to v$ it follows that $L_{x_n} v_n \to v$.

Lemma II.2. Let $m \in \mathbb{Z}^+$ and $E_n \in G^+(m)$, $n \in \mathbb{Z}^+$ be a sequence such that there exists $c > \gamma$ satisfying $\|Lv\| \geq c\|v\|$ for all $v \in E_n$, $n \in \mathbb{Z}^+$. Then there exists a subsequence E_{n_j}, $j \in \mathbb{Z}^+$ such that $L(E_{n_j})$, $j \in \mathbb{Z}^+$, is convergent in $G^+(m)$.

Proof. Suppose that $E_n \subset F_{x_n}$ and that $x_n \to x$. Let $\{v_1^{(n)},\ldots,v_m^{(n)}\}$ be a basis of $L(E_n)$ with $\|v_j^{(n)}\| = 1$ for all j and n and such that if $S_{n,k}$ is the subspace spanned by $\{v_1^{(n)},\ldots,v_k^{(n)}\}$ then:

$$\|v_{k+1}^{(n)} - w\| \geq 1/2 \qquad (1)$$

for all $n \in \mathbb{Z}^+$, $1 \leq k < m$ and $w \in S_{n,k}$. Set $w_j^{(n)} = L^{-1}v_j^{(n)}$. Then $\sup\{\|w_j^{(n)}\| \mid 1 \leq j < m,\ n \subset Z^+\} \leq c^{-1}$. By Lemma II.1 we can assume that for every $1 \leq j \leq m$ the sequence $v_j^{(n)} = Lw_j^{(n)}$ converges to some $v_j \in G_x$. We claim that the set $\{v_1,\ldots,v_n\}$ is linearly independent. To see this we shall prove by induction that $\{v_1,\ldots,v_k\}$ is linearly independent and that if S_k is the space spanned by $\{v_1,\ldots,v_k\}$ then $\lim\limits_{n\to+\infty} S_{n,k} = S_k$. Clearly the first property holds for k=1. If it is true for $k = \ell$ then to show that holds for $k = \ell+1$, it is sufficient to observe that $v_{\ell+1} \notin S_{n,\ell}$ would imply by the induction hypothesis that $v_{\ell+1} = \lim w_n$ with $w_n \in S_{n,\ell}$. Then: $0 = \|v_{\ell+1}-v_{\ell+1}\| = \lim\limits_{n\to+\infty} \|v_{\ell+1}^{(n)}-w_n\|$. But since $w_n \in S_{n,\ell}$, property (1) implies that $\|v_{\ell+1}^{(n)} - w_n\| \geq 1/2$ for all n thus contradicting its convergence to 0. This complete the proof of the claim. Then, since $v_j = \lim\limits_{n\to+\infty} v_j^{(n)}$ for all $1 \leq j \leq m$ it follows that $L(E_n)$ converges to the subspace spanned by $\{v_1,\ldots,v_m\}$.

In the statement of the next lemma we shall use the following definition of angle between subspaces of a Banach space. Let S_1, S_2 be closed subspaces of a Banach space F_o such that $S_1 \cap S_2 = \{0\}$.

We define the angle $\alpha(S_1,S_2)$ by:

$$\alpha(S_1,S_2) = \inf\{\|\pi v\|/\|v\| \mid 0 \neq v \in S_2\}$$

where $\pi: F_o \to F_o/S_1$ is the canonical homomorphism.

Lemma II.3. Let $m \in \mathbb{Z}^+$, $\{x_n \mid n \in \mathbb{Z}^+\}$ a sequence in K and at each x_n a splitting $F_{x_n} = E_n \oplus T_n$ with $\dim E_n = m$. Suppose that there exists $c > 0$ satisfying

$$\|Lv\| \geq c\|v\|$$

$$\alpha(L(E_n),\overline{L(T_n)}) \geq c$$

for all $n \in \mathbb{Z}^+$, $v \in E_n$. Then the sequence $\{T_n \mid n \in \mathbb{Z}^+\} \subset G^-(m)$ has a convergent subsequence.

Proof. We can assume that $x_n \to x$ and (by Lemma I.2) that $L(E_n) \to S \subset G_{\varphi(x)}$. As in the proof of Lemma I.1 we shall suppose that the vector bundles F and G are trivial, say $F = F_o \times K$, $G = G_o \times H$. Take a closed subspace $S' \subset G_o$ such that $S \oplus S' = G_o$. Then if n is large, say $n \geq N$ there exists a closed subspace $S'_n \subset S'$ and a linear continuous map $A'_n: S'_n \to S$ such that:

$$\sup_n \|A'_n\| < +\infty$$

$$\overline{L(T_n)} = \{x + A'_n x \mid x \in S'_n\}.$$

We leave the proof of these properties to the reader just remarking that the subspaces S'_n are given by:

$$S'_n = \pi(L(T_n))$$

where π is the projection of G onto S associated to the splitting $S \oplus S' = G$ and the maps A'_n by:

$$A'_n x = -x + (\pi/\overline{L(T_n)})^{-1} x.$$

By the Hahn-Banach theorem there exist continuous linear maps $A_n: S' \to S$ such that

$$A_n/S'_n = A'_n$$

$$\sup\|A_n\| < +\infty .$$

Set $R_n = \{x + A_n x \mid x \in S'\}$. We claim that

$$L_{x_n}^{-1}(R_n) = T_n$$

for large values of n. Obviously $L_{x_n}^{-1}(R_n) \supset T_n$. If $L_{x_n}^{-1}(R_n) \not\subset T_n$ there exists $v \notin T_n$ such that $Lv_{x_n} \in R_n$. Write $v = v_1+v_2$ with $v_1 \in T_n$ and $v_2 \in E_n$. Since $v \notin T_n$ we must have $v_2 \neq 0$. Then $L_{x_n} v = L_{x_n} v_1 + L_{x_n} v_2$ and since $L_{x_n} v_1 \in L(T_n) \subset R_n$ and $L_{x_n} v \in R_n$ it follows that $L_{x_n} v_2 \in R_n$. Then $L_{x_n} v_2 \in R_n \cap L(E_n)$. But clearly $R_n \oplus S = G_o$ and since $L_{x_n}(E_n) \to S$ it follows that for large values of n $R_n \cap L_{x_n}(E_n) = \{0\}$. Then $L_{x_n} v_2 = 0$. But L_{x_n}/E_n is injective. Hence $v_2 = 0$ contradicting the property $v_2 \neq 0$ implied by $v \in T_n$. This completes the proof of the claim. Now define B_n: $L_{x_n}^{-1}(S') \to L_{x_n}^{-1}(S)$ by:

$$B_n = L_{x_n}^{-1} A_n(L_{x_n}/L_{x_n}^{-1}(S')).$$

It is easy to see that $L_{x_n}^{-1}(R_n) = \{x + B_n x \mid x \in L_{x_n}^{-1}(S')\}$. Then, for large values of n:

$$T_n = L_{x_n}^{-1}(R_n) = \{x + B_n x \mid x \in L_{x_n}^{-1}(S')\}.$$

But $L_{x_n}^{-1}(S') \to L_x^{-1}(S')$ and $L_{x_n}^{-1}(S) \to L_x^{-1}(S)$. Hence there exist continuous injective linear maps i_n: $L_x^{-1}(S') \to F_o$, j_n: $L_x^{-1}(S) \to F_o$, such that $i_n(L_x^{-1}(S')) = L_{x_n}^{-1}(S')$, $j_n(L_x^{-1}(S)) = L_{x_n}^{-1}(S)$ and if i: $L_x^{-1}(S') \to F_o$, j: $L_x^{-1}(S) \to F_o$ are the inclusion maps, $\lim_{n\to+\infty} \|i_n-i\| = \lim_{n\to+\infty} \|j_n-j\| = 0$. We claim that the sequence $j_n^{-1} B_n i_n$: $L_x^{-1}(S') \to L_x^{-1}(S)$ has a convergent subsequence. If the claim is true, the lemma is proved because the sequence of subspaces

$$T'_n = \{x + (j_n^{-1} B_n i_n)x \mid x \in L_x^{-1}(S')\}$$

converges in $G^-(m)$ and the property $\lim_{n\to+\infty} \|i_n - i\| = \lim_{n\to+\infty} \|j_n - j\| = 0$ implies that the distances between T_n and T'_n converge to zero. Hence the convergence of T'_n implies that of T_n. To prove the clai take the closed linear hull $S'' \subset S'$ of $\bigcup_{n>0} L_{x_n}(L_{x_n}^{-1}(S'))$. It is a Banach separable space because every L_{x_n} is compact. Then the uniformely bounded sequence of linear maps $j_n^{-1}L_{x_n}^{-1} A_n/S'' \to L_x^{-1}(S)$ has a pointwise convergent subsequence and, to simplify the notation, we shall assume that the sequence itself is convergent. But if B is the unitary ball of $L_x^{-1}(S')$ the set $\bigcup_{n>0} L_{x_n} i_n(B) = L(\bigcup_{n>0} i_n(B))$ is relatively compact by Lemma II.1. Then the sequence $j_n^{-1}L_{x_n}^{-1} A_n/\bigcup_{n>0} L_{x_n} i_n(B)$ converges uniformely. Therefore so it does the sequence $j_n^{-1}L_{x_n}^{-1} A_n L_{x_n} i_n/B$ and the claim is proved.

Lemma II.4. For every $c > 0$ there exists $N > 0$ such that if $x \in K$ and $v_i \in F_x$, $i = 1,\ldots,n$ are unitary vectors satisfying $\|L_x v_i - L_x v_j\| \geq c$ for all $1 \leq i \leq j \leq n$, then $n \leq N$.

Proof. If this property is false, for some $c > 0$ there exists a sequence of points $x_n \in K$ and in each F_{x_n} a set of unitary vectors $\{v_1^{(n)},\ldots,v_n^{(n)}\}$ such that

$$\|L_{x_n} v_j^{(n)} = L_{x_n} v_j^{(n)}\| \geq c$$

if $1 \leq i < j \leq n$. Suppose that $x_n \to x$ and as we did in the proofs of II.3 and II.1 that $F = F_o \times K$, $G = G_o \times K$. By II.1 we can suppose that for each j the sequence $L_{x_n} v_j^{(n)}$ converges to some w_j. Then $\|w_j - w_i\| \geq c$ if $i < j$. For each j take n_j satisfying:

$$\|L_{x_n} v_j^{(n_j)} - w_j\| \leq c/3.$$

By II.1 the sequence $L_{x_n} v_n^{(n_j)}$ must have a convergent subsequence. However this is impossible because:

$$\|Lv_k^{(n_k)} - Lv_\ell^{(n_\ell)}\| \geq \|w_k - w_\ell\| - \|Lv_k^{(n_k)} - w_k\| -$$
$$- \|Lv_\ell^{(n_\ell)} - w_\ell\| \geq c - \frac{1}{3}c - \frac{1}{3}c = \frac{1}{3}c$$

for all k and ℓ.

III - Proof of Theorems A and C

Let K, F and L be as in the statement of Theorem A. By the ergodic decomposition theorem in order to prove Theorems A and B it is sufficient to show that if μ is a φ-invariant ergodic probability on K then $\mu(\Sigma(L)) = 1$ and that if $\mu(\bar{\Sigma}_2^{(k)} \cup \bar{\Sigma}_3) \neq 0$ then the equality of Theorem B holds μ-a.e. (observe that by the ergodicity of μ the property $\mu(\bar{\Sigma}_2^{(k)} \cup \bar{\Sigma}_3) \neq 0$ implies, by the φ-invariance of $\bar{\Sigma}_2^{(k)}$ and $\bar{\Sigma}_3$ and Theorem A, that $\mu(\Sigma_2^{(k)}) = 1$ or $\mu(\Sigma_3) = 1$). Then fix any φ-invariant ergodic probability measure μ and we shall prove these properties. For this purpose we shall need measurable versions of the concepts of Banach vector bundles and homomorphisms. We say that G is a μ-measurable Banach vector bundle on K if there exists a sequence of compact sets K_n, $n \in \mathbb{Z}^+$, called a normal sequence, such that $\mu(\bigcup_n K_n) = 1$, a map $\pi: G \to K$ (the canonical projection) and a function $\|\cdot\|: G \to \mathbb{R}$ such that $(\pi/\pi^{-1}(K_n), \|\cdot\|/K_n)$ is a Banach vector bundle on K for all n. A map $T: G \circlearrowleft$ is a μ-measurable homomorphism covering a continuous map $\varphi: K \circlearrowleft$ if the normal sequence K_n can be taken such that for all $m \in \mathbb{Z}^+$ there exists n satisfying $\varphi(K_m) \subset K_n$ and $T/\pi^{-1}(K_n)$ is a vector bundle homomorphism of $(\pi^{-1}(K_m), \pi/\pi^{-1}(K_m), \|\cdot\|/\pi^{-1}(K_m))$ in $(\pi^{-1}(K_n), \pi/\pi^{-1}(K_n), \|\cdot\|/\pi^{-1}(K_n))$. To simplify the notation we define $G/K_m = (\pi^{-1}(K_m) \cdot \pi/\pi^{-1}(K_m), \|\cdot\|/\pi^{-1}(K_m))$.

Before going into the proof of the property $\mu(\Sigma(L)) = 1$ we shall prove some lemmas, in whose statements K will be a compact metric space, $\varphi: K \circlearrowleft$ a homeomorphism, μ a φ-invariant ergodic probability on the Borel σ-algebra of K, G a μ-measurable Banach vector bundle on K and $T: G \circlearrowleft$ a μ-measurable homomorphism covering φ such that T/G_x is injective and compact for a.e. $x \in K$

and $\sup\|T/G_x\| < \infty$. If $\lambda \in \mathbb{R}$ and $x \in K$ define $E_\lambda(x)$ as the set of vectors $v \in G_x$ such that $T^{-n}v$ exists for all $n \geq 0$ and $\limsup n^{-1} \log\|T^{-n}v\| \leq \lambda$.

Lemma III.1. For every $\lambda \in \mathbb{R}$, $\dim E_\lambda(x) < \infty$ for a.e. $x \in K$. Moreover the set $\bigcup_x E_\lambda(x)$ is a μ-measurable Banach vector bundle on K.

Define:

$$k(T) = -\limsup_{n\to+\infty} \frac{1}{n} \log\|T^n/G_x\|.$$

Since μ is ergodic and $\sup\|T/G_x\| < +\infty$ this lim sup is constant a.e. and $> -\infty$.

Lemma III.2. $E_{k(T)}(x) \neq \{0\}$ a.e. if $k(T) < +\infty$,

Lemma III.3. $\lim_{n\to\pm\infty} \frac{1}{n} \log\|T^n v\| = -k(T)$ for a.e. x and every $0 \neq v \in E_{k(T)}(x)$.

Lemma III.4. For every $\lambda \in \mathbb{R}$ there exists a μ-measurable Banach vector bundle $G_\lambda \subset G$ such that if $G_\lambda(x)$ denotes its fiber at x the following properties hold for a.e. x:

a) $T(G_\lambda(x) \subset G_\lambda(\varphi(x))$

b) $E_\lambda(x) \oplus G_\lambda(x) = G_x$

c) $\limsup_{n\to+\infty} \frac{1}{n} \log\|T^n/G_\lambda(x)\| < -\lambda$

d) $\liminf_{n\to+\infty} \frac{1}{n} \log\|T^{-n}v\| \geq \lambda$ for every $0 \neq v \in G_\lambda(x)$ such that $T^{-n}v$ exists for all $n \geq 0$.

e) If $\pi(x)\colon G_x \to G_\lambda(x)$ is the projection associated to the splitting $G_x = E_\lambda(x) \oplus G_\lambda(x)$ then $\lim_{n\to\pm\infty} \frac{1}{n} \log \|\pi(\varphi^n(x))\| = 0$ a.e.

Before proving these lemmas we shall show how they imply the property $\mu(\Sigma(L)) = 1$. If $k(L) = +\infty$ then $\mu(\Sigma_1) = 1$ and we are done. If it is $< +\infty$, putting $\lambda_1 = -k(L)$ we have, by Lemmas III.4 and III.2, a μ-measurable bundle $G_{\lambda_1} \subset G$ such that for a.e. x

satisfies:

$$F_x = E_{\lambda_1}(x) \oplus G_{\lambda_1}(x)$$

$$0 < \dim E_{\lambda_1}(x) < +\infty$$

$$L(G_{\lambda_1}(x)) \subset G_{\lambda_1}(\varphi(x)).$$

By Lemma III.3:

$$\lim_{n\to\pm\infty} \frac{1}{n} \log\|T^n v\| = \lambda_1$$

for a.e. x and every $0 \neq v \in E_{\lambda_1}(x)$. Define $E_1(x) = E_{\lambda_1}(x)$. Moreover, Lemma III.4 implies:

$$k(T/G_{\lambda_1}) > -\lambda_1 .$$

If $k(T/G_{\lambda_1}) = +\infty$ we define $F_\infty(x) = G_{\lambda_1}(x)$ and we have proved that $\mu(\Sigma_1^{(2)}) = 1$ and that, by part (e) of III.4, that the equality of Theorem C holds a.e. in $\Sigma_1^{(2)}$. If $k(T/G_{\lambda_1}) < +\infty$, we put $F_1(x) = G_{\lambda_1}(x)$, and we have obtained the first subspaces of the families $E_1(x),\dots,F_1(x),\dots$ that will satisfy property II or III of the definition of regular points. We continue this process by applying Lemma III.4 and III.2 again, but this time to the μ-measurable homorphism $T/G_{\lambda_1} = T/F_1$ and putting $\lambda_2 = -k(T/F_1)$ we obtain a μ-measurable bundle $G_{\lambda_2} \subset F_1$ such that for a.e. x satisfies:

$$F_1(x) = G_{\lambda_2}(x) \oplus E_{\lambda_2}(x)$$

$$0 < \dim E_{\lambda_2}(x) < \infty$$

$$L(G_{\lambda_2}(x)) \subset G_{\lambda_2}(\varphi(x))$$

and by Lemma III.3:

$$\lim_{n\to\pm\infty} \frac{1}{n} \log\|T^n v\| = \lambda_2$$

for a.e. x and every $0 \neq v \in E_{\lambda_2}(x)$. Again, if $k(T/G_{\lambda_2}) = +\infty$, we proved $\mu(\Sigma_2^{(2)}) = 1$ and by part (e) of III.4 the equality of Theorem C holds a.e. If it is $< +\infty$ we put $E_2(x) = E_\lambda(x)$,

$F_2(x) = G_{\lambda_2}(x)$ and we have obtained the second elements in the sequences $E_1(x), E_2(x), \ldots, F_1(x), F_2(x), \ldots$ required to satisfy properties II or III of the definition of regular point. Continuing this process either we arrive to some λ_m such that $k(T/G_{\lambda_m}) = +\infty$, and then $\mu(E_2^{(m)}) = 1$ (and again, by (e) of III.4, the equality of Theorem C holds a.e.), or the process continues indefinitely and then we have proved $\mu(\Sigma_3) = 1$ and that Theorem C holds a.e. In this last case we still have to show that $\lim_{n\to+\infty} \lambda_n = -\infty$. If there exists λ satisfying $\lambda < \lambda_n$ for all n, we must have $E_\lambda(x) \supset E_1(x) \oplus \ldots \oplus E_n(x)$ a.e. for all n. But this contradicts the property $\dim E_\lambda(x) < +\infty$ a.e. given by III.1. This proves Theorem A.

Proof of Lemma III.1. Let $K_o \subset K$ be a compact set such that for all $n \geq 0$ $G/K_o \cup \varphi^n(K_o)$ is a Banach vector bundle and $T: G/K_o \to G/\varphi^n(K_o)$ is continuous. Let K' be the set of points $x \in K$ where:

$$\mu(K_o) = \lim_{n\to+\infty} \frac{1}{n} \#\{1 \leq j \leq n \mid \varphi^{-j}(x) \in K_o\}.$$

By the ergodicity of μ, $\mu(K') = 1$. We shall prove that $\dim E_\lambda(x) < \infty$ for all $x \in K'$. Suppose that $\dim E_\lambda(x) = +\infty$ for some $x \in K'$. We shall show this contradicts Lemma II.4.
Let $\{v_j^{(n)} \mid j \in \mathbb{Z}^+\} \subset E_\lambda(\varphi^{-n}(x))$, $n \in \mathbb{Z}^+$, be a sequence of unitary vectors such that if $G_j^{(n)}$ is the space spanned by $\{v_1^{(n)}, \ldots, v_j^{(n)}\}$ then:

$$T^{-1}(G_j^{(n)}) = G_j^{(n+1)} \tag{1}$$

and:

$$\|v_j^{(n)} - w\| \geq 1/2 \tag{2}$$

for all $j > 1$, $n \geq 1$, $w \in G_{j-1}^{(n)}$. This sequences can be easily constructed by induction on j as follows: we take $v_1^{(1)} \in E_\lambda(\varphi^{-1}(x))$ with $\|v_1^{(1)}\| = 1$ and then we define $v_1^{(n)} = T^{-(n-1)}v_1^{(1)}/\|T^{-(n-1)}v_1^{(1)}\|$. Suppose that we have constructed $\{v_1^{(n)}, \ldots, v_k^{(n)}\}$ for all $n \in \mathbb{Z}^+$ satisfying (1) and (2) for $1 \leq j \leq k$. Then we take $v_{k+1}^{(1)} \in E_\lambda(\varphi^{-1}(x))$

with norm 1 and such that (2) holds for $n=1$, $j = k+1$ and every $w \in G_k^{(1)}$. Set $G_{k+1}^{(n)} = T^{-(n-1)}(G_{k+1}^{(1)})$ and take $v_{k+1}^{(n)}$ with $\|v_{k+1}^{(n)}\| = 1$ and $\|v_{k+1}^{(n)} - w\| \geq 1/2$ for all $w \in G_k^{(n)}$. Then the set $\{v_1^{(n)},\dots,v_{k+1}^{(n)}\}$ satisfy (1), (2) for all $1 \leq j \leq k+1$, $n \in Z^+$. Now let $E_j^{(n)}$ be the one dimensional space generated by $v_j^{(n)}$ and let $\pi_j^{(n)}: G_j^{(n)} \to E_j^{(n)}$ be the projection associated to the splitting $G_j^{(n)} = E_j^{(n)} \oplus G_k^{(n)}$. Then the sets $\{v_1^{(n)},\dots,v_{k+1}^{(n)}\}$ satisfy (1) and (2) for all $1 \leq j \leq k+1$, $n \in Z^+$. Now let $E_j^{(n)}$ be the one dimensional space generated by $v_j^{(n)}$ and let $\pi_j^{(n)}: G_j^{(n)} \to E_j^{(n)}$ be the projection associated to the splitting $G_j^{(n)} = E_j^{(n)} \oplus G_{j-1}^{(n)}$. By (2)

$$\|\pi_j^{(n)}\| \leq 2 \tag{3}$$

for all $j \geq 1$, $n \in \mathbb{Z}^+$. Define $P_j^{(n)} = \|\pi_j^{(n+1)}T^{-1}/E_j^{(n)}\|$. By (3)

$$\frac{1}{N}\log\prod_{n=1}^{N} P_j^{(n)} = \frac{1}{N}\log\|\pi_j^{(n+1)}T^{-N}/E_j^{(1)}\| \leq$$

$$\leq \frac{1}{N}\log 2 + \frac{1}{N}\log\|T^{-N}/E_j^{(1)}\| =$$

$$= \frac{1}{N}\log 2 + \frac{1}{N}\log\|T^{-N}v_j^{(1)}\| .$$

Hence:

$$\limsup_{N\to+\infty}\frac{1}{N}\sum_{n=1}^{N}\log P_j^{(n)} \leq \lambda .$$

Denote $S_N = \{1 \leq n \leq N \mid \varphi^{-n}(x) \in K_o\}$. Since $x \in K'$:

$$\lim \frac{1}{N}\# S_N = \mu(K_o).$$

Take $0 < b < 1$ and N_o such that:

$$\# S_N \geq b\mu(K_o)N$$

if $N \geq N_o$. Take $c > 0$ such that:

$$\log P_j^{(n)} \geq -c$$

for all $j \geq 1$, $n \in \mathbb{Z}^+$. This number exists because the norm of T

is by hypothesis bounded on the fibers of G. Define:

$$S_N^{(j)} = \{n \in S_N \mid \log P_j^{(n)} \le c\}$$

$$\alpha_N^{(j)} = \frac{\# S_N^{(j)}}{\# S_N}.$$

Take N_j such that $N \ge N_j$ implies:

$$\sum_1^N \log P_j^{(n)} \le (\lambda+1)N.$$

Hence, for every $j \ge 1$:

$$(\lambda+1)N \ge c\,\#(S_N \setminus S_N^{(j)}) - c(N - \#(S_N \setminus S_N^{(j)})) \ge$$

$$\ge c(1-\alpha_N^{(j)})\,\#\,(S_N \setminus S_N^{(j)}) - c(N-(1-\alpha_N^{(j)})\# S_N) \ge$$

$$\ge c(1-\alpha_N^{(j)})b\mu(K_o)N - cN.$$

Then:

$$\alpha_N^{(j)} \ge 1 - \frac{\lambda+1+c}{2cb\mu(K_o)}$$

Then

$$\alpha_N^{(j)} \ge 1/3 \qquad (4)$$

for all j, if $\mu(K_o)$ and b are taken near to 1 and c is large enough. Now we shall use the following elementary combinatorial lemma:

<u>Lemma III.5</u>. If A is a finite set and $\mathcal{G} = \{A_1,\ldots,A_n\}$ a family of subsets there exists a point in A that belongs to $\sum_{j=1}^{n} (\#A_j/\#A)$ sets of $\mathcal{G}$.

<u>Proof</u>. Suppose that k is the maximum number of elements of $\mathcal{G}$ with non empty intersection. If $p \in A$ and $1 \le j \le n$ define $m(p,j) = \#\{1 \le i \le j \mid p \in A_i\}$. Define subsets $\tilde{A}_j$ of $A\times\{1,\ldots,k\}$ as follows:

$$\tilde{A}_j = \{(p,m(p,j)) \mid p \in A_j\}.$$

These sets are disjoint because if $(p,q) \in \tilde{A}_{j_1} \cap \tilde{A}_{j_2}$ with $j_1 > j_2$, then

$p \in A_{j_1}$, $p \in A_{j_2}$ and $q = m(p,j_1) = m(p,j_2)$. But since $p \in A_{j_2}$ we must have $m(p,j_2) \ge 1 + m(p,j_1)$.

Then:

$$k \# A = \#(A\times\{1,\dots,k\}) \ge \# \bigcup_{j=1}^{n} \tilde{A}_j = \sum_{j=1}^{n} \# \tilde{A}_j =$$

$$= \sum_{j=1}^{n} \# A_j = \left(\sum_{j=1}^{n} (\#A_j/\#A)\right)\#A.$$

Hence:

$$k \ge \left[\sum_{j=1}^{n} (\#A_j/\#A)\right]$$

that is what we wanted to prove.

Now apply this lemma to $A = \{1,\dots,N\}$ and $\mathcal{C} = \{S_N^{(1)},\dots, S_N^{(n)}$ By (4) and the lemma there exists a point m that belongs to $S_N^{(j_1)},\dots,S_N^{(j_q)}$, with $q \ge [n/4]$. Let $y = \varphi^{-m}(x)$. Then

$$\log P^{(m)}_{j_i} \le c$$

for $i \le i \le q$. This means:

$$\|\pi^{(r)}_{j_i} T\ v^{(r+1)}_{j_i}\| \exp(-c).$$

If $1 \le k < i \le q$ we can write:

$$\|Tv^{(r+1)}_{j_i} - Tv^{(r+1)}_{j_k}\| = \|\pi^{(r)}_{j_i}\ Tv^{(r+1)}_{j_k} - \pi^{(r+1)}_{j_k}\ Tv^{(r+1)}_{j_i} - Tv^{(r+1)}_{j_i}\| =$$

$$= \|\pi^{(r)}_{j_i} T_{v_{j_i}}^{(r+1)}\| \cdot \|v^{(r)}_{j_i} + \|\pi^{(r)}_{j_i} Tv^{(r+1)}_{j_i}\|^{-1}(Tv^{(r+1)}_{j_i} - \pi^{(r)}_{j_i} Tv^{(r+1)}_{j_i} - Tv^{(r+1)}_{j_k})\|$$

But the expression between brackets is a vector in $G^{(r)}_{j-1}$ because $Tv^{(r+1)}_{j_k} \in TG^{(r+1)}_{j_k} \subset G^{(r)}_{j_k} \subset G^{(r)}_{j_i-1}$ (recall that $k < i$, hence $j_k < j_i$) and the range of $(I - \pi^{(r)}_{j_i})$ is $G^{(r)}_{j_i-1}$. Using (4) and (2) we obtain

$$\|Tv^{(r+1)}_{j_i} - Tv^{(r+1)}_{j_k}\| \ge \exp(-c)\cdot\tfrac{1}{2}$$

for all $1 \le i < k \le q$. But Lemma II.4 applied to the homomorphisms

$T: G/\varphi^{-1}(K_o) \to G/K_o$ gives a constant that bounds q as a function of $(1/2)\exp(-c)$. On the other hand $q \geq [n/4]$ and n is arbitrary. This contradiction completes the proof of $\dim E_\lambda(x) < +\infty$ for $x \in K'$.

To complete the proof of III.1 we still have to show that $\bigcup \{E_\lambda(x) \mid x \in K\}$ is a μ-measurable Banach vector bundle on K. Since $E_{\lambda+\varepsilon}(x)$ is finite dimensional for all $\varepsilon > 0$ and a.e. $x \in K$ it follows that we can take $\varepsilon > 0$ so small that $E_{\lambda+\varepsilon}(x) = E_\lambda(x)$ a.e. Let $K(C)$ be a compact subset of K_o such that for all $n \in \mathbb{Z}^+$:

$$\|T^{-n}/E_\lambda(x)\| \leq C \exp(\lambda+\varepsilon)n.$$

Denote K_1 the set of points where $E_{\lambda+\varepsilon}(x) = E_\lambda(x)$ and the dimensions of these subspaces is constant, say m. We claim that the function $K_1 \cap K(C) \to G^+(m)$ is continuous. Let $\{x_n \mid n \in \mathbb{Z}^+\} \subset \subset K_1 \cap K(C)$ be a sequence converging to $x \in K_1 \cap K(C)$. Suppose that it is not true that $E_\lambda(x_n) \to E_\lambda(x)$. Then it is not true that $T^{-1}(E_\lambda(x_n)) \to T^{-1}(E_\lambda(x))$. But $\|Tv\| \geq C^{-1}\exp(-(\lambda+\varepsilon)n)$ for all $v \in \bigcup_n T^{-1}(E_\lambda(x_n))$. Then by II.6 there exists a convergent subsequence $T^{-1}(E_\lambda(x_{n_j})) \to E \subset G_{\varphi^{-1}(x)}$, $E \neq T^{-1}(E_\lambda(x))$. Then $E_\lambda(x_{n_j}) \to T(E)$. But then, since $x \in K(C) \subset K_o$,

$$\|T^{-n}/T(E)\| \leq C \exp(\lambda+\varepsilon)n$$

for all $n \in \mathbb{Z}^+$. Therefore $T(E) \subset E_{\lambda+\varepsilon}(x)$. But, since $x \in K_1$, $E_{\lambda+\varepsilon}(x)$ and $E_\lambda(x)$ coincide. Hence $T(E) = E_\lambda(x)$ and $E = T^{-1}(E_\lambda(x))$. This completes the proof of the claim. It is clear that we can choose $K(n)$, $n \in \mathbb{Z}^+$, satisfying:

$$\bigcup_{n\in\mathbb{Z}^+} K(n) \cap K_1 = K_o \cap K_1 \ \mathrm{mod}(0).$$

But the definition of μ-measurable vector bundle implies that K_o can be taken with measure arbitrarily near to 1. Since $\mu(K_1) = 1$ the lemma is proved.

Proof of Lemma III.2. We shall prove $\dim E_\lambda(x) \geq 1$ a.e. for every $\lambda > k(T)$. Observing that

$$E_{k(T)}(x) = \bigcap_{\lambda > k(T)} E_\lambda(x)$$

this will complete the proof of the theorem. Fix any $\lambda > k(T)$. Let A_n be the set of points $x \in K$ where there exists $0 \neq v \in G_x$ satisfying $\|T^{-m}v\| \leq \lambda^m \|v\|$ for all $1 \leq m \leq n$. We claim that there exists $c > 0$ such that $\mu(A_n) \geq c$ for all $n \geq 0$. Before proving this claim we shall show how it implies that $\dim E_\lambda(x) \geq 1$ a.e. Take a compact set $K_o \subset K$ such that for all n $G/\varphi^{-n}(K_o)$ is a Banach vector bundle and $T/(G/\varphi^{-n}(K_o))$ is continuous. As we mentioned before, K_o can be chose with measure arbitrarily near to 1. Suppose then that $\mu(K_o) \geq 1 - c/2$. Set $K_n = K_o \cap A_n$. Then $\mu(K_n) = \mu(K_n) = \mu(K_o \cap A_n) \geq \mu(A_n) - \mu(K_o^c) \geq c/2$. This implies $\mu(\bigcap_{m=1}^{\infty} \bigcup_{n=m}^{\infty} A_n) \geq c/2$. If we prove that $\dim E_\lambda(x) > 0$ for $x \in \bigcap_{m=1}^{\infty} \bigcup_{n=m}^{\infty} A_n$ we are done because as we just proved the measureof this set is >0 and if $\dim E_\lambda(x) > 0$ then $\dim E_\lambda(\varphi^n(x)) > 0$ for all n. Then $\dim E_\lambda(x) > 0$ holds for all x in the φ-invariant set $\bigcup_j \varphi^j(\bigcap_{m=1}^{\infty} \bigcup_{n=m}^{\infty} A_n)$. Since this set has positive measure (because it contains $\bigcap_{m=1}^{\infty} \bigcup_{n=m}^{\infty} A_n$), the ergodicity of μ implies that $\dim E_\lambda(x) > 0$ a.e. Now, if $x \in \bigcap_{m=1}^{\infty} \bigcup_{n=m}^{\infty} A_n$ there exists a sequence of positive integers $n_1, n_2, \ldots$ such that $x \in A_{n_j}$ for all j. Let $v_j \in G_x$, with $\|v_j\| = 1$, such that $x \in A_{n_j}$ for all j. Let Let $v_j \in G_x$, with $\|v_j\| = 1$, such that $\|T^{-n}v_j\| \leq \lambda^n$ for all $0 \leq n \leq n_j$. Then the set $\{T^{-1}v_j \mid j \in \mathbb{Z}^+\}$ is bounded. Hence we can assume that $v_j = T(T^{-1}v_j)$ converges to some vector $v \in G_x$. It is clear that $\|v\| = 1$ and $\|T^{-n}v\| \leq \lambda^n$ for all $n \geq 0$ and then $v \in E_\lambda(x)$.

Now let us prove the claim. We shall use the following lemma due to V.I. Pliss.

<u>Lemma III.6</u>. Given $H > \lambda$ and $\varepsilon > 0$ there exist $N_o = N_o(H,\lambda,\varepsilon)$ and $\delta = \delta(H,\lambda,\varepsilon)$ such that if $a_i \in \mathbb{R}$, $i = 0,\dots,N$ satisfy:

$$N \geq N_o$$

$$a_n \geq H \tag{1}$$

$$\sum_{0}^{N} a_n \leq (\lambda-\varepsilon)(N+1) \tag{2}$$

then there exist $0 \leq n_1 < n_2 < \dots < n_\ell \leq N$ such that:

$$\ell \geq \delta N \tag{3}$$

and

$$\sum_{n=n_j+1} a_n \leq (m-n_j)\lambda \tag{4}$$

for all $n_j < m \leq N$, $1 \leq j \leq \ell$.

We shall apply this lemma taking $0 < \varepsilon < \lambda - k(T)$ and N such that:

$$\log \frac{\|Tv\|}{\|v\|} \leq -H$$

for all $0 \neq v \in G$. We shall prove that for all n:

$$\mu(A_n) = \delta = \delta(H,\lambda,\varepsilon).$$

For this define:

$$\tau_N(x,A_n) = \frac{1}{N}\#\{0 < j \leq N \mid \varphi^j(x) \in A_n\}.$$

Chose x such that

$$\lim \tau_N(x,A_n) = \mu(A_n)$$

for all n. Since $-(\lambda-\varepsilon) < -k(T)$, for every large N there exists $0 \neq v \in G_x$ satisfying

$$\log \frac{\|T^{N+1}v\|}{\|v\|} \geq -(\lambda-\varepsilon)(N+1).$$

Hence, defining

$$a_k = \log \frac{\|T^{N-k}v\|}{\|T^{N-k+1}v\|}$$

for $0 \leq k \leq N$, we obtain:

$$\sum_{k=0}^{N} a_k = \log \frac{\|v\|}{\|T^{N+1}v\|} \leq (\lambda-\varepsilon)(N+1)$$

and, by the definition of H,

$$a_n \geq H.$$

Without loss of generality we can suppose:

$$a_n \geq N_0(H,\lambda,\varepsilon).$$

Then Lemma III.6 gives integers $0 < n_1 < n_2 < \ldots < n_\ell \leq N$ satisfying (1) and (2). But (2) means:

$$(m-n_j)\lambda \geq \sum_{n=n_j+1}^{m} a_n = \sum_{m=n_j+1}^{m} \log \frac{\|T^{N-n}v\|}{\|T^{N-n+1}v\|} = \log \frac{\|T^{N-m}v\|}{\|T^{N-n_j}v\|}.$$

Set

$$u_j = \frac{T^{N-n_j}v}{\|T^{N-n_j}v\|}.$$

Then, for all $n_j \leq m \leq N$

$$\|T^{-(m-n_j)}u_j\| \leq \exp \lambda(m-n_j)$$

or:

$$\|T^{-k}u_j\| \leq \exp \lambda k$$

for all $0 \leq k \leq N-n_j$. Since $N - n_j \geq n$ for $j > n$, it follows that $\varphi^{N-n_j}(x) \in A_n$ for $n < j \leq \ell$ (because $u_j \in G_{\varphi^{N-n_j}(x)}$ and satisfies the required conditions). But

$$\ell-n \geq \delta N-n$$

and then:

$$\frac{1}{N}\#\{0 < i \leq N \mid \varphi^i(x) \in A_n\} \geq \frac{1}{N}\#\{0 < i \leq N \mid i = N-n_j, n < j \leq \ell\} \frac{1}{N}(\ell-n) \geq \delta - \frac{n}{N}.$$

If for a fixed value of x we take limits for $N \to +\infty$ we conclude

$$\mu(A_n) = \lim_{N\to+\infty} \tau_N(x,A_n) \geq \delta.$$

<u>Proof of Lemma III.6</u>. Define $b_n = a_n - \lambda$ and $S_n = \sum_{0}^{m} b_n$. These numbers satisfy:

$$S_N \leq -\varepsilon(N+1) \tag{3}$$

$$b_n \geq H-\lambda \ . \tag{4}$$

Define $0 \leq n_1 < \ldots < n_\ell \leq N$ as follows:

$$n_1 = \min\{m \mid S_m = \max_{0 \leq k \leq N} S_k\}$$

$$n_{j+1} = \min\{m \mid S_m = \max_{n_j < k \leq N} S_k\}.$$

Then:

$$S_{n_1} \geq H-\lambda$$

$$S_{n_{j+1}} \leq S_{n_j} \ .$$

To continue this argument we shall need that the set $n_1,\ldots,n_\ell$ contains at least two elements, i.e. $\ell > 1$. But since obviously $n_\ell = N$ we have only to show $n_1 < N$. But $n_1 = N$ would imply $S_N \geq S_o = b_o$, that means by (3) and (4):

$$-\varepsilon(N+1) \geq S_N \geq S_o = b_o \geq H-\lambda$$

and

$$N+1 \leq \frac{1}{\varepsilon}(H-\lambda). \tag{5}$$

Then if we impose on $N_o(H,\lambda,\varepsilon)$ the requirement

$$N_o(H,\lambda,\varepsilon) > \frac{1}{\varepsilon}(H-\lambda) - 1 \tag{6}$$

we make (5) impossible and then $\ell > 1$ as we wanted.

Now observe that by the way we choose the n_j, we have:

$$S_m - S_{n_j} \leq 0$$

for $m \geq n_j$. But then:

$$\sum_{n=n_j+1}^{m} b_n = \sum_{0}^{m} b_n - \sum_{0}^{n_j} b_n = S_m - S_{n_j} \leq 0.$$

Recalling that $b_n = a_n - \lambda$ we obtain

$$\sum_{n=n_j+1}^{m} a_n \leq (m-n_j)\lambda$$

for all $1 < j \leq \ell$, $n_j < m \leq N$. This proves property (2) in III.6. To show (4) observe that hypothesis (1) and (2) of III.6 imply $H < \lambda$ and define

$$\delta(H,\lambda,\varepsilon) = \frac{\varepsilon}{\lambda - H}.$$

Then:

$$S_{n_{j+1}} \geq S_{n_j+1} = S_{n_j} + b_{n_j+1} \geq S_{n_j} + H-\lambda.$$

Applying this inequality for $j = 1,\ldots,\ell-1$ and adding:

$$S_{n_\ell} \geq S_{n_1} + (\ell-1)(H-\lambda).$$

But $S_{n_1} \geq H-\lambda$ and $S_{n_\ell} = S_N \leq -\varepsilon(N+1)$. Hence:

$$-\varepsilon(N+1) \geq (H-\lambda)(\ell-1).$$

Therefore:

$$\ell \geq 1 + \frac{\varepsilon}{\lambda-H}(N+1) \geq \delta(H,\lambda,\varepsilon)N.$$

Proof of Lemma III.3. We shall prove first that

$$\liminf_{n\to+\infty} \frac{1}{n} \log\|T^n v\| \geq -k(T) \tag{1}$$

for a.e. x and every $0 \neq v \in E_{k(T)}$. Together with the definition of $k(T)$ this proves III.3 for $n \to +\infty$. Instead of (1) we shall prove the following more general property that will be useful later:

Lemma III.7. If $\lambda > 0$, for a.e. $x \in K$ the inequality:

$$\liminf_{n\to+\infty} \frac{1}{n} \log\|T^n v\| \geq -\lambda$$

holds for all $v \in E_\lambda(x)$.

Proof. Take $\varepsilon > 0$ and define:

$$C_\varepsilon(x) = \sup \frac{\|T^{-n}/E_\lambda(x)\|}{\exp(\lambda+\varepsilon)n}.$$

Then if $x \in K$ and $v \in E_\lambda(x)$:

$$\|T^n v\| \geq \|(T/E_\lambda(x))^{-n}\|^{-1}\,\|v\| \geq C_\varepsilon^{-1}(x)\,\exp(-(\lambda+\varepsilon)n)\|v\|.$$

Hence:

$$\liminf_{n\to+\infty} \frac{1}{n} \log\|T^n v\| \geq -(\lambda+\varepsilon) - \liminf_{n\to+\infty} \frac{1}{n} \log C_\varepsilon(\varphi^n(x)).$$

Since ε is arbitrary our problem is reduced to show that for all $\varepsilon > 0$:

$$\liminf_{n\to+\infty} \frac{1}{n} C_\varepsilon(\varphi^n(x)) \leq 0 \qquad \text{a.e.}$$

This will follow from the following lemma. In fact it is the easy part of it, the others conclusions in it will be very useful later.

Lemma III.8. Let $(X,\mathcal{G},\mu)$ be a probability space and $\varphi: X \circlearrowleft$ a measure preserving transformation. If $A: X \to \mathbb{R}$ is a measurable function, then:

$$\liminf_{n\to+\infty} \frac{1}{n} A(\varphi^n(x)) = 0 \qquad \text{a.e.} \qquad (1)$$

Moreover, if there exists $F \in \mathcal{L}^1(X,\mathcal{G},\mu)$ such that

$$A(\varphi(x)) - A(x) \leq F(x) \qquad \text{a.e.} \qquad (2)$$

or if there exists $G \in \mathcal{L}^1(X,\mathcal{G},\mu)$ such that

$$A(\varphi(x)) - A(x) \geq G(x) \qquad \text{a.e.} \qquad (3)$$

Then $A\circ\varphi - A \in \mathcal{L}^1(X,\mathcal{G},\mu)$ and

$$\lim_{n\to+\infty} \frac{1}{n} A(\varphi^n(x)) = 0 \qquad \text{a.e.} \qquad (4)$$

Finally, if φ is invertible,

$$\lim_{n\to-\infty} \frac{1}{n} A(\varphi^n(x)) = 0 \qquad \text{a.e.} \qquad (5)$$

Proof. Suppose that (1) is false. Then there exist $S \in \mathcal{G}$ with $\mu(S) > 0$ and $c > 0$ such that

$$\liminf_{n\to+\infty} \frac{1}{n} A(\varphi(x)) > c$$

for all $x \in S$. Since A is measurable, there exists $S_o \in \mathcal{G}$ contained in S with $\mu(S_o) > 0$ and $K > 0$ such that $\varphi(x) \leq K$ for all $x \in S_o$. By Poincaré recurrence theorem we can take $x \in S_o$ and a sequence $0 < n_1 < n_2 < \ldots$ satisfying $\varphi^{n_j}(X) \in S_o$

for all $j \geq 1$. Then $A(\varphi^{n_j}(x)) \geq cn_j$ if j is large enough. But $\varphi^{n_j}(x) \in S_o$. Then $K \geq A(\varphi^{n_j}(x)) \geq cn_j$ for every large value of j. This contradiction proves (1). The integrability of $A\circ\varphi - A$ and (4) will be proved under hypothesis (2). When hypothesis (3) holds, the proof of these properties are reduced to the case of hypothesis (2) because

$$(-A)(\varphi(x)) - (-A(x)) = A(x) - A(\varphi(x)) \leq -G(x).$$

Then $(-A)\circ\varphi - (-A)$ is integrable thus implying the integrability of $A\circ\varphi - A = -((-A)\circ\varphi - (-A))$ and

$$\lim_{n\to+\infty} \frac{1}{n} A(\varphi^n(x)) = -\lim \frac{1}{n} (-A)(\varphi^n(x)) = 0 \qquad \text{a.e.}$$

So let us assume that (2) holds. If $N \in \mathbb{Z}^+$ define $F_N: X \to \mathbb{R}$ by

$$F_N(x) = A(\varphi(x)) - A(x)$$

if $A(\varphi(x)) - A(x) \geq -N$, and

$$F_N(x) = -N$$

otherwise. Then the sequence $\{F_N \mid N \in \mathbb{Z}^+\}$ is monotone and converges a.e. to $A\circ\varphi - \varphi$. Hence to prove the integrability of $A\circ\varphi - \varphi$ it suffices to show that

$$\inf \int_X F_N \, d\mu > -\infty.$$

We have:

$$\int_X F_N \, d\mu = \int_X \left(\lim_{n\to+\infty} \frac{1}{n} \sum_{j=0}^{n-1} F_N\circ\varphi^j\right) d\mu \geq$$

$$\geq \int_X \left(\limsup_{n\to+\infty} \frac{1}{n} \sum_{j=0}^{n-1} (A\circ\varphi^{j+1} - A\circ\varphi^j)\right) d\mu =$$

$$= \int_X \limsup_{n\to+\infty} \frac{1}{n} (A\circ\varphi^n - A) d\mu =$$

$$= \int_X \limsup_{n\to+\infty} \frac{1}{n} (A\circ\varphi^n) d\mu.$$

But

$$\limsup_{n\to+\infty} \frac{1}{n} A(\varphi^n(x)) = -\limsup_{n\to+\infty} \frac{1}{n} (-A)(\varphi^n(x)) \geq 0. \tag{6}$$

Then:

$$\int_X F_N \, d\mu \geq -\infty$$

completing the proof of the integrability of $A\circ\varphi - A$. Now let us show (4). From (6) and (1) it follow that it suffices to show that the sequence $n^{-1} A(\varphi^n(x))$ converges a.e. Write:

$$\frac{1}{n} A(\varphi^n(x)) = \frac{1}{n} \sum_{j=0}^{n-1} (A\circ\varphi - A)(\varphi^j(x)) + \frac{1}{n} A(x).$$

By Birkhoff's theorem the sequence

$$\frac{1}{n} \sum_{j=0}^{n-1} (A\circ\varphi - A)(\varphi^j(x))$$

converges a.e. Since $n^{-1} A(x)$ converges to 0 the proof of (4) is completed. To prove (5) it suffices to apply (4) to A and φ^{-1} and observe that if $A(\varphi(x)) - A(x) \leq F(x)$ a.e. then

$$A(x) - A(\varphi^{-1}(x)) \leq F(\varphi^{-1}(x)) \quad \text{a.e.}$$

that implies

$$A(\varphi^{-1}(x)) - A(x) \geq -(F\circ\varphi^{-1})(x)$$

so that A and φ^{-1} satisfy hypothesis (3). In the same way it is proved that if A and φ satisfy hypothesis (3), then A and φ^{-1} satisfy hypothesis (2).

It remains to prove III.3 when $n \to -\infty$. By the definition of $E_{k(T)}(x)$ it suffices to show that:

$$\liminf_{n\to+\infty} \frac{1}{n} \|T^{-n}v\| \geq k(T) \tag{4}$$

for a.e. x and every $0 \neq v \in E_{k(T)}(x)$. Given $\varepsilon > 0$ define:

$$C_\varepsilon(x) = \sup_{n\geq 0} \frac{\|T^n/E_{k(T)}(x)\|}{\exp(-k(T)+\varepsilon)^n}.$$

The number $C_\varepsilon(x)$ is finite by the definition of $k(T)$ and obviously C_ε is a measurable function. Then, if $v \in E_\lambda(x)$

$$\|v\| = \|T^n T^{-n} v\| \leq C_\varepsilon(\varphi^{-n}(x)) \exp(-k(T)+\varepsilon)n \ \|T^{-n}v\|.$$

Hence:

$$\liminf_{n\to+\infty} \frac{1}{n} \|T^{-n}v\| \geq k(T)-\varepsilon - \limsup_{n\to+\infty} \frac{1}{n} \log C_\varepsilon(\varphi^{-n}(x)).$$

Therefore, since ε is arbitrary, (4) will follow if we show:

$$\lim_{n\to+\infty} \frac{1}{n} \log C_\varepsilon(\varphi^{-n}(x)) \qquad \text{a.e.}$$

for all $\varepsilon > 0$. By III.8 we have only to show that $\log C_\varepsilon \circ \varphi^{-1} - \log C_\varepsilon$ is bounded by an integrable function. But if $n \geq 1$

$$\|T^n/E_{k(T)}(\varphi^{-1}(x)\| \leq \|T/E_{k(T)}(\varphi^{-1}(x))\| \|T^{n-1}/E_{k(T)}(x)\| \leq$$
$$\leq \|T/E_{k(T)}(\varphi^{-1}(x))\| C_\varepsilon(x)\exp(-k(T)+\varepsilon)(n-1) =$$
$$= \|T/E_{k(T)}(\varphi^{-1}(x))\| \ \exp(k(T)-\varepsilon)\exp(-k(T)+\varepsilon)n \ C_\varepsilon(x).$$

Therefore:

$$C_\varepsilon(\varphi^{-1}(x)) \leq \sup\{1, \sup_{n\geq 1} \frac{\|T^n/E_{k(T)}(\varphi^{-1}(x))\|}{\exp(-k(T)+\varepsilon)n}\}$$
$$\leq \sup\{1, \|T/E_{k(T)}(\varphi^{-1}(x))\| \exp(k(T)-\varepsilon) \ C_\varepsilon(x)\}.$$

But $C_\varepsilon(x) \geq 1$ for all x. Then the last expression is less than or equal to:

$$C_\varepsilon(x)\sup\{1, \|T/E_{k(T)}(\varphi^{-1}(x))\| \exp(k(T)-\varepsilon)\}$$

and then

$$\log C_\varepsilon(\varphi^{-1}(x)) - \log C_\varepsilon(x) \leq \sup\{0, \log\|T/E_{k(T)}(\varphi^{-1}(x))\| + k(T)-\varepsilon\}$$

that is integrable.

<u>Proof of Lemma III.4</u>. Let $F \subset G$ be a μ-measurable vector bundle such that $F_x \oplus E_\lambda(x) = G_x$ a.e. and such that if $\pi_x\colon G_x \to F_x$, $P_x\colon G_x \to E_\lambda(x)$ are the projections associated with this splitting

there exists $P > 0$ such that $\|p_x\| \le P$, $\|\pi_x\| \le P$ a.e. Define a μ-measurable homomorphism $\tilde{T}: F \circlearrowleft$ by $\tilde{T}/F_x = \pi_{\varphi(x)} T/F_x$. We claim that $k(\tilde{T}) > \lambda$. If $k(\tilde{T}) \le \lambda$ the previous lemmas imply that there exists a μ-measurable vector bundle $E = E_{k(\tilde{T})} \subset F$ such that a.e.:

$$\tilde{T}(E_x) = E_{\varphi(x)}$$

$$\lim_{n\to\pm\infty} \frac{1}{n} \log\|\tilde{T}^n/E_x\| = -k(\tilde{T}) \ge -\lambda .$$

Take $\hat{E}_x = \pi_x^{-1}(E_x)$. Then $\hat{E}_x \supset E_\lambda(x)$ and $\hat{E}_x \ne E_\lambda(x)$.

By III.7:

$$\liminf_{n\to+\infty} \frac{1}{n} \log\|T^n v\| \ge -\lambda$$

for a.e. x and every $0 \ne v \in E_\lambda(x)$. Then:

$$\liminf_{n\to+\infty} \frac{1}{n} \log\|T^n v\| \ge -\lambda \tag{1}$$

for a.e. x and every $0 \ne v \in \hat{E}_\lambda$. On the other hand

$$k(T^{-1}/\hat{E}) = - \limsup_{n\to+\infty} \frac{1}{n} \log\|T^{-n}/\hat{E}_x\| < -\lambda$$

because otherwise we should have $\hat{E} \subset E_\lambda$. Let $\tilde{E} \subset \hat{E}$ be the μ-measurable vector bundle $E_{k(T^{-1}/\hat{E})}$ associated to T^{-1}. Then

$$\lim_{n\to\pm\infty} \frac{1}{n} \log\|T^{-n}/\tilde{E}_x\| = -k(T^{-1}/\hat{E}) > \lambda .$$

In particular

$$\lim_{n\to+\infty} \frac{1}{n} \log\|T^n/\tilde{E}_x\| = k(T^{-1}/\hat{E}) < -\lambda$$

contradicting (1) and completing the proof of the claim. Now let M be the μ-measurable vector bundle on K whose fibers M_x are the spaces of continuous linear maps $L: F_x \to E_\lambda(x)$. Define a μ-measurable homomorphism $\Phi: M \circlearrowleft$ by:

$$\Phi(L) = T \circ L \circ (\tilde{T}^{-1}/F_{\varphi(x)}).$$

Let $\Gamma(M)$ be the space of measurable sections of M, i.e. measurable maps that to a.e. $x \in K$ associate a linear map $L_x \in M_x$.

To each $\Phi \in \Gamma(M)$ we can associate a μ-measurable vector bundle $G(\Phi)$ whose fibers are given by:

$$G_x(\Phi) = \{v + \Phi_x v \mid v \in F_x\}.$$

This μ-measurable vector bundle is T-invariant, i.e.

$$T(G_x(\Phi)) = G_{\varphi(x)}(\Phi) \qquad \text{a.e.}$$

if and only if:

$$(T^{-1} \Phi_{\varphi(x)}(\tilde{T}/F_x)) - \Phi_x = P_{\varphi(x)}(T/F_x) \quad \text{a.e.} \tag{1}$$

Then, to find Φ satisfying this condition, we define

$$A_x = -P_{\varphi(x)}(T/F_x)$$

and

$$\Phi_x - \sum_{n=0}^{\infty} T^{-n} A_{\varphi^n(x)}(\tilde{T}^n/F_x). \tag{2}$$

If we show that a.e. the sequence:

$$\sum_{n=0}^{\infty} \# \, T^{-n}/E_\lambda(\varphi^n(x))\| \cdot \| \, \|\tilde{T}^n/F_n\| \tag{3}$$

converges, then (2) will define an element of $\Gamma(M)$ satisfying (1). Then we shall define $G_\lambda = G(\Phi)$ and show that satisfies the required properties. To prove the convergence a.e. of the series (3) take $\varepsilon > 0$ so small that:

$$2\varepsilon < k(\tilde{T}) - \lambda. \tag{4}$$

Define

$$C(x) = \sup_{n\geq 0} \frac{\|T^{-n}/E_\lambda(x)\|}{\exp(\lambda+\varepsilon)n}$$

$$C_1(x) = \sup \frac{\|\tilde{T}^n/F_x\|}{\exp(-k(\tilde{T})+\varepsilon}.$$

Then:

$$\|\Phi_x\| \leq P \sum_{n=0}^{\infty} \|T^{-n}/E_\lambda(\varphi^n(x))\| \cdot \|\tilde{T}(x)\| \cdot \|\tilde{T}^n/F_x\| \leq$$

$$\leq P \sum_{n=0}^{\infty} C(\varphi^n(x))C_1(x)\exp(\lambda - k(\tilde{T})+2\varepsilon)n.$$

But it is not difficult to show, using methods similar to those applied in the proof of Lemma III.3, that $\log C\circ\varphi - \log C$ is bounded by an integrable function. Then Lemma III.8 proves that given $0 < \delta < k(\tilde{T}) - \lambda$, the number:

$$C_2(x) = \sup_{n\geq 0} \frac{C(\varphi^n(x))}{\exp \delta n}$$

is finite. Then:

$$\|\Phi_x\| \leq P \sum_{n=0}^{\infty} C_2(x)C_1(x) \exp(\lambda - k(\tilde{T}) + 2\varepsilon + \delta)n =$$

$$= P\ C_2(x)C_1(x)(1-\exp(\lambda - k(\tilde{T}) + 2\varepsilon + \delta)^{-1} = \bar{P}\ C_2(x)C_1(x)$$

where $\bar{P} = P(1-\exp(\lambda - k(\tilde{T}) + 2\varepsilon + \delta)^{-1})$. Now define $G_\lambda = G(\Phi)$. The previous considerations show that satisfy part (a) of III.4. Part (b) is clear. To prove part (c) set $B_x: F_x \to G_\lambda(x)$ defined by:

$$B_x v = v + \Phi_x v.$$

Then Φ_x is an isomorphism with

$$\|B_x\| \leq 1 + \|\Phi_x\|$$

$$\|B_x^{-1}\| \leq p'$$

a.e., where p' depends only on P. Moreover

$$T^n/G_\lambda(x) = B_{\varphi^n(x)}\ \tilde{T}^n B_x^{-1}\ . \tag{5}$$

Then:

$$\|T^n/G_{\lambda,x}\| \leq \|\tilde{T}^n\|\ P'(1 + \|\Phi_{\varphi^n(x)}\|) \leq$$

$$\leq \|\tilde{T}^n\|\ p'(1+\bar{P}C_2(\varphi^n(x))C_1(\varphi^n(x)))$$

and by III.8:

$$\limsup_{n\to+\infty} \frac{1}{n} \log\|T^n/G_\lambda(x)\| \leq \limsup_{n\to+\infty} \frac{1}{n} \log\|\tilde{T}^n\| +$$

$$+ \limsup_{n\to+\infty} \frac{1}{n} \log C_2(\varphi^n(x)) + \limsup_{n\to+\infty} \frac{1}{n} \log C_1(\varphi^n(x)) \leq$$

$$\leq \limsup_{n\to+\infty} \frac{1}{n} \log\|\tilde{T}^n\| = k(\tilde{T}) < -\lambda$$

proving part (c) of III.4. Here we used only, the fact of $\log C_1$,

$\log C_2$ being measurable. But in fact it is not difficult to check that they satisfy also the property that $\log C_i\circ\varphi - \log C_i$ is bounded by an integrable function. Then $\lim_{n\to+\infty} n^{-1}C_i(\varphi^n(x)) = 0$ for $i = 1,2$. From this property together with (5), part (d) of III.4 follows immediately.

To prove part (e) of III.4 we start observing that there exist constants $0 < k_1 < k_2$, depending only on P, such that

$$k_1\|\Phi(x)\| \le \|\pi(x)\| \le k_2\|\Phi(x)\|.$$

We leave to the reader the verification of the existence of k_1 and k_2. This reduces the problem to prove

$$\lim \frac{1}{n}\log\|\Phi(\varphi^n(x))\| = 0 \quad \text{a.e.}$$

But since $\|\Phi(x)\|$ is a measurable function of x we have by Lemma III.8 that:

$$\limsup_{n\to\pm\infty} \frac{1}{|n|}\log\|\Phi(\varphi^n(x))\| = -\liminf_{n\to\pm\infty} \frac{1}{|n|}(-\log\|\Phi(\varphi^n(x))) \ge 0.$$

On the other hand:

$$\liminf_{n\to\pm\infty} \frac{1}{|n|}\log\|\Phi(\varphi^n(x))\| \le \liminf_{n\to\pm\infty} \frac{1}{|n|}\bar{P}\, C_2(\varphi^n(x))C_1(\varphi^n(x)) = 0$$

thus completing the proof.

IV - Proof of Theorem B

Of the statement of Theorem B we shall only prove the properties concerning the sets $G_3(n_1,\ldots,n_k)$. The corresponding properties for $G_2(n_1,\ldots,n_k)$ follow applying the same methods with minor modifications.

Let $\hat{G}(n_1,\ldots,n_k)$ be the set of points $x \in K$ such that there exists a splitting

$$F_x = E_1(x) \oplus\ldots\oplus E_k(x) \oplus F_k(x)$$

and numbers $\lambda_1(x) >\ldots> \lambda_{k+1}(x)$ such that $\dim E_j(x) = n_j$ and:

$$\lim_{n\to+\infty} \frac{1}{n} \log\| L^n/F_k(x)\| = \lambda_{k+1}(x)$$

$$\lim_{n\to-\infty} \frac{1}{n} \log\| L^{-n}v\| = -\lambda_{k+1}(x)$$

for all $0 \neq v \in F_k(x)$ such that $L^{-n}v$ exists for all $n \geq 0$, and

$$\lim_{n\to\pm\infty} \frac{1}{n} \log\| L^n/E_j(x)\| = \lambda_j(x)$$

for all $1 \leq j \leq k$, $0 \neq v \in E_j(x)$. Let $\hat{G}_k$ be the union of the sets $\hat{G}(n_1,\ldots,n_k)$ when $n_1,\ldots,n_k$ varies over all the possible k-uples of positive integers. Then:

$$\Sigma_3 = \bigcap_{k\geq 1} \hat{G}_k \ .$$

Moreover if $x \in \Sigma_3 \cap \hat{G}(n_1,\ldots,n_k)$ the spaces $E_j(x)$, $F_k(x)$ and the numbers $\lambda_1(x) >\ldots> \lambda_k(x)$ coincide with those given in the definition of Σ_3 and:

$$\Sigma_3 \cap \hat{G}(n_1,\ldots,n_k) = G_3(n_1,\ldots,n_k). \tag{2}$$

Then if prove that the functions:

$$\hat{G}(n_1,\ldots,n_k) \ni x \mapsto E_j(x) \in G^+(n_j)$$

$$\hat{G}(n_1,\ldots,n_k) \ni x \mapsto F_k(x) \in G^-(\Sigma n_j)$$

are strongly measurable, that the functions:

$$\hat{G}(n_1,\ldots,n_k) \ni x \mapsto \lambda_j(x)$$

are measurable and that $\hat{G}(n_1,\ldots,n_k)$ is a Borel set, the proof of Theorem B will be complete. The measurability of the functions λ_j follows from (1). To prove the other properties we introduce the set $\hat{G}_m(n_1,\ldots,n_k)$ of points $x \in K$ such that there exists a splitting $F_x = E_1(x) \oplus\ldots\oplus E_k(x) \oplus F_k(x)$ and numbers $\lambda_1(x) >\ldots> \lambda_{k+1}(x)$ satisfying:

a) $\lambda_j(x) - \frac{1}{m} > \lambda_{j+1}(x) + \frac{1}{m}$, $0 \le j \le k$

b) $\|v\| \frac{1}{m} \exp(\lambda_j(x) - \frac{1}{m})n \le \|L^n v\| \le \|v\| m \exp(\lambda_j(x) + \frac{1}{m})n$

for all $1 \le j \le k$, $n \ge 0$, $v \in E_j(x)$

c) For all $n \ge 0$:

$$\frac{1}{m} \exp(\lambda_{k+1}(x) - \frac{1}{m})n \le \|L^n/F_k(x)\| \le m \exp(\lambda_{k+1}(x) + \frac{1}{m})n$$

d) For all $n \ge 0$ and every $0 \neq v \in F_k(x)$ such that $L^{-n}v$ exists for all $n \ge 0$:

$$\|L^{-n}v\| \ge \|v\| \frac{1}{m} \exp(\lambda_{k+1}(x) - \frac{1}{m})n$$

e) If $\alpha(\cdot,\cdot)$ is the angle defined as in Section II, then, for all $1 \le j < k$,

$$\alpha(E_1(x) \oplus \dots \oplus E_j(x), E_{j+1}(x) \oplus \dots \oplus E_k(x) \oplus F_k(x)) \ge \frac{1}{m}$$

and

$$\alpha(E_1(x) \oplus \dots \oplus E_k(x), F_k(x)) \ge \frac{1}{m} .$$

It is clear that:

$$\hat{G}(n_1,\dots,n_k) = \bigcup_{N=1}^{\infty} \bigcap_{m=N}^{\infty} \hat{G}_m(n_1,\dots,n_k) \tag{3}$$

and at points $x \in \bigcap_{m=N}^{\infty} \cap \hat{G}_m(n_1,\dots,n_k)$ the spaces $E_j(x)$, $F_k(x)$ and the numbers $\lambda_1(x) > \dots > \lambda_{k+1}(x)$ given by the definition of $\hat{G}(n_1,\dots,n_k)$ and $\hat{G}_m(n_1,\dots,n_k)$ concide. Therefore (3) reduces our problem to show that each set $\hat{G}_m(n_1,\dots,n_k)$ is <u>closed</u> and that the functions:

$$\hat{G}_m(n_1,\dots,n_k) \ni x \mapsto E_j(x) \tag{4}$$

$$\hat{G}_m(n_1,\dots,n_k) \ni x \mapsto F_k(x) \tag{5}$$

are <u>continuous</u>. To show these properties take a sequence $\{x_n\} \subset \hat{G}_m(n_1,\dots,n_k)$ converging to $x \in K$. By Lemmas II.2 and II.3 we can suppose that the numbers $\lambda_j(x_n)$, $j=1,\dots,k+1$ and the subspaces $E_j(x_n)$, $j=1,\dots,k$ and $F_k(x_n)$ converge to numbers λ_j,

$j=1,\dots,k+1$, and subspaces E_j, $j=1,\dots,k$, F_k. The uniformity of properties (a)-(e) imply that they are satisfied putting $\lambda_j(x) = \lambda_j$, $E_j(x) = E_j$, $F_k(x) = F_k$. Therefore $x \in \hat{G}_m(n_1,\dots,n_k)$. To show the continuity of the functions λ_j, E_j, F_k at x we proceed by contradiction. Suppose for instance that $F_k(x_n)$ doesn't converge to $F_k(x)$. Then we can assume (by Lemma II.3) that the sequences $\lambda_j(x_n)$, $E_j(x_n)$, $F_k(x_n)$ converge to λ_j, E_j, F_k and that $F_k \neq F_k(x)$. The same argument used before shows that $\lambda_j = \lambda_j(x)$, $E_j = E_j(x)$ and $F_k = F_k(x)$ contradicting that $F_k \neq F_k(x)$. This concludes the proof of Theorem B.

V - Proof of Theorem D

Let K, F, φ and L as in the statement of Theorem D. First we shall prove the theorem under the assumption that L/F_x is injective for all x. Suppose that μ is a φ-invariant probability on K and define $K_1 = \bigcap_{n\geq 0} \varphi^n(K)$. Then:

$$\varphi(K_1) = K_1. \tag{1}$$

Moreover:

$$1 = \mu(K) = \mu(\varphi^{-n}(\varphi^n(K))) = \mu(\varphi^n(K)).$$

Then $\mu(K_1) = 1$. Let $\tilde{K} = \{\theta: \mathbb{Z} \to K_1 \mid \varphi(\theta(n)) = \theta(n+1),\ n \in \mathbb{Z}\}$. Define $\tilde{\varphi}: \tilde{K} \circlearrowleft$ by $\tilde{\varphi}(\theta)(n) = \varphi(\theta(n))$. This map is a homeomorphism with inverse given by $\tilde{\varphi}^{-1}(\theta)(n) = \theta(n-1)$. Let $p: \tilde{K} \to K_1$ be defined by $p(\theta) = \theta(0)$. By (1), p is surjective and $\varphi p = p\tilde{\varphi}$. Then there exists a $\tilde{\varphi}$-invariant probability $\tilde{\mu}$ on $\tilde{K}$ such that $\tilde{\mu}(p^{-1}(A)) = \mu(A)$ for every Borel set $A \subset K$. It is not clear that in general p maps Borel sets onto Borel sets. Consider the Banach vector bundle $\tilde{F}$ on $\tilde{K}$ with fibers $\tilde{F}_\theta = F_{\theta(0)}$ and the homomorphism $\tilde{L}: \tilde{F} \circlearrowleft$ covering $\tilde{\varphi}$ given by $\tilde{L}/\tilde{F}_\theta = L/F_{\theta(0)}$. To $\tilde{L}$ we can apply Theorem A because all the injectivity assumptions are satisfied and it is easy to see that $p(\Sigma(\tilde{L})) \subset \Lambda(L)$ because if $\theta \in \Sigma(\tilde{L})$ and

$\lambda_1(\theta),\dots,E_1(\theta),\dots,F_1(\theta),\dots$ are given by the definition of $\Sigma(\tilde{L})$ then the numbers $\lambda_1(\theta),\dots$ and the spaces $F_{\theta(0)} \supset F_1(\theta) \supset F_2(\theta) \supset\dots$ satisfy the definition of $\Lambda(L)$. Now we are tempted to write $\mu(\Lambda(L)) = \tilde{\mu}(p^{-1}(\Lambda(L))) \geq \tilde{\mu}(\Sigma(\tilde{L})) = 1$. But we dont know whether $\Lambda(L)$ is a Borel set. Anyhow there exists a sequence $U_1 \subset U_2 \subset \dots$ of <u>compact</u> sets contained in $\Sigma(\tilde{L})^c$ such that

$$\lim_{n\to+\infty} \tilde{\mu}(U_n) = 1.$$

Then

$$\Lambda(L)^c \subset p(\Sigma(\tilde{L}))^c \subset p(\bigcup_{n\geq 1} U_n)^c \subset (\bigcup_{n\geq 1} p(U_n))^c = \bigcap_{n\geq 1} p(U_n)^c.$$

Since U_n is compact, $p(U_n)$ is compact and a fortiori a Borel set. Then:

$$\mu(p(U_n)) = \tilde{\mu}(p^{-1}(p(U_n))) \geq \tilde{\mu}(U_n)$$

and

$$\mu(p(U_n)^c) \leq 1 - \tilde{\mu}(U_n).$$

Hence $\Lambda(L)^c$ is a measure zero set with respect to μ.

Now let us consider the general case. Define a new Banach vector bundle $\tilde{F}$ on K whose fiber $\tilde{F}_x$ on x is the space of bounded sequences $\theta : \mathbb{Z}^+ \to F_x$ endowed with the norm $\|\theta\|_\infty = \sup\|\theta(n)\|$. On $\tilde{F}$ define a homomorphism $\tilde{L}: \tilde{F} \circlearrowleft$ covering φ by:

$$(\tilde{L}\theta)(0) = L\theta(0)$$

$$(\tilde{L}\theta)(m) = \frac{1}{a^m}\,\theta(m-1) \qquad m=1,2,\dots$$

where $a > \max\{1, \sup_x \|L/F_x\|\}$. Then:

$$(\tilde{L}^n\theta)(0) = L^n\theta(0)$$

$$(\tilde{L}^n\theta)(m) = \frac{1}{a^{mn}}\,L^{n-m}\,\theta(0) \qquad m=1,2,\dots,n-1$$

$$(\tilde{L}^n\theta)(m) = \frac{1}{a^{mn}}\,\theta(m-n) \qquad m=n,\dots$$

Using this expression it is easy to check that for any $x \in K$ and

$v \in F_x$, the limit $\lim_{n\to+\infty} n^{-1} \log\|\tilde{L}^n\theta\|$ exists if and only if $\lim_{n\to+\infty} n^{-1} \log\|L^n\theta(0)\|$ exists and in that case:

$$\lim_{n\to+\infty} \frac{1}{n} \log\|L^n\theta(0)\| = \lim \frac{1}{n} \log\|\tilde{L}^n\theta\| \tag{2}$$

(where we agree in setting $\log 0 = -\infty$). Now define $\pi_x: \tilde{F}_x \to F_x$ by $\pi_x\theta = \theta(0)$. Since $\tilde{L}/\tilde{F}_x$ is injective for all x we can apply to $\tilde{L}$ the restricted version of Theorem D we proved above. Then $\Lambda(\tilde{L})$ is a total probability set. Therefore (2) implies that $\Lambda(L) \supset \Lambda(\tilde{L})$ because if $x \in \Lambda(\tilde{L})$ and $\lambda_1(x) > \lambda_2(x) > \ldots$, $F_1(x) \supset \supset F_2(x) \supset \ldots$ are given by the definition of $\Lambda(\tilde{L})$, then (2) implies that the numbers $\lambda_1(x) > \lambda_2(x) > \ldots$ and the subspaces $\pi_x F_1(x) \supset \pi_x F_2(x) \supset \ldots$ satisfy all the properties required to grant $x \in \Lambda(L)$.

VI - Proof of Theorem E

We shall consider only the invertible case, i.e. when φ is a homeomorphism and L is injective on each fiber. In this case the statement of the theorem is equivalent to show that for all $\lambda > 0$ there exists $N = N(\lambda)$ such that if $x \in \Sigma(L)$ and $\lambda_j(x) > -\lambda$ then $\dim E_j(x) \le N(\lambda)$. Once this is proved the general case is reduced to invertible case using the method employed in Section V to prove Theorem D. The proof of the statement above for the invertible case is contained in the proof of Lemma III.1. In fact the argument given in the proof of that lemma bounds $\dim E_\lambda(x)$ by a constant depending only on a number c on which the only conditions imposed is that $\log P_j^{(n)} \ge -c$ for all n and j and that b and $\mu(K_o)$ can be chosen near to one in order to have

$$\frac{\lambda + 1 + c}{2cb_\mu(K_o)} \ge \frac{2}{3} .$$

For the first requirement, recalling the definition of $P_j^{(n)}$ and that $\|\pi_j^{(n)}\| \le 2$, it is sufficient that

$$c \geq \exp 2(\sup\{\|L/F_x\| \mid x \in K\})^{-1}.$$

For the second it suffices to take c such that:

$$\frac{\lambda+1+c}{2c} > \frac{2}{3}.$$

So c can be taken depending only on λ.

VII - Proof of Theorems F and H.

The main tool for the proof of Theorem F is the following lemma:

Lemma VII.1. Let E be a Banach space, $U \subset E$ an open set, $f: U \to E$ a $C^{k,\gamma}$ map and $K_o \subset U$ a subset satisfying the following properties:

a) $f^n(K_o) \subset U$ for all $n \geq 0$

b) For all $n \geq 0$ and $x \in f^n(K_o)$ there exists a splitting $E = H_x^{(n)} \oplus G_x^{(n)}$ such that:

b_1) $H_x^{(0)}$ and $G_x^{(0)}$ depend continuously on x.

b_2) For all $n \geq 0$ and $x \in f^n(K_o)$:

$$(D_x f)H_x^{(n)} = H_{f(x)}^{(n+1)}$$

$$(D_x f)G_x^{(n)} \subset G_{f(x)}^{(n+1)}$$

b_3) $(D_x f)/H_x^{(n)}$ is injective for all $n \geq 0$ and $x \in f^n(K_o)$

b_4) There exist $C > 0$, $\mu > 0$ and $0 < \varepsilon < \gamma\mu/3$ satisfying:

$$\|(D_{f^n(x)} f^m)/G_{f^n(x)}^{(n)}\| \leq C \exp \varepsilon n \exp(-\mu m)$$

for all $x \in K_o$, $n \geq 0$, $m \geq 0$. Moreover, if $n-m \geq 0$:

$$\|((D_{f^{n-m}(x)} f^m)/H_{f^{n-m}(x)}^{(n-m)})^{-1}\| \leq C \exp \varepsilon n \exp(\mu-3\varepsilon)m.$$

b_5) If $\pi_x^{(n)}: E \to G_x^{(n)}$ is the projection associated to the splitting $E = H_x^{(n)} \oplus G_x^{(n)}$, then:

$$\|\pi^{(n)}_{f^n(x)}\| \le C \exp \epsilon n$$

for all $x \in K_o$, $n \ge 0$.

c) There exists $A > 0$ such that:

$$d(f^n(K_o), E \setminus U) \ge A \exp(\mu+\epsilon)n$$

for all $n \ge 0$.

Then there exist $r > 0$ and $\delta > 0$ arbitrarily small such that if $\Gamma_r = \{(x,v) \mid x \in K_o,\ v \in G_x^{(0)},\ \|v\| < r\}$ there exists a map $\varphi: \Gamma_r \to E$ satisfying the following properties:

I) $\pi_x^{(0)} \varphi(x,v) = v$ for all $(x,v) \in \Gamma_r$

II) For all $(x,v) \in \Gamma_r$, $\varphi(x,v)$ is the unique vector in E with $\pi_x^{(0)}(\varphi(x,v)-x) = v$ and:

$$\|f^n(\varphi(x,v)) - f^n(x)\| \le \delta \exp(-n\mu)$$

for all $n \ge 0$. In particular this implies $\varphi(x,0) = x$.

III) φ is continuous and for every x the map $\varphi(x,\cdot)$ is $C^{k,\gamma}$ and its derivatives depend continuously on (x,v). Moreover its first derivative at $(x,0)$ is the inclusion map.

For the proof of Theorem G we shall need the dual version of this lemma:

<u>Lemma VII.2</u>. Let E, U and f be as in Lemma VII.1. Let $K_o \subset U$ be a subset satisfying the following properties:

a) $f^{-n}(K_o) \subset U$ and $f/f^{-n}(K_o)$ is injective for all $n \ge 0$.

b) For all $n \ge 0$ and $x \in f^{-n}(K_o)$ there exists a splitting $E = H_x^{(n)} \oplus G_x^{(n)}$ such that:

b_1) $H_x^{(0)}$, $G_x^{(0)}$ depend continuously on $x \in K_o$.

b_2) $(D_x f)H_x^{(n)} = H_{f(x)}^{(n-1)}$

$(D_x f)G_x^{(n)} \subset G_{f(x)}^{(n-1)}$

for all $n \geq 1$ and $x \in f^{-n}(K_o)$.

b_3) $(D_x f)/H_x^{(n)}$ is injective for all $n \geq 1$, $x \in f^{-n}(K_o)$.

b_4) There exist $C > 0$, $\mu > 0$ and $0 < \epsilon < \mu\gamma/3$ such that:

$$\|(D_x f^m)/G_x^{(n)}\| \leq C \exp \epsilon n \exp(\mu - 3\epsilon)m$$

for all $x \in f^{-n}(K_o)$ and $0 \leq m \leq n$. Moreover:

$$\|((D_{f^{-m}(x)} f^m)/H_{f^{-m}(x)}^{(n+m)})^{-1}\| \leq C \exp \epsilon n \exp(-\mu m)$$

for all $x \in f^{-n}(K_o)$, $n \geq 0$, $m \geq 0$.

b_5) If $\pi_x^{(n)}: E \to H_x^{(n)}$ is the projection associated to the splitting $E = G_x^{(n)} \oplus H_x^{(n)}$, then:

$$\|\pi_{f^{-n}(x)}^{(n)}\| \leq C \exp \epsilon n$$

for all $x \in K_o$, $n \geq 0$.

c) There exists $A > 0$ such that:

$$d(f^{-n}(K_o), E \setminus U) \geq A \exp(\mu+\epsilon)n$$

for all $n \geq 0$.

Then there exists $r > 0$ and $\delta > 0$ arbitrarily small such that if $\Gamma_r = \{(x,v) \mid x \in K_o, v \in H_x^{(0)}, \|v\| < r\}$, there exists a map $\varphi: \Gamma_r \to E$ satisfying the following properties:

I) $\pi_x^{(0)} \varphi(x,v) = v$ for all $(x,v) \in \Gamma_r$.

II) For all $(x,v) \in \Gamma_r$, $\varphi(x,v)$ is the unique vector in E with $\pi_x^{(0)}(\varphi(x,v)-x) = v$ and such that there exists a sequence $v_o, v_1, \ldots$ with $v_o = \varphi(x,v)$, $f(v_j) = v_{j-1}$ for all $j \geq 1$ and $\|v_n\| \leq \delta \exp(-\mu n)$. In particular $\varphi(x,0) = x$.

III) φ is continuous and for every x the map $\varphi(x,\cdot)$ is $C^{k,\gamma}$ and its derivatives depend continuously on (x,v). Moreover its first derivative at $(x,0)$ is the inclusion map.

Remark. In the proof of both lemmas the hypothesis of f being $C^{k,\gamma}$ will be used via the following property: set

$$p(x,v) = f(x+v) - f(x) - (D_x f)v$$

and let $(D_2^j p)(x,v)$ be the j-th derivative of p at (x,v) with respect to the second variable. Then there exist $C_1 > 0$

$$\|p(x,v)\| \leq C_1\|v\|^{1+\gamma} \tag{*}$$

$$\|p(x,v+w) - p(x,v) - \sum_{j=2}^{k} \frac{1}{j} (D_2^j p)(x,v)w^j\| \leq C_1\|v\|^{k+\gamma} \tag{**}$$

for all $n \geq 0$, $x \in f^n(K_o)$ in the case of VII.1, $x \in f^{-n}(K_o)$ in the case of VII.2, $\|v\| \leq 2^{-1} A \exp(\mu+\epsilon)n$, $\|w\| \leq 2^{-1} A \exp(\mu+\epsilon)n$.

Let us see how Theorems F and H follow from this lemmas. To prove Theorem F we have to show that for any φ-invariant ergodic measure ν on K, the properties of Theorem F hold for the points of a set $K' \subset K$ with $\nu(K') = 1$. For this we need the following corollary of VII.1:

Lemma VII.3. Suppose that the Lyapounov exponents of $Df: E\times K ⟲$ are $\lambda_1 > \lambda_2 > \ldots$ ν-a.e. If $k \geq 1$, $\mu > 0$ and $\lambda_k > -\mu > \lambda_{k+1} \geq -\infty$ there exist a compact set K_μ with measure arbitrarily near to 1 and $\delta(\mu) > 0$, $r(\mu) > 0$ arbitrarily small such that the maps $x \to E_j(x)$, $x \to F_k(x)$, $1 \leq j \leq k$, are continuous on K_μ and if $\Gamma_{r(\mu)} = \{(x,v) \mid x \in K_\mu, v \in F_k(x), \|v\| < r(\mu)\}$, there exists a map $\varphi_\mu = \Gamma_{r(\mu)} \to E$ such that for all $(x,v) \in \Gamma_{r(\mu)}, \varphi_\mu(x,v)$ is the unique vector in E of the form $\varphi_\mu(x,v) = x + v + w$ where $w \in E_1(x) \oplus \ldots \oplus E_k(x)$, such that:

$$\|f^n(\varphi_\mu(x,v)) - f^n(x)\| \leq \delta(\mu)\exp(-\mu n).$$

The map φ_μ is continuous and for every $x \in K_\mu$, the map $\varphi_\mu(x,\cdot)$ is $C^{k,\gamma}$ and its derivatives vary continuously with (x,v).

The set $\{\varphi_\mu(x,v) \mid \|v\| < r(\mu)\}$ will be denoted $W^s_{\mu,loc}(x)$. Lemma VII.3 states that $W^s_{\mu,loc}(x)$ is a $C^{k,\gamma}$ submanifold characterized as the set of vectors $y \in E$ of the form $y = x + v + w$ with $v \in F_k(x)$, $\|v\| < r(\mu)$, $w \in E_1(x) \oplus \ldots \oplus E_k(x)$, and satisfying:

$$\|f^n(y)-f^n(x)\| \leq \delta(\mu)\exp(-\mu n)$$

for all $n \geq 0$. This characterization implies that if $x \in K_\mu$ and $f^n(x) \in K_\mu$, then $f^n(W^s_{\mu,loc}(x)) \subset W^s_{\mu,loc}(x)) \subset W^s_{\mu,loc}(f^n(x))$. If $x \in K$ define

$$\tilde{W}^s_\mu(x) = \bigcup \{f^{-k}(W^s_{\mu,loc}(f^k(x))) \mid f^k(x) \in K_\mu,\ k \in \mathbb{Z}^+\}.$$

If f satisfies the hypothesis of Theorem F it follows that $f^{-k}(W^s_{\mu,loc}(x))$ is a $C^{r,\gamma}$ submanifold because the density of the range of $D_y f$ at every $y \in U$ implies that the map f^k is transversal to the finite codimensional $W^s_{\mu,loc}(x)$. Moreover the sets $f^{-k}(W^s_{\mu,loc}(f^k(x)))$ with $f^k(x) \in K_\mu$ increase with k because of the characterization of the sets $W^s_{\mu,loc}(f^k(x)))$ with $f^k(x) \in K_\mu$. increase with k because of the characterization of the sets Then $\tilde{W}^s_\mu(x)$ is an embedded submanifold. Now let $\Lambda \subset K$ be the set of points x such that for every positive rational μ the property $f^n(x) \in K_\mu$ holds for infinitely many positive values of n. We shall prove that all the properties in Theorem F are valid at this points. Since $\nu(\Lambda) = 1$ (because of the ergodicity of ν) this will prove that those properties hold ν-a.e. and then that they hold for a total probability set. Let $\lambda > 0$ and $x \in \Lambda$. Take $k \geq 1$ such that $\lambda_k > -\lambda \geq \lambda_{k+1}$. Such k exists if the sequence $\lambda_1 > \lambda_t > \ldots$ contains positive elements, and we can always assume that it does because otherwise we could replace $f\colon U \times \mathbb{R} \to E \to \mathbb{R}$ defined by $f(x,t) = (f(x),2t)$, that contains 2 in the sequence. Now we claim that $W^s_\lambda(x) = \tilde{W}^s_\mu(x)$ for every $\mu > 0$ with $\lambda_k > -\mu > \lambda_{k+1}$. Clearly this claim and the previous remarks on $\tilde{W}^s_\mu(x)$ prove that $W^s_\lambda(x)$ is an embedded submanifold and that $W^s_{\lambda'}(x) = W^s_{\lambda''}(x)$ if the interval $(-\lambda',\lambda'']$ doesn't contain elements of the sequence $\{\lambda_j\}$. To prove the claim assume first that $\lambda > \mu$. To show $\tilde{W}^s_\mu(x) \supset W^s_\lambda(x)$ if $x \in \Lambda$, take $p \in W^s_\lambda(x)$ $A > 0$ such that $\|f^n(p)-f^n(x)\| \le A\exp(-\mu n)$ for all $n \ge 0$. Such A exists because of the definition of $W^s_\lambda(x)$ and the inequality $-\mu > -\lambda$. Let $\pi_y: E \to F_k(y)$ be the projection associated to the splitting $E = F_k(y) \oplus (E_1(y) \oplus \ldots \oplus E_k(y))$ and set $B = \sup\{\|\pi_y\| \mid y \in K_\mu\}$. Fix $n_o \ge 0$ such that $f^{n_o}(x) \in K_\mu$, $BA\exp(-\mu n_o) < r(\mu)$ and $A\exp(-\mu n_o) \le A\exp(-\mu n_o) \le \delta(\mu)$. Then:

$$\|\pi_{f^{n_o}(x)}(f^{n_o}(p) - f^{n_o}(x))\| \le$$

$$\le BA\exp(-\mu n_o) < r(\mu)$$

and:

$$\|f^n(f^{n_o}(p)) - f^n(f^{n_o}(x))\| \le$$

$$\le A\exp(-\mu(n+n_o)) \le \delta(\mu(\exp(-\mu n)).$$

By the characterization of the submanifolds $W^s_{\mu,loc}(y)$, these inequalities imply $f^{n_o}(p) \in W^s_{\mu,loc}(f^{n_o}(x))$. Then $p \in f^{-n}(W^s_{\mu,loc}(f^{n_o}(x))) \subset \tilde{W}^s_\mu(x)$. To show $W^s_\lambda(x) \supset \tilde{W}^s_\mu(x)$ take $\epsilon > 0$ and a rational $\mu_1 > 0$ satisfying $\min(\mu,-\lambda+\epsilon) > -\mu_1 > \lambda_{k+1}$. Without lose of generalizty we can assume that the constant $\delta(\mu)$ were taken satisfying $\delta(\mu') < \delta(\mu'')$, if $\mu' > \mu''$. Then $\delta(\mu_1) <$ $< \delta(\mu)$. If $p \in \tilde{W}^s_\mu(x)$ and $x \in \Lambda$ there exists $A > 0$ satisfying $\|f^n(p)-f^n(x)\| \le \exp(-\mu n)$ for all $n \ge 0$. Take $N > 0$ such that:

$$BA\exp(-\mu N) \le \min(r(\mu),r(\mu_1))$$

$$f^N(x) \in K_{\mu_1} \cap K_\mu$$

$$A\exp(-\mu N) \le \delta(\mu).$$

Then:

$$\|f^n(f^N(p)) - f^n(f^N(x))\| \le$$
$$\le A \exp(-(n+N)\mu) \le \delta(\mu)\exp(-\mu n) \qquad (1)$$

and:

$$\|\pi_{f^N(x)}(f^N(p) - f^N(x))\| \le$$
$$\le B\|f^N(p) - f^N(x)\| \le BA \exp(-\mu N) \le \min(r(\mu), r(\mu_1)).$$

Since $f^N(x) \in K_\mu$ these inequalitites imply $f^N(p) \in W^s_{\mu,loc}(f^N(x))$. The uniqueness properties of VII.3 grant that if some $q \in E$ satisfies:

$$\pi_{f^N(x)}(q-f^N(x)) = \pi_{f^N(x)}(f^N(p) - f^N(x))$$
$$\|f^n(q) - f^n(f^N(x))\| \le \delta(\mu)\exp(-\mu n)$$

then $q = f^N(p)$. Moreover

$$(f^N(x), \pi_{f^N(x)}(f^N(p)-f^N(x)) \in \Gamma_{r(\mu_1)}$$

because $f^N(x) \in K_{\mu_1}$. Then there exists $\varphi_{\mu_1}(f^N(x), \pi_{f^N(x)}(f^N(p)-f^N(x)))$ satisfying:

$$\pi_{f^N(x)}(\varphi_{\mu_1}(f^N(x), \pi_{f^N(x)}(f^N(p)-f^N(x)))) =$$
$$= \pi_{f^N(x)}(f^N(p) - f^N(x))$$

and

$$\|f^n(\varphi_{\mu_1}(f^N(x), \pi_{f^N(x)}(f^N(p)-f^N(x)))) - f^n(f^N(x))\| \le$$
$$\le \delta(\mu_1)\exp(-n\mu_1) \le \delta(\mu_1)\exp(-n\mu_1) \le \delta(\mu)\exp(-n\mu).$$

Therefore $f^N(x) + \varphi_{\mu_1}(f^N(x), \pi_{f^N(x)}(f^N(p)-f^N(x)))$ satisfies the properties of q and then it coincides with $f^N(p)$. But:

$$\limsup_{n\to+\infty} \frac{1}{n} \log\|f^n(p)-f^n(x)\| =$$
$$= \limsup_{n\to+\infty} \frac{1}{n} \log\|f^n(f^N(p))-f^n(f^N(x))\|$$

and since $f^N(p) = f^N(x) + \varphi_{\mu_1}(f^N(x), \pi_{f^N(x)}(f^N(p)-f^N(x)))$ it follows that the last limit is $\leq \mu_1 \leq -\lambda+\varepsilon$. This inequality holds for all $\varepsilon > 0$. Then $p \in W^s_\lambda(x)$. This completes the proof of $W^s_\lambda(x) = \tilde{W}^s_\mu(x)$ when $\lambda > \mu$. With minor charges the same methods can be applied to show the same property for $\mu \geq \lambda$.

Similar techniques can be used to prove Theorem G. Instead of VII.3 we shall use the following lemma:

Lemma VII.4. Let ν be a ergodic φ-invariant probability on K. Suppose that the Lyapounov exponents of $Df\colon E \times K \circlearrowleft$ are $\lambda_1 > \lambda_2 > \ldots$ ν-a.e. If $k \geq 1$, $\mu > 0$ and $\lambda_{k+1} < \mu < \lambda_k$ there exists a compact set K_μ with measure arbitrarily near to 1 and $\delta(\mu) > 0$, $r(\mu) > 0$ arbitrarily small such that the maps $x \mapsto F_k(x)$, $x \mapsto H_k(x) = E_1(x) \oplus \ldots \oplus E_k(x)$ are continuous on K_μ and if $\Gamma_{r(\mu)} = \{(x,v) \mid x \in K_\mu,\ v \in H_k(x),\ \|v\| < r(\mu)\}$, there exists $\varphi_\mu\colon \Gamma_{r(\mu)} \to E$ such that for all $(x,v) \in \Gamma_{r(\mu)}$, $\varphi_\mu(x,v)$ is the unique vector of the form $x + v + w$ with $w \in F_k(x)$, such that $f^{-n}(x)$ exists for all $n \geq 0$ and satisfies:

$$\|f^{-n}(\varphi_\mu(x,v)) - f^{-n}(x)\| \leq \delta(\mu)\exp(-\mu n).$$

Moreover φ_μ is continuous and has continuous derivative with respect to the second variable.

To prove Theorem G using this lemma we proceed as we did to deduce Theorem F from Lemma VII.3. We have to show that for any φ-invariant ergodic probability ν the properties of Theorem H hold ν-a.e. For this we take any $\mu > 0$ such that there exists $k \geq 1$ satisfying $\lambda_{k+1} < \mu < \lambda_k$. Then we define $W^u_{\mu,loc}(x) = \{\varphi_\mu(x,v) \mid v \in H_k(x),\ \|v\| < r(\mu)\}$ if $x \in K_\mu$. Define $\tilde{W}^u_\mu(x) = \bigcup \{f^n(W^u_{\mu,loc}(f^{-n}(x))) \mid n \geq 0,\ f^{-n}(x) \in K_\mu\}$. By Lemma VII.4, $W^u_{\mu,loc}(y)$ is a $C^{r,\gamma}$ finite dimensional submanifold. Since f and its first derivative at every point of U are injective then $f^n(W^u_{\mu,loc}(f^{-n}(x)))$ is, for all $n \geq 0$, a $C^{r,\gamma}$ finite dimensional submanifold. As in the proof of Theorem F, we prove that $f^{n_1}(W^u_{\mu,loc}(f^{-n_1}(x))) \supset f^{n_2}(W^u_{\mu,loc}(f^{-n_2}(x)))$ if $n_1 > n_2$ and $f^{n_1}(x) \in K_\mu$, $f^{n_2}(x) \in K_\mu$. Then $\tilde{W}^u_\mu(x)$ is a finite dimensional $C^{r,\gamma}$ immerse submanifold. Now consider the set $\Lambda \subset K$ of points x such that $f^{-n}(x) \in K_\mu$ for infinitely many values of $n \geq 0$. This set satisfies $\nu(\Lambda) = 1$ because of the ergodicity of ν. As in the proof of Theorem F we show that if $\lambda > 0$ and $\mu > 0$ satisfy $\lambda_{k+1} < \lambda \leq \lambda_k$ and $\lambda_{k+1} < \mu < \lambda_k$ then $W^u_\lambda(x) = \tilde{W}^u_\lambda(x)$ for all $x \in \Lambda$. From this follows that the properties of Theorem H hold for every point in Λ. Then they hold ν-a.e. as we wished to prove.

It remains to show how VII.3 and VII.4 follow from VII.1 and VII.2 respectively. In fact we shall only prove VII.3. The proof of VII.4 is basically the same with some obvious modifications.

Take $\varepsilon > 0$ such that $\lambda_k > -\mu + 3\varepsilon > \lambda_{k+1}$ and define:

$$C_1(x) = \sup_{n\geq 0} \|(D_x f^n)/F_k(x)\| \exp \mu n$$

$$C_2(x) = \sup_{n\geq 0} \|(D_{f^{-n}(x)} f^n)/E_1(x) \oplus \dots \oplus E_k(x)\| \exp(-\mu+3\varepsilon)n.$$

The basic properties of the regular points stated in its definition imply that $C_1(x) < +\infty$ ν-a.e. Let $\pi_y\colon E \to F_k(y)$ be the projection associated to the splitting $E = (E_1(x) \oplus \dots \oplus E_k(x)) \oplus F_k(x)$. Define

$$C_3(x) = \sup_{n\geq 0} \|\pi_{f^n(x)}\| \exp(-\varepsilon n).$$

By Theorem B we have $C_3(x) < +\infty$ ν-a.e. Moreover it is easy to check that $\log(C_1(\varphi(x))/C_1(x))$ and $\log(C_2(x)/C_2(\varphi(x)))$ are bounded by integrable functions. Then we can apply Lemma III.8 to define the function:

$$C_4(x) = \sup_{n\geq 0} \exp(-n\varepsilon)\max(C_1(\varphi^n(x)),C_2(\varphi^n(x))).$$

By Lemma III.8, $C_4(x) < +\infty$ ν-a.e. Then we can take a compact set K_μ with arbitrarily large measure such that C_4 is bounded on K_μ by a constant $C > 0$. We can assume that also C_3 is bounded by C on K_μ. Then we can apply Lemma VII.1 to the compact set K_μ setting $H_x^{(n)} = E_1(x) \oplus \ldots \oplus E_k(x)$, $G_x^{(n)} = F_k(x)$ for $x \in f^n(K_\mu)$. Obviously $\epsilon > 0$ can be taken so small that $2\epsilon < \gamma\mu$ as required by condition (b) of VII.1. Applying VII.1 we obtain $r = r(\mu)$, $\delta = \delta(\mu)$ and $\varphi = \varphi_\mu : \Gamma_{r(\mu)} \to E$ satisfying the properties required by VII.3.

Now let us prove VII.1. Let Σ be the space of sequences $\xi : Z^+ \to E$ such that:

$$\sup_{n\geq 0} \|\xi(n)\| \exp \mu n < +\infty$$

endowed with the norm:

$$\|\xi\| = \sup_{n\geq 0} \|\xi(n)\| \exp \mu n$$

Σ is a Banach space. Set:

$$\Sigma_\rho = \{\xi \in \Sigma \mid \|\xi\| < \rho\}$$

$$\Gamma_r = \{(x,v) \mid x \in K_o, \ v \in G_x^{(0)}, \ \|v\| < r\}$$

and $\tilde{\pi}_x^{(n)} = I - \pi_x^{(n)}$ for $x \in f^n(K_o)$. Take $0 < \rho < d(K_o, E\ U)$ and define a map $F: \Gamma_\rho \times \Gamma_\rho \to \Sigma$ by:

$$F(x,v,\xi)(n) =$$

$$= (D_x f^n)v + \sum_{j=0}^{n-1} (D_{f^{j+1}(x)} f^{(n-1)-j})\pi_{f^{j+1}(x)}^{(j+1)} p(f^{j+1}(x), \xi(j)) -$$

$$- \sum_{j=n}^{\infty} (D_{f^n(x)} f^{j-(n-1)})^{-1} \tilde{\pi}_{f^{j+1}(x)}^{(j+1)} p(f^j(x), \xi(j))$$

for $n \geq 1$ and

$$F(x,v,\xi)(0) = v - \sum_{j=0}^{\infty} (D_x f^{j+1})^{-1} \tilde{\pi}_{f^{j+1}(x)}^{(j+1)} p(f^j(x), \xi(j)).$$

The series that appear in this definition and the fact that the

Lemma VI.5. Let $x \in K_o$, $0 < \delta < \rho$ and $v \in G_x^{(0)}$ with $\|v\| \le \rho$. Then $y \in E$ satisfies:

$$\pi_x^{(0)}(y-x) = v \tag{1}$$

$$\|f^n(y)-f^n(x)\| \le \rho \exp(-\mu n) \tag{2}$$

for all $n \ge 0$ if and only if there exists $\xi \in \Sigma_\rho$ such that

$$\xi(0) = y-x \tag{3}$$

$$F(x,v,\xi) = \xi. \tag{4}$$

Proof. Suppose y satisfies (1), (2). It is easy to see that if $w_n = f^n(y) - f^n(x)$ then:

$$f^n(y)-f^n(x) = (D_x f^n)w_o + \sum_{j=0}^{n-1} (D_{f^{j+1}(x)} f^{(n-1)-j}) p(f^j(x),w_j).$$

and

$$\tilde{\pi}^{(n)}_{f^n(x)} (f^n(y)-f^n(x)) = (D_x f^n)\tilde{\pi}_x^{(0)} w_o +$$

$$+ \sum_{j=0}^{n-1} (D_{f^{j+1}(x)} f^{(n-1)-j}) \tilde{\pi}^{(j+1)}_{f^{j+1}(x)} p(f^j(x),w_j) =$$

$$= (D_x f^n)(\tilde{\pi}_x^{(0)} w_o + \sum_{j=0}^{n-1} (D_x f^{j+1})^{-1} \tilde{\pi}^{(j+1)}_{f^{j+1}(x)} p(f^j(x),w_j).$$

Hence:

$$\rho \exp(-\mu n) C \exp \varepsilon n \ge \|\tilde{\pi}^{(n)}_{f^n(x)} (f^n(y)-f^n(x))\| \ge$$

$$\ge \frac{1}{C \exp(\mu-3\varepsilon)n} \|\tilde{\pi}_x^{(0)} w_o + \sum_{j=a}^{n-1} (D_x f^{j+1})^{-1} \tilde{\pi}^{(j+1)}_{f^{j+1}(x)} p(f^j(x),w_j)\|.$$

Finally we obtain that for all $n \ge 0$:

$$C^2 \rho \exp(-\mu n)\exp(-2\varepsilon n) \ge \|\tilde{\pi}_x^{(0)} w_o + \sum_{j=0}^{n-1} (D_x f^{j+1})^{-1} \tilde{\pi}^{(j+1)}_{f^{j+1}(x)} p(f^j(x),w_j)\|.$$

This implies:

$$\tilde{\pi}_x^{(0)} w_o + \sum_{j=0}^{\infty} (D_x f^{j+1})^{-1} \tilde{\pi}^{(j+1)}_{f^{j+1}(x)} p(f^j(x),w_j) = 0. \tag{5}$$

Therefore for $n > 0$ we can write:

sequence $n \to F(x,v,\xi)(n)$ belongs to Σ follow from the inequality $\|(D_x f^n)v\| \le C\exp(-\mu n)$ and the inequality (*) in the remark after the statement of VII.1. In fact:

$$\|\sum_{j=0}^{n-1} (D_{f^{j+1}(x)} f^{(n-1)-j})\pi_{f^{j+1}(x)}^{(j+1)} p(f^j(x),\xi(j))\| \le$$

$$\le \sum_{j=0}^{n-1} C_1C^2 \exp 2(j+1)\epsilon \exp(-\mu(n-1-j))\|\xi\|^{1+\gamma} \exp(-(1+\gamma)\mu n) \le$$

$$\le C_1C^2\|\xi\|^{1+\gamma} \exp(-(2+\gamma)\mu n) \sum_{j=0}^{n-1} \exp(2(j+1)\epsilon + \mu(j+1)) =$$

$$= C_1C^2\|\xi\|^{1+\gamma} \exp(-(2+\gamma)\mu n)\exp(2\epsilon+\mu) \frac{1-\exp(2\epsilon+\mu)n}{1-\exp(2\epsilon+\mu)} \le$$

$$\le \frac{C_1C^2\|\xi\|^{1+\gamma} \exp(-(2+\gamma)\mu n)\exp(2\epsilon+\mu)}{\exp(2\epsilon+\mu)-1}$$

$$\frac{1}{\exp(2\epsilon+\mu)-1} C_1C^2\|\xi\|^{1+\gamma} \exp(2\epsilon+\mu)\exp(-(\mu+\gamma\mu-2\epsilon)n).$$

Since $2\epsilon \lesssim \gamma\mu$ the last exponential is dominated by $\exp(-\mu n)$. Finally:

$$\|\sum_{j=n}^{\infty} (D_{f^n(x)} f^{j-(n-1)})^{-1} \tilde{\pi}_{f^{j+1}(x)}^{(j+1)} p(f^j(x),\xi(j))\| \le$$

$$\le \sum_{j=n}^{\infty} C_1C^2 \exp 2\epsilon(j+1)\exp(\mu-3\epsilon)(j-n+1)\|\xi\|^{1+\gamma} \exp(-(1+\gamma)\mu j) \le$$

$$\le C_1C^2\|\xi\|^{1+\gamma} \exp(\mu-\epsilon)\exp(2\epsilon-(1+\gamma)\mu)n \sum_{j=n}^{\infty} \exp(-\epsilon-\gamma\mu)(j-n) \le$$

$$\le C_1C^2\|\xi\|^{1+\gamma} \exp(\mu-\epsilon)\exp(-\mu n) \frac{1}{1-\exp(-\epsilon-\gamma\mu)}.$$

Similar computations show that F is continuous and, using inequality (**) of the remark after Lemma VII.2, that for each x the function $F(x,\cdot,\ )$ is $C^{r,\gamma}$ and its derivatives depend continuously on (x,v,ξ).

The main property of F is stated in the next lemma:

$$w_n = f^n(y) - f^n(x) =$$

$$= (D_x f^n)v + \sum_{j=0}^{n-1} (D_{f^{j+1}(x)} f^{(n-1)-j})\, \pi^{(j+1)}_{f^{j+1}(x)}\, p(f^{j+1}(x), w_j) +$$

$$+ (D_x f^n)\tilde{\pi}^{(0)}_x w_o + \sum_{j=0}^{n-1} (D_{f^{j+1}(x)} f^{(n-1)-j})\, \tilde{\pi}^{(j+1)}_{f^{j+1}(x)}\, p(f^{j+1}(x), w_j) =$$

$$= (D_x f^n)v + \sum_{j=0}^{n-1} (D_{f^{j+1}(x)} f^{(n-1)-j})\pi^{(j+1)}_{f^{n+1}(x)}\, p(f^{j+1}(x), w_j) +$$

$$+ (D_x f^n)(\tilde{\pi}^{(0)}_x w_o + \sum_{j=0}^{n-1} (D_x f^{j+1})^{-1}\, \tilde{\pi}^{(j+1)}_{f^{j+1}(x)}\, p(f^{j+1}(x), w_j)) =$$

$$= (D_x f^n)v + \sum_{j=0}^{n-1} (D_{f^{j+1}(x)} f^{(n-1)-j})\pi^{(j+1)}_{f^{j+1}(x)}\, p(f^{j+1}(x), w_j) +$$

$$- (D_x f^n)(\sum_{j=0}^{\infty} (D_x f^{j+1})^{-1}\, \tilde{\pi}^{(j+1)}_{f^{j+1}(x)}\, p(f^{j+1}(x), w_j) = F(x,v,w)(n). \qquad (6)$$

Moreover, (5) implies:

$$v = \pi^{(0)}_x w_o = w_o - \tilde{\pi}^{(0)}_x w_o =$$

$$= w_o + \sum_{j=0}^{\infty} (D_x f^{j+1})^{-1}\, \tilde{\pi}^{(j+1)}_{f^{j+1}(x)}\, p(f^{j+1}(x), w_j) \qquad (7)$$

Then (6) and (7) imply:

$$F(x,v,w) = w.$$

Where $w \in \Sigma_\rho$ is the sequence $w_o, w_1, \ldots$ Clearly this sequence satisfies $y-x = \pi^{(0)}_x w_o$ so that (3) and (4) are satisfied. Conversely suppose that there exists $\xi \in \Sigma_\rho$ satisfying (3) and (4). By induction on n it is easy to prove that:

$$f^n(y) - f^n(x) = F(x,v,\xi)(n).$$

Then $\|f^n(y)-f^n(x)\| = \|F(x,v,\xi)(n)\| = \|\xi(n)\| \le \rho \exp(-\mu n)$. Moreover $\pi^{(0)}_x (y-x) = \pi^{(0)}_x \xi(0) = \pi^{(0)}_x F(x,v,\xi)(0) = v$.

Then $y \in E$ satisfies (1), (2) if and only if $y = x + \xi$ where ξ satisfies

$$\Phi(x,v,\xi) = 0 \qquad (8)$$

where $\Phi: \Gamma_\rho \times \Sigma_\rho \to \Sigma$ is just $\Phi(x,v,\xi) = \xi - F(x,v,\xi)$.

Moreover the derivative of G respect to the third variable at $(a,0,0)$ is the identity and $G(x,0,0) = 0$. Then we can apply the implicit function theorem to solve (8) and we obtain $\rho > r > 0$ and $\rho > \delta > 0$ arbitrarily small such that if $(x,v) \in \Gamma_r$ there exists a unique $\xi \in \Sigma_\delta$ such that $G(x,v,\xi) = 0$ and defining $\xi = \Psi(x,v)$ the function $\Psi: \Gamma_r \to \Sigma_\delta$ satisfies that for all $x \in K_o$ the map $\varphi(x,\cdot)$ is $C^{r,\gamma}$ and its derivatives depend continuously on (x,v). By VI.4 it follows that for all $(x,v) \in \Gamma_r$ there exists a unique $\gamma \in E$ satisfying (1), (2) and this y is given by $y = x + \Psi(x,v)$. Then the function $\varphi: \Gamma_r \to \Sigma_\delta$ defined by $\varphi(x,v) = x + \Psi(x,v)$ satisfies all the required properties.

REFERENCES

[1] J. Hale - Theory of functional differential equations. Springer-Verlag, 1977.

[2] D. Henry - Geometric Theory of Semilinear Parabolic Equations. Lecture Notes in Mathematics (1980) 840

[3] V.I. Oseledec - A multiplicative ergodic theorem. Trans. Moscow Math. Soc. 19 (1968) 197-231.

[4] Ya. Pesin - Invariant manifolds which correspond to non vanishing exponents. Math. USSR 10, (1976).

[5] V.A. Pliss - On a conjecture due to Smale, Diff. Uravnenija, 8 (1972).

[6] D. Ruelle - Ergodic theory of differentiable systems. Publications IHES 50 (1979).

[7] D. Ruelle - Characteristic exponents and invariant manifolds in Hilbert space. Annals of Mathematics, 115 (1982).

Commun. Math. Phys. 100, 495–524 (1985)

Communications in
Mathematical
Physics

Hyperbolicity, Sinks and Measure in One Dimensional Dynamics

Ricardo Mañé
Institute de Matematica Pura e Aplicada, Estrada Dona Castorina 110, 20460 Rio de Janeiro-RJ, Brasil

Abstract. Let f be a C^2 map of the circle or the interval and let $\Sigma(f)$ denote the complement of the basins of attraction of the attracting periodic orbits. We prove that $\Sigma(f)$ is a hyperbolic expanding set if (and obviously only if) every periodic point is hyperbolic and $\Sigma(f)$ doesn't contain the critical point. This is the real one dimensional version of Fatou's hyperbolicity criteria for holomorphic endomorphisms of the Riemann sphere. We also explore other applications of the techniques used for the result above, proving, for instance, that for every C^2 immersion f of the circle (i.e. a map of the circle onto itself without critical points), either its Julia set has measure zero or it is the whole circle and then f is ergodic, i.e. positively invariant Borel sets have zero or full measure.

Introduction

The subject of this paper is the dynamics of C^2 maps of the circle or the interval, on regions bounded away from the critical points. The aspects of the dynamics that we shall consider, and the corresponding results that we shall prove, can be summarized as follows:

Hyperbolicity—If the map is not topologically equivalent to an irrational rotation of the circle, every compact invariant set not containing critical points, sinks or non-hyperbolic periodic points is hyperbolic.

Stability—Structural stability is generic in the space of C^r immersions of the circle and is characterized by the hyperbolicity of the non-wandering set.

Ergodicity—Transitive C^2 immersions of the circle are ergodic, i.e. every invariant Borel set has either zero or full Lebesgue measure.

Measure—If Γ is a compact invariant set with empty interior not containing critical points, then either the Lebesgue measure of Γ is zero or there exists an interval U that is mapped diffeomorphically into itself by some power of the map and such that $\Gamma \cap U$ has positive Lebesgue measure.

Sinks—For every compact set K that doesn't contain critical points, the periods of the sinks or non-hyperbolic periodic orbits contained in K are bounded.

Originally published in: *Comm. Math. Phys. 100 (1985), no. 4, 495–524,*

Before entering into the precise statements of these results, let us recall the definition and basic properties of the concepts they involve.

Let N denote either the circle $S^1=\{z\in\mathbb{C}\,|\,|z|=1\}$ or the interval $[0,1]$. Let $\mathrm{End}^r(N)$ be the space of C^r maps $f:N\circlearrowleft$ (endomorphisms) endowed with the C^r topology. As usual we say that $x\in N$ is a *periodic point* of $f\in\mathrm{End}^r(N)$ if $f^n(x)=x$ for some $n\geqq 1$. In this case we say that it is *hyperbolic* if $|(f^n)'(x)|\neq 1$, a *sink* if $|(f^n)'(x)|<1$ and a *source* if $|(f^n)'(x)|>1$. The *basin* of a sink x is defined as the set of points y such that $\lim_{n\to+\infty} d(f^n)(x), f^n(y))=0$. It is an open set containing x. We say that a set $\Lambda\subset N$ is an *invariant* set of $f\in\mathrm{End}^r(N)$ if $f(\Lambda)\subset\Lambda$. If there exists a neighborhood U of Λ such that $\Lambda=\bigcap_n f^n(U)$, we say that Λ is *isolated* and that U is an *isolating block* Λ. A *hyperbolic set* of $f\in\mathrm{End}^r(N)$ is a compact invariant set Λ such that there exists constants $K>0$ and $\lambda>1$ satisfying $|(f^n)'(x)|\geqq K\lambda^n$ for all $x\in\Lambda$, $n\geqq 0$. Hyperbolic sets have remarkable properties, the most outstanding being its stability and the reducibility of its dynamics to certain simpler models.

What follows is a short glossary of these properties:

I) If N is a hyperbolic set of $f\in\mathrm{End}^r(N)$, then $N=S^1$, and f *is topologically equilvalent to a map* $f_d:S^1\circlearrowleft$ given by $f_d(z)=z^d$, where d is an integer $\neq 1$ or -1. Recall that topologically equivalent means that there exists a homeomorphism $h:S^1\circlearrowleft$ such that $hf=f_dh$.

II) If $\Lambda\subset N-\partial N$ is an isolated hyperbolic set of $f\in\mathrm{End}^r(N)$ and $\Lambda\neq N$, then $f|\Lambda$ is topologically equivalent to a subshift of finite type. A subshift of finite type is a map $\partial:\Sigma^+(A)\circlearrowleft$, where A is a $m\times m$ matrix whose entries a (i,j) are 0 or 1; $\Sigma^+(A)$ is the space of the sequences $\theta:\mathbb{Z}^+\to\{1,\dots,m\}$ that satisfy the transition law a $(\theta(n),\theta(n+1))=1$ for all $n\geqq 0$, endowed with the pointwise convergence topology, and $\sigma:\Sigma^+(A)\circlearrowleft$ is defined by $\sigma(\theta)(n)=\theta(n+1)$.

III) If $\Lambda\subset N-\partial N$ is an isolated hyperbolic set of $f\in\mathrm{End}^1(N)$, then for every isolating block U of Λ, there exists a neighborhood $\mathscr{U}$ of f such that if $g\in\mathscr{U}$, $f|\Lambda$ is topologically equivalent to $g/\bigcap_n g^n(U)$.

IV) For every hyperbolic set $\Lambda\subset N-\partial N$ of $f\in\mathrm{End}^r(N)$ and every neighborhood V of Λ there exists an isolated hyperbolic contained in V and containing Λ.

Property (I) is the one dimensional case of a result of Shub [6]. Property (II) was proved by Jacobson [1] and Nitecki [5] (for a certain Λ, but their techniques with minor modifications prove (II)). Property (III) is nowadays a minor application of the stability theory of hyperbolic sets, but proofs can be found in [5]. (IV) is folklorical. We shall give a simple proof of it in the Appendix.

In [1], Jacobson introduced the set $\Sigma(f)$ of an endomorphism $f\in\mathrm{End}^r(N)$ defining it as the complement of the union of the basins of the sinks and, attempting to characterize the C^r structurally stable endormorphisms (i.e. those that are topoligically equivalent to every C^r nearby endomorphism), he considered the class of maps $f\in\mathrm{End}^r(N)$ such that $\Sigma(f)$ is hyperbolic and contained in $N-\partial N$. This class will be denoted $\mathscr{S}^r(N)$. He proved the following results:

V) $\mathscr{S}^1(N)$ is open and dense in $\mathrm{End}^1(N)$.

VI) Every $f\in\mathscr{S}^1(N)$ is Σ-stable, i.e. if $g\in\mathrm{End}^1(N)$ is C^1 near to f, $g/\Sigma(g)$ is topologically equivalent to $f/\Sigma(f)$.

VII) If $f \in \mathscr{S}^1(N)$ and $\Sigma(f) = N$, then $N = S^1$ and $f: S^1 \supsetarrow$ is topologically equivalent to a map $f_d: S^1 \supsetarrow$ (defined as in property (I), of which this property is an immediate corollary).

VIII) If $f \in \mathscr{S}^1(N)$ and $\Sigma(f) \neq N$, then $f/\Sigma(f)$ is topologically equivalent to a finite type subshift. This is a corollary of (II).

IX) If $f \in \mathscr{S}^1(N)$ and has not critical points, then f is C^1-structurally stable.

X) If $f \in \mathscr{S}^2(N)$ and satisfies:

a) Every critical point is non-degenerate, i.e. $f'(x) \Rightarrow f''(x) \neq 0$.

b) If $\tilde{C}(f)$ denotes the set of critical points of f, plus, in the case $N = [0, 1]$, the points 0 and 1, then $f^n(x) \neq f^m(y)$ for all x and y in $\tilde{C}(f)$ and every $n \geqq 0, m \geqq 1$.

c) If $N = [0, 1]$, 0 and 1 are not critical points of f, then f is C^2-structurally stable.

These properties pose two questions:
Problem I. Is $\mathscr{S}^r(N)$ dense in $\mathrm{End}^r(N)$ for $r \geqq 2$?
Problem II. Is $\Sigma(f)$ hyperbolic for every C^r-structurally stable $f \in \mathrm{End}^r(N)$?

An affirmative answer to Problem II would yield a complete characterization of structural stability, namely that $f \in \mathrm{End}^1(N)$ is structurally stable if and only if $C(f) = \phi$ and $f \in \mathscr{S}^1(N)$, and that $f \in \mathrm{End}^r(N)(r \geqq 2)$ is C^r-structurally stable if and only if $f \in \mathscr{S}^r(N)$ and satisfies (a), (b) and (c). We shall give positive answers to both questions when restricted to the space of C^r immersions (that is an open subset of $\mathrm{End}^r(N)$) with $r \geqq 2$. This will be based on the following hyperbolicity criteria that also has an independent intrinsic interest.

Theorem A. *If $f \in \mathrm{End}^2(N)$ and $\Lambda \subset N$ is a compact invariant set that doesn't contain critical points, sinks or non-hyperbolic periodic points, then either $\Lambda = N = S^1$ and f is topologically equivalent to an irrational rotation or Λ is a hyperbolic set.*

Corollary I. *If all the periodic points of $f \in \mathrm{End}^2(N)$ are hyperbolic and $\Sigma(f)$ doesn't contain critical points, then either $N = S^1$ and f is topologically equivalent to an irrational rotation or $\Sigma(f)$ is a hyperbolic set.*

Corollary II. *For all $r \geqq 1$, every C^r immersion $f: S^1 \supsetarrow$ can be approximated in the C^r topology by an immersion $g: S^1 \supsetarrow$ such that $\Sigma(g)$ is a hyperbolic set.*

Corollary III. *For all $r \geqq 2$, a C^r immersion $f: S^1 \supsetarrow$ is C^r structurally stable if and only if $\Sigma(f)$ is hyperbolic.*

Corollary I is an easy consequence of Theorem A. Corollary II also follows easily from Theorem A recalling that for all $r \geqq 1$, the hyperbolicity of all the periodic points is a generic property in $\mathrm{End}^r(N)$. Corollary III follows from Theorem A, property (IX) and the fact that all the periodic points of a C^r-structurally stable endomorphism are hyperbolic.

We were not able to solve Problems I and II, but using Theorem A we can reduce its solution to a problem related to links of critical points. To define this notion first recall that given $f \in \mathrm{End}^r(N)$, the α-limit set of a point $x \in N$ is defined by:

$$\alpha(x) = \bigcap_{n \geqq 0} \overline{\bigcup_{m \geqq n} f^{-m}(\{x\})}.$$

Denote $\tilde{C}(f)$ the set of critical points of f plus, in the case $N = [0, 1]$, the points 0 and 1. A *link* of f is a pair (x, y) of points in $\tilde{C}(f)$ such that y is in the forward orbit of x. A *fake link* is a pair (x, y) of points in $\tilde{C}(f)$ such that it is not a link and $x \in \alpha(y)$. Denote $v(f)$ the number of links of f.

Conjecture. If all the critical points of $f \in \mathrm{End}^r(N)$ $(r \geqq 2)$ are non-degenerate and all its periodic points are hyperbolic, then, if f has a fake link, there exists $g \in \mathrm{End}^r(N)$ arbitrarily near to f in the C^r topology and satisfying $v(g) > v(f)$.

In other words, using the fake link a new link is created by a small perturbation without destroying those already existing. To see how this conjecture implies an affirmative answer to Problems I and II we need the following approximation theorem.

Theorem B. *Suppose that $f \in \mathrm{End}^r(N)$, $r \geqq 2$, satisfies:*

a) *all the periodic points of f are hyperbolic,*
b) *all the critical points are non-degenerate,*
c) *f has no fake links.*

Then f can be approximated in the C^r topology by an endomorphism $g \in \mathscr{S}^r(N)$.

Now let $\mathscr{S}_0^r(N)$ be the set of maps $f \in \mathrm{End}^r(N)$ such that all its periodic points are hyperbolic and all its critical points are nondegenerate. If $f_0 \in \mathscr{S}_0^r(N)$ take an open neighborhood $\mathscr{U}$ of f_0 such that the number of critical points of every $f \in \mathscr{U}$ is the same, say N. Then $v(f) \leqq (N+2)^2$ for all $f \in \mathscr{U}$. Take $f_1 \in \mathscr{U}$ such that $v(f_1) \geqq v(f)$ for all $f \in \mathscr{U}$. It is easy to see that we can construct an endomorphism $f_2 \in \mathscr{U}$ such that all its periodic points are hyperbolic and having the same links of f_1. Then $v(f_2) = v(f_1)$. Suppose that f_2 has a fake link. If the conjecture is true there exists $f_3 \in \mathscr{U}$ with $v(f_3) > v(f_2)$. But then $v(f_3) > v(f_1)$ contradicting the way we choose f_1. Therefore f_2 has no fake links. By Theorem B we can approximate f_2 by $g \in \mathscr{U}$ such that $g \in \mathscr{S}^r(N)$. Since $\mathscr{U}$ is arbitrary, we have proved that, if the conjecture is true, every element of $\mathscr{S}_0^r(N)$ can be approximated by an element in $\mathscr{S}^r(N)$. Since $\mathscr{S}_0^r(N)$ is obviously dense in $\mathrm{End}^r(N)$ this yields an affirmative answer to Problem I. Therefore, if $f \in \mathrm{End}^r(N)$ is C^r-structurally stable $(r \geqq 2)$ we can approximate it by $g \in \mathscr{S}^r(N)$, and because of the structural stability of f, g and f are topologically equivalent. Moreover, it is easy to verify that if $r \geqq 2$ all the critical points of a C^r-structurally stable endomorphism are non-degenerate and that a topological equivalence transforms non-degenerate critical points in critical points. Then, the fact that $\Sigma(g)$ contains no critical points implies that $\Sigma(f)$ contains no critical points. Since all the periodic points of f are hyperbolic (because f is structurally stable) it follows from Corollary I that $\Sigma(f)$ is hyperbolic.

This approach is obviously motivated by Peixoto's proof [6] of the density of Morse–Smale vectorfields in two dimensional compact orientable manifolds, where the role of the critical points is played by the saddles and saddle connections correspond to links. The phenomena corresponding to fake links is a separatrix of a saddle accumulating in a saddle. In Peixoto's proof there is a step, corresponding to Theorem B, where it is proved that a Kupka–Smale vectorfield without fake links can be approximated by a Morse–Smale vectorfield. But the crucial point is to show that when a separatrix of a saddle accumulates in another saddle then a new saddle connection can be created without destroying those already existing. However this step doesn't work in the non-orientable case because of the reversing of orientation of certain Poincaré maps. The same problem appears here, but instead of being produced by the non-orientability of the manifold its cause is the existence of critical points.

Now let us consider the problem of whether there exists a one dimensional version of the infinitely many sinks phenomena found by Newhouse for diffeomorphisms in dimension $\geqq 2$. To produce examples of endomorphisms $f:N \circlearrowleft$ with infinitely many sinks is trivial. For instance one can easily construct C^∞ diffeomorphisms $f:[0,1] \circlearrowleft$ having infinitely many sinks. However the following theorem states that the only way to produce C^2 endomorphisms with infinitely many sinks bounded away from the critical points is by inserting intervals where some power of f acts as a diffeomorphism with infinitely many sinks. To give the precise statement of this property, let us first introduce a definition. Given if $\in \mathrm{End}^r(N)$ we shall say that two periodic orbits γ_1 and γ_2 are homologous if they have the same period n and there exist $x_1 \in \gamma_1$ and $x_2 \in \gamma_2$ and an interval J with x_1 and $x_2 \in \gamma_2$ as endopoints such that f^n/J maps J diffeomorphically onto itself. Obviously this is an equivalence relation.

Theorem C. *If $f \in \mathrm{End}^2(N)$, every compact set that doesn't contain critical points contains only finitely many non-homologous orbits of sinks or non-hyperbolic periodic points.*

Corollary I. *If $f \in \mathrm{End}^2(N)$ and K is a compact set not containing critical points, then the periods of the sinks or non-hyperbolic periodic points whose orbit is contained in K are bounded.*

Corollary II.*If $f:N \circlearrowleft$ is real analytic and K is a compact set not containing critical points then the set of sinks or non-hyperbolic periodic points whose orbits are contained in K is finite.*

Both Corollaries follow immediately from Theorem C. Another interesting application is the finiteness of the set of periodic plateaus of an immersion. Let us recall the basic dynamical properties of immersions. Given a C^1 endomorphism $f:S^1 \circlearrowleft$ with degree $d \neq 1$ or -1 there exists a continuous map $h:S^1 \circlearrowleft$ of degree 1 such that $hf = f_d h$, where $f_d:S_1 \circlearrowleft$ is defined by $f_d(z) = z^d$. Moreover, if f is an immersion, h is monotone, i.e. for every $z \in S^1$, $h^{-1}(\{z\})$ is either a unique point or an interval $[a,b]$ with $a \neq b$. In the last case we say that (a,b) is a *plateau* of f. Denote $J(f)$ the complement of the union of the plateaux of f. Using the map h it is easy to check the following properties:

XI) Two plateaux are either disjoint or coincide.

XII) f maps diffeomorphically plateaux onto plateaux.

XIII) Every plateau U is either periodic (i.e. $f^N(U) = U$ for some $N \geqq 1$), eventually periodic (i.e. $f^m(U)$ is periodic for some $m \geqq 1$) or wandering (i.e. $f^n(U) \cap f^m(U) = l$ for all $n \geqq 1$).

XIV) If f is C^2, every plateau is periodic or eventually periodic.

XV) $x \in J(f)$ if and only if for every neighborhood W of x there exists $n \geqq 0$ such that $f^n(U) = S^1$. This implies that $F/J(f)$ is transitive.

XVI) $J(f)$ is either a Cantor set or coincides with S^1. In this case f is topologically equivalent to $z \mapsto z^a$.

XVII) $J(f)$ contains a dense subset of sources.

Property (XIV) is not a corollary of the existence of the semiconjugacy h but follows easily from an adaptation of Denjoy's theorem (see Lemma I.4 below).

Each periodic plateau is mapped by h in a periodic point of f_d with the same period as the plateau. Moreover different plateaux are mapped in different points. Therefore if an immersion $f: S^1 \circlearrowleft$ has infinitely many periodic plateaux, the periods of these periodic plateaux is an unbounded set. But it is clear that every periodic plateau contains a sink or a non-hyperbolic periodic points. Then, by Corollary *I* of Theorem C, there can be only finitely many periodic plateaux. We have thus proved:

Corollary III. *The set of periodic plateaux of a C^2 immersion $f: S^1 \circlearrowleft$ is finite*

The question of the Lebesgue measure of compact invariant sets not containing critical points has an answer similar to that of the finiteness of the set of sinks. Every diffeomorphism of the interval has compact invariant sets with positive measure, but the next theorem shows that this is essentially the only way of producing examples of a compact invariant set without critical points and with positive measure.

Theorem D. *If $f \in \mathrm{End}^2(N)$ and $\Lambda \subsetneq N$ is a compact invariant set not containing critical points, then, either the Lebesgue measure of Λ is a zero of there exist an interval $J \subset N$ and an integer $n \geqq 1$ such that $f^n(J) \subset J$, f^n/J has no critical points and $J \cap \Lambda$ has positive Lebesgue measure.*

Corollary I. *If $f: S^1 \circlearrowleft$ is a C^2 immersion, either $J(f) = S^1$ or the Lebesgue measure of $J(f)$ is zero.*

Now observe that if all the periodic points of a C^2 immersion are hyperbolic, then by Theorem *A*, f is expanding, i.e. there exists $N \geqq 1$ such that $|(f^N)'(x)| > 1$ for all $x \in S^1$. This class of transformations have a very developed ergodic theory. For instance, they have a unique f-invariant probability absolutely continuous with respect to the Lebesgue measure and this f-invariant probability is ergodic and its Radon–Nikodym derivative with respect to the Lebesgue measure is a Holder continuous positive function. For simple reasons this result is false if there exists a non-hyperbolic periodic point x_0, even if for some $N \geqq 1$ the inequality $|(f^N)'(x)| > 1$ holds for all $x \neq x_0$. However, immersions $f: S^1 \circlearrowleft$ with $J(f) = S^1$ are ergodic:

Theorem E. *Every C^2 immersion $f: S^1 \circlearrowleft$ with $J(f) = S^1$ is ergodic,* i.e., *every invariant Borel set has either zero or full Lebesgue measure.*

The proof of these theorems, with the exception of Theorem B, will be based on Lemma I.3, stated in Sect. I and proved in Sect. II. Essentially it states that given a compact invariant set $\Lambda \subset N$ of a C^2 endomorphism, then if Λ doesn't contain critical points and $\bigcap_{n \geqq 0} f^n(\Lambda)$ is not a union of periodic orbits, there exists an open interval J having non-empty intersection with $\bigcap_{n \geqq 0} f^n(\Lambda)$, where the backward dynamics of $f / \bigcap_{n \geqq 0} f^n(\Lambda)$ is reasonably hyperbolic. More precisely, if $a \in J \cap (\bigcap_{n \geqq 0} f^n(\Lambda))$

and $a = a_0, a_1, \ldots,$ are points in $\bigcap_{n \geqq 0} f^n(\Lambda)$ satisfying $f(a_{n+1}) = a_n$ for all $n \geqq 0$, there exists maps $\varphi_n: J \to N, n = 1, 2, \ldots$ such that:

$$\varphi_n(a) = a_n,$$

$$f^n \varphi_n(x) = x \quad \text{for all } x \in J$$

for all $n \geqq 1$, and also satisfying the following three properties:

a) $\varphi_n(J) \cap J = \phi$ or $\varphi_n(J) \subset J$,

b) $\lim_{n \to +\infty} |\varphi_n(y)| = 0$ for all $y \in J$,

c) There exists $K > 0$ such that

$$\frac{1}{K} \leqq \frac{|\varphi_n'(z)|}{|\varphi_n'(y)|} \leqq K$$

for all $n \geqq 1, z \in J, y \in J$.

The reader familiar with the ergodic theory of hyperbolic sets will recognize in (a), (b), (c) an analog of the properties on which this theory is founded. In fact, once the existence of J satisfying all these properties is proved, the proofs of Theorems A, C, D and E requires little or very standard work. Theorem B is a corollary of Theorem A and a folklorical extension of Denjoy's theorem (Lemma I.4). It has no intrinsic value. Its interest may reside in motivating research on the link conjecture. This conjecture has a Closing Lemma flavour and as such it can prompt the remark that it can be harder than the problems it attempts to solve. However one must not forget that in the interval, Closing Lemma problems have been successfully handled (L. S. Young [7], I. Malta [4]) and that what Jacobson does in his paper is (essentially) to solve the conjecture exploiting the fact that the points involved are critical.

We suggest to the readers to (at least in the first lecture) follow the proofs, reasoning in the case of f being an immersion of S^1. By doing this they will reach the core of the proofs faster and avoid tedious technicalities.

A natural question is whether Theorems C and D survive without the hypothesis that keeps the critical points away. A concise simple question that exposes the lack of good general techniques to analyse the dynamics near critical points is the following.

Problem III. Does there exist a real analytic endomorphism $f: N \supsetneq$ with infinitely many sinks?

I. Proof of the Theorems

To prove the hyperbolicity of a set, instead of directly finding the constants K and λ required by the definition, it is easier to check the formally weaker condition required by the next lemma (whose very easy proof is left to the reader). We shall use the following notation: If $f \in \text{End}^1(N)$ and Λ is a subset of N, we denote $\mathscr{S}(\Lambda, x)$ the set of sequences $\theta: \mathbb{Z}^+ \to \Lambda$ such that $\theta(0) = x$ and $f(\theta(n)) = \theta(n-1)$ for all $n \geqq 1$.

Lemma I.1. *If $f \in \text{End}^1(N)$ and Λ is a compact invariant set such that*

$$\lim_{n \to +\infty} |(f^n)'(\theta(n))| = +\infty \tag{1}$$

for all $x \in \Lambda$ and $\theta \in \mathscr{S}(\Lambda, x)$, then Λ is hyperbolic.

To verify the hypothesis of this lemma we shall use the concept of coherent sequence of branches. If $J \subset N$ is an open interval we say that $\varphi: J \to N$ is a branch of f^{-n}/J if φ is C^1 and $f^n\varphi(x) = x$ for all $x \in J$. A *coherent sequence of branches* is a couple $(J, \{\varphi_n\})$, where J is an open interval and $\varphi_n: J \to N$ is a branch of f^{-n}/J, $n = 1, 2, 3, \ldots$, satisfying $f\varphi_{n+1} = \varphi_n$ for all $n \geqq 1$.

Given $f \in \mathrm{End}^1(N)$ and a compact invariant set $\Lambda \subset N$, we say that an open interval J is *adapted* to Λ if there exists $\delta > 0$ such that for every $x \in J \cap \Lambda$ and $\theta \in \mathscr{S}(\Lambda, x)$ there exists a coherent sequence $(J, \{\varphi_n\})$ satisfying:

a) $\varphi_n(x) = \theta(n)$,

b) $d(\varphi_n(J),\ C(f)) > \delta$,

c) $\varphi_n(J) \subset J$ or $\varphi_n(J) \cap J = \phi$,

for all $n \geqq 1$. Let us say that a coherent sequence $(J, \{\varphi_n\})$ is *associated* to Λ if there exists $x \in \Lambda \cap J$ such that $\varphi_n(x) \in \Lambda$ for all $n \geqq 1$.

Lemma I.2. *If $f \in \mathrm{End}^2(N)$ is not topologically equivalent to an irrational rotation and $\Lambda \subset N - \partial N$ is a compact invariant set not containing critical points, then, for all non-periodic points $x \in \bigcap_{n \geqq 0} f^n(\Lambda)$ there exists an interval $J \supset \{x\}$ adapted to Λ.*

This lemma and the next one will be proved in Sect. II. Given $f \in \mathrm{End}^1(N)$, a compact invariant set Λ and an interval J adapted to Λ, we say that a map $\psi: J \circlearrowleft$ is a *return map* of Λ if there exist $m \geqq 1$ such that ψ is a branch of f^{-m}/J satisfying $f^j(\psi(J)) \cap J = \phi$ for all $0 < j < m$ and there exists $x \in J \cap (\bigcap_{n \geqq 0} f^n(\Lambda))$ such that $\psi(x) \in \bigcap_{n \geqq 0} f^n(\Lambda)$. It is clear that $f^m(\psi(J)) = J$. Denote $F(\Lambda, J)$ the set of return maps $\psi: J \circlearrowleft$ of Λ. The next lemma is the fundamental step of the proof of the theorems.

Lemma I.3. *If $f \in \mathrm{End}^2(N)$ is not topologically equivalent to an irrational rotation and $\Lambda \subset N - \partial N$ is a compact invariant set not containing critical points, then, if $\bigcap_{n \geqq 0} f^n(\Lambda)$ contains non-periodic points, there exists an interval J adapted to Λ and constants $K_1 > 0$, $0 < \lambda < 1$ such that every coherent sequence $(J, \{\varphi_n\})$ associated to Λ satisfies*

$$\sum_{m=1}^{\infty} |\varphi_m'(x)| \leqq K_1, \tag{2}$$

$$|\varphi_n'(x)| \leqq K_1 |\varphi_n'(y)|, \tag{3}$$

for all $x \in J$, $y \in J$, $n \geqq 1$, and

$$|\psi'(x)| \leqq \lambda \tag{4}$$

for all $x \in J$, $\psi \in F(J, \Lambda)$.

Now let us prove Theorem A. First observe that *given $f \in \mathrm{End}^r(N)$ and a compact invariant set Λ, we can assume without loss of generality that $\Lambda \subset N - \partial N$.* In fact, $\Lambda \cap \partial N \neq \phi$ means that $N = [0, 1]$ and Λ contains either 0 or 1. Then we can extend f to a C^r map g of a bigger interval, say $[-1, 2]$, and now Λ will be a compact invariant set of $g \in \mathrm{End}^r([-1, 2])$ such that $\Lambda \subset [-1, 2] - \partial[-1, 2]$. Now we work with g and $N = [-1, 2]$ (where we can apply I.2 and I.3) and the conclusion on Λ will hold also for

Λ as an invariant set of $f:[0,1] \circlearrowleft$. For this reason, *in the proofs of* A, B, C *and* D, *we shall assume that the compact invariant sets that we shall handle, are contained in* $N - \partial N$.

To prove Theorem A suppose by contradiction that f is not topologically equivalent to an irrational rotation and that Λ is not hyperbolic. Denote $\mathscr{F}$ the family of non-hyperbolic compact invariant subsets of Λ. Order $\mathscr{F}$ by inclusion. $\mathscr{F} \neq \phi$ because $\Lambda \in \mathscr{F}$. Moreover, if $\mathscr{F}_0 \subset \mathscr{F}$ is a totally ordered subfamily, the set $\cap\{\Gamma | \Gamma \in \mathscr{F}_0\}$ belongs to $\mathscr{F}_0$ because otherwise it would be hyperbolic and then it would have a compact neighborhood U whose maximal invariant $\bigcap_n f^n(U)$ is hyperbolic. But $\Gamma \subset U$ for some $\Gamma \in \mathscr{F}_0$, and then $\Gamma \subset \bigcap_n f^n(U)$, thus implying that Γ is hyperbolic and contradicting $\Gamma \in \mathscr{F}_0 \subset \mathscr{F}$. Therefore, by Zorn's Lemma, there exists a minimal $\Lambda_0 \in \mathscr{F}_0$. Observe that $\bigcap_{n \geq 0} f^n(\Lambda_0)$ is again non-hyperbolic (because if it were, Λ_0 would be also hyperbolic; this follows easily from Lemma I.1). Then the minimality of Λ_0 implies $\Lambda_0 = \bigcap_{n \geq 0} f^n(\Lambda_0)$. Hence $f(\Lambda_0) = \Lambda_0$. Moreover Λ_0 can't be a union of periodic orbits because every periodic orbit is hyperbolic and then Λ_0 would be hyperbolic (a compact union of hyperbolic sets is hyperbolic, this is another simple corollary of I.1). Then we can apply Lemma I.3. Let J be the interval given by I.3. Take any $\theta \in \mathscr{S}(\Lambda_0, x)$, $x \in \Lambda_0$. Let Γ be the limit set of θ (i.e. the set of points p such that $\liminf_{n \to +\infty} d(\theta(n), p) = 0$). Γ is compact and invariant. If $\Gamma \neq \Lambda_0$ then Γ is hyperbolic by the minimality of Λ in $\mathscr{F}$. Using the hyperbolicity of Γ it is easy to prove that

$$\lim_{n \to +\infty} |(f^n)'(\theta(n))| = \infty. \tag{5}$$

If $\Gamma = \Lambda_0$ there exists n_0 such that $\theta(n_0) \in J \cap \Lambda_0$. Define $\hat{\theta} \in \mathscr{S}(\Lambda_0, \theta(n_0))$ by $\hat{\theta}(n) = \theta(n + n_0)$. Since J is adapted to Λ_0 there exists a coherent sequence $(J, \{\varphi_n\})$ satisfying $\varphi_n(\theta(n_0)) = \hat{\theta}(n)$ for all $n \geq 1$. Moreover (2) implies

$$\sum_{n=1}^{\infty} |\varphi_n'(\hat{\theta}(n))| < +\infty.$$

In particular $\lim_{n \to +\infty} |\varphi_n'(\hat{\theta}(n))| = 0$. Then $\lim_{n \to +\infty} |(f^n)'(\theta(n))| = \infty$ because $(f^n)'(\hat{\theta}(n)) = (\varphi_n'(\hat{\theta}(n)))^{-1}$. Hence:

$$\begin{aligned}
\lim_{n \to +\infty} |(f^n)'(\hat{\theta}(n))| &= \lim_{n \to +\infty} |(f^n)'(\hat{\theta}(n - n_0))| \\
&= \lim_{n \to +\infty} |(f^{n_0})'(\theta(n_0))| \, |(f^{n-n_0})'(\hat{\theta}(n - n_0))| \\
&= +\infty.
\end{aligned}$$

Therefore (5) holds for all $x \in \Lambda_0$ and $\theta \in \mathscr{S}(\Lambda_0, x)$ thus proving, by Lemma I.1, that Λ_0 is hyperbolic and contradicting $\Lambda_0 \in \mathscr{F}$.

To prove Theorem B we need one more lemma. It is a property essentially contained in Schwarz's proof of Denjoy's theorem. Given $f \in \mathrm{End}^1(N)$ we say that an open interval $J \subset N$ is a *d-interval* if f^n/J is injective and has no critical points for all $n \geq 1$. We say that a d-interval J is *eventually periodic* if there exist a d-interval J_1 and integers $n \geqq 0$, $m \geqq 1$ satisfying

$$f^m(J_1) \subset J_1, \tag{6}$$

$$f^n(J) \subset J_1. \tag{7}$$

Lemma I.4. *If $f \in \mathrm{End}^2(N)$ is not topologically equilvalent to an irrational rotation, every d-interval J is either eventually periodic or satisfies:*

$$\liminf_{n \to +\infty} d(f^n(J); C(f)) = 0,$$

where $C(f)$ is the set of critical points of f.

Now suppose that $f \in \mathrm{End}^r(N)$ satisfies the hypothesis of Theorem B. Denote $\hat{C}(f)$ the set of critical points of f contained in $\Sigma(f)$. Order the points in $\hat{C}(f)$ by the relation $y \geqq x$ if there exists $n \geqq 1$ such that $y = f^n(x)$. This order has no cycles because a cycle would be a periodic critical point, hence a sink that cannot be contained in $\Sigma(f)$. Then there exist points $x_1, \dots, x_m$ in $\hat{C}(f)$ such that every $x \in \hat{C}(F)$ satisfies $x \leqq x_i$ for some $1 \leqq i \leqq m$ and if $x_i \leqq x$ then $x = x_i$. Denote S_i the set of points $x \in \hat{C}(f)$ such that $x \leqq x_i$. Let us say that a d-interval J is *maximal* if every open interval properly containing J is not a d-interval. It is clear that if f is not a diffeomorphism of the circle every d-interval is contained in a unique maximal d-interval and two maximal d-intervals are either disjoint or coincide. Since for the proof of Theorem B the case of f being a diffeomorphism of the circle is trivial, we can assume that this property holds and then the set of maximal d-intervals is a countable family $J_1, J_2, \dots$ of disjoint intervals. Moreover, since f maps d-intervals in d-intervals it follows that for all $i \geqq 1$ there exists $j \geqq 1$ such that $f(J_i) \subset J_j$. We shall say that J_i is *periodic* if $f^n(J_i) \subset J_i$ for some $n \geqq 1$, and that it is *eventually periodic* if $f^m(J_i)$ is contained in a periodic maximal d-interval for some $m \geqq 0$. The absence of fake links easily implies that for every critical point x there exist intervals (a, x) and (x, b) (only one if $N = [0, 1]$ and x is an endpoint) not intersecting the backward orbits of the critical points. Hence (a, x) and (x, b) are d-intervals. Therefore to every critical point x we can associate maximal d-intervals $J^-(x) = (a(x), x)$, $J^+(x) = (x, b(x))$ (again, only one when $N = [0, 1]$ and x is an endpoint). If we apply Lemma I.4 to $J^\sigma(x)$ ($\sigma = +$ or $-$) we get that either $J^\sigma(x)$ is eventually periodic or

$$\liminf_{n \geqq 0} d(f^n(J^\sigma(x)), C(f)) = 0. \tag{8}$$

If (8) holds it means that there exists $y \in C(f)$ such that for suitable values of n, $f^n(J^\sigma(x))$ is arbitrarily near to y. Hence, for $\alpha = +$ or $-$, we have $f^n(J^\sigma(x)) \cap J^\alpha(y) \neq \phi$. But since $f^n(J^\sigma(x))$ is a d-interval it follows that

$$f^n(J^\sigma(x)) \subset J^\alpha(y). \tag{9}$$

Hence every $J^\sigma(x)$ is either eventually periodic or satisfies (9) for some $n \geqq 0$, $y \in C(f)$ and $\alpha = +$ or $-$. If $J^\alpha(y)$ is eventually periodic then so is $J^\sigma(x)$. If it is not, we have $f^k(J^\alpha(y)) \subset J^\beta(z)$ for some $k \geqq 1$, $y \in C(f)$, $\beta = +$ or $-$. Since the number of critical points is finite, an interval must appear twice in this process, thus proving that the initial $J^\sigma(x)$ is eventually periodic. Therefore every $J^\sigma(x)$ is eventually periodic. Then to every x_i we can associate a maximal d-interval A_i and integers $n_i \geqq 0$, $m_i \geqq 1$

satisfying

$$f^{n_i}(J^+(x_i)) \subset A_i, \quad f^{m_i}(A_i) \subset A_i.$$

Without loss of generality we can assume (replacing m_i by $2m_i$ if necessary) that f^{m_i}/A_i is order preserving. Since every periodic point of f^{m_i} is hyperbolic it follows that every $x \in \bar{A}_i$ is either a source or belongs to the basin of a sink (recall that f^{m_i}/A_i has no critical points). Then $f^{n_i}(x_i) \in \bar{A}_i$ is either a source or belongs to the basin of a sink. The second possibility cannot hold because $x_i \in \Sigma(f)$. Then $f^{n_i}(x_i)$ is a source for all $1 \leqq i \leqq m$.

Using that f^{m_i}/A_i has no critical points, it is easy to find a closed interval $B_i \subset \bar{A}_i$ such that one of its endpoints is $f^{n_i}(x_i)$ and the other endpoint is a sink whose basin contains Int B_i. Now we take $g \in \mathrm{End}^r(N)$ near to f, having all its periodic points hyperbolic and satisfying

a) f and g coincide in a neighborhood of $\bigcup_{i=1}^{m} B_i$,

b) $C(f) = C(g)$,

c) $g^{n_i}(x_i) \in \mathrm{Int}\, B_i$,

d) g and f coincide in $\bigcup_{i=1}^{m} S_i$.

From (c) and (d) follows that all the critical points in $\hat{C}(f)$ eventually under g will fall in $\bigcup_{i=1}^{m} \mathrm{Int}(B_i)$. By (a), every Int B_i is in the basin of a sink of g. Hence

$$\hat{C}(f) \subset \Sigma(g)^c. \tag{10}$$

On the other hand, points in $C(f) - \hat{C}(f)$ are in the basins of sinks of f, and if g is sufficiently near to f, are in the basins of sinks of g. Therefore:

$$C(f) - \hat{C}(f) \subset \Sigma(g)^c. \tag{11}$$

From (10), (11) and (b) we get

$$C(g) = C(f) \subset \Sigma(g)^c. \tag{12}$$

Since every periodic point of g is hyperbolic and $\Sigma(g)$ by definition doesn't contain sinks and by (12) doesn't contain critical points, we can apply Theorem A to $\Sigma(g)$ proving that it is a hyperbolic set

To prove Theorem C suppose by contradiction that there exists an infinite set S of sinks and non-hyperbolic periodic points such that any pair of points $x, y \in S$ are not homologous. First we shall prove that $\bar{S}$ contains a non-periodic point. If it doesn't there exists a periodic point $p \in \bar{S}$ and a sequence $\{p_n\} \subset S$ converging to p. Without loss of generality we can assume that $p_1 < p_2 < \cdots < p$ and take N such that $f^N(p) = p$ and $(f^N)'(p) > 0$. Then in a neighborhood of p, f^N is order preserving. Therefore we can assume that

$$p_0 \leqq x \leqq y \leqq p \Rightarrow f^N(x) < f^N(y). \tag{13}$$

In particular $p_0 < p_n < p$ implies

$$f^N(p_n) < f^N(p) = p.$$

Then, if

$$p_n \leqq f^N(p_n),$$

we have $p_0 \leqq p_n \leqq f^N(p_n) < p$, and applying (13) we get $p_0 \leqq f^N(p_n) \leqq f^{2N}(p_n) < f^{N'}(p) = p$. Applying again (13) we get $p_0 \leqq f^{2N}(p_n) \leqq f^{3n}(p_n) \subset f^N(p) = p$. Continuing with this method we obtain:

$$p_n \leqq f^N(p_n) \leqq f^{2N}(p_n) \leqq \cdots \leqq f^{m_n N}(p_n) = p_n$$

if $m_n N$ is a period of p_n. Hence $f^N(p_n) = p_n$. Then the integer N and the interval $[p_n, p]$ prove that p_n is homologous to p. Since there can be at most one point in S homologous to p, it follows that we can assume that no p_n is homologous to p and then

$$f^N(p_n) < p_n \tag{14}$$

for all $n \geqq 0$. Using the fact that f^N is injective in a neighborhood of p, we can take $\delta > 0$ such that if $x \in (p_0, p)$ and $y \neq x$ satisfy $f^N(x) = f^N(y)$, then

$$d(x, y) > \delta. \tag{15}$$

Now take $n \geqq 0$. If $f^{jm_n}(p_n) \in (p_{n-1}, p)$ for all $0 \leqq j \leqq N$, then p_n is a periodic point of the injective order preserving map $f^N/(p_{n-1}, p)$. Hence p_n is actually a fixed point, and arguing as before, p_n is homologous to p. But we have assumed that no p_n is homologous to p. Hence for all $n \geqq 0$ there exists $0 < j_n < m_n$ such that

$$f^{j_n N}(p_n) \notin (p_{n-1}, p), \tag{16}$$

$$f^{(j_n+1)N}(p_n) \in (p_{n-1}, p). \tag{17}$$

Since by (14) $f^N((p_{n-1}, p)) \supset (p_{n-1}, p)$, there exists $q_n \in (p_{n-1}, p)$ such that

$$f^N(q_n) = f^{(j_n+1)N}(p_n). \tag{18}$$

By (16) $f^{j_n N}(p_n) \neq q_n$ and by (18) $f^N(f^{j_n N}(p_n)) = f^N(q_n)$. Hence (15) implies:

$$d(q_n, f^{j_n N}(p_n)) > \delta. \tag{19}$$

Moreover

$$p - q_n < p - p_{n-1} \tag{20}$$

because $q_n \in (p_{n-1}, p_n)$. Taking limit when $n \to +\infty$ (on a subsequence if necessary) we have that $q_n \to p$ and the sequence $f^{j_n N}(p_n)$ converges to a point $\bar{p}$ that is in $\bar{S}$ but that by (19) is different from p, and by (17) $f^N(\bar{p}) = p = f^N(p)$. Since p is periodic and $\bar{p} \neq p$, the relation $f^N(\bar{p}) = p$ shows that $\bar{p} \in \bar{S}$ is not periodic. Now that we know that $\bar{S}$ contains non-periodic points we can apply Lemma I.3 to $\Lambda = \bar{S}$. Observe that $f(\bar{S}) = \bar{S}$ and then $\bigcap_{n \geqq 0} f^n(\bar{S}) = \bar{S}$. Then by Lemma I.3 there exists an interval J adapted to Λ satisfying (2) for every coherent sequence $(J, \{\varphi_n\})$ associated to $\bar{S}$. Take $x \in J \cap S$ and choose the periodic $\theta \in \mathscr{S}(\Lambda, x)$ (that exists because

x is periodic). By the definition of adapted interval there exists a coherent sequence $(J,\{\varphi_n\})$ such that $\varphi_n(x)=\theta(n)$ for all $n \geqq 1$. From (2) it follows that $|(f^n)'(\theta(n))|^{-1} = |\varphi_n'(x)| \to 0$ when $n \to +\infty$ and if N is the period of x we have $\theta(nN)=\theta(0)$ for all $n \geqq 1$. Then:

$$|(f^{nN})'(x)| = |(f^{nN})'(\theta(0))| = |(f^{nN})'(\theta(nN))|.$$

Then $|(f^{nN})'(x)| \to \infty$ when $n \to +\infty$ thus implying that x is a source and contradicting the definition of S.

To prove Theorem D we can assume that f is not topologically equivalent to an irrational rotation because in that case the property follows from a stronger property proved by Herman [2]. Suppose also that Λ doesn't satisfy the second option of Theorem D. We have to show that $\lambda(\Lambda)=0$, where $\lambda(\cdot)$ denotes the Lebesgue measure. Let $\mathscr{F}$ be the family of compact invariant sets $\Gamma \subset \Lambda$ such that

$$\lambda(\Lambda - \{x \in \Lambda \,|\, \omega(x) \subset \Gamma\}) = 0.$$

Order $\mathscr{F}$ by inclusion. Let $\mathscr{F}_0 \subset \mathscr{F}$ be a totally ordered subfamily. We claim that $\cap\{\Gamma \,|\, \Gamma \in \mathscr{F}_0\} \in \mathscr{F}_0$. Denote $\Lambda_0 = \cap\{\Gamma \,|\, \Gamma \in \mathscr{F}_0\}$. Take a neighborhood V of Λ_0. Then $\Gamma \subset V$ for some $\Gamma \in \mathscr{F}_0$ and

$$\lambda(\Lambda - \{x \in \Lambda \,|\, \omega(x) \subset V\} \leqq \lambda(\Lambda - \{x \in \Lambda \,|\, \omega(x) \subset \Gamma\}) = 0.$$

Since this is true for every neighborhood V of Λ_0 it follows that $\lambda(\Lambda - \{x \in \Lambda \,|\, \omega(x) \subset \Lambda_0\}) = 0$. Hence $\Lambda_0 \in \mathscr{F}_0$ proving the claim. We can now apply Zorn's Lemma to $\mathscr{F}$ and obtain a *minimal* $\Lambda_0 \in \mathscr{F}$. We claim that Λ_0 has the following property: for every open interval J such that $j \cap \Lambda_0 \neq \phi$, the set

$$\tilde{J} = \{x \in J \cap \Lambda \,|\, f^n(x) \in J \text{ for infinitely many } n\text{'s}\}$$

has positive measure. Suppose by contradiction that $\lambda(\tilde{J})=0$ and define

$$\Lambda_1 = \{x \in \Lambda \,|\, f^n(x) \notin J \quad \text{for all } n \geqq 0\}.$$

Clearly Λ_1 is compact and invariant. Moreover:

$$\{x \in \Lambda \,|\, \omega(x) \subset \Lambda_0\} = \left(\bigcup_{n \geqq 0} f^n(\tilde{J})\right) \cup \{x \in \Lambda \,|\, \omega(x) \subset \Lambda_1\}.$$

Since $\lambda\left(\bigcup_{n \geqq 0} f^{-n}(\tilde{J})\right) = 0$ (because $\lambda(\tilde{J})=0$) it follows that:

$$\lambda(\Lambda - \{x \in \Lambda \,|\, \omega(x) \subset \Lambda_0\}) = \lambda(\Lambda - \{x \in \Lambda \,|\, \omega(x) \subset \Lambda_1\}).$$

Hence $\Lambda_1 \in \mathscr{F}$. But then the minimality of Λ_0 implies $\Lambda_1 = \Lambda_0$. This means $J \cap \Lambda_0 = \phi$ contradicting our hypothesis on J. Now we claim that if $\lambda(\{x \in \Lambda \,|\, \omega(x) \subset \Lambda_0\}) \neq 0$ there exists an open interval J with $J \cap \Lambda_0 \neq \phi$ and $\lambda(\tilde{J})=0$. Since such interval cannot exist, it will follow that $\lambda(\{x \in \Lambda \,|\, \omega(x) \subset \Lambda_0\})=0$ and then:

$$\lambda(\Lambda) = \lambda(\Lambda - \{x \in \Lambda \,|\, \omega(x) \subset \Lambda_0\}).$$

But the measure at right is zero because $\Lambda_0 \in \mathscr{F}$, and then $\lambda(\Lambda)=0$ as we wished to prove.

To prove the claim observe that $f(\Lambda) \in \mathscr{F}$, and then the minimality of Λ_0 implies

$\Lambda_0 = f(\Lambda_0)$. Hence $\Lambda_0 = \bigcap_{n\geq 0} f^n(\Lambda_0)$. If Λ_0 is finite, every point in Λ_0 is periodic. Then $\lambda(\{x\in\Lambda \,|\, \omega(x)\subset\Lambda_0\})=0$ because if it is $\neq 0$ there exists a periodic orbit $\gamma\subset\Lambda_0$ such that $\lambda(x\in\Lambda\,|\,\omega(x)=\gamma\})\neq 0$. This inequality easily implies the existence of an interval J satisfying the second option in Theorem D. Since we are assuming that this option doesn't hold it follows that $\lambda(\{x\in\Lambda\,|\,\omega(x)\subset\Lambda_0\})=0$. If Λ_0 is an infinite union of periodic orbits, it is easy to prove, using Theorem C, that there exists an open interval J such that f^N/J is a diffeomorphism of J for some $N\geqq 1$ and containing infinitely many periodic orbit of Λ_0. Since we are assuming that the second option of Theorem D doesn't hold, it follows that $\lambda(J\cap\Lambda_0)=0$. But on the other hand $J\cap\Lambda_0=\tilde{J}$ because every point in Λ_0 is periodic. It remains the case when $\Lambda_0=\bigcap_{n\geq 0} f^n(\Lambda_0)$ contains non-periodic points. We can apply Lemma I.3 and obtain an interval J adapted to Λ_0 and $K_1>0$ satisfying (4). We shall prove that $\lambda(\tilde{J})=0$. Denote $F_n(J,\Lambda)$ the set of maps $\psi\colon\circlearrowleft$ that can be written as a composition of n elements of $F(J,\Lambda)$. By Lemma I.3 there exists $\sigma<\lambda<1$ such that:

$$|\psi'(x)|\leqq\lambda^n \tag{21}$$

for all $x\in F_n(J,\Lambda)$. In particular, (21) implies:

$$\operatorname{diam}\psi(J)\leqq\lambda^n\operatorname{diam}(J) \tag{22}$$

for all $\psi\in F_n(J,\Lambda)$. Denote $\tilde{F}_n(J,\Lambda)$ the set of maps ψ in $F_n(J,\Lambda)$ such that there exists a coherent sequence $(J,\{\varphi_n\})$ satisfying $\psi=\varphi_n$. By I.3 there exists $K_1>0$ such that

$$|\psi'(x)|\leqq K_1|\psi'(y)|$$

for all $x\in J$, $y\in J$ and $\psi\in\bigcup_{n\geq 0}\tilde{F}_n(J,\Lambda)$. It follows that

$$\frac{\lambda(\psi(A))}{\lambda(\psi(J))}\leqq K_2\frac{\lambda(S)}{\lambda(J)}$$

for all $\psi\in\bigcup_{n\geq 0}\tilde{F}_n(J,\Lambda)$ and every Borel set $A\subset J$. Then:

$$\begin{aligned}\frac{\lambda(\tilde{J})}{\lambda(J)}&=1-\frac{\lambda(J-\tilde{J})}{\lambda(J)}\geqq 1-K_2\frac{\lambda(\psi(J-\tilde{J}))}{\lambda(\psi(J))}\\&=1-K_2\left(1-\frac{\lambda(\psi(\tilde{J}))}{\lambda(\psi(J))}\right),\end{aligned}$$

and since $\psi(\tilde{J})\supset\psi(J)\cap\tilde{J}$,

$$\frac{\lambda(\tilde{J})}{\lambda(J)}\geqq 1-K_2\left(1-\frac{\lambda(\psi(J)\cap\tilde{J})}{\lambda(\psi(J))}\right). \tag{23}$$

If $\lambda(\tilde{J})\neq 0$ we can take a density point $x\in\tilde{J}$. There exists a sequence $\psi_{n_j}\in\tilde{F}_{n_j}(\Lambda,J)$, with $n_j\to+\infty$ such that $x\in\psi_{n_j}(J)$ for all n_j. By (22), $\operatorname{diam}\psi_{n_j}(J)\to 0$ when $j\to+\infty$. The definition of density point grants:

$$\lim_{j\to+\infty}\frac{\lambda(\psi_{n_j}(J)\cap\tilde{J})}{\lambda(\psi_{n_j}(J))}=1.$$

Applying (23) to $\psi = \psi_{n_j}$ and taking limits we obtain $\lambda(\tilde{J}) \geqq \lambda(J)$. Hence $\lambda(\tilde{J}) \neq 0$ implies that $\tilde{J}$ is dense in J. In particular Λ contains J. But Λ doesn't contain critical points. Therefore f^n/J has no critical points for all $n \geqq 0$. Moreover $f^n(J) \subset \Lambda \neq N$ for all $n \geqq 1$. This excludes the possibility $f^n(J) = N = S^1$ and shows that every power of f is injective in J and has no critical points. In other words J is a d-interval. But by Lemma I.4 this means that there exists an open interval J_1 such that $f^n(J_1) \subset J_1$, f^n has no critical points in J_1 and $f^N(J) \subset J_1$ for some $N \geqq 1$. Therefore the interval $f^n(J) \subset \Lambda \cap J_1$ proves that f satisfies the second option of Theorem D. Since we are assuming that this option doesn't hold, it follows that $\tilde{J}$ is not dense in J and $\lambda(\tilde{J}) = 0$.

To prove Theorem E we take the interval J obtained applying Lemma I.3 to $\Lambda = S^1$. We shall first prove that $\lambda(\Gamma \cap J) \neq 0$ implies $\lambda(\Gamma \cap J) = \lambda(J)$. Denote Γ_0 the set of points $x \in S^1$ whose forward orbit doesn't intersect J. Then Γ_0 is compact and invariant. By Theorem D, $\lambda(\Gamma_0) = 0$. Denote J_0 the set of points $x \in J$ such that there exists $N > 0$ satisfying $f^n(x) \notin J$ for all $n \geqq N$. Then $J_0 \subset \bigcup_{n \geqq 0} f^{-n}(\Gamma_0)$ and $\lambda(J_0)$

Define $J_1 = J - J_0$. Since $\lambda(J_0) = 0$, to prove that $\lambda(\Gamma \cap J) = J$ it suffices to show that $\lambda(\Gamma \cap J_1) = 0$. Denote $F_n(J, S^1)$ the set of maps $\psi: J \circlearrowleft$ that can be written as a composition of n elements of $F(J, S^1)$. Then for every $x \in J_1$ there exists a sequence $\psi_{n_i} \in F_n(J, S^1)$ such that

$$x \in \psi_{n_i}(J) \tag{24}$$

for all $i \geqq 1$. Moreover, I.3 implies that:

$$|\psi'(x)| \leqq \lambda^n$$

for all $\psi \in F_n(J, S^1)$ and $n \geqq 1$. In particular:

$$\operatorname{diam} \psi(J) \leqq \lambda^n \operatorname{diam} J \tag{25}$$

for all $n \geqq 1$ and $\psi \in F_n(J, S^1)$. Moreover, by Lemma I.3

$$|\psi'(x)| \leqq K_2 |\psi'(y)|$$

for all $n \geqq 1$, $\psi \in F_n(J, S^1)$, $x \in J$, $y \in J$. Then

$$\frac{\lambda(\psi(A))}{\lambda(\psi(J))} \geqq K_1^{-1} \frac{\lambda(A)}{\lambda(J)}$$

for all $\psi \in F_n(J, S^1)$, $n \geqq 1$ and every Borel set $A \subset J$. Therefore, if $x \in \Gamma \cap J$ and the sequence $\{\psi_{n_i}\}$ is chosen satisfying (24) for all n_i, we obtain:

$$\frac{\lambda(\psi_{n_i}(\Gamma^c \cap J_1))}{\lambda(\psi_{n_i}(J))} \geqq K^{-1} \frac{\lambda(\Gamma^c \cap J_1)}{\lambda(J)}.$$

But $\psi_{n_i}(\Gamma^c) \subset f^{-n_i}(\Gamma^c) \subset \Gamma^c$ and $\psi_{n_i}(J_1) \subset J_1$. Then

$$\Gamma^c \cap J_1 \cap \psi_{n_i}(J) \supset \psi_{n_i}(\Gamma^c \cap J_1).$$

Hence:

$$\frac{\lambda(\Gamma^c \cap J_1 \cap \psi_{n_i}(J))}{\lambda(\psi_{n_i}(J))} \geqq K_1^{-1} \frac{\lambda(\Gamma^c \cap J_1)}{\lambda(J)}. \tag{26}$$

But if x is a density point of $\Gamma \cap J_1$, it follows from (25) and (24) that:

$$\lim_{i \to +\infty} \frac{\lambda(\Gamma^c \cap J_1 \cap t_{n_i}(J))}{\lambda(\psi_{n_i}(J))} = 0.$$

Then, by (26), $\lambda(\Gamma^c \cap J_1) = 0$ and $\lambda(\Gamma \cap J_1) = \lambda(J_1)$. Now we know that $\lambda(\Gamma \cap J) \neq 0$ implies $\lambda(\Gamma \cap J) = \lambda(J)$. Suppose that $\lambda(\Gamma) = 0$. Observe that $\bigcup_{n \geqq 0} f^{-n}(J)$ has full measure because its complement is a proper compact invariant set. Then:

$$0 < \lambda(\Gamma) = \lambda\left(\Gamma \cap \left(\bigcup_{n \geqq 0} f^{-n}(J)\right)\right) \leqq \lambda\left(\bigcup_{n \geqq 0} f^{-n}(\Gamma \cap J)\right).$$

This inequality shows that $\lambda(\Gamma \cap J) > 0$. But we proved that $\lambda(\Gamma \cap J) > 0$ implies $\lambda(\Gamma \cap J) = \lambda(J)$. Hence $\lambda(\Gamma^c \cap J) = 0$ and then $\lambda(f^n(\Gamma^c \cap J)) = 0$ for all $n \geqq 0$. Taking n such that $f^n(J) = S^1$ we obtain:

$$\lambda(\Gamma^c) = \lambda(\Gamma^c \cap f^n(J)) \leqq \lambda(f^n(\Gamma^c \cap J)) = 0.$$

II. Proof of the Lemmas

We shall begin by proving Lemma I.4 because it is in fact previous to the other Lemmas. It is essentially an easy reformulation of Denjoy's theorem and therefore we shall only outline those parts of the proof that are only straightforward modifications of the proof of Denjoy's theorem.

We shall use the concept of maximal interval introduced in Sect. I to prove Theorem B. Given a d-interval J of $f \in \mathrm{End}^{1+\varepsilon}(N)$ we can take (because f is not a differmorphism of the circle), unique maximal d-intervals $J_i \supset f^i(J)$, $i = 0, 1, \ldots$. Suppose that there exist $m > n \geqq 0$ such that $J_m \cap J_n \neq \phi$. Then $J_m = J_n$ and

$$f^{m-n}(J_m) \cap J_m \supset f^{m-n}(f^n(J_0)) \cap J_m = f^m(J_0) \cap J_m = f^m(J_0) \neq \phi.$$

But $f^{m-n}(J_m)$ is a d-interval. Hence

$$f^{m-n}(J_m) \subset J_m,$$

and the theorem is proved. Now suppose that

$$J_m \cap J_n = \phi \tag{1}$$

for all $m > n \geqq 0$. Then

$$\sum_{n=0}^{\infty} \operatorname{diam} J_n < +\infty. \tag{2}$$

Suppose that

$$\liminf_{n \to +\infty} d(J_n, C(f)) = 0. \tag{3}$$

We also have:

$$d(f^n(J_0), C(f)) \leqq d(J_n, C(f)) + \operatorname{diam} J_n. \tag{4}$$

But (2) implies $\operatorname{diam} J_n \to 0$. This together with (3) and (4) implies

$$\liminf_{n\to+\infty} d(f^n(J_0), C(f)) = 0,$$

and the Lemma is proved. It remains to consider the case

$$\liminf_{n\to+\infty} d(J_n, C(f)) > 0.$$

This means that there exists $n_0 \geqq 0$ such that

$$\inf_{n\geqq n_0} d(J_n, C(f)) > 0.$$

Since $f^n(J_{n_0}) \subset J_{n+n_0}$, we obtain:

$$\inf_{n\geqq n_0} d(f^n(J_{n_0}), C(f)) > 0, \tag{5}$$

$$\sum_{n=0}^{\infty} \operatorname{diam} f^n(J_{n_0}) < +\infty. \tag{6}$$

We shall prove that (5) and (6) together contradict the maximality of J_{n_0}. From (5) and (6), as in the proof of Denjoy's theorem, we get

$$\sum_{n=0}^{\infty} |(f^n)'(x)| < +\infty \tag{7}$$

for all $x \in \bar{J}_{n_0}$.

As in the proof of Denjoy's theorem, we can find, using (5) and (7), an open interval $A \supset \{a\}$, where a is an endpoint of $\bar{J}_{n_0}$, such that

$$\lim_{n\to+\infty} \operatorname{diam} f^n(A) = 0. \tag{8}$$

Observe that the case $\bar{J}_{n_0} = N = [0,1]$ doesn't arise because of (1). The case $\bar{J}_{n_0} = N = S^1$ is also impossible for the same reason. Then we can take an open interval $A \supset A' \supset \{a\}$ so small that

$$A' \cup J_{n_0} \neq N. \tag{9}$$

Moreover, since we cannot have $J_{n_0} = N = [0,1]$, we can take the endpoint a different from 0 and 1. Hence:

$$A' \cup J_{n_0} \neq J_{n_0}. \tag{10}$$

Finally (8) implies that we can take A' so small that

$$\inf_{n\geqq 0} \operatorname{diam} f^n(A') < \tfrac{1}{2} \inf_{n\geqq 0} d(f^n(J_{n_0}), C(f)),$$

and then

$$\inf_{n\geqq 0} d(f^n(A' \cup J_{n_0}), C(f)) > 0. \tag{11}$$

By (9), $A' \cup J_{n_0}$ is an interval. By (11) it is a d-interval and by (16) it properly contains J_{n_0}. Hence J_{n_0} is not a maximal d-interval and this contradiction completes the proof of I.4.

The proof of Lemma I.2 requires a preliminary result:

Lemma II.1. *If $f \in \mathrm{End}^2(N)$ and $\Lambda \subset N - \partial N$ is a compact invariant set not containing critical points, then, for every non-periodic point $a \in \bigcap_{n \geq 0} f^n(\Lambda)$ and $\delta > 0$, there exists $\varepsilon > 0$ such that if $J \supset \{a\}$ is an open interval with $\mathrm{diam}(J) \leq \varepsilon$ then, for all $x \in J \cap \Lambda$ and $\theta \in \mathscr{S}(\Lambda, x)$ there exists a coherent sequence $(J, \{\varphi_n\})$ satisfying*

$$\varphi_n(x) = \theta(n), \tag{12}$$

$$\mathrm{diam}\, \varphi_n(J) \leq \delta \tag{13}$$

for all $n \geq 0$.

Proof. Take a non-periodic point $a \in \bigcap_{n \geq 0} f^n(\Lambda)$ and $\delta > 0$. Let $\varepsilon_n > 0$ be the maximum positive number such that if $J_n = (a - \varepsilon_n, a + \varepsilon_n)$ then, for every $x \in \Lambda \cap J_n$ and every $\theta \in \mathscr{S}(\Lambda, x)$ there exists branches $\varphi_j \colon J_n \to N$ of $f^{-j}/J, j = 1, \ldots, n$ such that

$$\varphi_j(x) = \theta(j),$$

$$\mathrm{diam}\, \varphi_j(J_n) \leq \delta$$

for all $1 \leq j \leq n$ and

$$f \varphi_{j+1} = \varphi_j$$

for all $1 \leq j \leq n$. Without loss of generality we can suppose

$$0 < \delta < \tfrac{1}{2} d(\Lambda, C(f) \cup f(N)).$$

This implies that there exists $0 \leq j_n \leq n$ such that

$$\mathrm{diam}\, \varphi_{j_n}(J_n) = \delta.$$

If we prove that $\liminf_{n \to +\infty} \varepsilon_n = \varepsilon > 0$, the Lemma is proved just taking $J = (a - \varepsilon, a + \varepsilon)$. Suppose that $\liminf_{n \to +\infty} \varepsilon_n = 0$. Then $j_n \to +\infty$. Now define $U_n = \varphi_{j_n}(J_n)$. A subsequence of the sequence of intervals $\{U_n\}$ converges to an open interval U that satisfies diam $U = \delta$. We shall assume to simplify the notation that $U_n \to U$. Then U is a d-interval. By I.4 there exists a d-interval V and integers $n \geq 1$, $N \geq 1$ such that

$$f^m(U) \subset V, \tag{14}$$

$$f^N(V) \subset V, \tag{15}$$

$$f^N/V \text{ has no critical points.} \tag{16}$$

Take a point $q \in U$ such that, $q \in U_n$ for every large value of n, say for all $n \geq n_0$. By (14), (15), (16) the ω-limit set of q is a periodic orbit γ. Then

$$d(a, \gamma) \leq \varepsilon_n + d((a - \varepsilon_n, a + \varepsilon_n), \gamma) = \varepsilon_n + d(f^{j_n}(U_n), \gamma) \leq \varepsilon_n + d(f^{j_n}(q), \gamma). \tag{17}$$

But $j_n \to +\infty$ when $n \to +\infty$. Then

$$\lim_{n\to+\infty} d(f^{j_n}(q), \gamma) = 0.$$

Moreover $\liminf_{n\to+\infty} \varepsilon_n = 0$. Hence (17) proves that $a \in \gamma$, contradicting that a is non-periodic.

To prove I.2 we shall consider first the case when f is an immersion of S^1. If its degree is 1 or -1, it is a diffeormorphism. Moreover it has periodic points because it is not topologically equivalent to an irrational rotation. Then, using that x is not a periodic point it is trivial to construct the adapted interval J (in fact any open interval $J \supset \{x\}$ such that $f^{-n}(J) \cap J = \phi$ for all $n > 0$ works). If the degree d of f is not 1 or -1 we define $g: S^1 \circlearrowleft$ by $g(z) = z^d$ and, as we explained in the Introduction, there exists a monotone map $h: S^1 \circlearrowleft$ satisfying $gh = hf$. Now observe that if J is an interval whose endpoints are fixed points of some power g^n of g (i.e. roots of the equation $z^{d^n - 1} - 1 = 0$) and doesn't contain fixed points of g^n, then J satisfies (c) (with respect to g). Condition (b) is obvious since $C(g) = \phi$ and the existence condition (condition (a)) is easy to check (even for $\Lambda = S^1$). Therefore J is an interval adapted to S^1, in particular adapted to Λ. Now, given $x \in \Lambda$, we take $h(x)$ and an interval $J \supset \{h(x)\}$ as described above. Using h it is easy to show that $h^{-1}(J)$ is an interval adapted to S^1 (in particular to Λ).

Now suppose that f is not an immersion of S^1. Given the non-periodic point $x \in \bigcap_{n \geq 0} f^n(\Lambda)$, let $\varepsilon > 0$ be given by Lemma II.1, taking as $\delta > 0$ a number satisfying

$$0 < \delta < \tfrac{1}{2} d(\Lambda, C(f)). \tag{18}$$

If $J \supset \{x\}$ is an open interval with $\operatorname{diam} J < \varepsilon$ then, by Lemma II.1 the first condition of the definition of interval adapted to Λ is satisfied. Condition (b) is also satisfied because if $(J, \{\varphi_n\})$ is a coherent sequence associated to Λ, every $\varphi_n(J)$ contains a point in Λ and, by Lemma II.1, $\operatorname{diam} \varphi_n(J) < \delta$. Then

$$d(\varphi_n(J), C(f)) \leqq d(\Lambda, C(f)) - \operatorname{diam} \varphi_n(J) \leqq d(\Lambda, C(f)) - \delta \leqq \delta.$$

Therefore conditions (a) and (b) are granted just by taking J having diameter $\leqq \varepsilon$. The problem is condition (c). If J is an open interval denote $\mathscr{C}(J)$ the set of all the coherent sequences $(J, \{\varphi_n\})$ associated to J and denote $\hat{J}$ the connected component containing x of the open set:

$$J \cup (U(\{\varphi_n(J) | n \geqq 1, \quad (J, \{\varphi_n\}) \in \mathscr{C}(J)\}).$$

Take a sequence $J_1 \supset J_2 \supset \cdots \supset \{x\}$ of open intervals with $\operatorname{diam}(J_n) \to 0$. Then $\hat{J}_1 \supset \hat{J}_2 \supset \cdots \{x\}$. If $\operatorname{diam}(\hat{J}_m) \to 0$ we are done because it is clear from the definition of $\hat{J}$ that if $(\hat{J}, \{\varphi_n\}) \in \mathscr{C}(\hat{J})$, then $\varphi_n(\hat{J}) \cap \hat{J} \neq \phi$ implies $\varphi_n(\hat{J}) \subset \hat{J}$. Hence any $\hat{J}_m$ with diameter less than ε will satisfy condition (c) (and, as we observed before (a) and (b) are implied by $\operatorname{diam}(\hat{J}_m) < \varepsilon$). Then suppose that $\bigcap_{m \geq 0} \hat{J}_m$ is not $\{x\}$. Then its interior is an open interval U. We claim that $U \subset \Lambda$. Given $y \in U$ and $\varepsilon_0 > 0$, take m so large that $\operatorname{diam}(J_m) < \varepsilon_0$ and

$$\sup\{\operatorname{diam} \varphi_n(J_m) | n \geqq 1, (J_m \{\varphi_n\}) \in \mathscr{C}(J_m)) < \varepsilon_0.$$

We can take such m by Lemma II.1 By the definition of $\hat{J}_m$ and since $y \in \hat{J}_m$, either $y \in J_m$ or $y \in \varphi_n(J_m)$ for some $n \geqq 1$ and $(J_m, \{\varphi_n\}) \in \mathscr{C}(J_m)$. In the first case $d(y,x) \leqq \operatorname{diam} J_m \leqq \varepsilon_0$. In the second case, observing that $\varphi_n(J_m)$ contains points of Λ because $(J_m\{\varphi_n\})$ is associated to Λ, we obtain again

$$d(y,\Lambda) \leqq d(y, \varphi_n(J_m)) \leqq \operatorname{diam} \varphi_n(J_m) < \varepsilon_0 .$$

Hence $d(y,\Lambda) \leqq \varepsilon_0$ for all ε_0 and then $y \in \Lambda$. This completes the proof of the claim. But $U \subset \Lambda$ implies $f^n(U) \subset \Lambda$ for all $n > 0$ and then f^n/U has no critical points for all $n > 0$. Since f is not an immersion of S^1 it follows that U is a d-interval. Let V be the maximal d-interval containing U. By Lemma I.4 it is eventually periodic. Moreover the construction of U shows that $x \in \bar{U}$. We have now two cases to consider

I) $x \in V$. From the fact that V is eventually periodic follows the existence of an open interval $\{x\} \subset J \subset V$ satisfying

$$f^n(J) \cap J = \phi$$

for all $n \geqq 1$. This implies that

$$\varphi_n(J) \cap J = \phi$$

for all $n \geqq 1$, $(J, \{\varphi_n\}) \in \mathscr{C}(J)$. Hence condition (c) is satisfied and we are done.

II) $x \in \partial V$. An endpoint of an eventually periodic interval is either eventually periodic or its forward orbit contains a critical point (maybe both). The second possibility cannot hold for x because x is contained in Λ that doesn't contain critical points. Then x is eventually periodic. But it cannot be periodic. Therefore V itself is eventually periodic but not periodic and:

$$\liminf_{n \to +\infty} d(f^n(V), V) > 0. \tag{19}$$

On the other hand for every m there exists $n_m \geqq 1$ and $(J_m, \{\varphi_n\}) \in \mathscr{C}(J_m)$ such that

$$U \cap \varphi_{n_m}(J_m) \cap J_m \neq \phi. \tag{20}$$

In particular

$$d(x, \varphi_{n_m}(x)) \leqq \operatorname{diam} J_m + \operatorname{diam} \varphi_{n_m}(J_n).$$

But $\operatorname{diam} J_m \to 0$ when $m \to +\infty$ and, by Lemma II.1.

$$\operatorname{Sup}\{\operatorname{diam} \varphi_n(J_m) | n \geqq 1, (J_m, \{\varphi_n\}) \in \mathscr{C}(J_m)\}$$

converges to zero when $m \to +\infty$. Hence $d(x, \varphi_{n_m}(x)) \to 0$ when $m \to +\infty$. If the sequence $\{n_m\}$ is bounded, this implies that x is periodic, contradicting our hypothesis. If it is unbounded we use (20) to obtain

$$\phi \neq f^{n_m}(U \cap \varphi_{n_m}(J_m) \cap J_n) \subset f^{n_m}(U) \cap J_m \cap f^{n_m}(J_m),$$

which implies:

$$f^{n_m}(V) \cap J_m \neq \phi.$$

Then,

$$d(f^{n_m}(V), V) \leqq d(f^{n_m}(V), x) \leqq \operatorname{diam} J_m,$$

contradicting (19).

The proof of Lemma I.3 will require different methods according to whether $\Lambda = N, \Lambda \neq N$. To prove it in the case when $\Lambda \neq N$ we need the three following lemmas. In its statements Λ and f will be as in the statement of I.3 and the notation $\mathscr{C}(J_0, \Lambda)$ will have the same meaning as that of the proof of II.1.

Lemma II.2. *If* $x \in \bigcap_{n \geqq 0} f^n(\Lambda)$ *is non-periodic there exists an interval* $J_0 \supset \{x\}$ *adapted to* Λ *such that for all* $r \geqq 0$ *there exists an arbitrarily small interval* $\{x\} \subset J \subset J_0$ *adapted to* Λ *satisfying*

$$\operatorname{diam} \varphi_n(J_0) \leqq r$$

for all $n \geqq N$ *and* $(J_0, \{\varphi_n\}) \in \mathscr{C}(J_0, \Lambda)$ *such that*

$$\varphi_n(J) \cap J \neq \phi.$$

Proof. By Lemma I.2 there exist arbitrarily small intervals $J_0 \supset \{x\}$ adapted to Λ. If the ω-limit set of x is a periodic orbit, γ, take J_0 satisfying

$$\bar{J}_0 \cap \gamma = \phi. \tag{21}$$

This can be done because since x is not periodic $x \notin \gamma$. If the ω-limit set of x is not a periodic orbit, choose any $J_0 \supset \{x\}$ adapted to Λ. Suppose that J_0 doesn't satisfy the required property. Then there exists $r > 0$, a sequence of coherent sequences $(J_0, \{\varphi_n^{(i)}\}) \in \mathscr{C}(J_0, \Lambda)$ $i = 1, 2, 3, \ldots$, a sequence of integers $n_1 < n_2 < \ldots$ and a sequence of intervals $J_1 \supset J_2 \supset \cdots \supset \{x\}$, with $\bigcap_{n \geqq 1} J_n = \{x\}$, such that

$$\varphi_{n_i}^{(i)}(J_i) \cap J_i \neq \phi \tag{22}$$

$$\operatorname{diam} \varphi_{n_i}^{(i)}(J_0) > r \tag{23}$$

for all $i \geqq 1$. Also from the fact that J_0 is adapted to Λ it follows that for any pair of coherent sequences $(J_0, \{\varphi_n\}) \in \mathscr{C}(J_0, \Lambda)$ and $(J_0, \{\psi_n\}) \in \mathscr{C}(J_0, \Lambda)$, and every $n \geqq l \geqq 1$, one of the following relations holds:

$$\varphi_n(J_0) \cap \psi_l(J_0) = \phi,$$
$$\varphi_n(J_0) \subset \psi_l(J_0) \quad \text{or} \quad \psi_l(J_0) \subset \varphi_n(J_0).$$

Using this property it is easy to see that the sequence above can be chosen satisfying

$$\varphi_{n_i}^{(i)}(J_0) \subset \varphi_{n_j}^{(j)}(J_0) \subset J_0 \tag{24}$$

for all $i \geqq j \geqq 1$. Therefore (23) and (24) imply that the set

$$U = \operatorname{Int} \bigcap_{i \geqq 1} \varphi_{n_i}^{(i)}(J_0) \tag{24}$$

is a non-empty open interval contained in J_0. It is easy to check that it is a d-interval. Since f is not topologically equivalent to an irrational rotation, U is eventually periodic by Lemma I.4. On the other hand

$$\lim_{i\to+\infty} \operatorname{diam} \varphi_{n_i}^{(i)}(J_i)=0 \tag{25}$$

by Lemma II.1. Moreover, by (22) and (24):

$$d(x, U)=\lim_i d(x, \varphi_{n_i}^{(i)}(J_0)) \leqq \lim_i d(x, \varphi_{n_i}^{(i)}(J_i))$$

$$\leqq \lim_i (\operatorname{diam} J_i + \operatorname{diam} \varphi_{n_i}^{(i)}(J_i)).$$

Since $\lim_i \operatorname{diam} J_i=0$, this inequality together with (25) implies that $x\in\bar{U}$. Moreover from the definition of U it follows that $f^{n_i}(U)\subset J_0$. Hence $\omega(x)\cap\bar{J}_0\neq\phi$. But since $x\in\bar{U}$ and U is eventually periodic it follows that $\omega(x)$ is a periodic orbit. Then the relation $\omega(x)\cap\bar{J}_0\neq\phi$ contradicts the way we choose J_0.

Lemma II.3. *If $f\in\operatorname{End}^2(N)$ for all $\delta>0$ there exists $K_0=K_0(\delta, f)>0$ such that $(J,\{\varphi_n\})$ is a coherent sequence satisfying:*

$$\inf_{n\geqq 0} d(\varphi_n(J), C(f))>0,$$

then

$$\frac{|\varphi_n'(x)|}{|\varphi_n'(y)|}\leqq \exp K_0 \sum_{j=0}^{n-1} \operatorname{diam} \varphi_j(J),$$

$$|\varphi_n'(x)|\leqq \frac{\operatorname{diam} \varphi_n(J)}{\operatorname{diam} J} \exp K_0 \sum_{j=0}^{n-1} \operatorname{diam} \varphi_j(J)$$

for all $n\geqq 0$, $x\in J$.

This Lemma is proved using a trivial adaptation of the method of Schwarz, proof of Denjoy's theorem. For a proof see Jacobson [1] Lemma Ia, Ib.

Lemma II.4. *If J is an interval adapted to Λ such that either $F(J,\Lambda)=\phi$ or there exists $0<\lambda<1$ satisfying:*

$$|\psi'(x)|<\lambda$$

for all $x\in J$ and $\psi\in F(J,\Lambda)$, then there exists $K_1>0$ such that

$$|\varphi_n'(x)|\leqq K_1|\varphi_n'(y)|,$$

$$\sum_{n=1}^{\infty} |\varphi_n'(x)|\leqq K_1$$

for all x, y in J and $(J,\{\varphi_n\})\in\mathscr{C}(J,\Lambda)$.

Proof. We shall first prove that there exists $K>0$ such that:

$$\sum_{n=0}^{\infty} \operatorname{diam} \varphi_n(J)\leqq K \tag{26}$$

for all $(J,\{\varphi_n\})\in\mathscr{C}(J,\Lambda)$. We note that here φ_0 stands for the identity and the same notation will be used in all this proof. Take $(J,\{\varphi_n\})\in\mathscr{C}(J,\Lambda)$. Suppose that

$$\varphi_n(J)\cap J=\phi$$

for all $n\geqq 1$ (this is the only possibility if $F(J,\Lambda)=\phi$). Then:

$$\varphi_n(J)\cap\varphi_m(J)=\phi$$

for all $1\leqq n\leqq m$, and then

$$\sum_{n=0}^{\infty}\operatorname{diam}\varphi_n(J)\leqq 2\operatorname{diam}N. \tag{27}$$

Now suppose that $\varphi_n(J)\cap J\neq\phi$ for infinitely many values of n, and let $n_1<n_2<\cdots$ be those integers for which $\varphi_{n_i}(J)\cap J\neq\phi$. Since J is adapted to Λ,

$$\varphi_{n_i}(J)\subset J$$

for all $i\geqq 1$. Clearly $\varphi_n\in F(\Lambda,J)$. Moreover it is easy to see (using again that J is adapted to Λ) that for all $i>1$ there exists $\psi_i\in F(J,\Lambda)$ such that

$$\psi_i/\varphi_{n_{i-1}}(J)=\varphi_{n_i}f^{n_{i-1}}$$

Set $\psi_1=\varphi_{n_1}$. Define $m_1=n_1$ and $m_i=n_i-n_{i-1}$ if $i>1$. The maps ψ_i can be described as follows. There exist branches $\varphi_n^{(i)}$ of f^{-n}/J, where $1\leqq n\leqq m_i$ satisfying the following properties:

a) $\varphi_n^{(i)}=f\varphi_{n+1}^{(i)}$ for all $i\geqq 1$, $1\leqq n<m_i$,
b) $\varphi_{m_i}^{(i)}=\psi_i$ for all $i\geqq 1$.

Now observe that the definition of the sequence $\{n_i\}$ plus the fact of J being adapted to Λ imply:

$$\varphi_n^{(i)}(J)\cap J=\phi$$

for all $1\leqq n<m_i$ and $i\geqq 1$. Then

$$\varphi_n^{(i)}(J)\cap\varphi_m^{(i)}(J)=\phi$$

for all $1\leqq n<m_i$. Hence

$$\sum_{n=0}^{m_i-1}\varphi_n^{(i)}(J)\leqq 2\operatorname{diam}N$$

for all $i\geqq 1$. By II.3 there exists a constant K_0 depending only on J such that

$$\frac{|(\varphi_n^{(i)})'(x)|}{|(\varphi_n^{(i)})'(y)|}\leqq\exp K_0\sum_{j=0}^{n-1}\operatorname{diam}\varphi_j^{(i)}(J)\leqq\exp 2K_0\operatorname{diam}N. \tag{28}$$

Set $K_1=\exp 2K_0\operatorname{diam}N$. From (28) it follows easily that if $A\subset J$ is any subset:

$$\frac{\operatorname{diam}\varphi_n^{(i)}(A)}{\operatorname{diam}A}\leqq K_1\frac{\operatorname{diam}\varphi_n^{(i)}(J)}{\operatorname{diam}J},$$

and then

$$\sum_{n=0}^{m_i-1}\operatorname{diam}\varphi_n^{(i)}(A)\leqq\operatorname{diam}A\frac{K_1}{\operatorname{diam}J}\sum_{n=0}^{m_i-1}\operatorname{diam}\varphi^{(i)}(J)$$

$$\leqq \operatorname{diam} A \frac{K_1}{\operatorname{diam} J} 2 \operatorname{diam} N. \tag{29}$$

Set

$$K_2 = 2K_1 \operatorname{diam} N.$$

If $i > 1$, applying (29) to $A = \varphi_{n_i - 1}(J)$, we obtain:

$$\sum_{n=0}^{m_i - 1} \operatorname{diam} \varphi_n^{(i)}(\varphi_{n_{i-1}}(J)) < K_2 \operatorname{diam} \varphi_{n_{i-1}}(J) \operatorname{diam}(J).$$

But $\varphi_{n_{i-1}}$ is a composition of $i-1$ maps in $F(J,\Lambda)$, namely the maps $\psi_1, \ldots, \psi_{i-1}$. Then:

$$|\varphi_{n_{i-1}}(x)| \leqq \lambda^{i-1}$$

for all $x \in J$. Hence

$$\operatorname{diam} \varphi_{n_{i-1}}(J) \leqq \operatorname{diam}(J) \lambda^{i-1},$$

and then

$$\sum_{n=0}^{m_i - 1} \operatorname{diam} \varphi_n^{(i)}(\varphi_{n_{i-1}}(J)) \leqq K_2 \lambda^{i-1}.$$

In a similar way we obtain:

$$\sum_{n=0}^{m_1 - 1} \operatorname{diam} \varphi^{(1)}(J) \leqq K_2.$$

Then

$$\sum_{n=0}^{\infty} \operatorname{diam} \varphi_n(J) \leqq \sum_{i=1}^{\infty} \sum_{n=0}^{m_i - 1} \operatorname{diam} \varphi_n^{(1)}(\varphi_{n_i - 1}(J))$$

$$\leqq \sum_{i=1}^{\infty} K_2 \lambda^i. \tag{30}$$

With minor modifications the same methods can be applied to the case when $\varphi_n(J) \cap J \neq \phi$ holds for a finite non-empty set of values of n, and the result is

$$\sum_{n=0}^{\infty} \varphi_n(J) \leqq 2 \operatorname{diam} N + \sum_{i=0}^{\infty} K_2 \lambda^i. \tag{31}$$

Then (27), (30) and (31) prove (26). The proof of II.4 is now an easy corollary of II.3. In fact, applying II.3 to $(J, \{\varphi_n\}) \in \mathscr{C}(J,\Lambda)$, we obtain

$$\frac{|\varphi_n'(x)|}{|\varphi_n'(y)|} \leqq \exp K_0 \sum_{j=0}^{n-1} \operatorname{diam} \varphi_j(J) \leqq \exp K_0 K \tag{32}$$

for all $x, y \in J$ and $n \geqq 1$. Moreover

$$|\varphi_n'(x)| \leqq \frac{\operatorname{diam} \varphi_n(J)}{\operatorname{diam} J} \exp K_0 \sum_{j=0}^{n-1} \operatorname{diam} \varphi_j(J) \leqq \frac{\operatorname{diam} \varphi_n(J)}{\operatorname{diam} J} \exp K_0 K.$$

Hence:

$$\sum_{n=0}^{\infty} |\varphi_n'(x)| \leqq \frac{\exp K_0 K}{\operatorname{diam} J} \sum_{n=0}^{\infty} \operatorname{diam} \varphi_n(J) \leqq \frac{K \exp K_0 K}{\operatorname{diam}(J)}.$$

This inequality and (32) prove the Lemma.

Now we are ready to prove Lemma I.3 in the case $\Lambda \neq N$. This property ensures the existence of a point $x \in \Lambda$ that is an endpoint of an open interval U contained in Λ^c. Let $J_0 \supset \{x\}$ be the interval adapted to Λ given by II.2. Suppose that $U = (x, b)$. Choose J_0 so small that $b \notin \bar{J}_0$. Since J_0 is adapted to Λ, there exists $\delta > 0$ such that $d(\varphi_n(J_0), C(f)) > \delta$ for all $n \geqq 1$ and $(J_0, \{\varphi_n\}) \in \mathscr{C}(J_0, \Lambda)$. Let $K_0 = K_0(\delta, f)$ be the constant given by II.3. Let J_0 be the interval (b^-, b^+) and take $r > 0$ satisfying:

$$\exp 4K_0 \operatorname{diam} N \cdot \frac{r}{b^+ - x} < \frac{1}{2}. \tag{33}$$

By Lemma II.2 there exists an interval $\{x\} \subset J \subset J_0$ adapted to Λ and such that if $(J_0, \{\varphi_n\}) \in \mathscr{C}(J_0, \Lambda)$ and

$$\varphi_n(J) \cap J \neq \phi$$

for some $n \geqq 1$, then

$$\operatorname{diam} \varphi_n(J_0) \leqq r.$$

Suppose that $J = (a^-, a^+)$. Define:

$$J_1 = (x, b^+).$$

Let ψ be an element of $F(J, \Lambda)$. By definition of $F(J, \Lambda)$ there exists $x_0 \in J \cap (\bigcap_{n \geqq 0} f^n(\Lambda))$ such that $\psi(x_0) \in \bigcap_{n \geqq 0} f^n(\Lambda)$. Take $k \geqq 1$ satisying $f^k(\psi(x_0)) = x_0$, and $\theta \in \mathscr{S}(\Lambda, x_0)$ such that $\theta(k) = \psi(x_0)$. Take $(J_0, \{\varphi_n\}) \in \mathscr{C}(J_0, \Lambda)$ satisfying $\theta(n) = \varphi_n(x_0)$ for all $n \leqq 1$. Then:

$$\varphi_k / J = \psi, \tag{34}$$

and

$$\varphi_n(J) \cap J = \phi \tag{35}$$

for all $1 \leqq n < k$. We claim that for all $n \geqq 1$

$$\varphi_n(J) \cap J = \phi \Rightarrow \varphi_n(J_1) \cap J_1 = \phi. \tag{36}$$

Suppose that $\varphi_n(J) \cap J = \phi$ and $\varphi_n(J_1) \cap J_1 \neq \phi$. The last property implies $\varphi_n(J_0) \cap J_0 \neq \phi$, and then $\varphi_n(J_0) \subset J_0$. In particular $\varphi_n(J) \subset J_0$, and since $\varphi_n(J) \cap J = \phi$, either $\varphi_n(J) \subset (b^-, a^-)$ or $\varphi_n(J) \subset (a^+, b^+)$. The last relation is not possible because $\varphi_n(J)$ contains $\varphi_n(x_0)$, and $\varphi_n(x_0) = \lambda(n)$ is contained in Λ. But $(a^+, b^+) \subset (x, b^+)$ doesn't contain points in Λ. Then $\varphi_n(J) \subset (b^-, a^-)$. One of the endpoints of $\varphi_n(J_1)$ is contained in $\overline{\varphi_n(J)}$. Since $\varphi_n(J) \subset (b^-, a^-)$ it follows that one endpoint of

$\varphi_n(J_1)$ is contained in $[b^-, a^-]$. Moreover $\varphi_n(J_1) \subset J_0$ (because $\varphi_n(J_1) \subset \varphi_n(J_0)$ and, as we observed above, $\varphi_n(J_0) \subset J_0$). Hence $\varphi_n(J_1)$ is an interval contained in (b^-, b^+) with an endpoint in $[b^-, a^-]$ and non-empty intersection with $(x, b^+) = J_1$. It follows that $\varphi_n(J_1) \supset (x, a^+)$. But then $x_0 \in \varphi_n(J_1)$. This implies $f^n(x_0) \in J_1$. Then $\Lambda \cap J_1 \neq \phi$ because $f^n(x_0)$ belongs to Λ and J_1. On the other hand $J_1 \subset (x, b)$. Therefore it doesn't intersect Λ. This contradiction completes the proof of (36). From (35) and (36) follows that $\varphi_n(J_1) \cap J_1 = \phi$ for all $1 \leqq n < k$. Hence $\varphi_n(J_1) \cap \varphi_m(J_1) = \phi$ for all $1 \leqq n < m < k$ and then

$$\sum_{n=0}^{k-1} \operatorname{diam} \varphi_n(J_1) \leqq 2 \operatorname{diam} N.$$

By Lemma II.3:

$$|\varphi_k'(y)| \leqq \frac{\operatorname{diam} \varphi_k(J_1)}{\operatorname{diam} J_1} \cdot \exp K_0 \sum_{n=0}^{k-1} \operatorname{diam} \varphi_n(J_1)$$

$$\leqq \exp 2K_0 \operatorname{diam} N \cdot \frac{\operatorname{diam} \varphi_k(J_0)}{b^+ - x} \tag{37}$$

for all $y \in \bar{J}_1$. Moreover (35) implies $\varphi_n(J) \cap \varphi_m(J) = \phi$ for all $1 \leqq n < m$. Then, by II.3:

$$\frac{|\varphi_k'(z)|}{|\varphi_k'(y)|} \leqq \exp K_0 \sum_{n=0}^{k-1} \operatorname{diam} \varphi_n(J) \leqq \exp 2K_0 \operatorname{diam} N \tag{38}$$

for all y and z in $\bar{J}$. Given $z \in J$ and using (33), (37) and (38):

$$|\psi'(z)| = |\varphi_k'(z)| = |\varphi_k'(a^+)| \cdot \frac{|\varphi_k'(z)|}{|\varphi_k'(a^+)|} \leqq \exp 4K_0 \operatorname{diam} N \cdot \frac{r}{b^+ - x} < \frac{1}{2}.$$

This completes the proof in the case $\Lambda \neq N$. Now let us consider the case $\Lambda = N$. In the case f has no critical points. If it is also injective, it will be a diffeormorphism. Since f is not topologically equivalent to a rotation, it will be a diffeomorphism with periodic points. Moreover by hypothesis not every point in $N = \bigcap_{n \geqq 0} f^n(N) = \bigcap_{n \geqq 0} f^n(\Lambda)$ will be periodic. Under this condition it is trivial to find an open interval $J \subset N$ such that $J \cap f^{-n}(J) = \phi$ for all $n \geqq 1$. This means that J is adapted and $F(J, N) = \phi$. By II.4, J satisfies the properties required in Lemma I.3:

If f in not injective, then $N = S^1$ and $f: S^1 \circlearrowleft$ is an immersion with degree $d \neq 1$ or -1. If $J(f) \neq S^1$ (where $J(f)$ is defined as in the introduction) there exists a plateau J satisfying $f^{-n}(J) \cap J = \phi$ for all $n \geqq 1$ (any non-periodic plateau satisfies this property). Then J is adapted and $F(J, \Lambda) = \phi$. Hence, by Lemma II.4, J satisfies the thesis of I.3. It remains to consider the case $J(f) = S^1$. This means that f is topologically equivalent to the map $z \to z^d$. We claim that there exists an interval J_0 adapted to S^1 such that for all $\varepsilon > 0$ there exists $\psi_0 \in F(J_0, S^1)$ satisfying

$$\operatorname{diam} \psi_0(J_0) < \varepsilon \tag{39}$$

and

$$\operatorname{diam} \varphi\psi_0(J_0) < \varepsilon \tag{40}$$

for all $n \geqq 1$ and every branch φ of f^{-n}/J_0. First let us prove this property for $f(z)=z^d$. Take as J_0 an open interval whose endpoints are fixed points of some power f^m and not containing fixed points of f^m. It is easy to verify that if $m|d| \geqq 3$ then $F(J_0, S^1)$ contains infinitely many maps. Moreover $\psi_1(J_0) \cap \psi_2(J_0) = \phi$ if ψ_1 and ψ_2 are different maps in $F(J_0, \Lambda)$. Therefore there exists $\psi_0 \in F(J_0, \Lambda)$ satisfying (39). Property (40) also holds because any branch of f^{-n}/J_0 is a contraction. This completes the proof of the claim for the map $z \to z^d$. The general case of an immersion $f: S^1 \circlearrowleft$ topologically equivalent to $z \to z^d$ follows easily from this case using the conjugacy between f and $z \mapsto z^d$.

On the other hand, since f is a C^2 immersion, we can apply Lemma II.3 to obtain a constant K_0 such that the inequalities of Lemma II.3 hold for every open interval $J \subset S^1$ and every coherent sequence $(J, \{\varphi_n\}) \in \mathscr{C}(J, S^1)$.

Fix $\varepsilon > 0$ satisfying:

$$\varepsilon \frac{\exp 4K_0 \operatorname{diam} N}{\operatorname{diam} J} < \frac{1}{2}. \tag{41}$$

By the claim there exist an adapted interval J_0 and $\psi_0 \in F(J_0, \Lambda)$ satisfying (39) and (40). Set $J = \psi_0(J_0)$. We claim that:

$$|\psi'(x)| < \tfrac{1}{2} \tag{42}$$

for all $\psi \in F(J, \Lambda)$ and every $x \in J$. First we shall prove that:

$$|\psi'(x)| \leqq \exp 2K_0 \operatorname{diam} N \frac{\operatorname{diam} \psi(J_0)}{\operatorname{diam} J_0} \tag{43}$$

for all $\psi \in F(J_0, \Lambda)$. Take a coherent sequence $(J_0, \{\varphi_n\})$ and $k \geqq 1$ that satisfies:

$$\psi = \varphi_k$$

and

$$\varphi_n(J_0) \cap J_0 = \phi$$

for all $1 \leqq n < k$. Then $\varphi_n(J_0) \cap \varphi_m(J_0) = \phi$ for all $1 \leqq n < m \leqq k$. Hence

$$\sum_{n=1}^{k} \operatorname{diam} \varphi_n(J_0) < 2 \operatorname{diam} N,$$

and by II.3:

$$\begin{aligned} |\psi'(x)| = |\varphi_k'(x)| &\leqq \frac{\operatorname{diam} \psi(J_0)}{\operatorname{diam} J_0} \exp K_0 \sum_{j=0}^{n-1} \operatorname{diam} \varphi_j(J_0) \\ &\leqq \frac{\operatorname{diam} \psi(J_0)}{\operatorname{diam} J_0} \exp 2K_0 \operatorname{diam} N, \end{aligned}$$

completing the proof of (43). Now take a coherent sequence $(J_0, \{\varphi_n\})$ and $k \geqq 1$ satisfying $\psi = \varphi_k/J$ and $\varphi_n(J) \cap J = \phi$ for $1 \leqq n < k$. Suppose first that $\varphi_n(J_0) \cap$

$J_0=\phi$ for all $1\leqq n<k$. Then $\varphi_n\in F(J_0,S^1)$ and by (43), (39) and (41):

$$|\psi'(x)|=|\varphi_k'(x)|\leqq \exp 4K_0 \operatorname{diam} N \frac{\operatorname{diam}\psi(J_0)}{\operatorname{diam} J_0}$$

$$\leqq \exp 2K_0 \operatorname{diam} N\cdot\frac{\varepsilon}{\operatorname{diam} J_0}<\frac{1}{2} \tag{44}$$

for all $x\in J$. Now suppose that there exist integers $1\leqq n<k$ such that $\varphi_n(J_0)\cap J_0\neq\phi$. Let n_0 be the maximum of such integers. Consider the maps:

$$\psi_n=\varphi_{n_0+n}f^{n_0}/\varphi_{n_0}(J_0) \tag{45}$$

with $n\geqq 1$. Then $(\varphi_{n_0}(J_0),\{\psi_n\})$ is a coherent sequence. We can extend it to a coherent sequence $(J_0,\{\psi_n\})$. Moreover

$$\psi_n(J_0)\cap J_0=\phi \tag{46}$$

for $1\leqq n<k-n_0$ because $\psi_n(J_0)\cap J_0\neq\phi$ implies $\psi_n(J_0)\subset J_0$ because J_0 is adapted. and then

$$\varphi_{n_0+n}(J_0)\cap J_0=\psi_n(\varphi_{n_0}(J_0))\cap J_0\neq\phi.$$

By the way we choose n_0, this implies $n_0+n\geqq k$, hence $n\geqq k-n_0$ and (46) is proved. Property (46) implies that $\psi_{k-n_0}\in F(J_0,S^1)$. Then, by (43),

$$|\psi_{k-n_0}'(x)|\leqq \exp 2K_0 \operatorname{diam} N\frac{\operatorname{diam}\psi_{k-n_0}(J_0)}{\operatorname{diam} J_0}$$

for all $x\in J_0$. But by (45)

$$\psi_{k-n_0}(J_0)\subset J.$$

Hence

$$|\psi_{k-n_0}'(x)|\leqq \exp 2K_0 \operatorname{diam} N\frac{\operatorname{diam} J}{\operatorname{diam} J_0}. \tag{47}$$

Moreover:

$$\begin{gathered}\psi=\psi_{k-n_0}\varphi_{n_0},\\ \varphi_n(J)\cap\varphi_m(J)=\phi\end{gathered} \tag{48}$$

for all $1\leqq n<m<n_0$. Then:

$$\sum_{n=1}^{n_0-1}\operatorname{diam}\varphi_n(J)\leqq 2\operatorname{diam} N.$$

Applying II.3:

$$|\varphi_{n_0}'(x)|\leqq \exp 2K_0 \operatorname{diam} N\frac{\operatorname{diam}\varphi_{n_0}(J)}{\operatorname{diam} J}. \tag{49}$$

Then, using (48), (49) and (47) we obtain:

$$|\psi'(x)| \leqq |\psi'_{k-no}(\varphi_{no}(x))| \, |\varphi'_{no}(x))|$$

$$\leqq \exp 4K_0 \operatorname{diam} N \frac{\operatorname{diam} \varphi_{no}(J)}{\operatorname{diam} J_0}$$

$$\leqq \exp 4K_0 \operatorname{diam} N \frac{\varepsilon}{\operatorname{diam} J_0} \leqq \frac{1}{2}.$$

We have thus proved that J satisfies the hypothesis of Lemma II.4. Then J satisfies the properties required in Lemma I.3.

III. Appendix. Isolated Hyperbolic Sets

Here we shall prove property (IV) of the Introduction. Let $\Lambda \subset N - \partial N$ be a hyperbolic set of $f \in \operatorname{End}^r(N)$. Let V be a neighborhood of Λ. Define

$$U(\lambda) = \{x \in N | d(x, \Lambda) < \lambda\}, \quad \Lambda(\lambda) = \bigcap_{n \geqq 0} f^n(\overline{U(\lambda)}).$$

Then $\bigcap_{\lambda > 0} \overline{U(\lambda)} = \Lambda$ and $\bigcap_{\lambda > 0} \Lambda(\lambda) = \Lambda$. It follows that there exists $\varepsilon > 0$ such that

$$\Lambda \subset \Lambda(\lambda) \subset V, \tag{1}$$

$$\Lambda(\lambda) \text{ is hyperbolic} \tag{2}$$

for every $0 < \lambda < \varepsilon$. Since $U(\lambda)$ is open it can be written as a union of intervals $(a^{2i}(\lambda), a^{2i+1}(\lambda))$, $i = 0, \ldots, m(\lambda)$. Clearly $m(\lambda'') \geqq m(\lambda')$ when $0 < \lambda'' < \lambda'$. Then it is easy to see that there exist $0 < a < b < \varepsilon$ such that $m(\lambda)$ is constant for $a < \lambda < b$. Call m this constant. It follows that the functions $a^{(i)}(\lambda)$ are continuous and monotone for $1 \leqq i \leqq m$ and $\lambda \in (a, b)$. Suppose that $\overline{U(\lambda)}$ is a neighborhood of $\Lambda(\lambda)$, is hyperbolic, contains Λ, is contained in V and is isolated (with $\overline{U(\lambda)}$ as an isolating block). Then property (IV) is proved. Now suppose that $\overline{U(\lambda)}$ is not a neighborhood of $\Lambda(\lambda)$ for all $\lambda \in (a, b)$. This means that for all $\lambda \in (a, b)$ there exists $1 \leqq i \leqq m$ such that $a^{(i)}(\lambda) \in \Lambda(\lambda)$. Therefore

$$(a, b) = \bigcup_{i=1}^{m} S_i, \tag{3}$$

where S_i is define by

$$S_i = \{\lambda \in (a, b) | a^{(i)}(\lambda) \in \Lambda(\lambda)\}.$$

Since every S_i is closed, (3) implies that there exists $1 \leqq i \leqq m$ such that S_i has non-empty interior. Let (c, d) be an interval contained in S_i. Then

$$a^{(i)}(\lambda) \in \Lambda(\lambda) \subset \Lambda(d)$$

for all $\lambda \in (c, d)$, and it follows that the interval $J = a^{(i)}((c, d))$ is contained in $\Lambda(d)$. If f^n/J is injective for all $n \geqq 1$, the hyperbolicity of $\Lambda(d)$ implies that diam $f^n(J) \to +\infty$ when $n \to +\infty$. But since $\Lambda(d) \supset f^n(J)$ for all $n \geqq 0$, this is impossible. Therefore

f^n/J is not injective for some $n \geqq 0$. Since for all $n \geqq 1$ f^n/J has no critical points, because for all $n \geqq 0$, $f^n(J) \subset \Lambda(d)$, and $\Lambda(d)$ contains no critical points because it is hyperbolic, the non-injectivity of f^n/J implies that $N = S^1$ and $f^n(J) = S^1$. Then $\Lambda(d) \supset f^n(J) = S^1$. If $\Lambda = S^1$ without loss of generality we can take the neighborhood V satisfying $V \neq S^1$. But then (1) implies $\Lambda(d) \subset V$. On the other hand we have proved that $S^1 < \Lambda(d)$. Then $V = S^1$. This contradiction shows that $\overline{U(\lambda)}$ must be a neighborhood of $\Lambda(\lambda)$ for some $\lambda \in (a, \beta)$ and completes the proof of property (IV).

References

1. Jacobson, M. V.: On smooth mappings of the circle into itself. Mat. U.S.S.R. Sb. **14**, 161–185 (1971)
2. Herman, M.: Sur la conjugaison différentiable des difféomorphisms du cercle à des rotations, Publications Mathématiques IHES
3. Lasota, A., Yorke, J.: On the existence of invariant measures for piecewise monotonic transformations, Trans. Am. Math. Soc. **186**, 481–448 (1973)
4. Malta, I.: On Denjoy theorem for endomorphisms. (preprint, 1983)
5. Nitecki, Z.: Factorization of nonsingular circle endomorphisms, salvador symposium on dynamical systems. Peixoto, M. ed. New York: Academic Press, 1973
6. Peixoto, M.: Structural stability on two dimensional manifolds. Topology **1**, 101–120 (1962)
7. Shub, M.: Endomorphisms of compact manifolds. Am. J. Math. **91**, 175–199 (1969)
8. Young, L. S.: A closing Lemma on the interval. Inv. Math. **54**, 179–187 (1970)

Communicated by A. Jaffe

Received December 6, 1984

Commun. Math. Phys. 112, 721–724 (1987)

Communications in Mathematical Physics

Erratum

Hyperbolicity, Sinks and Measure in One Dimensional Dynamics

Ricardo Mañé

Instituto de Matematica Pura e Aplicada, Edificio Lelio Gama, Estrada Dona Castorina, 110, CEP 22, 460, Rio de Janeiro, Brasil

Commun. Math. Phys. **100**, 495–524 (1985)

In our paper "Hyperbolicity, sinks and measure in one dimensional dynamics" (Commun. Math. Phys. **100**, 495–524 (1985)), the proof of the crucial Lemma I.3 is not complete. This proof is divided in two cases depending on whether $\Lambda = N$ or $\Lambda \neq N$. The case $\Lambda = N$ is correct. The proof of the case $\Lambda \neq N$ begins by taking an open interval $U \subset \Lambda^c$ having an endpoint x contained in Λ, and almost immediately we apply to x a Lemma (Lemma II.2) that requires x to be non-periodic. But a point x chosen as we did can be periodic, and, even more important, it is easy to produce examples of sets Λ satisfying the hypothesis of I.3 such that every $x \in \Lambda$ that is an endpoint of an open interval contained in Λ^c, is periodic. Therefore the proof must be extended in order to cover these cases. This can be done as follows, by a suitable combination of tools already developed in the paper.

We have to prove Lemma I.3 when $\Lambda \neq N$. We shall do it through the following steps.

We have to prove Lemma I.3 when $\Lambda \neq N$. First observe that it follows trivially from the definition that $(J, \{\varphi_n\})$ is a coherent sequence associated to Λ if and only if it is a coherent sequence associated to $\bigcap_{n \geqq 0} f^n(\Lambda)$. Follows from this remark that an interval J is adapted to Λ if and only if it is adapted to $\bigcap_{n \geqq 0} f^n(J)$ because the definition of interval adapted to a set Λ only involves the family of coherent sequences associated to Λ, that as we said, coincides with the family of coherent sequences associated to $\bigcap_{n \geqq 0} f^n(\Lambda)$.

Therefore it suffices to prove Lemma I.3 in the surjective case, i.e. when $f(\Lambda) = \Lambda$, because, once proved in this case, if we want to prove it in the general case of a compact set $\Lambda \neq N$ satisfying only $f(\Lambda) \subset \Lambda$, we apply the surjective case to $\bigcap_{n \geqq 0} f^n(\Lambda)$ (that obviously satisfies $f(\bigcap_{n \geqq 0} f^n(\Lambda)) = \bigcap_{n \geqq 0} f^n(\Lambda)$). This yields an interval J adapted to $\bigcap_{n \geqq 0} f^n(\Lambda))$ (and then, as explained above, adapted to Λ) satisfying the properties in Lemma I.3 for $\bigcap_{n \geqq 0} f^n(\Lambda)$. Since these properties only involve the family of coherent

Originally published in: *Comm. Math. Phys. 112 (1987), no. 4, 721–724,*

sequences associated to $\bigcap_{n\geq 0} f^n(\Lambda)$, that coincides with the family of coherent sequences associated to Λ, it follows that J satisfies Lemma I.3 for Λ.

Then we have to prove I.3 when $\Lambda \neq N$ and $f(\Lambda)=\Lambda$. We shall do it through the following steps.

Step I. If an interval J adapted to Λ satisfies the following property:

(*) Either $F(J,\Lambda)=\emptyset$ or there exists $0<\lambda<1$ that $|\psi'(x)|<\lambda$ for all $x\in J$ and $\psi\in F(J,\Lambda)$, then J satisfies Lemma I.3.

This step is just Lemma II.4. We include it here because it reduces the proof of I.3 to finding an interval J adapted to Λ and satisfying (*).

Step II. If there exists an open interval $U\subset\Lambda^c$ having an endpoint $x\in\Lambda$ that is not a periodic point, then there exists an interval J adapted to Λ satisfying (*).

This step is what is correctly and explicitly proved in the proof of I.3 when $\Lambda\neq N$ (page 519).

Step III. Let J_0 be an interval adapted to Λ such that for all $\varepsilon>0$ there exists $\psi_0\in F(J_0,\Lambda)$ satisfying $\operatorname{diam}\psi_0(J_0)<\varepsilon$, then there exists an interval J adapted to Λ satisfying (*).

The proof of this property is contained in the proof of I.3 in the case $\Lambda=N$. In fact, to prove I.3 when $\Lambda=N$, we use $\Lambda=N$ to show that there exists an interval J_0 satisfying the hypothesis of Step III, and then (without using again the hypothesis $\Lambda=N$) we prove that this hypothesis implies the existence of J satisfying (*).

Step IV. If there exists an interval J_0 adapted to Λ such that $F(J_0,\Lambda)$ contains infinitely many maps, then there exists an interval J adapted to Λ satisfying (*).

To prove this property observe that $\Sigma\operatorname{diam}\psi(J_0)<\operatorname{diam}(J_0)$, where the sum is taken over all the maps $\psi\in F(J_0,\Lambda)$. Then, if $\#F(J_0,\Lambda)=\infty$, there exists $\psi_0\in F(J_0,\Lambda)$ with $\operatorname{diam}\psi(J_0)$ arbitrarily small, thus proving that J_0 satisfies the hypothesis of Step III.

Given a set $J\subset N$ and $x\in J$ define $N(x,J)$ as the minimum $n>0$ such that $f^n(x)\in J$. If $f^n(x)\notin J$ for all $n>0$ set $N(x,J)=+\infty$.

Step V. If J_0 is an interval adapted to Λ such that $\sup\{N(x,J_0)|x\in\Lambda\cap J_0\}=\infty$, then there exists J satisfying (*).

First suppose that there exists a sequence $\{x_i\}\subset J_0\cap\Lambda$ such that $n_i=N(x_i,J_0)<\infty$ for all i and $\lim_{i\to+\infty} N(x_i,J_0)=\infty$. We can assume that $n_i\neq n_j$ if $i\neq j$. Then the maps $\psi_i\in F(J_0,\Lambda)$ with $\psi_i(f^{n_i}(x_i))=x_i$ are all different. Hence $\#F(J_0,\Lambda)=\infty$ and the existence of J follows from Step IV. Now suppose that there exists $x\in J_0\cap\Lambda$ with $N(x,J_0)=\infty$. When x is accumulated by points $y\in J_0\cap\Lambda$ with $N(y,J_0)<\infty$, then, since $N(x,J)=\infty$, there exists a sequence $\{x_i\}$ satisfying the hypothesis of the previous case. When every $y\in J_0\cap\Lambda$ nearby x satisfies $N(y,J_0)=\infty$, then there exists an interval $J\subset J_0$ adapted to Λ, containing x, such that $N(y,J)=\infty$ for all $y\in\Lambda\cap J$. This means that $F(J,\Lambda)=\emptyset$ and then (*) is satisfied.

Step VI. If there exists $x\in\Lambda$ such that $x\notin\omega(x)$, then there exists J satisfying (*).

Suppose that $x\in\Lambda$ satisfies $x\notin\omega(x)$. Take (by II.1) an interval J_0 adapted to Λ

containing x and so small that $f^n(x) \notin J_0$ for all $n > 0$. Then $N(x, J_0) = \infty$ and Step VI follows from V.

Step VII. If there exists a sequence of periodic points $x_i \in \Lambda$ with periods $n_i \to +\infty$ and $\{x_i\}$ converging to a periodic point x, then there exists J satisfying (*).

Let γ be the orbit of x. The existence of the sequence $\{x_i\}$ with periods $n_i \to +\infty$ implies that there exists $y \notin \gamma$ such that $f(y) \in \gamma$ and the orbits of the points x_i accumulate in y. Hence $y \in \Lambda$. Obviously $\omega(y) = \gamma$. Since $y \notin \gamma$, we obtain $y \notin \omega(y)$ that by Step VI implies the existence of J satisfying (*).

Step VIII. Let Λ_1 be the closure of the non-periodic points in Λ. If there exists a source $x \in \Lambda - \Lambda_1$, then there exists J satisfying (*).

Suppose first that $x \notin \overline{\Lambda - \{x\}}$. Then we can take an open interval U containing x and such that, if n is the period of x, f^n maps U diffeomorphically onto $f^n(U) \supset U$ and $f^n(U) \cap \Lambda = \{x\}$. Then $J = f^n(U)$ is obviously adapted to Λ and $F(J, \Lambda)$ contains only one element, namely the map $\psi: J \to U$ such that $\psi(x) = x$ and $f^n \psi(y) = y$ for all $y \in J$. Then J satisfies (*). Now suppose that x is accumulated by points in $\Lambda - \{x\}$. Since $x \notin \Lambda_1$ these points must be periodic, and since x is a source their periods must go to ∞ when they approach x. Then we fall in the hypothesis of Step VII.

Step IX. If there exists a periodic point $x \in \Lambda_1$, then there exists J satisfying (*).

Suppose that γ is the orbit of x. If there exists $y \in \Lambda$ such that $f(y) \in \gamma$ and $y \notin \gamma$, then $y \notin \omega(y) = \gamma$ and by Step VI we are done. Now suppose that there is no such y. There exists an arbitrarily small neighborhood U of γ and a continuous map $\varphi: U \to N$ such that $\varphi/\gamma = (f/\gamma)^{-1}$ and $f\varphi(x) = x$ for all $x \in U$. The absence of points $y \in \Lambda - \gamma$ with $f(y) \in \gamma$ implies that if U is small enough then

$$f^{-1}(\{p\}) \cap \Lambda = \varphi(p) \tag{**}$$

for all $p \in U \cap \Lambda$. Now take a non-periodic point $z \in \Lambda$ so close to x that for some $m > 0$ the interval (x, z) is mapped diffeomorphically onto $(x = f^m(x), f^m(z))$ and $(f^j(x), f^j(z)) \subset U$ for all $0 \leqq j \leqq m$. When $f^m(z) < z$ the ω-limit of z is a periodic orbit in U, and since z is not periodic we have $z \notin \omega(z)$, and by VI we are done. When $f(z) > z$ we claim that z is not recurrent. Again by VI, this proves what we want. To prove the claim, suppose that z is recurrent. Then we can take $N > n$ such that $f^N(z) \in (x, f^m(z))$. Define $V = \bigcup_0^m (f^j(x), f^j(z))$. Then $V \subset U$, and it is easy to check (using that $f^m(z) > z$) that

$$\varphi(V) \subset V. \tag{***}$$

Moreover $f^N(z) \in V \cap \Lambda$. By (***) and (**):

$$f^{N-1}(z) = \varphi f^N(z) \in \varphi(V).$$

Continuing this argument we conclude that $f^j(z) \in V$ for all $0 \leqq j \leqq N$. But then z has all its orbit contained in V because N can be taken arbitrarily large. Then z cannot be recurrent, thus completing the proof of the claim and Step IX.

Now let us prove Lemma I.3 when $\Lambda \neq N$. Take an open interval $U \subset \Lambda_1^c$ having an endpoint $x_0 \in \Lambda_1$. When x_0 is periodic, Lemma I.3 follows from Step IX and I. When x_0 is not periodic we apply Step II to Λ_1 and we obtain an interval J adapted to Λ_1 such that either $F(J, \Lambda_1) = \emptyset$ or there exists $0 < \lambda < 1$ satisfying $|\psi'(x)| < \lambda$ for all $x \in J$ and $\psi \in F(J, \Lambda_1)$. Take a non-periodic point $x_1 \in \Lambda_1 \cap J$ (recall that the non-periodic points are dense in Λ_1). Suppose first that $\Lambda - \Lambda_1$ doesn't accumulate at x_1. Then (by II.2) we can take an interval $J_0 \subset J$ adapted to Λ_1 containing x_1, and so small that $J_0 \cap (\Lambda - \Lambda_1) = \emptyset$. This means that $J_0 \cap \Lambda = J_0 \cap \Lambda_1$ and then $F(J_0, \Lambda_1) = F(J_0, \Lambda)$. Then, if $\psi \in F(J_0, \Lambda)$, it follows that $\psi \in F(J_0, \Lambda_1)$. It is easy to see that this implies that ψ can be written as a composition of elements of $F(J, \Lambda_1)$. Hence $|\psi'(x)| < \lambda$ for all $x \in J_0$. We have thus checked that either $F(J_0, \Lambda) = \emptyset$ or $|\psi'(x)| < \lambda$ for all $x \in J_0$ and $\psi \in F(J_0, \Lambda)$. According to Step I, this proves Lemma I.3. Now suppose that $\Lambda - \Lambda_1$ accumulates at x_1. Take a sequence of points $x_i \in \Lambda - \Lambda_1$ (that by definition of Λ_1 must be periodic) converging to x_1. We claim that if γ_i is the orbit of x_i, then for all $\varepsilon > 0$ there exists i_0 such that $\gamma_i \subset B_\varepsilon(\Lambda_1) = \{z \mid d(z, \Lambda_1) < \varepsilon\}$ for all $i \geqq i_0$. To prove this first observe that the periods of x_i must go to ∞ when $i \to +\infty$. because they converge to a non-periodic point x_1. Then, if the claim were false, there would exist points $y_i \in \gamma_i$, for arbitrarily large values of i, converging to a point $y \notin \Lambda_1$. Then y is periodic (because it doesn't belong to Λ_1) and it is a limit of a sequence of periodic points whose periods go to infinity. Hence, by Step VII, Lemma I.3 is proved. Therefore we have only to prove Lemma I.3 when the claim is true. Now take an interval $J_1 \subset \bar{J}_1 \subset J$ adapted to Λ and containing x_1 (whose existence is granted by II.2). Set $\delta = d(J_1, J^c)$. We can assume that $N_0 = \sup \{N(x, J_0) \mid x \in J_0 \cap \Lambda\} < \infty$, because otherwise we are done by Step V. Choose $0 < \varepsilon < \delta/2$ such that if $a \in \Lambda_1$ and $y \in N$ satisfy $d(y, a) < \varepsilon$, then $d(f^j(a), f^j(y)) < \delta/2$ and the interval $(f^j(a), f^j(y))$ doesn't contain critical points for all $0 \leqq j \leqq N_0$. Now take a point x_i such that its orbit is contained in $B_\varepsilon(\Lambda_1)$. We shall prove that for every $x \in \gamma_i \cap J_1$ there exists $n > 0$ satisfying $f^n(x) \in J_1$ and $|(f^n)'(x)| > 1/\lambda$. This easily implies that γ_i is a source, and since this source is not contained in Λ_1, we are done by Step VIII. Suppose that $x \in \gamma_i \cap J_1$. There exists $0 < n \leqq N_0$ such that $f^n(x) \in J_1$. Since $\gamma_i \in B_\varepsilon(\Lambda_1)$ and $0 < \varepsilon < \delta/2$, there exists $a \in \Lambda_1$ such that $d(a, x) < \varepsilon < \delta/2$. Hence $a \in \Lambda_1 \cap J$. Moreover $d(f^n(a), f^n(x)) < \delta/2$. Since $f^n(x) \in J_0$, it follows that $f^n(a) \in J$, and then $f^n(a) \in J \cap \Lambda_1$. Denote $F'(J, \Lambda_1)$ the set of maps obtained as compositions of maps in $F(J, \Lambda_1)$. Then there exists $\psi \in F'(J, \Lambda_1)$ with $\psi(f^n(a)) = a$. Since the interval $(f^j(a), f^j(x))$ doesn't contain critical points for all $0 \leqq j \leqq n$, it is easy to check $\psi(f^n(x)) = x$. Hence $|(f^n)'(x)| = |\psi'(f^n(x))|^{-1} > 1/\lambda$.

Communicated by J.-P. Eckmann

Received April 22, 1987

Invent. math. 81, 459–471 (1985)

Inventiones mathematicae

On the instability of Herman rings

Ricardo Mañé

I.M.P.A., Estrada Dona Castorina 110, 22460 Rio de Janeiro, RJ, Brasil

Introduction

Let $\bar{\mathbb{C}} = \mathbb{C} \cup \{\infty\}$ be the Riemann sphere and let $\mathrm{End}(\bar{\mathbb{C}})$ be the space of analytic maps (endomorphisms) of $\bar{\mathbb{C}}$ into itself with degree >1 endowed with the C^0 topology (analytic in this paper will always mean complex analytic). It is well known that these maps are given by rational functions. Denote by $A(r_1, r_2)$ the annulus

$$A(r_1, r_2) = \{z \in \mathbb{C} \mid r_1 < |z| < r_2\}.$$

A *Herman ring* of $f \in \mathrm{End}(\bar{\mathbb{C}})$ is a connected component U of the complement of the Julia set $J(f)$ of f homeomorphic to an annulus and such that there exists $n \geqq 1$ satisfying $f^n(U) = U$. In this case it can be proved that if φ: $A(r_1, r_2) \to U$ is a conformal representation, the map $\varphi^{-1} f^n \varphi$: $A(r_1, r_2) \hookleftarrow$ is a rotation i.e. there exists $\theta \in \mathbb{R}$ such that

$$(\varphi^{-1} f^n \varphi)(x) = e^{2\pi\theta i} z$$

for all $z \in A(r_1, r_2)$. By elementary reasons, the number θ is independent of the conformal representation. It is called the rotation, number of the Herman ring. The minimum $n \geqq 1$ such that $f^n(U) = U$ is called the period of U.

Herman rings are one of the five possible dynamical behaviours in the complement of the Julia set (Sullivan [3]). It can be easily shown that of the other four, three of them don't appear for maps in an open and dense subset of $\mathrm{End}(\bar{\mathbb{C}})$. The purpose of this paper is to prove the same property for Herman rings.

Theorem. *The set of maps* $f \in \mathrm{End}(\bar{\mathbb{C}})$ *without Herman rings contains an open and dense subset of* $\mathrm{End}(\bar{\mathbb{C}})$.

This result is a step in the approach proposed in [2] to solve the fundamental conjecture on the generic dynamics of rational maps.

Originally published in: *Invent. Math. 81 (1985), no. 3, 459–471,*

Reprint by Springer International Publishing 2017

Conjecture I. *For a dense set of maps $f \in \operatorname{End}(\bar{\mathbb{C}})$, the Julia set of f is hyperbolic*, i.e. *there exist $K>0$, $\lambda>1$ such that*

$$|(f^n)'(z)| \geqq K\lambda^n$$

for all $z \in J(f)$ and $n \geqq 1$.

The results in [2], together with the theorem above, show that this conjecture is implied by the following one:

Conjecture II. *For a dense set of maps $f \in \operatorname{End}(\bar{\mathbb{C}})$ the Julia set of f has no invariant line fields.*

Recall that an *invariant line field* on the Julia set of f is a pair (A, L) where A (called the support of the line field) is an invariant Borel set (i.e. $f^{-1}(A)=A$) with *positive* Lebesgue measure contained in $J(f)$, and L is a map that to every $z \in A$ associates a one dimensional subspace $L(z) \subset T_z\bar{\mathbb{C}}$ such that $L(f(z)) = f'(z)\,L(z)$ for a.e. $z \in A$. In particular, if the Lebesgue measure of $J(f)$ is zero, there are no invariant line fields on $J(f)$. Since when $J(f)$ is hyperbolic its Lebesgue measure is zero, it follows that Conjectures I and II are equivalent. Obviously, a proof that $J(f)$ has Lebesgue measure zero for a dense set of maps $f \in \operatorname{End}(\bar{\mathbb{C}})$, would imply Conjecture II. However, to prove Conjecture II could be easier. One may start with an invariant line field (A, L) and, using some dense hypothesis on the map, prove that the Lebesgue measure of A is zero, thus reaching a contradiction. The existence of the line field should play a role in obtaining the distortion inequalities on which proofs that invariant sets have measure zero usually rely. This approach is motivated by a successful similar idea introduced by Sullivan to study the Ahlfors conjecture, that states that the limit set of a finitely generated Kleinian group has measure zero. The conjecture remains open but in [3] Sullivan proved that there are no invariant line fields on the limit set; a fact that suffices to replace the Ahlfors conjecture in several relevant applications.

The results in [3], plus the theorem above, show that the following conjecture also implies Conjecture I:

Conjecture III. *For every $f \in \operatorname{End}(\bar{\mathbb{C}})$, either the Lebesgue measure of $J(f)$ is zero or $J(f)=\bar{\mathbb{C}}$ and f is ergodic*, i.e. *for every Borel set $A \subset \bar{\mathbb{C}}$ such that $f^{-1}(A) \subset A$, either A or A^c has measure zero.*

However this conjecture is stronger than the first one and has an intrinsic interest that goes far beyond its application to prove Conjecture I.

There also exists a more direct approach to prove Conjecture I. It is based on the fact that when every periodic orbit is hyperbolic and there are no critical points in the Julia set, then this set is hyperbolic. Since the hyperbolicity of the periodic orbits is a generic property, it remains to prove that the absence of critical points in the Julia set is also a generic property. Suppose that $J(f)$ contains a critical point p. By elementary properties of $J(f)$, the set $\bigcup_{n \geqq 0} f^{-n}(\{p\})$ is dense in $J(f)$. In particular, it accumulates at p. Then one could try to find $g \in \operatorname{End}(\bar{\mathbb{C}})$ arbitrarily near to f, such that p is a critical point of g and $p \in \bigcup_{n \geqq 0} g^{-n}(\{p\})$. This means that p is a critical periodic point of g. In particular it is an attractive periodic point. Since the attractive periodic points of f survive the perturbation and appear (moving its positions a little) as attractive periodic points of g, the net result of the perturbation is to increase the number of attractive periodic points. However, in an open and dense subset of $\mathrm{End}(\bar{\mathbb{C}})$, this number is locally constant [2]. Then, the conclusion would be that elements of this open and dense set cannot have critical points in the Julia set.

The proof of the theorem is given in the next section. It is based on the fact that an open and dense set of maps $f \in \mathrm{End}(\bar{\mathbb{C}})$ is structurally stable [2, 3]. We shall show that the structural stability of f implies that f has a certain property that we shall call infinitesimal stability. This means that given an infinitesimal variation G of f (i.e. G is the derivative at $w=0$ of a map $H: D \times \bar{\mathbb{C}} \to \bar{\mathbb{C}}$, where D is the unit disk, such that $H(0,z)=f(z)$ for all $z \in \bar{\mathbb{C}}$) there exists a continuous vectorfield V on $\bar{\mathbb{C}}$ such that

$$f'(z)\,V(z) - V(f(z)) = G(z) \tag{1}$$

for all $z \in \bar{\mathbb{C}}$. What we shall do is to prove that a map with a Herman ring is not structurally stable. As an example of the method we shall use, consider a map $f \in \mathrm{End}(\bar{\mathbb{C}})$ with a fixed Herman ring U that contains the unit circle $S^1 = \{z \in \mathbb{C} \,|\, |z|=1\}$ as an invariant circle where f acts as the rotation $z \to e^{i\theta} z$. Such a behaviour is impossible because it would force f to be a rotation but it is a good heuristical example. Then, at points of S^1 Eq. (1) becomes:

$$e^{i\theta}\,V(z) - V(e^{i\theta} z) = G(z). \tag{2}$$

Now take $J: D \times \bar{\mathbb{C}}$ defined by $H(w,z)=(l+w)f(z)$. Then $G(z)=f(z)$ is an infinitesimal variation of f. Since $f(z)=e^{i\theta} z$, Eq. (2) implies that there exists a continuous vectorfield V on $\bar{\mathbb{C}}$ such that for all $z \in S^1$

$$e^{i\theta}\,V(z) - V(e^{i\theta} z) = e^{i\theta} z. \tag{3}$$

However it is quite easy to show (intergrating this equation on S^1) that there is no continuous $V: S^1 \to \mathbb{C}$ satisfying (3). Then, the map f is not infinitesimally stable. In the general case the basic idea is the same but first we have to transform the map, using a special change of coordinates, in a map that has a Herman ring that contains an invariant circle that almost coincides with the unit circle and where f acts almost as a rotation. More precisely, if $f \in \mathrm{End}(\bar{\mathbb{C}})$ has a Herman ring U that contains an invariant circle γ, we shall construct a quasiconformal homeomorphism $g: \mathbb{C} \hookleftarrow$ such that $gfg^{-1} \in \mathrm{End}(\bar{\mathbb{C}})$ and $g(\gamma)$, which is an invariant circle of gfg^{-1}, is near to S^1. By a result of Sullivan [3], if f is structurally stable, then gfg^{-1} is structurally stable. Using the invariant circle $g(\gamma)$ we shall prove that gfg^{-1} is not infinitesimally stable. Hence it is not structurally stable and then f is not structurally stable.

This idea seems essentially analytical. However it evolved from a formally different geometric idea that can be used to prove the theorem for *fixed* Herman rings (i.e. having period $n=1$) but that we couldn't extend to the general case. Since this geometric approach throws light on the nature of the proof of the theorem, we shall now briefly explain it. Suppose that $f \in \mathrm{End}(\bar{\mathbb{C}})$

has a Herman ring U with period $n \geqq 1$ and let $\varphi: A(r_1, r_2) \to U$ be a conformal representation. It is easy to prove that the curves $\gamma = \varphi(\{z \mid |z| = r\})$, for $r_1 < r < r_2$ are invariant under f^n and independent of the conformal representation. This family of curves is called the canonical foliation of U. Moreover observe that if α is a simple closed curve contained in U, then $f_n(\alpha) \cap \alpha \neq \emptyset$. Now we claim that a structurally stable $f \in \mathrm{End}(\bar{\mathbb{C}})$ cannot have a Herman ring whose canonical foliation contains a convex curve $\gamma \subset \mathbb{C}$. In fact, if f is structurally stable every nearby $g \in \mathrm{End}(\bar{\mathbb{C}})$ contains a fixed Herman ring U_g satisfying $\gamma \subset U_g$ (but obviously γ doesn't need to belong to the canonical foliation of U_g). Without lose of generality we can suppose that the origin of $\mathbb{C}$ is in the interior of γ. Take $g = (1-\varepsilon) f$ with $0 < \varepsilon < 1$. Then $g(\gamma) = (1-\varepsilon) f(\gamma) = (1-\varepsilon)\gamma$. Then, since γ is convex and contains the origin in its interior, we obtain $\gamma \cap g(\gamma) = \emptyset$ contradicting the intersection property stated above. Since g can be taken arbitrarily near to f, taking ε small, this proves that f is not structurally stable thus completing a proof by contradiction of our claim. However this proof is not immediatly extendable to rings with period $n > 1$ because the calculation of $g^n(\gamma)$ is not straightforward as in the previous case. Anyhow, in the case of a fixed Herman ring, the assumption of the existence of a convex closed curve γ in its canonical foliation can be avoided. The idea is to produce a quasiconformal homeomorphism $h = \bar{\mathbb{C}} \hookleftarrow$ such hfh^{-1} is a rational map and the ring $h(U)$ has large modulus. This means that there is a conformal representation $\varphi: A(r_1, r_2) \to h(U)$ with r_1 very small and r_2 very large. Without lose of generality we can assume that $\varphi(1) = 1$ and that 0 and ∞ belong to different connected components of the complement of $h(U)$. According to Theorem I.6, that we shall prove in Section II, such a conformal representation is very near to the identity map or the inversion $z \to z^{-1}$ in the annulus $A(1/2, 2)$ if r_1 is very small and r_2 very large. Then $\varphi(\{z \mid |z| = 1\})$ is convex. Hence hfh^{-1} is not structurally stable and then f is not structurally stable.

Proof of the theorem. First we shall recall some basic results about quasiconformal structures and mappings, analytic motions, structural stability and quasiconformal conjugacies necessary for the proof of the theorem.

If $U \subset \bar{\mathbb{C}}$ is an open set, a *quasiconformal structural* on U is a function $\mu: U \to \mathbb{C}$ with

$$\|\mu\|_\infty < 1.$$

Associated to μ we have a field of ellipses that to a.e. $z \in U$ associates the ellipse

$$E_\mu(z) = \{e^{i\theta} + \mu e^{-i\theta} \mid \theta \in \mathbb{R}\}.$$

The set of quasiconformal structures on U is denoted $Qc(U)$.

If $U \subset \bar{\mathbb{C}}$ and $V \subset \bar{\mathbb{C}}$ are open sets, we say that a continuous map $f: U \to V$ is *quasiconformal* if it satisfies the following conditions:

a) For a.e. $z \in U$, f is differentiable at z and its derivative is an orientation preserving isomorphism.

b) The function $K(z) = \|f'(z)\| \cdot \|(f')^{-1}(z)\|$ satisfies $\|K\|_\infty < +\infty$.

c) f is absolutely continuous on lines, i.e. the real and imaginary parts of the restriction of f to a line parallel to the real or imaginary axis are absolutely continuous. This, together with (a) and (b), implies that the Lebesgue measure of a Borel set $A \subset U$ is zero if and only if the measure of $f(A)$ is zero [1].

Every quasiconformal map $f: U \to V$ induces a map $f_*: Qc(V) \to Qc(U)$ defined as follows. Suppose that $\mu \in Qc(V)$. Using (a), (b) and (c) it is easy to prove that there exists a unique $\nu \in Qc(U)$ such that the ellipse $f'(z) E\ (z)$ is homothetic to $E_\mu(f(z))$. Define $f_*(\mu) = \nu$.

The following is the well known measurable Riemann mapping theorem due to Morrey. For a proof see [1].

Theorem I.1. [1] *For every $\mu \in Qc(\bar{\mathbb{C}})$ there exists a quasiconformal homeomorphism $g: \bar{\mathbb{C}} \hookleftarrow$ such that $g_*(\mu) = 0$.*

An *analytic motion* of $\bar{\mathbb{C}}$ is a map $\varphi: W \times \bar{\mathbb{C}} \to \bar{\mathbb{C}}$, where W is a connected complex manifold, such that, denoting $\varphi_\lambda(\cdot) = \varphi(\lambda, \cdot)$, the following properties are satisfied:

a) $\varphi_\lambda: \bar{\mathbb{C}} \hookleftarrow$ is injective for all $\lambda \in W$.

b) For all $z \in \mathbb{C}$, the map $W \ni w \to \varphi_\lambda(w) \in \bar{\mathbb{C}}$ is analytic.

c) There exists $\lambda_0 \in W$ such that φ_{λ_0} is the identity.

The next theorem is the λ-lemma of [2]:

Theorem I.2. *If $\varphi: W \times \bar{\mathbb{C}} \to \bar{\mathbb{C}}$ is an analytic motion, then φ is continuous and for all $\lambda \in W$, $\varphi_\lambda: \bar{\mathbb{C}} \hookleftarrow$ is a quasiconformal homeomorphism.*

We shall also need the following easy corollary:

Corollary. *If $\varphi: W \times \bar{\mathbb{C}} \to \bar{\mathbb{C}}$ is an analytic motion, then, for every $\lambda_1 \in W$, the vectorfield*

$$\bar{\mathbb{C}} \ni z \to \frac{\partial \varphi}{\partial \lambda}(\lambda_1, z) \in T_z \bar{\mathbb{C}}$$

Proof. Take $\lambda_1 \in W$ and consider the vectorfield $V(z) = (\partial\varphi/\partial\lambda)(\lambda_1, z)$. To prove the continuity of V take a point $z \in \bar{\mathbb{C}}$ and a sequence $z_n \to z$. Define a sequence of analytic functions $F_n: W \to \bar{\mathbb{C}}$ by $F_n(\lambda) = \varphi_\lambda(z_n)$ and define $F: W \to \bar{\mathbb{C}}$ by $F(z) = \varphi_\lambda(z)$. By I.2 φ is continuous. This obviously implies that $F_n \to F$ when $n \to +\infty$ uniformely on compact sets. Then $F'(\lambda_1) = \lim_{n \to +\infty} F_n'(\lambda_1)$. Since $F'(\lambda_1) = V(z)$ and $F_n'(\lambda_1) = V(z_n)$ this concludes the proof of the corollary.

We say that $f \in \mathrm{End}(\bar{\mathbb{C}})$ is *structurally stable* if there exists a neighborhood W of f and an analytic motion $\varphi: W \times \bar{\mathbb{C}} \to \bar{\mathbb{C}}$ such that

$$\varphi_f = \text{Identity}$$

and

$$\varphi_g^{-1} f \varphi_g = g$$

for all $g \in W$.

A *variation* of $f \in \mathrm{End}(\bar{\mathbb{C}})$ is an analytic map $F: D \times \bar{\mathbb{C}} \to \bar{\mathbb{C}}$, where $D = \{z \in C \mid |z| < 1\}$ and

$$F(0, z) = f(z)$$

for all $z \in D$. The analytic map $G: \bar{\mathbb{C}} \to T\bar{\mathbb{C}}$ defined as

$$G(z) = \frac{\partial F}{\partial \lambda}(0, z) \tag{1}$$

is called an *infinitesimal variation* of f. We say that $f \in \mathrm{End}(\bar{\mathbb{C}})$ is *infinitesimally stable* if for every infinitesimal variation G of f there exists a continuous vectorfield V on $\bar{\mathbb{C}}$ such that:

$$f'(z)\,V(z) - V(f(z)) = G(z). \tag{2}$$

Theorem I.3. *Structurally stable maps are infinitesimally stable.*

To prove this theorem take a structurally stable $f \in \mathrm{End}(\bar{\mathbb{C}})$ and a variation $F: D \times \bar{\mathbb{C}} \to \bar{\mathbb{C}}$ of the map f as above. By the definition of structural stability there exists an analytic motion $\varphi: D \times \bar{\mathbb{C}} \to \bar{\mathbb{C}}$ such that:

$$\varphi_{\lambda_0} = \text{Identity}$$

and

$$\varphi_\lambda(f(z)) = F(\lambda, \varphi_\lambda(z)) \tag{3}$$

for all $\lambda \in D$ and $z \in \bar{\mathbb{C}}$. Defining the vectorfield V on $\bar{\mathbb{C}}$ as

$$V(z) = \frac{\partial \varphi}{\partial \lambda}(\lambda_0, z) \tag{4}$$

and derivating (3) with respect to λ at $\lambda = \lambda_0$, we obtain:

$$f'(z)\,V(z)\,V(z) = V(f(z)) + \frac{\partial F}{\partial \lambda}(\lambda_0, z).$$

Then, if G is the infinitesimal variation given by (1), it follows that Eq. (2) is satisfied by V given by (4), which by Theorem I.2 is continuous.

Theorem I.4. [2, 3] *The set of structurally stable maps* $f \in \mathrm{End}(\bar{\mathbb{C}})$ *is an open and dense subset of* $\mathrm{End}(\bar{\mathbb{C}})$.

Theorem I.5. [3] *If* $f \in \mathrm{End}(\bar{\mathbb{C}})$ *is structurally stable and* $g \in \mathrm{End}(\bar{\mathbb{C}})$ *is quasiconformally equivalent to* f (i.e. *there exists a quasiconformal homeomorphism* $h: \bar{\mathbb{C}} \hookleftarrow$ *such that* $h^{-1} f h = g$) *then* g *is structurally stable.*

The following theorem will be proved in the next section.

Theorem I.6. *Given two sequences:*

$$\dots r_2 < r_1 < 1 < R_1 < R_2 < \dots$$

such that $R_n \to +\infty$ *and* $r_n \to 0$ *and a sequence of injective analytic maps* $\varphi_n: A(r_n, R_n) \to \mathbb{C}$ *such that:*

a) $\varphi_n(1) = 1$ *for all* n;

b) *For all* n, *the origin is contained in the bounded connected component of the complement of* $\varphi_n(A(r_n, R_n))$;

c) *For all* n, *the winding number with respect to the origin of the curve* $\theta \to \varphi_n(e^{i\theta})$ *is* 1, *then* $\{\varphi_n\}$ *converges uniformely to the identity on compact subsets of* $\mathbb{C} - \{0\}$.

To prove the theorem, we shall show that if $f \in \mathrm{End}(\bar{\mathbb{C}})$ has a Herman ring, then there exists $g \in \mathrm{End}(\bar{\mathbb{C}})$ quasiconformally equivalent to f, such that g is not infinitesimally stable. By Theorem I.3, g is not structurally stable. Hence, by I.5, f is not structurally stable. Therefore structurally stable maps don't have Herman rings. By I.4, this proves the theorem.

Suppose that $f \in \mathrm{End}(\bar{\mathbb{C}})$ has a Herman ring U with period m and rotation number θ. Denote:

$$C = \bigcup_{j=0}^{m} U_j, \quad U_j = f^j(U), \; j = 0, \ldots, m.$$

Observe that $U_0 = U_m = U$. The property stated in the next lemma is technical but will play a fundamental role at the end of the proof. However it is unnecessary when $m = 1$. Given a point $x \in \bar{\mathbb{C}}$ *and* $0 \leqq j < m$ define the number $\tau_j(x)$ as 1 or -1 according to whether the map $f \mid U_j$: $U_j \to U_{j+1}$ preserves or reverses orientation of the invariant ourves of the annulus U_j in the *plane* $\bar{\mathbb{C}} - \{x\}$, i.e., whether the clockwise orientation of the invariant curves in U_j in the plane $\bar{\mathbb{C}} - \{x\}$ is mapped in the clockwise orientation or not. Define

$$v_j(x) = \prod_{i=j}^{m-1} \tau_i(x).$$

Lemma I.7. *There exists* $x \in \bar{\mathbb{C}} - C$ *such that* $\sum_{j=0}^{m-1} v_j(x) \neq 0$.

Proof. We say that two points x' and x'' in $\bar{\mathbb{C}} - C$ are related by a j-crossing if x' and x'' are in different connected components of $\bar{\mathbb{C}} - U_j$ but in the same connected component of $\bar{\mathbb{C}} - U_i$ for all $i \neq j$, $0 \leqq i < m$. Suppose that x' and x'' are in this situation and fix in each U_i an invariant curve γ_i endowed with the clockwise orientation of the plane $\bar{\mathbb{C}} - \{x'\}$. Then in the plane $\bar{\mathbb{C}} - \{x''\}$ all the curves γ_i, $i \neq j$ are clockwise oriented and γ_j is oriented counterclockwise. Using this remark, it follows that, if $0 < j \leqq m-1$, we have:

$$\begin{aligned} \tau_j(x') &= -\tau_j(x'') \\ \tau_{j-1}(x') &= -\tau_j(x'') \\ \tau_i(x') &= \tau_i(x''), \quad 0 \leqq i \leqq m-1, \; i \neq j, \; i \neq j-1. \end{aligned}$$

This condition and the definition of the v_j's imply:

$$\begin{aligned} v_j(x') &= -v_j(x'') \\ v_i(x') &= v_i(x'') \quad 0 \leqq i \leqq m-1, \; i \neq j. \end{aligned}$$

Now choose two points x' and x'' in $\bar{\mathbb{C}} - C$ related by a 1-crossing and suppose that x' doesn't satisfy the required condition, i.e.:

$$\sum_{j=0}^{m-1} v_j(x') = 0.$$

By the relations above

$$\begin{aligned} \sum_{j=0}^{m-1} v_j(x'') &= v_1(x'') + \sum_{j \neq 1} v^j(x'') = -v_1(x') + \sum_{j \neq 1} v_j(x') \\ &= -2v_1(x') + \sum_{j=0}^{m-1} v_j(x') = -2v_1(x') \neq 0. \end{aligned}$$

This completes the proof of the lemma. Using it, choose x in the complement of C such that

$$\sum_{j=0}^{m-1} v_j(x) \neq 0.$$

Without loss of generality we can suppose that $x = \infty$. Hence:

$$\sum_{j=0}^{m-1} v_j(\infty) \neq 0.$$

Take a conformal representation:

$$\varphi : A(r, R) \to U$$

that maps clockwise oriented circles centered at the origin onto clockwise oriented invariant curves of U. We can suppose

$$1 \in U$$

and

$$\varphi(1) = 1.$$

Take a sequence of quasiconformal homeomorphisms:

$$k_n : A(r/n, Rn) \to A(r, R)$$

satisfying:

$$k_n(1) = 1$$

and commuting with the rotation of angle θ, i.e.

$$k_n(e^{i\theta} z) = e^{i\theta} k_n(z).$$

Now define a sequence of quasiconformal structures

$$\mu_n \in Qc(\bar{\mathbb{C}})$$

by:

$$\mu_n/U = (\varphi k_n^{-1})_*(0)$$

$$\mu_n/f^{-j}(U) = (f^j)_*(\mu_n), \quad j \geqq 0$$

$$\mu_n(z) = 0, \quad z \notin \bigcup_{j \geqq 0} f^{-j}(U).$$

It is easy to check (using that k_n commutes with the rotation $z \to e^{i\cdot}z$) that μ_n is f-invariant, i.e.:

$$f_*(\mu_n) = \mu_n. \tag{1}$$

Now take a quasiconformal homeomorphism $h_n : \bar{\mathbb{C}} \hookleftarrow$, whose existence is granted by Theorem I.1, such that

$$(h_n)_*(0) = \mu_n \tag{2}$$

$$h_n(0) = 0 \tag{3}$$

$$h_n(1) = 1 \tag{4}$$

$$h_n(\infty) = \infty. \tag{5}$$

Then, defining $g_n = h_n f h_n^{-1}$, we have

$$(g_n)_*(0) = 0 \tag{6}$$

because, by (1) and (2):

$$\begin{aligned}(g_n)_*(0) &= (h_n^{-1})_* f_*(h_n)_*(0) = (h_n^{-1})_* f_*(\mu_n) \\ &= (h_n^{-1})_*(\mu_n) = 0.\end{aligned}$$

But (6) implies that $g_n\colon \bar{\mathbb{C}} \hookleftarrow$ is analytic [1]. Then, to complete the proof of the theorem, it suffices to show that g_n is not infinitesimally stable for large values of n. Observe that $U_n = h_n(U)$ is a Herman ring of g_n with period m and rotation number θ. Moreover, by (5), the annuli $g_n^j(U_n)$, $j = 0, 1, \ldots, m-1$, are contained in $\mathbb{C}$. Suppose that G is an infinitesimal variation of g_n and consider the infinitesimal stability equation:

$$g_n'(z)\, V(z) - V(g_n(z)) = G(z). \tag{7}$$

We shall work with this equation restricted to $C_n = \bigcup_{j=0}^{m-1} g_n^j(U_n)$, and restricted to this set, g_n is invertible. Denote $g_n^{-1}\colon C_n \hookleftarrow$ the inverse of $g_n/C_n\colon C_n \hookleftarrow$. Define the vectorfield $\tilde{G}$ on C_n by:

$$\tilde{G} = G g_n^{-1}.$$

Then, on C_n, Eq. (7) can be written as:

$$g_n'(g_n^{-1}(z))\, V(g_n^{-1}(z)) - V(z) = \tilde{G}(z).$$

From this equation follows that

$$(g_n^m)'(g_n^{-m}(z))\, V(g_n^{-m}(z)) - V(z) = \sum_{j=0}^{m-1} (g_n^j)'(g_n^{-j}(z))\, \tilde{G}(g_n^{-j}(z)). \tag{8}$$

Now consider the map

$$\psi_n = h_n \varphi k_n\colon A(r/n, Rn) \to U_n.$$

This map is a bijection. Moreover it is analytic because

$$(\psi_n)_*(0) = (\psi k_n)_*(h_n)_*(0)$$

and, by (2) and the definition of μ_n on U:

$$(\psi_n)_*(0) = (\varphi k_n)_*(\mu_n) = 0.$$

Now define the vectorfield $\hat{V}$ on $A(r/n, Rn)$ by

$$\hat{V}(z) = (\psi_n^{-1})'(\psi_n(z))\, V(\psi_n(z))$$

and the conformal representation:

$$\Phi_n^{(j)}\colon A(r/n, Rn) \to g_n^{-j}(U_n)$$

as:

$$\Phi_n^{(j)} = g_n^{-j}\psi_n.$$

Then, applying $(\psi_n^{-1})'(\psi_n(z))$ to Eq. (8) evaluated at the point $\psi_n(z)$, we obtain:

$$(\psi_n^{-1} g_n^m \psi_n)'((\psi_n^{-1} g_n^{-m}\psi_n)(z))\,\hat{V}((\psi_n^{-1} g_n^{-m}\psi_n)(z)) - \hat{V}(z)$$
$$= \sum_{j=0}^{m-1} ((\Phi_n^{(j)})^{-1})'(\Phi_n^{(j)}(z))\,\tilde{G}(\Phi_n^{(j)}(z)). \tag{9}$$

Recalling that $(\psi_n^{-1} g_n^m \psi_n)(z) = e^{i\theta} z$ for $z \in A(r/n, Rn)$, Eq. (9) becomes:

$$e^{i\theta}\hat{V}(e^{-i\theta}z) - \hat{V}(z) = \sum_{j=0}^{m-1} ((\Phi_n^{(j)})^{-1})'(\Phi_n^{(j)}(z))\,\tilde{G}(\Phi_n^{(j)}(z)) \tag{10}$$

for all $z \in A(r/n, Rn)$. It is easy to see, using Fourier series, that:

$$\int_{S^1} \frac{1}{z^2}(e^{i\theta} H(e^{-i\theta}z) - H(z))\,dz = 0$$

for every continuous function $H: S^1 \to \mathbb{C}$. Hence, for all n:

$$\sum_{j=0}^{m-1} \int_{S^1} \frac{1}{z^2}((\Phi_n^{(j)})^{-1}(\Phi_n^{(j)}(z))\,\tilde{G}(\Phi_n^{(j)}(z))\,dz = 0. \tag{11}$$

Now observe that the functions $\Phi_n^{(j)}$, after being composed with suitable affine maps $z \to az + b$ and inversions $z \to z^{-1}$, satisfy the hypothesis of I.6. More precisely, we can choose numbers $\hat{\lambda}_n^{(j)}$, $\hat{\gamma}_n^{(j)}$ and maps $\hat{L}_n^{(j)}$ that are either the identity or inversions $z \to z^{-1}$, such that for all j the sequence

$$\varphi_n^{(j)} = \hat{\lambda}_n^{(j)} \hat{L}_n^{(j)}(\Phi_n^{(j)} + \hat{\gamma}_n^{(j)}) : A(r/n, Rn) \to \mathbb{C}$$

satisfies the hypothesis of I.6. Then, for all j:

$$\varphi_n^{(j)} \to I \tag{12}$$

when $n \to +\infty$, uniformely on compact subsets of $\mathbb{C} - \{0\}$. Setting $\lambda_n^{(j)} = (\hat{\lambda}_n^{(j)})^{-1}$, $L_n^{(j)} = (\hat{L}_n^{(j)})^{-1}$ and $\gamma_n^{(j)} = (\hat{L}_n^{(j)})^{-1}(-\hat{\gamma}_n^{(j)}/\hat{\lambda}_n^{(j)})$, we can write:

$$\Phi_n^{(j)} = \lambda_n^{(j)} L_n^{(j)} \varphi_n^{(j)} + \gamma_n^{(j)}. \tag{13}$$

Using (12) and (13), we can compute $((\Phi_n^{(j)})^{-1}(\Phi_n^{(j)}(z))$ and show it can be written in a neighborhood N of S^1 in the form:

$$((\Phi_n^{(j)})^{-1})'(\Phi_n^{(j)}(z)) = \frac{\varepsilon_n^{(j)}}{\lambda_n^{(j)}} S_n^{(j)}(z)\, z^{\delta(j,n)}$$

where for all j:

$$S_n^{(j)} \to 1 \tag{14}$$

uniformely on N when $n \to +\infty$, and $\varepsilon_n^{(j)}$ and $\delta(j,n)$ are respectively -1 and 2 or 1 and 0 according to whether $L_n^{(j)}$ is an inversion or the identity. Then each term of (11) is

$$\int_{S^1} \frac{1}{z^2}((\Phi_n^{(j)})^{-1})'(\Phi_n^{(j)}(z))\,\tilde{G}(\Phi_n^{(j)}(z))\,dz$$

$$= \varepsilon_n^{(j)} \int_{S^1} \frac{1}{z^2}\cdot\frac{1}{\lambda_n^{(j)}}\,S_n^{(j)}(z)\,z^{\delta(j,n)}\,\tilde{G}(\lambda_n^{(j)} L_n^{(j)} \varphi_n^{(j)}(z)+\gamma_n^{(j)})\,dz. \qquad (15)$$

Now take the variation $F\colon D \times \overline{\mathbb{C}} \to \overline{\mathbb{C}}$, where D is the unit disk, defined by:

$$f(w,z) = (1+w)\,g_n(z).$$

The corresponding infinitesimal variation is:

$$G(z) = \frac{\partial F}{\partial w}(0,z) = g_n(z)$$

and then

$$\tilde{G}(z) = z.$$

Replacing this $\tilde{G}$ in (15) we obtain:

$$\varepsilon_n^{(j)} \int_{S^1} \frac{1}{z^2}\,S_n^{(j)}(z)\,z^{\delta(j,n)}\left(L_n^{(j)}\varphi_n^{(j)}(z)+\frac{\gamma_n^{(j)}}{\lambda_n^{(j)}}\right)dz.$$

By (14), when $n \to +\infty$, this integral is approximated by:

$$\varepsilon_n^{(j)} \int_{S^1} \frac{1}{z^2}\,z^{\delta(j,n)}\left(L_n^{(j)} z+\frac{\gamma^{(j)}}{\lambda_n^{(j)}}\right)dz$$

$$= \varepsilon_n^{(j)} \int_{S^1} \frac{1}{z^2}\,z^{\delta(j,n)}\,L_n^{(j)}\,z\,dz + \varepsilon_n^{(j)} \int_{S^1} \frac{1}{z^2}\,z^{\delta(j,n)}\,\frac{\gamma_n^{(j)}}{\lambda_n^{(j)}}\,dz.$$

When $L_n^{(j)}$ is the identity, $\delta(j,n)$ and $\varepsilon_n^{(j)}$ are 0 and 1 and the value of this expression is $2\pi i$. When $L_n^{(j)}$ is an inversion, $\delta(j,n)$ and $\varepsilon_n^{(j)}$ are 2 and -1 and then the value is $-2\pi i$. Now observe that $L_n^{(j)}$ is the identity or an inversion according to whether

$$g_n^j\colon g_n^{-j}(U_n) \to U_n$$

preserves or reverses orientation of the invariant curves of the Herman ring (in the sense of the statement of Lemma I.7), and then $L_n^{(j)}$ is the identity or an inversion according to whether the number $v_j(\infty)$ (defined as in the statement of I.7) is 1 or -1. Therefore the sum at left in (11) is

$$2\pi i \sum_{j=0}^{m-1} v_j(\infty),$$

which it is $\neq 0$ by the way we placed the point ∞. Therefore the sum at left in (11) converges to a number $\neq 0$ when $n \to +\infty$ thus contradicting (11).

Proof of Theorem I.6. Given two points p and q in $\overline{\mathbb{C}}$, we say that an analytic map $\varphi: (A(r,R) \to \overline{\mathbb{C}}$ *separates* p and q if they belong to different connected components of the complement of $\varphi(A(r,R))$. The proof of I.6 is an application of the following lemma.

Lemma. *For all* $0<r<R<\infty$ *and* $p \in \overline{\mathbb{C}}$, $q \in \overline{\mathbb{C}}$, *the family of analytic injective maps* $\varphi: A(r,R) \to \overline{\mathbb{C}}$ *that separate* p *and* q *is normal.*

Proof. We can suppose that $p=0$ and $q=\infty$. Let $\varphi_n: A(r,R) \to \mathbb{C}$ be a sequence of analytic injective maps separating 0 and ∞. Take

$$r_1 < r < R < R_1. \tag{1}$$

Then $\varphi_n(A(r,R))$ cannot contain $A(r_1, R_1)$ because this would imply that $\varphi_n^{-1}/A(r_1,R_1)$ is an analytic injective map of $A(r_1,R_1)$ into $A(r,R)$ such that the bounded connected component of the complement of $\varphi_n^{-1}(A(r_1,R_1))$ contains the disk $\{z \mid |z|<r\}$, and it is well known that this is impossible under Condition (1). Then for all n there exists $\alpha_n \in \mathbb{C}$ with

$$|\alpha_n| = r_1 \quad \text{or} \quad |\alpha_n| = R_1 \tag{2}$$

and

$$\alpha_n \notin \varphi_n(A(r,R)). \tag{3}$$

Then, by (3), the sequence of functions

$$\alpha_n^{-1}\varphi_n:\ A(r,R) \to \mathbb{C}$$

doesn't take the value 1 and, since φ_n doesn't take the value 0, the same property holds for $\alpha_n^{-1}\varphi_n$. Therefore there exists a subsequence $\{\alpha_{n_j}^{-1}\varphi_{n_j}\}$ that converges uniformely on compacts sets of $A(r,R)$ either to ∞ or to an analytic map $\varphi: A(r,R) \to \mathbb{C}$. In the first case, since $|\alpha_n| \geqq r_1$ by (2), it follows that $\varphi_{n_j} \to \infty$ on compact subsets of $A(r,R)$. In the second case, since by (2) we can suppose that the sequence $\{\alpha_{n_j}\}$ converges to some $\alpha \in \mathbb{C}$, it follows that $\varphi_{n_j} \to \alpha\varphi$.

Now let us prove Theorem I.6. Let $\{\varphi_n\}$ be a sequence satisfying its hypothesis. Suppose that it is not true that $\varphi_n \to$ Identity uniformely on compact subsets of $\mathbb{C} - \{0\}$. Then there exist $m_1 < m_2 < m_3 < \dots$ such that for all m_i the sequence

$$\varphi_n: A(r_{m_i}, R_{m_i}) \to \mathbb{C}, \qquad n \geqq m_i,$$

doesn't converge uniformely on compact subsets of $A(r_{m_i}, R_{m_i})$ to the identity. But by the hypothesis of I.6 every φ_n separates 0 and ∞. By the lemma this is a normal sequence for each m_i. Therefore, using the normality of these sequences and a standard diagonal procedure, there exists an analytic function

$$\varphi: \mathbb{C} - \{0\} = \bigcup_i A(r_{m_i}, R_{m_i}) \to \mathbb{C}$$

and a subsequence $\{\varphi_{n_j}\}$ such that for all i

$$\varphi_{n_j}/A(r_{m_i}, R_{m_i}) \to \varphi$$

when $j \to +\infty$, uniformely on compact subsets of $A(r_{m_i}, R_{m_i})$. Since $\varphi_n(1)=1$ for all n it follows that $\varphi(1)=1$, and since φ_n is injective for all n, the function φ is also injective or it is constant. But φ cannot be constant because by hypothesis (c) of Theorem I.6 the winding number with respect to zero of the arc $\theta \to \varphi(e^{i\theta})$ is $\neq 0$. Hence φ is injective. Then the origin and ∞ cannot be essential singularities. Hence φ is a rational injective function with $\varphi(1)=1$. This means that φ is either the identity or $z \to z^{-1}$. But hypothesis (c) of Theorem I.6 implies that the winding number with respect to the origin of the are $\theta \to \varphi(e^{i\theta})$ is one. Therefore φ is the identity.

References

1. Ahlfors, L.: Lectures on quasiconformal mappings. Van Nostrand Co. 1966
2. Mañé, R., Sad, P., Sullivan, D.: On the dynamics of rational maps. Ann. Sci. Éc. Norm. Super., **16**, 193–217 (1983)
3. Sullivan, D.: Quasi conformal homeomorphisms and dynamics I, II, III. (To appear in Ann. Math.)

Oblatum 8-IV-1984 & 19-II-1985

Ricardo Mañé
I.M.P.A.,
Estrada Dona Castorina, 110,
22460 - Rio de Janeiro - RJ,
Brasil

R MAÑÉ

On a theorem of Klingenberg

I - Introduction

Let M be a complete Riemannian manifold and let UM be its unit tangent bundle endowed with its standard Riemannian structure. The geodesic flow $\varphi_t: UM \circlearrowleft$, $t \in \mathbb{R}$, is defined by $\varphi_t(p,v) = (\gamma(t),\dot{\gamma}(t))$ where $\gamma: \mathbb{R} \to M$ is the geodesic with initial condition $\gamma(0) = p$, $\dot{\gamma}(0) = v$. This and all the geodesics in this work will be parametrized by arc lenght. When the curvature of M is $\leq -a < 0$, its geodesic flow is an Anosov flow, i.e., there exists a continuous splitting $T(UM) = E^s \oplus E^u \oplus E^{\varphi}$, where E^{φ} is the one dimensional subbundle of $T(UM)$ tangent to the orbits of φ, such that there exist constants $C > 0$, $0 < \lambda < 1$ satisfying

$$\|(D\varphi_t)/E^s_\theta\| \leq C\lambda^t$$

$$\|(D\varphi_{-t})/E^u_\theta\| \leq C\lambda^t$$

for all $\theta \in UM$ and $t \geq 0$.

This strong property was discovered by Hadamard [6]. Later, it was successfully exploited by many authors; most notably by Hopf [8] (in the two-dimensional case) and Anosov [1] (in the general n-dimensional case), whose works are the foundation of the nowadays remarkably rich theory of the geodesic flow of negatively curved manifolds.

Originally published in: *Dynamical systems and bifurcation theory (Rio de Janeiro, 1985), 319–345, Pitman Res. Notes Math. Ser., 160, Longman Sci. Tech., Harlow, 1987,*

On the other hand there are manifolds whose geodesic flow is of Anosov type and whose curvature is strictly positive in certain regions. Examples of this phenomena and a study of sufficient conditions that grant the Anosov property to the geodesic flow were developed by Eberlein [3].

Riemannian manifolds whose geodesic flow is Anosov share several of the main properties of negatively curved manifolds. Some of them, like (in the compact case) the ergodicity of the geodesic flow and the density of the closed geodesics, because they follow from a theorem that holds for any Anosov volume preserving flow. But there also other shared properties, of a more strictly geometric nature. Here is a list of the main ones in the case when M is compact:

I) M has no conjugate points.

II) There exist $c_0 > 0$ and $\mu > 0$ such that every perpendicular Jacobi vector field J with $J(0) = 0$ satisfies $\|J(t)\| \geq c_0 \|\dot{J}(0)\| \sinh \mu t$ for all t.

III) If $\tilde{M}$ is the universal covering of M, then two different geodesics $\gamma_i : \mathbb{R} \to \tilde{M}$, $i = 1,2$ with $\gamma_1(0) = \gamma_2(0)$, satisfy

$$\lim_{t \to \pm\infty} d(\gamma_1(t), \gamma_2(t)) = +\infty.$$

IV) Given a geodesic $\gamma : \mathbb{R} \to \tilde{M}$ and a point $p \in \tilde{M}$ there exists a unique geodesic γ_1 with $\gamma_1(0) = p$ and ω-asymptotic with γ i.e. satisfying

$$\sup_{t \geq 0} d(\gamma_1(t), \gamma(\mathbb{R}^+)) < \infty.$$

In this case there exists a such that

$$\lim_{t \to +\infty} d(\gamma_1(t), \gamma(t+a)) = 0.$$

In a similar way, but replacing $t \geq 0$ and $t \to +\infty$ by $t \leq 0$ and $t \to -\infty$, we define α-asymptotic geodesics.

V) If two geodesics of $\tilde{M}$ are α and ω-asymptotic then they coincide up to a translation of the parametrization.

VI) Given two geodesics γ_1 and γ_2 such that γ_1 is not ω-asymptotic with $t \to \gamma_2(-t)$, there exists a geodesic γ in $\tilde{M}$ that is ω-asymptotic with γ_1 and α-asymptotic with γ_2. Moreover γ is unique up to reparametrizations by a translation of the parameter.

Of these properties the crucial one is the expanding property of the Jacobi vector fields (property II); that in particular implies that there are no conjugate points. The other properties follow from (II) as reasonably easy corollaries adapting the proofs of the negatively curved case. That property was proved by Klingenberg in [9] assuming M to be compact.

The objective of this paper is to extend this result to complete non compact manifolds with curvature bounded from below. The rest of the list of properties then follow by strict repetition of their proofs in the compact case. Our proof will contain a proof for the compact case different from that given by Klingenberg in [9].

When M is compact we shall show that if the derivative of the geodesic flow leaves invariant a continuous Lagrangian subbundle, then M has no conjugate points. When the geodesic flow is Anosov, both the stable and unstable subbundles are Lagrangian.

To state our results let us first recall the basic definitions and properties of the unit tangent bundle and the geodesic flow.

If $\theta = (p,v) \in UM$, denote $N(\theta)$ the space of vectors $w \in T_pM$ such that $\langle w,v \rangle = 0$. Let π: $UM \to M$ be the canonical projection and define $S(\theta) = (D_\theta\pi)^{-1}N(\theta)$. $S(\theta)$ is orthogonal to E^φ_θ and $(D_\theta\pi)E^\varphi_\theta$ is the one dimensional space generated by v. Moreover $S(\theta)$ is invariant under $D\varphi$, i.e., $(D\varphi)S(\theta) = S(\varphi_t(\theta))$ for all θ and t. The vertical subspace of $S(\theta)$ is defined as $V(\theta) = (D_\theta\pi)^{-1}(\{0\})$. The horizontal subspace $H(\theta)$ is the orthogonal complement of $V(\theta)$ in $S(\theta)$. The map $(D_\theta\pi)/H(\theta)$ is an isometry between $H(\theta)$ and $N(\theta)$. Moreover there exists an isometry $J_\theta: S(\theta)\circlearrowleft$ such that $J^2_\theta = -I$, $J_\theta V(\theta) = H(\theta)$, $J_\theta H(\theta) = V(\theta)$ and the symplectic form $\Omega_\theta: S(\theta)\times S(\theta) \to \mathbb{R}$ defined by $\Omega_\theta(v,w) = \langle v,J_\theta w\rangle$ is invariant under $D\varphi$, i.e., $\Omega_\theta(v,w) = \Omega_{\varphi_t(\theta)}((D\varphi_t)v,(D\varphi_t)w)$ for all θ, t, w and v. For proofs of this basic properties see [4]. Denoting $n = \dim M$, it is clear that $\dim H(\theta) = \dim V(\theta) = \dim N(\theta) = n-1$. A subspace $E \subset S(\theta)$ is Lagrangian if $\dim E = n-1$ and $\Omega_\theta(v,w) = 0$ for all $v \in E$, $w \in E$ (or, what is the same, if $S(\theta)$ is the orthogonal sum of E and $J_\theta E$. The angle between two subspaces $E \subset S(\theta)$, $F \subset S(\theta)$ with $\dim E = \dim F = n-1$, is defined by $\alpha(E,F) = \infty$ if $E \cap F^\perp \neq \{0\}$ and, when $E \cap F^\perp = \{0\}$, by $\alpha(E,F) = \|L\|$ where $L: F \to F^\perp$ is the unique linear map such that E is the graph of L, i.e., $E = \{x+Lx \mid x\in F\}$.

When the geodesic flow is Anosov, E^s_θ and E^u_θ are Lagrangian subspaces of $S(\theta)$. To prove this property let us first show that $E^s_\theta \subset S(\theta)$, $E^u_\theta \subset S(\theta)$ for all θ. Suppose that $v \in T_\theta(UM)$ and $v \notin S(\theta)$. Let w be its orthogonal projection on E^φ_θ. Then, since $(D\varphi_t)S(\theta) = S(\varphi_t(\theta))$ it follows that $(D\varphi_t)w$ is the orthogonal projection of $(D\varphi_t)v$ on $E^\varphi_{\varphi_t(\theta)}$. Moreover $\|(D\varphi_t)w\| = \|w\|$. Then $\|(D\varphi_t)v\| \geq \|(D\varphi_t)w\| \geq \|w\|$. This means that a vector not contained in $S(\theta)$ cannot be contracted by $(D\varphi_t)$. Hence $E^s_\theta \subset S(\theta)$. Moreover, if x and y belong to E^s_θ, we have $\Omega_\theta(x,y) = \Omega_{\varphi_t(\theta)}((D\varphi_t)x,(D\varphi_t)y)$ for all t. Taking the limit for $t \to +\infty$

322

we obtain $\Omega_\theta(x,y) = 0$. Hence E^s_θ is a Lagrangian subspace of $S(\theta)$. A similar argument proves the same property for E^u_θ.

Theorem A. If M is a complete Riemannian manifold whose geodesic flow is Anosov and with curvature $\geq -c$, $c > 0$, then M has no conjugate points and for all $\theta \in UM$ the stable and unstable spaces E^s_θ, E^u_θ, are Lagrangian subspaces of $S(\theta)$ satisfying

$$\alpha(E^s_\theta, H(\theta)) \leq \sqrt{c}$$

$$\alpha(E^u_\theta, H(\theta)) \leq \sqrt{c}.$$

Corollary. Under the assumptions of Theorem A, there exist constants $C_0 > 0$ and $\mu > 1$ such that every perpendicular Jacobi vector field J, with $J(0) = 0$, satisfies

$$\|J(t)\| \geq C_0 \|\dot{J}(0)\| \sinh \mu t$$

for all $t \geq 0$.

The proof of Theorem A and its corollary will be given in the next section.

For the statement of the next theorem we need the following definition: we say that the geodesic flow of a complete Riemannian manifold M has an invariant Lagrangian subbundle if there exists a continuous subbundle F of $T(UM)$ such that for all $\theta \in UM$ the fiber F_θ is a Lagrangian subspace of $S(\theta)$ and $F_{\varphi_t(\theta)} = (D\varphi_t)F_\theta$ for all $t \in R$.

Theorem B. A complete, finite volume Riemannian manifold whose geodesic flow has an invariant Lagrangian subbundle, has no conjugate points.

The proof of Theorem B is completely different from that of Theorem A and can be read independently. It uses an argument involving a Maslov index, that we shall introduce through a definition slightly different from the standard one exposed by Arnold in [1], that makes possible its calculation in our case as a direct application of the Ricatti equation.

II - Proof of Theorem A

In this section we shall prove Theorem A using the following propositions that will be proved after completing the proof of Theorem A.

Proposition II.1. Let M be a Riemannian manifold and $\gamma: [0,a] \to M$ a geodesic arc. If there exists a Lagrangian subspace $E \subset S(\gamma(0),\dot{\gamma}(0))$ such that $V(\gamma(t),\dot{\gamma}(t)) \cap (D\varphi_t)E = \{0\}$ for all $0 \leq t \leq a$ then the geodesic arc γ doesn't contain conjugate points.

Proposition II.2. Let M be a complete Riemannian manifold with curvature $\geq -c$, $c \geq 0$, and let $\gamma: \mathbb{R} \to M$ be a geodesic such that there exists a Lagrangian subspace $E \subset S(\gamma(0),\dot{\gamma}(0))$ such that $E \cap V(\gamma(0),\dot{\gamma}(0)) \neq \{0\}$ and constants $C > 0$, $0 < \lambda < 1$ satisfying

$$\|(D\varphi_t)/(D\varphi_T)E\| \leq C\lambda^t$$

for all $t \geq 0$, $T \geq 0$. Then there exists in $\gamma(\mathbb{R}^+)$ points conjugate to $\gamma(0)$ and moreover there exists $\tau = \tau(c,\lambda,C)$ such that every interval $[a,b] \subset \mathbb{R}^+$ with $b-a > \tau$ contains a point t such that

$$V(\gamma(t),\dot{\gamma}(t)) \cap (D\varphi_t)E \neq \{0\}.$$

Proposition II.2 also holds reversing the time, i.e., if $\|(D\varphi_t)/(D\varphi_T)E\| \leq C\lambda^{|t|}$ for all $t \leq 0$, $T \leq 0$ then there exist in $\gamma((-\infty,0])$ points conjugate to $\gamma(0)$ and every interval $[a,b] \subset$ $\subset (-\infty,0]$ with $b-a > \tau$ contains a point t such that

$V(\gamma(t),\dot{\gamma}(t)) \cap (D\varphi_t)E \neq \{0\}$.

To prove Theorem A let us first recall the formalism through which the dynamics of $D\varphi$ is related to Jacobi vector fields and Lagrangian subspaces to symmetric linear maps. Given $\theta \in UM$ define a linear map $L_\theta: S(\theta) \to N(\theta)\times N(\theta)$ by $L_\theta(v) = ((D_\theta\pi)v,0)$ if $v \in H(\theta)$ and $L_\theta v = (0,(D_\theta\pi)J_\theta v)$ if $v \in V(\theta)$. By the properties explained in the Introduction L_θ is an isometry. Identifying $S(\theta)$ with $N(\theta)\times N(\theta)$ through this isometry the vertical subspace becomes $\{0\}\times N(\theta)$ and the horizontal becomes $N(\theta)\times\{0\}$. Any subspace $E \subset S(\theta)$ with $\dim E = n-1$ (where $\dim M = n$) such that $E \cap (\{0\}\times N(\theta)) = \{0\}$ can be written as the graph of a linear map $T: N(\theta)\circlearrowleft$ i.e. $E = \{(v,Tv) \mid v \in N(\theta)\}$ and E is Lagrangian if and only if T is symmetric. Moreover for every $(p,v) = \theta \in UM$ we have the curvature symmetric map $K(\theta): N(\theta)\circlearrowleft$ given by $K(\theta)w = R_p(v,w)v$. If E is a Lagrangian subspace of $S(\theta)$ such that $(D\varphi_t)E \cap V(\varphi_t(\theta)) = \{0\}$ for all t in an interval $(-\epsilon,\epsilon)$, then, as explained above, we can write $(D\varphi_t)E = \text{graph } U(t)$ where $U(t): N(\varphi_t(\theta))\circlearrowleft$ is a symmetric map. This family of maps satisfies the Ricatti equation

$$\dot{U}(t) + U^2(t) + K(t) = 0 \tag{1}$$

where $K(t) = K(\varphi_t(\theta))$. Finally, if $v \in S(\theta)$, we can write, using the identifications of $S(\varphi_t(\theta))$ with $N(\varphi_t(\theta)) \times N(\varphi_t(\theta))$, $(D\varphi_t)v = (J(t),Y(t))$, where $J(t)$ and $Y(t)$ belong to $N(\varphi_t(\theta))$ for all t. Then $Y(t) = \dot{J}(t)$ and $J(t)$ satisfies the Jacobi equation

$$\ddot{J}(t) = -K(t)J(t).$$

Now suppose that M satisfies the hypothesis of Theorem A. Define Λ as the set of points $\theta \in UM$ such that

$$E^s_{\varphi_t(\theta)} \cap V(\varphi_t(\theta)) = \{0\}$$

for all t. If $\theta \in \Lambda$ we can write

$$E^s_{\varphi_t(\theta)} = \text{graph } U(t)$$

where $U(t): N(\varphi_t(\theta)) \circlearrowleft$ is a symmetric linear map that satisfies the Ricatti equation (1) for all $t \in \mathbb{R}$. By a result of Green, [5], a symmetric solution of (1) defined for all $t \in \mathbb{R}$ satisfies

$$\sup_t \|U(t)\| \leq \left(\sup_{v \neq 0, t} (\langle -K(t)v,v\rangle/\|v\|^2)\right)^{1/2}$$

Hence, in our case:

$$\sup_t \|U(t)\| \leq \sqrt{c}.$$

Therefore, $\theta \in \Lambda$ implies

$$\sup_t \alpha(E^s_{\varphi_t(\theta)}, H(\theta)) \leq \sqrt{c}. \tag{2}$$

This proves that Λ is closed. Now let us study the complement Λ^c of Λ. We claim that there exists $\tau > 0$ such that if $\theta \in \Lambda^c$ then every interval $[a,b]$ with $b-a > \tau$ contains a point t such that $E^s_{\varphi_t(\theta)} \cap V(\varphi_t(\theta)) \neq \{0\}$. Let $C > 0$ and $0 < \lambda < 1$ be the contraction constants of the Anosov flow φ. Let $\tau = \tau(c,\lambda,C) > 0$ be given by II.2. Suppose that $\theta \in \Lambda^c$. Hence $E^s_{\varphi_t(\theta)} \cap V(\varphi_t(\theta)) \neq$ $\neq \{0\}$ for some $t \in \mathbb{R}$. Suppose that θ doesn't satisfy the property of the claim for $\tau = \tau(c,\lambda,C)$. Then there exists an interval $[a,b]$, with $b-a > \tau$, such that $E^s_{\varphi_t(\theta)} \cap V(\varphi_t(\theta)) = \{0\}$ for all $t \in [a,b]$. But then $E^s_{\varphi_t(\theta)} \cap V(\varphi_t(\theta)) = \{0\}$ for all $t < a$ because if $E^s_{\varphi_{\bar{t}}(\theta)} \cap V(\varphi_{\bar{t}}(\theta)) \neq \{0\}$ for some $\bar{t} \leq a$ then, by II.2 (applied with the origin of the parametrization placed at $\bar{t}$) there exists $t \in [a,b]$ satisfying $E^s_{\varphi_t(\theta)} \cap V(\varphi_t(\theta)) \neq \{0\}$. This contradiction shows that $E^s_{\varphi_t(\theta)} \cap V(\varphi_t(\theta)) = \{0\}$ for all $t < a$. Then, by II.1, there no conjugate points in $\gamma/(-\infty,a]$. But then

$E^u_{\varphi_t(\theta)} \cap V(\varphi_t(\theta)) = \{0\}$ for all $t \le a$ because if $E^u_{\varphi_{\bar t}(\theta)} \cap$ $\cap V(\varphi_{\bar t}(\theta)) \ne \{0\}$ for some $\bar t \le a$, then, applying II.2 (reversing time and placing the origin of the parametrization at $\bar t$) we obtain points in $(-\infty, \bar t]$ conjugate with $\bar t$. This contradiction shows that $E^u_{\varphi_t(\theta)} \cap V(\varphi_t(\theta)) = \{0\}$ for all $t \le a$. But then $E^u_{\varphi_t(\theta)} \cap$ $\cap V(\varphi_t(\theta)) = \{0\}$ for all $t \in R$ because if this property fails at $\bar t \ge a$, applying II.2 (reversing time and placing the origin at $\bar t$) we should have in any interval $[c,d] \subset (-\infty, \bar t]$ with $d-c > \tau$ points t where $E^s_{\varphi_t(\theta)} \ne \{0\}$. We have thus proved $E^s_{\varphi_t(\theta)} \cap V(\varphi_t(\theta)) =$ $= \{0\}$ for all t. By II.1 it follows that γ doesn't contain conjugate points. But $E^s_{\varphi_r(\theta)} \cap V(\varphi_r(\theta)) \ne \{0\}$ and this, by II.2, implies the existence of conjugate points in $[r, +\infty)$. This contradiction completes the proof of the claim.

Now observe that the claim easily implies that Λ^c is closed. Since Λ is also closed one of them must be empty. When M is not compact there exists $\theta \in UM$ such that the geodesic $\gamma: \mathbb{R} \to M$ with initial condition $(\gamma(0), \dot\gamma(0)) = \theta$ is a ray i.e. $\gamma/\mathbb{R}^+$ doesn't contain conjugate points. In particular, by II.2 $E^s_{\varphi_t(\theta)} \cap V(\varphi_t(\theta)) =$ $= 0$ for all $t \ge 0$. Hence $\theta \notin \Lambda^c$ because it doesn't satisfy the property of the claim. Hence $\Lambda \ne \emptyset$ i.e. $\Lambda = UM$. By the definition of Λ and Proposition II.1 this implies that M has no conjugate points. Moreover by (1) we have $\alpha(E^s_\theta, H(\theta)) \le \sqrt{c}$ for all $\theta \in \Lambda$ i.e. for all $\theta \in UM$. The same argument replacing E^s by E^u shows that $\alpha(E^u_\theta, H(\theta)) \le \sqrt{c}$ for all $\theta \in UM$. When M is compact there are no conjugate points (by Theorem B). Hence $\Lambda^c = \emptyset$ because II.2 the geodesic determined by an element $\theta \in \Lambda^c$ should contain conjugate points. Then $\Lambda = UM$ and the proof of Theorem A is completed as before.

Now let us prove Proposition II.1. Take symmetric linear maps $S(t)\colon N(\varphi_t(\theta))\circlearrowleft$ such that:

$$\operatorname{graph} S(t) = (D\varphi_t)E \tag{2}$$

for all $0 \le t \le a$. Let $J(t)$ be a perpendicular Jacobi vector field along γ with $J(c) = 0$ and $\dot{J}(c) \ne 0$ for some $0 < c < a$. We want to prove that $J(t) \ne 0$ for all $0 < t < a$. For each $0 \le t \le a$ let $S(t)\colon N(\varphi_t(\theta))\circlearrowleft$ be a symmetric map whose graph is $(D\varphi_t)E$. Let $Y(t)\colon N(\theta)) \to N(\varphi_t(\theta))$ be a family of isomorphisms satisfying $Y(c) = I$ and

$$\dot{Y}(t) = S(t)Y(t)$$

for $0 \le t \le a$. If $v \in E$, we can write

$$(D\varphi_t)v = (v(t), S(t)v(t))$$

where $v(t) \in N(\varphi_t(\theta))$. But, as we explained above, we have $S(t)v(t) = \dot{v}(t)$. Hence $v(t) = Y(t)w$, where $w = v(c)$, and then

$$(D\varphi_t)v = (Y(t)w, S(t)Y(t)w)$$

for all $0 \le t \le a$ and $J_1(t) = Y(t)w$ is a perpendicular Jacobi vector field. Hence, for all $0 \le t \le a$

$$\langle J(t), \dot{J}_1(t)\rangle - \langle \dot{J}(t), J_1(t)\rangle = \langle J(c), \dot{J}_1(c)\rangle - \langle \dot{J}(c), J_1(c)\rangle = -\langle \dot{J}(c), J_1(c)\rangle.$$

Then, for all $0 \le t \le a$

$$\langle J(t), S(t)Y(t)w\rangle - \langle \dot{J}(t), Y(t)w\rangle = -\langle \dot{J}(c), w\rangle.$$

Therefore

$$\langle Y^*(t)(S(t)J(t) - \dot{J}(t)), w\rangle = -\langle \dot{J}(c), w\rangle$$

for all $0 \le t \le a$ and $w \in N(\theta)$. This implies

$$\dot{J}(t) = S(t)J(t) + Y^*(t)^{-1}\,\dot{J}(c).$$

The solution of this equation with $J(c) = 0$ is

$$J(t) = Y(t)\int_c^t Y(s)^{-1}Y^*(s)^{-1}\dot{J}(c)ds.$$

Hence

$$\langle Y^{-1}(t)J(t),\dot{J}(c)\rangle = \int_c^t \|Y^*(s)^{-1}\dot{J}(c)\|^2\, ds.$$

Then $J(t) \neq 0$ for all $0 \leq t \leq a$ different from c.

Finally let us prove Proposition II.2. We claim that under the hypothesis of II.2, there exists $\tau(c,\lambda,C) > 0$ such that $\gamma/[0,\tau]$ contains points conjugate to $\gamma(0)$ and there exists $t_o \in (0,\tau]$ such that:

$$(D\varphi_{t_o})E \cap V(\gamma(t_o),\dot{\gamma}(t_o)) \neq \{0\}.$$

Clearly this claim implies II.2 because it grants the existence of points in $\gamma/(0,+\infty)$ conjugate to $\gamma(0)$. Moreover, applying the claim to the geodesic $\gamma_1(t) = \gamma(t+t_o)$ and the space $(D\varphi_{t_o})E$, it follows the existence of $0 < t_1 \leq \tau$ such that

$$(D\varphi_{t_1})(D\varphi_{t_o})E \cap V(\gamma(t_o+t_1),\dot{\gamma}(t_o+t_1)) \neq \{0\}.$$

Continuing this method we obtain a sequence $0 < t_n < \tau$ such that denoting $s_n = t_o +\ldots+ t_n$ we have

$$(D\varphi_{s_n})E \cap V(\gamma(s_n),\dot{\gamma}(s_n)) \neq \{0\}$$

thus completing the proof of II.2. To prove the claim it suffices to find $\tau_o = \tau_o(c,\lambda,C)$ such that there exists a point in $\gamma/(0,\tau_o)$ conjugate to $\gamma(0)$. Once this proved it follows that $\gamma(\varepsilon,\tau_o+\varepsilon)$ contains a point conjugate to $\gamma(\varepsilon)$ if ε is small enough and then there exists $\varepsilon \leq t \leq \tau_o + \varepsilon$ such that $(D\varphi_t)E \cap V(\gamma(t),\dot{\gamma}(t)) \neq \{0\}$ because otherwise, by II.1, there would be no conjugate points in $\gamma/[\varepsilon,\tau_o+\varepsilon]$. Hence, taking, let us say, $\tau = \tau_o+1$ the claim is proved. To find $\tau_0 = \tau_0(c,\lambda,C)$ we shall use the following lemma due to Green [5].

Lemma II.3. Let M be a Riemannian manifold with curvature $\geq -c$, $c > 0$. Then there exists $A = A(c) > 0$ such that if $\gamma: [0,a] \to M$ is a geodesic arc without conjugate points and $J(t)$, $0 \leq t \leq a$ is a perpendicular Jacobi vector field on γ with $J(0) = 0$, then

$$\|\dot{J}(t)\| \leq A\|J(t)\|$$

for all $1 \leq t \leq a$.

Take a C^∞ function $\psi: \mathbb{R} \circlearrowleft$ satisfying $\psi(t) = 0$ if $|t| > 1$ and $\psi(0) = 1$. Set

$$B = \int_{-1}^{1} (|\psi''\psi| + c\psi^2)\,dt.$$

Define $D = ((c/3)+1+A)\exp A$ where A is given by the previous lemma. Take $T > 0$ such that

$$C\lambda^T < (DB)^{-1/2} \tag{4}$$

and let $\tau_0 = T+1$. We shall prove that $\gamma/(0,\tau_0)$ contains a point conjugate to $\gamma(0)$. Suppose that it doesn't. Let $J(t)$, $t \geq 0$, be a perpendicular Jacobi vector field on γ such that $J(0) = 0$, $\|\dot{J}(0)\| = 1$, $\dot{J}(0) \in E$. Since $\gamma(t)$ is not conjugate to $\gamma(0)$ for all $0 < t < \tau_0 = T+1$ there exists a perpendicular Jacobi vector field $\tilde{Z}(t)$ along γ with $\tilde{Z}(T) = 0$ and $\tilde{Z}(0) = \dot{J}(0)$. We have

$$1 = \langle \dot{J}(0), \tilde{Z}(0)\rangle = -\langle J(T), \dot{\tilde{Z}}(T)\rangle.$$

Define $Z(t) = \tilde{Z}(t)/\|\dot{\tilde{Z}}(T)\|$. Then $\|\dot{Z}(T)\| = 1$ and

$$\|Z(0)\| = \frac{\|\tilde{Z}(0)\|}{\|\dot{\tilde{Z}}(T)\|} = \frac{1}{\|\dot{\tilde{Z}}(T)\|} = -\frac{\langle J(T), \dot{\tilde{Z}}(T)\rangle}{\|\dot{\tilde{Z}}(T)\|} \leq$$

$$\leq \|J(T)\| \leq C\lambda^T.$$

330

Now consider the bilinear form I on the space Ω of piecewise C^2 perpendicular vector fields along $\gamma/[0,T+1]$ defined by

$$I(V,W) = -\int_0^{T+1} \langle \ddot{V}(t)+K(\gamma(t),\dot{\gamma}(t))V,W\rangle -$$

$$- \sum_i \langle (\Delta_i \dot{V})(t_i),W(t_i)\rangle$$

where $(\Delta_i \dot{V})(t_i)$ is the jump of $\dot{V}$ at the discontinuity t_i. Define $V \in \Omega$ as $V(t) = 0$ for $T \le t \le T+1$, $V(t) = Z(t)$ for $1 \le t \le T$ and $V(t) = t\Phi_1^t Z(1)$ for $0 \le t \le 1$, where $\Phi_1^t: T_{\gamma(1)}M \to T_{\gamma(t)}M$ denotes the parallel transport along γ. Define $W \in \Omega$ by $W(t) = \psi(t-T)\Phi_T^t \dot{Z}(T)$. Then

$$I(V,V) = -\int_0^1 t^2 \langle K(\gamma(t),\dot{\gamma}(t))\Phi_1^t Z(1),\Phi_1^t Z(1)\rangle +$$

$$+ \langle Z(1)-\dot{Z}(1),Z(1)\rangle.$$

Using Green's lemma:

$$I(V,V) \le \frac{1}{3} c\|Z(1)\|^2 + A\|Z(1)\|^2 + \|Z(1)\|^2.$$

But also from Green's Lemma and Gronwall's inequality we obtain:

$$\|Z(1)\| \le \|Z(0)\| \exp A.$$

Hence

$$I(V,V) \le (\tfrac{1}{3} c+A+1)(\exp A)\|Z(0)\|^2 = D\|Z(0)\|^2 .$$

Moreover

$$I(V,W) = -\|\dot{Z}(T)\| = 1$$

$$I(W,W) = -\int_{T-1}^{T+1} (\psi''\psi + \psi^2\langle K(\gamma(t),\dot{\gamma}(t))\Phi_T^t\dot{Z}(T),\Phi_T^t\dot{Z}(T)\rangle dt \le$$

$$\le \int_{-1}^{1} (|\psi''\psi| + c\psi^2)dt = B.$$

Therefore:

$$I(V+\lambda W,V+\lambda W) \le D\|Z(0)\|^2 + 2\lambda + B\lambda^2 .$$

But

$$4-4DB\|Z(0)\|^2 > 0$$

because by (3) and (4)

$$DB\|Z(0)\|^2 \leq DB(C\lambda^T)^2 < 1.$$

Hence $I(V+\lambda W, V+\lambda W) < 0$ for certain values of λ thus proving that the interval $(0,T+1)$ contains points conjugate to $\gamma(0)$.

To prove the corollary of Theorem A consider a perpendicular Jacobi vector field J along the geodesic γ with initial condition $(\gamma(0),\dot{\gamma}(0)) = \theta \in UM$ such that $J(0) = 0$. Let $U(t):N(\varphi_t(\theta))\circlearrowleft$ be the symmetric map whose graph is $E^u_{\varphi_t(\theta)}$. Let $Y(t): N(\theta) \to$ $\to N(\varphi_t(\theta))$ be a family of linear maps satisfying $Y(0) = I$ and

$$\dot{Y}(t) = U(t)Y(t) \tag{1}$$

for all $t \in \mathbb{R}$. Then if $v \in E^u_\theta$ is written as $v = (w,u)$ with $w \in N(\theta)$, $u \in N(\theta)$ we have

$$(D\varphi_t)v = (Y(t)w, U(t)Y(t)w)$$

for all t. Then

$$\|(D\varphi_t)v\|^2 = \|Y(t)w\|^2 + \|U(t)Y(t)w\|^2 \leq$$
$$\leq \|Y(t)w\|^2 + c\|Y(t)w\|^2 = (1+c)\|Y(t)w\|^2.$$

Moreover

$$\|(D\varphi_t)v\|^2 \geq C^2\lambda^{-2t}\|v\|^2 = C^2\lambda^{-2t}(\|w\|^2+\|u\|^2) \geq C^2\lambda^{-2t}\|w\|^2.$$

Hence

$$\|Y(t)w\| \geq C\lambda^{-t}(1+c)^{-1/2}\|w\|$$

for all $t \geq 0$ and w. This can be written as

$$\|Y(t)w\| \geq Ae^{at}\|w\| \tag{2}$$

for all $t \geq 0$, and w, for some $A > 0$, $a > 0$ depending only on

332

C, λ and c. Now define a perpendicular Jacobi vector field J_1 as $J_1(t) = Y(t)w$. Then

$$\langle J(t),\dot{J}_1(t)\rangle - \langle\dot{J}(t),J_1(t)\rangle =$$
$$= \langle J(0),\dot{J}_1(0)\rangle - \langle\dot{J}(0),J_1(0)\rangle = -\langle\dot{J}(0),J_1(0)$$

for all t. Then

$$\langle J(t),U(t)Y(t)w\rangle - \langle\dot{J}(t),Y(t)w\rangle = -\langle\dot{J}(0),w\rangle$$

for all t. Hence

$$\langle Y^*(t)(U(t)J(t)-\dot{J}(t)),w\rangle = -\langle\dot{J}(0),w\rangle$$

for all t and w. Then

$$\dot{J}(t) = U(t)J(t) + Y^*(t)^{-1}\,\dot{J}(0).$$

The solution of this equation with $J(0) = 0$ is given by

$$J(t) = Y(t)\int_0^t Y(s)^{-1}\,Y^*(s)^{-1}\,\dot{J}(0)ds.$$

Denote

$$u(t) = \int_0^t Y(s)^{-1}\,Y^*(s)^{-1}\,\dot{J}(0)ds.$$

Then

$$\langle u(t),\dot{J}(0)\rangle' = \langle\dot{u}(t),\dot{J}(0)\rangle = \|Y^*(t)^{-1}\,\dot{J}(0)\|^2.$$

But since Y satisfies equation (1) and $\|U(t)\| \le \sqrt{c}$ for all t, it follows that

$$\|Y(t)\| \le e^{t\sqrt{c}} \tag{3}$$

for all $t \ge 0$. Hence

$$\langle u(t),\dot{J}(0)\rangle \ge \|\dot{J}(0)\|^2\int_0^t e^{-2s\sqrt{c}}\,ds =$$
$$= \|\dot{J}(0)\|^2\frac{1}{2\sqrt{c}}(1-e^{-2t\sqrt{c}})ds.$$

Therefore

$$\|u(t)\| \geq \|\dot{J}(0)\|^{-1} \langle u(t), J(0)\rangle \geq \|\dot{J}(0)\| \frac{1}{2\sqrt{c}}(1-e^{-2t\sqrt{c}})$$

and then, by (2):

$$\|J(t)\| \geq A \frac{1}{2\sqrt{c}}(e^{at}-e^{(a-2\sqrt{c})t})\|\dot{J}(0)\|.$$

Comparing (2) and (3) we obtain $\sqrt{c} > a$. Then $2\sqrt{c} - a > a$ and then, defining $\mu = 2\sqrt{c} - a$, we obtain that

$$\|J(t)\| \geq \|\dot{J}(0)\| A \frac{1}{\sqrt{c}} \sin h\, \mu\, t$$

for all $t \geq 0$.

III - Proof of Theorem B

If M is a Riemannian manifold and $\theta \in UM$ denote $\Lambda(\theta)$ the set of Lagrangian subspaces of $S(\theta)$. $\Lambda(\theta)$ has a natural manifold structure (see [1]). Denote $\Lambda(UM)$ the bundle on UM whose fiber on θ is $\Lambda(\theta)$. Let $\Lambda_k(\theta)$ be the set of Lagrangian subspaces $E \in \Lambda(\theta)$ such that $\dim E \cap V(\theta) = k$. Then $\Lambda_k(\theta)$ is a submanifold of $\Lambda(\theta)$ with codimension $k(k+1)/2$ ([1]). Then, denoting $\Lambda_k(UM)$ the set of elements $(\theta, E) \in \Lambda(UM)$ such that $E \in \Lambda_k(\theta)$, it follows that $\Lambda_k(UM)$ is a submanifold of $\Lambda(UM)$ with codimension $k(k+1)/2$. Define

$$\Gamma = \bigcup_{k \geq 2} \Lambda_k(UM)$$

and let Σ be the complement of Γ. Clearly Σ is an open and dense subset of $\Lambda(UM)$. Let S^1 be the unit circle of the complex plane. Define the function $m: \Sigma \to S^1$ as follows: if $(\theta, E) \in$ $\in \Lambda_o(UM)$, take a symmetric map $L: N(\theta) \circlearrowleft$ such that graph $(L) = E$ and define

$$m(\theta, E) = \frac{1-i\,tr\, L}{1+i\,tr\, L}.$$

334

When $(\theta,E) \in \Lambda_1(UM)$, set

$$m(\theta,E) = -1.$$

Let us prove that m is continuous. Suppose that $(\theta,E) \in \Lambda_1(UM)$ and that $(\theta_n,E_n) \in \Lambda_o(UM)$ is a sequence converging to (θ,E). We can write $E_n = \text{graph}(L_n)$ where $L_n\colon N(\theta_n) \circlearrowleft$ is a symmetric linear map. We claim that there exists $C > 0$ such that for all n the number of eigenvalues of L_n (counted with multiplicity) larger in modulus than C is ≤ 1. If this property is false there exist integers $n_1 < n_2 < \ldots$, vectors x_j, y_j, and two real sequences $\{\lambda_j\}$ and $\{\mu_j\}$ such that $|\lambda_j| \to +\infty$, $|\mu_j| \to +\infty$ and:

$$\langle x_j,y_j\rangle = 0 \qquad \|x_j\| = \|y_j\| = 1$$

$$L_{n_j}x_j = \lambda_j x_j \qquad L_{n_j}y_j = \mu_j y_j$$

for all j. If S_j is the space spanned by $(x_j,\lambda_j x_j)$ and $(y_j,\mu_j y_j)$, the property $|\lambda_j| \to +\infty$ implies that the sequence $\{S_j\}$ converges to a two dimensional subspace of $V(\theta)$ and, since $S_j \subset E_{n_j}$, this two dimensional subspace is contained in E. This contradicts the property $E \in \Lambda_1(\theta)$ and proves the claim. Moreover the spectral radius of L_n goes to $+\infty$ when $n \to +\infty$ because otherwise, since the maps L_n are symmetric, we would have a sequence of integers $n_1 < n_2 < \ldots$ such that

$$\sup_j \|L_{n_j}\| < +\infty$$

and then E would be also a graph of a linear map $L\colon N(\theta) \circlearrowleft$ (obtained as the limit of a subsequence of $\{L_{n_j}\}$) thus implying $E \in \Lambda_o(\theta)$. Then every L_n has exactly one eigenvalue λ_n with $|\lambda_n| > C$ and $\lim|\lambda_n| = +\infty$. Hence $\lim\limits_{n\to+\infty} |\text{tr } L_n| = +\infty$ and this implies $\lim\limits_{n\to+\infty} m(\theta_n,E_n) = -1$.

Observe that m cannot be continuously extended to Γ because if $(\theta,E) \in \Gamma$ it is easy to see that given any $\lambda \in \mathbb{R}$ there exist, arbitrarily near to E, spaces $E' \in \Lambda_o(\theta)$ such that $E' = \text{graph}(L)$ and $\text{tr}\, L = \lambda$.

Now we shall define an index that to every continuous map $\gamma: S^1 \to \Lambda(UM)$ associates an integer $\text{Ind}(\gamma) \in Z$.

If $\gamma: S^1 \to \Sigma$ is a continuous map, define $\text{Ind}(\gamma)$ as the degree of $m \circ \gamma: S^1 \circlearrowleft$. Observe that if $\gamma_1: S^1 \to \Sigma$ and $\gamma_2: S^1 \to \Sigma$ are homotopic, then $\text{Ind}(\gamma_1) = \text{Ind}(\gamma_2)$. Given a continuous map $\gamma: S^1 \to \Lambda(UM)$ define $\text{Ind}(\gamma)$ as $\text{Ind}(\gamma) = \text{Ind}(\tilde{\gamma})$ where $\tilde{\gamma}: S^1 \to \Sigma$ is homotopic to γ. To show that this definition is correct we have to show that there exists $\tilde{\gamma}: S^1 \to \Sigma$ homotopic to γ and that $\text{Ind}(\tilde{\gamma})$ is independent of $\tilde{\gamma}$. The first property is proved taking $\tilde{\gamma}: S^1 \to \Sigma$ very near in the C^o topology (in particular homotopic) to γ. This can be done because Γ is a union of submanifolds with codimensions ≥ 3. The independence property is proved observing that if $\hat{\gamma}: S^1 \to \Sigma$ is also homotopic to γ, then $\hat{\gamma}$ and $\tilde{\gamma}$ are homotopic and since Γ is a union of submanifolds with codimensions ≥ 3, the homotopy between $\hat{\gamma}$ and $\tilde{\gamma}$ can be taken avoiding Γ i.e. $\tilde{\gamma}$ and $\hat{\gamma}$ are homotopic as maps into Σ, and then, as we observed above, $\text{Ind}(\tilde{\gamma}) = \text{Ind}(\hat{\gamma})$.

Now suppose that there exists an invariant Lagrangian subbundle E on UM i.e. for all $\theta \in UM$ the fiber E_θ is Lagrangian and

$$(D\varphi_t)E_\theta = E_{\varphi_t(\theta)}$$

for all t. Given a continuous map $\alpha: S^1 \to UM$ we define $\hat{\alpha}: S^1 \to \Lambda(UM)$ by

$$\hat{\alpha}(t) = (\alpha(t), E_{\alpha(t)})$$

and we introduce an index $\text{ind}(\alpha)$ by

336

$$\operatorname{ind}(\alpha) = \operatorname{Ind}(\hat{\alpha}).$$

Every map $\alpha: S^1 \to UM$ can be written in the form $\alpha(t) = (\gamma(t), v(t))$ with $\gamma(t) \in M$ and $v(t) \in U_{\gamma(t)}M$. We shall denote α^* the map of S^1 into UM defined by $\alpha^*(t) = (\gamma(t), -v(t))$. Observe that α and α^* are always homotopic. Hence

$$\operatorname{ind}(\alpha^*) = \operatorname{ind}(\alpha). \tag{1}$$

We shall say that a continuous map $\alpha: S^1 \to UM$ is a pseudo geodesic if for all $s \in S^1$ where

$$E_{\alpha(s)} \cap V(\alpha(s)) \neq \{0\} \quad \text{or} \quad E_{\alpha^*(s)} \cap V(\alpha^*(s)) \neq \{0\} \tag{2}$$

there exists $\epsilon > 0$ and $k > 0$ such that

$$\varphi_{kt}(\alpha(o)) = \alpha(e^{it}s)$$

for all $-\epsilon \leq t \leq \epsilon$. In other words, writing α in the form $\alpha = (\gamma, v)$ as above, α is a pseudo geodesic if every $s \in S^1$ where (2) holds has a neighborhood V such that γ/V is a geodesic arc (parametrized by a multiple of arc lenght) and $v(t) = \dot{\gamma}(t)/\|\dot{\gamma}(t)\|$ for all $t \in V$.

The following is the key lemma in the proof of Theorem B.

<u>Lemma III.1</u>. If $\alpha: S^1 \to UM$ is a pseudo geodesic such that $E_{\alpha(s)} \cap V(\alpha(s)) \neq \{0\}$ for some $s \in S^1$, then $\operatorname{ind}(\alpha) > 0$.

Before proving III.1 we shall complete the proof of Theorem B. First we shall prove the following corollary of III.1.

<u>Corollary</u>. Every pseudo geodesic $\alpha: S^1 \to UM$ satisfies

$$E_{\alpha(t)} \cap V(\alpha(t)) = \{0\}$$

for all $t \in S^1$.

337

Proof: Suppose that $\alpha\colon S^1 \to UM$ is a pseudo geodesic and there exists $s \in S^1$ where $E_{\alpha(s)} \cap V(\alpha(s)) \neq \{0\}$ holds. By III.1, $ind(\alpha) > 0$. Define $\bar{\alpha}\colon S^1 \to UM$ by $\bar{\alpha}(z) = \alpha(\bar{z})$. Then

$$ind(\bar{\alpha}) = -ind\,\alpha < 0 \tag{3}$$

because $\bar{\alpha}$ is just the composition of α with the map $S^1 \ni z \to \bar{z} \in S^1$ whose degree is -1. Write α in the form $\alpha(t) = (\gamma(t), v(t))$ where $\gamma\colon S^1 \to M$ is a continuous map and $v(t) \in U_{\gamma(t)}M$ for all $t \in S^1$. Define $\beta\colon S^1 \to UM$ by

$$\beta(t) = (\gamma(\bar{t}), -v(\bar{t})).$$

This map is a pseudo geodesic because if $s \in S^1$ satisfies

$$E_{\beta(s)} \cap V(\beta(s)) \neq \{0\} \quad \text{or} \quad E_{\beta^*(s)} \cap V(\beta^*(s)) \neq \{0\}$$

then, from the definition of β follows that (2) is satisfied at $\bar{s}$ and then there exists a neighborhood V of s such that γ/V is a geodesic and $v(t) = \dot{\gamma}(t)/\|\dot{\gamma}(t)\|$ for t in V. But then the map $\bar{V} \ni t \to \gamma(\bar{t})$ is also a geodesic. Moreover, the property $v(t) = \dot{\gamma}(t)/\|\gamma(t)\|$ for all $t \in V$ implies that the derivative of the map $V \ni t \to \gamma(\bar{t})$ is collinear with $-v(\bar{t})$ for all $t \in \bar{V}$. Hence β is a pseudo geodesic. But observe that $\beta = \bar{\alpha}^*$. Hence, using (1) and (2)

$$ind(\beta) = ind(\bar{\alpha}^*) = ind(\bar{\alpha}) = -ind(\alpha) < 0.$$

Since β is a pseudo geodesic this contradicts III.1 and proves the corollary.

The final step in the proof of Theorem B is to show that $E_\theta \cap V(\theta) = \{0\}$ for all $\theta \in UM$. By II.1 this implies that M has no conjugate points and completes the proof of Theorem B. Let us suppose that $E_\theta \cap V(\theta) \neq \{0\}$. Using this property we shall construct a pseudo geodesic $\alpha\colon S^1 \to UM$ with $\alpha(0) = \theta$. Since

$E_\theta \cap V(\theta) \neq \{0\}$ this pseudo geodesic contradicts the corollary of III.1 and shows that we must have $E_\theta \cap V(\theta) = \{0\}$ for all θ.

To construct α take $a > 0$ such that, denoting $(p_1, v_1) = \varphi_{-a}(\theta)$ and $(p_2, v_2) = \varphi_a(\theta)$, we have

$$E_{(p_1,v_1)} \cap V(p_1,v_1) = \{0\}, \quad E_{(p_1,-v_1)} \cap V(p_1,-v_1) = \{0\}$$

$$E_{(p_2,v_2)} \cap V(p_2,v_2) = \{0\}, \quad E_{(p_2,-v_2)} \cap V(p_2,-v_2) = \{0\}.$$

The existence of a is not immediate. It requires the following lemma, that will be proved after finishing the construction of α.

<u>Lemma III.2</u>. If $\theta \in UM$ and $E \subset S(\theta)$ is a Lagrangian subspace, then the set of $t \in \mathbb{R}$ where $(D\varphi_t)E \cap V(\varphi_t(\theta)) \neq \{0\}$ is discrete.

Apply this lemma to θ and E_θ. Then the set $S = \{t \in \mathbb{R} \mid E_{\varphi_t(\theta)} \cap V(\varphi_t(\theta)) \neq \{0\}\}$ is discrete. Moreover, if $\theta = (p,v)$ and we define $\xi = (p,-v)$, then, III.2 applied to ξ and E_ξ proves that the set $S' = \{t \in \mathbb{R} \mid E_{\varphi_t(\xi)} \cap V(\varphi_t(\xi)) \neq \{0\}\}$ is also discrete. Then, taking $a > 0$ such that a and $-a$ don't belong to $S \cup S'$, the desired properties are satisfied. Now take neighborhoods V_1, V_2 of (p_1,v_1) and (p_2,v_2) respectively such that

$$E_{(q,w)} \cap V(q,w) = \{0\} \tag{4}$$

and

$$E_{(q,-w)} \cap V(q,-w) = \{0\} \tag{5}$$

for all $(q,w) \in V_1 \cup V_2$. Since every point of UM is a non wandering point of φ (because the volume of UM is finite), we can take $\eta \in V_2$ and $T > 0$ such that $\varphi_T(\eta) \in V_1$. Now set $\omega_1 = \pi/4a$, $\omega_2 = \pi/T$ and take $\alpha\colon S^1 \to UM$ satisfying

$$\alpha(\exp \omega_1 ti) = \varphi_t(\theta)$$

339

for $-a \le t \le a$,

$$\alpha(\exp(\omega_2 t + \frac{\pi}{2})i) = \varphi_t(\eta)$$

for $0 \le t \le T$, and $\alpha(\text{esp } ti) \in V_2$ when $\pi/4 \le t \le \pi/2$ and $\alpha(\exp ti) \in V_1$ for $-\pi/2 \le t \le -\pi/4$. Obviously α satisfies the condition $\alpha(0) = \theta$. Moreover it is a pseudo geodesic because if at $s \in S^1$ condition (2) is satisfied, then by properties (4) and (5), the point $\alpha(s)$ cannot belong to $V_1 \cup V_2$. Hence the argument of s must be contained in the interval $(-\pi/4, \pi/4)$ or in the interval $(\pi/2, 3\pi/2)$. Since α restricted to these intervals is an arc of trajectory of the geodesic flow parametrized by a constant multiple of time, then the condition required in the definition of pseudo geodesics is satisfied.

To prove Lemma III.2 it suffices to show that if E is a Lagrangian subspace of $S(\theta)$ such that $E \cap V(\theta) \neq \{0\}$ there exists a neighborhood V of $t = 0$ such that $(D\varphi_t)E \cap V(\theta) = \{0\}$ for all $0 \neq t \in V$. Let $p: S(\theta) \to H(\theta)$ be the orthogonal projection. We claim that $p(E)$ is the orthogonal complement of $J_\theta(E \cap V(\theta))$ in $H(\theta)$. In fact

$$\begin{aligned}\dim p(E) + \dim J_\theta(E \cap V(\theta)) &= \dim E - \dim(E \cap V(\theta)) + \\ &+ \dim(E \cap V(\theta)) = \dim E = \dim H(\theta).\end{aligned}$$

Moreover, if $x \in p(E)$, we can write $x = y+z$ with $v \in E$ and $z \in E \cap V(\theta)$. Then, if $w \in E \cap V(\theta)$ we have $\langle y, J_\theta w\rangle = 0$ because y and w belong to the Lagrangian subspace E, and $\langle z, J_\theta w\rangle = 0$ because $z \in V(\theta)$ and $J_\theta w \in J_\theta(E \cap V(\theta)) \subset H(\theta)$. Hence $\langle x, J_\theta w\rangle = 0$ completing the proof of the claim. Take a basis $\{v_1, \dots, v_n\}$ of $p(E)$ and let $p_t: S(\varphi_t(\theta)) \to H(\varphi_t(\theta))$ be the orthogonal projection. Then if t is near to 0 there exists a set of m linearly independent vectors $\{v_1(t), \dots, v_m(t)\} \subset p_t((D\varphi_t)E)$.

340

Let $\{w_1,\ldots,w_k\}$ be a basis of unitary vectors of $E \cap V(\theta)$. Then $k+m = \dim E = \dim H(\theta)$. Take perpendicular Jacobi vector fields $Y_1(t),\ldots,Y_k(t)$, with $Y_i(0) = 0$ and $\dot{Y}_i(0) = J_\theta w_i$ for $1 \leq i \leq k$. Define $w_i(t) = Y_i(t)/\|Y_i(t)\|$ for $t \neq 0$. Then

$$\lim_{t\to 0} w_i(t) = \lim_{t\to 0} \frac{Y_i(t)/t}{\|Y_i(t)/t\|} = \frac{\dot{Y}_i(0)}{\|\dot{Y}_i(0)\|} = J_\theta w_i .$$

Moreover

$$p_t((D\varphi_t)w_i) = Y_i(t).$$

Hence

$$p_t((D\varphi_t)E) \supset \{v_1(t),\ldots,v_m(t),w_1(t),\ldots,w_k(t)\}.$$

When $|t|$ is small, $\{v_1(t),\ldots,v_m(t),w_1(t),\ldots,w_k(t)\}$ is near to $\{v_1,\ldots,v_m,J_\theta w_1,\ldots,J_\theta w_k\}$ that is a basis of $H(\theta)$ because $H(\theta) = p(E) \oplus J_\theta(E\cap V)$. Then $p_t((D\varphi_t)E) = H(\varphi_t(\theta))$ for $|t|$ small and $\neq 0$. Obviously $p_t((D\varphi_t)E) = H(\varphi_t(\theta))$ implies $(D\varphi_t)E \cap$ $\cap\ V(\varphi_t(\theta)) = \{0\}$.

Finally, we shall prove Lemma III.1. Given a continuous map $\beta: S^1 \to \Lambda(UM)$ we shall denote $P(\beta,c) = \{t \mid (m\circ\beta)(t) = c\}$ and, if β is differentiable at $t \in S^1$, we shall denote $\sigma(\beta,t)$ the sign of the derivativo $(m\circ\beta)'(t)$ i.e. $\sigma(t) = 1$ or -1 according to whether $(m\circ\beta)'(t)$ preserves or reverses orientation. Now let $\alpha: S^1 \to UM$ be a pseudo geodesic and define $\tilde{\alpha}: S^1 \to \Lambda(UM)$ by $\tilde{\alpha}(t) = (\alpha(t),E_{\alpha(t)})$. We have to prove that $\mathrm{Ind}(\tilde{\alpha}) > 0$. For this it suffices to show (as we explained above) that for all C^1 map $\beta: S^1 \to \Lambda(UM)$ that is a C^1 approximation of $\tilde{\alpha}$ (and then homotopic to $\tilde{\alpha}$) and satisfies $\beta(S^1) \cap \Gamma = \emptyset$ the degree of the map $m\circ\beta: S^1 \circlearrowleft$ (that is by definition $\mathrm{Ind}(\beta)$), is strictly positive. To prove this we shall show that there exists $\bar{c} > 0$ such that for all $c > \bar{c}$

341

$$\sum_{t\in P(\tilde{\alpha},c)} \sigma(\tilde{\alpha},t) > 0. \tag{6}$$

This implies that the degree of $m\circ\beta$ is >0 because this degree can be calculated as

$$\sum_{t\in P(\beta,c)} \sigma(\beta,t)$$

and an elementary application of the implicit function theorem shows that

$$\sum_{t\in P(\beta,c)} \sigma(\beta,t) \geq \sum_{t\in P(\tilde{\alpha},c)} \sigma(\beta,t).$$

Hence (6) suffices to prove III.1. To prove (6) we first transform $\tilde{\alpha}$ in a map of $[0,2\pi]$ in $\Lambda(UM)$ defining $\hat{\alpha}: [0,2\pi] \to \Lambda(UM)$ by $\tilde{\alpha}(t) = \tilde{\alpha}(e^{it})$. Then (6) is equivalent to

$$\sum_{t\in P(\hat{\alpha},c)} \sigma(\hat{\alpha},t) > 0. \tag{7}$$

But since $m\circ\hat{\alpha}$ takes values in S^1, its derivative satisfies:

$$\frac{(m\circ\hat{\alpha})'(t)}{i(m\circ\hat{\alpha})(t)} \in \mathbb{R}$$

and

$$\sigma(\hat{\alpha},t) = \operatorname{sgn} \frac{(m\circ\hat{\alpha})'(t)}{i(m\circ\hat{\alpha})(t)}.$$

Hence (7) can be written as:

$$\sum_{t\in P(\hat{\alpha},c)} \operatorname{sgn} \frac{(m\circ\hat{\alpha})'(t)}{i(m\circ\hat{\alpha})(t)} > 0. \tag{8}$$

By Lemma III.2 the set of points $t \in [0,2\pi]$ where $(m\circ\hat{\alpha})(t) = -1$ is finite. Let $t_1,\dots,t_r$ be this set. Without lose of generality we can assume it doesn't contain 0 or 2π. Each t_j has a neighborhood V_j such that $\hat{\alpha}/V_j$ can be written as

$$\hat{\alpha}(t) = (\alpha_j(t), E_{\alpha_j(t)})$$

342

where $\alpha_j: V_j \to UM$ is an arc of trajectory of the geodesic flow parametrized by a constant positive multiple of time i.e. there exists $k > 0$ such that

$$\alpha_j(t_j+s) = \varphi_{ks}(\alpha_j(t_j)).$$

Write, for $t \in V_j$

$$E_{\alpha_j(t)} = \text{graph } U_j(t)$$

where $U_j(t): N(\alpha_j(t))\circlearrowleft$ is a linear symmetric map. Then, the invariance of the bundle E implies that U_j satisfies the Ricatti equation (see Section II):

$$k\dot{U}_j(t) + U_j^2(t) + K(t) = 0.$$

Now take $-1 < c < 0$ so near to -1 that

$$P(\hat{\alpha},c) \subset \bigcup_{j=1}^{r} V_j .$$

Let t be a point in $P(\hat{\alpha},c)$. Suppose that it belongs to V_j. Then

$$(m\circ\hat{\alpha})'(t) = \left(\frac{1-i\,tr\,U_j}{1+i\,tr\,U_j}\right)'(t) = -\frac{2i\,tr\,\dot{U}_j(t)}{(1+i\,tr\,U_j(t))^2}.$$

Then:

$$\frac{(m\circ\hat{\alpha})'(t)}{i(m\circ\hat{\alpha})(t)} = -\frac{2tr\,\dot{U}_j(t)}{1+tr^2U_j(t)} = \frac{2}{k}\,\frac{tr\,U_j^2(t) + tr\,K(t)}{1 + tr^2\,U_j(t)} .$$

Since U_j is symmetric we have

$$tr\,U_j \le (n-1)^{1/2}\,(tr\,U_j^2)^{1/2}.$$

Hence

$$\frac{(m\circ\hat{\alpha})'(t)}{i(m\circ\hat{\alpha})(t)} \ge \frac{2}{k}\,\frac{(n-1)^{-1}\,tr^2\,U_j(t) + tr\,K(t)}{1 + tr^2\,U_j(t)} . \qquad (9)$$

343

Moreover $(m\circ\hat{\alpha})(t) = c$ implies

$$\operatorname{tr} U_j(t) = -i\,\frac{1-c}{1+c}\,.$$

Therefore, if we take c very near to -1, the condition $(m\circ\alpha)(t) = c$ implies that $|\operatorname{tr} U_j(t)|$ is very large. Then the quotient at left becomes near to $\frac{2}{k}(n-1)^{-1}$. Hence, if c is taken close enough to -1, it follows that

$$\frac{(m\circ\hat{\alpha})'(t)}{i(m\circ\hat{\alpha})(t)} \geq \frac{1}{(n-1)k}$$

for all $t \in P(\hat{\alpha}, c)$. This obviously implies (8).

REFERENCES

[1] D.V. Anosov, Geodesic flows on closed Riemannian manifolds with negative curvature, Proc. Steklow Inst. Math. 90 (1967). Amer. Math. Soc.

[2] V.I. Arnold, Une classe charactéristique intervenant dans les conditions de quantification, Appendix I in Théorie des perturbations et méthodes assymptotiques, V.P. Maslov, Dunod, Paris (1972).

[3] P. Eberlein, Geodesic flows on negative curved manifolds I, Ann. of Math. 95 (1972), 492-410.

[4] P. Eberlein, When is a geodesic flow of Anosov type? I, J. Differential Geometry 8, (1973), 437-463.

[5] L. Green, A theorem of E. Hopf, Michigan Math. J. 5 (1958), 31-34.

[6] J. Hadamard, Les surfaces a courbures apposées et leur lignes geodesiques, J. Math. Pures et appl. (1898), 27-73.

[7] M. Hirsch, C. Pugh, M. Shub, Invariant manifolds, Springer Verlag Lecture Notes in Mathematics 583 (1976).

THE HAUSDORFF DIMENSION OF INVARIANT PROBABILITIES OF RATIONAL MAPS

by

Ricardo Mañé

INTRODUCTION.

The first objective of this note is to prove the following result:

THEOREM A. If f is an analytic endomorphism of the Riemann sphere $\overline{\mathbb{C}} = \mathbb{C} \cup \{\infty\}$ (i.e. a rational function) and μ is an ergodic invariant probability of f with entropy $h_\mu(f) > 0$, then

$$\int \log|f'|\,d\mu > 0$$

and

$$\lim_{r\to 0} \frac{\log \mu(B_r(x))}{\log r} = h_\mu(f)\left(\int \log|f'|\,d\mu\right)^{-1}$$

for μ-a.e. x, where $B_r(x)$ denotes the disk of radius r centered at x.

This result can be restated in terms of the Hausdorff dimension of μ that is defined (following Young [11]) as the infimum of the Hausdorff dimensions of the sets X with $\mu(X) = 1$. As Young observed in [11], when the limit in Theorem A exists and is constant for μ-a.e. x, then it is equal to $HD(\mu)$. Hence we can write

$$(1) \qquad h_\mu(f) = HD(\mu) \int \log|f'|\,d\mu$$

for every ergodic μ with positive entropy.

This formula is the version for rational maps of Young's formula for diffeomorphisms of surfaces [11]. When f satisfies Axiom A, it was proved by Manning in [7].

Originally published in: *Dynamical systems, Valparaiso 1986, 86–117, Lecture Notes in Math. 1331*,
Reprint by Springer International Publishing 2017

We shall prove Theorem A using a formula (Proposition I.1) for the calculation of the entropy of continuous maps, that unifies, in an easy but useful version, the Brin-Katok formula ([1]) with the lower bound that we proved in [8].

Our second objective is to apply (1) to the study of the Hausdorff dimension of the maximal measure. Recall that every rational map f of degree d has topological entropy $\log d$ ([3], [10]) and there exists a unique f-invariant probability μ_f (the maximal measure) such that $h_{\mu_f}(f) = \log d$ ([9], [2], [5]). Moreover μ_f has the following interesting property. Let $C^\circ(\bar{\mathbb{C}})$ be the space of real continuous functions on $\bar{\mathbb{C}}$ endowed with the C° topology. Define the linear map $\mathcal{L}_f : C^\circ(\bar{\mathbb{C}}) \circlearrowleft$ by

$$\mathcal{L}_f\, \phi(x) = \frac{1}{d} \sum_{y \in f^{-1}(x)} \phi(y)$$

where the preimages of x are repeated according with their multiplicities. On the other hand recall that the exceptional points of f are defined as those points p such that $f^{-1}(\{p\}) = \{p\}$. These points are at most 2 and always attracting periodic points. Then, for every $\phi \in C^\circ(\bar{\mathbb{C}})$ and every compact set K not containing exceptional points, the sequence $\mathcal{L}_f^n\, \phi / K$ converges to the μ_f-integral of ϕ.

Let $M(\bar{\mathbb{C}})$ be the space of probabilities on the Borel σ-algebra of $\bar{\mathbb{C}}$ endowed with the weak topology and let $\mathrm{End}(\bar{\mathbb{C}})$ be the space of rational maps of $\bar{\mathbb{C}}$ endowed with the C° topology. Let $d(\cdot,\cdot)$ be a metric on $\bar{\mathbb{C}}$ generated by a Riemannian inner product.

THEOREM B. The functions

$$\mathrm{End}(\bar{\mathbb{C}}) \ni f \to \mu_f \in M(\bar{\mathbb{C}})$$

$$\mathrm{End}(\bar{\mathbb{C}}) \ni f \to HD(\mu_f)$$

$$\mathrm{End}(\bar{\mathbb{C}}) \times \bar{\mathbb{C}} \ni (f,w) \to \int \log d(z,w)\, d\mu_f(z)$$

are continuous.

This result will be proved in Section II. When f varies in the space of Axiom A maps, Lopes proved in [6] that $HD(\mu_f)^{-1}$ is an harmonic function of f.

The continuity of μ_f will be a very easy consequence of the properties of μ_f and $\mathcal{L}_f$ recalled above. We include it here because it is an unavoidable step in the proof of the continuity of $HD(\mu_f)$. The continuity of the last function in Theorem B will be useful to prove the following Corollary that, on its turn, will be used to give a formula for the Hausdorff dimension of the maximal measure of polynomials.

COROLLARY 1. If $f(z) = a_d z^d + \ldots + a_o$, $a_d \neq 0$, is a polynomial then

$$\int \log|z - c| d\mu_f(z) = -\frac{1}{d-1} \log|a_d|$$

for all c in the Julia set of f.

To deduce this Corollary from Theorem B, first recall that the Julia set of f is the boundary of the basin of ∞, that will be denoted $B(\infty)$ and is defined as the set of points z such that $|f^n(z)| \to \infty$ when $n \to +\infty$. Then, if c belongs to the Julia set of f, it can be written as the limit of a sequence $\{c_n | n \geq 1\} \subset B(\infty)$. The third continuity property in Theorem B implies that

$$(2) \qquad \lim_{n\to+\infty} \int \log|z - c_n| d\mu_f(z) = \int \log|z - c| d\mu_f(z) .$$

Moreover, when $w \in B(\infty)$, the integral

$$\int \log|z - w| d\mu_f(z)$$

can be calculated through the following easy and well known method.

Take a continuous function $\phi \in C^\circ(\bar{\mathbb{C}})$ that coincides with $z \to \log|z - w|$ when z is in the Julia set. Take a point x_o in the Julia set. Then, if $x_i^{(n)}$, $i = 1, \ldots, d^n$ are the roots of $f^n(x) - x_o = 0$ (repeated according to their multiplicities), we have

$$\int \log|z - w| d\mu_f(z) = \int \phi \, d\mu_f = \lim_{n\to+\infty} \mathcal{L}_f^n \phi(x_o) =$$

$$= \lim_{n\to+\infty} \frac{1}{d^n} \sum_{i=1}^{d^n} \phi\left(x_i^{(n)}\right) = \lim_{n\to+\infty} \frac{1}{d^n} \sum_{i=1}^{d^n} \log|w - x_i^{(n)}| =$$

$$= \lim_{n\to+\infty} \frac{1}{d^n} \log \prod_{i=1}^{d^n} |w - x_i^{(n)}| .$$

But since f^n is a polynomial we can factorize $f^n - x_o$ as follows

$$f^{(n)}(z) - x_o = a_d(n) \prod_{i=1}^{d^n} |z - x_i^{(n)}|$$

where $a_d(n)$ is the coefficient of the term of degree d^n of f^n . Then

$$\int \log|z - w| d\mu_f(z) = \lim_{n\to+\infty} \frac{1}{d^n} \log \frac{1}{|a_d(n)|} |f^n(w) - x_o|$$

But $a_d(n) = a_d^{s_n}$ where $s_n = (d^n - 1) / (d - 1)$. Hence

$$(3) \qquad \int \log|z - w| d\mu_f(z) = - \frac{1}{d - 1} \log|a_d| + \lim_{n\to+\infty} \frac{1}{d^n} \log|f^n(w) - x_o|$$

When $w \in B(\infty)$, $|f^n(w) - x_o|$ can be replaced by $|f^n(w)|$ because $\lim_{n\to+\infty} |f^n(w)| = +\infty$. Then, if $w \in B(\infty)$

$$(4) \qquad \int \log|z - w| d\mu_f(z) = - \frac{1}{d - 1} \log|a_d| + \lim_{n\to+\infty} \frac{1}{d^n} \log|f^n(w)| .$$

Therefore, to prove Corollary I, it suffices to show that

$$(5) \qquad \lim_{w\to\partial B(\infty)} \lim_{n\to+\infty} \frac{1}{d^n} \log|f^n(w)| = 0$$

because if this property is true, we can apply (4) to $w = c_n$ and, taking limit when $n\to+\infty$, (5) implies

$$\lim_{n\to\infty} \int |\log|z - c_n| d\mu_f(z) = - \frac{1}{d - 1} \log|a_d|$$

and together with (2) proves the Corollary. To prove (5) define $V : B(\infty) \to \mathbb{R}$ by

$$V(w) = \lim_{n\to+\infty} \frac{1}{d^n} \log|f^n(w)|$$

and observe that $V(f^m(w)) = d^m V(w)$ for all $w \in B(\infty)$. Take $R > 0$ such that

$$\overline{f(B_R(\infty))} \subset B_R(\infty)$$

$$B_R(\infty) \subset B(\infty) .$$

Define $U = B_R(\infty) - f(B_R(\infty))$. Then, for all $z \in B(\infty) - B_R(\infty)$ there exists $m(z) > 0$ such that $f^{m(z)}(z) \in B_R(\infty)$ and $f^n(z) \notin B_R(\infty)$ for $0 < n < m(z)$. Hence

$$f^{m(z)}(z) \in U .$$

Moreover it is easy to check that $m(z) \to +\infty$ if $z \to \partial B(\infty)$. Then

$$\lim_{w \to \partial B(\infty)} \frac{1}{d^n} \log|f^n(w)| = \lim_{w \to \partial B(\infty)} V(w) =$$

$$= \lim_{w \to \partial B(\infty)} \frac{1}{d^{m(w)}} V(f^{m(w)}(w)) \leq$$

$$\leq \limsup_{w \to \partial B(\infty)} \frac{1}{d^{m(w)}} \sup_{z \in U} V(z) = 0$$

thus proving (5).

Now we shall apply Corollary 1 to prove a formula for $HD(\mu_f)$.

COROLLARY 2. If $f(z) = a_d z^d + \ldots + a_o$, $a_d \neq 0$, is a polynomial and c_j, $1 \leq j \leq d - 1$, are the zeroes of f', then:

$$HD(\mu_f) = \lim_{r \to 0} \frac{\log \mu_f(B_r(x))}{\log r} =$$

$$= \log d \Big(\log d + \sum_{c_j \in B(\infty)} \lim_{n \to \infty} \frac{1}{d^n} \log|f^n(c_j)|\Big)^{-1} .$$

By Theorem A, to prove this formula is equivalent to showing that

$$\int \log|f'| d\mu_f = \log d + \sum_{c_j \in B(\infty)} \lim_{n \to \infty} \frac{1}{d^n} \log|f^n(c_j)| . \tag{6}$$

But

$$f'(z) = da_d \prod_{j=1}^{d-1} (z - c_j) .$$

Hence

(7) $$\int \log|f'|d\mu_f = \log|a_d| + \log d + \sum_{j=1}^{d-1} \int \log|z - c_j|d\mu_f(z) .$$

When $c_j \in B(\infty)$ it follows from (4) that

(8) $$\int \log|z - c_j|d\mu_f(z) = -\frac{1}{d-1}\log|a_d| + \lim_{n\to+\infty} \frac{1}{d^n}\log|f^n(c_j)|$$

When $c_j \in \partial B(\infty)$, Corollary 1 says that

(9) $$\int \log|z - c_j|d\mu_f(z) = -\frac{1}{d-1}\log|a_d| .$$

Finally, when $c_j \notin B(\infty)$, the sequence $f^n(c_j)$ is bounded. Then (3) implies that (9) also holds for these points. Putting together (7), (8) and (9) we obtain

$$\int \log|f'|d\mu_f = \log|a_d| + \log d - \frac{N}{d-1}\log|a_d| +$$

$$+ \sum_{c_j\in B(\infty)} \left(-\frac{1}{d-1}\log|a_d| + \lim_{n\to+\infty}\frac{1}{d^n}\log|f^n(c_j)|\right)$$

where N is the number of c_j's not contained in $B(\infty)$. Therefore since the number of c_j's is $d - 1$, we obtain (6).

I. PROOF OF THEOREM A.

First let us prove a general formula for the entropy of continuous map that, as we said in the Introduction, is a unification of the Brin-Katok formula [1] and the inequality that we proved in [8].

Let K be a metric space, $f : K \circlearrowleft$ a continuous map and $\rho : K \to (0,1]$ a measurable function. Define:

$$S(x,n,\rho) = \{y | d(f^j(y),f^j(x)) \le \rho(f^j(x)) , \quad 0 \le j \le n\} .$$

Let μ be a probability on K . Define

$$h_\mu^+(x,f,\rho) = \limsup_{n\to+\infty} -\frac{1}{n}\log \mu(S(x,n,\rho))$$

$$h_\mu^-(x,f,\rho) = \liminf_{n\to+\infty} -\frac{1}{n}\log \mu(S(x,n,\rho))$$

When μ is f-invariant it is easy to check that

$$h_\mu^+(x,f,\rho) \leq h_\mu^+(f(x),f,\rho)$$

$$h_\mu^-(x,f,\rho) \leq h_\mu^-(f(x),f,\rho)$$

for all x . Since μ is f-invariant this implies

$$h_\mu^+(x,f,\rho) = h_\mu^+(f(x),f,\rho)$$

$$h_\mu^-(x,f,\rho) = h_\mu^-(f(x),f,\rho)$$

for μ-a.e. x . Hence, if μ is ergodic, this implies that $h_\mu^+(x,f,\rho)$ and $h_\mu^-(x,f,\rho)$ have constant values $h_\mu^+(f,\rho)$ and $h_\mu^-(f,\rho)$ for μ-a.e. x

Recall that the capacity $c(K)$ of a compact metric space K is defined as

$$c(K) = \limsup_{r\to 0} \frac{\log N(r)}{\log (1/r)}$$

where $N(r)$ denotes the minimum number of balls of radius r equired to cover K .

PROPOSITION I.1. Let K be a compact metric space with finite capacity, $f : K \circlearrowleft$ a continuous map, μ and ergodic f-invariant probability and $\rho : K \to (0,1]$ a measurable function such that $\log \rho$ is μ-integrable. Then

$$h_\mu(f) = \lim_{\varepsilon\to 0} h_\mu^+(f,\varepsilon\rho) = \lim_{\varepsilon\to 0} h_\mu^-(f,\varepsilon\rho) .$$

PROOF. By the Brin-Katok formula [1]:

$$h_\mu(f) = \lim_{\varepsilon\to 0} h_\mu^-(f,\varepsilon)$$

and by the inequality that we proved in [8]:

(1) $$h_\mu(f) \geq \lim_{\varepsilon\to 0} h^+_\mu(f,\varepsilon\rho)$$

Hence

$$\lim_{\varepsilon\to 0} h^+_\mu(f,\varepsilon\rho) \leq h_\mu(f) = \lim_{\varepsilon\to 0} h^-_\mu(f,\varepsilon) \leq \lim_{\varepsilon\to 0} h^-_\mu(f,\varepsilon\rho) \leq \lim_{\varepsilon\to 0} h^+_\mu(f,\varepsilon\rho)$$

where the second inequality follows immediately from the property $\varepsilon\rho \leq \varepsilon$. This concludes the proof. Observe that in [8] we proved (1) for the case when K is a manifold. However the proof obviously holds when K is a compact metric space with finite capacity because the fact that K is a manifold is used in [8] only to grant that there exists m such that $N(r)$ is bounded for every small r by $(1/r)^m$. This is precisely equivalent to $c(K) < \infty$.

Now let us prove Theorem A. Observe that $\log|f'|$ is measurable and bounded from above. Then, by Birkhoff's theorem:

$$\lim_{n\to+\infty} \frac{1}{n} \log|(f^n)'(x)| = \lim_{n\to+\infty} \frac{1}{n} \sum_{j=0}^{n-1} \log|f'(f^j(x))| =$$

$$= \int \log|f'|d\mu .$$

Clearly the first limit is the unique Lyapounov exponent of f and we shall denote it λ. By Ruelle's inequality:

$$h_\mu(f) \leq 2 \max \{0,\lambda\} .$$

But, by hypothesis, $h_\mu(f) > 0$. Then

$$0 < h_\mu(f) \leq 2\lambda = 2 \int \log|f'|d\mu$$

thus proving the first part of Theorem A. The proof of the second part requires several lemmas.

LEMMA I.2. Suppose that there exist two sequences $0 < \beta_n \leq \alpha_n < 1$ both converging to zero, $x \in \bar{\mathbb{C}}$ and numbers α,β,a and b such that:

(1) $$\limsup_{n\to+\infty} \frac{1}{n} \log \alpha_n \le -\alpha$$

(2) $$\liminf_{n\to+\infty} \frac{1}{n} \log \beta_n \ge -\beta$$

(3) $$\limsup_{n\to+\infty} \frac{\log \mu(B_{\alpha_n}(x))}{\log \alpha_n} \le a$$

(4) $$\liminf_{n\to+\infty} \frac{\log \mu(B_{\beta_n}(x))}{\log \beta_n} \ge b$$

Then:

(5) $$\limsup_{r\to 0} \frac{\log \mu(B_r(x))}{\log r} \le \left(\frac{\beta}{\alpha}\right) a$$

(6) $$\liminf_{r\to 0} \frac{\log \mu(B_r(x))}{\log r} \ge \left(\frac{\alpha}{\beta}\right) b$$

PROOF. Define $\tilde{\alpha} = \exp(-\alpha + \varepsilon)$, $\tilde{\beta} = \exp(-\beta - \varepsilon)$ for $\varepsilon > 0$. Then by (1) and (2) there exists n_o such that for $n \ge n_o$:

(7) $$\tilde{\alpha}^n \ge \alpha_n \ge \beta_n \ge \tilde{\beta}^n .$$

Then, for $n \ge n_o$:

$$-\log \mu(B_{\alpha_n}(x)) \ge -\log \mu(B_{\tilde{\alpha}^n}(x))$$

and

$$\frac{\log \mu(B_{\alpha_n}(x))}{\log \alpha_n} \ge \frac{\log \mu(B_{\tilde{\alpha}^n}(x))}{\log \tilde{\alpha}^n} \cdot \frac{\log \tilde{\alpha}^n}{\log \alpha_n} .$$

Taking lim sup in both sides and using (3) and (7):

$$a \geq \limsup_{n\to+\infty} \frac{\log \mu(B_{\alpha_n}(x))}{\log \alpha_n} \geq \limsup_{n\to+\infty} \frac{\log \mu(B_{\tilde{\alpha}^n}(x))}{\log \tilde{\alpha}^n} \cdot \frac{\log \tilde{\alpha}^n}{\log \alpha_n} \geq$$

$$\geq \limsup_{n\to+\infty} \frac{\log \mu(B_{\tilde{\alpha}^n}(x))}{\log \tilde{\alpha}^n} \frac{\log \tilde{\alpha}^n}{\log \tilde{\beta}^n} .$$

Hence

$$\limsup_{n\to+\infty} \frac{\log \mu(B_{\tilde{\alpha}^n}(x))}{\log \tilde{\alpha}^n} \leq a \frac{\log \tilde{\beta}}{\log \tilde{\alpha}} = a \frac{-\beta - \varepsilon}{-\alpha + \varepsilon} .$$

For each $r > 0$ take $n \in \mathbb{Z}^+$ such that $\tilde{\alpha}^n \leq r \leq \tilde{\alpha}^{n-1}$. Then

$$\frac{\log \mu(B_r(x))}{\log r} = \frac{-\log \mu(B_r(x))}{-\log r} \leq$$

$$\leq \frac{-\log \mu(B_{\tilde{\alpha}^n}(x))}{-\log \tilde{\alpha}^{n-1}} = \frac{\log \mu(B_{\tilde{\alpha}^n}(x))}{\log \tilde{\alpha}^n} \frac{\log \tilde{\alpha}^n}{\log \tilde{\alpha}^{n-1}} .$$

Therefore

$$\limsup_{r\to 0} \frac{\log \mu(B_r(x))}{\log r} \leq \frac{-\beta - \varepsilon}{-\alpha + \varepsilon} \cdot a$$

Since ε is arbitrary

$$\limsup_{r\to 0} \frac{\log \mu(B_r(x))}{\log r} \leq \frac{\beta}{\alpha} \cdot a$$

thus proving (5). Inequality (6) is proved using the same methods.

Let $C(f)$ be the set of critical points of f . Define $\rho : \bar{\mathbb{C}} \to (0,1]$ by

$$\rho(z) = \frac{1}{2} \min \{d(z,C(f)),1\} .$$

Let us see that $\log \rho$ is μ-integrable. Obviously it suffices to check it in a neighbourhood of each critical point since except for these points $\log \rho$ is continuous. Take $z_o \in C(f)$. There exist $r > 0$, $A > 0$ and $m \geq 1$ such that

$$|f'(z)| \leq A|z - z_o|^m$$

for all $z \in B_r(z_o)$. If r is small enough $\rho(z) = (1/2)|z - z_o|$ for all $z \in B_r(z_o)$. Hence

$$|f'(z)| \leq A\ \rho(z)^m$$

and then

$$\log|f'(z)| \leq \log A + m \log \rho(z)$$

for all $z \in B_r(z_o)$. Hence

$$0 < -\log \rho / B_r(z_o) < (\log A) / m - \frac{1}{m} \log|f'|/ B_r(z_o) \ .$$

Since we have proved that $\log|f'|$ is μ-integrable it follows that so is $-\log \rho / B_r(z_o)$ concluding the proof of the μ-integrability of $\log \rho$.

LEMMA I.3. For all $\delta > 0$ there exists $\varepsilon(\delta) > 0$ such that

$$e^{-\delta}|f'(a)|d(a,b) \leq d(f(a),f(b)) \leq e^{\delta}|f'(a)|d(a,b)$$

for all $a,b \notin C(f)$ such that $d(a,b) \leq \varepsilon(\delta)d(a,C(f))$ and, setting $\rho = \varepsilon(\delta)d(a,C(f))$, $f / B_\rho(a)$ is inyective.

The proof of this property is a corollary of Koebe's distortion theorem ([4]).

Set:

$$\lambda = \int \log|f'|d\mu$$

and fix $\delta > 0$ satisfying $0 < \delta < \lambda$ and $\varepsilon = \min \{1,\varepsilon(\delta)\}$. Now define

$$\alpha_n(x) = \min \{r|B_r(x) \supset S(x,n,\varepsilon\rho)\}$$

$$\beta_n(x) = \max \{r|B_r(x) \subset S(x,n,\varepsilon\rho)\}$$

$$r_n(x) = \max \{r|B_r(f^n(x)) \subset f^n(S(x,n,\varepsilon\rho))\} \ .$$

LEMMA I.4. If $x \notin \bigcup_{j=0}^{n} f^{-j}(C(f))$ and $y \in S(x,n,\varepsilon\rho)$; then:

$$e^{-\delta j}|(f^j)'(x)|d(x,y) \leq d(f^j(x),f^j(y)) \leq e^{\delta j}|(f^j)'(x)|d(x,y)$$

and $f^j / S(x,n,\varepsilon\rho)$ is injective for all $0 < j \leq n$.

PROOF. This Lemma follows easily by induction on j using I.3, the definition of $S(x,n,\varepsilon\rho)$ and the obvious inequality $0 < \varepsilon\rho < \varepsilon = \varepsilon(\delta)$.

LEMMA I.5. If $x \notin \bigcup_{j=0}^{\infty} f^{-j}(C(f))$,

$$\alpha_n(x) \leq e^{\delta n}|(f^n)'(x)|^{-1}$$

for all $n \geq 1$.

PROOF. If $x \notin \bigcup_{j=0}^{\infty} f^{-j}(C(f))$, $n \geq 1$ and $y \in S(x,n,\varepsilon\rho)$, then from Lemma I.4 it follows that

$$1 \geq \varepsilon\rho(f^n(x)) \geq d(f^n(x),f^n(y)) \geq e^{-\delta n}|(f^n)'(x)|d(y,x) .$$

Hence

$$d(x,y) \leq e^{\delta n}|(f^n)'(x)|^{-1}$$

for all $y \in S(x,n,\varepsilon\rho)$. Clearly this implies

$$\alpha_n(x) \leq e^{\delta n}|(f^n)'(x)|^{-1}$$

for all $n \geq 1$ and $x \notin \bigcup_{j=0}^{\infty} f^{-j}(C(f))$ thus proving the Lemma.

COROLLARY I.6. If $x \notin \bigcup_{j=0}^{\infty} f^{-j}(C(f))$:

$$\limsup_{n\to+\infty} \frac{1}{n} \log \alpha_n(x) \leq -\lambda + \delta .$$

PROOF. If $x \notin \bigcup_{j=0}^{\infty} f^{-j}(C(f))$ we have, by I.5, that

$$\alpha_n(x) \leq e^{\delta n} |(f^n)'(x)|^{-1}$$

for all $n \geq 1$. Then

$$\frac{1}{n} \log \alpha_n(x) \leq -\frac{1}{n} \log|(f^n)'(x)| + \delta \ .$$

Hence

$$\limsup_{n\to+\infty} \frac{1}{n} \log \alpha_n(x) \leq -\lambda + \delta \ .$$

COROLLARY I.7. For μ-a.e. x :

$$\limsup_{n\to+\infty} \frac{\log \mu(B_{\alpha_n(x)}(x))}{\log \alpha_n(x)} \leq \frac{h^+_\mu(f,\varepsilon\rho)}{\lambda - \delta}$$

PROOF. Since $S(x,n,\varepsilon\rho) \subset B_{\alpha_n(x)}(x))$:

$$-\log \mu(S(x,n,\varepsilon\rho)) \geq -\log \mu(B_{\alpha_n(x)}(x)) \ .$$

Then

$$\frac{-(1/n) \log \mu(S(x,n,\varepsilon\rho))}{-(1/n) \log \alpha_n(x)} \geq \frac{\log \mu(B_{\alpha_n(x)}(x))}{\log \alpha_n(x)} \ .$$

Moreover, for μ-a.e. x :

$$\limsup_{n\to+\infty} -\frac{1}{n} \log \mu(S(x,n,\varepsilon\rho)) = h^+_\mu(f,\varepsilon\rho) \ .$$

Applying this property, the previous inequality and I.6 (that holds in a full measure set) we obtain that for μ-a.e. x :

$$\limsup_{n\to+\infty} \frac{-(1/n)\log \mu(S(x,n,\varepsilon\rho))}{-(1/n)\log \alpha_n(x)} \le \frac{\limsup_{n\to+\infty}(-1/n)\log \mu(S(x,n,\varepsilon\rho))}{\liminf_{n\to+\infty}(-1/n)\log \alpha_n(x)} \le$$

$$\le \frac{h_\mu^+(f,\varepsilon\rho)}{-\limsup_{n\to+\infty}(1/n)\log \alpha_n(x)} \le \frac{h_\mu^+(f,\varepsilon\rho)}{\lambda-\delta}$$

LEMMA I.8. If $x \notin \bigcup_{j=0}^{\infty} f^{-j}(C(f))$ and $n \geq 1$:

$$\beta_n(x) \geq e^{-\delta n}|(f^n)'(x)|^{-1}\, r_n(x) .$$

PROOF. By Lemma I.4, given $x \notin \bigcup_{j=0}^{\infty} f^{-j}(C(f))$, $n \geq 1$ and $y \in S(x,n,\varepsilon\rho)$ we have

$$d(f^n(y),f^n(x)) \leq e^{\delta n}|(f^n)'(x)|d(y,x)$$

and moreover $f^n / S(x,n,\varepsilon\rho)$ is injective. Therefore the boundary of $f^n(S(x,n,\varepsilon\rho))$ is the image under f^n of the boundary of $S(x,n,\varepsilon\rho)$. Since, by definition of $\beta_n(x)$ there exists a point in this boundary at distance $\beta_n(x)$ of x , it follows by the inequality above that the boundary of $f^n(S(x,n,\varepsilon\rho))$ contains a point at distance $\leq \exp(\delta n)|(f^n)'(x)|\beta_n(x)$ from $f^n(x)$, thus implying

$$r_n(x) \leq e^{\delta n}|(f^n)'(x)|\beta_n(x) .$$

LEMMA I.9. If $x \notin \bigcup_{j=0}^{\infty} f^{-j}(C(f))$ and $n \geq 1$:

$$r_n(x) \geq \min\{\varepsilon\rho(f^n(x)) , e^{-\delta}|f'(f^{n-1}(x))|r_{n-1}(x)\} .$$

PROOF. Given $x \notin \bigcup_{j=0}^{\infty} f^{-j}(C(f))$ and $n \geq 1$ set

(1) $$R = e^{-\delta}|f'(f^{n-1}(x))|r_{n-1}(x) .$$

Then, Lemma I.4 implies

(2) $$f\left(B_{r_{n-1}(x)}(f^{n-1}(x))\right) \supset B_R(f^n(x))$$

because $r_{n-1}(x) \leq \varepsilon\rho(f^{n-1}(x))$. Suppose first that

(3) $$R \leq \varepsilon\rho(f^n(x)) .$$

Let us prove that this inequality implies:

(4) $$R \leq r_n(x)$$

Observe that by (2) there exists $S \subset B_{r_{n-1}(x)}(f^{n-1}(x))$ such that

(5) $$f(S) = B_R(f^n(x)) .$$

Moreover, since

$$S \subset B_{r_{n-1}(x)}(f^{n-1}(x)) \subset f^{n-1}(S(x,n-1,\varepsilon\rho))$$

there exists $S_o \subset S(x,n-1,\varepsilon\rho)$ such that

$$f^{n-1}(S_o) = S$$

that together with (5) implies

$$f^n(S_o) = B_R(f^n(x))) .$$

Then $S_o \subset S(x,n,\varepsilon\rho)$ because for $0 \leq j \leq n-1$

$$f^j(S_o) \subset f^j(S(x,n-1,\varepsilon\rho)) \subset B_{\varepsilon\rho(f^j(x))}(f^j(x))$$

and, for $j = n$,

$$f^n(S_o) = B_R(f^n(x)) \subset B_{\varepsilon\rho(f^n(x))}(f^n(x))$$

where the last inclusion follows immediatly from (3). Then $S_o \subset S(x,n,\varepsilon\rho)$ thus implying

$$B_R(f^n(x)) = f^n(S_o) \subset f^n(S(x,n,\varepsilon\rho)) .$$

This proves (4). By equation (1), (4) is equivalent to

(6) $$r_n(x) \geq e^{-\delta}|f'(f^{n-1}(x))|r_{n-1}(x) \; .$$

Now consider the case

(7) $$R > \varepsilon\rho(f^n(x)) \; .$$

Let us prove that in this case

(8) $$r_n(x) \geq \varepsilon\rho(f^n(x)) \; .$$

From (2) and (7) it follows that

$$f\big(B_{r_{n-1}(x)}(f^{n-1}(x))\big) \supset B_R(f^n(x)) \supset B_{\varepsilon\rho(f^n(x))}(f^n(x)) \; .$$

Then there exists $S \subset B_{r_{n-1}(x)}(f^{n-1}(x))$ such that

$$f(S) = B_{\varepsilon\rho(f^n(x))}(f^n(x)) \; .$$

Take $S_o \subset S(x,n-1,\varepsilon\rho)$ such that

$$f^{n-1}(S_o) = S \; .$$

Therefore

(9) $$f^n(S_o) = B_{\varepsilon\rho(f^n(x))}(f^n(x)) \; .$$

Arguing as before we prove that

$$S_o \subset S(x,n,\varepsilon\rho)$$

and then (9) implies

$$B_{\varepsilon\rho(f^n(x))}(f^n(x)) \subset f^n(S(x,n,\varepsilon\rho))$$

and this implies (8). This completes de proof of I.9 because (6) and (8) say that $r_n(x)$ is $\geq$ than the minimum between $\varepsilon\rho(f^n(x))$ and R .

COROLLARY I.10. For μ-a.e. x and every $n \geq 1$ there exists $0 \leq m \leq n$ such that

$$r_n(x) \geq \varepsilon\rho(f^m(x))\ e^{-\delta(n-m)}|(f^{n-m})'(f^m(x))|$$

and $m \to +\infty$ when $n \to +\infty$.

PROOF. By I.9, if $x \notin \bigcup_{j=0}^{\infty} f^{-j}(C(f))$ and $n \geq 1$:

$$r_n(x) \geq \min\ \{\varepsilon\rho(f^n(x)), e^{-\delta}|f'(f^{n-1}(x))|r_{n-1}(x)\}\ .$$

Then either $r_n(x) \geq \varepsilon\rho(f^n(x))$, and in this case I.10 is proved taking $m = n$, or

(1) $$r_n(x) \geq e^{-\delta}|f'(f^{n-1}(x))|r_{n-1}(x)\ .$$

Now apply I.9 to $r_{n-1}(x)$. Either $r_{n-1}(x) \geq \varepsilon\rho(f^{n-1}(x))$, and then

$$r_n(x) \geq e^{-\delta}|f'(f^{n-1}(x))|\varepsilon\rho(f^{n-1}(x))$$

that proves I.10 with $m = n - 1$, or

(2) $$r_{n-1}(x) \geq e^{-\delta}|f'(f^{n-2}(x))|r_{n-2}(x)\ .$$

Then, (1) and (2) imply

$$r_n(x) \geq e^{-2\delta}|(f^2)'(f^{n-2}(x))|r_{n-2}(x)\ .$$

Continuing with this method either we find $0 \leq m \leq n$ with the desired property or for all $0 \leq m \leq n$ we have

$$r_n(x) \geq e^{-\delta(n-m)}|(f^{n-m})'(f^m(x))|r_m(x)\ .$$

When $m = 0$

(3) $$r_n(x) \geq e^{-\delta n}|(f^n)'(x)|r_o(x)\ .$$

But for μ-a.e. x we have

$$\lim_{n\to+\infty} \frac{1}{n} \log|(f^n)'(x)| = \lambda > \delta\ .$$

Hence $\lim_{n\to+\infty} e^{-\delta n}|(f^n)'(x)| = +\infty$ for μ-a.e. x. This and (3) imply $\lim_{n\to+\infty} r_n(x) =$ $= +\infty$ contradicting the obvious inequality $r_n(x) \leq \varepsilon\rho(f^n(x)) \leq \varepsilon$. This proves the existence of m satisfying I.10. Let us see that $m \to +\infty$ when $n \to +\infty$. The argument is similar to that used to prove the existence of m. If the property is false there exist $n_1 < n_2 < \dots$ and $N \geq 0$ such that for each n_j there exists $0 \leq m_j \leq N$ satisfying

$$r_n(x) \geq e^{-\delta(n_j-m_j)} |(f^{n_j-m_j})'(f^{m_j}(x))| r_{m_j}(x) .$$

Then

$$r_n(x) \geq e^{-\delta n_j} |(f^{n_j})'(x)| C \tag{4}$$

where

$$C = \min \{ |(f^m)'(x)|^{-1} r_m(x) | 0 \leq m \leq N \} .$$

Arguing as above, (4) implies $\lim_{n\to+\infty} r_n(x) = +\infty$ that again contradicts $r_n(x) < \varepsilon$.

COROLLARY I.11. For μ-a.e. x and all $n \geq 1$ there exists $0 \leq m \leq n$ such that:

$$\beta_n(x) \geq e^{-2\delta n} e^{\delta m} |(f^m)'(x)|^{-1} \varepsilon\rho(f^m(x))$$

and $m \to +\infty$ when $n \to +\infty$.

PROOF. This is an immediate consequence of I.8 and I.10.

COROLLARY I.12. For μ-a.e. x.

$$\liminf_{n\to+\infty} \frac{1}{n} \log \beta_n(x) \geq -(\lambda + 2\delta) .$$

PROOF. By I.11, for μ-a.e. x and every $n \geq 1$ there exists $0 \leq m \leq n$ such that

$$\frac{1}{n} \log \beta_n(x) \geq \frac{1}{n} \log(e^{-2\delta n} e^{\delta m} |(f^m)'(x)|^{-1} \varepsilon\rho(f^m(x))) =$$

$$= -2\delta + \delta \frac{m}{n} - \frac{1}{n} \log|(f^m)'(x)| + \frac{1}{n} \log \varepsilon + \frac{1}{n} \log \rho(f^m(x)) \geq$$

$$\geq -2\delta - \frac{m}{n}\cdot\frac{1}{m} \log|(f^m)'(x)| + \frac{1}{n}\log\varepsilon + \frac{1}{m}\log\rho(f^m(x)) .$$

Since $(1/m)|\log(f^m)'(x)|$ is positive when $m \to +\infty$ for μ-a.e. x and $m \to +\infty$ if $n \to +\infty$, it follows that

$$(1) \qquad \frac{1}{n}\log\beta_n(x) \geq -2\delta - \frac{1}{m}\log|(f^m)'(x)| + \frac{1}{n}\log\varepsilon + {}$$

$$+ \frac{1}{m}\log\rho(f^m(x)) .$$

But observe that for μ-a.e. x.

$$(2) \qquad \lim_{m\to+\infty} \frac{1}{m}\log\rho(f^m(x)) = 0$$

because setting $F = \log\rho$ we have

$$\frac{1}{m}\log\rho(f^m(x)) = \frac{1}{m}\sum_{j=0}^{m-1} (F\circ f - F)(f^j(x)) + \frac{1}{m}F(x) .$$

Then, by Birkhoff's Theorem

$$\lim_{m\to+\infty}\frac{1}{m}\log\rho(f^m(x)) = \int(F\circ f - F)d\mu =$$

$$= \int F\circ f d\mu - \int F d\mu = 0$$

because μ is ergodic and f-invariant. Then, recalling that by I.10 $m \to +\infty$ when $n \to +\infty$, (1) and (2) imply:

$$\liminf_{n\to+\infty}\frac{1}{n}\log\beta_n(x) \geq -2\delta - \lambda .$$

COROLLARY I.13. For μ-a.e. x

$$\liminf_{n\to+\infty}\frac{\log\mu\left(B_{\beta_n}(x)\right)}{\log\beta_n(x)} \geq \frac{h_\mu^-(f,\varepsilon\rho)}{\lambda + 2\delta}$$

PROOF. Clearly

$$\frac{(-1/n)\log\mu(S(x,n,\varepsilon\rho))}{(-1/n)\log\beta_n(x)} \leq \frac{\log\mu\left(B_{\beta_n(x)}(x)\right)}{\log\beta_n(x)}$$

because

$$\mu(S(x,n,\varepsilon\rho)) \geq \mu\left(B_{\beta_n(x)}(x)\right) .$$

Then

$$\liminf_{n\to+\infty} \frac{\log \mu\left(B_{\beta_n(x)}(x)\right)}{\log \beta_n(x)} \geq \frac{\liminf_{n\to+\infty} (-1/n) \log \mu(S(x,n,\varepsilon\rho))}{\limsup_{n\to+\infty} (-1/n) \log \beta_n(x)} =$$

$$= \frac{\liminf_{n\to+\infty} (-1/n) \log \mu(S(x,n,\varepsilon\rho))}{-\liminf_{n\to+\infty} (1/n) \log \beta_n(x)} \geq \frac{h_\mu^-(f,\varepsilon\rho)}{\lambda + 2\delta} .$$

Now, for μ-a.e. x, we shall apply I.2 to

$$\alpha_n = \alpha_n(x)$$

$$\beta_n = \beta_n(x)$$

$$a = h_\mu^+(f,\varepsilon\rho) / (\lambda - \delta)$$

$$b = h_\mu^-(f,\varepsilon\rho) / (\lambda + 2\delta)$$

$$\alpha = \lambda - \delta$$

$$\beta = \lambda + 2\delta .$$

Hypothesis (1), (2), (3) and (4) of Lemma I.2 follow from I.6, I.12, I.7 and I.13 respectively. Then, I.2 implies:

(*) $$\limsup_{r\to 0} \frac{\log \mu(B_r(x))}{\log r} \leq \frac{\beta}{\alpha}\cdot a = \frac{\lambda + 2\delta}{\lambda - \delta}\cdot\frac{h_\mu^+(f,\varepsilon\rho)}{\lambda - \delta}$$

(**) $$\liminf_{r\to 0} \frac{\log \mu(B_r(x))}{\log r} \geq \frac{\alpha}{\beta}\cdot b = \frac{\lambda - \delta}{\lambda + 2\delta}\cdot\frac{h_\mu^-(f,\varepsilon\rho)}{\lambda + 2\delta}$$

But ε and δ are arbitrarily small and by I.1

$$\lim_{\varepsilon\to 0} h_\mu^+(f,\varepsilon\rho) = \lim_{\varepsilon\to 0} h_\mu^-(f,\varepsilon\rho) = h_\mu(f) .$$

Then (*) and (**) imply

$$\lim_{r\to 0} \frac{\log \mu(B_r(x))}{\log r} = \frac{1}{\lambda} h_\mu(f) = h_\mu(f)\left(\int \log|f'|\,d\mu\right)^{-1} .$$

II. PROOF OF THEOREM B.

Let us first prove the continuity of $\mu_f \in M(\bar{\mathbb{C}})$, that is, as we said in the Introduction, the easiest part of Theorem B.

Given $f \in \mathrm{End}(\bar{\mathbb{C}})$ denote by $\mathrm{Exc}(f)$ the set of its exceptional points. Let $C^\circ(\bar{\mathbb{C}})$ and $\mathcal{L}_f : C^\circ(\bar{\mathbb{C}}) \circlearrowleft$ be defined as in the Introduction. Denote $\mathcal{L}_f^* : M(\bar{\mathbb{C}}) \circlearrowleft$ the adjoint map of $\mathcal{L}_f$. We claim that if $\mu \in M(\bar{\mathbb{C}})$ satisfies $\mathcal{L}_f^*\mu = \mu$ and $\mathrm{supp}(\mu) \cap \mathrm{Exc}(f) = \Phi$ then $\mu = \mu_f$. To prove this property take $\phi \in C^\circ(\bar{\mathbb{C}})$ and $\mu \in M(\bar{\mathbb{C}})$ with support contained in $\mathrm{Exc}(f)^c$ and observe that:

$$\int \phi \, d((\mathcal{L}_f^*)^n \mu) = \int (\mathcal{L}_f^n \phi) d\mu .$$

Since $\mathcal{L}_f^n \phi$ converges to the μ_f-integral of ϕ uniformely on compact sets of $\mathrm{Exc}(f)^c$ and the support of μ is a compact set of $\mathrm{Exc}(f)^c$, we conclude that

$$\lim_{n\to+\infty} \int \phi \, d((\mathcal{L}_f^*)^n \mu) = \int \left(\int \phi \, d\mu_f \right) d\mu = \int \phi \, d\mu_f .$$

This means that

$$\lim_{n\to+\infty} (\mathcal{L}_f^*)^n \mu = \mu_f$$

for all $\mu \in M(\bar{\mathbb{C}})$ whose support is contained in $\mathrm{Exc}(f)^c$. This obviously implies that $\mathcal{L}_f^* \mu_f = \mu_f$ and that $\mathcal{L}_f^* \mu = \mu$ implies $\mu = \mu_f$ if the support of μ is contained in $\mathrm{Exc}(f)^c$.

To prove the continuity of μ_f as a function of f we have to show that if f_n , $n \geq 1$ is a sequence of rational maps converging to a rational map f and the sequence μ_{f_n} , $n \geq 1$, converges to $\mu \in M(\bar{\mathbb{C}})$, then $\mu = \mu_f$. By the claim it suffices to prove that $\mathcal{L}_f^* \mu = \mu$ and that the support of μ is contained in $\mathrm{Exc}(f)^c$. The first property follows trivially from:

$$\mathcal{L}_f^* \mu = \lim_{n\to+\infty} \mathcal{L}_{f_n}^* \mu_{f_n} = \lim_{n\to+\infty} \mu_{f_n} = \mu .$$

The second property follows from the fact the points in $\mathrm{Exc}(f)$ are attracting periodic points. Then there exists an open neighbourhood U of $\mathrm{Exc}(f)$ and a neighbourhood $\mathcal{U}$ of f such that, denoting $J(\cdot)$ the Julia set,

$$J(g) \cap U = \Phi$$

for all $g \in \mathcal{U}$. In particular the support $\mathrm{supp}(\mu_g)$ of μ_g, that coincides with $J(g)$, satisfies $\mathrm{supp}(\mu_g) \cap U = \Phi$ for all $g \in \mathcal{U}$. Hence

$$\mathrm{supp}\left(\mu_{f_n}\right) \cap U = \Phi$$

when n is large. Therefore $\mathrm{supp}(\mu) \cap U = \Phi$ as we wished to prove.

The proof of parts (b) and (c) of Theorem B rely on the following Lemma

LEMMA II.1. If K is a compact subset of $\mathrm{End}(\bar{\mathbb{C}})$, there exist $\rho > 0$ and $\gamma > 0$ such that

$$\mu_f(B_r(x)) \leq r^\gamma$$

for all $0 < r < \rho$, $f \in K$ and $x \in \bar{\mathbb{C}}$.

Before proving this Lemma, we shall show how it implies part (b) and (c) of Theorem B. The first step is to deduce the following Corollary

COROLLARY II.2. If K is a compact subset of $\mathrm{End}(\bar{\mathbb{C}})$, for all $\varepsilon > 0$ there exists $r > 0$ such that

$$\int_{B_r(a)} |\log \|f'\|| \, d\mu_f < \varepsilon$$

for all $f \in K$ and $a \in \bar{\mathbb{C}}$.

PROOF. The compacity of K implies that there exist $d > 0$ and $A > 0$ s.t. every $f \in K$ has degree $\leq d$ and satisfies

$$(1) \qquad |\log \|f'(z)\|| \leq A + A \sum_{i=1}^{k} |\log d(z,c_i)|$$

where $c_1, \ldots, c_k$ are the critical points of f. Take $r > 0$ satisfying $4r < \rho$, where ρ is given by II.1 and

(2) $$Ar^{\gamma} < \varepsilon / (2d + 1)$$

(3) $$\sum_{n=-2}^{\infty} (-\log r\, 2^{-n})(r\, 2^{-n})^{\gamma} \leq \frac{\varepsilon}{A(2d + 1)}$$

(4) $$r^{\gamma} \max \{\log \frac{1}{r} , \operatorname{diam} \bar{\mathbb{C}}\} \leq \varepsilon / A(2d + 1)$$

where γ is the constant given by II.1. Given $1 \leq i \leq k$ define

$$S_n(c_i) = \{z | r\, 2^{-(n+1)} \leq d(z,c_i) < r\, 2^{-n}\} .$$

Then

$$\int_{B_{3r}(c_i)} |\log d(z,c_i)| d\mu_f(z) \leq \sum_{n=-2}^{\infty} \int_{S_n(c_i)} |\log d(z,c_i)| d\mu_f(z) \leq$$

$$\leq \sum_{n=-2}^{\infty} (-\log r\, 2^{-n}) \mu_f(S_n(c_i)) \leq \sum_{n=-2}^{\infty} (-\log r\, 2^{-n}) \mu_f(B_{r2^{-n}}(c_i)) \leq$$

$$\leq \sum_{n=-2}^{\infty} (-\log r\, 2^{-n})(r\, 2^{-n})^{\gamma} .$$

Applying (3) we obtain

(6) $$\int_{B_{3r}(c_i)} |\log d(z,c_i)| d\mu_f(z) \leq \frac{\varepsilon}{A(2d + 1)} .$$

Applying (1) and II.1 we get

$$\int_{B_r(w)} |\log \|f'\| | d\mu_f \leq \int_{B_r(w)} (A + A \sum_{i=1}^{k} |\log d(z,c_i)|) d\mu_f(z) \leq$$

$$\leq Ar^{\gamma} + A \sum_{i=1}^{k} \int_{B_r(w)} |\log d(z,c_i)| d\mu_f(z)$$

When $d(w,c_i) \leq 2r$ we have

$$\int_{B_r(w)} |\log d(z,c_i)|\,d\mu_f(z) \le \int_{B_{3r}(c_i)} |\log d(z,c_i)|\,d\mu_f(z) \le$$

$$\le \frac{\varepsilon}{A(2d+1)}$$

When $d(w,c_i) > 2r$, then $d(z,c_i) > r$ for all $z \in B_r(w)$ and

$$|\log d(z,c_i)| \le \max\{\log\frac{1}{r}, \operatorname{diam}\bar{\mathbb{C}}\}.$$

Then, by II.1 and (4):

$$\int_{B_r(w)} |\log d(z,c_i)|\,d\mu_f(z) \le r^\gamma \max\{\log\frac{1}{r}, \operatorname{diam}\bar{\mathbb{C}}\} \le \frac{\varepsilon}{A(2d+1)}.$$

Hence

$$\int_{B_r(w)} |\log|f'||\,d\mu_f \le Ar^\gamma + A\cdot\frac{\varepsilon}{A(2d+1)}k \le$$

$$\le Ar^\gamma + \frac{\varepsilon}{2d+1}\cdot 2d \le \frac{\varepsilon}{2d+1} + \frac{\varepsilon}{2d+1}\cdot 2d = \varepsilon.$$

To prove part (b) of Theorem B first observe that by formula (1) in the Introduction:

$$HD(\mu_f) = \log d\Big(\int \log\|f'\|\,d\mu_f\Big)^{-1}.$$

Therefore, to prove part (b) of Theorem B is equivalent to show the continuity of the function

$$\text{(1)} \qquad \operatorname{End}(\bar{\mathbb{C}}) \ni f \to \int \log|f'|\,d\mu_f.$$

From the continuity of μ_f as a function of f follows that if $f = \lim_{n\to+\infty} f_n$ and $U \subset \bar{\mathbb{C}}$ is an open set with $\mu_f(\partial U) = 0$, then

$$\text{(2)} \qquad \lim_{n\to+\infty} \int_U \phi_n\,d\mu_{f_n} = \int_U \phi\,d\mu_f$$

for any sequence of continuous functions $\phi_n : \bar{U} \to \mathbb{R}$ converging uniformely to $\phi : \bar{U} \to \bar{\mathbb{C}}$. Now choose an arbitrarily small value of $r > 0$ such that if $c_1, \ldots, c_k$ are the critical points of f then $\mu_f(\partial B_r(c_i)) = 0$ for all i. Then, setting

$$U = \Big(\bigcup_i B_r(c_i)\Big)^c ,$$

follows from (2) that

(3) $$\lim_{n\to+\infty} \int_U \log|f_n'|\,d\mu_{f_n} = \int_U \log|f'|\,d\mu_f .$$

But by II.2, given any $\varepsilon > 0$ we can choose $r > 0$ so small that

(4) $$\int_{U^c} |\log|f_n'||\,d\mu_{f_n} < \varepsilon$$

for all n, and

(5) $$\int_{U^c} |\log|f'||\,d\mu_f < \varepsilon .$$

From (3), (4) and (5) follows that

$$\limsup_{n\to+\infty} \Big| \int \log|f_n'|\,d\mu_{f_n} - \int \log|f'|\,d\mu_f \Big| < 2\varepsilon .$$

Since ε is arbitrary, this proves the continuity of (1).

Part (c) of Theorem B is proved following the same method of combining the continuity of μ_f with Lemma II.1.

Now let us prove II.1. This requires the following two Lemmas. In their statements, K will be, as in II.1, a compact set in $\mathrm{End}(\bar{\mathbb{C}})$. The set of critical points of a rational map f will be denoted $C(f)$. Using the compacity of K it is easy to see that there exist $A > 0$ and $m \in \mathbb{Z}^+$ such that

(1) $$\|f'(z)\| \le A$$

(2) $$\big|\|f'(z)\| - \|f'(y)\|\big| \le A\,d(z,y)$$

(3) $$\|f'(z)\| \ge A^{-1} d(z,C(f))^m$$

for all $z,y \in \bar{\mathbb{C}}$ and $f \in K$. Obviously (2) implies

(4) $\|f'(z)\| \leq Ad(z,C(f))$

for all $z \in \bar{\mathbb{C}}$ and $f \in K$.

LEMMA II.3. If $f \in K$, $w \in C(f)$, $z \in \bar{\mathbb{C}}$, $n > 0$ and $r_o > 0$ satisfy

(5) $d(z,w) < r_o$

(6) $d(f^n(z),w) < r_o$

(7) $4A^n r_o < 1/4$

then $B_{3r_o}(z)$ is contained in the basin of an attracting periodic orbit.

PROOF. If $y \in B_{3r_o}(z)$, then $y \in B_{4r_o}(w)$ and, by (4) ,

$$\|f'(y)\| \leq 4r_o A .$$

Hence $y \in B_{3r_o}(z)$ implies (using (1) and (7))

(8) $\|(f^n)'(y)\| \leq \|(f^{n-1})'(f(y))\|\|f'(y)\| \leq A^{n-1}\cdot 4r_o A < 1/4 .$

Then

$$\operatorname{diam} f^n\left(B_{3r_o}(z)\right) \leq 3r_o\cdot\sup\{\|(f^n)'(y)\| \mid y \in B_{3r_o}(z)\} \leq (3/4)r_o .$$

This means that

(9) $f^n\left(B_{3r_o}(z)\right) \subset B_{3r_o}(z)$,

because if $y \in B_{3r_o}(z)$ it follows from (8), (5) and (6) that:

$$d(f^n(y),z) \leq d(f^n(y),f^n(z)) + d(f^n(z),w) + d(w,z) \leq$$

$$\leq \frac{1}{4} d(y,z) + r_o + r_o \leq \frac{3}{4} r_o + 2r_o < 3r_o .$$

From (8) and (9) follows that there exists $p \in B_{3r_o}(z)$ that is an attracting fixed point of f^n and whose basin contains $B_{3r_o}(z)$. Therefore, $B_{3r_o}(z)$ is contained in the basin of the attracting periodic orbit $\{p, \ldots, f^n(p) = p\}$.

LEMMA II.4. There exist $r_1 > 0$ and $B > 0$ such that if $f \in K$, $z \in \bar{\mathbb{C}}$ and $y \in \bar{\mathbb{C}}$ satisfy

$$d(y,z) < r_1$$

$$d(y,z) < Bd(z,C(f))^m$$

then $f(z) \neq f(y)$.

PROOF. Take $r_1 > 0$ so small that there exists $A_1 > 0$ such that for all $f \in K$ and $z \in \bar{\mathbb{C}}$, if we move z to the origin of $\mathbb{C}$, then

$$f\big(B_{r_1}(0)\big) \subset \mathbb{C}$$

$$|f'(w) - f'(y)| \leq A_1|w - y|$$

$$|f'(y)| \geq A_1^{-1} d(y,C(f))^m \tag{10}$$

for all y and w in $B_{r_1}(0)$. Then we can write $f / B_{r_1}(0) \to \mathbb{C}$ as

$$f(x) = f(0) + f'(0)x + p(x)$$

and

$$|p'(x)| \leq |f'(x) - f'(0)| \leq A_1|x| .$$

Since $p(0) = 0$, this implies

$$|p(y)| = |p(y) - p(0)| \leq A_1|y|^2 .$$

Hence

$$|f(y) - f(0)| \geq |f'(0)||y| - |p(y)| \geq |f'(0)||y| - A_1|y|^2 =$$

$$= (|f'(0)| - A_1|y|)|y| .$$

Therefore, $f(y) \neq f(0)$ if $|f'(0)| > A_1|y|$. But by (10)

$$|f'(0)| \geq A_1^{-1} d(0,C(f))^m .$$

Then, when

$$(11) \qquad A_1|y| < A_1^{-1} d(0,C(f))^m ,$$

the condition $|f'(0)| > A_1|y|$ is satisfied and then $f(y) \neq f(0)$. But (11) can be written as

$$(12) \qquad |y| < Bd(0,C(f))^m$$

with $B = A_1^{-2}$. Therefore (12) implies $f(y) \neq f(0)$ and proves the Lemma.

Now we shall prove II.1. Take $\rho > 0$ such that:

$$\rho A < \min \{r_1, 1\}$$

$$4A(1 + B^{-1/m})(A\rho)^{1/m} < 1/4 .$$

For $0 < r < \rho$ define $n = n(r)$ as the maximum integer such that

$$(13) \qquad rA^n < \min \{r_1, 1\}$$

$$(14) \qquad 4A^n(1 + B^{-1/m})(A^n r)^{\frac{1}{m}} < 1/4 .$$

There exists $\alpha > 0$ such that for all $0 < r < \rho$:

$$n(r) > \alpha \log\left(\frac{1}{r}\right) .$$

Given a map $g : \bar{\mathbb{C}} \circlearrowleft$ and a set $S \subset \bar{\mathbb{C}}$, denote $\nu(g / S)$ the maximum number of preimages in S of the points in $g(S)$. Let d be the maximum degree of the maps in K.

We claim that for all $z \in \bar{\mathbb{C}}$ and $0 < r < \rho$, either $B_r(z)$ is contained in the basin of a sink or

$$(15) \qquad \nu(f^{n(r)} / B_r(z)) < 2^{2d-1} .$$

Observe that

$$\nu(f^{n(r)} / B_r(z)) \leq \prod_{i=0}^{n(r)-1} \nu(f / f^i(B_r(z))) .$$

Hence, if (15) doesn't hold there exist $2d - 1$ values of i between 0 and $n(r) - 1$, say $0 \leq n_i \leq n(r) - 1$, $i = 1, \ldots, 2d - 1$, such that

$$\nu(f / f^{n_i}(B_r(z))) \geq 2 . \tag{16}$$

But (1) implies

$$\operatorname{diam} f^{n_i}(B_r(z)) \leq A^{n_i} r \leq A^{n(r)} r$$

and, by (13)

$$\operatorname{diam} f^{n_i}(B_r(z))) < r_1 .$$

Suppose that:

$$d\left(f^{n_i}(z), C(f)\right) > (1 + B^{-1/m})(A^{n(r)} r)^{1/m}$$

Then any couple of points y_1, y_2 in $f^{n_i}(B_r(z))$ satisfy

$$d(y_1, y_2) < r_1 \tag{17}$$

and (recalling that, by (13), $A^{n(r)} r < 1$)

$$\begin{aligned} d(y_1,y_2) &\leq \operatorname{diam} f^{n_i}(B_r(z)) \leq A^{n(r)} r = B\left|(1 + B^{-1/m})(A^{n(r)} r)^{1/m} - \right. \\ &\left. - (A^{n(r)} r)^{1/m}\right|^m \leq B\left|d(f^{n_i}(z), C(f)) - (A^{n(r)} r)^{1/m}\right|^m \leq \\ &\leq B\left|d(f^{n_i}(z), C(f)) - d(y_2, f^{n_i}(z))\right|^m \leq B\, d(y_2, C(f))^m . \end{aligned} \tag{18}$$

Then, by properties (17) and (18) we can apply II.4 to conclude that $f(y_1) \neq f(y_2)$ for all y_1, y_2 in $f^{n_i}(B_r(z))$. This means that

$$\nu(f / f^{n_i}(B_r(z))) = 1$$

contradicting (16). Hence, we have proved that (16) implies

$$d\left(f^{n_i}(z), C(f)\right) \leq (1 + B^{-1/m})(A^{n(r)}\ r)^{1/m}$$

for all $1 \leq i \leq 2d - 1$. Then there exist points $p_i \in C(f)$, $1 \leq i \leq 2d - 1$ such that

$$d\left(f^{n_i}(z), p_i\right) \leq (1 + B^{-1/m})(A^{n(r)}\ r)^{1/m}$$

Since $C(f)$ contains at most $2d - 2$ elements, there exist $1 \leq i < j \leq 2d - 1$ such that $p_i = p_j = p$. Hence

$$d\left(f^{n_i}(z), p\right) \leq (1 + B^{-1/m})(A^{n(r)}\ r)^{1/m}$$

$$d\left(f^{n_j}(z), p\right) \leq (1 + B^{-1/m})(A^{n(r)}\ r)^{1/m}$$

Let us apply II.3 to $f^{n_i}(z)$, $p \in C(f)$, $n = n_j - n_i$ and

$$r_o = (1 + B^{-1/m})(A^{n(r)}\ r)^{1/m}$$

From (14) follows that

$$4\ A^{n_j - n_i}\ r_o \leq 4\ A^{n(r)}(1 + B^{-1/m})(A^{n(r)}\ r)^{1/m} < 1/4\ .$$

Hence II.3 can be applied and we conclude that $B_{3r_o}\left(f^{n_i}(z)\right)$ is contained in the basin of an attracting periodic orbit. But

$$\operatorname{diam} f^{n_i}(B_r(z)) \leq A^{n_i}\ r \leq A^{n(r)}\ r \leq 3(1 + B^{-1/m})A^{n(r)}\ , \quad r = 3r_o$$

Hence $f^{n_i}(B_r(z)) \subset B_{3r_o}\left(f^{n_i}(z)\right)$ and then $B_r(z)$ is contained in the basin of an attracting periodic orbit. This completes the proof of the claim. To complete the proof of II.1 recall that for every Borel set S such that f^n / S is injective and d_1 is the degree of f , then

$$\mu_f(S) = \frac{1}{d_1^n}\ \mu_f(f^n(S))\ .$$

From this property follows easily that for every Borel set B :

$$\mu_f(B) \leq \frac{\nu(f^n / B)}{d_1^n}\mu_f(f^n(B)) .$$

Then, if $0 < r < \rho$ and $z \in \bar{\mathbb{C}}$ and d_o is the minimum degree of the maps in K :

$$\mu_f(B_r(z)) \leq \frac{\nu(f^n / B_r(z))}{d_1^n} \mu_f(f^n(B_r(z)) \leq \frac{\nu(f^n / B_r(z))}{d_1^n} \leq \frac{1}{d_o^n} \nu(f^n / B_r(z))$$

Taking $n = n(r)$ and applying the claim:

$$\mu_f(B_r(z)) \leq \frac{1}{d_o^{n(r)}} 2^{2d-1}$$

Since $n(r) \geq \alpha \log(1 / r)$:

$$\mu_f(B_r(z)) \leq 2^{2d-1} r^{\alpha \log d_o} .$$

Diminish ρ so that:

$$2^{2d-1} \rho^{(1/2)\alpha \log d_o} < 1 .$$

Then, if $0 < r < \rho$:

$$\mu_f(B_r(z)) \leq 2^{2d-1} r^{\alpha \log d_o} < r^{(1/2)\alpha \log d_o} .$$

Hence, setting $\gamma = (1/2)\alpha \log d_o$, the Lemma is proved.

REFERENCES.

[1] M. Brin, A. Katok., On local entropy Geometric Dynamics Lecture Notes in Mathematics, 1007 (1983) 30-38.

[2] A. Freire, A. Lopes, R. Mañé., An invariant measure for rational maps., Bol. Soc. Bras. Mat., 14 (1983) 45-62.

[3] M. Gromov., On the entropy of holomorphic maps IHES Publ. Math., 53 (1981), 53-78.

[4] E. Hille., Analytic Function Theory Ginn and Co. 1959.

[5] V. Ljubich., Entropy properties for rational endomorphisms of the Riemann sphere Ergod. Th. and Dyn. Syst. 3 (1983), 351-385.

[6] A. Lopes., Orthogonality and the Hausdorff dimension of the maximal measure Proc. Amer. Math. Soc. 98 (1986), 51-55.

[7] A. Manning., The dimension of the maximal measure for polynomial maps., Ann. of Math., 119 (1984) 425-430.

[8] R. Mañé., A proof of Pesin's formula Ergod. Th. and Dyn. Syst. 1 (1981) 95-102.

[9] R. Mañé., On the uniqueness of the maximizing measure for rational maps Bol. Soc. Bras. Mat. 14 (1983) 27-42.

[10] M. Misiurewicz, F. Przytycki., Topological entropy and degree of smooth mappings Bull. Acad. Polon. Sci. 25 (1977) 573-574.

[11] L.S. Young, Dimension, entropy and Lyapounov exponents Ergod. Th. and Dyn. Syst. 2 (1982) 109-124.

IMPA
Estrada Dona Castorina 110
22.460 Rio de Janeiro, RJ
BRAZIL.

ON THE CREATION OF HOMOCLINIC POINTS

by RICARDO MAÑÉ

INTRODUCTION

Let M be a compact boundaryless smooth manifold, $f: M \circlearrowleft$ a C^r diffeomorphism and p a hyperbolic fixed point of f (i.e. all the eigenvalues of $D_p f: T_p M \circlearrowleft$ have eigenvalues with modulus $\neq 1$). It is well known that the stable and unstable sets of p, denoted and defined respectively by

$$W^s(p) = \{ x \mid \lim_{n \to +\infty} d(f^n(x), p) = 0 \}$$

and

$$W^u(p) = \{ x \mid \lim_{n \to +\infty} d(f^{-n}(x), p) = 0 \},$$

are C^r injectively immersed submanifolds of M. The points of intersection of $W^s(p)$ with $W^u(p)$ different from p (i.e. the points in $W^s(p) \cap (W^u(p) - \{p\})$) are called *homoclinic points* associated to p. The points of intersection of the closure of $W^s(p)$ with $W^u(p)$ or of the closure of $W^u(p)$ with $W^s(p)$, different from p, will be called *almost homoclinic points* associated to p. In other words the set of almost homoclinic points associated to p is:

$$(\overline{W^s(p)} \cap W^u(p)) \cup (\overline{W^u(p)} \cap W^s(p)) - \{p\}.$$

The purpose of this paper is to study the well known problem of whether, when there exist almost homoclinic points, it is possible to create homoclinic points by a small perturbation of the diffeomorphism. To state this question more precisely, recall that if a C^r diffeomorphism $f: M \circlearrowleft$ has a hyperbolic fixed point p, there exist a neighborhood U of p and a C^r neighborhood $\mathcal{U}$ of f such that every $g \in \mathcal{U}$ has a unique fixed point $p(g)$ in U and moreover this fixed point is hyperbolic and depends continuously on g. Obviously $p(f) = p$.

Problem. — Suppose that a C^r diffeomorphism $f: M \circlearrowleft$ has almost homoclinic points associated to the hyperbolic fixed point p. Does there exist a C^r diffeormophism $g: M \circlearrowleft$, arbitrarily near to f in the C^r topology, having homoclinic points associated to $p(g)$?

Originally published in: *Publ. Math. Inst. Hautes Etudes Sci. No. 66 (1988), 139–159,*

A stronger form of this problem would be to specify an almost homoclinic point q of f and require g to have an homoclinic point nearby q. Observe that if this problem has an affirmative answer, then, by very soft arguments, it can be proved that for C^r generic diffeomorphisms, the set of transversal homoclinic points associated to a hyperbolic fixed point is dense in the set of almost homoclinic points. This conclusion has obvious interest since almost homoclinic points carry little further knowledge. On the other hand, every transversal homoclinic point is known to be an accumulation point of a sequence of invariant compact sets where the diffeomorphism acts as a transitive finite type subshift (Smale [4]). But we shall restrict our discussion of the problem to the version stated above because it seems idle to strenghten a question that even in its weaker form remains, in its full generality, completely unanswered.

However, positive answers have been obtained under supplementary hypotheses. Robinson [3] and Pixton [2] solved affirmatively the problem for diffeomorphisms of the two dimensional sphere. In fact they solve also the strong form of the problem. Takens [5] solved the problem for Hamiltonian diffeomorphisms, but only in the case $r = 1$. His solution too covers the strong version of the problem and as a corollary follows that for C^1 generic Hamiltonian diffeomorphisms the set of homoclinic points associated to a hyperbolic point p is dense in $W^s(p)$ and $W^u(p)$.

The purpose of this paper is to answer the problem under a different type of supplementary hypothesis. To explain it we need some preliminary definitions. Let $\mathcal{M}(M)$ denote the set of probabilities on the Borel σ-algebra of M endowed with its usual topology, i.e. the unique metrizable topology such that $\mu_n \to \mu$ if and only if $\int \varphi \, d\mu_n \to \int \varphi \, d\mu$ for every continuous function $\varphi : M \to \mathbf{R}$. Given $f \in \mathrm{Diff}^r(M)$, denote by $\mathcal{M}(f)$ the set of f-invariant elements of $\mathcal{M}(M)$. Associated to every $x \in M$ and $n \in \mathbf{Z}^+$, define a probability $\mu(x, n) \in \mathcal{M}(M)$ by

$$\mu(x, n) = \frac{1}{n} \sum_{j=1}^{n} \delta_{f^j(x)}.$$

Denote by $\mathcal{M}(x)$ the set of $\mu \in \mathcal{M}(M)$ such that there exists a sequence $n_1 < n_2 < \ldots$ of positive integers satisfying

$$\mu = \lim_{j \to +\infty} \mu(x, n_j).$$

Observe that $\mathcal{M}(x) \subset \mathcal{M}(f)$.

Theorem A. — Suppose that $f \in \mathrm{Diff}^r(M)$, $r = 1$ *or* 2, *has a hyperbolic fixed point* p *such that there exist* $x \in W^u(p) - \{p\}$ *and* $\mu \in \mathcal{M}(x)$ *satisfying* $\mu(\{p\}) > 0$. *Then in every* C^r *neighborhood of* f *there exists a diffeomorphism* g *that coincides with* f *in a neighborhood of* p *and has homoclinic points associated to* p.

This theorem will follow as a corollary of a stronger result that in particular implies that Theorem A also holds replacing the hyperbolic fixed point by an isolated hyperbolic set.

Let us recall some definitions that will appear in the statement of our main result. We say that $\Lambda \subset M$ is a *hyperbolic set* of $f \in \mathrm{Diff}^r(M)$ if it is compact, f-invariant and there exist a continuous splitting $TM/\Lambda = E^s \oplus E^u$, invariant under Df, and constants $C > 0$, $0 < \lambda < 1$ such that

$$\|(D_x f^n)/E^s_x\| \leqslant C\lambda^n,$$

$$\|(D_x f^{-n})/E^u_x\| \leqslant C\lambda^n$$

for all $x \in \Lambda$, $n > 0$. If there exists a compact neighborhood U of Λ such that

$$\bigcap_n f^n(U) = \Lambda,$$

we say that Λ is *isolated* and U is an *isolating block* of Λ. Observe that every compact neighborhood of Λ contained in an isolating block is also an isolating block. Define the stable and unstable manifolds of Λ by

$$W^s(\Lambda) = \{ x \mid \lim_{n\to+\infty} d(f^n(x), \Lambda) = 0 \},$$

$$W^u(\Lambda) = \{ x \mid \lim_{n\to+\infty} d(f^{-n}(x), \Lambda) = 0 \}.$$

When U is an isolating block of Λ, it is known ([1]) that

$$\bigcap_{n\geqslant 0} f^{-n}(U) = W^s(\Lambda) \cap U$$

and

$$\bigcap_{n\geqslant 0} f^n(U) = W^u(\Lambda) \cap U.$$

We say that p is an *homoclinic point* associated to Λ if

$$p \in W^s(\Lambda) \cap W^u(\Lambda) - \Lambda.$$

The next theorem is a generalization of Theorem A to hyperbolic sets and more general probabilities.

Theorem B. — Let f: M $\circlearrowleft$ be a C^r diffeomorphism ($r = 1$ or 2) and Λ an isolated hyperbolic set of f. Let $\{x_n\} \subset W^u(\Lambda)$ be a sequence converging to a point $x \notin \Lambda$ and $m_1 < m_2 < \ldots$ be a sequence of integers such that the probabilities $\mu(x_n, m_n)$ converge to a probability μ with $\mu(\Lambda) > 0$. Then every C^r neighborhood of f contains a diffeomorphism g coinciding with f in a neighborhood of Λ and having a homoclinic point associated to Λ.

Our next result says something about the case when the hypothesis $\{x_n\} \subset W^u(\Lambda)$ is dropped. We need another definition. Let us say that $\Lambda_0 \subset M$ is a *basic set* of $f \in \mathrm{Diff}^r(M)$ if it is hyperbolic, isolated and f/Λ_0 is transitive (i.e. there exists $p \in \Lambda_0$ with $\omega(p) = \Lambda_0$). Recall that if Λ is an isolated hyperbolic set of f: M $\circlearrowleft$ and $\Omega(f/\Lambda) = \Lambda$ (where $\Omega(f/\Lambda)$ denotes the set of nonwandering points of f/Λ) then Λ can be uniquely decomposed as a union $\Lambda = \Lambda_0 \cup \ldots \cup \Lambda_m$ of disjoint basic sets that we shall call *basic components* of Λ. This property is an obvious adaptation of Smale's Spectral Decomposition Theorem.

Theorem C. — *Let $f: M \circlearrowleft$ be a C^r diffeomorphism ($r = 1$ or 2) and Λ an isolated hyperbolic set of f such that $\Omega(f/\Lambda) = \Lambda$. Suppose that $\{x_n\} \subset M$ is a sequence converging to a point $x \notin \Lambda$ and $m_1 < m_2 < \ldots$ is a sequence of integers such that the probabilities $\mu(x_n, m_n)$ converge to a probability μ with $\mu(\Lambda) > 0$. Then given a C^r neighborhood $\mathcal{U}$ of f one of the following properties holds:*

I) *There exists $g \in \mathcal{U}$ coinciding with f in a neighborhood of Λ and having homoclinic points associated to a basic component of Λ.*

II) *For every neighborhood U of Λ there exist a neighborhood $V \subset U$, a diffeomorphism $g \in \mathcal{U}$ whose inverse coincides with f^{-1} in $V \cup U^c$, $k > 0$ and $0 < m \leqslant m_k$ such that*

$$g^j(x_k) = f^j(x_k)$$

for all $0 \leqslant j \leqslant m - 2$ and

$$g^m(x_k) \in \bigcap_{n \geqslant 0} g^{-n}(V) = \bigcap_{n \geqslant 0} f^{-n}(V).$$

Our final result shows a special case of Theorem C when property (I) always holds.

Theorem D. — *Let $f: M \circlearrowleft$ be a C^r diffeomorphism ($r = 1$ or 2) and Λ an isolated hyperbolic set of f with $\Omega(f/\Lambda) = \Lambda$. If there exists $x \notin W^s(\Lambda)$ such that $\mu(\Lambda) > 0$ for all $\mu \in \mathcal{M}(x)$ then every C^r neighborhood of f contains a diffeomorphism that coincides with f in a neighborhood of Λ and has an homoclinic point associated to a basic component of Λ.*

Theorems B and C can be stated in a unified (but less clear) form. In fact the next result (Theorem I.1) gives more information about the position of the homoclinic orbit that is created by a perturbation. Theorem D is proved through a variation of part of Theorem I.1.

We advise the reader to follow the proofs in the simpler case when Λ is a fixed point p, M is two dimensional and f is linear in a neighborhood of p. This simplifies the technicalities and exposes directly the idea of the proof.

I. — Proof of Theorems B and C

Here we shall prove the following result from which Theorems B and C follow immediately.

Theorem **I.1.** — *Let $f: M \circlearrowleft$ be a C^r diffeomorphism ($r = 1$ or 2) and Λ an isolated hyperbolic set of f. Let $\bar{x} \notin \Lambda$ be a point such that there exist a sequence $\{x_n\} \subset M$ converging to $\bar{x}$ and a sequence of integers $0 < m_1 < m_2 < \ldots$ such that the sequence of probabilities $\mu(x_n, m_n)$ converges to a probability μ that satisfies $\mu(\Lambda) > 0$. Then, for every neighborhood $\mathcal{U}$ of f in* $\mathrm{Diff}^r(M)$ *and every neighborhood U of Λ in M, there exist $g \in \mathcal{U}$ and a neighborhood $V \subset U$ of Λ such that*

$$f/(V \cup U^c) = g/(V \cup U^c),$$
$$f^{-1}/(V \cup U^c) = g^{-1}/(V \cup U^c)$$

and moreover satisfy one of the following properties:

I) *There exist* $k > 0$ *and* $0 < s < m_k$ *such that:*

$$g^j(x_k) = f^j(x_k)$$

for all $0 \leqslant j < s$ *and*

$$g^{s+1}(x_k) \in \bigcap_{n \geqslant 0} g^{-n}(V) = \bigcap_{n \geqslant 0} f^{-n}(V).$$

Moreover when there exists an isolating block U_0 *of* Λ *such that* $\{x_n\} \subset \bigcap_{n \geqslant 0} f^m(U_0)$, *then g can be chosen also satisfying* $g^{-j}(x_n) = f^{-j}(x_n)$ *for all* $j \geqslant 0$, $n \geqslant 0$.

II) *There exist* $k > 0$ *and* $0 < s_0 < s_1 < m_k$ *such that*

(i) $$g^j(f^{s_0}(x_k)) = f^j(f^{s_0}(x_k))$$

for $0 \leqslant j < s_1 - s_0$ *and*

(ii) $$g^{-2}(f^{s_0}(x_k)) \in \bigcap_{n \geqslant 0} g^n(V) = \bigcap_{n \geqslant 0} f^n(V),$$

(iii) $$g^{(s_1 - s_0)+1}(f^{s_0}(x_k)) \in \bigcap_{n \geqslant 0} g^{-n}(V) = \bigcap_{n \geqslant 0} f^{-n}(V).$$

The proof of Theorem I.1 relies on four lemmas. Ths first one is quite simple and we leavc its proof to the reader.

Lemma 1. — *Given* $c > 0$ *and a neighborhood* $\mathcal{N}$ *of the identity in* $\mathrm{Diff}^k(M)$, *there exists* $R > 0$ *such that for every* $0 < r < R$ *and every pair of points* $a \in M$ *and* $b \in M$ *satisfying*

$$d(a, b) \leqslant r^{k+c}$$

there exists $h \in \mathcal{N}$ *such that*

$$h(a) = b$$

and

$$h(x) = x$$

for all x outside the ball $B_r(a)$.

For the statement of the second lemma suppose that f, Λ and U satisfy the hypotheses of Theorem I.1. Suppose also M endowed with a Riemannian metric such that if $TM/\Lambda = E^s \oplus E^u$ is the hyperbolic splitting of TM/Λ, there exists $0 < \lambda_0 < 1$ such that $\|(D_x f)/E^s_x\| < \lambda_0$, $\|(D_x f^{-1})/E^u_x\| < \lambda_0$ for all $x \in \Lambda$.

Lemma 2. — *There exist compact sets* $V^+ \subset U$, $V^- \subset U$, $0 < \varepsilon_0 < 1$ *and* $0 < \gamma < \lambda < 1$ *satisfying:*

a) $f(V^+) \subset V^+$, $f^{-1}(V^-) \subset V^-$;

b) $\Lambda = V^+ \cap V^-$;

c) $\gamma\, d(x, V^-) \leqslant d(f(x), V^-) \leqslant \lambda\, d(x, V^-)$

and

$$\gamma\, d(x, V^+) \leqslant d(f^{-1}(x), V^+) \leqslant \lambda\, d(x, V^+),$$

for all x satisfying $d(x, V^+) < \varepsilon_0$ *and* $d(x, V^-) < \varepsilon_0$.

If the reader, following the suggestion given at the end of the Introduction, is keeping track of the proof in the case there described, then this lemma becomes completely trivial just taking a square W contained in U and defining V^+ and V^- as the connected components (that are segments) of $W^s(p) \cap W$ and $W^u(p) \cap W$, as illustrated in figure I.

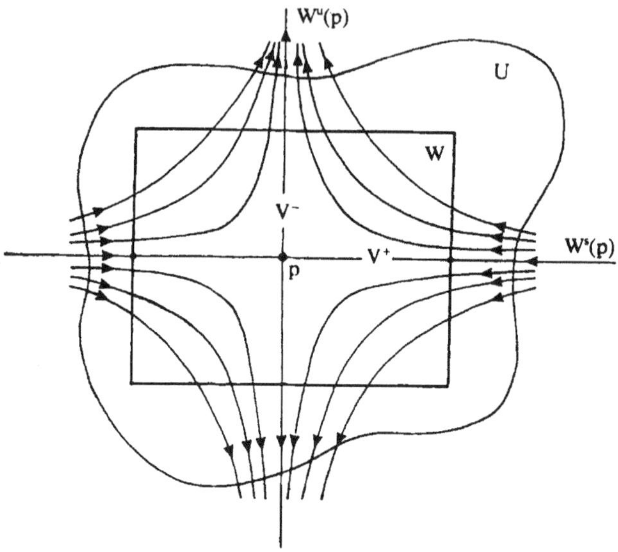

FIG. I

The proof of Lemma 2 will be given after the proof of Theorem I.1.

Let $\bar{x} \notin \Lambda$ be given as in the statement of the theorem and denote $\mu_k = \mu(x_k, m_k)$ for $k > 0$.

To prove Theorem I.1 we start by taking a compact neighborhood $W \subset U$ of Λ such that

$$\bigcap_n f^n(W) = \Lambda,$$

$$\bar{x} \notin W$$

and $$f(\bar{x}) \notin W.$$

Define $$V(r) = \{ x \mid d(x, V^+) \leqslant r, d(x, V^-) \leqslant r \}.$$

Choose $0 < \varepsilon_1 < \varepsilon_0$, where ε_0 is given by Lemma 2, such that

$$V(\varepsilon_1) \subset W.$$

Now take $0 < \delta < 1$ and a sequence $r_0, r_1, \ldots$ satisfying

(1) $$0 < r_0 < \varepsilon_1,$$

(2) $$r_{n+1} = r_n^{1+\delta}$$

and

(3) $$\mu_k(\partial V(r_n)) = \mu(\partial V(r_n)) = 0$$

for all $n \geqslant 0$ and $k > 0$. Set $V_n = V(r_n)$ and define S_n as the set of points $x \in V_0$ that can be written as $x = f^m(y_n)$, $m \in \mathbf{Z}$, with $y_n \in V_n$ and $f^j(y_n) \in V_0$ for all $0 \leqslant j \leqslant m$ if $m \geqslant 0$, or for all $m \leqslant j \leqslant 0$ when $m \leqslant 0$. In the simplified case the sets V_n and S_n look as in figure II where V_n is a square of side $2r_n$ centered at p and S_n is the subset of V_0 bounded by hyperbolas passing through the vertices of V_n.

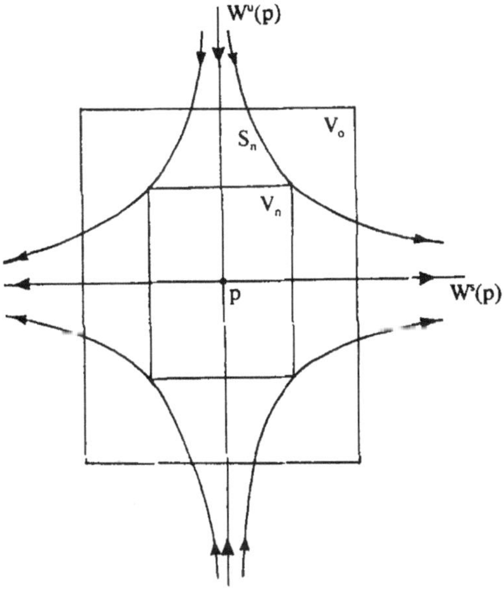

FIG. II

For $x \in V_0$ define:

$$T(x) := \sup\{j \mid f^i(x) \in V_0 \text{ for all } 0 \leqslant i \leqslant j\} + \sup\{j \mid f^{-i}(x) \in V_0 \text{ for all } 0 < i \leqslant j\}.$$

Lemma 3. — There exist $C_2 > C_1 > 0$ *such that, for* δ *as above and for all* $n > 0$,

$$T(x) \leqslant C_2(1+\delta)^n$$

if $x \in V_0 - S_n$ *and*

$$C_1(1+\delta)^n \leqslant T(x)$$

if $x \in S^n$.

19

Proof. — Observe that by (1), $V_0 = V(r_0) \subset V(\varepsilon_1) \subset V(\varepsilon_0)$. Then we can apply the inequalities of part c) of Lemma 2 to the points in V_0. Hence, if $f^j(x) \in V_0$ for all $-n_1 \leqslant j \leqslant n_2$, it follows that

$$r_0 \geqslant d(f^{n_2}(x), V^+) \geqslant \lambda^{-n_2} d(x, V^+),$$
$$r_0 \geqslant d(f^{-n_1}(x), V^-) \geqslant \lambda^{-n_1} d(x, V^-).$$

Since $x \notin V_n$ either $d(x, V^+) \geqslant r_n$ or $d(x, V^-) \geqslant r_n$. Then

$$r_n \leqslant \max\{\lambda^{n_2} r_0, \lambda^{n_1} r_0\}.$$

But by (2) we have $\log r_n = (1+\delta)^n \log r_0$, implying

$$(1+\delta)^n \log r_0 \leqslant \log r_0 + \log \lambda(\max\{n_1, n_2\}).$$

Hence
$$n_1 + n_2 \leqslant 2\,\frac{\log r_0}{\log \lambda}\,(1+\delta)^n.$$

From this inequality the upper bound of the lemma follows easily. The lower bound is obtained by the same method applying the lower bounds of part c) of Lemma 2.

Define a (n, k)-string as a set σ of the form $\sigma = \{f^j(x_k), \ldots, f^{j+\ell}(x_k)\} \subset V_n$, $1 \leqslant j \leqslant j + \ell \leqslant m_k$, such that $f^{j-1}(x_k) \notin V_n$ and $f^{j+\ell+1}(x_n) \notin V_n$. Observe that there is a natural order relation among (n, k)-strings, defined by $\sigma_1 < \sigma_2$ if the first element of σ_2 is a strictly positive iterate of the last member of σ_1. To simplify the notation, $(0, k)$-strings will be called k-strings. Most of the time we shall deal only with k-strings. More general strings will appear only once in our arguments.

Now we are ready to state the fundamental lemma of the proof of Theorem I.1.

Lemma 4. — For all $n_1 > 0$, one of the following properties holds:

a) *There exist $n \geqslant n_1$, $k > 0$ and two k-strings $\sigma_1 \subset S_{n+1}$, $\sigma_1 < \sigma_2 \subset S_{n+1}$ such that $\sigma \cap (S_n - S_{n+1}) = \varnothing$ for every k-string $\sigma_1 < \sigma < \sigma_2$.*

b) *There exist $n \geqslant n_1$, $k > 0$ and a k-string $\sigma_1 \subset S_{n+1}$ such that $\sigma \cap (S_n - S_{n+1}) = \varnothing$ for every k-string $\sigma < \sigma_1$.*

Proof. — Take $1 + \delta < \xi < 2$ and $s \in \mathbf{Z}^+$ such that $2s - 1 > \xi s$. Define $\nu_k(S_n)$ as the number of k-strings contained in S_n. Now we shall prove three claims.

Claim 1. — Suppose that property a) *of Lemma 4 does not hold for a certain $n_1 > 0$. Then*

$$\nu_k(S_n) \leqslant (1/\xi)\, \nu_k(S_{n-1})$$

for all $k > 0$ and $n > n_1$ such that $\nu_n(S_n) > s$.

Proof. — Suppose that $\sigma_1, \ldots, \sigma_p$, $p = \nu_k(S_n)$, are the k-strings contained in S_n. If $n > n_1$, the hypothesis of the claim implies that between any two consecutive k-strings σ_i, σ_{i+1} in S_n there exists a k-string $\hat{\sigma}_i$ such that $\hat{\sigma}_i \cap S_{n-1} \neq \varnothing$, therefore $\hat{\sigma}_i \subset S_{n+1}$ (by the definition of S_{n-1}) and then S_{n-1} contains the strings $\sigma_1, \ldots, \sigma_p$ and $\hat{\sigma}_1, \ldots, \hat{\sigma}_{p-1}$. Hence

$$\nu_k(S_{n-1}) \geqslant 2p - 1 = 2\nu_k(S_n) - 1 \geqslant \xi\nu_k(S_n),$$

because $\nu_k(S_n) > s$.

Claim 2. — Suppose that property a) *of Lemma 4 does not hold for a certain* $n_1 > 0$. *Then*

$$\mu_k(S_n - S_{n+1}) \leqslant C_2 \left(\frac{1+\delta}{\xi}\right)^n (1+\delta)\xi^{n_1},$$

for all $k > 0$ *and* $n > n_1$ *such that* $\nu_k(S_n) > s$.

Proof. — The first claim proves that

$$\nu_k(S_n) \leqslant (1/\xi)^{n-n_1} \nu_k(S_{n_1})$$

for all $k > 0$ and $n > n_1$ such that $\nu_k(S_n) > s$. Denote by T the supremum of the lenghts of the k-strings contained in $S_n - S_{n+1}$; then

$$\begin{aligned}\mu_k(S_n - S_{n+1}) &= (1/m_k)\,\#\{1 \leqslant j \leqslant m_k \mid f^j(x_k) \in S_n - S_{n+1}\}\\ &\leqslant (1/m_k)\,T(\nu_k(S_n) - \nu_k(S_{n+1})) \leqslant (1/m_k)\,T\nu_k(S_n).\end{aligned}$$

But by Lemma 3,

$$T \leqslant C_2(1+\delta)^{n+1}.$$

Hence
$$\begin{aligned}\mu_k(S_n - S_{n+1}) &\leqslant C_2(1+\delta)^{n+1}(1/m_k)\,\nu_k(S_n)\\ &\leqslant C_2(1+\delta)^{n+1}(1/m_k)(1/\xi)^{n-n_1}\nu_k(S_{n_1}).\end{aligned}$$

But
$$(1/m_k)\,\nu_k(S_{n_1}) \leqslant (1/m_k)\,\#\{1 \leqslant j \leqslant m_k \mid f^j(x_k) \in S_{n_1}\} \leqslant 1.$$

Then
$$\mu_k(S_n - S_{n+1}) \leqslant C_2\left(\frac{1+\delta}{\xi}\right)^n (1+\delta)\,\xi^{n_1}.$$

Claim 3. — Suppose that there exists $n_1 > 0$ *such that property* b) *of Lemma 4 does not hold. Then*

$$\mu_k(S_n) \leqslant C_2\, C_1^{-1}(1+\delta)^{s+2}\, s\mu_k(S_{n-2} - S_n)$$

for all $n > n_1 + 2$ *and* $k > 0$ *such that* $\nu_k(S_n) \leqslant s$.

Proof. — If property b) of Lemma 4 does not hold for n_1, then, for all $m \geqslant n_1$, $k > 0$ and every S_j with $j > m$ that contains a k-string σ_1, it follows that $S_{j-1} - S_j$ also contains k-strings. This means that the family of sets $S_{j-1} - S_j, j > m$, that contain a k-string has the form $\{S_m - S_{m+1}, \ldots, S_{m+\ell} - S_{m+\ell+1}\}$. Now suppose that $n \geqslant n_1 + 3$. Put $m = n - 2$. Then $m > n_1$. Hence $S_m - S_{m+1} = S_{n-2} - S_{n-1}$ and $S_{m+1} - S_{m+2} = S_{n-1} - S_n$ contain k-strings. At least one of these two strings has the form $\sigma = \{f^j(x_k), \ldots, f^{j+\ell}(x_k)\}$ with $1 \leqslant j \leqslant j + \ell < m_k$ (remember that the definition of k-strings only requires $j \leqslant m_k$; but there exists at most *one* k-string whose last point is an iterate larger than m_k). By Lemma 3 the lenght of σ is larger than or equal to $C_1(1+\delta)^{n-2}$. Then

$$\mu_k(S_{n-2} - S_n) \geqslant (1/m_k)\, C_1(1+\delta)^{n-2}.$$

Moreover, denoting by T the supremum of the lenghts of the k-strings contained in S_n, we have

$$\mu_k(S_n) \leqslant (1/m_k)\, T\nu_k(S_n).$$

But since $n > n_1$, the above argument shows that there exists ℓ such that every set $S_j - S_{j+1}$, $n \leqslant j \leqslant n+\ell$, contains a k-string and conversely every k-string contained in S_n is contained in one of these sets. Therefore $\nu_k(S_n) \geqslant \ell + 1$ and $\nu_k(S_n) \leqslant s$ implies $\ell < s$. Now, by Lemma 3, we have

$$T \leqslant C_2(1+\delta)^{n+s},$$

hence $$\mu_k(S_n) \leqslant (1/m_k)\, C_2(1+\delta)^{n+s} s.$$

Therefore $$\mu_k(S_n) \leqslant C_2\, C_1^{-1}(1+\delta)^{s+2} s\mu_k(S_{n-2} - S_n).$$

Now we are ready to prove Lemma 4. Suppose it is false. Then there exists $n_1 > 0$ such that a) and b) do not hold for n_1. Define

$$r(k) = \min\{j \mid \nu_k(S_j) \leqslant s\}.$$

Then $$\mu_k(S_n) = \mu_k(S_{r(k)}) + \sum_{n \leqslant j < r(k)} \mu_k(S_j - S_{j+1}).$$

Moreover it is easy to see that

$$\lim_{k \to +\infty} r(k) = \infty.$$

Hence $r(k) > n_1 + 2$ if k is large enough. Applying Claim 3, we obtain:

$$\mu_k(S_{r(k)}) \leqslant C_2\, C_1^{-1}(1+\delta)^{s+2} s\mu_k(S_{r(k)-2} - S_{r(k)}).$$

Now $r(k) - 2 > n_1$ and $\nu_k(S_{r(k)-1}) > s$; therefore Claim 2 implies

$$\mu_k(S_{r(k)-2} - S_{r(k)}) = \mu_k(S_{r(k)-2} - S_{r(k)-1}) + \mu_k(S_{r(k)-1} - S_{r(k)})$$
$$\leqslant C_3\left(\frac{1+\delta}{\xi}\right)^{r(k)-2} + C_3\left(\frac{1+\delta}{\xi}\right)^{r(k)-1} \leqslant 2C_3\left(\frac{1+\delta}{\xi}\right)^{r(k)-2},$$

where $C_3 = C_2(1+\delta)\,\xi^{n_1}$. Moreover, Claim 2 implies

$$\mu_k(S_j - S_{j+1}) \leqslant C_3\left(\frac{1+\delta}{\xi}\right)^j$$

for every $n_1 < j < r(k)$. Hence, setting $C_4 = 2sC_1^{-1}\, C_2\, C_3(1+\delta)^{s+2}$ we obtain

$$\mu_k(S_n) \leqslant C_4\left(\frac{1+\delta}{\xi}\right)^{r(k)-2} + C_3 \sum_{n \leqslant j < r(k)} \left(\frac{1+\delta}{\xi}\right)^j$$

for $n > n_1$ and k large. Taking limits when $k \to +\infty$, and recalling that $(1+\delta)/\xi < 1$ and $\lim_{k\to+\infty} r(k) = \infty$, we obtain

$$\lim_{n\to\infty} \lim_{k\to+\infty} \mu_k(S_n) = 0.$$

But (3) implies

$$\mu(S_n) = \lim_{k\to+\infty} \mu_k(S_n).$$

Hence
$$\mu(\Lambda) = \lim_{n\to+\infty} \mu(S_n) = 0,$$

contradicting the hypothesis $\mu(\Lambda) > 0$ and proving Lemma 4.

Now let us complete the proof of Theorem I.1. By Lemma 4 there exist n, arbitrarily large, and $k > 0$ satisfying property a) or property b). Suppose that property a) is satisfied; in other words there exist consecutive k-strings $\sigma_1 < \sigma_2$ contained in S_{n+1}, such that $\sigma \cap S_n = \emptyset$ for all k-strings $\sigma_1 < \sigma < \sigma_2$. Now observe that by the definition of strings, the intersections $\sigma_1 \cap V_n$, $\sigma_2 \cap V_n$ are (n, k)-strings. Let q_1 be the last point of the (n, k)-string $\sigma_1 \cap V_n$ and q_2 the first point of the (n, k)-string $\sigma_2 \cap V_n$. In the simplified case, the situation is described in figure III.

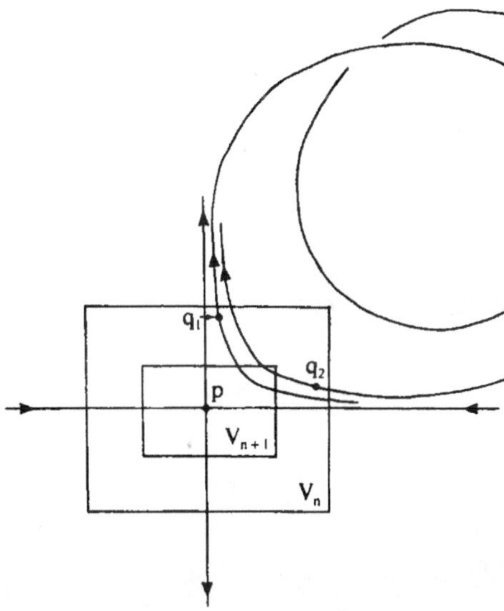

Fig. III

Since q_1 is the last point of $\sigma_1 \cap V_n$, $f(q_1)$ does not belong to V_n. Hence either $d(f(q_1), V^+) > r_n$ or $d(f(q_1), V^-) > r_n$. But Lemma 2 implies

$$d(f(q_1), V^-) \leqslant \lambda\, d(q_1, V^-) \leqslant \lambda r_n < r_n,$$

thus

$$d(f(q_1), V^+) > r_n. \tag{4}$$

Take a point $p_1 \in \sigma_1 \cap V_{n+1}$ and write $q_1 = f^a(p_1)$. Then (4) and Lemma 2 imply

$$r_n < d(f^{a+1}(p_1), V^+) \leqslant \gamma^{-(a+1)} d(p_1, V^+) \leqslant \gamma^{-(a+1)} r_{n+1} = \gamma^{-(a+1)} r_n^{1+\delta},$$

and therefore

$$r_n^\delta \geqslant \gamma^{a+1}. \tag{5}$$

Using again Lemma 2, we have

$$d(q_1, V^-) = d(f^a(p_1), V^-) \leqslant \lambda^a d(p_1, V^-) \leqslant \lambda^a r_{n+1} = \lambda^a r_n^{1+\delta}.$$

Write $\lambda = \gamma^\alpha$ with $0 < \alpha < 1$. From (5) it follows that

$$d(q_1, V^-) \leqslant \gamma^{\alpha a} r_n^{1+\delta} \leqslant (r_n^\delta/\gamma)^\alpha . r_n^{1+\delta} = \gamma^{-\alpha} r_n^{1+\delta+\alpha\delta}.$$

The only restriction we used about δ was $0 < \delta < 1$. Suppose we take it so near to 1 that $\delta(1 + \alpha) > 1$ and take $0 < \beta < \delta(1 + \alpha) - 1$. Then the last inequality, if n is large, yields

$$d(q_1, V^-) < r_n^{2+\beta}.$$

Then we can choose $y_1 \in V^-$ such that

$$q_1 \in B(r_n^{2+\beta}, y_1). \tag{6}$$

Now we shall prove that if n is large enough, then

$$f^{-j}(y_1) \notin B(r_n^{1+(\beta/3)}, y_1) \tag{7}$$

for all $j > 0$. From (4) and Lemma 2 it follows that

$$d(q_1, V^+) \geqslant \gamma\, d(f(q_1), V^+) > \gamma r_n,$$

and therefore (6) implies

$$d(y_1, V^+) \geqslant d(q_1, V^+) - d(y_1, q_1) \geqslant \gamma r_n - r_n^{2+\beta}. \tag{8}$$

Also, for all $j > 0$, Lemma 2 implies that

$$d(f^{-j}(y_1), V^+) \leqslant \lambda^j d(y_1, V^+) < \lambda\, d(y_1, V^+),$$

hence from (8) it follows that

$$d(f^{-j}(y_1), y_1) \geqslant (1 - \lambda)\, d(y_1, V^+) \geqslant (1 - \lambda)\, (\gamma r_n - r_n^{2+\beta}).$$

But if n is large,

$$(1 - \lambda)\, (\gamma r_n - r_n^{2+\beta}) > r_n^{1+(\beta/3)}.$$

Hence (7) holds for all $j > 0$, if n is large enough. Now write $q_1 = f^{t_1}(x_k)$, $q_2 = f^{t_2}(x_k)$ and define $N_0 = t_2 - t_1$. We claim that if n is large enough, then

$$f^j(q_1) \notin B(r_n^{1+(\beta/3)}, y_1) \tag{9}$$

for all $0 < j \leqslant N_0$. First we prove (9) for $j = 1$. Take $A > 0$ such that $d(f(z), f(w)) \leqslant A\, d(z, w)$ for all $z, w \in M$. Then (6) implies

$$d(f(q_1), y_1) \geqslant d(y_1, f(y_1)) - d(f(q_1), f(y_1)) \geqslant d(y_1, f(y_1)) - A r_n^{2+\beta}.$$

Now from Lemma 2 and (8) it follows that

$$d(y_1,f(y_1)) \geqslant d(f(y_1), V^+) - d(y_1, V^+) \geqslant \left(\frac{1}{\lambda} - 1\right)(\gamma r_n - r_n^{2+\beta}),$$

hence
$$d(f(q_1),y_1) \geqslant \left(\frac{1}{\lambda} - 1\right)\gamma r_n - \left(\frac{1}{\lambda} - 1 + A\right) r_n^{2+\beta}.$$

But if n is large, the right hand side of that inequality is larger than $r_n^{1+(\beta/3)}$, and (9) holds for $j = 1$. Now suppose that

$$d(f^j(q_1),y_1) \leqslant r_n^{1+(\beta/3)}$$

for some $2 \leqslant j \leqslant N_0$ and take $B > 0$ such that $d(f^{-1}(z),f^{-1}(w)) \leqslant B\, d(z, w)$ for all $z, w \in M$. Since $y_1 \in V^-$, Lemma 2 implies

$$d(f^{j-1}(q_1), V^-) \leqslant d(f^{j-1}(q_1),f^{-1}(y_1)) \leqslant Br_n^{1+(\beta/3)}$$

and therefore, if n is large,

(10) $$d(f^{j-1}(q_1^-), V^-) \leqslant r_n.$$

On the other hand, (6) and Lemma 2 imply

$$d(f^{-1}(y_1), V^+) \leqslant d(f^{-1}(y_1),f^{-1}(q_1)) + d(f^{-1}(q_1), V^+)$$
$$\leqslant B\, d(y_1, q_1) + \lambda\, d(q_1, V^+) \leqslant Br_n^{2+\beta} + \lambda r_n$$

because $q_1 \in V_n$ and therefore

$$d(f^{j-1}(q_1), V^+) \leqslant d(f^{j-1}(q_1),f^{-1}(y_1)) + d(f^{-1}(y_1), V^+)$$
$$\leqslant (Br_n^{\beta/3} + Br_n^{1+\beta} + \lambda)\, r_n.$$

When n is large, this means

(11) $$d(f^{j-1}(q_1), V^+) \leqslant r_n.$$

From (10) and (11) it follows that $f^{j-1}(q_1) \in V_n$. Let σ be the k-string containing $f^{j-1}(q_1)$. Then $\sigma \cap S_n \neq \varnothing$. Also $\sigma_1 \leqslant \sigma$ because $f^{j-1}(q_1) \in \sigma$, $j - 1 \geqslant 1$ and $q_1 \in \sigma_1$. Moreover $\sigma \leqslant \sigma_2$ because

$$f^{N_0-j+1}(f^{j-1}(q_1)) = f^{N_0}(q_1) = q_2,$$

$N_0 - j + 1 \geqslant 1$ and $f^{j-1}(q_1) \in \sigma$. Now $\sigma_1 \leqslant \sigma \leqslant \sigma_2$ and $\sigma \cap S_n = \varnothing$ imply $\sigma_1 = \sigma$ or $\sigma = \sigma_2$. Suppose $\sigma_1 = \sigma$; then $f^{j-1}(q_1) \in \sigma_1 \cap V_n$ and now $j - 1 \geqslant 1$ means that q_1 is not the last point of the k-string σ_1 in V_n, which is a contradiction. Suppose $\sigma = \sigma_2$. Then $f^{-(N_0-j+1)}(q_2) = f^{j-1}(q_1) \in \sigma_2 \cap V_n$ and $N_0 - j + 1 \geqslant 1$ imply that q_2 is not the first point of the k-string σ_2 in V_n, which also is a contradiction. This completes the proof of (9).

Similarly to (6) and (7) we prove that if n is large, there exists $y_2 \in V^+$ such that

(12) $$q_2 \in B(r_n^{2+\beta}, y_2)$$

and

(13) $$f^j(y_2) \notin B(r_n^{1+(\beta/3)}, y_2)$$

for all $j > 0$. Moreover

(14) $$f^j(y_2) \notin B(r_n^{1+(\beta/3)}, y_1)$$

for all $j \geqslant 0$. To prove (14), observe that $f^j(y_2) \in V^+$ for all $j \geqslant 0$. Hence, using (8) we obtain

$$d(y_1, f^j(y_2)) \geqslant d(y_1, V^+) \geqslant \gamma r_n - r_n^{2+\beta}.$$

But if n is large, $\gamma r_n - r_n^{2+\beta} > r_n^{1+(\beta/3)}$, proving (14).

In a similar way we prove the property

(15) $$f^{-j}(y_1) \notin B(r_n^{1+(\beta/3)}, y_2)$$

for all $j \geqslant 0$. Moreover, the proof of property (9) can be adapted to show that

(16) $$f^j(q_1) \notin B(r_n^{1+(\beta/3)}, y_2)$$

for all $0 \leqslant j < N_0$. The case $j = N_0 - 1$ is treated separately and if (16) is false for some $0 \leqslant j \leqslant N_0 - 2$, proceeding similarly to the proof of (9), we find a string σ satisfying $\sigma_1 \leqslant \sigma \leqslant \sigma_2$ and $\sigma \cap S_n \neq \emptyset$ and then, by the way σ_1 and σ_2 were selected, it must satisfy $\sigma_1 = \sigma$ or $\sigma = \sigma_2$. As in the proof of (9), both possibilities immediately lead to contradictions.

Now apply Lemma 1 to $\mathcal{N} = \mathcal{U} \circ f^{-1}$ (where $\mathcal{U}$ is the neighborhood in the statement of Theorem I.1) and

$$c = \frac{2+\beta}{1+(\beta/3)} - 2$$

when $r = 2$, or

$$c = \frac{2+\beta}{1+(\beta/3)} - 1$$

when $r = 1$. This gives a number $R > 0$, and we can suppose, taking n large enough, that $r_n < R$. Then, by Lemma 1, there exists $h \in \mathcal{N}$ such that

$$h(q_2) = y_2,$$

$$h(y_1) = q_1$$

and $$h(z) = z \quad \text{if } z \notin B(r_n^{1+(\beta/3)}, y_1) \cup B(r_n^{1+(\beta/3)}, y_2).$$

Define $g = hf$ and

$$V = V_n - S - f^{-1}(S),$$

where $$S = B(r_n^{1+(\beta/3)}, y_1) \cup B(r_n^{1+(\beta/3)}, y_2).$$

Then $$g \in \mathcal{N} \circ f = \mathcal{U}.$$

Clearly $V \subset V_n$ and then, by (1),

$$V \subset U.$$

Moreover it is easy to check that

$$f/(V \cup U^c) = g/(V \cup U^c) \quad \text{and} \quad f^{-1}/(V \cup U^c) = g^{-1}/(V \cup U^c).$$

Now let us verify that V is a neighborhood of Λ if n is large. First we shall show that

$$(17) \qquad d(\Lambda, y_1) > r_n^{1+(\beta/8)},$$

$$(18) \qquad d(\Lambda, y_2) > r_n^{1+(\beta/8)}.$$

Using (8), we obtain

$$d(\Lambda, y_1) \geqslant d(V^+, y_1) \geqslant \gamma r_n - r_n^{2+\beta}.$$

When n is large, $\gamma r_n - r_n^{2+\beta} > r_n^{1+(\beta/8)}$, thus proving (17). Inequality (18) is proved similarly. Properties (17) and (18) imply that the closed set S does not intersect Λ. Since Λ is invariant, $\Lambda \cap f^{-1}(S) = \emptyset$. Then $V_n - (S \cup f^{-1}(S))$ is a neighborhood of Λ.

Now we shall check that option (II) of Theorem I.1 is satisfied for g, V and an appropriate choice of k and $0 < s_0 < s_1 < m_k$. As k we obviously take the k in the construction of σ_1 and σ_2. Choose $0 < s_0 < s_1 < m_k$ satisfying

$$f^{s_0}(x_k) = q_1, \quad f^{s_1}(x_k) = q_2.$$

Observe that s_0 and s_1 coincide with the numbers t_1 and t_2 used to define the number $N_0 = t_2 - t_1$ of property (9). Then $s_1 = s_0 + N_0$. We shall prove property (i) of Theorem I.1 by induction on j.

For $j = 1$,

$$g(f^{s_0}(x_k)) = g(q_1) = hf(q_1).$$

But by (9) and (16), $f(q_1) \notin S$. Then $g(q_1) = f(q_1)$ and

$$g(f^{s_0}(x_k)) = f(q_1) = f(f^{s_0}(x_k)).$$

Now assume that $f^i(q_1) = g^i(q_1)$, $1 \leqslant i < (s_1 - s_0) - 1$. Then

$$g^{i+1}(f^{s_0}(x_k)) = g^{i+1}(q_1) = g(g^i(q_1)) = g(f^i(q_1)).$$

But $f^{i+1}(q_1) \notin S$ because $i + 1 < s_1 - s_0 = N_0$ and, according to (9) and (16), $f^j(q_1) \notin S$ for all $0 < j < N_0$. Hence $g(f^i(q_1)) = hf(f^i(q_1)) = h(f^{i+1}(q_1)) = f^{i+1}(q_1)$ and then

$$(19) \qquad f^{i+1}(f^{s_0}(x_k)) = g^{i+1}(f^{s_0}(x_k)),$$

completing the proof of (i). To prove property (ii) of Theorem I.1, we first prove, by induction on j, that

$$(20) \qquad g^{-j}(f^{s_0}(x_k)) = f^{-j}(y_1)$$

for all $j \geqslant 1$. When $j = 1$,

$$g^{-1}(f^{s_0}(x_k)) = f^{-1}h^{-1}(q_1) = f^{-1}(y_1).$$

Assume that (20) holds for $j = i \geqslant 1$. Then

$$g^{-(i+1)}(f^{s_0}(x_k)) = g^{-1}g^{-i}(f^{s_0}(x_k)) = g^{-1}f^{-i}(y_1) = f^{-1}h^{-1}f^{-i}(y_1).$$

20

But by (7) and (15), $f^{-i}(y_1) \notin S$. Then

$$g^{-(i+1)}(f^{s_0}(x_k)) = f^{-1}f^{-i}(y_1) = f^{-(i+1)}(y_1).$$

This completes the proof of (20). It follows that

$$(21) \qquad g^{-j}(f^{s_0}(x_k)) \in V_n$$

for all $j \geqslant 1$ because, by Lemma 2, $f^{-j}(V_n \cap V^-) \subset V_n \cap V^-$ for all $j \geqslant 0$. Since $y_1 \in V_n \cap V^-$, this implies that $f^{-j}(y_1) \in V_n \cap V^- \subset V_n$ for all $j \geqslant 0$. This property together with (20) implies (21). Moreover, property (20), together with (7) and (15), implies

$$(22) \qquad g^{-j}(f^{s_0}(x_k)) = f^{-1}(y_1) \notin S$$

for all $j \geqslant 1$. Hence

$$(23) \qquad g^{-j}(f^{s_0}(x_k)) \notin f^{-1}(S)$$

for all $j \geqslant 2$. Clearly, property (ii) of Theorem I.1 follows from (21), (22) and (23). The equality

$$\bigcap_{n \geqslant 0} f^n(V) = \bigcap_{n \geqslant 0} g^n(V),$$

also contained in that property, is a trivial consequence of $g/V = f/V$. Property (iii) o Theorem I.1 is proved using similar methods, observing that (19) implies

$$\begin{aligned} g^{s_1 - s_0}(f^{s_0}(x_k)) &= gg^{s_1 - s_0 - 1}(f^{s_0}(x_k)) = hff^{s_1 - s_0 - 1}(f^{s_0}(x_k)) \\ &= hf^{s_1 - s_0}(f^{s_0}(x_k)) = hf^{s_1}(x_k) = h(q_2) = y_2. \end{aligned}$$

Then we have to prove that $g^j(y_2) = f^j(y_2) \in V$ for all $j > 0$. We shall do it by induction. For $j = 1$

$$g(y_2) = hf(y_2).$$

But by (13) and (14), $f(y_2) \notin S$. Then $g(y_2) = f(y_2)$. Moreover $y_2 \in V^+ \cap V_n$. Then $f(y_2) \in f(V^+ \cap V_n) \subset V^+ \cap V_n \subset V_n$. Finally observe that $f(y_2) \notin f^{-1}(S)$ because $f(y_2) \in f^{-1}(S)$ implies $f^2(y_2) \in S$ contradicting (13) or (14). Then

$$f(y_2) \in V_n - S - f^{-1}(S) = V.$$

Now suppose that $f^i(y_2) \in V$. Then $g^{i+1}(y_2) = g(g^i(y_2)) = g(f^i(y_2)) = hf^{i+1}(y_2)$. By (13) and (14), $f^{i+1}(y_2) \notin S$. Then $g^{i+1}(y_2) = f^{i+1}(y_2)$ and, by the argument already used in the case $j = 1$, $f^{i+1}(y_2) \in V_n$. Hence $g^{i+1}(y_2) \in V_n$. Moreover $g^{i+1}(y_2) \notin S \cup f^{-1}(S)$ because otherwise we have $g^{i+1}(y_2) = f^{i+1}(y_2) \in S \cup f^{-1}(S)$ and then either $f^{i+1}(y_2) \in S$ or $f^{i+2}(y_2) \in S$, violating properties (13) or (14). Hence $g^{i+1}(y_2) \in V_n - (S \cup f^{-1}(S)) = V$, thus concluding the proof or property (II) of Theorem I.1.

Finally, let us show that when in the application of Lemma 4 it is option b) that holds, then we can find $g \in \mathcal{U}$ and V satisfying property (I) of Theorem I.1. Let $\sigma_1 = \{f^{s_0}(x_k), \ldots, f^{s_0+\ell}(x_k)\}$ be the string given by option b) of Lemma 4. As in the previous case, we prove that, if n is large enough, there exists $y \in V^+$ such that, setting $q = f^{s_0}(x_k)$,

$$q \in B(r_n^{2+\beta}, y),$$
$$f^j(y) \notin B(r_n^{1+(\beta/3)}, y)$$

for all $j > 0$, and

$$f^j(x_k) \notin B(r_n^{1+(\beta/3)}, y)$$

for all $0 \leqslant j < s_0$. Now take $h \in \mathrm{Diff}^r(M)$ satisfying

$$h(q) = y$$

and $$h(z) = z \quad \text{if } z \notin B(r_n^{1+(\beta/3)}, y).$$

Using Lemma 1 as before, we show that if n is very large then $g = hf \in \mathscr{U}$. Define

$$S = B(r_n^{1+(\beta/3)}, y)$$

and $$V = V_n - S - f^{-1}(S).$$

Using the same methods as in the previous case it is easy to check that g, k, V and $f^{s_0}(x_k)$ satisfy property (I) of Theorem I.1. Moreover, in the particular case when there exists an isolating block U_0 of Λ such that $\{x_n\} \subset \Lambda^s$, where Λ^s is defined as $\Lambda^s = \bigcap_{m \geqslant 0} f^m(U_0)$, we have $g^{-j}(x_n) = f^{-j}(x_n)$ for all $j \geqslant 0$, $n \geqslant 0$, as property (I) of Theorem I.1 requires. To see this, observe that f^{-1} and g^{-1} differ only on S. Therefore we have only to show that $f^{-j}(x_n) \in S$ for all $j \geqslant 0$ and $n \geqslant 0$. But there exists $N_0 > 0$ such that $f^{-j}(x_n) \in V^-$ for all $j > N_0$, $n \geqslant 0$. Since $S \cap V^- = \varnothing$, it remains to show that $f^{-j}(x_n) \notin S$ for all $0 \leqslant j \leqslant N_0$, $n \geqslant 0$. But the set $\{f^{-j}(x_n) \mid n \geqslant 0, 0 \leqslant j \leqslant N_0\}$ is compact and disjoint from Λ; since S can be taken in an arbitrarily small neighborhood of Λ, it follows that S does not contain points of that set. This completes the proof of Theorem I.1 but for Lemma 2.

Finally, let us prove Lemma 2. For all $x \in \Lambda$ define

$$W_\varepsilon^u(x) = \{y \mid d(f^n(x), f^n(y)) \leqslant \varepsilon \text{ for all } n \geqslant 0\},$$
$$W_\varepsilon^s(x) = \{y \mid d(f^{-n}(x), f^{-n}(y)) \leqslant \varepsilon \text{ for all } n \geqslant 0\}.$$

When ε is small, these sets are C^1 disks tanget at x to the subspaces E_x^s, E_x^u. Take a compact neighborhood U_0 of Λ such that

$$\bigcap_n f^n(U_0) = \Lambda.$$

Choose $\varepsilon > 0$ so small that

(24) $$W_\varepsilon^u(x) \cup W_\varepsilon^s(x) \subset U_0$$

for all $x \in \Lambda$ and define $V^+ = \bigcup_{x \in \Lambda} W_\varepsilon^s(x)$, $V^- = \bigcup_{x \in \Lambda} W_\varepsilon^u(x)$. Clearly these sets are compact and satisfy

(25) $$f(V^+) \subset V^+$$

(26) $$f^{-1}(V^-) \subset V^-.$$

Moreover property (24) implies that $V^+ \subset U_0 \subset U$ and $V^- \subset U_0 \subset U$. From (25) and (26) follows that if $y \in V^+ \cap V^-$ then $f^n(y) \in f^n(V^+) \subset V^+ \subset U_0$, for all $n \geqslant 0$, and $f^{-n}(y) \in f^{-n}(V^-) \subset V^- \subset U_0$, for all $n \geqslant 0$. Hence $y \in \bigcap_n f^n(U_0) = \Lambda$. Then

$$(27) \qquad V^+ \cap V^- = \Lambda.$$

To complete the proof of Lemma 2 it remains to show the inequalities of c). To prove them we take ε so small that

$$(28) \qquad \begin{aligned} \overline{V^+ - f(V^+)} \cap \Lambda &= \emptyset \\ \overline{V^- - f^{-1}(V^-)} \cap \Lambda &= \emptyset. \end{aligned}$$

That such an ε exists is proved in [1]. Besides this non trivial property we shall need two easy lemmas.

Lemma 5. — *There exist* $\delta > 0$ *and* $0 < \gamma < \lambda < 1$ *such that if* $p \in \Lambda$, $x \in M$ *and* $d(x, p) \leqslant \delta$, *then*

$$\gamma\, d(x, W^u_\varepsilon(p)) \leqslant d(f(x), W^u_\varepsilon(f(p))) \leqslant \lambda\, d(x, W^u_\varepsilon(p))$$

and

$$\gamma\, d(x, W^s_\varepsilon(p)) \leqslant d(f^{-1}(x), W^s_\varepsilon(f^{-1}(p))) \leqslant \lambda\, d(x, W^s_\varepsilon(p)).$$

This lemma is an easy application of the basic properties of hyperbolic sets. We leave its proof to the reader.

Lemma 6. — *For all* $N > 0$ *there exists* $c(N) = c > 0$ *such that if* $d(x, \Lambda) \leqslant c$ *and* $p \in V^-$ *satisfies* $d(x, p) = d(x, V^-)$ *then* $p \in f^{-N}(V^-)$.

Proof. — If the lemma is false, there exist $N > 0$ and sequences $\{x_n\}$ and $\{p_n\} \subset V^-$ such that

$$(29) \qquad d(x_n, \Lambda) \leqslant 1/n,$$

$$(30) \qquad p_n \in V^-,$$

$$(31) \qquad d(x_n, p_n) = d(x_n, V^-)$$

and

$$(32) \qquad p_n \notin f^{-N}(V^-)$$

for all $n \geqslant 0$. We can suppose that the sequences $\{p_n\}$ and $\{x_n\}$ converge to p and x respectively. From (29) it follows that $x \in \Lambda$. Moreover, by (31),

$$d(x_n, p_n) = d(x_n, V^-) \leqslant d(x_n, \Lambda) \leqslant 1/n.$$

Hence $p = x$. But (30) and (32) imply that

$$p_n \in \overline{V^- - f^{-N}(V^-)}$$

for all $n \geqslant 0$, hence

$$x = p \in \overline{V^-} - \overline{f^{-N}(V^-)}.$$

Then $$\overline{V^- - f^{-N}(\overline{V^-})} \cap \Lambda \neq \emptyset,$$

which is easily seen to contradict (28), thus proving Lemma 6.

To complete the proof of Lemma 2 take $0 < \delta_1 < \delta$ (where δ is given by Lemma 5) such that $d(a, b) \leqslant \delta_1$ implies $d(f^{-1}(a), f^{-1}(b)) \leqslant \delta$. Choose $N > 0$ such that

$$\operatorname{diam} f^{-N}(W^u_\varepsilon(p)) \leqslant \delta_1/2 \tag{33}$$

for all $p \in \Lambda$. By (27) there exists $\varepsilon_1 > 0$ that if $d(x, V^+) \leqslant \varepsilon_1$ and $d(x, V^-) \leqslant \varepsilon_1$ then $d(x, \Lambda) \leqslant c$ and $d(f(x), \Lambda) \leqslant c$, where $c = c(N)$ is given by Lemma 6. Define $\varepsilon_0 = \min\{\varepsilon_1, \delta_1/2\}$. Let us show that this choice of ε_0 satisfies property c) of Lemma 2. If $d(x, V^+) \leqslant \varepsilon_0$ and $d(x, V^-) \leqslant \varepsilon_0$, then

$$d(x, \Lambda) \leqslant c$$

and then, by Lemma 6, there exists $y \in f^{-N}(V^-)$ such that

$$d(x, y) = d(x, V^-).$$

Since $y \in f^{-N}(V^-)$ it follows that

$$y \in f^{-N}(W^u_\varepsilon(q)) \subset W^u_\varepsilon(f^{-N}(q))$$

for some $q \in \Lambda$. Define $p = f^{-N}(q)$. By (33), we have

$$d(y, p) \leqslant \delta_1/2$$

and then

$$d(x, p) \leqslant d(x, y) + d(y, p) \leqslant d(x, V^-) + \delta_1/2 \leqslant \varepsilon_0 + \delta_1/2 \leqslant \delta_1/2 + \delta_1/2 = \delta_1 \leqslant \delta.$$

Now we apply Lemma 5 to obtain

$$d(f(x), V^-) \leqslant d(f(x), W^u_\varepsilon(f(p))) \leqslant d(x, W^u_\varepsilon(p)) \leqslant \lambda\, d(x, y) = \lambda\, d(x, V^-),$$

thus proving one of the desired inequalities. Moreover, by the way we chose ε_0 and ε_1, we know that $d(x, V^-) \leqslant \varepsilon_0$ and $d(x, V^+) \leqslant \varepsilon_0$ imply

$$d(f(x), \Lambda) \leqslant c.$$

Then, by Lemma 6, there exists $z \in f^{-N}(V^-)$ such that

$$d(f(x), z) = d(f(x), V^-).$$

Since $z \in f^{-N}(V^-)$ we can write

$$z \in f^{-N}(W^u_\varepsilon(q)) \subset W^u_\varepsilon(f^{-N}(q))$$

for some $q \in \Lambda$. Define $p = f^{-N}(q)$. By (33), $d(z, p) \leqslant \delta_1/2$. Hence, arguing as above, we obtain

$$d(f(x), p) \leqslant d(f(x), z) + d(z, p) \leqslant d(f(x), V^-) + \delta_1/2 \leqslant \varepsilon_0 + \delta_1/2 \leqslant \delta_1$$

and, by the way we chose δ_1, this inequality implies

$$d(x, f^{-1}(p)) \leqslant \delta.$$

Applying Lemma 5 to x and $f^{-1}(p)$, we have

$$\gamma\, d(x, V^-) \leqslant \gamma\, d(x, W^u_\varepsilon(f^{-1}(p))) \leqslant d(f(x), W^u_\varepsilon(p)) \leqslant d(f(x), z) = d(f(x), V^-).$$

This concludes the proof of the inequalities of Lemma 2 involving $x, f(x)$ and V^-. The proof of those involving $x, f^{-1}(x)$ and V^+ is the same with some obvious modifications.

II. — Proof of Theorem D

To prove Theorem D we shall use all the auxiliary sets and notation of the proof of Theorem I.1. Suppose that $x \notin W^s(\Lambda)$ and $\mu(\Lambda) > 0$ for all $\mu \in \mathcal{M}(x)$. Since $x \notin W^s(\Lambda)$ there exists $N > 0$ such that $f^N(x) \notin V_0$. Obviously $\omega(x) \cap \Lambda \neq \emptyset$. Then there exists $N_1 > N$ such that $f^{N_1}(x) \in V_0$. If $j \geqslant N_1$ define

$$A(j) = \sup\{n \mid S_n \cap \{f^N(x), \ldots, f^j(x)\} \neq \emptyset\}.$$

Observe that $A(j) < +\infty$ because otherwise we would have:

$$\{f^N(x), \ldots, f^j(x)\} \cap (V^+ \cup V^-) = \bigcap_{n \geqslant 0} (\{f^N(x), \ldots, f^j(x)\} \cap S_n) \neq \emptyset$$

and then either $f^N(x) \in V^-$ (that is impossible since $f^N(x) \notin V_0$) or $f^j(x) \in V^+$ (that is impossible because $x \notin W^s(\Lambda)$). Hence $A(j) < +\infty$ and then in its definition the supremum is actually a maximum. Let $a_1 < a_2 < \ldots$ be the set of values of $A(j)$ for $j \geqslant N_1$. Define

$$\tilde{m}_n = \max\{j \mid A(j) = a_n\}.$$

Then $\tilde{m}_1 < \tilde{m}_2 < \ldots$. Take a subsequence $\tilde{m}_{n_1} < \tilde{m}_{n_2} < \ldots$ such that, denoting $m_k = \tilde{m}_{n_k}$, the probabilities $\mu(x, m_k)$ converge to $\mu \in \mathcal{M}(x)$. By hypothesis, $\mu(\Lambda) > 0$; hence $\mu(\Lambda_0) > 0$ for some basic component Λ_0 of Λ.

Let us apply Lemma 4 of Section I to $\bar{x} = x$, the isolated hyperbolic set Λ_0 and the sequences $x_k = x$ and $\mu(x, m_k)$, $k \in \mathbf{Z}^+$. If property a) of the conclusion of that lemma holds, then the proof of Theorem I.1 shows that it implies the validity of property I.1 (II). This means that in every neighborhood of f there exists a diffeomorphism g that coincides with f in a neighborhood of Λ and has an homoclinic point associated to Λ_0, thus proving Theorem D. Therefore the proof of Theorem D will be completed if we can show that when property b) of the conclusion of Lemma 4 holds, then property a) also holds. More specifically we shall prove that if for some $k > 0$ and $n > 0$ there exists a k-string $\sigma_1 \subset S_{n+1}$ such that $\sigma \cap (S_n - S_{n+1}) = \emptyset$ for every k-string, then there exist two $(k+1)$-strings $\sigma'_1 < \sigma'_2$ contained in S_{n+1} and such that $\sigma \cap (S_n - S_{n+1}) = \emptyset$ for every $(k+1)$-string $\sigma'_1 < \sigma < \sigma'_2$. To prove this observe that since σ_1 is a k-string and is contained in S_{n+1}, we have

$$a_{n_k} = A(\widetilde{m}_{n_k}) = A(m_k) \geqslant n+1.$$

Let σ_2 be the first $(k+1)$-string that is not a k-string. The definition of m_k implies that $\sigma_2 \subset S_\ell$ for some $\ell > a_{n_k}$, thus $\ell > n+1$. Then σ_1 and σ_2 are $(k+1)$-strings contained in S_{n+1} and $\sigma_2 > \sigma_1$. Moreover, if there exists a $(k+1)$-string σ such that $\sigma_1 < \sigma < \sigma_2$, then, by our choice of σ_2, necessarily σ is also a k-string and therefore $\sigma \cap (S_n - S_{n+1}) = \emptyset$ by hypothesis. Setting $\sigma'_1 = \sigma_1$ and $\sigma'_2 = \sigma_2$, the proof of Theorem D is concluded.

BIBLIOGRAPHY

[1] M. Hirsch, J. Palis, C. Pugh, M. Shub, Neighborhoods of hyperbolic sets, *Invent. Math.*, **9** (1970), 121-134.

[2] D. Pixton, Planar homoclinic points, *Journal of Differential Equations*, **44** (1982), 365-382.

[3] C. Robinson, Closing stable and unstable manifolds on the two-sphere, *Proc. Amer. Math. Soc.*, **41** (1973), 299-303.

[4] S. Smale, Diffeomorphisms with many periodic points, *Differential and Combinatorial Topology*, 63-70, Princeton Univ. Press, 1965.

[5] F. Takens, Homoclinic points in conservative systems, *Invent. Math.*, **18** (1972), 267-292.

Instituto de Matemática Pura e Aplicada (IMPA)
Estrada Dona Castorina, 110
Jardim Botânico
Rio de Janeiro - RJ

Manuscrit reçu le 12 mars 1987.

A PROOF OF THE C^1 STABILITY CONJECTURE

by RICARDO MAÑÉ

INTRODUCTION

Two continuous maps, $f_1 : X_1 \circlearrowleft$ and $f_2 : X_2 \circlearrowleft$ are *topologically equivalent* if there exists a homeomorphism $h : X_1 \to X_2$ such that $h^{-1} f_2 h = f_1$. A C^r diffeomorphism f of a closed manifold M is C^r *structurally stable* if it has a C^r neighborhood $\mathcal{U}$ such that every $g \in \mathcal{U}$ is topologically equivalent to f. This concept was introduced in the thirties by Andronov and Pontrjagin [1], in the limited (when compared with its present range) framework of flows on the two dimensional disk. The turning point of its development that connected it with much richer possibilities, came in the early sixties, through the work of Smale who, as a consequence of his improved version of a classical result of Birkhoff about homoclinic points, showed that structural stability can coexist with highly developed forms of recurrence [24].

Immediatly afterwards, the understanding of the mechanisms that grant structural stability grew substantially through the papers of Anosov [2], Smale [25] and Palis and Smale [16], that proved several new classes of dynamical systems to be structurally stable. On the light of these results, and intending to unify them, Palis and Smale conjectured in their joint paper that the two conditions known as Axiom A and the Strong Transversality Condition (whose definitions we shall recall below) are necessary and sufficient for a C^r diffeomorphism to be C^r structurally stable. Their sufficiency was proved in the well known papers of Robbin [20] for $(r \geqslant 2)$ and Robinson [22] (for $r = 1$). The question of their necessity was reduced to prove that C^r structural stability implies Axiom A (Robinson [21]). This problem became known as the Stability Conjecture, and it is the objective of this paper to prove it in the C^1 case.

Theorem A. — *Every* C^1 *structurally stable diffeomorphism of a closed manifold satisfies Axiom A.*

In the next section we shall prove this result. The proof will be supported on six theorems. Three of them were already known; the other three will be proved in the remaining sections.

Originally published in: *Publ. Math. Inst. Hautes Etudes Sci. No. 66 (1988), 161–210,*

Several relevant problems closely connected with the Stability Conjecture remain open; notably the C^r case (that looks beyond the scope of the available techniques) and, even in the C^1 case, the characterization of the more flexible form of stability known as Ω-stability as well as the corresponding problems for flows, for which the methods we use here open realistic possibilities. The case of flows on compact manifolds with boundary that are tangent to the boundary pose a different type of problem. Recent examples show that Axiom A is *not* necessary for structural stability [7].

Before developing the discussion of these questions, we shall first recall the definition and main virtues of Axiom A dynamics.

From now on M will denote a closed manifold and $\mathrm{Diff}^r(M)$ will be the space of C^r diffeomorphisms of M endowed with the C^r topology. We say that Λ is a *hyperbolic set* of $f \in \mathrm{Diff}^r(M)$ if it is compact, invariant (i.e. $f(\Lambda) = \Lambda$) and there exists a continuous splitting $TM/\Lambda = E^s \oplus E^u$ (where TM/Λ is the tangent bundle restricted to Λ) that is invariant (i.e. $(Df)\,E^s = E^s$, $(Df)\,E^u = E^u$) and there exist constants $C > 0$, $0 < \lambda < 1$ such that

$$\|(Df^n)/E^s(x)\| \leq C\lambda^n,$$
$$\|(Df^{-n})/E^u(x)\| \leq C\lambda^n$$

for all $x \in \Lambda$ and $n \geq 0$. Expositions of the rich theory of hyperbolic sets can be found in the books of Bowen [3], Newhouse [14] and Shub [23]. Given $f \in \mathrm{Diff}^r(M)$ and $x \in M$ define the *stable* and *unstable manifolds* of x as:

$$W^s_f(x) = \{y \in M \mid \lim_{n\to+\infty} d(f^n(x), f^n(y)) = 0\}$$
$$W^u_f(x) = \{y \in M \mid \lim_{n\to+\infty} d(f^{-n}(x), f^{-n}(y)) = 0\}.$$

When dealing with only one diffeomorphism, as will be the case in this section, we shall denote these sets as $W^s(x)$ and $W^u(x)$.

When x belongs to a hyperbolic set, then $W^s(x)$ and $W^u(x)$ are C^r injectively immersed submanifolds ([25], [6]).

The nonwandering set $\Omega(f)$ of f is defined as the set of points $x \in M$ such that for every neighborhood U of x there exists $n \geq 1$ satisfying $f^n(U) \cap U \neq \emptyset$. When $\Omega(f)$ is a hyperbolic set and the periodic points are dense in $\Omega(f)$, we say that f satisfies *Axiom A*. In this case it is known [25] that

$$M = \bigcup_{x\in\Omega(f)} W^s(x) = \bigcup_{x\in\Omega(f)} W^u(x). \tag{1}$$

Using this property it is easy to see that $W^s(y)$ and $W^u(y)$ are C^r injectively immersed manifolds for all $y \in M$, because by (1), for all $y \in M$, there exists $x \in \Omega(f)$ such that $y \in W^s(x)$ and then $W^s(x) = W^s(y)$. Since $W^s(x)$ is a C^r injectively immersed manifold, the property is proved.

We say that an Axiom A diffeomorphism f satisfies the *Strong Transversality Condition* when

$$T_x W^s(x) + T_x W^u(x) = T_x M$$

for all $x \in M$, or, what is equivalent by (1), if for all p and q in $\Omega(f)$, $W^s(p)$ and $W^u(q)$ intersect transversally. There are several characterizations of diffeomorphisms satisfying Axiom A and the Strong Transversality Condition. For instance, $f \in \text{Diff}^r(M)$ satisfies Axiom A and the Strong Transversality Condition if and only if every tangent vector $v \in TM$ can be decomposed as $v = v^+ + v^-$, where v^+ and v^- satisfy

$$\liminf_{n\to+\infty} \|(Df^n)\, v^+\| = \liminf_{n\to-\infty} \|(Df^n)\, v^-\| = 0.$$

For this and other characterizations, see [11].

Let us now discuss the open problems related to Theorem A. The first one must be the C^r case of Theorem A with $r > 1$. Unfortunately there is little to say about this question. Not being even known whether a C^2 structurally stable diffeomorphism has at least one periodic point it seems, to say the least, difficult to prove that they are dense in the nonwandering set as the definition of Axiom A requires. Even if this density property is proved and unless the method used to achieve this feat sheds new light on these questions, other disturbingly simple unanswered questions remain (see the Introduction of [12]).

Turning to more feasible questions, we have the problem of characterizing Ω-stability, that is defined as follows: f is C^r *Ω-stable* if it has a C^r neighborhood $\mathcal{U}$ such that $g/\Omega(g)$ is topologically equivalent to $f/\Omega(f)$ for all $g \in \mathcal{U}$. Smale proved that if f satisfies Axiom A plus the so called no cycles condition then f is Ω-stable [26]. The converse problem has been reduced to proving that C^r Ω-stability implies Axiom A (Palis [15]). When $r > 1$ this problem runs into the same (or worse) stumbling blocks than the Stability Conjecture. When $r = 1$ we think, as we say above, that the techniques developed here make of it a realistic target. Similar comments hold for the corresponding problems for flows on boundaryless compact manifolds. But in the quite natural attempt to study structural stability in the space of flows on a compact manifold with boundary that are tangent to the boundary, new and different problems arise. Labarca and Pacifico [7] have found examples that show that in this framework there exist structurally stable flows that do not satisfy Axiom A. The conjecture itself, then, must be reformulated in terms that so far have not been proposed.

Returning to the case of diffeomorphisms of a closed manifold M, define $\mathcal{F}^r(M)$ as the set of diffeomorphisms $f: M \circlearrowleft$ having a C^r neighborhood $\mathcal{U}$ such that all the periodic points of every $g \in \mathcal{U}$ are hyperbolic. It is easy to see [4] that C^r structurally or Ω-stable diffeomorphisms belong to $\mathcal{F}^r(M)$. Moreover most of the steps toward proving that structural or Ω-stability imply Axiom A use only the weaker fact that such diffeomorphisms belong to $\mathcal{F}^r(M)$. For this reason we conjectured in [12] that every element of $\mathcal{F}^r(M)$ satisfies Axiom A. For the reasons we have just explained, this conjecture contains the questions of whether structural or Ω-stability imply Axiom A. Once more, and for the same reasons than in the previous problems, let us leave aside the case $r > 1$. When $\dim M = 2$ (and $r = 1$) we proved this conjecture in [12]. Even if the techniques developed here fall short of extending this result to the n-dimensional case, it is interesting, and promising, that most of the steps of the proof of Theorem A require only the hypothesis $f \in \mathscr{F}^1(M)$. It is only in the last step where we need the whole weight of the structural stability of f.

On the other hand, if we define $\mathscr{F}^1(M)$ for flows in the obvious, analogous form to that used for diffeomorphisms, it is not true that flows in $\mathscr{F}^1(M)$ satisfy Axiom A. An exemple is the Guckenheimer-Lorenz attractor [5], that also plays the key role in the construction of the example of Labarca and Pacifico mentioned above.

I wish to thank Jacob Palis for several important corrections and to Claus Doering for his deep and exhaustive revision of the first version of this work.

I. — Proof of Theorem A

As we explained in the Introduction, in this section we shall prove Theorem A, using for this purpose six theorems that either have been already proved elsewhere or will be proved in the next sections.

Let M be a closed manifold and let $\mathscr{F}^1(M)$ be defined as in the Introduction. Let $P(f)$ denote the set of periodic points of the diffeomorphism f and, if $x \in P(f)$, let $E^s(x)$ and $E^u(x)$ be the stable and unstable subspaces of $T_x M$, i.e. the subspaces associated to the eigenvalues of $Df^n : T_x M \circlearrowleft$ (where n is the period of x) that have respectively modulus < 1 and > 1. Clearly $(Df)\, E^s(x) = E^s(f(x))$, $(Df)\, E^u(x) = E^u(f(x))$ and, if x is hyperbolic, $T_x M = E^s(x) \oplus E^u(x)$. Denote by $\overline{P}(f)$ the closure of $P(f)$.

The first step of the proof of Theorem A is the following corollary of Pugh's Closing Lemma [19] proved in the Introduction of [12].

Theorem **I.1.** — *If* $f \in \mathscr{F}^1(M)$, *then* $\Omega(f) = \overline{P}(f)$.

Now define $P_i(f)$ as the set of points $x \in P(f)$ such that $\dim E^s(x) = i$. By I.1

$$\Omega(f) = \bigcup_{i=0}^{\dim M} \overline{P}_i(f)$$

when $f \in \mathscr{F}^1(M)$. Then, if $f \in \mathscr{F}^1(M)$, it is sufficient to show that $\overline{P}_i(f)$ is a hyperbolic set for all $0 \leqslant i \leqslant \dim M$. The cases $i = 0$ and $i = \dim M$ follow from a theorem due to Pliss.

Theorem **I.2** (Pliss [18]). — *If* $f \in \mathscr{F}^1(M)$, *then* $P_0(f)$ *and* $P_{\dim M}(f)$ *are finite.*

Obviously this implies that $\overline{P}_0(f) = P_0(f)$ and $P_{\dim M}(f) = \overline{P}_{\dim M}(f)$ are hyperbolic sets when $f \in \mathscr{F}^1(M)$. To prove the hyperbolicity of the sets $\overline{P}_i(f)$ for $1 \leqslant i \leqslant \dim M$ the basic strategy is the obvious one: to start with the splittings $T_x M = E^s(x) \oplus E^u(x)$ that we have when $x \in P(f)$ to show that this splitting of $TM/P_i(f)$ extends to a splitting of $TM/\overline{P}_i(f)$ satisfying the definition of hyperbolicity. The next result provides the extension and some indications of its hyperbolicity. Its statement uses the concept of *dominated splitting*, that will appear also in several results of this section and is defined as follows. Given a compact invariant set Λ of a diffeomorphism f we say that a splitting $TM/\Lambda = E \oplus F$ is a dominated splitting if it is continuous, invariant and there exists $C > 0$ and $0 < \lambda < 1$ such that

$$\|(Df^n)/E(x)\| . \|(Df^{-n})/F(f^n(x))\| \leqslant C\lambda^n$$

for all $x \in \Lambda$ and $n \geqslant 0$. In geometric terms this is equivalent to say that for every one-dimensional subspace $L \subset T_x M$, $x \in \Lambda$, not contained in $E(x)$, the angle between $(Df^n) L$ and $F(f^n(x))$ converge exponentially to zero as $n \to +\infty$.

Theorem **I.3.** — *If* $f \in \mathscr{F}^1(M)$ *there exist* $C > 0$, $0 < \lambda < 1$, $m > 0$ *and a* C^1 *neighborhood* $\mathscr{U}$ *of* f *such that for all* $g \in \mathscr{U}$ *and* $0 < i < \dim M$ *there exists a dominated splitting* $TM/\overline{P}_i(g) = \widetilde{E}_i^s \oplus \widetilde{E}_i^u$ *satisfying:*

a) $\|(Dg^m)/\widetilde{E}_i^s(x)\| . \|(Dg^{-m})/\widetilde{E}_i^u(g^m(x))\| \leqslant \lambda$ for all $x \in \overline{P}_i(g)$,

b) $\widetilde{E}_i^s(x) = E^s(x)$ *and* $\widetilde{E}_i^u(x) = E^u(x)$ *if* $x \in \overline{P}_i(g)$

c) *If* $x \in P_i(g)$ *and has period* $n > m$, *then*

$$\prod_{j=0}^{[n/m]-1} \|(Dg^m)/E^s(g^{mj}(x))\| \leqslant C\lambda^{[n/m]}$$

$$\prod_{j=1}^{[n/m]} \|(Dg^{-m})/E^u(g^{mj}(x))\| \leqslant C\lambda^{[n/m]},$$

d) *For all* $x \in P_i(g)$

$$\lim_{n\to+\infty} \frac{1}{n} \sum_{j=0}^{n-1} \log\|(Dg^m)/E^s(g^{mj}(x))\| \leqslant \log \lambda$$

$$\lim_{n\to+\infty} \frac{1}{n} \sum_{j=0}^{n-1} \log\|(Dg^{-m})/E^u(g^{mj}(x))\| \leqslant \log \lambda.$$

Observe that d) is interesting only when the period of x is $\leqslant m$. Otherwise it is just a corollary of c).

Theorem I.3 was independently proved in [17], [10] and [8]. The statement used above is taken from [12], where there is also a proof of I.3.

After Theorem I.3 the problem becomes to show that the splitting $TM/\overline{P}_i(f) = \widetilde{E}_i^s \oplus \widetilde{E}_i^u$ is hyperbolic for all $1 \leqslant i < \dim M$. If there is a hyperbolic splitting it must be this. The following, and fundamental, step is a theorem saying that to prove the hyperbolicity of the splitting $TM/\overline{P}_i(f) = \widetilde{E}_i^s \oplus \widetilde{E}_i^u$ it suffices to show only that Df contracts the subbundle $\widetilde{E}_i^s$. To state this result it is convenient to introduce a definition: given a compact invariant set Λ of $f: M \circlearrowleft$, we say that a subbundle $E \subset TM/\Lambda$ is *contracting* if it is continuous, invariant and there exist $C > 0$ and $0 < \lambda < 1$ such that

$$\|(Df^n)/E(x)\| \leqslant C\lambda^n$$

for all $n \geqslant 0$ and $x \in \Lambda$. We say that it is *expanding* if there exist $C > 0$ and $0 < \lambda < 1$ such that

$$\|(Df^{-n})/E(x)\| \leqslant C\lambda^n$$

for all $x \in \Lambda$ and $n \geqslant 0$.

Theorem **I.4.** — *If* $f \in \mathscr{F}^1(M)$, $0 < i < \dim M$ *and* $\tilde{E}_i^s$ *is contracting, then* $\tilde{E}_i^u$ *is expanding.*

This theorem will be proved in Section II as a corollary of a slightly more general result.

Now our problem is reduced to show that $f \in \mathscr{F}^1(M)$ implies that $\tilde{E}_i^s$ is contracting for all $0 < i < \dim M$. To recognize the contracting property the following easy lemma is extremely useful because it translates this property into averages with respect to ergodic measures.

Denote by $\mathscr{M}(f/\Lambda)$ the set of invariant probabilities on the Borel σ-algebra of Λ endowed with the weak topology, i.e. the unique metrizable topology such that

$$\mu_n \to \mu \Leftrightarrow \int \varphi \, d\mu_n \to \int \varphi \, d\mu$$

for every continuous $\varphi : \Lambda \to \mathbf{R}$.

Lemma **I.5.** — *Let* Λ *be a compact invariant set of* $f \in \mathrm{Diff}^1(M)$ *and* $E \subset TM/\Lambda$ *be a continous invariant subbundle. If there exists* $m > 0$ *such that*

$$\int \log \|(Df^m)/E\| \, d\mu < 0$$

for every ergodic $\mu \in \mathscr{M}(f^m/\Lambda)$, *then* E *is contracting.*

Proof. — It is easy to see that if for each $x \in \Lambda$ there exists $n > 0$ such that

$$\|(Df^n)/E(x))\| < 1,$$

then E is contracting. Stronger than this is to say that for each $x \in \Lambda$ there exists $n > 0$ satisfying

$$\prod_{j=0}^{n-1} \|(Df^m)/E(f^{mj}(x))\| < 1.$$

Suppose this property is false. Then there exists $x \in \Lambda$ such that

$$\prod_{j=0}^{n-1} \|(Df^m)/E(f^{mj}(x))\| \geqslant 1$$

for all n. Hence, for all n,

$$\frac{1}{n}\sum_{j=0}^{n-1} \log \|(Df^m)/E(f^{mj}(x))\| \geqslant 0.$$

Define a probability μ_n by

$$\mu_n = \frac{1}{n}\sum_{j=0}^{n-1} \delta_{f^{mj}(x)}$$

and let $\{ \mu_{n_k} \mid k \geqslant 0 \}$ be a convergent subsequence. Its limit μ_0 belongs to $\mathscr{M}(f^m/\Lambda)$ and

$$\int \log \|(\mathrm{D}f^m)/\mathrm{E}\| \, d\mu_0 = \lim_{k\to+\infty} \int \log \|(\mathrm{D}f^m)/\mathrm{E}\| \, d\mu_{n_k}$$
$$= \lim_{k\to+\infty} \frac{1}{n_k}\sum_{j=0}^{n_k-1} \log \|(\mathrm{D}f^m)/\mathrm{E}(f^{mj}(x))\| \geqslant 0.$$

But if the integral with respect to μ_0 of $\|(\mathrm{D}f^m)/\mathrm{E}\|$ is $\geqslant 0$, by the Ergodic Decomposition Theorem there exists an ergodic $\mu \in \mathscr{M}(f^m/\Lambda)$ with the same property and the lemma is proved.

Now suppose that $f \in \mathscr{F}^1(\mathrm{M})$ and let us try to prove that f satisfies Axiom A, which, as we explained above, is reduced to the contracting property of $\widetilde{\mathrm{E}}_i^s$ for all $0 < i < \dim \mathrm{M}$, and we shall try to do it by induction on i and using Lemma I.5. If $f \in \mathscr{F}^1(\mathrm{M})$, $\mathrm{P}_0(f)$ is hyperbolic by Theorem I.2. Now suppose that $\overline{\mathrm{P}}_k(f)$ is hyperbolic for $0 \leqslant k \leqslant j$ and let us try to prove the hyperbolicity of $\overline{\mathrm{P}}_{j+1}(f)$. For this purpose we need the following result, that was implicitly proved in [12] and will be explicitly proved in Section III.

Theorem **I.6.** — *If* $f \in \mathscr{F}^1(\mathrm{M})$ *and* $m > 0$ *is given by* I.3, *there exists* $0 < \lambda_0 < 1$ *such that if* $\overline{\mathrm{P}}_k(f)$ *is hyperbolic for all* $0 \leqslant k < i$ *and* $\mu \in \mathscr{M}(f^m/\overline{\mathrm{P}}_i(f))$ *satisfies*

(1) $$\int \log \|(\mathrm{D}f^m)/\widetilde{\mathrm{E}}_i^s\| \, d\mu \geqslant \log \lambda_0,$$

then

(2) $$\mu\Big(\bigcup_{0\leqslant k<i} \overline{\mathrm{P}}_k(f)\Big) > 0.$$

To complete the induction step, it suffices to show that

(3) $$\mu\Big(\bigcup_{0\leqslant k\leqslant j} \overline{\mathrm{P}}_k(f)\Big) = 0$$

for all $\mu \in \mathscr{M}(f^m/\overline{\mathrm{P}}_{j+1}(f))$ because, by Theorem I.6, this implies that there are no measures $\mu \in \mathscr{M}(f^m/\overline{\mathrm{P}}_{j+1}(f))$ satisfying (1). Hence

$$\int \log \|(\mathrm{D}f^m)/\widetilde{\mathrm{E}}_{j-1}^s\| \, d\mu < \log \lambda_0 < 0$$

for all $\mu \in \mathscr{M}(f^m/\overline{\mathrm{P}}_{j+1}(f))$ and then, by Lemma I.5, $\widetilde{\mathrm{E}}_{j+1}^s$ is contracting and, by I.4, $\overline{\mathrm{P}}_{j+1}(f)$ is hyperbolic. This would complete the induction step and also the proof of the Axiom A property for f. However we are not able to prove that (3) holds for every $\mu \in \mathscr{M}(f^m/\overline{\mathrm{P}}_{j+1}(f))$ using only the hypothesis $f \in \mathscr{F}^1(\mathrm{M})$. We shall do it when f is C¹ structurally stable (thus proving Theorem A). For this we need the following theorem, for whose statement we shall recall the definition of a basic set. A *basic set* of $f \in \mathrm{Diff}^1(\mathrm{M})$

is a hyperbolic set Λ that is transitive (i.e. there exists $x \in \Lambda$ whose ω-limit set is Λ) and isolated, i.e. it has a compact neighborhood U satisfying

$$\bigcap_n f^n(U) = \Lambda.$$

The transitivity implies that the dimension of the fibers of the stable subspace of the hyperbolic splitting of TM/Λ is constant and we shall call it the index of Λ and denote it $\mathrm{Ind}(\Lambda)$. The stable and unstable sets of Λ are defined by

$$W^s_f(\Lambda) = \{y \mid \lim_{n\to+\infty} d(f^n(y), \Lambda) = 0\}$$

$$W^u_f(\Lambda) = \{y \mid \lim_{n\to+\infty} d(f^{-n}(y), \Lambda) = 0\}.$$

When it is clear with respect to which diffeomorphism we are considering $W^s_f(\Lambda)$ and $W^s_f(\Lambda)$, we shall denote them by $W^s(\Lambda)$ and $W^u(\Lambda)$. The following theorem will be proved in Section V.

Theorem **I.7.** — *Let* Λ *be a compact invariant set of* $f \in \mathrm{Diff}^1(M)$ *such that* $\Omega(f/\Lambda) = \Lambda$ *and having a dominated splitting* $TM/\Lambda = E \oplus F$. *Suppose that there exist basic sets* $\Lambda_1, \ldots, \Lambda_s$ *of* f *and constants* $m > 0$, $c > 0$ *and* $0 < \lambda < 1$ *satisfying:*

I) $\mathrm{Ind}(\Lambda_i) < \dim E(x)$ *for all* $1 \leqslant i \leqslant s$ *and* $x \in \Lambda$.

II) *There exists a* C^1 *neighborhood* $\mathcal{U}$ *of* f *such that if* $g \in \mathcal{U}$ *coincides with* f *in a neighborhood of* $\bigcup_1^s \Lambda_k$ *then*

$$W^s_g(\Lambda_i) \cap W^u_g(\Lambda_i) = \Lambda_i$$

for all $1 \leqslant i \leqslant s$.

III) *If* $\mu \in \mathcal{M}(f^m/\Lambda)$ *satisfies*

$$\int \log \|(Df^m)/E\| \, d\mu \geqslant -c$$

then

$$\mu(\bigcup_1^s)\Lambda_k > 0.$$

IV) $\|(Df^m)/E(x)\| \cdot \|(Df^{-m})/F(f^m(x))\| \leqslant \lambda$ *for all* $x \in \Lambda$.

Then, given $1 \leqslant i \leqslant s$ *such that* $\Lambda - \Lambda_i$ *is not closed, there exist* $g \in \mathrm{Diff}^1(M)$ *arbitrarily* C^1 *near to* f, *coinciding with* f *in a neighborhood of* $\bigcup_1^s \Lambda_k$ *and* $1 \leqslant r \leqslant s$, $r \neq i$, *such that* $\Lambda - \Lambda_r$ *is not closed and*

$$W^s_g(\Lambda_i) \cap W^u_g(\Lambda_r) \neq \emptyset.$$

Besides this theorem, we shall need the following minor remark. If $g \in \mathcal{F}^1(M)$, denote by $N(i, n, g)$ the number of fixed points of g^n contained in $P_i(g)$.

Lemma **I.8.** — *If* $f \in \mathcal{F}^1(M)$ *there exists a* C^1 *neighborhood* $\mathcal{U}$ *of* f *such that*

a) $N(i, n, g_1) = N(i, n, g_2)$ *for all* $g_1 \in \mathcal{U}$, $g_2 \in \mathcal{U}$, $n > 0$ *and* $0 \leqslant i \leqslant \dim M$;

b) *if* $g \in \mathcal{U}$ *and* g *coincides with* f *in a neighborhood of* $\overline{P}_i(f)$ *for some* $0 \leqslant i \leqslant \dim M$, *then* $\overline{P}_i(g) = \overline{P}_i(f)$.

Proof. — Let $\mathcal{U} \subset \mathcal{F}^1(M)$ be an open connected neighborhood of f. To prove a) it suffices to show that $N(i, n, g_2) \geqslant N(i, n, g_1)$ because then, reversing the roles of g_1 and g_2, it follows that $N(i, n, g_2) \leqslant N(i, n, g_1)$ and then $N(i, n, g_1) = N(i, n, g_2)$. Let $g(t) \in \mathcal{U}$, $0 \leqslant t \leqslant 1$, be a continuous arc of diffeomorphisms with $g(0) = g_1$, $g(1) = g_2$. For every fixed point x of g_1^n there exists an arc $x(t) \in M$, $0 \leqslant t \leqslant 1$, such that $g(t)^n(x(t)) = x(t)$ and $x(0 = x$. The existence of this arc follows from the implicit function theorem recalling that, since $g(t) \in \mathcal{U} \subset \mathcal{F}^1(M)$ for all $0 \leqslant t \leqslant 1$, then if $g(t)^n(p) = p$ it follows that $D(g(t))(p) - I : T_p M \circlearrowleft$ is an isomorphism. Moreover observe that if $x \in P_i(g_1)$ then $x(t) \in P_i(g(t))$ for all $0 \leqslant t \leqslant 1$ (again because $g(t) \in \mathcal{F}^1(M)$ for all $0 \leqslant t \leqslant 1$). Then, for each fixed point x of g_1^n in $P_i(g_1)$ we have found a fixed point $x(1)$ of g_2^n in $P_i(g_2)$ and obviously the correspondance $x \mapsto x(1)$ is injective. This proves that $N(i, n, g_2) \geqslant N(i, n, g_1)$. To prove b), suppose that $g \in \mathcal{U}$ coincides with f in a neighborhood of $\overline{P}_i(f)$. Clearly every periodic point of f in $P_i(f)$ is also a periodic point of g in $P_i(g)$. Then $P_i(g) \supset P_i(f)$. But since $N(i, n, g) = N(i, n, f)$ for all $n > 0$, we have $P_i(g) = P_i(f)$ and then $\overline{P}_i(g) = \overline{P}_i(f)$ completing the proof of b).

Now let us return to the problem to which we had reduced the proof of Theorem A. The problem was to show that if f is C^1 structurally stable (and then $f \in \mathcal{F}^1(M)$) and $\overline{P}_k(f)$ is hyperbolic for all $0 \leqslant k \leqslant j$, then $\overline{P}_{j+1}(f)$ is hyperbolic. As we explained above, the hyperbolicity of $\overline{P}_{j+1}(f)$ is reduced to show that (3) holds for all $\mu \in \mathcal{M}(f^m/\overline{P}_{j+1}(f))$. Suppose that there exists $\mu_0 \in \mathcal{M}(f^m/\overline{P}_{j+1}(f))$ which does not satisfy (3), i.e.:

$$\mu_0\Big(\bigcup_{0 \leqslant k \leqslant j} \overline{P}_k(f)\Big) > 0. \tag{4}$$

To exhibit a contradiction between the existence of μ_0 and the structural stability of f we shall use I.7 and I.8. First observe that the hyperbolic set $\bigcup_{0 \leqslant k \leqslant j} \overline{P}_k(f)$ can be decomposed as

$$\bigcup_{0 \leqslant k \leqslant j} \overline{P}_k(f) = \Lambda_1 \cup \ldots \cup \Lambda_s,$$

where $\Lambda_1, \ldots, \Lambda_s$ are disjoint basic sets. This follows from a straighforward adaptation of Smale's Spectral Decomposition Theorem [25]. Moreover, let us show that there exist sets Λ_i such that $\overline{P}_{j+1}(f) - \Lambda_i$ is not closed. This will follow from the next lemma.

Lemma **I.9.** — *If* $\Lambda_i \cap \overline{P}_{j+1}(f) \neq \emptyset$ *then* $\overline{P}_{j+1}(f) - \Lambda_i$ *is not closed.*

Proof. — Suppose that there exists Λ_i such that $\Lambda_i \cap \overline{P}_{j+1}(f) \neq \emptyset$ and $\overline{P}_{j+1}(f) - \Lambda_i$ is closed. Then we can decompose $\overline{P}_{j+1}(f)$ as

$$\overline{P}_{j+1}(f) = (\overline{P}_{j+1}(f) \cap \Lambda_i) \cup (\overline{P}_{j+1}(f) - \Lambda_i)$$

and both sets in the union at right are compact and obviously disjoint. Moreover $\overline{P}_{j+1}(f) - \Lambda_i$ is not empty because if it were, then $\overline{P}_{j+1}(f)$ would be a subset of the hyperbolic set Λ_i that has index $\leqslant j$ and this is impossible because $\overline{P}_{j+1}(f)$ contains hyperbolic periodic points whose stable subspace has dimension $j+1$. Now take neighborhoods U and V of $\Lambda_i \cap \overline{P}_{j+1}(f)$ and $\overline{P}_{j+1}(f) - \Lambda_i$ respectively, such that

$$(5) \qquad f(U) \cap V = \emptyset.$$

Take a sequence of points $\{x_n\} \subset \overline{P}_{j+1}(f)$ converging to a point of $\overline{P}_{j+1}(f) \cap \Lambda_i$. Let γ_n be the orbit of x_n. We claim that for n sufficiently large, $\gamma_n \subset U$. If this is false there exist arbitrarily large values of n with $\gamma_n - U \neq \emptyset$. On the other hand, since $x_n \in \gamma_n$ converges to a point in $\overline{P}_{j+1}(f) \cap \Lambda_i$, for large values of n we have $\gamma_n \cap U \neq \emptyset$. Then for infinitely many values of n the orbit γ_n contains points both in U and U^c. By (5), an orbit γ_n that intersects U cannot be contained in $U \cup V$. Then there are infinitely many values of n such that γ_n contains points in the complement of $U \cup V$. Therefore, since every γ_n is contained in $\overline{P}_{j+1}(f)$, this contradicts the fact that $\overline{P}_{j+1}(f)$ is contained in $U \cup V$ and proves the claim, i.e. that for n large, $\gamma_n \subset U$. Then

$$(6) \qquad \gamma_n \subset \bigcap_j f^j(U).$$

Taking U very small, the intersection at right is a hyperbolic set close to the hyperbolic set Λ_i. Then its stable fibers have dimensions $\leqslant \operatorname{Ind}(\Lambda_i) \leqslant j$. Then by (6) the stable subspaces of the points of γ_n have dimensions $\leqslant j$, contradicting the property $\gamma_n \subset P_{j+1}(f)$. This contradiction completes the proof of the lemma.

Corollary **I.10**. — *There exist values of i such that $\overline{P}_{j+1}(f) - \Lambda_i$ is not closed.*

Proof. — If $\overline{P}_{j+1}(f) - \Lambda_i$ is closed for all $1 \leqslant i \leqslant s$ then, by Lemma I.9, the intersections $\overline{P}_{j+1}(f) \cap \Lambda_i$ are empty for all $1 \leqslant i \leqslant s$. But then

$$\mu_0(\bigcup_{0 \leqslant k \leqslant j} \overline{P}_k(f)) = \mu_0(\bigcup_1^s \Lambda_i) = 0$$

because the support of μ_0 is contained in $\overline{P}_{j+1}(f)$, thus contradicting (4).

Now let us show that we can apply Theorem I.7 to the set $\Lambda = \overline{P}_{j+1}(f)$, the dominated splitting $TM/\overline{P}_{j+1}(f) = \widetilde{E}^s_{j+1} \oplus \widetilde{E}^u_{j+1}$, the basic sets $\Lambda_1, \ldots, \Lambda_s$, $m > 0$ and $0 < \lambda < 1$ given by I.3 and $c = -\log \lambda_0$ given by Theorem I.6. Since $\operatorname{Ind}(\Lambda_i) \leqslant j$ for all i and $\dim \widetilde{E}^s_{j+1}(x) = j+1$ for all $x \in \overline{P}_{j+1}(f)$, hypothesis (I) is satisfied. Clearly $\Omega(f/\overline{P}_{j+1}(f)) = \overline{P}_{j+1}(f)$ because of the density of the periodic points in $\overline{P}_{j+1}(f)$; also IV) follows from I.3. Moreover, Theorem I.6 says that every $\mu \in \mathcal{M}(f^m/\overline{P}_{j+1}(f))$ satisfying

$$\int \log \|(Df^m)/\widetilde{E}^s_{j+1}\| \, d\mu \geqslant -c = \log \lambda_0$$

must also satisfy

$$\mu(\Lambda_1 \cup \ldots \cup \Lambda_s) = \mu(\bigcup_{0 \leqslant k \leqslant j} \overline{P}_k(f)) > 0,$$

thus proving hypothesis III). It remains to check hypothesis II). Suppose that a diffeomorphism g is so close to f that it belongs to the neighborhood $\mathcal{U}$ of f given by Lemma I.8. Suppose that g coincides with f in a neighborhood of $\bigcup_1^s \Lambda_k$ and by contradiction suppose that

$$W_g^s(\Lambda_q) \cap W_g^u(\Lambda_q) \neq \Lambda_q$$

for some $1 \leqslant q \leqslant s$. Without loss of generality we can assume that there exists $p \in W_g^s(\Lambda_q) \cap W_g^u(\Lambda_q) - \Lambda_q$ that is a transversal homoclinic point associated to Λ_q (perturbing g a little if necessary). Then there exist periodic points z, arbitrarily close to p, and having a stable subspace with dimension equal to $\mathrm{Ind}(\Lambda_q)$. Denote by ℓ this index. Then $z \in P_\ell(g)$. Moreover observe that p not only does not belong to Λ_q but also that $p \notin \bigcup_1^s \Lambda_k$. In fact, if $p \in \Lambda_k$ for some $1 \leqslant k \leqslant s$, it follows that its whole orbit is contained in Λ_k. Since this orbit accumulates in Λ_q, it follows that $\Lambda_k \cap \Lambda_q \neq \emptyset$ thus implying $k = q$ and contradicting $p \notin \Lambda_q$. Then we can assume that $z \notin \bigcup_1^s \Lambda_k$, because it can be taken arbitrarily close to $p \notin \bigcup_1^s \Lambda_k$. Let n be the period of z. Then $N(\ell, n, g) > N(\ell, n, f)$ because the fixed points of g^n in $P_\ell(g)$ include all the fixed points of f^n in $P_\ell(f)$ (because g coincides with f in a neighborhood of $P_\ell(f)$) and also z (that is not an element of $P_\ell(f)$ because $z \notin \bigcup_1^s \Lambda_k$, which contains $P_\ell(f)$). This contradiction with Lemma I.8 completes the proof of hypothesis II) of Theorem I.7.

Now let us apply I.7 to $\Lambda_1, \ldots, \Lambda_s$ and $\overline{P}_{j+1}(f)$. We take Λ_i such that $\overline{P}_{j+1}(f) - \Lambda_i$ is not closed (that exists by Corollary I.10), and I.7 yields a diffeomorphism g arbitrarily C^1 near to f, coinciding with f in a neighborhood of $\bigcup_1^s \Lambda_k$, and Λ_r such that the set $\overline{P}_{j+1}(f) - \Lambda_r$ is not closed, $r \neq i$ and

$$W_g^s(\Lambda_i) \cap W_g^u(\Lambda_r) \neq \emptyset. \tag{7}$$

But (7) is not enough, as far as we can see, to contradict the structural stability of f, unless we pick Λ_i with some extra properties that will yield that contradiction. Let us explain how to choose Λ_i. Let t be the minimum of the indexes of the sets Λ_k such that $\overline{P}_{j+1}(f) - \Lambda_k$ is not closed. Take Λ_i such that $\overline{P}_{j+1}(f) - \Lambda_i$ is not closed, $\mathrm{Ind}(\Lambda_i) = t$ and there do not exists sets Λ_k, with $k \neq i$, such that $\overline{P}_{j+1}(f) - \Lambda_k$ is not closed and

$$W_f^s(\Lambda_i) \cap W_f^u(\Lambda_k) \neq \emptyset.$$

Let us show that there exists such a Λ_i. If it does not exist, there is a family of different basic sets $\Lambda_{i_1}, \ldots, \Lambda_{i_p}$ such that their indexes are all t and

$$W_f^s(\Lambda_{i_n}) \cap W_f^u(\Lambda_{i_{n+1}}) \neq \emptyset$$

for $1 \leqslant n \leqslant p$ and

$$W_f^s(\Lambda_{i_p}) \cap W_f^u(\Lambda_{i_1}) \neq \emptyset.$$

Moreover all these intersections are transversal because of the structural stability of f. Then it is well known that a point z in, for instance, $W^s_f(\Lambda_{i_1}) \cap W^u_f(\Lambda_{i_2})$, belongs to $\overline{P}_i(f)$. Therefore z belongs to some set Λ_q. Hence the orbit of z is contained in Λ_q. This implies that Λ_q intersects Λ_{i_1} and Λ_{i_2} thus implying $q = i_1 = i_2$. This contradiction with the fact that all the sets $\Lambda_{i_1}, \ldots, \Lambda_{i_p}$ are different, completes the proof of the existence of Λ_i exhibiting the properties described above.

Now let $\mathcal{U}$ be the neighborhood of f given by Lemma I.8 and take $g \in \mathcal{U}$ as above, coinciding with f in a neighborhood of $\bigcup_1^s \Lambda_k$, and such that there exists Λ_r, with $r \neq i$, satisfying (7) and such that $\overline{P}_{j+1}(f) - \Lambda_r$ is not closed. Observe that since the intersection in (7) must be transversal, because of the structural stability of f (Robinson [21]), then $\mathrm{Ind}(\Lambda_i) \geqslant \mathrm{Ind}(\Lambda_r)$. But since $\mathrm{Ind}(\Lambda_i) = t$, the definition of t implies $\mathrm{Ind}(\Lambda_r) = \mathrm{Ind}(\Lambda_i) = t$. Moreover without loss of generality we can assume that g is topologically equivalent to f. Let $h : M \circlearrowleft$ be a homeomorphism such that $gh = hf$. Clearly $h(P_i(f)) = P_i(g)$ for all $0 \leqslant i \leqslant \dim M$ and then $h(\overline{P}_i(f)) = \overline{P}_i(g)$ for all $0 \leqslant i \leqslant \dim M$. Hence

(8) $$h\Big(\bigcup_1^s \Lambda_k\Big) = h\Big(\bigcup_{0 \leqslant k \leqslant j} \overline{P}_k(f)\Big) = \bigcup_{0 \leqslant k \leqslant j} \overline{P}_k(g).$$

But, by part b) of Lemma I.8, $\overline{P}_k(g) = \overline{P}_k(f)$ for all $0 \leqslant k \leqslant j$ because g and f coincide in a neighborhood of the union of the sets $\overline{P}_k(f)$, $0 \leqslant k \leqslant j$. Then (8) implies

$$h\Big(\bigcup_1^s \Lambda_k\Big) = \bigcup_1^s \Lambda_k$$

and it is easy to check that for all $1 \leqslant k \leqslant s$, $h(\Lambda_k)$ is another set of the family $\Lambda_1, \ldots, \Lambda_s$ with the same index as Λ_k. Define $T(f)$ as the set of pairs (n, q) such that $n \neq q$, $\mathrm{Ind}(\Lambda_n) = \mathrm{Ind}(\Lambda_q) = t$ and

(9) $$W^s_f(\Lambda_n) \cap W^u_f(\Lambda_q) \neq \emptyset.$$

Define $T(g)$ exactly in the same way replacing f by g. From the fact that h maps every set of the family $\Lambda_1, \ldots, \Lambda_s$ onto another set of the family with the same index, it follows that

(10) $$\# T(g) = \# T(f).$$

Moreover, all the intersections in (9) are transversal by the structural stability of f. Hence, when g is sufficiently close to f, if (9) holds for certain values n and q, it holds also replacing f by g. Then $T(g) \supset T(f)$. But by (10) this implies

(11) $$T(g) = T(f).$$

Now observe that $(i, r) \notin T(f)$ because by the way we chose i, no pair with i in the first entry belongs to $T(f)$. But, on the other hand, $(i, r) \in T(g)$ because of (7) and the property proved above ($\mathrm{Ind}(\Lambda_r) = t$). This contradicts (11), concluding the proof of Theorem A.

II. — Proof of Theorem I.4

Theorem I.4 will follow as a corollary of the following more general result. If Λ is a compact invariant set of $f \in \mathrm{Diff}^1(M)$, we say that a dominated splitting $TM/\Lambda = E \oplus F$ is *homogeneous* if the dimension of the subspaces $E(x)$, $x \in \Lambda$, is constant. We say that a compact neighborhood U of Λ is an *admissible neighborhood* if $TM/\bigcap_n f^n(U)$ has one and exactly one homogeneous dominated splitting $TM/\bigcap_n f^n(U) = \hat{E} \oplus \hat{F}$ extending the splitting $TM/\Lambda = E \oplus F$. It is known, and not difficult to show, that if TM/Λ has a homogeneous dominated splitting, then Λ has an admissible neighborhood U (see [6] for instance). Moreover it is clear that every compact neighborhood of Λ contained in U is another admissible neighborhood. To simplify the notation, in what follows we shall write $\bigcap_{n \in \mathbb{Z}} f^n(U) = M(f, U)$.

Theorem **II.1.** — *Let Λ be a compact invariant set of $g \in \mathrm{Diff}^1(M)$ such that $\Omega(g/\Lambda) = \Lambda$, let $TM/\Lambda = E \oplus F$ be a homogeneous dominated splitting such that E is contracting and suppose $c > 0$ is such that the inequality*

$$\liminf_{n \to +\infty} \frac{1}{n} \sum_{j=1}^{n} \log \|(Dg^{-1})/F(g^j(x))\| \leqslant -c$$

holds for a dense set of points $x \in \Lambda$. Then either F is expanding (and therefore Λ is hyperbolic) or for every admissible neighborhood V of Λ and every $0 < \gamma < 1$ there exists a periodic point $p \in M(g, V)$ with arbitrarily large period N and satisfying

$$\gamma^N \leqslant \prod_{j=1}^{N} \|(Dg^{-1})/\hat{F}(g^j(p))\| < 1,$$

where $\hat{F}$ is given by the unique homogeneous dominated splitting $TM/M(g, V) = \hat{E} \oplus \hat{F}$ that extends $TM/\Lambda = E \oplus F$.

Let us see how Theorem I.4 follows from II.1. Suppose that $f \in \mathscr{F}^1(M)$ and $\tilde{E}_i^s$ is contracting. Let m be given by Theorem I.3 and apply II.1 to $g = f^m$, $\Lambda = \overline{P}_i(f)$ and the splitting $TM/\overline{P}_i(f) = \tilde{E}_i^s \oplus \tilde{E}_i^u$. Then, by Theorem II.1, either $\tilde{E}_i^u$ is expanding (and then Theorem I.4 is proved), or, given an admissible neighborhood V and $0 < \gamma < 1$, there exists a periodic point $p \in M(f^m, V)$, with arbitrarily large period N, such that:

$$\gamma^N \leqslant \prod_{j=1}^{N} \|(Dg^{-1})/\hat{F}(g^j(p))\| < 1.$$

The fact that this product is < 1 implies that $\hat{F}(p)$ is contained in the unstable subspace $E_u(p)$. Then, by Theorem I.3

$$\prod_{j=1}^{N} \|(Dg^{-1})/\hat{F}(g^j(p))\| \leqslant \prod_{j=1}^{N} \|(Df^{-m})/E^u(f^{mj}(p))\| \leqslant C\lambda^N.$$

Thus $\gamma^N \leqslant C\lambda^N$. But if we pick γ satisfying $\lambda < \gamma < 1$ and N sufficiently large this is a contradiction.

To prove Theorem II.1 we shall use several lemmas that will be useful also in Section IV to develop the tools that in Section V will be used to prove Theorem I.7.

From now on, g, Λ and $TM/\Lambda = E \oplus F$ will be as in the statement of II.1. We shall use the following definitions. We say that a pair of points $(x, g^n(x))$ contained in Λ, $n > 0$, is a *γ-string* if

$$\prod_{j=1}^{n} \|(Dg^{-1})/F(g^j(x))\| \leqslant \gamma^n$$

and we say that it is a *uniform γ-string* if $(g^k(x), g^n(x))$ is a γ-string for all $0 \leqslant k < n$.

The first step in the proof of II.1 is the following lemma, that is a sophisticated modification of the Shadowing Lemma ([23], [14]), (or Anosov Closing Lemma) and can be proved with similar methods. An explicit proof was given by Liao [8] (only for $k = 1$, but obviously the proof applies also to the general case).

Lemma **II.2.** — *Given $0 < \hat{\gamma} < 1$ and $\delta > 0$, there exists $\varepsilon = \varepsilon(\hat{\gamma}, \delta) > 0$ such that if $(x_i, g^{n_i}(x_i))$, $i = 1, \ldots, k$, are uniform $\hat{\gamma}$-strings satisfying $d(g^{n_i}(x_i), x_{i+1}) < \varepsilon$ for all $1 \leqslant i < k$ and $d(g^{n_k}(x_k), x_1) < \varepsilon$, then there exists a periodic point x of g with period $N = n_1 + \ldots + n_k$ such that*

$$d(g^n(x), g^n(x_1)) < \delta$$

for $0 \leqslant n \leqslant n_1$ and, setting $N_i = n_1 + \ldots + n_i$,

$$d(g^{N_i+n}(x), g^n(x_{i+1})) < \delta$$

for $0 \leqslant n \leqslant n_{i+1}$, $1 \leqslant i < k$.

Before continuing with the proof of II.1, let us first give a rough outline of it. Suppose that F is not expanding. Then to prove II.1 we have to show that given $0 < \gamma < 1$ and an admissible neighborhood V of Λ, there exists a periodic point $p \in M(g, V)$, with arbitrarily large period N, such that

$$\gamma^N < \prod_{j=1}^{N} \|(Dg^{-1})/\hat{F}(g^j(p))\| < 1. \tag{1}$$

Choose $\gamma < \hat{\gamma} < 1$. The periodic point p satisfying (1) will be the point $p = x$ obtained applying II.2 to $\hat{\gamma}$ and a suitable choice of uniform $\hat{\gamma}$-strings $(x_i, g^{n_i}(x_i))$. Since by II.2 the points of the orbit of x are δ-near to the points $g^{\ell}(x_i)$, $1 \leqslant i \leqslant k$, $0 \leqslant \ell \leqslant n_i$, then the condition $x \in M(g, V)$ will be satisfied if we work with a sufficiently small δ. To check (1) let us analyze the product in (1). Define α by

$$\alpha^N = \prod_{j=1}^{N} \|(Dg^{-1})/\hat{F}(g^j(x))\|$$

and γ_i by

$$\gamma_i^{n_i} = \prod_{j=1}^{n_i} \|(Dg^{-1})/F(g^j(x_i))\|.$$

What we want to prove is $\gamma^N < \alpha^N < 1$. But observe that, again by the δ-condition that x satisfies, α^N is such that $\alpha/(\prod_i \gamma_i^{n_i})^{1/N}$ becomes arbitrarily close to 1 if δ is sufficiently small. For the purpose of this informal outline we shall assume

$$\alpha^N = \prod_i \gamma_i^{n_i}.$$

Then $\alpha^N < 1$ is satisfied. The problem is to check $\gamma^N < \alpha^N$. Take $\gamma < \tilde{\gamma} < \hat{\gamma}$. We shall show that we can choose $x_1, \ldots, x_k$ and $n_1, \ldots, n_k$, with a value of k, say $k = 2n + 1$, such that when i is even we have $\gamma_i > \tilde{\gamma}$. Moreover every γ_i satisfies $\gamma_i \geqslant C$ where C is the minimum of the norms $\|(Dg^{-1})/T_x M\|$, $x \in M$. Then

$$\alpha^N = \prod_j \gamma_i^{n_i} \geqslant \prod_{j=1}^{n} \tilde{\gamma}^{n_{2j}} \prod_{j=0}^{n} C^{n_{2j+1}}.$$

Let N_1 be the sum of the n_i's for even values of i and N_2 the corresponding sum for odd values of i. Then

$$\alpha^N \geqslant \tilde{\gamma}^{N_1} C^{N_2}.$$

Since $\tilde{\gamma} > \gamma$, if N_1 is sufficiently larger than N_2, then $\alpha^N > \gamma^N$. Therefore we shall choose n_i much larger than n_{i+1} for every even i. The selection of the points x_i and the integers n_i requires the hypothesis of II.1 about the existence of a dense set of points z where a subsequence of the products

$$\prod_{j=1}^{n} \|(Dg^{-1})/F(g^j(z))\|$$

converges exponentially to 0, together with the fact that since F is not expanding, there exists a value of z such that this property does not hold. Then, we shall carefully pick the points x_i in such a way that $(x_i, g^{n_i}(x_i))$ is a uniform $\hat{\gamma}$-string (thus implying $\gamma_i < \hat{\gamma}$) but with γ_i not too small (that is, $\gamma < \gamma_i$) and also satisfying all the properties that we used in our sketch.

The selection of the points x_i requires several lemmas that in Section IV will be also useful to prepare the proof of Theorem I.7.

Lemma **II.3.** — *For all* $0 < \gamma_0 < \gamma_3 < 1$ *there exist* $N(\gamma_0, \gamma_3) > 0$ *and* $0 < c(\gamma_0, \gamma_3) < 1$ *such that if* $(x, g^n(x))$ *is a* γ_0*-string and* $n \geqslant N(\gamma_0, \gamma_3)$, *then there exist* $0 < n_1 < \ldots < n_k \leqslant n$, $k > 1$, *such that* $k \geqslant nc(\gamma_0, \gamma_3)$ *and* $(x, g^{n_i}(x))$ *is a uniform* γ_3*-string for all* $1 \leqslant i \leqslant k$.

We shall not prove this lemma because it is an immediate reformulation of a result of Pliss ([17], [18]).

Let us say that $(x, g^n(x))$ is an (N, γ)-*obstruction*, $0 < \gamma < 1$, $0 \leqslant N < n$, if $(x, g^m(x))$ is not a γ-string for all $N \leqslant m \leqslant n$.

Lemma II.4. — *Take* $0 < \gamma_0 < \gamma_3 < 1$, $0 < \gamma_2 < \gamma_3$ *and a* γ_0*-string* $(x, g^n(x))$. *Let* $0 < n_1 < \ldots < n_k \leqslant n$ *be the set of integers such that* $(x, g^{n_i}(x))$ *is a uniform* γ_3*-string, and let* $N = N(\gamma_2, \gamma_3)$. *Then, for all* $1 \leqslant i < k$, *either* $n_{i+1} - n_i \leqslant N$ *or* $(g^{n_i}(x), g^{n_{i+1}}(x))$ *is a* (N, γ_2)*-obstruction. Moreover, either* $n_1 \leqslant N$ *or* $(x, g^{n_1}(x))$ *is* (N, γ_2)*-obstruction.*

Proof. — *If* $n_{i+1} - n_i > N$ and $(g^{n_i}(x), g^{n_{i+1}}(x))$ is not a (N, γ_2)-obstruction, there exists $n_i + N \leqslant m \leqslant n_{i+1}$ such that $(g^{n_i}(x), g^m(x))$ is a γ_2-string. Hence, by II.3, that can be applied because $m - n_i \geqslant N = N(\gamma_2, \gamma_3)$, there exists $n_i < r < m$ such that $(g^{n_i}(x), g^r(x))$ is a uniform γ_3-string. Hence $(x, g^r(x))$ is a uniform γ_3-string and r should be in the sequence $n_1 < \ldots < n_k$. But on the other hand $n_i < r < m \leqslant n_{i+1}$.

Lemma II.5. — *If* $0 < \gamma_0 < \gamma_3 < 1$, $0 < \gamma_1 < \gamma_2 < \gamma_3$, *and* $(x, g^n(x))$ *is a* γ_0*-string containing a* (N, γ_2)*-obstruction* $(g^r(x), g^{r+\ell}(x))$ *such that*

a) $n \geqslant N(\gamma_0, \gamma_3)$,

b) $nc(\gamma_0, \gamma_3) > r + \ell$,

c) $r + \ell \geqslant N(\gamma_1, \gamma_2)$ *and*

d) $(r + \ell)\, c(\gamma_1, \gamma_2) > r + N$,

then there exists a uniform γ_3*-string* $(x, g^m(x))$, $r + \ell \leqslant m \leqslant n$, *that is not a* γ_1*-string.*

In a more informal language these conditions require n to be large with respect to $r + \ell$ and $r + \ell$ to be large with respect to $r + N$.

Proof. — Let $0 < n_1 < \ldots < n_k \leqslant n$ be the integers such that $(x, g^{n_i}(x))$ is a uniform γ_3-string. By a), we can apply II.3 that implies

$$k \geqslant nc(\gamma_0, \gamma_3).$$

Then, by b)

$$k > r + \ell.$$

Hence $n_k > r + \ell$ because obviously $n_k \geqslant k$. Let j be the smallest integer such that

$$n_j \geqslant r + \ell. \tag{2}$$

Let us prove that $(x, g^{n_j}(x))$ (that is a uniform γ_3-string) is not a γ_1-string, thus completing the proof of the lemma. Suppose that $(x, g^{n_j}(x))$ is a γ_1-string. By c)

$$n_j \geqslant r + \ell \geqslant N(\gamma_1, \gamma_2),$$

then we can apply II.3 to the γ_1-string $(x, g^{n_j}(x))$ and $0 < \gamma_1 < \gamma_2$. It yields a family $0 < m_1 < \ldots < m_s \leqslant n_j$ such that $(x, g^{m_j}(x))$ is a uniform γ_2-string for all j (hence all the numbers m_i belong to $\{n_1, \ldots, n_j\}$, since $\gamma_2 < \gamma_3$) and

$$s \geqslant n_j\, c(\gamma_1, \gamma_2) \geqslant (r + \ell)\, c(\gamma_1, \gamma_2).$$

Applying d) we obtain

$$m_{s-1} \geqslant s - 1 \geqslant (r + \ell)\, c(\gamma_1, \gamma_2) - 1 \geqslant r + N.$$

But $m_{s-1} = n_i$ for some $1 \leqslant i < j$. Hence

$$m_{s-1} = n_i < r + \ell$$

because j was the minimum index such that (2) holds. Then $r + \mathrm{N} \leqslant n_i < r + \ell$. Since $(g^r(x), g^{r+\ell}(x))$ is a (N, γ_2)-obstruction, $(g^r(x), g^{n_i}(x))$ is not a γ_2-string. But this contradicts the fact that $(x, g^{n_i}(x))$ is a uniform γ_2-string.

If $x \in \Lambda$, denote by $\mathrm{J}(x, \Lambda)$ the set of points $y \in \Lambda$ that can be written as

$$y = \lim_{n \to +\infty} g^{m_n}(x_n),$$

where $\{ x_n \mid n \geqslant 0 \}$ is a sequence converging to x and $\lim_{n \to +\infty} m_n = +\infty$. Clearly to obtain $\mathrm{J}(x, \Lambda)$, it is sufficient to use sequences $\{ x_n \mid n \geqslant 0 \}$ contained in some dense subset of Λ. Moreover the hypothesis $\Omega(g/\Lambda) = \Lambda$ implies

$$x \in \mathrm{J}(x, \Lambda)$$

for all $x \in \Lambda$.

Let us say that a compact invariant set $\Sigma \subset \Lambda$ is a *(t, γ)-set* ($t \in \mathbf{Z}^+$, $0 < \gamma < 1$) if for every $x \in \Sigma$ there exists $-t < m < t$ such that $(g^{m-n}(x), g^m(x))$ is a γ-string for all $n > 0$. Clearly this implies that $\mathrm{F} \mid \Sigma$ is an expanding subbundle.

Take γ_0 such that $\exp(-c) < \gamma_0 < 1$, where c is as in the statement of Theorem II.1. Then, by hypothesis, there exists a dense set $\Lambda_0 \subset \Lambda$ such that if $x \in \Lambda_0$ then there are infinitely many values of n satisfying

$$\prod_{j=1}^{n} \|(\mathrm{D}g^{-1})/\mathrm{F}(g^j(x))\| < \gamma_0^n.$$

Take $\gamma_1, \gamma_2, \bar{\gamma}_2, \gamma_3$ with

$$(3) \qquad 0 < \gamma_0 < \gamma_1 < \gamma_2 < \bar{\gamma}_2 < \gamma_3 < 1.$$

Lemma **II.6.** — *For every* $\varepsilon > 0$ *there exists* $\mathrm{N}(\varepsilon)$ *such that for all* $x \in \Lambda$, *either* $\mathrm{J}(x, \Lambda)$ *is an* $(\mathrm{N}(\varepsilon), \gamma_3)$*-set or there exists* $y \in \mathrm{J}(x, \Lambda)$ *such that* $(y, g^n(y))$ *is an* (N, γ_2)*-obstruction for all* $n > \mathrm{N}(\varepsilon)$, *where* $\mathrm{N} = \mathrm{N}(\bar{\gamma}_2, \gamma_3)$ *is given by* II.3, *and moreover* y *satisfies one of the following properties:*

a) $d(x, y) \leqslant \varepsilon$;

b) *there exists* $z_0 \in \Lambda$ *arbitrarily near to* x *and* $m > 0$ *such that* $d(g^m(z_0), y) < \varepsilon$ *and* $(z_0, g^m(z_0))$ *is a uniform* γ_3*-string.*

Proof. — Denote by $\Lambda(\mathrm{N})$ the set of points $y \in \Lambda$ such that $(y, g^n(y))$ is an (N, γ_2)-obstruction for all $n > \mathrm{N}$. It is easy to check that given $\varepsilon > 0$ there exists $\mathrm{N}(\varepsilon) > \mathrm{N}$ such that when $(y, g^n(y))$ is an $(\mathrm{N}, \bar{\gamma}_2)$-obstruction and $n > \mathrm{N}(\varepsilon)$, then $d(y, \Lambda(\mathrm{N})) < \varepsilon$ (here we use that $\bar{\gamma}_2 > \gamma_2$). Given $x \in \Lambda$ and $z \in \mathrm{J}(x, \Lambda)$ there exists a sequence $\{ x_n \mid n \geqslant 0 \} \subset \Lambda_0$ converging to x and satisfying $z = \lim_{n \to +\infty} g^{m_n}(x_n)$ and $\lim_{n \to +\infty} m_n = +\infty$. For $n \geqslant 0$ define

$$\mathscr{S}(n) = \{ m > 0 \mid (x_n, g^m(x_n)) \text{ is a uniform } \gamma_3\text{-string} \} \cup \{ 0 \}.$$

By II.3 it is easy to see that $\mathcal{S}(n)$ is unbounded (since $\gamma_0 < y_3$ and $x_n \in \Lambda_0$). Set

$$k_n^+ = \min \mathcal{S}(n) \cap [m_n, +\infty)$$

and
$$k_n^- = \max \mathcal{S}(n) \cap [0, m_n).$$

Suppose that $\liminf(k_n^+ - k_n^-) \leqslant N(\varepsilon)$. Then there exists $0 \leqslant m \leqslant N(\varepsilon)$ such that $g^m(z)$ is the limit of a subsequence of $\{ g^{k_n^+}(x_n) \mid n \geqslant 0 \}$. Hence, if $r > 0$, $(g^{m-r}(z), g^m(z))$ is a γ_3-string because it is the limit of a sequence of γ_3-strings $(g^{k_n^+ - r}(x_n), g^{k_n^+}(x_n))$ (that indeed are γ_3-strings for $r \leqslant k_n^+$ because $(x_n, g^{k_n^+}(x_n))$ is a uniform γ_3-string for all n). Therefore, for some $0 \leqslant m \leqslant N(\varepsilon)$, $(g^{m-r}(z), g^m(z))$ is a γ_3-string for all $r > 0$. If this holds for all $z \in J(x, \Lambda)$ then $J(x, \Lambda)$ is a $(N(\varepsilon), \gamma_3)$-set. If it does not hold for all $z \in J(x, \Lambda)$ this argument shows that we can pick z such that for many n, $k_n^+ - k_n^- > N(\varepsilon)$. Hence $k_n^+ - k_n^- > N$ because $N(\varepsilon) > N$. Then, by Lemma II.4, $(g^{k_n^-}(x_n), g^{k_n^+}(x_n))$ is an $(N, \bar{\gamma}_2)$-obstruction. Therefore $d(g^{k_n^-}(x_n), \Lambda(N)) < \varepsilon$ for infinitely many values of n. If for an unbounded set of these we have $k_n^- > 0$, we take $y \in \Lambda(N)$ such that $d(g^{k_n^-}(x_n), y) < \varepsilon$ and then this point y, the point $z_0 = x_n$ and $m = k_n^-$ satisfy the requirements of Lemma II.6 and option b). If $k_n^- = 0$ for all sufficiently large values of n that satisfy $d(g^{k_n^-}(x_n), \Lambda(N)) < \varepsilon$, then $d(x_n, \Lambda(N)) < \varepsilon$ and since $x_n \to x$ we obtain $d(x, \Lambda(N)) \leqslant \varepsilon$. Taking $y \in \Lambda(N)$ such that $d(x, y) \leqslant \varepsilon$ it follows that y satisfies II.6 and option a).

Lemma **II.7**. — *If* F *is not expanding, for all* $\varepsilon > 0$ *there exists a compact invariant set* $\Lambda(\varepsilon) \subset \Lambda$ *such that every* $x \in \Lambda(\varepsilon)$ *has the following property: there exist* x_0 *arbitrarily near to* x, $n_0 \geqslant 0$ *and* $y \in \Lambda(\varepsilon)$ *such that* $d(g^{n_0}(x_0), y) < \varepsilon$, $(y, g^n(y))$ *is an* (N, γ_2)*-obstruction for all* $n > N(\varepsilon)$, *and, if* $n_0 > 0$, $(x_0, g^{n_0}(x_0))$ *is a uniform* γ_3*-string. Moreover* $\Lambda(\varepsilon)$ *is the closure of its interior. (One has* $N = N(\bar{\gamma}_2, \gamma_3)$ *and* $N(\varepsilon)$ *is given by* II.6).

Proof. — Let Σ be the union of all the $(N(\varepsilon), \lambda_3)$-sets. Then its closure $\bar{\Sigma}$ is an $(N(\varepsilon), \lambda_3)$-set. Since F is not expanding, $\bar{\Sigma} \neq \Lambda$. Define $\Lambda(\varepsilon)$ as the closure of the open set $\Lambda - \bar{\Sigma}$. Given $x \in \Lambda(\varepsilon)$ take $\bar{x} \in \Lambda - \bar{\Sigma}$ near to x. Since $\bar{x} \notin \Sigma$, the set $J(\bar{x}, \Lambda)$ (that contains $\bar{x}$ because $\Omega(g/\Lambda) = \Lambda$) cannot be an $(N(\varepsilon), \lambda_3)$-set. Then by II.6, there exist a point $y \in J(\bar{x}, \Lambda)$ such that $(y, g^n(y))$ is an (N, γ_2)-obstruction for all $n > N(\varepsilon)$, and x_0 arbitrarily near to $\bar{x}$ (hence near to x) and $n_0 \geqslant 0$ such that $d(g^{n_0}(x_0), y) < \varepsilon$ and, if $n_0 > 0$, $(x_0, g^n(x_0))$ is a uniform γ_3-string. To complete the requirements of Lemma II.7 we have only to show that $y \in \Lambda(\varepsilon)$. But $\bar{x} \in \Lambda - \bar{\Sigma}$. Hence there exists a neighborhood U of $\bar{x}$ with $U \cap \bar{\Sigma} = \emptyset$. Therefore if a point $z \in J(\bar{x}, \Lambda)$ is given by $z = \lim_{n \to +\infty} g^{m_n}(x_n)$, where $\{ x_n \mid n \geqslant 0 \} \subset \Lambda$ is a sequence contained in Λ converging to $\bar{x}$ and $m_n \to +\infty$, it follows that $x_n \in U$ for large values of n. This means that $x_n \in \Lambda - \bar{\Sigma} \subset \Lambda(\varepsilon)$, implying $g^{m_n}(x_n) \in \Lambda(\varepsilon)$ for n large enough; thus $z \in \Lambda(\varepsilon)$. This proves $J(\bar{x}, \Lambda) \subset \Lambda(\varepsilon)$. In particular $y \in \Lambda(\varepsilon)$ because $y \in J(\bar{x}, \Lambda)$.

Now let V be an admissible neighborhood of Λ, let $TM/M(g, V) = \hat{E} \oplus \hat{F}$ be the homogeneous dominated splitting extending $TM/\Lambda = E \oplus F$ and let $0 < \gamma < 1$

be given. Take $\gamma_0, \gamma_1, \gamma_2, \bar{\gamma}_2, \gamma_3$ such that (3) and $\gamma < \gamma_0$ and choose $0 < k_0 < 1$ such that

(4) $$\gamma < k_0^2 \gamma_1$$

(5) $$k_0^{-1} \gamma_3 < 1.$$

Take $\delta > 0$ that such if $a, b \in M(g, V)$ and $d(a, b) < \delta$ then

(6) $$\|(Dg^{-1})/\hat{F}(a)\| \geqslant k_0 \|(Dg^{-1})/\hat{F}(b)\|.$$

Let $\varepsilon = \varepsilon(\delta, \gamma_3)$ be given by Lemma II.2. We claim that if F is not expanding there exists a sequence $\{ x_i \mid i \geqslant 1 \} \subset \Lambda(\varepsilon/4)$ and a sequence of integers $n_i \geqslant 0$ such that:

I) $d(g^{n_i}(x_i), x_{i+1}) < \varepsilon/2$ for all $i \geqslant 1$.

II) If $n_i > 0$, then $(x_i, g^{n_i}(x_i))$ is a uniform γ_3-string. For all even values of i, $n_i > 0$ and $(x_i, g^{n_i}(x_i))$ is not a γ_1-string.

III) If $K = \min\{\|(Dg^{-1})/F(x)\| \mid x \in \Lambda\}$, then

$$\gamma_1^{n_i} K^{n_{i-1}} \geqslant (k_0 \gamma_1)^{n_i + n_{i-1}}$$

for every even value of i.

We shall construct this sequence by induction. We should begin by the cases $i = 1$ and $i = 2$, but we shall proceed directly to the induction step that is sufficiently illuminating about the construction of the first two terms of the sequence. Suppose then $(x_i, g^{n_i}(x_i))$ constructed for $1 \leqslant i \leqslant j$, j even. Since $g^{n_j}(x_j) \in \Lambda(\varepsilon/4)$ we can apply II.7 that gives a point x_{j+1} arbitrarily near to $g^{n_j}(x_j)$ (in particular we can assume $d(g^{n_j}(x_j), x_{j+1}) < \varepsilon/2$) and $n_0 \geqslant 0$ such that $g^{n_0}(x_{j+1})$ is $\varepsilon/4$-near to a point $y \in \Lambda(\varepsilon/4)$ such that $(y, g^n(y))$ is an (N, γ_2)-obstruction for all $n > N(\varepsilon/4)$. Moreover, if $n_0 > 0$, then $(x_{j+1}, g^{n_0}(x_{j+1}))$ is a uniform γ_3-string. Since $\Lambda(\varepsilon/4)$ is the closure of its interior, and in its interior there is a dense set of values of x such that $(x, g^n(x))$ is a γ_0-string for infinitely many values of n, there exists $x_{j+2} \in \Lambda(\varepsilon/4)$ so near to y that

$$d(g^{n_0}(x_{j+1}), x_{j+2}) < \varepsilon/2,$$

and such that $(x_{j+2}, g^n(x_{j+2}))$ is a γ_0-string for infinitely many values of n. Take $N_1 > N(\varepsilon/4)$; taking x_{j+2} sufficiently near to y, we obtain that $(x_{j+2}, g^{N_1}(x_{j+2}))$ is an (N, γ_2)-obstruction. Taking N_1 large with respect to N, and a value of n large with respect to N_1 and such that $(x_{j+2}, g^n(x_{j+2}))$ is a γ_0-string, we can apply Lemma II.5 (with $r = 0$, $\ell = N_1$), and obtain $N_1 \leqslant n_{j+2} \leqslant n$ such that $(x_{j+2}, g^{n_{j+2}}(x_{j+2}))$ is a uniform γ_3-string but is not a γ_1-string. Then $x_{j+1}, x_{j+2}, n_{j+1} = n_0$ and n_{j+2} satisfy conditions I), II). Condition III) holds if n_{j+2} is large with respect to n_{j+1}. Then to satisfy it, take in the previous construction N_1 so large that

$$\gamma_1^n K^{n_{j+1}} \geqslant (k_0 \gamma_1)^{n + n_{j+1}}$$

for all $n \geqslant N_1$. It will hold for $n = n_{j+2}$.

Now take two points x_ℓ, x_k, with ℓ and k odd, $k > \ell$ such that $d(x_\ell, x_k) < \varepsilon/2$. To simplify the notation translate the indexes in order to have $\ell = 1$, $k = 2t + 1$. Then

$$d(g^{n_k}(x_{k-1}), x_1) = d(g^{n_k}(x_{k-1}), x_k) + d(x_k, x_1) \leqslant \varepsilon/2 + \varepsilon/2 = \varepsilon.$$

Now we apply Lemma II.2 to $\hat{\gamma} = \gamma_3$, the sequence $x_1, \ldots, x_k$ and the sequence of integers $n_1, \ldots, n_k$. However Lemma II.2 requires every $(x_i, g^{n_i}(x_i))$ to be a uniform γ_3-string (thus, in particular, $n_i > 0$) and we have satisfied this condition only when $n_i > 0$. Since this holds for every even value of i, we can expurgate the values of i such that $n_i = 0$; in other words we apply Lemma II.2 to the set S of points x_i with $n_i > 0$. If a certain x_j is not among these points, then j must be odd, x_{j-1} and x_{j+1} are in S and

$$\begin{aligned} d(g^{n_{j-1}}(x_{j-1}), x_{j+1}) &\leqslant d(g^{n_{j-1}}(x_{j-1}), x_j) + d(x_j, x_{j+1}) \\ &= d(g^{n_{j-1}}(x_{j-1}), x_j) + d(g^{n_j}(x_j), x_{j+1}) \leqslant \varepsilon/2 + \varepsilon/2 = \varepsilon. \end{aligned}$$

Hence, Lemma II.2 gives a periodic point x, with period $N = n_1 + \ldots + n_k$, such that

$$d(g^n(x), g^n(x_1)) \leqslant \delta$$

for $0 \leqslant n \leqslant n_1$ and, setting $N_i = n_1 \ldots + n_i$,

$$d(g^{N_i+n}(x), g^n(x_{i+1})) < \delta$$

for $1 \leqslant i < k$ and $0 \leqslant n \leqslant n_{i+1}$. In particular, every point of the orbit of x is δ-near to Λ. If δ is taken conveniently small, the above inequality implies that the orbit of x is contained in $M(g, V)$ as Theorem II.1 requires. Moreover it implies by (6) that

$$||(Dg^{-1})/\hat{F}(g^n(x))|| \geqslant k_0 \, ||(Dg^{-1})/F(g^n(x_1))||$$

for $0 \leqslant n \leqslant n_1$ and

$$||(Dg^{-1})/\hat{F}(g^{N_i+n}(x))|| \geqslant k_0 \, ||(Dg^{-1})/F(g^n(x_{i+1}))||$$

for $1 \leqslant i < k$, $0 \leqslant n \leqslant n_{i+1}$. Hence, setting $N_0 = 0$, we have

$$\prod_{n=1}^{n_{i+1}} ||(Dg^{-1})/\hat{F}(g^{N_i+n}(x))|| \geqslant k_0^{n_{i+1}} \prod_{n=1}^{n_{i+1}} ||(Dg^{-1})/F(g^n(x_{i+1}))||$$

for $0 \leqslant i < k$. Hence, if i is odd,

$$\prod_{n=1}^{n_{i+1}} ||(Dg^{-1})/\hat{F}(g^{N_i+n}(x))|| \geqslant k_0^{n_{i+1}} \gamma_1^{n_{i+1}}$$

and, for i even,

$$\prod_{n=1}^{n_{i+1}} ||(Dg^{-1})/\hat{F}(g^{N_i+n}(x))|| \geqslant k_0^{n_{i+1}} K^{n_{i+1}}.$$

Consequently,

$$\prod_{n=1}^{N} ||(Dg^{-1})/\hat{F}(g^n(x))|| \geqslant \prod_j k_0^{n_j + n_{j+1}} \gamma_1^{n_{j+1}} K^{n_j},$$

where the last product is taken over all the odd values in $[1, k-1]$. Applying property (III) we obtain

$$\prod_j k_0^{n_j+n_{j+1}} \gamma_1^{n_{j+1}} K^{n_j} \geqslant \prod_j k_0^{2(n_j+n_{j+1})} \gamma_1^{n_j+n_{j+1}} = k_0^{2N} \gamma_1^N.$$

Since $k_0^2 \gamma_1 > \gamma$ by (4), we have proved

$$\prod_{n=1}^{N} \|(Dg^{-1})/\hat{F}(g^n(x))\| > \gamma^N.$$

Finally these same methods can be used to prove the desired upper estimate for the product on the left. Recalling that $(x_i, g^{n_i}(x))$ is a γ_3-string if $n_i \neq 0$, we obtain analogously that

$$\prod_{n=1}^{N} \|(Dg^{-1})/\hat{F}(g^n(x))\| \leqslant k_0^{-N} \gamma_3^N.$$

Since γ_3 and k_0 satisfy $k_0^{-1}\gamma_3 < 1$, by (5), this implies

$$\prod_{n=1}^{N} \|(Dg^{-1})/\hat{F}(g^n(x))\| < 1.$$

III. — Proof of Theorem I.6

Let $\mathcal{M}(M)$ be the space of probabilities on the Borel σ-algebra of M endowed with the weak topology. If $f \in \mathrm{Diff}^1(M)$, let $\mathcal{M}(f)$ be the set of f-invariant elements of $\mathcal{M}(M)$ and $\mathcal{M}_e(f)$ be the set of ergodic elements of $\mathcal{M}(f)$. Specially interesting for our purpose will be the f-invariant probabilities supported on a periodic orbit of f, i.e. probabilities of the form

$$\mu = \frac{1}{n}\sum_{j=0}^{n-1} \delta_{f^j(x)}$$

where x satisfies $f^n(x) = x$. Denote by $\mathcal{M}_p(f)$ the set of these probabilities. The following result is a corollary of the main Theorem in [12].

Theorem **III.1.** — *Suppose that $f \in \mathrm{Diff}^1(M)$ and $\mu \in \mathcal{M}_e(f^m)$ for some $m > 0$. Then, given a neighborhood V of μ and a compact set K disjoint from the support of μ, there exists a diffeomorphism g, arbitrarily C^1 close to f and coinciding with f on K, such that there exists a probability $\mu_0 \in \mathcal{M}_p(g^m)$ contained in V whose support is disjoint from K.*

Proof. — Recall that an invariant set Λ of f is said to have total probability if $\mu(\Lambda) = 1$ for all $\mu \in \mathcal{M}(f)$. Define $\Sigma(f)$ as the set of points $x \in M$ such that for all $\varepsilon > 0$, every compact set K disjoint from the closure of the orbit of x, and every C^1 neighborhood $\mathcal{U}$ of f, there exists $g \in \mathcal{U}$ which coincides with f on K and has a periodic point y such that, if n is its period, one has $d(f^j(x), g^j(y)) \leqslant \varepsilon$ for all $0 \leqslant j \leqslant n$. Theorem A of [12] states that $\Sigma(f)$ has total probability. It is easy to see that then $\mu(\Sigma(f)) = 1$

for all $\mu \in \mathcal{M}(f^m)$ and all $m \neq 0$. Suppose that μ, m, K and V are as in the statement of Theorem III.1. We can assume that there exist continuous functions $\varphi_i : M \to \mathbf{R}$, $1 \leqslant i \leqslant s$, such that V is the set of $\nu \in \mathcal{M}(M)$ satisfying

$$\left| \int \varphi_i \, d\nu - \int \varphi_i \, d\mu \right| \leqslant 1$$

for all $1 \leqslant i \leqslant s$. Take $x \in \Sigma(f)$ and $N > 0$ such that

$$\left| \int \varphi_i \, d\mu - \frac{1}{n} \sum_{j=0}^{n-1} \varphi_i(f^{jm}(f^k(x))) \right| \leqslant 1/4 \tag{1}$$

for all $1 \leqslant i \leqslant s$, $0 \leqslant k \leqslant m-1$, $n \geqslant N$. Such an x exists because $\mu \in \mathcal{M}_e(f^m)$ and $\mu(\Sigma(f)) = 1$. If x is a periodic point of f, take $n \geqslant N$ such that $f^{mn}(x) = x$ and define

$$\mu_0 = \frac{1}{n} \sum_{j=0}^{n-1} \delta_{f^{mj}(x)}.$$

Then, by (1), $\mu_0 \in V$. Hence the theorem is proved taking $g = f$ and μ_0. Suppose that x is not periodic. Take $\varepsilon > 0$ such that if $a, b \in M$ satisfy $d(a, b) \leqslant \varepsilon$ then $| \varphi_i(a) - \varphi_i(b) | \leqslant 1/4$ for all $1 \leqslant i \leqslant s$. By the definition of $\Sigma(f)$ there exists g arbitrarily C^1 near to f, coinciding with f on K and having a periodic point y such that if n is its period then $d(f^j(x), g^j(y)) \leqslant \varepsilon$ for all $0 \leqslant j \leqslant nm$. Define μ_0 as

$$\mu_0 = \frac{1}{n} \sum_{j=0}^{n-1} \delta_{g^{mj}(y)}.$$

Observe that since x is not a periodic point of f then, taking g very near to f and ε very small, the period of y becomes arbitrarily large. Then we can assume $n \geqslant m(N+1)$. Now observe that $\{jn \mid 0 \leqslant j \leqslant m-1\}$ partitions $\{jm \mid 0 \leqslant j \leqslant n-1\}$ into a disjoint union of m sets, each with approximately $[n/m]$ multiples of m. More precisely, write $\{jm \mid 0 \leqslant j \leqslant n-1\} = \bigcup_{r=1}^{m} \{N_r m, N_r m + m, \ldots, N_r m + (n_r - 1) m\}$, where $N_r m - m < (r-1) n \leqslant N_r m = (r-1) n + k_r$. The integers k_r, n_r and N_r are obtained inductively by $N_1 = k_1 = 0$ and, for $1 \leqslant r \leqslant m$, $n - k_r \leqslant n_r m < n - k_r + m$, $k_{r+1} = n_r m - n + k_r$ and $N_{r+1} = N_r + n_r$. It follows that $0 \leqslant k_r \leqslant m-1$ and $0 \leqslant k_r + (n_r - 1) m < n$, as well as $n_1 + n_2 + \ldots + n_m = N_{m+1} = n$. But

$$g^{N_r m + jm}(y) = g^{(r-1)n + k_r + jm}(y) = g^{k_r + jm}(y)$$

for each $1 \leqslant r \leqslant m$, $0 \leqslant j \leqslant n_r - 1$, and therefore

$$\int \varphi_i \, d\mu_0 = \frac{1}{n} \sum_{j=0}^{n-1} \varphi_i(g^{jm}(y)) = \frac{1}{n} \sum_{r=1}^{m} \sum_{j=0}^{n_r-1} \varphi_i(g^{k_r+jm}(y))$$

for each $1 \leqslant i \leqslant s$. Moreover, since $n_r > (n-m)/m \geqslant N$, (1) implies

$$\left| \int \varphi_i \, d\mu - \frac{1}{n} \sum_{r=1}^{m} \sum_{j=0}^{n_r-1} \varphi_i(f^{k_r+jm}(x)) \right|$$

$$\leqslant \frac{1}{n} \sum_{r=1}^{m} n_r \left| \int \varphi_i \, d\mu - \frac{1}{n_r} \sum_{j=0}^{n_r-1} \varphi_i(f^{jm}(f^{k_r}(x))) \right| \leqslant 1/4.$$

Hence, for all $1 \leqslant i \leqslant s$,

$$\left|\int \varphi_i \, d\mu - \int \varphi_i \, d\mu_0\right| \leqslant \frac{1}{4} + \left|\int \varphi_i \, d\mu_0 - \frac{1}{n} \sum_{r=1}^{m} \sum_{j=0}^{n_r-1} \varphi_i(f^{k_r+jm}(x))\right|$$

$$\leqslant \frac{1}{4} + \frac{1}{n} \sum_{r=1}^{m} \sum_{j=0}^{n_r-1} \left|\varphi_i(g^{k_r+jm}(y)) - \varphi_i(f^{k_r+jm}(x))\right| \leqslant 1/2.$$

Lemma **III.2.** — *Suppose that* $f \in \mathscr{F}^1(M)$ *and* $P_k(f)$ *is hyperbolic for* $0 \leqslant k < i$. *Then, for every sufficiently small neighborhood* U *of* $\bigcup_{0 \leqslant k < i} \ddot{P}_k(f)$ *there exists a* C^1 *neighborhood* $\mathscr{U}_0$ *of* f *such that*

$$\bigcup_{0 \leqslant k < i} \overline{P}_k(g) = \bigcap_n g^n(U)$$

for all $g \in \mathscr{U}_0$.

Proof. — By the hyperbolicity of $\overline{P}_k(f)$ for $0 \leqslant k < i$, there exists a neighborhood $\mathscr{U}_0$ of f, that we can and shall assume to be connected and contained in $\mathscr{F}^1(M)$, such that for each $g \in \mathscr{U}_0$ there exists a homeomorphism

$$h_g : \bigcup_{0 \leqslant k < i} \overline{P}_k(f) \to \bigcap_n g^n(U)$$

such that $gh_g(x) = h_g f(x)$ for all x in the union on the left, and, for all x in that union, $h_g(x)$ depends continuously on $g \in \mathscr{U}_0$. Moreover, $h_g(\bigcup_{0 \leqslant k < i} \overline{P}_k(f))$ is hyperbolic. Let us prove that

$$h_g(\bigcup_{0 \leqslant k < i} P_k(f)) \subset \bigcup_{0 \leqslant k < i} P_k(g). \tag{2}$$

Suppose that $x \in P_k(f)$ and $g \in \mathscr{U}_0$. Take a continuous arc of diffeomorphisms $g_t \in \mathscr{U}_0$, $0 \leqslant t \leqslant 1$, with $g_0 = f$, $g_1 = g$. Then $x_t = h_{g_t}(x)$ is periodic for g_t, with the same minimum period than x. Since every $g_t \in \mathscr{U}_0 \subset \mathscr{F}^1(M)$, x_t is a hyperbolic periodic point of g_t for all $0 \leqslant t \leqslant 1$. Then it easy to see that the dimensions of the stable manifolds of x_t are the same for all $0 \leqslant t \leqslant 1$. Hence $x_1 = h_g(x) \in P_k(g)$. This proves (2). Now let us show that

$$h_g(\bigcup_{0 \leqslant k < i} P_k(f)) = \bigcup_{0 \leqslant k < i} P_k(g).$$

Denote by $P_{k,\ell}(g)$ the set of points in $P_k(g)$ whose minimum period is ℓ. It is easy to see, using the fact that $\mathscr{U}_0$ is connected and contained in $\mathscr{F}^1(M)$, that, for all $0 \leqslant k \leqslant \dim M$ and $\ell \geqslant 1$, $\# P_{k,\ell}(g)$ is the same for all $g \in \mathscr{U}_0$. Moreover, the argument used to prove (2) shows that for all $0 \leqslant k \leqslant i$, $\ell \geqslant 1$ and $g \in \mathscr{U}_0$, $\# P_{k,\ell}(f)$ is equal to the number of points in $h_g(P_k(f))$ with period ℓ. Since $\# P_{k,\ell}(g) = \# P_{k,\ell}(f)$, this shows that $P_{k,\ell}(g) \subset h_g(P_k(f))$ for all $\ell \geqslant 1$. Hence

$$P_k(g) = \bigcup_{\ell \geqslant 1} P_{k,\ell}(g) \subset h_g(P_k(f)).$$

This implies

$$\bigcup_{0\leqslant k<i} \mathrm{P}_k(g) \subset h_g(\bigcup_{0\leqslant k<i} \mathrm{P}_k(f)),$$

thus proving the equality, which clearly implies

$$\bigcap_n g^n(\mathrm{U}) = h_g(\bigcup_{0\leqslant k<i} \overline{\mathrm{P}}_k(f)) = \bigcup_{0\leqslant k<i} \overline{\mathrm{P}}_k(g).$$

Now let us prove Theorem I.6.

Given $f \in \mathscr{F}^1(\mathrm{M})$ such that $\overline{\mathrm{P}}_k(f)$ is hyperbolic for $0 \leqslant k < i$, take $m > 0$ and $0 < \lambda < 1$ given by Theorem I.3. Take any $\lambda < \lambda_0 < 1$ and suppose that $\mu \in \mathscr{M}(f^m/\overline{\mathrm{P}}_i(f))$ satisfies

$$(3) \qquad \int \log \|(\mathrm{D}f^m)/\widetilde{\mathrm{E}}_i^s\|\, d\mu \geqslant \log \lambda_0$$

and let us prove that

$$(4) \qquad \mu(\bigcup_{0\leqslant k<i} \overline{\mathrm{P}}_k(f)) > 0.$$

First suppose that μ is ergodic but not supported on a periodic orbit. Then, if (4) does not hold, we have

$$\mu(\bigcup_{0\leqslant k<i} \overline{\mathrm{P}}_k(f)) = 0.$$

Take a neighborhood U of $\bigcup_{0\leqslant k<i} \overline{\mathrm{P}}_k(f)$ such that

$$\mu(\mathrm{U}) < 1/2$$

and

$$\mu(\partial \mathrm{U}) = 0.$$

These properties imply that there exists a neighborhood V of μ in $\mathscr{M}(\mathrm{M})$ such that

$$(5) \qquad \nu(\mathrm{U}) < 1/2$$

for all $\nu \in \mathrm{V}$. Take a neighborhood W of $\overline{\mathrm{P}}_i(f)$ such that there exists a dominated splitting $\mathrm{TM}/\mathrm{M}(f, \mathrm{W}) = \widetilde{\mathrm{E}}_i^s \oplus \widetilde{\mathrm{E}}_i^u$ extending $\mathrm{TM}/\overline{\mathrm{P}}_i(f) = \widetilde{\mathrm{E}}_i^s \oplus \widetilde{\mathrm{E}}_i^u$. See Section II for the definition of $\mathrm{M}(f, \mathrm{W})$ and the existence of W. Take a neighborhood W_0 of $\mathrm{M}(f, \mathrm{W})$ so small that there exists a continuous splitting $\mathrm{TM}/\mathrm{W}_0 = \hat{\mathrm{E}}^s \oplus \hat{\mathrm{E}}^u$ extending $\mathrm{TM}/\mathrm{M}(f, \mathrm{W}) = \widetilde{\mathrm{E}}_i^s \oplus \widetilde{\mathrm{E}}_i^u$. Then, by standard properties of dominated splittings ([6], [9]), there exists a neighborhood $\mathscr{U}_0$ of f such that for every $g \in \mathscr{U}_0$, $\mathrm{M}(g, \mathrm{W})$ is contained in W_0 and has a dominated splitting $\mathrm{TM}/\mathrm{M}(g, \mathrm{W}) = \widetilde{\mathrm{E}}_g^s \oplus \widetilde{\mathrm{E}}_g^u$ such that the number

$$\delta(g) = \sup\{d(\widetilde{\mathrm{E}}_g^s(x), \hat{\mathrm{E}}^s(x)) \mid x \in \mathrm{M}(g, \mathrm{W})\}$$

converges to zero when $g \to f$. Take a continuous function $\psi : \mathrm{M} \to \mathbf{R}$ such that $\psi(x) = \log \|(\mathrm{D}f^m)/\hat{\mathrm{E}}^s(x)\|$ for $x \in \mathrm{W}_0$. Then (3) can be written as

$$\int \psi\, d\mu \geqslant \log \lambda_0.$$

Take λ_1 and $\varepsilon > 0$ such that

$$\lambda_1 e^{\varepsilon} < \lambda_0$$

and

$$\lambda < \lambda_1 e^{-\varepsilon}$$

and suppose that the neighborhood V of μ is so small that

$$\int \psi \, d\nu \geqslant \log \lambda_1$$

for all $\nu \in$ V. Using Theorem III.1 we can take g, arbitrarily C^1 near to f, and a periodic point x of g^m such that the probability $\mu_0 \in \mathscr{M}_p(g^m)$ given by

$$\mu_0 = \frac{1}{n} \sum_{j=0}^{n-1} \delta_{g^{mj}(x)},$$

where n is the period of x, is in V. Then

$$\int \psi \, d\mu_0 \geqslant \log \lambda_1. \tag{6}$$

Suppose that g is so near to f that for all y in the g^m-orbit of x

$$\log \|(Dg^m)/\widetilde{E}^s(y)\| \geqslant \log \|(Df^m)/\hat{E}^s(y)\| - \varepsilon. \tag{7}$$

From (6) we obtain

$$\log \lambda_1 \leqslant \int \psi \, d\mu_0 = \frac{1}{n} \sum_{j=0}^{n-1} \psi(g^{mj}(x)) = \frac{1}{n} \sum_{j=0}^{n-1} \log \|(Df^m)/\hat{E}^s(g^{mj}(x)\|.$$

Using (7):

$$\frac{1}{n} \sum_{j=0}^{n-1} \log \|(Dg^m)/\widetilde{E}^s_g(g^{mj}(x))\| \geqslant \frac{1}{n} \sum_{j=0}^{n-1} \log \|(Df^m)/\hat{E}^s(g^{mj}(x))\| - \varepsilon$$
$$\geqslant \log(\lambda_1 e^{-\varepsilon}).$$

Then

$$\prod_{j=0}^{n-1} \|(Dg^m)/\widetilde{E}^s_g(g^{mj}(x))\| \geqslant \lambda_1^n e^{-n\varepsilon}. \tag{8}$$

Property a) of I.3 says that for all j

$$\|(Dg^m)/\widetilde{E}^s_g(g^{mj}(x))\| . \|(Dg^{-m})/\widetilde{E}^u_g(g^{m(j+1)}(x))\| < \lambda.$$

Together with (8) this yields

$$\prod_{j=1}^{n} \|(Dg^{-m})/\widetilde{E}^u_g(g^{mj}(x))\| \leqslant (\lambda\lambda_1^{-1} e^{\varepsilon})^n < 1.$$

Denoting as usual by $E^u(x)$ the unstable subspace of the periodic point x, this means that

$$E^u(x) \supset \widetilde{E}^u_g(x) \tag{9}$$

because $(Dg^{mn}) \widetilde{E}^u_g(x) = \widetilde{E}^u_g(x)$ and

$$\|(Dg^{-mn})/\widetilde{E}^u_g(x)\| \leqslant \prod_{j=1}^{n} \|(Dg^{-m})/\widetilde{E}^u_g(g^{mj}(x))\| < 1.$$

24

Then, by (9), $x \in P_k(g)$ for some $k \leq i$. Suppose that $k = i$. Observe that the hyperbolicity of the periodic point x easily implies that any subspace $E \subset T_x M$ satisfying $(Dg^{mn}) E = E$ and $E \cap E^u(x) = \{0\}$ must be contained in $E^s(x)$. When $k = i$, I.3 a) implies $E^u(x) = \widetilde{E}_g^u(x)$. Hence $\widetilde{E}_g^s(x) \cap E^u(x) = \{0\}$. Then $\widetilde{E}_g^s(x) \subset E^s(x)$. But $\widetilde{E}_g^s(x)$ and $E^s(x)$ have the same dimension, namely $\dim M - \dim E^u(x)$. Therefore $\widetilde{E}^s(x) = E^s(x)$. By c) of I.3

$$\prod_{j=0}^{n-1} \|(Dg^m)/\widetilde{E}^s(g^{mj}(x))\| \leq C\lambda^n.$$

This and (8) imply

$$\lambda_1^n e^{-\varepsilon n} \leq C\lambda^n. \tag{10}$$

But since $\lambda < \lambda_1 \exp(-\varepsilon)$, this inequality is impossible if n is very large. On the other hand, the fact that μ is not supported by a periodic orbit implies that the period of the periodic orbit that supports its aproximation μ_0 (i.e. the period n of x) is arbitrarily large if μ_0 is taken sufficiently near to μ. Hence, taking μ_0 sufficiently near to μ, (10) becomes impossible, thus showing that we cannot have $k = i$. Then $x \in P_k(g)$ for some $k < i$. But by Lemma III.2, if g is sufficiently near to f, we have

$$P_k(g) \subset U,$$

and then, by (5),

$$\mu_0(P_k(g)) \leq \mu_0(U) < 1/2.$$

But $x \in P_k(g)$ implies

$$\mu_0(P_k(g)) = 1.$$

This contradiction completes the proof of (4), when μ is ergodic and not supported in a periodic orbit of f^m. Observe that a μ satisfying (3) cannot be supported by a periodic point of f^m because if this were true, i.e. if μ had the form

$$\mu = \frac{1}{n} \sum_{j=0}^{n-1} \delta_{f^{mj}(y)}$$

and $f^{mn}(y) = y$, then (3) would mean

$$\log\lambda < \log\lambda_0 \leq \int \log \|(Df^m)/\widetilde{E}_i^s\| \, d\mu = \lim_{t\to+\infty} \frac{1}{t} \sum_{j=0}^{t-1} \log \|(Df^m)/\widetilde{E}_i^s(f^{mj}(y))\|,$$

thus contradicting part d) of I.3. Finally suppose that some $\mu \in \mathcal{M}(f^m/\overline{P}_i(f))$ satisfies (3). By the Ergodic Decomposition Theorem we can write

$$\int\left(\int \log \|(Df^m)/\widetilde{E}_i^s\| \, d\mu_x\right) d\mu(x) = \int \log \|(Df^m)/\widetilde{E}_i^s\| \, d\mu \geq \log\lambda_0.$$

This means that the set S of points x satisfying

$$\int \log \|(Df^m)/\widetilde{E}_i^s\| \, d\mu_x > \log\lambda \tag{11}$$

has positive measure with respect to μ. Moreover

$$\mu(\bigcup_{0\leqslant k<i} \overline{P}_k(f)) = \int \mu_x(\bigcup_{0\leqslant k<i} \overline{P}_k(f))\, d\mu(x) \geqslant \int_S \mu_x(\bigcup_{0\leqslant k<i} \overline{P}_k(f))\, d\mu(x). \tag{12}$$

But when $x \in S$, (11) holds, thus implying

$$\mu_x(\bigcup_{0\leqslant k<i} \overline{P}_k(f)) > 0,$$

because μ_x is ergodic. Then, since this union is invariant under f^m,

$$\mu_x(\bigcup_{0\leqslant k<i} \overline{P}_k(f)) = 1.$$

Then (12) implies

$$\mu(\bigcup_{0\leqslant k<i} \overline{P}_k(f)) \geqslant \mu(S) > 0.$$

IV. — Contracting sequences and attainability

In this section we shall develop a perturbation technique that will be used in the next section to prove Theorem I.7. But before entering into the somehow formally involved array of definitions and statements that form this method, we shall first expose the underlying ideas through the discussion of a simplified but closely related problem.

A loose description of the aim of Theorem I.7 is the creation of linkings between transitive hyperbolic sets that are bound together by orbits that accumulate in all of them. The problem is, using this loose linking, to create a real linking, meaning by this an intersection between a stable and an unstable manifold of these sets. A simplified version of this type of objective is the following old and still open question: Suppose that $f \in \text{Diff}^k(M)$ has a hyperbolic point p such that there exists $q \in W^s(p) - \{p\}$ whose α-limit set satisfies $\alpha(q) \cap (W^u(p) - \{p\}) \neq \emptyset$; then, is it possible to find a diffeomorphism g, C^k near to f such that it coincides with f in a neighborhood of p and satisfies $q \in W^s_g(p) \cap W^u_g(p)$? Even without the requirement that g coincide with f nearby p, this question admits an obvious formulation, as open and difficult as the above one but for simplicity we shall discuss this question as we stated it.

There are at least two possible approaches to this problem. The first is the local method that consists in taking $n \geqslant 0$ such that the point $f^{-n}(q)$ is very close to a point $z \in W^u(p) - \{p\}$ and trying to find a diffeomorphism ψ, C^k near to the identity, such that $\psi(f^{-n}(q)) = z$ and is the identity outside a ball $B_r(z)$ that does not contain p. Then if $B_r(z)$ does not contain points of the form $f^j(q)$ for $0 \geqslant j \geqslant -(n-1)$, the sequence $\{f^j(q) \mid 0 \geqslant j \geqslant -(n-1)\} \cup \{f^j(z) \mid j \geqslant 0\}$ is an orbit of the diffeomorphism $g = (\psi f^{-1})^{-1}$, obviously contained in $W^s_g(p) \cap W^u_g(p)$. But if we take r so small that $f^j(q) \notin B_r(z)$ for $0 \geqslant j \geqslant -(n-1)$, and on the other hand such that $f^{-n}(q) \in B_r(z)$, then the C^k distance between g and f becomes a function of r and $d(f^{-n}(q), z)$. More specifically, ψ can be taken C^k near to the identity if $d(f^{-n}(q), z)/r^k$ is small, a condition that requires a special choice of n which cannot be always satisfied. To make the local method work we can add supplementary hypotheses, as in the results in [13] that will be stated again in the next section before using them (together with the results of this section) to prove I.7.

The other approach is more global. It consists in taking a small neighborhood U of q and a diffeomorphism ψ that is the identity outside U and then defining the diffeomorphism $g = f\psi$ with the hope of finding ψ near to the identity and also satisfying the relation $g^{-n}(q) \in W^u_g(q)$ for some $n > 0$. Stated in this form it may seem outrageously naive, but the intention is to exploit the dynamics of f in such a way that the small perturbation introduced by ψ will be amplified under iteration in such a way that the orbit of q under f (that accumulated in $W^u(p)$) will move toward $W^u(p)$ and hit it. Again, to accomplish this project, we shall, in our situation, have supplementary hypotheses that grant certain expanding behaviour of f^{-1} through which the amplification of small perturbations will be obtained. In fact *the proof of Theorem I.7 exploits an alternative: either the local method works or there are enough expanding dynamics in f to make the global method work.* What we shall do now is to prepare the techniques of the second part of the alternative.

Let Λ be a compact invariant set of $f \in \text{Diff}^1(M)$ having a dominated splitting $TM/\Lambda = E \oplus F$. Given $m > 0$ and $0 < \gamma < 1$ we say that a pair of points $(x, f^{-mn}(x))$ in Λ, $n > 0$, is an *(m, γ)-string*, if

$$\prod_{j=1}^{n} \|(Df^m)/E(f^{-mj}(x))\| \leqslant \gamma^n,$$

and we say that $(x, f^{-mn}(x))$ is a *uniform (m, γ)-string* when $(f^{-mj}(x), f^{-mn}(x))$ is an (m, γ)-string for all $0 \leqslant j < n$. These two definitions are just repetitions of those introduced at the beginning of section II, applied now to $g = f^{-m}$ and the dominated splitting $TM/\Lambda = F \oplus E$.

A pair (S, ν), where $S = \{x_1, x_2, \dots\} \subset \Lambda$ is a sequence in Λ and $\nu : S \to \mathbf{Z}^+$ is a function satisfying $\lim_{n \to +\infty} \nu(x_n) = +\infty$, is an *$(m, \gamma)$-contracting sequence* if there exists $\hat{n}$ such that $(x, f^{-mj}(x))$ is an (m, γ)-string for all $\hat{n} < j \leqslant \nu(x)$ and $x \in S$. Moreover we say that (S, ν) is a *strongly (m, γ)-contracting sequence* if $(x, f^{-m\nu(x)}(x))$ is a uniform (m, γ)-string for all $x \in S$. The sequence

$$\hat{S} = \{f^{-m\nu(x_n)}(x_n) \mid n \geqslant 1\}$$

will be called the sequence of *endpoints* of (S, ν). If (S', ν') and (S'', ν'') are (m, γ)-contracting sequences, write $(S', \nu') \geqslant (S'', \nu'')$ if S'' is a subsequence of S' and $\nu'' \leqslant \nu'/S''$.

Theorem **IV.1.** — *If (S, ν) is an (m, γ)-contracting sequence then, given $\gamma < \gamma_1 < 1$, one of the following properties holds:*

a) *There exists a strongly (m, γ_1)-contracting sequence $(S', \nu') \leqslant (S, \nu)$ whose sequence of endpoints converges to a point y such that for every $0 < \gamma_2 < \gamma_1$ there exists $N > 0$ such that*

$$\prod_{j=1}^{n} \|(Df^m)/E(f^{-mj}(y))\| > \gamma_2^n$$

for all $n \geqslant N$.

b) *There exists a strongly* (m, γ_1)*-contracting sequence* $(S', \nu') \leqslant (S, \nu)$ *such that* $S - S'$ *is finite and* $\nu - \nu'$ *is bounded.*

Proof. — If $x \in S$, denote by $\mathscr{S}(x)$ the set of integers n in $(0, \nu(x)]$ such that $(x, f^{-mn}(x))$ is a uniform (m, γ_1)-string. By Lemma II.3 there exist $N_0 > 0$ and $c > 0$ such that

$$(1) \qquad \# \mathscr{S}(x) \geqslant c\nu(x)$$

when $\nu(x) \geqslant N_0$. Recalling that, by definition, if $S = \{x_1, x_2, \ldots\}$, then

$$\lim_{n \to +\infty} \nu(x_n) = +\infty,$$

it follows that (1) holds for all x_n with large n, say $n \geqslant n_0$. Define $S' = \{x_{n_0}, x_{n_0+1}, \ldots\}$ and, if $x \in S'$, let $\nu'(x)$ be the largest integer in $\mathscr{S}(x)$. Clearly (S', ν') is a strongly (m, γ_1)-contracting sequence and $S - S'$ is finite. Therefore, if $\nu - \nu'$ is bounded, (S', ν') has property b). Let us suppose $\nu - \nu'$ is unbounded and construct (S'', ν'') satisfying a). Let S'' be a subsequence of S' such that, setting $S'' = \{y_1, y_2, \ldots\}$, we have

$$\lim_{n \to +\infty} (\nu(y_n) - \nu'(y_n)) = +\infty$$

and the sequence of endpoints satisfies

$$\lim_{n \to +\infty} f^{-m\nu'(y_n)}(y_n) = y.$$

Let us prove that y satisfies the inequality required by property a), thus completing the proof of IV.1, since then the strongly (m, γ_1)-contracting sequence (S'', ν''), with $\nu'' = \nu'/S''$, has property a). Given $0 < \gamma_2 < \gamma_1$, take $\gamma_2 < \bar{\gamma}_2 < \gamma_1$. By Lemma II.3 there exists $N > 0$ such that if $(y, f^{-mn}(y))$ is a (m, γ_2)-string and $n > N$, there exists $0 < n_1 \leqslant n$ such that $(y, f^{-mn_1}(y))$ is a uniform $(m, \bar{\gamma}_2)$-string. Suppose then that for some $n > N$ the inequality in property a) does not hold. Then $(y, f^{-mn}(y))$ is a (m, γ_2)-string and since $n > N$ we have $0 < n_1 \leqslant n$ such that $(y, f^{-mn_1}(y))$ is a uniform $(m, \bar{\gamma}_2)$-string. Then there exists a neighborhood U of y such that $(z, f^{-mn_1}(z))$ is a uniform (m, γ_1)-string for every $z \in U$. Take j so large that

$$(2) \qquad f^{-m\nu'(y_j)}(y_j) \in U$$

and $\nu(y_j) = \nu'(y_j) > n_1$. Then $(y_j, f^{-m(\nu'(y_j)+n_1)}(y_j))$ is a uniform (m, γ_1)-string, because so are $(y_j, f^{-m\nu'(y_j)}(y_j))$ and (by (2)) $(f^{-m\nu'(y_j)}(y_j), f^{-m(\nu'(y_j)+n_1)}(y_j))$. Since $\nu'(y_j) + n_1 < \nu(y_j)$, it then follows that $\nu'(y_j) + n_1 \in \mathscr{S}(y_j)$, contradicting the definition of ν' and concluding the proof.

The important property of strongly contracting sequences, that in the next section we shall exploit to prove Theorem I.7, is given by the following definition and theorem.

Given a sequence $S = \{x_1, x_2, \ldots\}$ converging to a point x_0 and a set $\Sigma \subset M$, we say that Σ is *attainable* from S if given $\delta > 0$, a neighborhood U of x_0 and a C^1 neighborhood $\mathscr{U}$ of f, there exist $g \in \mathscr{U}$ and integers $k > 0$ and $\ell > 0$ such that

a) $x_k \in U$ and $g^{-\ell}(x_k) \in \Sigma$,
b) $g^{-1}(x) = f^{-1}(x)$ for all $x \notin U$, and
c) $d(f^{-n}(x_k), g^{-n}(x_k)) < \delta$ for all $0 \leqslant n \leqslant \ell$.

Recall that by [6], the domination property of the splitting $TM/\Lambda = E \oplus F$ implies that there exists a family of embedded C^1 disks $D(y)$, $y \in \Lambda$, such that:

1) $y \in D(y)$ and $T_y D(y) = F(y)$;
2) $f(D(y))$ contains a neighborhood of $f(y)$ in the disk $D(f(y))$;
3) $D(y)$ depends continuously on y.

Define $D_r(y)$ as the set of points in $D(y)$ whose distance in $D(y)$ to y is $\leq r$.

Theorem **IV.2.** — *Given* $r > 0$, $m \in \mathbf{Z}^+$ *and* $0 < \gamma < 1$, *there exists* $\varepsilon = \varepsilon(r, m, \gamma)$ *such that if* (S, ν) *is a strongly* (m, γ)*-contracting sequence and* S *converges to a non periodic point* x_0, *then, if* $y \in \Lambda$ *is* ε*-near to an accumulation point of the sequence of endpoints of* (S, ν), $D_r(y)$ *is attainable from* S.

Proof. — Take a neighborhood U_0 of $f^{-1}(x_0)$ and a C^∞ function $\psi : M \to [0, 1]$ satisfying $\psi(x) = 0$ if $x \notin U_0$ and $\psi(f^{-1}(x_0)) = 1$. For each $v \in T_{f^{-1}(x_0)} M$ define a C^∞ vector field on M setting $\xi_v(y) = 0$ if $y \notin U_0$ and

$$\xi_v(x) = \psi(x)\, \tau(f^{-1}(x_0), x)\, v$$

when $x \in U_0$, where $\tau(f^{-1}(x_0), x)$ is the linear map from $T_{f^{-1}(x_0)} M$ onto $T_x M$ given by the parallel translation along the minimizing geodesic that joins $f^{-1}(x_0)$ to x. Since this geodesic is unique when $f^{-1}(x_0)$ and x are sufficiently close, it follows that ξ_v is well defined U_0 sufficiently small. For v small, say $\|v\| < R$, define a diffeomorphism $f_v : M \circlearrowleft$ by

$$f_v^{-1}(x) = \exp_{f^{-1}(x)} \xi_v(x).$$

Observe that if $\|v\|$ is small, the map $x \mapsto \exp_{f^{-1}(x)} \xi_v(x)$ is a map C^1 close to f^{-1}. Hence it is a diffeomorphism and then f_v, that is the inverse of the map $x \mapsto \exp_{f^{-1}(x)} \xi_v(x)$, is well defined.

The idea of the proof consists in taking $\varepsilon_0 > 0$ and studying the sets $\Sigma(k)$ given for each $x_k \in S$ by

$$\Sigma(k) = \{ f_v^{-\nu(x_k)m}(x_k) \mid \|v\| < \varepsilon_0 \}.$$

We shall prove that there exists a constant $c > 0$ such that for k sufficiently large $\Sigma(k)$ contains a disk $D(k)$ tangent to $E(f^{-\nu(x_k)m}(x_k))$ at $f^{-\nu(x_k)m}(x_k)$ which (treating M as a Euclidean space) can be written as the graph of a C^1 map

$$\varphi_k : \{ w \in E(f^{-\nu(x_k)m}(x_k)) \mid \|w\| < c \} \to F(f^{-\nu(x_k)m}(x_k)).$$

This means that

$$D(k) = \{ w + \varphi_k(w) \mid w \in E(f^{-\nu(x_k)m}(x_k)), \|w\| < c \}.$$

Moreover the maps φ_k satisfy $\|(\varphi_k)'(w)\| \leq c$ for all k sufficiently large and every w in the domain of φ_k. These properties, plus a standard application of the proof of the implicit function theorem, imply that given $r > 0$, if y is near to a point $f^{-\nu(x_k)m}(x_k)$ with k sufficiently large, then $D_r(y) \cap D(k) \neq \emptyset$. This means that

$$f_v^{-\nu(x_k)m}(x_k) \in D_r(y).$$

Since ε_0 can be taken arbitrarily small and f_v^{-1} and f^{-1} coincide outside U_0, this proves two properties (a) and b)) required by the attainability of $D_r(y)$. The last property (property c)) will require more careful estimates.

To formalize this method we begin by taking a neighborhood W_0 of Λ such that there exists a continuous splitting $TM/W_0 = \hat{E} \oplus \hat{F}$ extending the splitting $TM/\Lambda = E \oplus F$. For $x \in W_0$ let $\pi_1 : T_x M \to \hat{E}(x)$ and $\pi_2 : T_x M \to \hat{F}(x)$ be the projections associated to this splitting and let $S_\varepsilon(x)$ be the cone

$$S_\varepsilon(x) = \{v \in T_x M \mid \| \pi_2 v \| < \varepsilon \| \pi_1 v \|\}.$$

Set

$$c_0 = \frac{1}{2} \min_{x \in M} \{\|(Df)/T_x M\|^{-1}, \|(Df^{-1})/T_x M\|^{-1}\}.$$

Using the domination condition we can choose $\gamma < \gamma_1 < 1$, an arbitrarily small $\bar{\varepsilon}_0 > 0$ and $0 < \varepsilon_1 < \varepsilon_0 < \bar{\varepsilon}_0$, $n_0 > 0$, $\ell > 0$, $A > 1$ such that for every x in a certain neighborhood $W_1 \subset W_0$ of Λ the following properties hold:

I) $(Df^{-mn_0}) \overline{S_{\bar{\varepsilon}_0}(x)} \subset S_{\varepsilon_0}(f^{-mn_0}(x))$;

II) $(Df^{-j}) \overline{S_{\varepsilon_0}(x)} \subset S_{\bar{\varepsilon}_0}(f^{-j}(x))$ for all $0 \leqslant j$;

III) $(Df^{-\ell mn_0}) \overline{S_{\bar{\varepsilon}_0}(x)} \subset S_{\varepsilon_1}(f^{-\ell mn_0}(x))$;

IV) if $x \in W_1$, $v \in S_{\varepsilon_1}(x)$ and $w \in T_x M$ satisfy $\| v \| \geqslant A \| W \|$, then $v + w \in S_{\varepsilon_0}(x)$ and $\| v + w \| \geqslant (1/2) \| v \|$;

V) ℓ is so large that

$$\left(\frac{1}{4} c_0^{4n_0 m}\right)^{k+1} \gamma_1^{-t} > A$$

when $tm > \ell k$. Moreover there exists $\beta > 1$ such that

$$\left(\frac{1}{4} c_0^{4n_0 m}\right)^{k} \gamma_1^{-t} > \beta^t$$

when $t > \ell k$.

Furthermore, recalling that x_0 is not periodic, we can choose U_0 so small that

VI) $f^{-j}(U_0) \cap U_0 = \emptyset$ for all $0 \leqslant j \leqslant 2\ell n_0 m$.

The rest of the proof consists in showing, following the method outlined above, that given $r > 0$ there exists $\varepsilon = \varepsilon(r, m, \gamma)$ such that if y is ε-near to an accumulation point of the sequence of endpoints of (S, ν), then for all $\delta > 0$ there exist an arbitrarily small $v \in T_{f^{-1}(x_0)} M$ and $x \in S \cap U_0$ such that $f_v^{-n}(x) \in D_r(y)$ for $n = m\nu(x)$ and $d(f_v^{-j}(x), f^{-j}(x)) < \delta$ for all $0 \leqslant j \leqslant n$. Clearly this suffices to prove the theorem.

Choose $R > 0$ so small that from properties I), II), III) and the definition of c_0 follows that, when $\| v \| < R$, then

I') $(Df_v^{-mn_0}) S_{\bar{\varepsilon}_0}(x) \subset S_{\varepsilon_0}(f_v^{-mn_0}(x))$ for all $x \in W_1$,

II') $(Df_v^{-j}) S_{\varepsilon_0}(x) \subset S_{\bar{\varepsilon}_0}(f_v^{-j}(x))$ for all $j \geqslant 0$ and x such that $f_v^{-i}(x) \in W_1$ for all $0 \leqslant i \leqslant j$,

III') $(Df_v^{-\ell m n_0}) S_{\bar{\varepsilon}_0}(x) \subset S_{\varepsilon_1}(f_v^{-\ell m n_0}(x))$ for all $x \in W_1$,

IV') $\|(Df_v^{-1})(x)\|^{-1} \geq c_0$ for all $x \in M$.

Observe that III') and I') imply:

V') $(Df_v^{-n n_0 m}) S_{\bar{\varepsilon}_0}(x) \subset S_{\varepsilon_1}(f^{-n n_0 m}(x))$ for all $x \in W_1$ and $n \geq \ell$ such that $f_v^{-j}(x) \in W_1$ for all $0 \leq j \leq n n_0 m$.

We will use the following notation for linear maps: given $T : E \to F$, $\operatorname{Im} T = T(E)$ denotes the image of T in F and $\nu \mid T \mid = \min_{\|w\|=1} \| Tw \|$ denotes the minimum norm of T.

Given $x \in W_1$ and $n > 0$, denote by

$$D_v f_v^{-n}(x) : T_{f^{-1}(x_0)} M \to T_{f_v^{-n}(x)} M$$

the derivative with respect to v of the map $v \mapsto f_v^{-n}(x)$. Assume that $R > 0$ is so small that $\| D_v f_v^{-1}(x) \| \leq 2$ for all $x \in M$, $\| v \| < R$ and also that there exists a neighborhood $U_1 \subset f(U_0)$ such that

$$\nu \mid D_v f_v^{-1}(x) \mid \geq 1/2, \tag{3}$$

$$\operatorname{Im} D_v f_v^{-1}(x) \subset S_{\varepsilon_0}(f_v^{-1}(x)) \quad \text{and} \quad f_v^{-1}(x) \in U_0 \tag{4}$$

for all $x \in U_1$, $\| v \| < R$. Take $\delta > 0$ and for $x \in S$, define $V(x)$ as the maximal star shaped open set in $T_{f^{-1}(x_0)} M$ such that if $v \in V(x)$ then $d(f_v^{-j}(x), f^{-j}(x)) \leq \delta$ for all $0 \leq j \leq m\nu(x)$. Take δ so small that $d(z, \Lambda) \leq \delta$ implies $z \in W_1$. Moreover, given any $0 < c < 1$, we can take δ, $\bar{\varepsilon}_0$ and R so small that

VI') $\|(Df_v^{-k}) w \| \geq c \|(Df^k)/E(f^{-k}(x))\|^{-1} \| w \|$ whenever $0 \leq k \leq m n_0$, $\| v \| < R$, $x \in \Lambda$, $w \in S_{\bar{\varepsilon}_0}(y)$, $d(y, x) \leq \delta$.

Take $0 < c < 1$ such that

$$c^{-1} \gamma < \gamma_1. \tag{5}$$

Lemma **IV.3.** — *Suppose that* $x \in S \cap U_1$, $v \in V(x)$ *and let* $0 < n_1 < n_2 < \ldots$ *be the sequence of integers such that* $f_v^{-n_k}(x) \in U_0$. *Then the following properties hold:*

a) $\operatorname{Im} D_v f_v^{-n}(x) \subset S_{\bar{\varepsilon}_0}(f_v^{-n}(x))$ *for all* $0 \leq n \leq m\nu(x)$.

b) *For all* $n_k \leq n < n_{k+1}$

$$\nu \mid D_v f_v^{-n}(x) \mid \geq \left(\frac{1}{4} c_0^{4 n_0 m}\right)^k \prod_{j=1}^{[n/n_0 m]} c \, \|(Df^{n_0 m})/E(f^{-n_0 m j}(x))\|^{-1}.$$

c) *For all* $k \geq 1$,

$$\operatorname{Im}(Df^{-1}) (D_v f_v^{-n_k+1}(x)) \subset S_{\varepsilon_1}(f_v^{-n_k}(x)).$$

d) *For all* $k \geq 1$,

$$\operatorname{Im} D_v f_v^{-n_k} \subset S_{\varepsilon_0}(f^{-n_k}(x)).$$

Proof. — We shall prove it by induction. Since $x \in U_1$, then properties (3) and (4) imply that the lemma holds for n_1. Let us suppose that it holds for every integer $n \leq n_k$ and let us prove that it holds for every integer $n \leq n_{k+1}$. First we shall show that a) holds for every $n < n_{k+1}$. That it holds also for $n = n_{k+1}$ will follow from d), which we shall have to prove for n_{k+1}. Observe that if $n_k < n < n_{k+1}$ then

$$\text{(6)} \qquad D_v f_v^{-n}(x) = (Df^{-(n-n_k)})\,(D_v f_v^{-n_k})\,(x)$$

because $f_v^{-j}(x) \notin U_0$ for $n_k < j \leqslant n$. Then (6), (V′) and the induction hypothesis imply:

$$\operatorname{Im} D_v f_v^{-n}(x) \subset (Df^{-(n-n_k)})\, S_{\varepsilon_0}(f_v^{-n_k}(x)) \subset S_{\tilde{\varepsilon}_0}(f_v^{-n}(x)),$$

thus completing the proof of a) for $n_k < n < n_{k+1}$. By the induction hypothesis it holds for $n \leqslant n_k$. Hence it holds for $n < n_{k+1}$, as we wished to prove. Now let us prove b) for $n_k < n < n_{k+1}$ (for $n \leqslant n_k$ it follows from the induction hypothesis). Immediatly afterwards we shall prove c) and d) for $k+1$ and b) for $n = n_{k+1}$. Given $n_k < n < n_{k+1}$, write it as $n = [n/n_0 m]\, n_0 m + r$ with $0 \leqslant r < n_0 m$, and write $p_1 = [n_k/n_0 m]$, $p_2 = [n_k/n_0 m] + 1$. Then, by the induction hypothesis applied to property c) we have

$$\text{(7)} \qquad \operatorname{Im}(Df^{-(n_k - p_1 n_0 m)})\,(D_v f_v^{-p_1 n_0 m}(x)) = \operatorname{Im}(Df^{-1})\,(D_v f_v^{-(n_k-1)}(x)) \subset S_{\varepsilon_1}(f^{n_k}(x)).$$

Applying b) and (5) we get

$$\begin{aligned} &\vee\,|\,(Df^{-(n_k - p_1 n_0 m)})\,(D_v f_v^{-p_1 n_0 m}(x))| \\ &\geqslant c_0^{(n_k - p_1 n_0 m)} \left(\frac{1}{4} c_0^{4 n_0 m}\right)^k \prod_{j=1}^{p_1} c\, \|(Df^{n_0 m})/E(f^{-n_0 m j}(x))\|^{-1} \\ &\geqslant \left(\frac{1}{4} c_0^{4 n_0 m}\right)^{k+1} (c^{-1}\gamma)^{-p_1 p_0} \geqslant \left(\frac{1}{4} c_0^{4 n_0 m}\right)^{k+1} \gamma_1^{-p_1 n_0}. \end{aligned}$$

Since $p_1 n_0 m > \ell k$ we can apply (V) to obtain

$$\text{(8)} \qquad \vee\,|(Df^{-(n_k - p_1 n_0 m)})\,(D_v f_v^{-p_1 n_0 m}(x))| \geqslant A$$

From IV), (3), (4), (7) and (8) it follows that

$$\begin{aligned} \vee\,|\,D_v f_v^{-n_k}(x)| &= \vee\,|(Df^{-(n_k p_1 n_0 m)})\,(D_v f_v^{-p_1 n_0 m}(x)) \\ &\quad + D_v f_v^{-1}(f^{-n_k+1}(x))| \geqslant \frac{1}{2} \vee\,|(Df^{-(n_k - p_1 n_0 m)})\,(D_v f_v^{-p_1 n_0 m}(x)|. \end{aligned}$$

Then

$$\vee\,|\,D_v f_v^{-n_k}(x)| \geqslant \frac{1}{2} c_0^{n_k - p_1 n_0 m} \vee\,|\,D_v f_v^{-p_1 n_0 m}(x)| \geqslant \frac{1}{2} c_0^{n_0 m} \vee\,|\,D_v f_v^{-p_1 n_0 m}(x)|.$$

Hence

$$\begin{aligned} \text{(9)} \qquad \vee\,|\,D_v f_v^{-p_2 n_0 m}(x)| &= \vee\,|(Df^{-(p_2 n_0 m - n_k)})\,(D_v f_v^{-n_k}(x))| \\ &\geqslant c_0^{n_0 m} \vee\,|\,D_v f_v^{-n_k}(x)| \geqslant \frac{1}{2} c_0^{2 n_0 m} \vee\,|\,D_v f_v^{-p_1 n_0 m}(x)|. \end{aligned}$$

Moreover, by V′), for all $n_k < j \leqslant n$

$$\begin{aligned} \operatorname{Im} D_v f_v^{-j}(x) &= \operatorname{Im}(Df^{-(j-n_k)}(D_v f_v^{-n_k}(x)) \\ &\subset (Df^{-(j-n_k)})\, S_{\varepsilon_0}(f_v^{-n_k}(x)) \subset S_{\varepsilon_0}(f_v^{-j}(x)). \end{aligned}$$

Hence, by VI') we obtain that, for all $n_k < j < j + mn_0 \leqslant n$,

$$\begin{aligned} \nu\,|\,D_v f_v^{-(j+mn_0)}(x)| &= \nu\,|(Df^{-mn_0})\,(D_v f_v^{-j})\,(x)| \\ &\geqslant c\,||(Df^{mn_0})/E(f^{-(mn_0+j)}(x))||^{-1}\,\nu\,|\,D_v f_v^{-j}(x)|. \end{aligned}$$

Then

$$\nu\,|\,D_v f_v^{-[n/mn_0]mn_0}(x)| \geqslant \prod_{j=1}^{[n/mn_0]} c\,||(Df^{mn_0})/E(f^{-(mn_0(j+p_2))}(x)||^{-1}.\nu\,|\,D_v f_v^{-p_2 n_0 m}(x)|. \tag{10}$$

Using (9) we get

$$\begin{aligned} \nu\,|D_v f_v^{-p_2 n_0 m}(x)| &\geqslant \frac{||(Df^{mn_0})/E(f^{-p_2 mn_0}(x))||^{-1}}{||(Df^{mn_0})/E(f^{-p_2 mn_0}(x))||^{-1}}\cdot\frac{1}{2}\,c_0^{2n_0 m}\,\nu\,|\,D_v f_v^{-p_1 n_0 m}(x)| \\ &\geqslant \frac{1}{2}\,c_0^{3n_0 m}\,||(Df^{mn_0})/E(f^{-p_2 mn_0}(x))||^{-1}\,\nu\,|\,D_v f_v^{-p_1 n_0 m}(x)|. \end{aligned}$$

Combining this with (10) we get

$$\begin{aligned} &\nu\,|\,D_v f_v^{-[n/mn_0]mn_0}(x)| \\ &\quad\geqslant \frac{1}{2}\,c_0^{3n_0 m}\prod_{j=0}^{[n/mn_0]} c\,||(Df^{mn_0})/E(f^{-(mn_0(j+p_2))}(x)||.\nu\,|\,D_v f_v^{-p_1 n_0 m}(x)|. \end{aligned}$$

Using the induction hypothesis

$$\begin{aligned} \nu\,|\,D_v f_v^{-[n/mn_0]mn_0}(x)| &\geqslant \frac{1}{2}\,c_0^{3n_0 m}\prod_{j=0}^{[n/mn_0]} c\,||(Df^{mn_0})/E(f^{-(mn_0(j+p_2))}(x))||^{-1} \\ &\quad\cdot\left(\frac{1}{4}\,c_0^{4n_0 m}\right)^k \prod_{j=1}^{P_1} c\,||(Df^{n_0 m})/E(f^{-n_0 mj}(x))||^{-1} \\ &\geqslant \frac{1}{2}\,c_0^{3n_0 m}\left(\frac{1}{4}\,c_0^{4n_0 m}\right)^k \prod_{j=1}^{[n/mn_0]} c\,||(Df^{mn_0})/E(f^{-mn_0 j}(x))||^{-1}. \end{aligned} \tag{11}$$

Then

$$\begin{aligned} \nu\,|\,D_v f_v^{-n}(x)\,| &= \nu\,|(Df^{-r})\,D_v f_v^{-[n/n_0 m]n_0 m}(x)| \geqslant c_0^r\,\nu\,|\,D_v f_v^{-[n/n_0 m]n_0 m}(x)| \\ &\geqslant c_0^{n_0 m}\,\frac{1}{2}\,c_0^{3n_0 m}\left(\frac{1}{4}\,c_0^{4n_0 m}\right)^k \prod_{j=1}^{[n/mn_0]} c\,||(Df^{mn_0})/E(f^{-mn_0 j}(x)||^{-1}. \end{aligned} \tag{12}$$

This completes the proof of b) for $n < n_{k+1}$. Now observe that

$$\begin{aligned} \mathrm{Im}(Df^{-1})\,(D_v f^{-n_{k+1}+1}(x)) &= \mathrm{Im}(Df^{-(n_{k+1}-n_k)})\,(D_v f_v^{-n_k}(x)) \\ &\subset (Df^{-(n_{k+1}-n_k)})\,S_{\varepsilon_0}(f_v^{-n_k}(x)) \subset S_{\varepsilon_1}(f_v^{-n_{k+1}}(x)) \end{aligned} \tag{13}$$

where the last inequality follows from VI) and V'). This proves c). Moreover

$$D_v f_v^{-n_{k+1}}(x) = (Df^{-1})\,(D_v f_v^{-n_{k+1}+1}(x)) + D_v f_v^{-n_{k+1}+1}(x).$$

Hence, as before, we can show that the minimum norm of the first term on the right is large, because $D_v f_v^{-n_{k+1}+1}(x)$ can be estimated using b), which we have proved to hold for $n < n_{k+1}$, and Df^{-1} contracts norms at most by a factor c_0. Then the first term on the right has a much larger minimum norm than the second term, and its image is contained in $S_{\varepsilon_1}(f_v^{-n_{k+1}}(x))$ by (13). Therefore, by IV) we have

$$\operatorname{Im} D_v f_v^{-n_{k+1}}(x) \subset S_{\varepsilon_0}(f_v^{-n_{k+1}}(x)),$$

thus proving d), and

$$\nu \,|\, D_v f_v^{-n_{k+1}}(x) | \geqslant \frac{1}{2} \nu \,|(Df^{-1})\,(D_v f_v^{-n_{k+1}+1}(x))|.$$

Write $n_{k+1} = nn_0\, m + r$ with $0 < r \leqslant n_0\, m$. Suppose $r < mn_0$. Then

$$\nu \,|\, D_v f_v^{-n_{k+1}}(x) | \geqslant \frac{1}{2} \nu \,|(Df^{-1})\,(D_v f_v^{-n_{k+1}+1}(x))|$$
$$= \frac{1}{2} \nu \,|(Df^{-r})\,(D_v f_v^{-nn_0 m}(x))| \geqslant \frac{1}{2} c_0^r \,\nu \,|\, D_v f_v^{-nn_0 m}(x)|.$$

Applying (11):

$$\nu \,|\, D_v f_v^{-n_{k+1}}(x) | \geqslant \frac{1}{2} c_0^r . \frac{1}{2} c_0^{3n_0 m} \left(\frac{1}{4} c_0^{4n_0 m}\right)^k . \prod_{j=1}^{n} c \,||(Df^{mn_0})/E(f^{-mn_0 j}(x))||^{-1}$$
$$\geqslant \left(\frac{1}{4} c_0^{4n_0 m}\right)^{k+1} \prod_{j=1}^{n} c \,||(Df^{mn_0})/E(f^{-mn_0 j}(x))||^{-1}.$$

From $r < mn_0$ it follows that $n = [n_{k+1}/mn_0]$ and this completes the proof of b) for $n = n_{k+1}$. When $r = mn_0$ we write

$$\nu \,|\, D_v f_v^{-n_{k+1}}(x) | \geqslant \frac{1}{2} \nu \,|(Df^{-1})\,(D_v f_v^{-n_{k+1}+1}(x))|$$
$$= \frac{1}{2} \nu \,|(Df^{-mn_0})\,(D_v f_v^{-nmn_0}(x))|.$$

But
$$\operatorname{Im} D_v f_v^{-nmn_0}(x) = \operatorname{Im}(Df^{-(nmn_0 - n_k)})\,(D_v f_v^{-n_k}(x))$$
$$\subset (Df^{-(nmn_0 - n_k)})\, S_{\varepsilon_0}(f_v^{-n_k}(x)).$$

Hence, by II′)

$$\operatorname{Im} D_v f_v^{-nmn_0}(x) \subset S_{\bar{\varepsilon}_0}(f_v^{-nmn_0}(x)).$$

Then, by VI′)

$$\nu \,|\, D_v f_v^{-n_{k+1}}(x) | \geqslant \frac{1}{2} c \,||(Df^{mn_0})/E(f^{-n_{k+1}}(x))||^{-1} . \nu \,|\, D_v f_v^{-mn_0 n}(x)|.$$

Using (12)

$$\nu \,|\, D_v f_v^{-n_{k+1}}(x) | \geqslant \frac{1}{2} c \,||(Df^{mn_0})/E(f^{-n_{k+1}}(x))||^{-1}$$
$$\cdot \frac{1}{2} c_0^{4n_0 m} \left(\frac{1}{4} c_0^{4n_0 m}\right)^k \prod_{j=1}^{n} c \,||(Df^{mn_0})/E(f^{-jmn_0}(x))^{-1}$$
$$= \left(\frac{1}{4} c_0^{4n_0 m}\right)^{k+1} \prod_{j=1}^{n+1} c \,||(Df^{mn_0})/E(f^{-jmn_0}(x))||^{-1}.$$

Since $n+1=[n_{k+1}/n_0 m]$, this concludes the proof of Lemma IV.3.

If $x \in S \cap U_1$ define, for all $0 \leqslant n \leqslant m\nu(x)$, $\Sigma_x(n)$ as the set

$$\Sigma_x(n) = \{f_v^{-n}(x) \mid v \in V(x)\}.$$

From now on we shall treat M as if it were a Euclidean space. All our arguments will be local and they can be exposed more clearly in that way instead of the formally necessary but cumbersome repeated use of local coordinates.

By IV.3 b), $\nu \mid D_v f_v^{-n}(x) \mid > 0$ for all $x \in U_1 \cap S$ and $0 \leqslant n \leqslant m\nu(x)$. By IV.3 a), $\operatorname{Im} D_v f_v^{-n}(x) \subset S_{\varepsilon_0} f_v^{-n}(x))$. Hence $\Sigma_x(n)$ is the graph of a C^1 map

$$(14) \qquad \psi_{n,x}: D(n, x) \to \hat{F}(f^{-n}(x))$$

where $D(n, x)$ is an open, simply connected subset of $\hat{E}(f_v^{-n}(x))$. Clearly the subspace

$$\{w + (D\psi_{n,x})(y)\, w \mid w \in \hat{E}(f_v^{-n}(x))\}$$

turns out to be the tangent space of $\Sigma_x(n)$ at the point $f_v^{-n}(x) + y + \psi_{n,x}(y)$. Hence, by IV.3 a), there exists $C > 0$ such that

$$(15) \qquad \|(D\psi_{n,x})(y)\| \leqslant C$$

for all $x \in U_1 \cap S$, $0 \leqslant n \leqslant m\nu(x)$, $y \in D(n, x)$.

Lemma **IV.4.** — a) *There exists* $\beta_0 > 1$ *such that*

$$\nu \mid D_v f_v^{-n}(x) \mid \geqslant \beta_0^n$$

for all $x \in S \cap U_1$, $\ell n_0 m < n \leqslant m\nu(x)$.

b) *Given* $\hat{c} > 1$ *there exists* $N(\hat{c})$ *such that*

$$\| D_v f_v^{-n}(x) \| \leqslant \prod_{j=1}^{t} \hat{c}c^{-1} \|(Df^m)/E(f^{-(n+jm)}(x))\| \cdot \| D_v f_v^{-m\nu(x)}(x) \|$$

for all $x \in U_1 \cap S$, $N(\hat{c}) < n < n + tm \leqslant m\nu(x)$.

Proof. — By IV.3 b) and (5), if $n > 0$ we have

$$\nu \mid D_v f_v^{-n}(x) \mid \geqslant \left(\frac{1}{4} c_0^{\ell n_0 m}\right)^k \gamma_1^{-[n/n_0 m] n_0}$$

where k is chosen by $n_k \leqslant n < n_{k+1}$. Then

$$k \leqslant [n/\ell n_0 m].$$

and, if $n > \ell n_0 m$,

$$\frac{[n/n_0 m]\, n_0}{[n/\ell n_0 m]} \geqslant \ell n_0.$$

Then, by V)

$$\nu \mid D_v f_v^{-n}(x) \mid \geqslant \beta^{[n/n_0 m] n_0}$$

for $n > \ell n_0 m$. From here it is easy to conclude the existence of $\beta_0 > 1$ as required by part a) of the lemma. Now let us prove b) by induction on n starting at $n = m\nu(x)$. Suppose the property proved for $n+m$, $n+2m$, $\ldots$, $n+tm \leqslant m\nu(x)$ and let us prove it for n. Suppose first that $f_v^{-(n+m)}(x) \notin U_0$. Then

$$D_v f_v^{-(n+m)}(x) = (Df^{-m})\,(D_v f_v^{-n}(x)).$$

Using IV.3 a) and VI') with $k = m$ and any $w \in T_{f^{-1}(x_0)} M$,

$$\begin{aligned} \| D_v f_v^{-(n+m)}(x).w \| &= \|(Df^{-m})\,(D_v f_v^{-n}(x).w)\| \\ &\geqslant c\,\|(Df^m)/E(f^{-(n+m)}(x))\|^{-1}\,\| D_v f_v^{-n}(x).w \|. \end{aligned}$$

From this inequality and the induction hypothesis follows

$$\begin{aligned} \| D_v f_v^{-n}(x)\| &\leqslant c^{-1}\,\|(Df^m)/E(f^{-(n+m)}(x))\| \\ &\qquad . \prod_{j=2}^{t} \hat{c}c^{-1}\,\|(Df^m)/E(f^{-(n+jm)}(x))\|\,\| D_v f_v^{-m\nu(x)}(x)\| \\ &\leqslant \prod_{j=1}^{t} \hat{c}c^{-1}\,\|(Df^m)/E(f^{-(n+jm)}(x))\|\,\| D_v J_v^{-m\nu(x)}(x)\|, \end{aligned}$$

thus proving b). Suppose now that $f_v^{-(n+m)}(x) \in U_0$. Then

$$D_v f_v^{-(n+m)}(x) = (Df^{-m})\,(D_v f_v^{-n}(x)) + D_v f_v^{-1}(f_v^{-(n+m)+1}(x)).$$

Since IV.4 a) says that $\nu \,|\, D_v f_v^{-(n+m)}(x)|$ is very large if n is large enough, we can write

$$\| D_v f_v^{-(n+m)}(x) - D_v f_v^{-1}(f_v^{-(n+m)+1}(x))\| \leqslant \hat{c}\,\| D_v f_v^{-(n+m)}(x)\|.$$

From IV.3 d)

$$\operatorname{Im} D_v f_v^{-(n+m)}(x) \subset S_{\varepsilon_1}(f_v^{-(n+m)}(x)).$$

Using property IV) we conclude

$$\operatorname{Im}(D_v f_v^{-(n+m)}(x) - D_v f_v^{-1}(f_v^{-(n+m)+1}(x))) \subset S_{\varepsilon_0}(f_v^{-(n+m)}(x)).$$

Hence, using VI') as above,

$$\begin{aligned} \| D_v f_v^{-n}(x)\| &= \|(Df^m)\,(D_v f_v^{-(n+m)}(x) - D_v f_v^{-1}(f_v^{-(n+m)+1}(x)))\| \\ &\leqslant c^{-1}\,\|(Df^m)/E(f^{-(n+m)}(x))\| \\ &\qquad .\| D_v f_v^{-(n+m)}(x) - D_v f_v^{-1}(f_v^{-(n+m)+1}(x))\| \\ &\leqslant c^{-1}\hat{c}\,\|(Df^m)/E(f^{-(n+m)}(x))\|.\| D_v f_v^{-(n+m)}(x)\| \\ &\leqslant c^{-1}\hat{c}\,\|(Df^m)/E(f^{-(n+m)}(x))\|.\| D_v f_v^{-m\nu(x)}(x)\| \\ &\qquad . \prod_{j=2}^{t} \hat{c}c^{-1}\,\|(Df^m)/E(f^{-(n+jm)}(x))\| \\ &\leqslant \prod_{j=1}^{t} \hat{c}c^{-1}\,\|(Df^m)/E(f^{-(n+jm)}(x))\|.\|(D_v f_v^{-m\nu(x)}(x))\|. \end{aligned}$$

This concludes the proof of IV.4. Take $\hat{c}$ such that $\hat{c}c^{-1}\gamma = \gamma_2 < 1$. Then b) implies

Corollary **IV.5.** — *For all* $x \in U_1 \cap S$, $N(\hat{c}) < n < n + tm \leqslant m\nu(x)$,

$$\| D_v f_v^{-m(\nu(x)-t)}(x) \| \leqslant \gamma_2^t \| D_v f_v^{-m\nu(x)}(x) \|.$$

Define $\Sigma_x(r, n)$ as the set of points in $\Sigma_x(n)$ whose distance to x in the manifold $\Sigma_x(n)$ is $\leqslant r$, and let $\Sigma_x^*(n)$ be the union of all the $\Sigma_x(r, n)$ that are complete when endowed with the metric of the submanifold $\Sigma_x(n)$. Then IV.5 implies:

Corollary **IV.6.** — *For all* $x \in U_1 \cap S$, $N(\hat{c}) < n < n + tm \leqslant m\nu(x)$,

$$\operatorname{diam} \Sigma_x^*(m\nu(x) - t) \leqslant \gamma_2^t \operatorname{diam} \Sigma_x^*(m\nu(x)).$$

Lemma **IV.7.** — *There exists* $\hat{\varepsilon}_0$ *such that*

$$\operatorname{diam} \Sigma_x^*(m\nu(x)) \geqslant \hat{\varepsilon}_0$$

for all $x \in S \cap U_1$ *with* $m\nu(x) \geqslant N(\hat{c})$.

Proof. — If the lemma is false there exist points x in $U_1 \cap S$ such that $\operatorname{diam} \Sigma_x^*(m\nu(x))$ is arbitrarily small. Set $\hat{\delta} = \operatorname{diam} \Sigma_x^*(m\nu(x))$. By the definition of $\Sigma_x^*(m\nu(x))$ there exists $p \notin \overline{\Sigma}_x(m\nu(x)) - \Sigma_x(m\nu(x))$ that is a limit of a sequence $\{p_n \mid n \geqslant 0\} \subset \Sigma_x^*(m\nu(x))$. Write

$$p_n = f_{v_n}^{-m\nu(x)}(x)$$

for some $v_n \in V(x)$. Then

$$f_{v_n}^{-j}(x) = f_{v_n}^{-(m\nu(x)-j)} f_{v_n}^{-m\nu(x)}(x) \in f_{v_n}^{-(m\nu(x)+j)} \Sigma_x(m\nu(x)) \subset \Sigma_x^*(j).$$

This means that

$$d(f_{v_n}^{-j}(x), f^j(x)) \leqslant \operatorname{diam} \Sigma_x^*(j).$$

Then, by IV.6, it follows that for all $0 \leqslant j \leqslant m\nu(x)$ and n, $d(f_{v_n}^{-j}(x), f^{-j}(x))$ remains arbitrarily small, say $\leqslant \hat{\varepsilon}$. Hence $d(f_v^{-j}(x), f^{-j}(x)) \leqslant \hat{\varepsilon}$ for all $0 \leqslant j \leqslant m\nu(x)$. Then, $\hat{\varepsilon} \lll \delta$ (where δ is the constant used in the definition of $V(x)$), there exists a disk D_0 in $T_{f^{-1}(x_0)} M$, centered at v such that $d(f_w^{-j}(x), f^{-j}(x)) < \delta$ for all $w \in D_0$. Then $V(x) \cup D_0$ is obviously open and, decreasing D_0 if necessary, star shaped because the center of D_0 is in the boundary of $V(x)$. Then $V(x) \cup D_0$ is open, star shaped and satisfies the condition required by the definition of $V(x)$. Moreover it contains v that is not in $V(x)$. This contradicts the maximality of $V(x)$ and completes the proof.

Using (14) we can now take an open subset $D^*(n, x) \subset D(n, x)$ such that

$$\Sigma_x^*(n) = \{ w + \psi_{n,x}(w) \mid w \in D^*(n, x) \}.$$

From the definition of $\Sigma_x^*(n)$, property (15) and Lemma IV.7 there exists $\rho > 0$ such that the disk $B_\rho(f^{-n}(x)) = \{ w \in \hat{E}(f^{-n}(x)) \mid \| w \| < \rho \}$ is contained in $D^*(n, x)$ for all $x \in S \cap U_1$, $0 < n \leqslant m\nu(x)$. To conclude the proof of Theorem IV.2 we shall use the following easy lemma.

Lemma **IV.8.** — *Let* E_1, E_2 *be Banach spaces and let* $B^i_r(p)$ *be the ball of radius* r *in* E_i *centered at* p. *Let* $C > 0$, $\rho_0 > 0$ *and* $\varepsilon > 0$ *be constants such that* ε *is so small that*

$$\varepsilon C < 1, \quad \frac{\varepsilon + \varepsilon^2}{1 - \varepsilon C} < \min\{(\rho_0 - \varepsilon)/C, \rho_0\}.$$

Suppose that

$$\varphi : B^2_{\rho_0}(0) \to E^1, \quad \psi : B^1_{\rho_0}(p) \to E^2$$

satisfy:

a) $\varphi(0) = 0$, $\| \varphi(w_2) - \varphi(w_1)\| \leqslant \varepsilon \| w_2 - w_1 \|$ *for all* $w_1, w_2 \in B^2_{\rho_0}(0)$;
b) $\| \psi(p)\| \leqslant \varepsilon$ *and* $\| \psi(w_2) - \psi(w_1)\| \leqslant C \| w_2 - w_1 \|$ *for all* $w_1, w_2 \in B^1_{\rho_0}(0)$;
c) $\| p \| \leqslant \varepsilon$.

Then $$\operatorname{graph}(\varphi) \cap \operatorname{graph}(\psi) \neq \emptyset.$$

Proof. — We have to find $x \in B^2_{\rho_0}(0)$ and $y \in B^1_{\rho_0}(p)$ such that

$$(x, \varphi(x)) = (\psi(y), y).$$

This is equivalent to finding $y \in B^1_{\rho_0}(p)$ such that

$$\psi(y) \in B^2_{\rho_0}(0)$$
$$\varphi\psi(y) = y.$$

Observe that

$$\| \psi(y)\| \leqslant \| \psi(p)\| + C \|y - p\| \leqslant \varepsilon + C \|y - p\|.$$

Then, for every $0 < \rho_1 < \min\{\rho_0, (\rho_0 - \varepsilon)/C\}$,

$$\psi(B^1_{\rho_1}(p)) \subset B^2_{\rho_2}(0).$$

Now we can consider

$$\varphi\psi : B^1_{\rho_1}(p) \to E^1$$

that satisfies

$$\| \varphi\psi(p)\| \leqslant \varepsilon \| \psi(p)\| \leqslant \varepsilon^2$$

and

(16) $$\| \varphi\psi(w_1) - \varphi\psi(w_2)\| \leqslant \varepsilon C \| w_1 - w_2 \|.$$

Hence

(17) $$\| \varphi\psi(w) - p \| \leqslant \| p \| + \varepsilon^2 + \varepsilon C \| w - p \| \leqslant \varepsilon + \varepsilon^2 + \varepsilon C \| w - p \|.$$

Since by hypothesis

$$\frac{\varepsilon + \varepsilon^2}{1 - \varepsilon C} < \min\{\rho_0, (\rho_0 - \varepsilon)/C\},$$

there exists

(18) $$\frac{\varepsilon + \varepsilon^2}{1 - \varepsilon C} < \rho_2 < \min\{\rho_0, (\rho_0 - \varepsilon)/C\}.$$

Thus $\rho_2 < \min\{\rho_0, (\rho_0 - \varepsilon)/C\}$ implies

$$\psi(B^1_{\rho_2}(p)) \subset B^2_{\rho_0}(0)$$

and then from (17) follows

$$(\varphi\psi)\,(B^1_{\rho_2}(p)) \subset B^1_{\rho_2}(p),$$

because $\varepsilon + \varepsilon^2 + \varepsilon C\rho_2 < \rho_2$ by (18).

Moreover, since $\varepsilon C < 1$, (16) implies that $\varphi\psi$ is a contraction $B^1_{\rho_2}(p)$ into itself. The fixed point of this contraction satisfies the required properties.

To complete the proof of Theorem IV.2, we shall apply Lemma IV.8 to the maps

$$\psi_{n,x} : B_\rho(f^{-n}(x)) \to \hat{F}(f^{-n}(x)),$$
$$\varphi : B_{r_1}(y) \to \hat{E}(y)$$

where $B_{r_1}(y)$ is the disk of radius r_1 centered at 0 of the fiber $\hat{F}(y)$ and φ is a C^1 map such that

$$\operatorname{graph}(\varphi) = D_r(y).$$

Then $\varphi(0) = 0$ and $(D\varphi)\,(0) = 0$. Diminishing r_1 we can satisfy the condition:

$$\|(D\varphi)\,(x)\| \leqslant \varepsilon$$

for all $x \in B_{r_1}(y)$, for any given $\varepsilon > 0$. Choose ε so small that $\varepsilon C \ll 1$. Then IV.8 says that if $f^{-n}(x)$ and y are ε-near, then

$$\operatorname{graph}(\varphi) \cap \operatorname{graph}(\psi_{n,x}) \neq \emptyset.$$

This means that

$$D_r(y) \cap \Sigma^s_x(n) \neq \emptyset$$

and then there exists $v \in V(x)$ such that

$$f_v^{-n}(x) \in D_r(y).$$

V. — Proof of Theorem I.7

We shall begin by recalling the statements of three theorems, proved in [13], about the creation of homoclinic points.

If $x \in M$ and $f \in \mathrm{Diff}^1(M)$, let $\mu_x(f, n)$ be the probability

$$\mu_x(f, n) = \frac{1}{n}\sum_{j=1}^{n} \delta_{f^{-j}(x)}.$$

Denote by $\mathcal{M}(f, x)$ the set of accumulation points of the sequence $\{\mu_x(f, n) \mid n > 0\}$. Clearly $\mathcal{M}(f, x) \subset \mathcal{M}(f)$.

Given a basic set Λ of a diffeomorphism f (see Section I for the definition of a basic set) we say that p is a homoclinic point associated to Λ if $p \in W^s(\Lambda) \cap W^u(\Lambda) - \Lambda$.

Recall that every hyperbolic set Σ of f that is isolated (i.e. $\bigcup_n f^n(U) = \Sigma$ for some compact neighborhood of Λ) and satisfies $\Omega(f/\Sigma) = \Sigma$, can be decomposed in a unique way in a union of disjoint basic sets

Theorem **V.1** ([13]). — *If Σ is an isolated hyperbolic set of $f \in \mathrm{Diff}^1(M)$, with $\Omega(f/\Sigma) = \Sigma$, and there exist $x \in W^s(\Sigma) - \Sigma$ and $\mu \in \mathcal{M}(f, x)$ such that $\mu(\Sigma) > 0$, then there exists $g \in \mathrm{Diff}^1(M)$ arbitrarily C^1 near to f, coinciding with f in a neighborhood of $\Sigma \cup \{f^n(x) \mid n \geqslant 0\}$ and such that either g has homoclinic points associated to a basic set of Σ or else,*

$$x \in W^s_g(\Sigma) \cap W^u_g(\Sigma).$$

Theorem **V.2** ([13]). — *If Σ is an isolated hyperbolic set of $f \in \mathrm{Diff}^1(M)$, with $\Omega(f/\Sigma) = \Sigma$, and there exists $x \notin W^u(\Sigma)$ such that $\mu(\Sigma) > 0$ for all $\mu \in \mathcal{M}(f, x)$, then there exists $g \in \mathrm{Diff}^1(M)$, arbitrarily near to f and coinciding with f in a neighborhood of Σ, having a homoclinic point associated to a basic set of Σ.*

Theorem **V.3** ([13]). — *Let Σ be an isolated hyperbolic set of $f \in \mathrm{Diff}^1(M)$ such that $\Omega(f/\Sigma) = \Sigma$. Suppose that $\{x_j\} \subset M$ is a sequence converging to a point $x \notin \Sigma$ and $n_1 < n_2 < \dots$ is a sequence of integers such that the probabilities $\mu_{x_j}(f, n_j)$ converge to a probability μ with $\mu(\Sigma) > 0$. Then, given a C^1-neighborhood $\mathcal{U}$ of f, one of the following properties holds:*

a) *There exists $g \in \mathcal{U}$ coinciding with f in a neighborhood of Σ and having a homoclinic point associated to a basic set of Σ.*

b) *For every neighborhood U of Σ, there exists another neighborhood $V \subset U$ of Σ and $g \in \mathcal{U}$, coinciding with f in $V \cup U^c$, such that for some $j > 0$ and $0 < n \leqslant n_j$, g satisfies*

$$g^{-i}(x_j) = f^{-i}(x_j)$$

for $0 \leqslant i \leqslant n - 2$ and

$$g^{-t}(x_j) \in V$$

for all $t \geqslant n$.

Observe that the last condition, together with the fact that g and f coincide in V, implies that

$$V \ni g^{-i}(g^{-n}(x_j)) = f^{-i}(g^{-n}(x_j))$$

for all $i \geqslant 0$. If U (and then V) is small enough, this implies

$$g^{-n}(x_j) \in W^u(\Sigma).$$

The proof of the following easy lemma is left to the reader.

Lemma **V.4.** — a) *If $A \subset M$ is a compact set and a sequence of probabilities $\mu_{x_j}(f^m, n_j)$, $j = 1, 2, \dots$, converges to $\mu \in \mathcal{M}(f^m)$ such that $\mu(A) > 0$, then every accumulation point ν of the sequence $\{\mu_{x_j}(f, mn_j) \mid j > 0\}$ satisfies $\nu(A) > 0$.*

b) *If $A \subset M$ is a compact set and $x \in M$ satisfies $\mu(A) > 0$ for all $\mu \in \mathcal{M}(f^m, x)$, then $\nu(A) > 0$ for all $\nu \in \mathcal{M}(f, x)$.*

Now let us prove Theorem I.7. We shall find a diffeomorphism g arbitrarily C^1 near to f, coinciding with f in a neighborhood of $\bigcup_{1}^{s} \Lambda_k$, and such that if $\Lambda - \Lambda_i$ is not closed then

$$W^s_g(\Lambda_i) \cap W^u_g(\Lambda_1 \cup \ldots \cup \Lambda_s) - (\Lambda_1 \cup \ldots \cup \Lambda_s) \neq \emptyset.$$

This proves I.7, because it implies the existence of $1 \leqslant r \leqslant s$ such that

$$W^s_g(\Lambda_i) \cap W^u_g(\Lambda_r) - (\Lambda_i \cup \Lambda_r) \neq \emptyset$$

and then, from hypothesis II) of I.7, it follows that $i \neq r$. Finally, it will be obvious from the construction of g that, as required by Theorem I.7, $\Lambda - \Lambda_r$ is not closed.

From now on, let $f, \Lambda, \Lambda_1, \ldots, \Lambda_s$, $m > 0$, $c > 0$ and $0 < \lambda < 1$ be given by the hypotheses of Theorem I.7. Choose $\gamma, \gamma_1, \gamma_2$ such that

$$\lambda < \gamma_2$$

and $$\exp(-c) < \gamma_2 < \gamma < \gamma_1 < 1.$$

Lemma **V.5.** — *If* $x \in \Lambda$ *satisfies*

$$\prod_{j=1}^{n} \|(Df^m)/E(f^{-mj}(x))\| \geqslant \gamma_2^n$$

for all n *sufficiently large, then* $x \in W^u(\Lambda_1 \cup \ldots \cup \Lambda_s)$.

Proof. — Suppose that the inequality holds for all $n \geqslant N$. Then, if $\mu \in \mathcal{M}(f^m, x)$ is the limit of a sequence $\mu_x(f^m, n_j)$, $j \geqslant 0$, we obtain

$$\begin{aligned}\int \log \|(Df^m)/E\|\, d\mu &= \lim_{j\to+\infty} \int \log \|(Df^m)/E\|\, d\mu_x(f^m, n_j)\\ &= \lim_{j\to+\infty} \frac{1}{n_j} \sum_{i=j}^{n_j} \log \|(Df^m)/E(f^{-mi}(x))\|\\ &= \lim_{j\to+\infty} \log\Big(\prod_{i=1}^{n_j} \|(Df^m)/E(f^{-mi}(x))\|\Big)^{1/n_j} \geqslant \log \gamma_2 > -c.\end{aligned}$$

Hence, by hypothesis III), for all $\mu \in \mathcal{M}(f^m, x)$ we have

$$\mu(\Lambda_1 \cup \ldots \cup \Lambda_s) > 0.$$

Using V.4 b), this inequality holds for all $\mu \in \mathcal{M}(f, x)$. Applying V.2 we conclude that either $x \in W^u(\Lambda_1 \cup \ldots \cup \Lambda_s)$ or that there exists $g \in \mathrm{Diff}^1(M)$ arbitrarily C^1 near to f and coinciding with f in a neighborhood of $\Lambda_1 \cup \ldots \cup \Lambda_s$ such that there exists a homoclinic point $p \notin \Lambda_1 \cup \ldots \cup \Lambda_s$ associated to one of the basic sets Λ_i. By hypothesis II) of I.7 this is impossible. Hence $x \in W^u(\Lambda_1 \cup \ldots \cup \Lambda_s)$, proving the lemma.

To construct g, let now Λ_i be such that $\Lambda - \Lambda_i$ is not closed. Suppose, to simplity the notation, that $i = 1$. Take an isolating block W of Λ_1 (i.e. $\bigcap_n f^n(W) = \Lambda_1$) and set

$$\Lambda_1^s = \bigcap_{n\geqslant 0} f^{-n}(W),$$

$$\Lambda_1^u = \bigcap_{n\geqslant 0} f^{n}(W).$$

Let us show that $(\Lambda_1^s - \Lambda_1) \cap \Lambda \neq \emptyset$. Suppose that $(\Lambda_1^s - \Lambda_1) \cap \Lambda = \emptyset$. Then

$$\emptyset = (\Lambda_1^s - \Lambda_1) \cap \Lambda = (\bigcap_{n \geqslant 0} f^{-n}(W) - \Lambda_1) \cap \Lambda = \bigcap_{n \geqslant 0} f^{-n}(W \cap \Lambda) - \Lambda_1.$$

Hence
$$\Lambda_1 = \bigcap_{n \geqslant 0} f^{-n}(W \cap \Lambda).$$

This implies ([26]) that there exists a compact neighborhood W_0 of Λ_1 in Λ such that $f^{-1}(W_0) \subset \operatorname{Int} W_0$ and

$$\bigcap_{n \geqslant 0} f^{-n}(W_0) = \Lambda_1.$$

When $\operatorname{Int} W_0 - f^{-1}(W_0) = \emptyset$, or, what is the same, if $W_0 = f^{-1}(W_0)$, this equality implies $W_0 = \Lambda_1$ because $f^{-n}(W_0) = W_0$ for all n. But this is impossible because a neighborhood of Λ_1 in Λ must contain points not in Λ_1, since $\Lambda - \Lambda_1$ is not closed. Suppose now that $\operatorname{Int} W_0 - f^{-1}(W_0) \neq \emptyset$. Clearly

$$f^{-n}(\operatorname{Int} W_0 - f^{-1}(W_0)) \subset f^{-n}(\operatorname{Int} W_0) = \operatorname{Int} f^{-n}(W_0) \subset f^{-n}(W_0) \subset f^{-1}(W_0).$$

Hence
$$f^{-n}(\operatorname{Int} W_0 - f^{-1}(W_0)) \cap (\operatorname{Int} W_0 - f^{-1}(W_0)) = \emptyset.$$

Therefore $\operatorname{Int} W_0 - f^{-1}(W_0)$ contains no nonwandering points of f/Λ. This contradicts $\Omega(f/\Lambda) = \Lambda$ and completes the proof of $(\Lambda_1^s - \Lambda_1) \cap \Lambda \neq \emptyset$.

Now, unfortunately, the proof divides in two cases. The first case is when there exists $p \in (\Lambda_1^s - \Lambda_1) \cap \Lambda$ such that $p \in \alpha(p)$. Then we can take a sequence of integers $n_1 < n_2 < \ldots$ such that the sequence $\{f^{-n_j m}(p) \mid j \geqslant 1\}$ converges to a point in Λ_1.

Lemma **V.6.** — *If* $(\{p\}, \{n_1, n_2, \ldots\})$ *is not an* (m, γ)*-contracting sequence, then Theorem I.7 is true.*

Proof. — By hypothesis, there exist $j_1 < j_2 < \ldots$ such that, setting $\bar{n}_i = n_{j_i}$, we have

$$\prod_{j=1}^{\bar{n}_i} \|(Df^m)/E(f^{-mj}(p))\| \geqslant \gamma^{\bar{n}_i}.$$

Hence
$$\int \log \|(Df^m)/E\| \, d\mu_p(f^m, \bar{n}_i) \geqslant \log \gamma > -c.$$

Then, if $\mu \in \mathscr{M}(f^m)$ is an accumulation point of the sequence $\{\mu_p(f^m, \bar{n}_i) \mid i \geqslant 1\}$,

$$\int \log \|(Df^m)/E\| \, d\mu \geqslant \log \gamma - c.$$

Hence, by hypothesis,

$$\mu(\Lambda_1 \cup \ldots \cup \Lambda_s) > 0.$$

Then, V.4 implies that there exists $\mu_0 \in \mathscr{M}(f, p)$ satisfying

$$\mu_0(\Lambda_1 \cup \ldots \cup \Lambda_s) > 0.$$

Applying V.1 to $\Sigma = \Lambda_1 \cup \ldots \cup \Lambda_s$ we obtain a diffeomorphism g, arbitrarily C^1 near to f, coinciding with f in a neighborhood of $\Sigma \cup \{f^n(p) \mid n \geqslant 0\}$ and having p as a homoclinic point associated to Σ, the other option being ruled out by hypothesis II) of I.7. Since $g^n(p) = f^n(p)$ for all $n \geqslant 0$, it follows that

$$p \in W^s_g(\Lambda_1) \cap W^u_g(\Lambda_1 \cup \ldots \cup \Lambda_s).$$

This proves I.7 and then also Lemma V.6.

By Lemma V.6 we can continue the proof of I.7 assuming that $(\{p\}, \{n_1, n_2, \ldots\})$ is an (m, γ)-contracting sequence. Let us apply IV.1 to $(\{p\}, \{n_1, \ldots\})$ and our choice of $\gamma < \gamma_1 < 1$. If property a) of IV.1 holds, there exists a subsequence $\{\bar{n}_1, \bar{n}_2, \ldots\}$ of $\{n_1, n_2, \ldots\}$ such that $(\{p\}, \{\bar{n}_1, \bar{n}_2, \ldots\})$ is strongly (m, γ)-contracting and $f^{-\bar{n}_j m}(p)$ converges to a point $y \in \Lambda$ satisfying

$$(1) \qquad \prod_{i=1}^{n} \|(Df^m)/E(f^{-mi}(y))\| \geqslant \gamma_2^n$$

for all n larger than a certain N. By Lemma V.5,

$$(2) \qquad y \in W^u(\Lambda_1 \cup \ldots \cup \Lambda_s).$$

Now hypothesis IV) of Theorem I.7 says that

$$\|(Df^m)/E(x)\| . \|(Df^{-m})/F(f^m(x))\| < \lambda$$

for all $x \in \Lambda$. Hence, this inequality and (1) imply

$$\prod_{i=0}^{n-1} \|(Df^{-m})/F(f^{-mi}(y))\|$$
$$= \prod_{i=0}^{n-1} \|(Df^m)/E(f^{-m(i+1)}(y)\| . \|(Df^{-m})/F(f^{-mi}(y))\|$$
$$. \prod_{i=0}^{n-1} \|(Df^m)/E(f^{-m(i+1)}(y))\|^{-1} \leqslant \lambda^n \gamma_2^{-n}.$$

Since we chose γ_2 satisfying $\lambda\gamma_2^{-1} = \lambda_0 < 1$, we have

$$(3) \qquad \prod_{i=0}^{n-1} \|(Df^{-m})/F(f^{-mi}(y))\| \leqslant \lambda_0^n$$

for all $n \geqslant$ N. Let $D_r(x)$, $x \in \Lambda$, be the family of disks tangent at x to $F(x)$, associated to the splitting $TM/\Lambda = E \oplus F$, as we explained in Section IV. From (3) it is easy to deduce

$$(4) \qquad \lim_{n \to +\infty} \operatorname{diam} f^{-n}(D_r(y)) = 0$$

when r is small enough; this, together with (2), implies

$$(5) \qquad D_r(y) \subset W^u(\Lambda_1 \cup \ldots \cup \Lambda_s).$$

Now let us apply the Attainability Theorem IV.2 to the strongly (m, γ_1)-contracting sequence $(\{p\}, \{\bar{n}_1, \bar{n}_2, \ldots\})$ and y. It yields a diffeomorphism g, arbitrarily C^1 near to f and such that g^{-1} and f^{-1} differ only in an arbitrarily small neighborhood U of p, and moreover satisfying

(6) $$g^{-n}(p) \in D_r(y)$$

for some $n > 0$. Suppose that U was chosen so small that

(7) $$U \cap (\{f^n(p) \mid n \geqslant 1\} \cup (\bigcup_{n \geqslant 0} f^{-n}(D_r(y))) \cup \Lambda_1) = \emptyset.$$

This can be done because

$$p \notin \bigcup_{n \geqslant 0} f^{-n}(D_r(y));$$

otherwise (5) would imply

$$p \in W^u(\Lambda_1 \cup \ldots \cup \Lambda_s)$$

and since $p \in W^s(\Lambda_1)$, Theorem I.7 would follow just taking $f = g$. Then p does not belong to the set

$$\{f^n(p) \mid n \geqslant 1\} \cup (\bigcup_{n \geqslant 0} f^{-n}(D_r(y)) \cup \Lambda_1,$$

that (using (5)) is easily seen to be closed. Then U can be chosen satisfying (7). From (7) it follows easily that

(8) $$D_r(y) \subset W^u_g(\Lambda_1 \cup \ldots \cup \Lambda_s).$$

Also from (7) follows that

$$g^k(p) = f^k(p)$$

for all $k \geqslant 0$. Hence

$$p \in W^s_g(\Lambda_1),$$

and this together with (6) and (8), implies

$$p \in W^s_g(\Lambda_1) \cap W^u_g(\Lambda_1 \cup \ldots \cup \Lambda_s)$$

which once more, proves Theorem I.7. Now consider the case when applying IV.1 to $(\{p\}, \{n_1, n_2, \ldots\})$ it is property b) that holds. Then there exists a sequence of positive integers $0 < \bar{n}_j \leqslant n_j$ such that $(\{p\}, \{\bar{n}_1, \bar{n}_2, \ldots\})$ is a strongly (m, γ_1)-contracting sequence and

$$\sup_j (n_j - \bar{n}_j) < \infty.$$

This last relation implies that without loss of generality we can assume (recalling that $f^{-n_j m}(p)$ converges to a point in Λ_1) that

$$\lim_{j \to +\infty} f^{-\bar{n}_j m}(p) = p_0 \in \Lambda_1.$$

Let $TM/\Lambda_1 = E^s \oplus E^u$ be the hyperbolic splitting of Λ_1. By hypothesis, $\dim E^u(x) > \dim F(x)$ for all $x \in \Lambda_1 \cap \Lambda$. Then we have two dominated splittings of $TM/\Lambda_1 \cap \Lambda$, namely $TM/\Lambda_1 = E^s \oplus E^u$ and $TM/\Lambda = E \oplus F$. Using that f/Λ_1 is transitive and well known (and easy) properties of dominated splittings (see [9], for instance) we obtain that $F(x) \subset E^u(x)$ for all $x \in \Lambda_1$. Then, by definition of hyperbolic splitting, $F(x)$ satisfies property (3), for all $x \in \Lambda_1$ (with y obviously replaced by x and suitable values of $0 < \lambda_0 < 1$ and $m > 0$). Take $x = p_0$. Then, as before, we have that

$$\lim_{n\to+\infty} \operatorname{diam} f^{-n}(D_r(p_0)) = 0$$

when r is small enough and then

$$D_r(p_0) \subset W^u(p_0) \subset W^u(\Lambda_1).$$

Now, arguing as before, take a neighborhood U of p such that

$$U \cap (\{f^n(p) \mid n > 0\} \cup (\bigcup_{n\geq 0} f^{-n}(D_r(p_0)))) = \emptyset.$$

Applying again the Attainability Theorem, now to the strongly (m, γ_1)-contracting sequence, we can find $g \in \mathrm{Diff}^1(M)$ arbitrarily C^1 near to f, such that g^{-1} and f^{-1} differ only in U, and satisfying for some $N > 0$

$$g^{-N}(p) \in D_r(p_0).$$

Since g^{-1} and f^{-1} coincide in the set

$$\{f^n(p) \mid n > 0\} \cup (\bigcup_{n\geq 0} f^{-n}(D_r(p_0))),$$

we have

$$f^n(p) = g^n(p)$$

for all $n \geq 0$ and

$$\lim_{n\to+\infty} \operatorname{diam} g^{-n}(D_r(p_0)) = 0.$$

Hence $\quad p \in W^s_g(\Lambda_1)$

and $\quad g^{-N}(p) \in D_r(p_0) \subset W^u_g(\Lambda_1).$

Therefore $\quad p \in W^s_g(\Lambda_1) \cap W^u_g(\Lambda_1) - \Lambda_1,$

a contradiction with hypothesis II) of I.7.

This completes the proof of Theorem I.7 when there exists $p \in (\Lambda_1^s - \Lambda_1) \cap \Lambda$ such that $p \in \alpha(p)$. Now let us suppose that

(9) $\quad p \notin \alpha(p)$ for all $p \in (\Lambda_1^s - \Lambda_1) \cap \Lambda.$

Take $q \in (\Lambda_1^s - \Lambda_1) \cap \Lambda$. Since $q \notin \alpha(q)$ there exists a ball $B_\rho(q)$ such that

(10) $\quad f^{-n}(q) \notin B_\rho(q)$

for all $n \geq 1$. Since $q \in \Lambda = \Omega(f/\Lambda)$, there exist a sequence of points $\{q_j \mid j \geq 1\} \subset \Lambda$ converging to q and integers $\bar{n}_j \to +\infty$ such that $\lim_{j\to+\infty} f^{-\bar{n}_j}(q_j) = q$. Given any $\varepsilon > 0$, we can take $\rho > 0$ so small that, using property (10), there exists a sequence of integers $0 < n_j < \bar{n}_j/m$ satisfying $\lim_{j\to+\infty} n_j = +\infty$ and for, all j,

(i) $f^{-n}(q_j) \notin B_\rho(q)$ for all $0 < n \leq mn_j$,

(ii) $d(f^{-mn_j}(q_j), \Lambda_1) < \varepsilon$.

The next lemma parallels Lemma V.6, in such a way that the whole proof, under assumption (9), parallels the proof of the case when we had $p \in (\Lambda_1^s - \Lambda_1) \cap \Lambda$ satisfying $p \in \alpha(p)$.

Lemma **V.7.** — *If* $(\{q_j\}, \{n_j\})$ *is not an* (m, γ)*-contracting sequence, then Theorem* I.7 *is true.*

Proof. — By hypothesis, there exist $j_1 < j_2 < \ldots$ such that, denoting $\hat{n}_i = n_{j_i}$ and $\hat{q}_i = q_{j_i}$, we have

$$\prod_{j=1}^{\hat{n}_i} \|(Df^m)/E(f^{-mj}(\hat{q}_i))\| \geqslant \gamma^{\hat{n}_i}.$$

Hence
$$\int \log \|(Df^m)/E\| \, d\mu_{\hat{q}_i}(f^m, \hat{n}_i) \geqslant \log \gamma > -c.$$

Then, if $\mu \in \mathcal{M}(f^m)$ is an accumulation point of the sequence $\{\mu_{\hat{q}_i}(f^m, \hat{n}_i) \mid i \geqslant 1\}$,

$$\int \log \|(Df^m)/E\| \, d\mu \geqslant \log \gamma > -c.$$

Hence, by hypothesis

$$\mu(\Lambda_1 \cup \ldots \cup \Lambda_s) > 0.$$

Then V.4 implies that if μ_0 is an accumulation point of the sequence $\{\mu_{\hat{q}_i}(f, m\hat{n}_i) \mid i \geqslant 1\}$, one has

$$\mu_0(\Lambda_1 \cup \ldots \cup \Lambda_s) > 0.$$

Now we can apply V.3 to obtain a diffeomorphism g arbitrarily C^1 near to f and coinciding with f in a neighborhood of $\Lambda_1 \cup \ldots \cup \Lambda_s$, such that (since option a) of V.3 is ruled out by hypothesis II) of I.7) there exists $\hat{q}_i$, with i arbitrarily large, such that

$$g^{-n}(\hat{q}_i) = f^{-n}(\hat{q}_i) \tag{11}$$

for all $n \geqslant 0$ less than a certain $N \leqslant m\hat{n}_j$, and

$$g^{-t}(\hat{q}_i) \in V \tag{12}$$

for all $t \geqslant N$, where V is an arbitrarily small neighborhood of $\Lambda_1 \cup \ldots \cup \Lambda_s$ where g and f coincide. Then, if V is taken being an isolating block of $\Lambda_1 \cup \ldots \cup \Lambda_s$, we conclude that

$$g^{-N}(\hat{q}_i) \in W^u(\Lambda_1 \cup \ldots \cup \Lambda_s).$$

Moreover (11) and (12) imply that

$$g^{-n}(\hat{q}_i) \notin B_\rho(q)$$

for all $n \geqslant 1$. Now observe that the forward orbit of q converges to Λ_1 and therefore, without loss of generality, we can assume that is does not intersect $B_\rho(q)$. By the local stability of the hyperbolic set Λ_1, there exists $\bar{q}$ nearby q whose forward orbit converges to Λ_1 without intersecting $B_{\rho/2}(q)$. Moreover the distance $d(q, \bar{q})$ is arbitrarily small if g is sufficiently C^1 near to f. Therefore the quotient $d(\hat{q}_i, \bar{q})/(\rho/2)$ can be obtained arbitrarily small. Then there exists a diffeomorphism $h \in \mathrm{Diff}^1(M)$ such that

$$h(\hat{q}_i) = \bar{q},$$

and $$h(x) = x \quad \text{if } x \notin B_{\rho/2}(q)$$

and whose C^1 distance to the identity goes to zero together with $d(\hat{q}_i, \bar{q})/(\rho/2)$. Consider $\bar{g} \in \mathrm{Diff}^1(M)$ defined by $\bar{g} = hg$. It is easy to check that

$$\{ \bar{g}^n(\bar{q}) \mid n \geqslant 0 \} \cup \{ \bar{g}^{-n}(\hat{q}_i) \mid n \geqslant 1 \}$$

has the property of being an orbit of $\bar{g}$, and clearly an orbit in

$$W^s_{\bar{g}}(\Lambda_1) \cap W^u_{\bar{g}}(\Lambda_1 \cup \ldots \cup \Lambda_s) - (\Lambda_1 \cup \ldots \cup \Lambda_s),$$

thus proving I.7 and also Lemma V.7.

As before, Lemma V.7 means that we have only to complete the proof of Theorem I.7 assuming that $(\{q_j\}, \{n_j\})$ is an (m, γ)-contracting sequence. Observe that we reached this conclusion independently of the ε used in the construction of the sequence $(\{q_j\}, \{n_j\})$. On the other hand we have already shown that if $r > 0$ is small enough then

$$D_r(x) \subset W^u(\Lambda_1)$$

for all $x \in \Lambda_1$. Let $0 < \gamma_2 < \gamma < \gamma_1 < 1$ be as chosen and let $\varepsilon > 0$ be smaller than the $\varepsilon(r, m, \gamma_1)$ given by Attainability Theorem IV.2. Now let us apply Theorem IV.1 to the (m, γ)-contracting sequence $(\{q_j\}, \{n_j\})$. If property a) of IV.1 holds, there exists $(\{q'_j\}, \{n'_j\}) \leqslant (\{q_j\}, \{n_j\})$ that is strongly (m, γ_1)-contracting and such that

$$\lim_{j \to +\infty} f^{-n'_j m}(q'_j) = y \in \Lambda$$

where for the same reasons as in the proof of (5), y satisfies

$$D_r(y) \subset W^u(\Lambda_1 \cup \ldots \cup \Lambda_s).$$

Take, as before, $0 < \rho_1 < \rho$ such that

$$B_{\rho_1}(q) \cap (\{f^n(q) \mid n \geqslant 1\} \cup (\bigcup_{n \geqslant 0} f^{-n}(D_r(y)))) = \varnothing.$$

Now, applying the Attainability Theorem to $(\{q'_j\}, \{n'_j\})$, we can take a diffeomorphism g, arbitrarily C^1 near to f, such that $g^{-1} = f^{-1}$ in the complement of $B_{\rho_1}(q)$ and such that for an arbitrarily large j there exists $N > 0$ satisfying

$$g^{-N}(q'_j) \in D_r(y). \tag{13}$$

The arguments used in the previous case now show that

$$D_r(y) \subset W^u_g(\Lambda_1 \cup \ldots \cup \Lambda_s)$$

and

$$f^n(q) = g^n(q) \quad \text{for all } n \geqslant 0. \tag{14}$$

Moreover, the Attainability Theorem also grants that g can be taken satisfying:

$$d(g^{-n}(q'_j), f^{-n}(q'_j)) \leqslant \rho - \rho_1$$

for all $0 \leqslant n \leqslant N$. Hence

$$\{ g^{-n}(q'_j) \mid 1 \leqslant n \leqslant N \} \cap B_{\rho_1}(q) = \emptyset.$$

Moreover by (13)

$$\{ g^{-n}(q'_j) \mid n \geqslant N \} \cap B_{\rho_1}(q) \subset (\bigcup_{n \geqslant 0} g^{-n}(D_r(y))) \cap B_{\rho_1}(q)$$
$$= (\bigcup_{n \geqslant 0} f^{-n}(D_r(y))) \cap B_{\rho_1}(q) = \emptyset.$$

Hence $$\{ g^{-n}(q'_j) \mid n \geqslant 1 \} \cap B_{\rho_1}(q) = \emptyset.$$

and from (14)

$$\{ g^n(q) \mid n > 0 \} \cap B_{\rho_1}(q) = \emptyset.$$

Now take $h \in \mathrm{Diff}^1(M)$ such that

$$h(q'_j) = q,$$

and $$h(x) = x \quad \text{if } x \notin B_{\rho_1}(q).$$

Then the set

$$\{ g^n(q) \mid n \geqslant 0 \} \cup \{ g^{-n}(q'_j) \mid n \geqslant 1 \}$$

has the property of being an orbit of $\bar{g} = hg$, and moreover, by (13) and (14), it is an orbit in

$$W^s_{\bar{g}}(\Lambda_1) \cap W^u_{\bar{g}}(\Lambda_1 \cup \ldots \cup \Lambda_s).$$

Hence

(15) $$q \in W^s_{\bar{g}}(\Lambda_1) \cap W^u_{\bar{g}}(\Lambda_1 \cup \ldots \cup \Lambda_s) - (\Lambda_1 \cup \ldots \cup \Lambda_s),$$

thus proving Theorem I.7. Finally, let us consider the case when in the application of Theorem IV.1 to $(\{q_j\}, \{n_j\})$ it is property b) that holds. This property means that without loss of generality we can suppose that $(\{q_j\}, \{n_j\})$ is a strongly (m, γ_1)-contracting sequence. Since $d(f^{-mn_j}(q_j), \Lambda_1) < \varepsilon$, there exists $y \in \Lambda_1$ ε-near to an accumulation point of the sequence of endpoints of $(\{q_j\}, \{n_j\})$. Then, given an arbitrarily small neighborhood U of q, we can apply the Attainability Theorem (observing that for this purpose we take ε smaller than the $\varepsilon(r, m, \gamma_1)$ of this theorem) and obtain g, arbitrarily C^1 near to f, such that $g^{-1} = f^{-1}$ in U^c, and satisfying

$$g^{-N}(q_j) \in D_r(y)$$

for some $N > 0$ and $j \geqslant 1$ that can be taken arbitrarily large. Now, repeating the method of the proof in the previous case (i.e. when it was option a) of IV.1 that held) we take a diffeomorphism h, C^1 close to the identity, satisfying $h(q_j) = q$ and $h(x) = x$ if $x \notin U$, and we define $\bar{g} = hg$ and show that q satisfies (15).

REFERENCES

[1] A. Andronov, L. Pontrjagin, Systèmes grossiers, *Dokl. Akad. Nauk. SSSR*, **14** (1937), 247-251.
[2] D. V. Anosov, Geodesic flows on closed Riemannian manifolds with negative curvature, *Proc. Steklov Inst. Math.*, **90** (1967), 1-235.
[3] R. Bowen, Equilibrium states and the ergodic theory of Anosov diffeomorphisms, *Springer Lecture Notes in Math.*, **470** (1975).
[4] J. Franks, Necessary conditions for stability of diffeomorphisms, *Trans. A.M.S.*, **158** (1971), 301-308.
[5] J. Guckenheimer, A strange, strange attractor, in *The Hopf bifurcation and its applications*, *Applied Mathematics Series*, **19** (1976), 165-178, Springer Verlag.
[6] M. Hirsch, C. Pugh, M. Shub, Invariant manifolds, *Springer Lecture Notes in Math.*, **583** (1977).
[7] R. Labarca, M. J. Pacifico, Stability of singular horseshoes, *Topology*, **25** (1986), 337-352.
[8] S. T. Liao, On the stability conjecture, *Chinese Ann. Math.*, **1** (1980), 9-30.
[9] R. Mañé, Persistent manifolds are normally hyperbolic, *Trans. A.M.S.*, **246** (1978), 261-283.
[10] R. Mañé, Expansive diffeomorphisms, in *Dynamical Systems-Warwick* 1974, *Springer Lecture Notes in Math.*, **468** (1975), 162-174.
[11] R. Mañé, Characterization of AS diffeomorphisms, *Proc. ELAM III*, *Springer Lecture Notes in Math.*, **597** (1977), 389-394.
[12] R. Mañé, An ergodic closing lemma, *Ann. of Math.*, **116** (1982), 503-540.
[13] R. Mañé, On the creation of homoclinic points, *Publ. Math. I.H.E.S.*, **66** (1987), 139-159.
[14] S. Newhouse, Lectures on dynamical systems, *Progr. in Math.*, **8** (1980), 1-114.
[15] J. Palis, A note on Ω-stability, in *Global Analysis, Proc. Sympos. Pure Math., A.M.S.*, **14** (1970), 221-222.
[16] J. Palis, S. Smale, Structural stability theorems, in *Global Analysis, Proc. Sympos. Pure Math., A.M.S.*, **14** (1970), 223-231.
[17] V. A. Pliss, Analysis of the necessity of the conditions of Smale and Robbin for structural stability of periodic systems of differential equations, *Diff. Uravnenija*, **8** (1972), 972-983.
[18] V. A. Pliss, On a conjecture due to Smale, *Diff. Uravnenija*, **8** (1972), 268-282.
[19] C. Pugh, The closing lemma, *Amer. J. Math.*, **89** (1967), 956-1009.
[20] J. Robbin, A structural stability theorem, *Ann. of Math.*, **94** (1971), 447-493.
[21] C. Robinson, C^r structural stability implies Kupka-Smale, in *Dynamical Systems, Salvador*, 1971, Academic Press 1973, 443-449.
[22] C. Robinson, Structural stability of C^1 diffeomorphisms, *J. Diff. Eq.*, **22** (1976), 28-73.
[23] M. Shub, Stabilité globale des systèmes dynamiques, *Astérisque*, **56** (1978).
[24] S. Smale, Diffeomorphisms with many periodic points, in *Differential and Combinatorial Topology*, Princeton Univ. Press, 1964, 63-80.
[25] S. Smale, Differentiable dynamical systems, *Bull. A.M.S.*, **73** (1967), 747-817.
[26] S. Smale, The Ω-stability theorem, in *Global Analysis, Proc. Sympos. Pure Math., A.M.S.*, **14** (1970), 289-297.

Instituto de Matemática Pura e Aplicada (IMPA)
Estrada Dona Castorina, 110
Jardim Botánico
Rio de Janeiro - RJ

Manuscrit reçu le 12 mars 1987.

THE EXPLOSION OF SINGULAR CYCLES

by RODRIGO BAMÓN (1), RAFAEL LABARCA (2)
RICARDO MAÑÉ, MARÍA JOSÉ PACÍFICO

Introduction

The objective of this work is to introduce and analyze a new natural mechanism through which a vector field depending on a parameter may evolve, when the parameter varies, from a vector field exhibiting a very simple dynamical nature(say, having a finite chain recurrent set), into one exhibiting non trivial forms of recurrence.

For diffeomorphisms, the study of such mechanisms goes back to the work of Newhouse and Palis [NP1], where they considered one-parameter families of diffeomorphims and analyzed the dynamics of the diffeomorphisms corresponding to values of the parameter close to the first bifurcation parameter (i.e. the first value of the parameter for which the diffeomorphism is not Morse-Smale). After avowing the difficulties of proving which is the generic dynamics at the first bifurcation, they focused their research on the case when at the first bifurcation value the chain recurrent set of the diffeomorphism is the union of a finite set of hyperbolic periodic orbits and a *cycle* i.e. a finite family of hyperbolic periodic orbits linked, in a cyclic way by orbits contained in the intersections of stable and unstable manifolds of different periodic orbits of the family [NP2]. They conjectured a certain genericity of this property, to be recalled below. In that paper and afterwards in a joint work with Takens [NPT], the authors describe how the cycle explodes when the parameter increases. Explosion here means, as usual in this context, a sudden increase of the size of a relevant dynamically defined set (say, the non wandering set) triggered by a small perturbation of the sytem. Essentially, in [NP1], [NP2] and [NPT], a perturbation of the system leads to the creation of homoclinic tangencies and then to the vast array of phenomena they carry on their wake (Newhouse wild horseshoes, persistent tangencies, non hyperbolic attractors, etc.). Their research then moves to the natural question of how large are the set of parameters for which each one of these phenomena arise, and their main and more accurate results are in the case of diffeomorphisms of two dimensional manifolds [NP2], [NPT], [PT].

(1) Partially supported by Fondecyt 1059/89, Chile.
(2) Partially supported by Fondecyt, Chile and FAPERJ, Brazil.

Originally published in: *Publ. Math. Inst. Hautes Etudes Sci. No. 78 (1993), 207–232 (1994)*,

For vector fields without singularities, a similarity with the case of diffeomorphisms of surfaces can be expected. But, when the cycle whose explosion gives birth to non trivial dynamical structures contains a singularity, new mechanisms, with no analogy with the case of diffeomorphisms of surfaces, arise.

The first such mechanism was studied by Afraimovic and Shilnikov in [AS], where they considered one-parameter families of vector fields on 3-dimensional manifolds that cross the boundary of the Morse-Smale region through the collision of two saddles producing a saddle-saddle singularity whose stable and unstable manifolds have transversal intersections. They analyse how, after crossing the boundary, this saddle-saddle self connection unfolds into non trivial hyperbolic sets that admit an accurate description in terms of Bernouilli shifts.

Our objective is the explosion of what we shall call *singular cycles*, i.e. cycles containing a *hyperbolic* singularity, and we shall describe how they explode in a way entirely different from that of the cycles of diffeomorphisms of surfaces or the Afraimovic-Shilnikov cycle. Through the explosion of such a cycle we enter a region largely filled by Axiom A flows, and in an important subcase, the non Axiom A flows that may appear are arranged in a codimension 1 lamination of the space of vector fields, having small Hausdorff dimension, and where each leaf of the lamination is characterized as a class of topologically equivalent vector fields, whose dynamics is a generalization of the Lorenz-Guckenheimer-Williams attractor [GW] and the Labarca-Pacífico [LP] singular horseshoe.

Let us now give the precise statements of our results. Let M be a compact and boundaryless 3-manifold and let $\mathscr{X}^r$ be the Banach space of C^r vector fields on M. If $X \in \mathscr{X}^r$ denote $\Gamma(X)$ its chain recurrent set. We say that $X \in \mathscr{X}^r$ is *simple* when $\Gamma(X)$ is a union of finitely many hyperbolic critical orbits. By a critical orbit we mean an orbit that is either periodic or singular. It is easy to see that the set S^r of simple C^r vector fields is an open subset of $\mathscr{X}^r$.

Denote by $C^k(I, \mathscr{X}^r)$ the space of families X_μ of C^r vector fields depending on a parameter $\mu \in [-1, 1]$ such that the map $[-1, 1] \ni \mu \mapsto X_\mu \in \mathscr{X}^r$ is C^k. Endow $C^k(I, \mathscr{X}^r)$ with the C^k topology.

Among the families $X_\mu \in C^k(I, \mathscr{X}^r)$, we distinguish those that start at a simple system, i.e.

$$(*) \qquad X_{-1} \in S^r,$$

and leave the region of simple systems, i.e.

$$(**) \qquad X_\mu \notin \overline{S}^r \quad \text{for some } \mu > -1.$$

The *crossing parameter value* of such a family is the supremum of the μ's such that $X_\mu \in S^r$ (or, what is the same, the minimum μ such that $X_\mu \notin S^r$). Denote by $C^k_*(I, \mathscr{X}^r)$ the set of such families. To simplify the notation, and without loss of generality we shall add to the definition of $C^k_*(I, \mathscr{X}^r)$ the requirement that *the crossing value of the parameter be zero*, i.e.

$$X_\mu \in C^k_*(I, \mathscr{X}^r) \Rightarrow \begin{cases} X_\mu \in S^r & \text{if } \mu < 0, \\ X_0 \notin S^r. \end{cases}$$

A *cycle* of a vector field $X \in \mathscr{X}^r$ is a compact invariant chain recurrent set of X consisting of a finite family of hyperbolic periodic orbits and orbits whose α and ω-limit sets are different hyperbolic periodic orbits of the family.

Translated into our context, what Newhouse and Palis conjectured in [NP] is that for a generic family $X_\mu \in C^k_*(I, \mathscr{X}^r)$, X_0 either has a non hyperbolic periodic orbit or a cycle. Even without the support of this conjecture, that remains widely open, cycles are a crucial concept for the understanding of how complex dynamical objects are born from very simple ones.

An orbit γ of a vector field X is nontransversal if $\alpha(\gamma)$ and $\omega(\gamma)$ are hyperbolic critical orbits and the stable and unstable manifolds of $\alpha(\gamma)$ and $\omega(\gamma)$ intersect non-transversally along γ.

Our object of study will be *simple singular cycles* defined as follows: a simple singular cycle Λ of a vector field $X \in \mathscr{X}^r$ is a cycle of X satisfying:

a) Λ contains a unique singularity σ_0.

b) The eigenvalues of $D_{\sigma_0} X : T_{\sigma_0} M \hookleftarrow$ are real and satisfy $-\lambda_3 < -\lambda_1 < 0 < \lambda_2$.

c) Λ contains a unique nontransversal orbit γ_0 which is contained in $W^u(\sigma_0)$ and $\omega(\gamma_0)$ is a periodic orbit σ_1.

d) For every $p \in \gamma_0$ and every invariant manifold $W(\sigma_0)$ of X, passing through σ_0 and tangent at σ_0 to the space spanned by the eigenvectors associated to $-\lambda_1$ and λ_2, we have

$$T_p W(\sigma_0) + T_p W^s(\sigma_1) = T_p M.$$

e) There is a neighborhood $\mathscr{U}$ of X such that if $Y \in \mathscr{U}$ the continuations $\sigma_i(Y)$, $0 \leqslant i \leqslant k$, of critical orbits σ_i of the cycle are well defined, the vector field Y is C^2-linearizable nearby $\sigma_0(Y)$ and the Poincaré maps of $\sigma_i(Y)$, $1 \leqslant i \leqslant k$, are C^2-linearizable.

f) Λ is isolated, i.e. it has an isolating block. Recall that an isolating block of an invariant set Λ of a vector field X is an open set U such that

$$\Lambda = \bigcap_t X^t(U),$$

where $X^t : M \hookleftarrow$ is the flow generated by X.

The motivation of this definition is closely related to the following property which is nowadays standard knowledge in bifurcation theory.

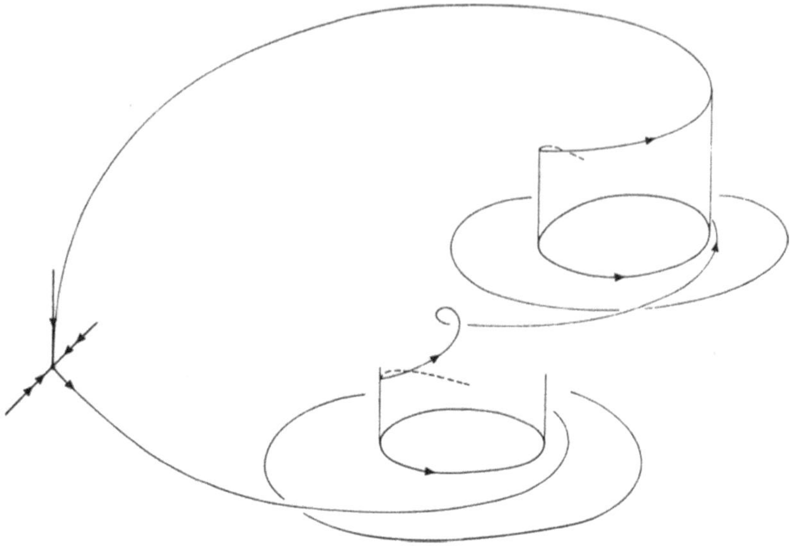

FIG. 1

Proposition. — *If* $r \geqslant 3$ *and* $k \geqslant 1$ *there exists an open and dense set* $\mathscr{A}$ *of* $C^k_*(I, \mathfrak{X}^r)$ *such that if* $X_\mu \in \mathscr{A}$ *and* X_0 *has a cycle* Λ *that contains a singularity, then* Λ *is a simple singular cycle and* $\Gamma(X_0) - \Lambda$ *is a finite union of hyperbolic critical orbits.*

The $r \geqslant 3$ condition is required because the C^3-topology is the weaker one where the openess and density of property (*e*) is granted by Sternberg's linearization Theorem [S].

The next result gives the basic elementary description of what happens nearby a simple singular cycle when you perturb the vector field. We shall use the following notation: given $Y \in \mathfrak{X}^r$ and $U \subset M$, set

$$\Lambda(Y, U) = \bigcap_t Y^t(U).$$

Theorem 1. — *Let* Λ *be a simple singular cycle of* $X \in \mathfrak{X}^r$ *and let* U *be an isolating block of* Λ. *Then there exists a neighborhood* $\mathscr{U}$ *of* X *and a codimension-one submanifold* $\mathscr{N} \subset \mathfrak{X}^r$ *containing* X *such that:*

a) $Y \in \mathscr{U} \cap \mathscr{N} \Rightarrow \Lambda(Y, U)$ *is a simple singular cycle topologically equivalent to* Λ.

b) $\mathscr{U} - \mathscr{N}$ *has two connected components, and one of them, that we shall denote by* $\mathscr{U}^-$, *is such that* $Y \in \mathscr{U}^-$ *implies that the chain recurrent set of* $Y/\Lambda(Y, U)$ *consists of the continuations* $\sigma_i(Y)$, $0 \leqslant i \leqslant k$, *of the critical orbits* σ_i *contained in* Λ.

This means that the cycle persists topologically unchanged in $\mathcal{N} \cap \mathcal{U}$ and is broken in $\mathcal{U}^-$, leaving behind a very simple dynamical object. The explosion will really take place in the other connected component of $\mathcal{U} - \mathcal{N}$, to be denoted $\mathcal{U}^+$. Define $\mathcal{U}_H^+$ as the set of $Y \in \mathcal{U}^+$ such that the chain recurrent set of $Y/\Lambda(Y, U)$ is $\sigma_0(Y)$ plus a transitive hyperbolic set. The set $\mathcal{U}_H^+$ fills a very large part of $\mathcal{U}^+$. This is the content of the next result, in whose statement $m(\cdot)$ denotes the Lebesgue measure of subsets of **R**.

Theorem 2. — If $X_\mu \in \mathcal{A}$ crosses transversally $\mathcal{N}$ at X_0 and X_0 has a simple singular cycle Λ, and U is an isolating block of Λ, then there exists $\delta > 1$ such that

$$\limsup_{t \to 0} \frac{1}{t^\delta} m\{0 \leqslant \mu \leqslant t \mid X_\mu \in \mathcal{U}^+ - \mathcal{U}_H^+\} = 0.$$

The study of $\mathcal{U}^+ - \mathcal{U}_H^+$ heavily depends on whether the eigenvalues

$$-\lambda_3 < -\lambda_1 < 0 < \lambda_2$$

at the singularity σ_0 satisfy $-\lambda_1 + \lambda_2 > 0$ or $-\lambda_1 + \lambda_2 < 0$. In the first case we say that the cycle is *expanding* and in the second that is *contracting*.

Denote by $c(\cdot)$ the upper limit capacity of subsets of **R**.

Theorem 3. — Let $X_\mu \in \mathcal{A}$, Λ and U be as in Theorem 2. Then:

a) If the cycle is expanding, the set

$$\{0 \leqslant \mu \leqslant t \mid X_\mu \in \mathcal{U}^+ - \mathcal{U}_H^+\},$$

for t sufficiently small, is a Cantor set whenever $X_t \in \mathcal{U}_H^+$, and

$$\lim_{t \to 0} c(\{0 \leqslant \mu \leqslant t \mid X_\mu \in \mathcal{U}^+ - \mathcal{U}_H^+\} = 0.$$

b) If the cycle is contracting then the set of parameters $\mu > 0$ for which the unstable manifold of the singularity $\sigma_0(X_\mu)$ converges to an attracting periodic orbit, accumulates on $\mu = 0$.

This leads naturally to introduce a new open subset $\mathcal{U}_{H'} \subset \mathcal{U}^+$ consisting of those vector fields $Y \in \mathcal{U}^+$ for which the chain recurrent set of $Y/\Lambda(Y, U)$ is the union of $\sigma_0(Y)$, a transitive hyperbolic set and a unique attracting periodic orbit. It is clear that this set is open. The natural question is:

Is $\mathcal{U}_H^+ \cup \mathcal{U}_{H'}^+$ dense in $\mathcal{U}$ in the contracting case?

In the expanding case, the next theorem will show that $\mathcal{U}_{H'}^+$ is empty and $\mathcal{U}_H^+$ is dense in $\mathcal{U}^+$. In a special contracting case, determined by supplementary conditions on the eigenvalues of the singularity, an affirmative answer to the question has been recently given by Pacífico and Rovella [PR].

Theorem 4. — *If* $X \in \mathfrak{X}^r$, Λ *and* U *are as in the statement of Theorem 2 and if* Λ *is expanding, then* $\mathcal{U} - \mathcal{U}_H^+$ *is laminated by codimension 1,* C^1 *submanifolds, such that for all the vector fields in each lamina, the dynamics of the maximal invariant set of* U *is the same up to topological equivalence. Moreover, when* $Y \in \mathcal{U} - \mathcal{U}_H^+$, $\Lambda(Y, U)$ *is a chain recurrent expansive set.*

This gives a complete description of $\mathcal{U} - \mathcal{U}_H^+$ in the expanding case. Our next and final result shows the stability of one-parameter families of vector fields crossing $\mathcal{N}$ transversally, in the expanding case.

We say that a one-parameter family $X_\mu \in C^k(I, \mathfrak{X}^r)$ is Γ-*stable* if for all $\varepsilon > 0$ there exists a neighborhood $\mathcal{U}$ of X_μ in $C^r(I, \mathfrak{X}^r)$ such that if $Y_\mu \in \mathcal{U}$ there exist a reparametrizing homeomorphism $\varphi : I \to I$, ε-near to the identity, and, for each $\mu \in I$, a topological equivalence h_μ between $X_\mu/\Gamma(X_\mu)$ and $Y_\mu/\Gamma(Y_\mu)$, ε-near to the identity, such that the map

$$(\mu, x) \in \{(\mu, x) ; \mu \in I, x \in \Gamma(X_\mu)\} \mapsto (\varphi(\mu), h_\mu(x)) \in \{(s, w) ; s \in I, \omega \in \Gamma(Y_\mu)\}$$

is continuous.

Theorem 5. — *If* $X_\mu \in \mathcal{A}$ *is as in the statement of Theorem 2 and* X_0 *has an expanding simple singular cycle, then there exists* $\delta > 0$ *such that* $\{ X_\mu \mid -1 \leqslant \mu \leqslant \delta \}$ *is* Γ*-stable.*

We are thankful to J. Palis for useful conversations and to M. Viana for the proof of the result in the Appendix. The first two authors are grateful to IMPA for its very kind hospitality.

Proof of the Theorems

Proof of Theorem 1.

Let $X \in \mathfrak{X}^r$ be a vector field having a simple singular cycle Λ with isolating block $U \subset M$. For simplicity we will assume that Λ contains a unique periodic orbit. So, Λ is the union of a singularity σ_0, a periodic orbit σ_1 and a unique nontransversal orbit γ such that $\lambda(\gamma) = \sigma_0$ and $w(\gamma) = \sigma_1$.

Let S be a cross section to the flow of X at $q \in \sigma_1$ parametrized by

$$\{(x, y) ; |x|, |y| \leqslant 1\}$$

and satisfying $W^s(\sigma_1) \supseteq \{(x, 0) ; |x| \leqslant 1\}$ and $W^u(\sigma_1) \supseteq \{(x, y) ; |y| \leqslant 1\}$.

We call a closed subset $C \subset S$ a *horizontal strip* if it is bounded (in S) by two disjoint continuous curves connecting the vertical sides of S, $\{(-1, y), |y| \leqslant 1\}$ and $\{(1, y), |y| \leqslant 1\}$.

Let p be the first intersection of γ with S. Then $p = (x_0, 0)$ and we assume $x_0 > 0$. Since $W^u(\sigma_1)$ intersects $W^s(\sigma_0)$ and γ has σ_0 as α-limit and σ_1 as ω-limit set, a first return map F is defined on a subset of S. Moreover, if $q_0 = (0, y_0) \in S$ is such that its ω-limit set is σ_0 then there exists a horizontal strip $R \ni q_0$ so that F is defined on R. As Λ is isolated, we have $\Lambda \cap S \subset \{(x, y) ; y \geqslant 0\}$ and $F(R) \subset \{(x, y), y \leqslant 0\}$. See figure 2.

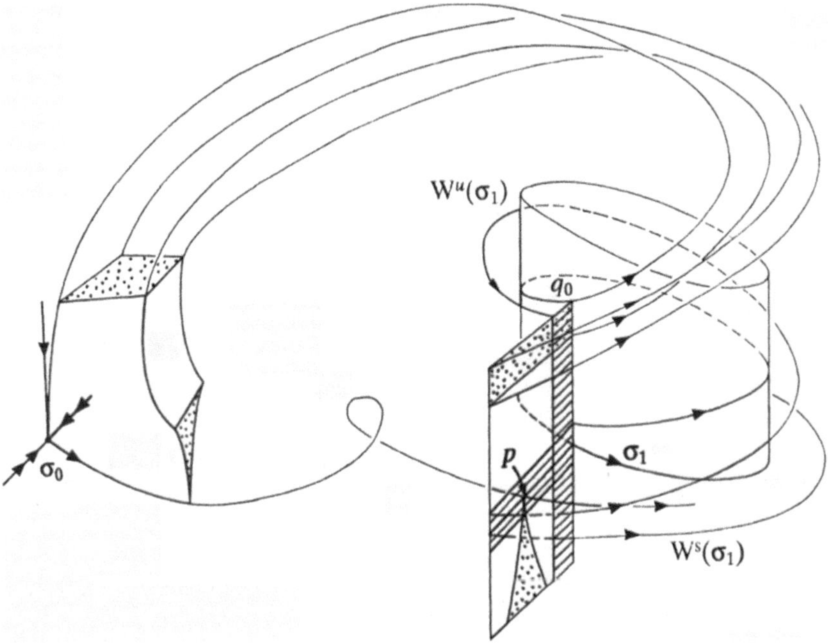

FIG. 2

If $Y \in \mathscr{X}^r$ is nearby X then $W^s(\sigma_1(Y))$ intersects S at a curve $c(Y)$ and the first intersection of $W^u(\sigma_0(Y))$ with S is a point $p(Y)$. Observe that both $c(Y)$ and $p(Y)$ vary smoothly with Y.

The implicit function theorem on Banach spaces implies that the condition $p(Y) \in c(Y)$ defines a C^1 local codimension-one submanifold $\mathscr{N}$ in a neighborhood $\mathscr{U} \subset \mathscr{X}^r$ of X such that $\mathscr{U} \backslash \mathscr{N}$ has two connected components; one of them, that we shall call $\mathscr{U}^-$, characterized by $p(Y) \in S$, lies below $c(Y)$. Let $\mathscr{U}^+$ be the other component.

Clearly if $Y \in \mathscr{U}^-$, then $\Gamma(Y/\Lambda(Y, U))$ is the union of $\sigma_0(Y)$ and $\sigma_1(Y)$ and, so, Y is simple.

If $Y \in \mathscr{U}^+$, then $\sigma_1(Y)$ has transversal homoclinic orbits and so Y is not simple. Thus, $\mathscr{U} \cap S^r$ coincides with $\mathscr{U}^-$, proving Theorem 1. ∎

Before going through the proof of Theorem 2, let us introduce some notation.

Let $X \in \mathscr{X}^r$ be a vector field having a simple singular cycle Λ with isolating block U. For simplicity we will assume that Λ contains a singularity σ_0, a unique periodic orbit σ_1 and a unique non transversal orbit $\gamma \subset W^u(\sigma_0)$ such that $\alpha(\gamma) = \sigma_0$ and $\omega(\gamma) = \sigma_1$.

Let S, $\mathscr{U}$, $\mathscr{N}$, $\mathscr{U}^+$ and $\mathscr{U}^-$ be as above. Taking $\mathscr{U}$ small enough, S is also a cross section for every $Y \in \mathscr{U}$ at $\sigma_1(Y)$, where $\sigma_1(Y)$ is the periodic orbit obtained by continuation of σ_1. As before, there exists a first return map F_Y defined on a subset of S for every $Y \in \mathcal{U}$.

Since $\Lambda(Y, U)$ is the closure of the saturation by Y^t of $\Lambda(Y, Y) \cap S$ and $\Lambda(Y, U) \cap S$ is the maximal invariant set of F_Y, it is necessary to describe F_Y to understand the dynamics of $\Lambda(Y, U)$. To do so we choose coordinates (x, y) on S depending smoothly on Y so that

(i) $Q = \{(x,y); |x|, |y| \leq 1\} \subset S$;
(ii) $\{(x, 0); |x| \leq 1\} \subseteq W^s(\sigma_1(Y))$;
(iii) $\{(0,y); |y| \leq 1\} \subseteq W^u(\sigma_1(Y))$;
(iv) $\Lambda \cap S \subseteq \{(x,y), y \geq 0\}$;
(v) the first intersection of $W^u(\sigma_0(Y))$ with S is a point $p_Y = (x_Y, y_Y)$ with $0 < x_Y < 1$.

Notice that $Y \in \mathcal{U}^+$ if and only if $y_Y > 0$; $\Gamma(Y/\Lambda(Y, U)) \neq \{\sigma_0(Y), \sigma_1(Y)\}$ if and only if $y_Y > 0$ and $\Lambda(Y, U) \cap S \subset Q^+ = \{(x, y) \in Q; x, y \geq 0\}$.

For $y_Y \geq 0$, $W^u(\sigma_0(Y))$ intersects $W^s(\sigma_1(Y))$, and since $W^u(\sigma_1(Y))$ intersects $W^s(\sigma_0(Y))$ (transversally!), we see that if $q_0(Y) = (0, y_0(Y)) \in Q^+$ is such that $\omega(q_0(Y)) = \sigma_0(Y)$ and $\alpha(q_0(Y)) = \sigma_1(Y)$, then there exists a horizontal strip $R_Y^0 \ni q_0(Y)$ so that the positive orbit of points at R_Y^0 pass first near $\sigma_0(Y)$ and then return to Q. On the other hand, the positive orbit of points at a horizontal strip R_Y containing $W^s(\sigma_1(Y)) \cap Q^+$ turns around the closed orbit $\sigma_1(Y)$ and then returns to Q. See figure 3.

So, F_Y is defined on $R_Y^0 \cup R_Y$ and the restriction of F_Y to R_Y coincides with the Poincaré map P_Y associated to $\sigma_1(Y)$. We further assume that P_Y is linear on Q. Let $\rho_Y > 1$ and $\tau_Y < 1$ be the eigenvalues of $DP_Y(0, 0)$. Thus,

$$R_Y^0 = \{(x, y); x \geq 0, \tilde{\theta} \leq y \leq \theta_Y(x)\}$$

where $\theta_Y(x) = \theta(Y, x)$ is a smooth real function satisfying

$$\{(x, \theta_Y(x)), 0 \leq x \leq 1\} \subseteq W^s(\sigma_0(Y))$$

and if $\delta_Y(x) = \delta(Y, x)$ is so that $\{(x, \theta_Y(x) - \delta_Y(x)), 0 \leq x \leq 1\} \subseteq W^s(\sigma_1(Y))$, then there is $\varepsilon > 0$ such that $\tilde{\theta} + \varepsilon < \theta_Y(x) - \delta_Y(x)$ for every x. Making a linear change of coordinates we can also assume (vi) $|\theta'_Y| < 1/100$, δ_Y goes to zero in the C^1 topology when Y approaches $\mathcal{N}$ and $\theta_Y(x_0) = 1$ for some $x_0 \in [0, 1]$.

Clearly $R_Y = \{(x, y); x \geq 0, \rho_Y^{-1} \leq y \leq \rho_Y^{-1}.\theta_Y(x)\}$ and $F_Y(x, y) = (\tau_Y.x, \rho_Y.y)$ for $(x, y) \in R_Y$.

To obtain the expression of $F_Y(x, y)$ for $(x, y) \in R_Y^0$ we proceed as follows.

Let $-\lambda_3(Y) < -\lambda_1(Y) < 0 < \lambda_2(Y)$ be the eigenvalues of $DY(\sigma_0(Y))$, $\alpha_Y = \dfrac{\lambda_1(Y)}{\lambda_2(Y)}$ and $\beta_Y = \dfrac{\lambda_3(Y)}{\lambda_2(Y)}$. Let (x_1, x_2, x_3) be C^2 linearizing coordinates for Y in a neighborhood $U_0 \ni \sigma_0(Y)$ depending smoothly on Y.

Let L, $\tilde{L}$ be the planes $x_1 = 1$ and $x_2 = 1$ respectively.

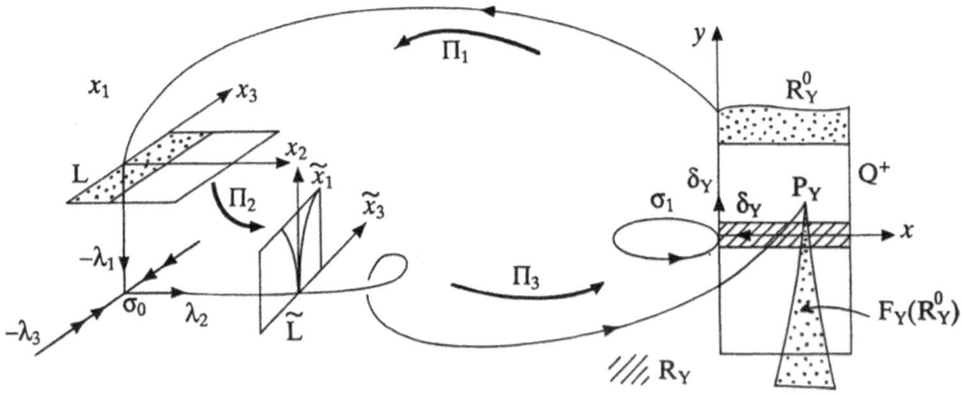

FIG. 3

For $(x,y) \in R_Y^0$ we have $F_Y(x,y) = \pi_3 \circ \pi_2 \circ \pi_1(x,y) = (f_Y(x,y), g_Y(x,y))$ where

a) $\pi_1 : Q^+ \to L$ is a diffeomorphism, $\pi_1(x, \theta_Y(x)) = (x_3, 0)$ for $0 \leqslant x \leqslant 1$,

$$D\pi_1(x,y) = \begin{pmatrix} a(x,y) & b(x,y) \\ c(x,y) & d(x,y) \end{pmatrix} \quad \text{with } k_1 \leqslant |a(x,y)|,\ |d(x,y)| \leqslant K_1,$$

for k_1 and K_1 two positive real constants. Up to replacing $\{(x, \theta_Y(x)), x \in [0,1]\}$ by some negative iterate of it (and shrinking $\mathcal{U}$ if necessary) we may assume that $|c(x,y)|/|d(x,y)| \leqslant \varepsilon$ for every $(x,y) \in R_Y^0$ and $Y \in \mathcal{U}^+$, $0 < \varepsilon \ll 1$.

b) $\pi_3 : \tilde{L} \to Q^+$ is a diffeomorphism, $D\pi_3(\tilde{x}_3, \tilde{x}_1) = \begin{pmatrix} \tilde{a}(\tilde{x}_3, \tilde{x}_1) & \tilde{b}(\tilde{x}_3, \tilde{x}_1) \\ \tilde{c}(\tilde{x}_3, \tilde{x}_1) & \tilde{d}(\tilde{x}_3, \tilde{x}_1) \end{pmatrix}$ with $k_2 \leqslant |\tilde{a}(\tilde{x}_3, \tilde{x}_1)|, |\tilde{d}(\tilde{x}_3, \tilde{x}_1)| \leqslant K_2$ for some positive constants k_2, K_2. Moreover, replacing $p(Y)$ by some positive iterate of it (also contained in $W^u(\sigma_0(Y) \cap S)$ has the effect of decreasing $|\tilde{b}|/|\tilde{d}|$ so that we may assume $|\tilde{b}|/|\tilde{d}| \leqslant \varepsilon$ for some small $\varepsilon > 0$.

c) $\pi_2 : L \to \tilde{L}$ is given by $\pi_2(x_3, x_2) = (\tilde{x}_3 = x_3 . x_2^{\beta_Y}, \tilde{x}_1 = x_2^{\alpha_Y})$.

From *a)*, *b)* and *c)* above follow

d)

$$\begin{aligned} \frac{\partial}{\partial x} f_Y(x,y) &= \alpha_Y . c(x,y) . \tilde{b}(\tilde{x}_3, \tilde{x}_1) . x_2^{\alpha_Y - 1} \\ &+ \beta_Y . c(x,y) . \tilde{a}(\tilde{x}_3, \tilde{x}_1) . x_3 . x_2^{\beta_Y - 1} \\ &+ a(x,y) . \tilde{a}(\tilde{x}_3, \tilde{x}_1) . x_2^{\beta_Y} + r_1(x,y), \quad |r_1(x,y)| \leqslant \text{constant} . x_2^{\beta_Y - 1}. \end{aligned}$$

$$\begin{aligned} \frac{\partial}{\partial y} f_Y(x,y) &= \alpha_y . d(x,y) . \tilde{b}(\tilde{x}_3, \tilde{x}_1) . x_2^{\alpha_Y - 1} \\ &+ \beta_Y . d(x,y) . \tilde{a}(\tilde{x}_3, \tilde{x}_1) . x_3 . x_2^{\beta_Y - 1} + r_2(x,y), \quad |r_2(x,y)| \leqslant \text{constant} . x_2^{\beta_Y}. \end{aligned}$$

$$\frac{\partial}{\partial x} g_Y(x,y) = \alpha_Y . c(x,y) . \tilde{d}(\tilde{x}_3, \tilde{x}_1) . x_2^{\alpha_Y - 1}$$
$$+ \beta_Y . c(x,y) . \tilde{c}(\tilde{x}_3, \tilde{x}_1) . x_3 . x_2^{\beta_Y - 1} + r_3(x,y), \quad |r_3(x,y)| \leqslant \text{constant} . x_2^{\beta_Y}.$$

$$\frac{\partial}{\partial y} g_Y(x,y) = \alpha_Y . d(x,y) . \tilde{d}(\tilde{x}_3, \tilde{x}_1) . x_2^{\alpha_y - 1}$$
$$+ \beta_Y . d(x,y) . \tilde{c}(\tilde{x}_3, \tilde{x}_1) . x_3 . x_2^{\beta_Y - 1} + r_4(x,y), \quad |r_4(x,y)| \leqslant \text{constant} . x_2^{\beta_Y}.$$

We now state a useful lemma establishing the existence of C^1 invariant stable foliations for F_Y. Its proof is in the Appendix.

Lemma 1. — For every $Y \in \mathcal{U} \backslash \mathcal{U}^-$ *there exists an invariant* C^1 *stable foliation* $\mathcal{F}_Y^s$ *for* F_Y *depending* C^1 *on* Y.

This lemma implies that after a C^1 change of coordinates, $\theta_Y(x)$ and $\delta_Y(x)$ do not depend on x and that $g_Y(x,y)$ does not depend on y. For simplicity we assume $\theta_Y(x) \equiv 1$. We also have $c(x,y) \equiv 0$. As π_1 is a diffeomorphism, $a(x,y) \neq 0$ and $d(x,y) \neq 0$ for every (x,y). Thus using *d)* we conclude that there are real positive constants C and K so that

e) $$0 < \left| \frac{\partial}{\partial x} f_Y(x,y) \right| \leqslant K . x_2^{\beta_y} + r_1(x,y),$$

$$\left| \frac{\partial}{\partial y} f_Y(x,y) \right| \leqslant K . x_2^{\alpha_y - 1} + r_2(x,y)$$

and $$\left| \frac{d}{dy} g_Y(y) \right| \leqslant C . x_2^{\alpha_Y - 1} + r_3(y), \quad |r_3(y)| \leqslant \text{constant} . x_2^{\alpha_Y}.$$

Moreover,

f) $$f_Y(x, 1) = x_Y \quad \text{for } x \in [0, 1], \quad g_Y(1) = y_Y.$$

g) $$f_Y(x, 1 - \delta_Y) \subseteq \{(x, 0), x \in (0, 1)\}, \quad g_Y(1 - \delta_Y) = 0.$$

Conditions *e)*, *f)* and *g)* imply $\delta_Y = A_Y . y_Y^{1/\alpha_Y}$, A_Y a positive constant. Finally, making another C^1 change of coordinates we obtain

$$F_Y(x,y) = (f_Y(x,y), g_Y(y))$$

with $$g_Y(y) = \begin{cases} \rho_Y . y & \text{for } y \in [0, \rho_Y^{-1}], \\ y_Y - A_Y^{-\alpha_Y} . (1-y)^{\alpha_Y} [1 + \varphi((1-y)^{\alpha_Y}, y_Y)] & \text{for } y \in [1 - A_Y . y_Y^{1/\alpha_Y}, 1], \end{cases}$$

where φ is continuous, C^1 on $[0, 1)$, $\varphi(0, y_Y) = \varphi(A_Y^{\alpha_Y} . y_Y, y_Y) = 0$.

Furthermore, using *e)*, *f)* and *g)* we obtain

h) $$\left|\frac{d}{dy} g_Y(y)\right| \leqslant C.|1-y|^{\alpha_Y - 1}.$$

$$0 < \left|\frac{\partial}{\partial x} f_Y(x,y)\right| \leqslant K.|1-y|^{\beta_Y} \quad \text{and} \quad \left|\frac{\partial}{\partial y} f_Y(x,y)\right| \leqslant K.|1-y|^{\alpha_y - 1}.$$

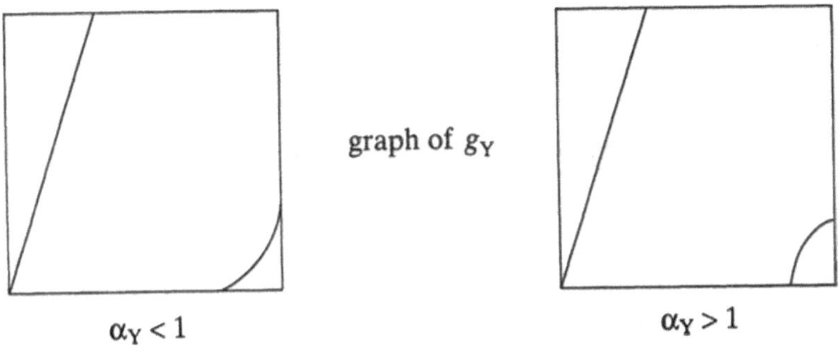

FIG. 4

We now come to the proof of Theorem 2.

Given $Y \in \mathcal{U}^+$, let $k_Y = 1 + y_Y^{\alpha_Y^{-1}}$ and

$$\mathcal{U}_0^+ = \{ Y \in \mathcal{U}^+ ; \exists n_0 \in \mathbf{N} \text{ so that } k_Y . \rho_Y^{-1} < \rho_Y^{n_0} . y_Y < 1 - \delta_Y \}.$$

The next two lemmas show that $\mathcal{U}_0^+ \subseteq \mathcal{U}_H^+$, that is, $\Gamma(Y/\Lambda(Y, U))$ is $\sigma_0(Y)$ plus a transive hyperbolic set for every $Y \in \mathcal{U}_0^+$.

To prove Theorem 2 we shall actually prove that there exist positive real constants M_1, M_2 and δ so that

(1) $$M_1 < \limsup_{\varepsilon \to 0} \frac{m(\mu \leqslant \varepsilon ; X_\mu \notin \mathcal{U}_0^+)}{\varepsilon^{1+\delta}} < M_2,$$

where X_μ is a one-parameter family as in the statement of Theorem 2. This obviously implies the result.

Lemma 2. — If $Y \in \mathcal{U}_0^+$ *then* $\Gamma(Y/\Lambda(Y, U))$ *is* $\sigma_0(Y)$ *plus a hyperbolic set.*

Proof. — Let $\Lambda(F_Y, Q^+) = \bigcap_{n \in \mathbf{Z}} F_Y^n(Q^+)$. For $Y \in \mathcal{U}_0^+$, $(x_Y, y_Y) \notin \Lambda(F_Y, Q^+)$ and then $\Lambda(Y, U)$ is the closure of the saturation of $\Lambda(F_Y, Q^+)$ by Y^t. Thus, to obtain the result it is enough to prove that $\Lambda(F_Y, Q^+)$ is a hyperbolic set.

28

Observe that for $Y \in \mathcal{U}_0^+$, $\Lambda(F_Y, Q^+) \subseteq \tilde{R}_Y^0 \cup R_Y$, where

$$\tilde{R}_Y^0 = \{(x, y) \in R_Y^0; y \leqslant 1 - A_Y . y_Y^{(\alpha_Y^{-2} + \alpha_Y^{-1})}\}.$$

The previous lemma implies the existence of a stable cone field for F_Y. So, to obtain the hyperbolicity of $\Lambda(F_Y, Q^+)$ we have only to prove the existence of an unstable cone field for F_Y. We will actually prove the existence of an unstable cone field for $G_Y = F_Y^{n_0+2}$ which easily implies the result. For this we claim that if $\mathcal{U}$ is small enough, then

(i) $DG_Y(x, y).(1, 1) = (u, v)$; $u = u(x, y)$, $v = v(x, y)$ and $|v| > |u|$ for every $(x, y) \in \Lambda(F_Y, Q^+)$;
(ii) $DG_Y(x, y).(1, 1) = (\bar{u}, \bar{v})$; $\bar{u} = \bar{u}(x, y)$, $\bar{v} = \bar{v}(x, y)$ and $|\bar{v}| > |\bar{u}|$ for every $(x, y) \in \Lambda(F_Y, Q^+)$;
(iii) there exists $\nu > 1$ such that

$$\| DG_Y(x, y).(u, v)\| \geqslant \nu . \|(u, v)\| \quad \text{for } |v| > |u|.$$

Notice that (iii) follows from (i), (ii) and the inequality

$$\| DG_Y(x, y).(\pm 1, 1)\| > \|(\pm 1, 1)\| \quad \text{for every } (x, y) \in \Lambda(F_Y, Q^+).$$

Suppose $(x, y) \in \tilde{R}_Y^0$. Then $G_Y = P_Y^{n_0+1} \circ F_Y$ where P_Y is the linear map $P_Y(x, y) = (\tau_Y . x, \rho_Y . y)$ and hence

$$DG_Y(x, y).(\pm 1, 1) = P_Y^{n_0+1} \circ \begin{pmatrix} \frac{\partial}{\partial x} f_Y(x, y) & \frac{\partial}{\partial y} f_Y(x, y) \\ 0 & \frac{\partial}{\partial y} g_Y(y) \end{pmatrix} \begin{pmatrix} \pm 1 \\ 1 \end{pmatrix} = \begin{pmatrix} u \\ v \end{pmatrix},$$

where
$$u = \tau_Y^{n_0+1} . \left(\pm \frac{\partial}{\partial x} f_Y(x, y) + \frac{\partial}{\partial y} f_Y(x, y)\right)$$

and
$$v = \rho_Y^{n_0+1} . \frac{\partial}{\partial y} g_Y(y).$$

Using *h)* above we obtain that there are positive real constants L and N such that

$$|u| < L . \tau_Y^{n_0+1} . |1 - y|^{\alpha_Y - 1}$$

and
$$|v| \geqslant N . \rho_Y^{n_0+1} . |1 - y|^{\alpha_Y - 1}.$$

Hence we have $|u| < |v|$ for $\mathcal{U}$ small enough.

For $(x, y) \in R_Y$, either $F_Y^j(x, y) \in R_Y$ for every $j = 0, 1, \ldots, n_0 + 1$ or $F_Y^j(x, y) \in R_Y$ for $j = 0, 1, \ldots, n_0$ and $F_Y^{n_0+1}(x, y) \in \tilde{R}_Y^0$. In the first case $G_Y(x, y) = P_Y^{n_0+2}(x, y)$ and the result follows trivially. In the second case $G_Y = F_Y \circ P_Y^{n_0+1}$ and we argue as before.

We now prove (iii). For $(x, y) \in \tilde{R}^0_Y$ we have

(2) $$\| DF_Y^{n_0+2}(x,y).(\pm 1, 1)\| = \|(u, v)\| \geqslant \| v \| \geqslant N.\rho_Y^{n_0+1}.|1-y|^{\alpha_Y - 1}.$$

If $\alpha_Y - 1 < 0$, choosing $\mathscr{U}$ small enough we are done. Otherwise, since $(x, y) \in \tilde{R}^0_Y$ we have $y < 1 - A.y_Y^{(\alpha_Y^{-2}+\alpha_Y^{1})}$ and hence $|1-y|^{\alpha_Y-1} \geqslant A^{\alpha_Y-1}.y_Y^{1-\alpha_Y^{-2}}$. So

(3) $$\| DF_Y^{n_0+2}(x,y).(\pm 1, 1)\| \geqslant \tilde{N}.\rho_Y^{n_0+1}.y_Y.y_Y^{-\alpha_Y^{-2}}.$$

But $Y \in \mathscr{U}_0^+$ implies $\rho_Y^{n_0+1}.y_Y \geqslant 1$ and thus

$$\| DF_Y^{n_0+2}(x,y).(\pm 1, 1)\| \geqslant \tilde{N}.y_Y^{-\alpha_Y^{-2}} > 10 \text{ for } \mathscr{U} \text{ small enough.}$$

If $(x, y) \in R_Y$ the result follows similarly. ■

Lemma 3. — If $Y \in \mathscr{U}_0^+$ *then* $\Lambda(F_Y, Q^+)$ *is a transitive set.*

Proof. — Let $F_Y(x,y) = (f_Y(x,y), g_Y(y))$ and $I_Y^0 = [0, \rho_Y^{-1}]$, $I_Y^1 = [1 - A_Y.y_Y^{1/\alpha_Y}, 1]$. Then $g_Y : I_Y^0 \cup I_Y^1 \to [0, 1]$. To each $x \in I_Y \cup I_Y^1$ we associate a 2-symbol sequence $\mathbf{x}$ defined by $x_i = 0$ if $g_Y^i(x) \in I_Y^0$, $x_i = 1$ if $g_Y^i(x) \in I_Y^1$. The sequence $\mathbf{x}$ is called the itinerary of x.

Given $Y \in \mathscr{U}_0^+$, let N_0 be so that $1 < y_Y.\rho_Y^{N_0} < \rho_Y$. We claim that given $(N_i)_{i \geqslant 1}$, $N_i \in \mathbf{N}$, $N_i \geqslant N_0$ for every i, the sequence $\mathbf{a} = 1\underbrace{0\ldots01}_{N_1}\underbrace{0\ldots01}_{N_2}\underbrace{0\ldots01}_{N_3}\ldots$ is realized as the itinerary of a unique point $x \in I_Y^0 \cup I_Y^1$. Indeed, for each $i \geqslant 1$, let $R_{N_i} = \rho_Y^{-N_i}[1 - A_Y.y_Y^{1/\alpha_Y}, 1]$. Then $g_Y^{N_i+1}(R_{N_i}) = I_Y^i$ for every i and defining inductively

$$C_1 = g_Y^{-1}(R_{N_i}) \cap I_Y^1,$$
$$C_k = g^{-(N_1+1)} \circ g^{-(N_2+1)} \circ \ldots \circ g^{-(N_{k-1}+1)}(g^{-1}(R_{n_k}) \cap I_Y^1) \cap I_Y^1$$

we have $C_1 \supset C_2 \supset \ldots$ Thus $\bigcap_{i \geqslant 1} C_i$ is nonempty and there exists $x \in Y_Y^0 \cup Y_Y^1$ so that its itinerary is $\mathbf{a}$. Now we observe that the proof we gave in Lemma 2 (actually inequality (3)) implies the unicity of such an x.

Denote by Λ_{g_Y} the maximal invariant set of g_Y. The result above implies that the pre-orbit of 0 by g_Y is dense in Λ_{g_Y} and so Λ_{g_y} is transitive. This implies the transitivity of $\Lambda(F_Y, Q^+)$. ■

To conclude the proof of Theorem 2 it remains to prove inequality (1) above. For this, let $X_\mu \in \mathscr{A}$ be as in the statement of the theorem. We start by reparametrizing X_μ in such way that $y_{X_\mu} = \mu$. As before $\alpha_0 = \lambda_1(0)/\lambda_2(0)$ where $-\lambda_3(0) < -\lambda_1(0) < 0 < \lambda_2(0)$ are the eigenvalues of $DX_0(\sigma_0)$, σ_0 the singularity contained in the simple singular cycle Λ_0 of X_0.

We will assume $\alpha_\mu = \alpha$, $A_\mu = A$ and $\rho_\mu = \rho$ for $0 \leqslant \mu \leqslant \varepsilon$. The general case, although more difficult can be done in a similar way.

Let $n_0 > 0$ be so that $\rho^{-n_0} < \varepsilon$. Call $\mu_0 = \rho^{-n_0}$ and for $n > n_0$ define

$$R_n^1 = \{ \mu; 1 - A_\mu^{\alpha^{-1}} \leqslant \rho^n . \mu \leqslant 1 \},$$
$$R_n^2 = \{ \mu; \rho^{-1} \leqslant \rho^n . \mu \leqslant k_\mu + \rho^{-1} \}, \quad \text{where } k_\mu = \mu^{\alpha^{-1}} . \rho^{-1}.$$

If $R^i = \bigcup_{n \geqslant n_0} R_n^i$, $i = 1, 2$, we have $\{ \mu; X_\mu \notin \mathscr{U}_0^+ \} = R^1 \cup R^2$. So, to obtain (1) it is enough to prove that

$$\frac{A + \rho^{-1}}{1 + \rho^{-(1+\alpha^{-1})}} \leqslant \frac{m(R^1 \cup R^2)}{\mu_0^{1+\alpha^{-1}}} \leqslant \frac{A + \rho^{-1}}{1 - \rho^{-(1+\alpha^{-1})}}.$$

Indeed, if $\mu \in R_n^1$, $n \geqslant n_0$, then $\mu \leqslant \rho^{-n}$ and so $\mu^{\alpha^{-1}} \leqslant \rho^{-n.\alpha^{-1}}$ which implies $1 - A_\mu^{\alpha^{-1}} \geqslant 1 - A.\rho^{-n.\alpha^{-1}}$. Hence $\mu \in \rho^{-n}[1 - A.\rho^{-n.\alpha^{-1}}, 1]$ and $m(R_n^1) \leqslant A.\rho^{-n(1+\alpha^{-1})}$.

If $\mu \in R_n^2$, $n \geqslant n_0$, then $\mu \leqslant \rho^{-n}$ which implies $k_\mu = \mu^{\alpha^{-1}}.\rho^{-1} \leqslant \rho^{-n(1+\alpha^{-1})}$. Hence $\rho^{-n-1} \leqslant \mu \leqslant \rho^{-n-1}(1 + \rho^{-n.\alpha^{-1}})$ and $m(R_n^2) \leqslant \rho^{-n(1+\alpha^{-1})}$. Thus

$$m(R^1 \cup R^2) \leqslant \sum_{n \geqslant n_0} (A.\rho^{-n(1+\alpha^{-1})} + \rho^{-n}.\rho^{-1}.\rho^{-n.\alpha^{-1}})$$
$$\leqslant (A + \rho^{-1}) \sum_{n \geqslant n_0} \rho^{-n(1+\alpha^{-1})} = (A + \rho^{-1}) . \frac{\rho^{n_0(1+\alpha^{-1})}}{1 - \rho^{-(1+\alpha^{-1})}}.$$

So
$$\frac{m(R^1 \cup R^2)}{\mu_0^{1+\alpha^{-1}}} = \frac{m(R^1 \cup R^2)}{\rho^{-n_0(1+\alpha^{-1})}} \leqslant \frac{A + \rho^{-1}}{1 - \rho^{-(1+\alpha^{-1})}}.$$

The other inequality of (1) is obtained in a similar way. This concludes the proof of Theorem 2. ∎

Proof of Theorem 3.

a) Let $X_\mu \in \mathscr{A}$ be as in the statement of Theorem 3, $\mathscr{F}_\mu^s$ be the stable foliation for F_μ given in Lemma 1 and $g_\mu : [0, 1] \to [0, 1]$ be the map induced by $\mathscr{F}_\mu^s$. Recall that after a C^1 change of coordinates g_μ is given by

$$g_\mu(y) = \begin{cases} \rho_\mu . y \quad \text{for } 0 \leqslant y \leqslant \rho_\mu^{-1}, \\ - A_\mu^{-\alpha_\mu} . (1 - y)^{\alpha_\mu}[1 + \varphi((1 - y)^{\alpha_\mu}, \mu)] + \mu \\ \qquad\qquad\qquad \text{for } 1 - A_\mu . \mu^{1/\alpha_\mu} \leqslant y \leqslant 1. \end{cases}$$

Define $T_\mu = \{ y \in \mathbf{R}; \rho_\mu^{-1} < y < 1 - A_\mu . \mu^{1/\alpha_\mu} \}$.

Given $\bar{\mu} > 0$ small and such that $X_{\bar{\mu}} \in \mathscr{U}_H^+$, Lemma 4 implies that

$$\{ \mu \leqslant \bar{\mu}; X_\mu \notin \mathscr{U}_H^+ \} = \{ \mu \in [0, \bar{\mu}]; g_\mu^n(1) \notin T_\mu \text{ for every } n \geqslant 0 \}.$$

Since the change of coordinates above is differentiable and depends C^1 on μ, the result is a consequence of following lemma.

Lemma 4. — *Given* $\bar{\mu}$ *as above*, $\{\mu \leqslant \bar{\mu};\ X_\mu \notin \mathcal{U}_H^+\}$ *is a Cantor set and its limit capacity satisfy*

$$\lim_{\bar{\mu}\to 0} c(\{\mu \leqslant \bar{\mu};\ X_\mu \notin \mathcal{U}_H^+\}) = 0.$$

Proof. — We will assume $\alpha_\mu = \alpha$, $\rho_\mu = \rho$ and $A_\mu = A$ for $0 \leqslant \mu \leqslant \bar{\mu}$. The general case, more difficult, can be done using the same ideas and similar calculations.

Given $n_0 \in \mathbf{N}$, let $\tilde{\mu} = \rho_{\tilde{\mu}}^{-n_0}$ and if $E \subset [0, \tilde{\mu}]$ denote by E^c the complement of E in $[0, \tilde{\mu}]$.

Observe that for $0 \leqslant \mu \leqslant \tilde{\mu}$ we have $g_\mu(1) = \mu$, $g_\mu^2(1) = \rho.\mu, \ldots, g_\mu^{n_0}(1) = \rho^{n_0}.\mu$ and $g_{\tilde{\mu}}^{n_0}(1) = g_{\tilde{\mu}}^{n_0}(\tilde{\mu}) = 1$.

Define, inductively, $G_0(\mu) = g_\mu^{n_0+1}(1) = g_\mu^{n_0}(\mu)$ and $G_k(\mu) = g_\mu(G_{k-1}(\mu))$ for $k \geqslant 1$.

We claim that $\min_\mu G'_k(\mu) \geqslant \rho.\min G'_{k-1}(\mu)$, $k \geqslant 1$.

Indeed

$$G'_k(\mu) = \frac{\partial}{\partial\mu} g_\mu(G_{k-1}(\mu)) + \frac{\partial}{\partial x} g_\mu(G_{k-1}(\mu)).G'_{k-1}(\mu).$$

As $\frac{\partial}{\partial\mu} g_\mu(G_{k-1}(\mu)) \geqslant 0$ and $\frac{\partial}{\partial x} g_\mu(G_{k-1}(\mu)) \geqslant \rho$ we get

(1) $$G'_k(\mu) \geqslant \rho.G'_{k-1}(\mu)$$

and so we have proved the claim.

Since $G'_0(\mu) \geqslant \rho^{n_0}$, (1) implies

(2) $$G'_k(\mu) \geqslant \rho^{n_0+k} \quad \text{for every } k \geqslant 0.$$

For $\mu \leqslant \tilde{\mu}$ define

$$\tilde{T}_\mu = T_\mu \cup g_\mu^{-1}(T_\mu) \cup \ldots \cup g_\mu^{-n_0+1}(T_\mu).$$

Given $k \geqslant 1$, let $E_k = \{\mu \in [0, \tilde{\mu}];\ \exists\, 0 \leqslant j < k;\ G_{n_0 j}(\mu) \in \tilde{T}_\mu\}$ and $N(k)$ be the number of connected components of E_k^c. Since $E_k = \{\mu \in [0, \tilde{\mu}]; \exists\, 0 \leqslant j \leqslant k; g_\mu^{n_0 j}(1) \in \tilde{T}_\mu\}$ and, for each $y \in [0, 1]$, the cardinality of $\{x \in [0, 1];\ g_\mu^{n_0}(x) = y\}$ has 2 as lower bound and $n_0 + 1$ as upper bound we obtain $N(k) \leqslant (n_0 + 1)^{k+1}$. So E_k^c is covered by at most $(n_0 + 1)^{k+1}$ intervals whose lenght is, by (2), bounded by $\rho^{-n_0.(k+1)}$. Moreover, if I_k^j, $0 \leqslant j \leqslant (n_0 + 1)^{k+1}$ is a connected component of E_k^c then the restriction of $G_{n_0.k}$ to I_k^j is an increasing function. Thus

$$\Lambda_{n_0} = \bigcap_{k \geqslant 0} E_k^c$$

is a Cantor set and its limit capacity satisfies

$$c(\Lambda_{n_0}) \leqslant \lim_{k\to\infty} \frac{\log(n_0 + 1)^{k+1}}{\log \rho^{n_0.(k+1)}}$$

$$= \lim_{k\to\infty} \frac{(k + 1).\log(n_0 + 1)}{(k + 1).n_0.\log \rho} = \frac{\log(n_0 + 1)}{n_0.\log \rho}.$$

But, for $\bar{\mu} \leqslant \rho^{-n_0} = \tilde{\mu}$, $I_B \cap [0, \bar{\mu}] = \Lambda_{n_0}$ which implies that $c(I_B \cap [0, \bar{\mu}]) = c(\Lambda_0)$ and since $\lim_{n_0 \to \infty} \dfrac{\log(n_0 + 1)}{n_0} = 0$ we get $\lim_{\bar{\mu} \to 0} c(I_B \cap [0, \bar{\mu}]) = 0$, proving the result. ∎

b) Let X_μ be as in the statement of Theorem 2 and suppose $\alpha_0 = \lambda_1/\lambda_2 > 1$. Then for $\mu_0 > 0$ small, $\alpha_\mu = \dfrac{\lambda_1(\mu)}{\lambda_2(\mu)} > 1$ for every $\mu \in [0, \mu_0]$. As before we can assume $y_\mu = \mu$.

Let $n_0 \in \mathbf{N}$ be such that $\rho_\mu^{-n_0} < \mu_0$ for every $\mu \in [0, \mu_0]$. To each $n > n_0$ let μ_n be so that $\rho_{\mu_n}^n \cdot \mu_n = 1$. Then $F_{\mu_n}^{n+1}(x, 1) = (\rho_{\mu_n}^n \cdot x_{\mu_n}, 1)$ and since $F_{\mu_n}(x, 1) = (x_{\mu_n}, \mu_n)$ for every $0 \leqslant x \leqslant 1$ we obtain $F_{\mu_n}^{n+1}(x_{\mu_n}, \mu_n) = (x_{\mu_n}, \mu_n)$. So, (x_{μ_n}, μ_n) is a periodic orbit of F_{μ_n} corresponding to a homoclinic orbit for X_{μ_n} associated to $\sigma_0(X_{\mu_n})$.

As $\alpha_{\mu_n} > 1$, F_{μ_n} is differentiable at $(x, 1)$ and $DF_{\mu_n}(x, 1) = 0$ for every $0 \leqslant x \leqslant 1$. Thus $DF_{\mu_n}^{n+1}(x_{\mu_n}, \mu_n) = 0$ and (x_{μ_n}, μ_n) is an attracting periodic orbit for F_{μ_n}. We also have $F_{\mu_n}^{n+1}(R_{\mu_n}^0) \subseteq R_{\mu_n}^0$, where $R_{\mu_n}^0 = R_{X_{\mu_n}}$ was defined before. As F_μ varies continuously with μ, we obtain that there is $0 < \tilde{\mu}_n < \mu_n$ such that $F_{\tilde{\mu}_n}^{n+1}(R_{\tilde{\mu}_n}^0) \subset \text{interior}\,(R_{\tilde{\mu}_n}^0)$. Hence, $F_{\tilde{\mu}_n}^{n+1}$ has an attracting fixed point in the *interior* of $R_{\tilde{\mu}_n}^0$ which is also an attracting periodic orbit of $X_{\tilde{\mu}_n}$. Clearly the ω-limit set of $\gamma_0(n)$ is this attracting periodic orbit, where $\gamma_0(n)$ is the separatrix of $\sigma_0(n)$ close, in compact parts, to γ_0. Letting n go to infinity we finish the prove. Notice that for every $\mu \in [\tilde{\mu}_n, \mu_n)$ the vector field X_μ has an attracting periodic orbit and the ω-limit set of $\gamma_0(\mu)$ is this orbit. ∎

From now on we assume that the simple singular cycle Λ of X is *expanding*, that is, $\alpha_0 = \lambda_1/\lambda_2 < 1$ where $-\lambda_3 < -\lambda_1 < 0 < \lambda_2$ are the eigenvalues of $DX(\sigma_0)$, σ_0 the singularity contained in Λ.

The next lemma characterizes the vector fields in $\mathcal{U}_H^+$ in this case.

Lemma 5. — The following conditions are equivalent:

(i) $Y \in \mathcal{U}_H^+$, *i.e.*, $\Gamma(Y/\Lambda(Y, U))$ *is* $\sigma_0(Y)$ *plus a transitive hyperbolic set;*
(ii) $\Lambda(Y, U)$ *is a hyperbolic set;*
(iii) $\exists\, n \in \mathbf{N}$ *such that* $F_Y^n(x_Y, y_Y) \notin D_Y = R_Y^0 \cup R_Y$, D_Y *is the domain of* F_Y;
(iv) $(x_Y, y_Y) \notin \Lambda(F_Y, Q^+)$.

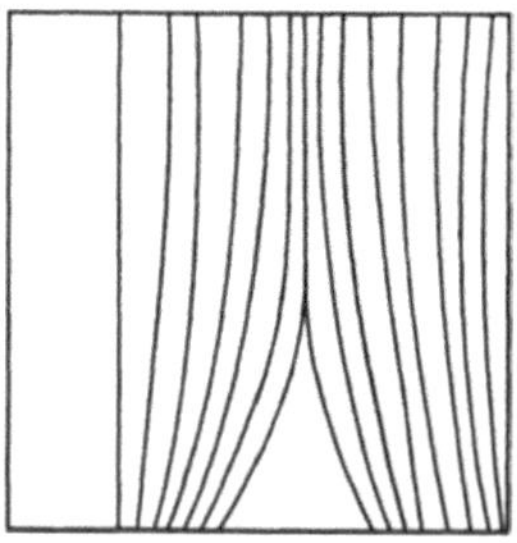

Fig. 5

The fundamental tool for the proof of this lemma is the existence of an invariant unstable cone field for F_Y, $Y \in \mathcal{U}\backslash\mathcal{U}^-$, and consequently, of an invariant unstable foliation $\mathcal{F}^u_Y$ for F_Y. This foliation is obtained by the F_Y-foward saturation of a foliation $\mathcal{G}_Y$ of the fundamental domain for F_Y which looks like the one on figure 5. The foliation $\mathcal{F}^u_Y$ is singular because through the F_Y-orbit of (x_Y, y_Y) there are infinitely many leaves.

The proof of the existence of such unstable cone field is contained in Lemma 2. We point out that for $\alpha_Y < 1$, the proof ends at inequality (2) of Lemma 2. These remarks together with Lemma 3 easily imply Lemma 5.

Proof of Theorems 4 and 5.

Let $\mathcal{U}$, $\mathcal{U}^-$, $\mathcal{U}^+$ and $\mathcal{N}$ be as in Theorem 1. By lemma 1, after a C^1 change of coordinates, we can assume that the horizontal lines $y =$ constant in Q^+ form a stable foliation $\mathcal{F}^s_Y$ for F_Y, for every $Y \in \mathcal{U}^+ \cup \mathcal{N}$. Recall that $\mathcal{U}^+_H$ is the set of $Y \in \mathcal{U}^+$ such that $\Lambda(Y, U)$ is hyperbolic and $\mathcal{U}^+_B = \mathcal{U}^+\backslash\mathcal{U}^+_H$. Given $X \in \mathcal{U}^+_B$, we shall prove that there exists a C^1 codimension-one submanifold $\mathcal{N}_X$ such that the dynamics of $\Lambda(Y, U)$ for every $Y \in \mathcal{N}_X$ is, up to a topological equivalence, the same. In order to prove this, let us first fix some notation.

To each $Y \in \mathcal{U}^+_B$ and $(x_0, y_0) \in \Lambda(F_Y, Q^+)$ define

$$W^s_Y(x_0, y_0) = \{(x, y) \in Q^+;\ \| F^n_Y(x, y) - F^n_Y(x_0, y_0)\| \to 0 \text{ as } n \to \infty\}.$$

Clearly $W^s_Y(x_0, y_0) \supseteq \{(x, y_0);\ 0 \leqslant x \leqslant 1\}$. Recall that $\{(x, y_0);\ 0 \leqslant x \leqslant 1\}$ is the stable leaf of $\mathcal{F}^s_Y$ at (x_0, y_0). Moreover, there exists $\delta_0 > 0$ such that

$$W^s_{Y,\delta_0}(x_0, y_0) = \{(x, y);\ \| F^n_Y(x, y) - F^n_Y(x_0, y_0)\| \leqslant \delta_0,\ \forall n \geqslant 0\} \subseteq \{(x, y_0);\ 0 \leqslant x \leqslant 1\}.$$

Here $\|(x, y)\| = \max\{|x|, |y|\}$ and $B((x, y), r)$ is the ball of radius r and center (x, y).

Set $p_0(Y) = (x_Y, y_Y)$ and $P_n(Y) = F^n_Y(P_0(Y)) = (x^n_Y, y^n_Y)$ for $n > 0$. Take $X \in \mathcal{U}^+_B$ so that $p_0(X) \in \Lambda(F_Y, Q^+)$. For $\delta > 0$ fixed and Y nearby X define

$$W_\delta(X, Y) = \{(x, y) \in Q^+;\ \| F^n_Y(x, y) - p_n(X)\| < \delta,\ \forall n \geqslant 0\},$$

that is, $W_\delta(X, Y)$ is the set of points in Q^+ whose F_Y-orbit δ-shadows the F_X-orbit of $p_0(X)$. The next lemma shows that if the F_X-orbit of $p_0(X)$ is at a distance δ from $\{(x, 1),\ 0 \leqslant x \leqslant 1\}$, then $W_\delta(X, Y)$ is nonempty for Y nearby X.

Lemma 6. — Suppose that there exists $\delta > 0$ such that $|y^n_X - 1| > \delta$ for every $n \geqslant 0$. Then $W_\delta(X, Y)$ is nonempty for Y nearby X and it depends smoothly on Y.

Proof. — Under the hypothesis of the lemma we have that $B(p_n(X), \delta) \subseteq Q$ for every $n \geqslant 0$.

Clearly $W_\delta(X, Y) = \bigcap_{n \geqslant 0} F_Y^{-n}(B(p_n(X), \delta))$.

But $\alpha_X < 1$ implies that F_X^{-1} contracts vertical segments and since it always expands

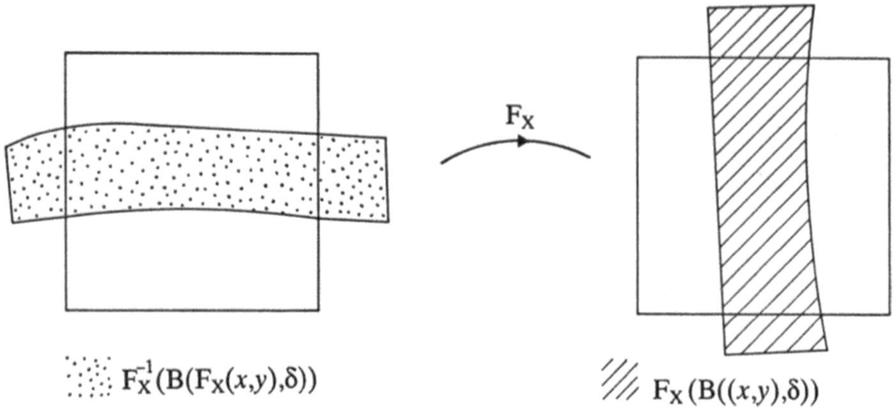

FIG. 6

horizontal segments, we obtain that for every $(x, y) \in \Lambda(F_X, Q^+)$, $F_X^{-1}(B(F_X(x, y), \delta))$ is a horizontal strip. See figure 6.

If $\mathscr{V}$ is a small neighborhood of X then for every $Y \in \mathscr{V}$, $F_Y^{-1}(B(F_X(x, y), \delta))$ is also a horizontal strip, near $F_X^{-1}(B(F_X(x, y), \delta))$. This implies that the sequence $B(p_0(X), \delta) \cap F_Y^{-n}(B(p_n(X), \delta))$ is a nested sequence of horizontal strips for $Y \in \mathscr{V}$. So $W_\delta(X, Y)$ is nonempty and it also depends smoothly on Y. ∎

Let us now suppose that $\sup\{y_X^n, n \geq 0\} = 1$. In this case it could happen that for every $\delta > 0$ there exists Y arbitrarily near X with $W_\delta(X, Y) = \Phi$. This is so because it could occur that $F_Y^{-n}(B(p_n(X), \delta)) \cap Q^+ = \Phi$ for some n. To bypass this difficulty we define a fake return map $\tilde{F}_Y$, much related to F_Y, in such way that it always induces a fake C^1 curve $\tilde{W}_\delta(X, Y)$ as above. We shall also prove that if $(x_Y, y_Y) = p_0(Y) \in \tilde{W}_\delta(X, Y)$ then $W_\delta(X, Y)$ exists and it coincides with $\tilde{W}_\delta(X, Y)$.

In order to define $\tilde{F}_Y$ recall that the domain D_Y of F_Y for $Y \in \mathscr{U}$ is $R_Y \cup R_Y^0$ where $R_Y = \{(x, y) \in Q; 0 \leq x \leq 1, -\rho_Y^{-1} \leq y \leq \rho_Y^{-1}\}$, $R_Y^0 = \{(x, y) \in Q; 0 \leq x \leq 1, 1 - \theta \leq y \leq 1\}$ and θ satisfies $\theta - \varepsilon \geq \delta_Y$ for some $\varepsilon > 0$, $\delta_Y = A_Y . y_Y^{1/\alpha_Y}$. Also recall that $F_Y(x, 1 - \delta_Y)$ is contained in $W^s(\sigma_1(Y))$ and $F_Y(x, 1) = (x_Y, y_Y)$ for every $0 \leq x \leq 1$.

Let $\tilde{R}_Y^0 = \{(x, y) \in Q;\ 0 \leq x \leq 1, 1 - \theta \leq y \leq 1 + \theta\}$ and consider

$$T_Y(x, y) = \begin{cases} (x, y) & \text{if } y \leq 1 \\ (x, 2 - y) & \text{if } y \geq 1, \end{cases}$$

$$T_Y^0(x, y) = \begin{cases} (x, y) & \text{if } y \geq y_Y \\ (x, 2y_Y - y) & \text{if } y \leq y_Y . \end{cases}$$

The fake return map $\tilde{F}_Y$ is defined by

$$\tilde{F}_Y(x, y) = \begin{cases} F_Y(x, y) & \text{if } (x, y) \in R_Y \cup R_Y^0 \\ T_Y^0 \circ F_Y \circ T_Y(x, y) & \text{if } (x, y) \in \tilde{R}_Y^0 \backslash R_Y^0 . \end{cases}$$

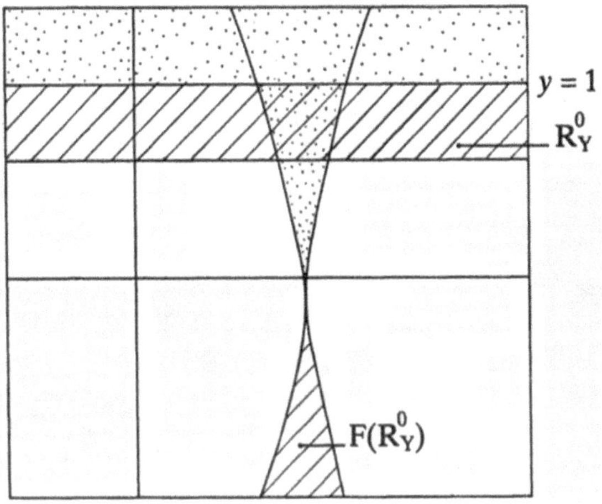

FIG. 7

Observe that the family of horizontal lines in $\tilde{D}_Y = R_Y \cup \tilde{R}^0_Y$ is an invariant stable foliation $\tilde{\mathcal{F}}^s_Y$ for $\tilde{F}_Y$, $\tilde{F}_Y$ is an extension of F_Y depending C^1 on Y and $\tilde{F}_Y$ also expands the vertical direction.

For $\delta > 0$ define

$$\tilde{W}_\delta(X, Y) = \{(x, y) \in Q;\ \|\tilde{F}^n_Y(x, y) - p_n(X)\| \leqslant \delta,\ n \geqslant 0\}.$$

Shrinking $\mathcal{U}$ if necessary we prove, as in lemma 6, the following.

Lemma 7. — *If* $\delta < \delta_0$ *then* $\tilde{W}_\delta(X, Y)$ *is nonempty for every* $Y \in \mathcal{U}$.

Notice that the curves $\tilde{W}_\delta(X, Y)$ *coincide with* $W_\delta(X, Y)$ *whenever* $(x, y) \in \tilde{W}_\delta(X, Y)$ *and* $\tilde{F}^n_Y(x, y) \in R_Y \cup R^0_Y$ *for every* $n \geqslant 0$. *Moreover*, $\tilde{W}_\delta(X, Y)$ *depends smoothly on* Y.

Let $\mathcal{N}_X = \{Y \in \mathcal{U};\ p_0(Y) \in \tilde{W}_\delta(X, Y)\}$. We claim that N_X is a codimension-one submanifold defined on a neighborhood $\mathcal{U}_X \subset \mathcal{U}$ of X. Furthermore, for every $\tilde{X} \in \mathcal{U}_X \cap \mathcal{U}^+_B$, the corresponding $\mathcal{N}_{\tilde{X}}$ is such that $\mathcal{U}_X \backslash \mathcal{N}_{\tilde{X}}$ has two connected components. The proof of these facts is a consequence of the implicit function theorem. Indeed, assuming $\tilde{W}_\delta(\tilde{X}, Y) = \{(x, y(\tilde{X}, Y)),\ 0 \leqslant x \leqslant 1\}$ and defining $H(\tilde{X}, Y) = y(\tilde{X}; Y) - y_Y$ one has $\mathcal{N}_{\tilde{X}} = \{Y; H(\tilde{X}, Y) = 0\}$. Since there exists a direction Y_0 along which $\frac{\partial}{\partial Y_0} H(X, X) = \frac{d}{ds} H(X, X + sY_0)\big|_{s=0} \neq 0$ the proof of the claim follows.

The next lemma provides a condition implying the coincidence between $\tilde{W}_\delta(X, Y)$ and $W_\delta(X, Y)$.

Lemma 8. — *If* $p_0(Y) \in \widetilde{W}_\delta(X, Y)$ *then* $\widetilde{W}_\delta(X, Y) = W_\delta(X, Y)$.

Proof. — Suppose first that $p_0(X)$ is an eventually periodic orbit of F_X. Then, the stable leaf through $p_0(X)$ is an eventually periodic leaf of the stable foliation $\mathscr{F}^s_X$ for F_X. This implies that there exists a hyperbolic periodic orbit q of F_X so that $p_0(X) \in W^s(q)$. Let q_Y be the continuation of q for Y in a small neighborhood of X. Then we clearly have $\widetilde{W}_\delta(X, Y) \subseteq W^s(q_Y)$ and $\widetilde{W}_\delta(X, Y) = W_\delta(X, Y)$. In this case $\mathscr{N}_X = \{ Y; p_0(Y) \in W^s(q_Y) \}$.

Now assume that $p_0(X)$ is not eventually periodic. Then $p_0(Y)$ is not eventually periodic for every $Y \in \mathscr{N}_X$. Suppose that there is $Y \in \mathscr{N}_X$ such that $p_j(Y) \in Q$ for $j = 0, 1, \ldots, n_0 - 1$ and $p_{n_0}(Y) = (x^{n_0}_Y, y^{n_0}_Y)$ with $y^{n_0}_Y > 1$. Let Y_t, $0 \leqslant t \leqslant 1$, be a C^1 arc contained in $\mathscr{N}_X$, $Y_0 = X$ and $Y_1 = Y$. Then there exists $t_0 \in (0, 1)$ so that $p_{n_0}(Y_{t_0}) \in \{(x, 1), |x| \leqslant 1\}$. Hence, $p_0(Y_{t_0})$ is eventually periodic, which is a contradiction.

Thus, $p_j(Y) \in Q$ for every $j \geqslant 0$ and reasoning as before we obtain

$$\widetilde{W}_\delta(X, Y) = W_\delta(X, Y). \quad \blacksquare$$

Lemmas 6, 7 and 8 prove that if $X_\mu \in \mathscr{A}$ is a C^1 arc such that X_0 has an expanding singular cycle, then $\mathscr{U}_B$ is laminated by codimension-one submanifolds. Moreover, for $\mathscr{U}$ small enough, if $X \in \mathscr{U}_B$ and $\mathscr{N}_X$ is the corresponding submanifold through X, then $\mathscr{U}^+ \backslash \mathscr{N}_X$ has two connected components. To conclude the proof of Theorem 4, it remains to prove that the dynamics of $\Lambda(Y, U)$, $Y \in \mathscr{N}_X$ is topologically equivalent to the dynamics of $\Lambda(X, U)$. To do so we proceed as follows.

Given $Y \in \mathscr{N}_X$, let $g_Y : [0, \rho_Y^{-1}] \cup [1 - \delta_Y, 1] \to [0, 1]$ be the map induced by $\mathscr{F}^s_Y$. As we already saw g_Y is an expanding map, C^1 on $J_Y - \{1\}$, where J_Y is the domain of g_Y. The itinerary $i(g_Y)$ of g_Y is the g_Y-orbit of 1. Since g_X is an expanding map, it follows from [MT] that the itinerary of g_X characterizes the dynamics of g_X, that is, an expanding map $g : J \to [0, 1]$, $J = [0, a] \cup [b, 1]$, $0 < a < b < 1$, is conjugate to g_X if and only if $i(g) = i(g_X)$. But for $Y \in \mathscr{N}_X$ we clearly have $i(g_Y) = i(g_X)$ and so g_Y and g_X are conjugate. This implies that the dynamics of the stable foliations $\mathscr{F}^s_X$ and $\mathscr{F}^s_Y$ for F_X and F_Y, respectively, are conjugate.

As in [LP], to obtain a conjugacy h between $\Lambda(F_X, Q^+)$ and $\Lambda(F_Y, Q^+)$ it remains to prove that the dynamics of the unstable foliations $\mathscr{F}^u_X$ and $\mathscr{F}^u_Y$ for F_X and F_Y, respectively, are conjugate. This follows from the fact that these dynamics are given by nearby expanding maps of the interval.

Since $\Lambda(X, U)$ is the closure of the forward saturation by the flow of X of $\Lambda(F_X, Q^+)$, standard methods (see [LP], [GW]) allow us to extend the homeomorphism already defined on Q^+ to a homeomorphism between $\Lambda(X, U)$ and $\Lambda(Y, U)$ sending orbits of $\Lambda(X, U)$ onto orbits of $\Lambda(Y, U)$, preserving their orientation. This completes the proof of Theorem 4. $\blacksquare$

The proof of Theorem 5 is easy and it is left to the reader.

APPENDIX

Here we shall prove Lemma 1. The stable foliation $\mathscr{F}^s_Y$ will be obtained as the integral curves of a C^1 vector field $\eta_Y : Q^+ \to [-1, 1]$, $\eta_Y(x, y) = (1, \varphi_Y(x, y))$, where φ_Y will be obtained as a fixed point of an appropriated graph transform. In order to define this we start by fixing, for each $Y \in \mathscr{U}\backslash\mathscr{U}^-$, a C^1 nearby horizontal foliation $\mathscr{G}_Y$ in $Q^+\backslash(R_Y \cup R^0_Y)$ containing $\{(x, P_Y^{-1}.\theta_Y(\tau_Y.x)); x \in [0, 1]\}$ and $\{(x, \theta_Y(x) - \delta_Y(x)); x \in [0, 1]\}$ as leaves, see Figure 8. We denote by $r(Y, (x, y))$ the inclination of the corresponding leaf passing through the point (x, y) and assume that $(Y, (x, y)) \mapsto r(Y, (x, y))$ is C^1.

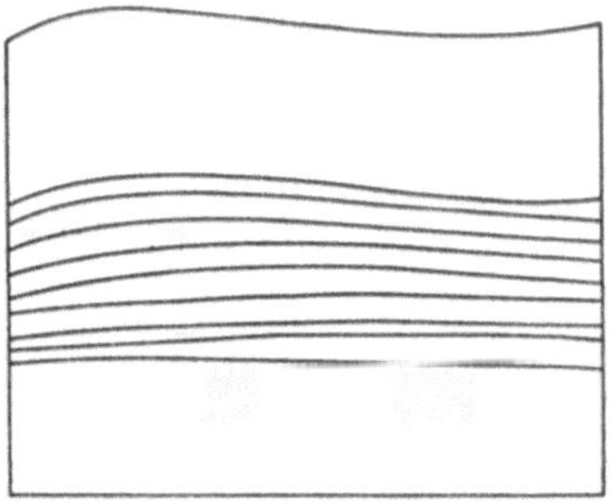

Fig. 8

Consider the space $\mathscr{A}$ of continuous functions $\varphi : (\mathscr{U}\backslash\mathscr{U}^-) \times Q^+ \to [-1, 1]$ satisfying

(i) $\varphi(Y, (x, y)) = r(Y, (x, y))$ if $(x, y) \in Q^+\backslash(R_Y \cup R^0_Y)$;

(ii) $\varphi(Y, (x, \theta_Y(x))) = \theta'_Y(x)$ and $(\varphi(Y, (x, 0)) = 0$ for every $x \in [0, 1]$.

To each $\varphi \in \mathscr{A}$ we associate a continuous vector field

$$\eta^\varphi(Y, (x, y)) = (0, (1, \varphi_Y(Y, (x, y))))$$

and by integrating η^φ we obtain a C^0 foliation $\mathscr{F}^\varphi$ of $(\mathscr{U}\backslash\mathscr{U}^-) \times Q^+$ such that

(i) each leaf of $\mathscr{F}^\varphi$ has the form $\{Y\} \times \mathscr{F}^\varphi_Y(x, y)$ where $\mathscr{F}^\varphi_Y(x, y)$ is a curve in Q^+ passing through (x, y);

(ii) for $(x, y) \in Q^+\backslash(R_Y \cup R^0_Y)$, $\mathscr{F}^\varphi_Y(x, y)$ coincides with the leaf $\mathscr{G}_Y(x, y)$ of $\mathscr{G}_Y$ passing through (x, y);

(iii) for each $Y \in (\mathscr{U}\backslash\mathscr{U}^-)$, $\{(x, \theta_Y(x)); x \in [0, 1]\}$, $\{(x, \rho_Y^{-1}.\theta_Y(\tau_Y.x)); x \in [0, 1]\}$, $\{(x, \theta_Y(x) - \delta_Y(x)); x \in [0, 1]\}$ and $\{(x, 0); x \in [0, 1]\}$ are leaves of $\mathscr{F}_Y$.

Now we define an operator $T : \mathscr{A} \to \mathscr{A}$ in such a way that having $T(\varphi) = \varphi$ is equivalent to $\mathscr{F}^\varphi$ being invariant under the map $F^{-1}(Y, (x, y)) = (Y, F_Y^{-1}(x, y))$. Since, for $(x, y) \in \mathrm{Dom}\, F_Y = R_Y \cup R_Y^0$,

$$DF_{(Y,(x,y))} = \begin{bmatrix} \mathrm{Id} & 0 & 0 \\ * & \frac{\partial}{\partial x} f_Y(x, y) & \frac{\partial}{\partial y} f_Y(x, y) \\ * & \frac{\partial g}{\partial x} g_Y(x, y) & \frac{\partial}{\partial y} g_Y(x, y) \end{bmatrix},$$

where $F_Y(x, y) = (f_Y(x, y), g_Y(x, y))$, we have

$$DF^{-1}_{(Y,(x,y))} \cdot \eta^\varphi(F(Y, (x, y)))$$

$$= \frac{1}{\Delta} \begin{bmatrix} \mathrm{Id} & 0 & 0 \\ * & \frac{\partial}{\partial y} g_Y(x, y) & -\frac{\partial}{\partial y} f_Y(x, y) \\ * & -\frac{\partial}{\partial x} g_Y(x, y) & \frac{\partial}{\partial x} f_Y(x, y) \end{bmatrix} \begin{bmatrix} 0 \\ 1 \\ \varphi(F(Y, (x, y))) \end{bmatrix},$$

with $\Delta = \det DF_{(Y,(x,y))}$. Hence $\mathscr{F}^\varphi$ is invariant if and only if for every $(x, y) \in \mathrm{Dom}\, F_Y$,

$$\varphi(Y, (x, y)) = \frac{-\frac{\partial}{\partial x} g_Y(x, y) + \frac{\partial}{\partial x} f_Y(x, y) \cdot \varphi(F(Y, (x, y)))}{\frac{\partial}{\partial y} g_Y(x, y) - \frac{\partial}{\partial y} f_Y(x, y) \cdot \varphi(F(Y, (x, y)))}.$$

We denote by $r^\varphi(Y, (x, y))$ the right-hand term in this equality and define $T : \mathscr{A} \to \mathscr{A}$ by

$$T(\varphi)(Y, (x, y)) = \begin{cases} r^\varphi(Y, (x, y)) & \text{if } (x, y) \in \mathrm{Dom}\, F_Y \\ r(Y, (x, y)) & \text{if } (x, y) \notin \mathrm{Dom}\, F_Y \end{cases}$$

In order to prove that T has a fixed point we endow $\mathscr{A}$ with the norm of the supremum and prove that T is a contraction with respect to this norm. Afterwards we shall prove that the fixed point of T is in fact C^1, completing the proof of Lemma 1.

(i) $T(\mathscr{A}) \subset \mathscr{A}$.

The continuity of $T(\varphi)(Y, (x, y))$ outside the graph θ_Y is clear. Along this curve it follows observing that *d)* of the expression for F_Y (see Section II) implies that $T(\varphi)(Y, (x, y))$ converges to $\frac{-c(x, y)}{d(x, y)}$ when (x, y) tends to graph θ_Y. As $\frac{-c(x, y)}{d(x, y)}$ is the inclination along graph θ_Y we obtain the result.

We now prove that $|r^{\varphi}(Y,(x,y))| \leqslant 1$ for every $(x,y) \in \operatorname{Dom} F_Y$, $Y \in \mathcal{U}\backslash\mathcal{U}^-$. For $(x,y) \in R_Y$,

$$|T(\varphi)(Y,(x,y))| = \frac{|\tau_Y . \varphi \circ F(Y,(x,y))|}{|\rho_Y|} \leqslant 1.$$

If $(x,y) \in R_Y^0$, using *d)* of the expression for $F_Y(x,y)$ given in Section II we obtain

$$|I| \equiv \left| -\frac{\partial}{\partial x} g_Y(x,y) + \varphi \circ F(Y,(x,y)) . \frac{\partial}{\partial x} f_Y(x,y) \right|$$

$$\leqslant |c| . |\tilde{d}| \left[1 + \frac{|\tilde{b}|}{|\tilde{d}|} \right] . x_2^{\alpha_Y - 1} + |r_I(x,y)|$$

$$|II| \equiv \left| -\frac{\partial}{\partial x} g_Y(x,y) + \varphi \circ F(Y,(x,y)) . \frac{\partial}{\partial x} f_Y(x,y) \right|$$

$$\leqslant |c| . |\tilde{d}| \left[1 - \frac{|\tilde{b}|}{|\tilde{d}|} \right] . x_2^{\alpha_Y - 1} + |r_{II}(x,y)|$$

where $|r_I(x,y)|$ and $|r_{II}(x,y)|$ are bounded by constant.$x_2^{\beta_Y - 1}$. Therefore,

$$\frac{|I|}{|II|} \leqslant \frac{|c|}{|d|} \frac{[1 + |\tilde{b}|/|\tilde{d}|] + \varepsilon}{[1 - |\tilde{b}|/|\tilde{d}|] - \varepsilon}, \qquad \varepsilon > 0 \text{ small.}$$

But, as explained at *a)* and *b)* in the expression for F_Y given in Section II, $|\tilde{b}|/|\tilde{d}|$ can be choosen as small as we want and $|c|/|d|$ were also fixed smaller than ε, with $\varepsilon \ll 1$. So, for $(x,y) \in R_Y^0$,

$$|r^{\varphi}(Y,(x,y))| \equiv \frac{|I|}{|II|} < 1.$$

Thus $|\varphi| \leqslant 1$ implies $|T(\varphi)| \leqslant 1$ as we claimed.

(ii) T is a contraction.

We have

$$(T(\varphi_1) - T(\varphi_2))(Y,(x,y))$$

$$= \frac{\left(\frac{\partial}{\partial x} f_Y . \frac{\partial}{\partial y} g_Y - \frac{\partial}{\partial y} f_Y . \frac{\partial}{\partial x} g_Y \right) (\varphi_1 \circ F - \varphi_2 \circ F)}{\left(\frac{\partial}{\partial y} f_Y - \varphi_1 \circ F . \frac{\partial}{\partial y} g_Y \right) . \left(\frac{\partial}{\partial y} g_Y - \varphi_2 \circ F . \frac{\partial}{\partial y} f_Y \right)} (x,y)$$

$$= \frac{\det DF_Y . (\varphi_1 \circ F - \varphi_2 \circ F)}{\left(\frac{\partial}{\partial y} g_Y - \varphi_1 \circ F . \frac{\partial}{\partial y} f_Y \right) . \left(\frac{\partial}{\partial y} g_Y - \varphi_2 \circ F . \frac{\partial}{\partial y} f_Y \right)} (x,y).$$

If $(x,y) \in R_Y$ we have $|(T(\varphi_1) - T(\varphi_2))(Y,(x,y))| = \frac{|\tau_Y|}{|\rho_Y|}|\varphi_1 \circ F - \varphi_2 \circ F|$ and $|\tau_Y| < |\rho_Y|$.

If $(x,y) \in R_Y^0$ then using *a)*, *b)*, *c)* and *d)* of the expression for F_Y one obtain

$$|\tilde{I}| \equiv |\det DF_Y(x,y)| \leqslant \text{constant}.x_2^{\beta_Y+\alpha_Y-1},$$

$$|\widetilde{II}| \equiv \left|\left(\frac{\partial}{\partial y}g_Y - \varphi_1 \circ F.\frac{\partial}{\partial y}f_Y\right).\left(\frac{\partial}{\partial y}g_Y - \varphi_2 \circ F.\frac{\partial}{\partial y}f_Y\right)(x,y)\right|$$

$$\geqslant \alpha_Y^2.|d|^2.|\tilde{d}|^2.\left(1 - \frac{|\tilde{b}|^2}{|\tilde{d}|^2}\right).x_2^{2\alpha_Y-2} - |s(x,y)|, |s(x,y)|$$

$$\leqslant \text{const}.x_2^{\beta_Y+\alpha_Y-2}.$$

Thus $\frac{|\tilde{I}|}{|\widetilde{II}|} \leqslant \text{constant}.x_2^{\beta_Y-\alpha_Y+1}$. As $\beta_Y - \alpha_Y > 0$ for every Y we obtain $\frac{|\tilde{I}|}{|\widetilde{II}|} \leqslant \frac{1}{2}$. This proves that T has a unique fixed point φ_0. In order to prove that φ_0 is of class C^1 we proceed as follows. Define

$$\tilde{\mathscr{A}} = \{A \in C^0((\mathscr{U}\backslash\mathscr{U}^-) \times Q^+, L(\mathbf{R}^2, \mathbf{R}));$$
$$A = Dr(Y,(x,y)) \text{ for } (x,y) \in Q^+\backslash R_Y \cup R_Y^0\}.$$

Here $Dr(Y,(x,y))$ means $\frac{\partial}{\partial(x,y)}r(Y,(x,y))$ where $r(Y,(x,y))$ is the inclination field of the foliation $\mathscr{G}_Y$ defined and fixed above.

We now introduce an operator $\tilde{T}: \mathscr{A} \times \tilde{\mathscr{A}} \to \mathscr{A} \times \tilde{\mathscr{A}}$ such that if f, g are of class C^2 and φ is of class C^1 then $\tilde{T}(\varphi, D\varphi) = (T\varphi, D(T\varphi))$. This operator is defined as $\tilde{T}(\varphi, A) = (T\varphi, S(\varphi, A))$ and the explicit form of $S(\varphi, A)$ is $\left(\text{denoting } f_x = \frac{\partial}{\partial x}f_Y, \text{ etc.}\right)$.

$$S(\varphi, A) = \begin{cases} \dfrac{\begin{array}{l}[(g_y - \varphi \circ F.f_y)(-Dg_x + f_x.A \circ F.DF + \varphi \circ F.Df_x) \\ \quad - (-g_x + \varphi \circ F.f_x)(Dg_y - f_y.A \circ F.DF - \varphi \circ F.Df_y)]\end{array}}{(g_y - \varphi \circ F.f_y)^2} \\ \text{for } (x,y) \in R_Y \cup R_Y^0 \\ Dr(Y,(x,y)) \quad \text{for } (x,y) \notin R_Y \cup R_Y^0, \end{cases}$$

where Dh means $\frac{\partial h}{\partial(x,y)}$ for $h = g_x, g_y, F$, etc.

Note that if $\tilde{T}$ has a globally attracting fixed point (φ_0, A_0), then choosing φ of class C^1 we obtain $\tilde{T}^n(\varphi, D\varphi) = (T^n\varphi, D(T^n\varphi)) \to (\varphi_0, D\varphi_0)$ as n goes to ∞. This implies that $A_0 = D\varphi_0$ and so φ_0 is of class C^1. Thus, to conclude the proof of the lemma it remains to show first that $\tilde{T}$ is a well defined operator and secondly that it has such an attracting fixed point. For this last point it is enough to show that each map $\tilde{T}_\varphi : \tilde{\mathscr{A}} \to \tilde{\mathscr{A}}$, $\tilde{T}_\varphi(A) = S(\varphi, A)$ is a contraction, with constant of contraction uniform on $\varphi \in \mathscr{A}$.

To see that $\tilde{T}$ is a well-defined operator involves a standard calculation and we will not do it here. But it is possible to prove, using *d)* of the expression for F_Y and the fact that $\beta_Y - \alpha_Y > 0$ that

(i) $| S(\varphi_1, A)(x,y) - S(\varphi_2, A)(x,y) | \leqslant \text{constant} | \varphi_1 - \varphi_2 |$ and

(ii) $S(\varphi, A)(x,y)$ converges to $D\left(\dfrac{-c(x,y)}{d(x,y)}\right)$ as (x,y) tends to graph θ_Y.

We now prove that $\tilde{T}$ is a contraction. Observe that

$$| \tilde{T}_\varphi(A) - \tilde{T}_\varphi(B) | = \frac{|f_x \cdot g_x - f_y \cdot g_x|}{|(g_y - \varphi \circ F . f_y)^2|} | (A \circ F - B \circ F) | . | DF |$$

$$= \frac{(\det DF) . | DF |}{| g_y - \varphi \circ F . f_y |^2} (A \circ F - B \circ F).$$

For $(x,y) \in R_Y$ we have

$$| \tilde{T}_\varphi(A) - \tilde{T}_\varphi(B) | = | \tau_Y | . | A \circ F - B \circ F | \quad \text{and} \quad | \tau_Y | < 1.$$

For $(x,y) \in R_Y^0$ we have

$$| \hat{I} | \equiv | \det DF | . | DF | \leqslant \text{constant}. x_2^{\beta_Y + \alpha_Y - 1} . x_2^{\alpha_Y - 1}$$

$$| \widehat{II} | \equiv | (g_y - \varphi \circ F . f_y) |^2 \geqslant \alpha_Y^2 . | d |^2 . | \tilde{a} |^2 (1 - K | \tilde{b} |^2 / | \tilde{a} |^2) . x_2^{2\alpha_Y - 2} - | t_Y(x,y) |$$

with $| t_Y(x,y) | \leqslant \text{constant}. x_2^{\beta_Y + \alpha_Y - 2}$.

It follows that $\dfrac{| \hat{I} |}{| \widehat{II} |} \leqslant \text{constant}. x_2^{\beta_Y} \leqslant \dfrac{1}{2}$. Thus $\tilde{T}_\varphi$ is a contraction with constant of contraction independent of φ. This proves the lemma. ■

REFERENCES

[AS] V. S. Afraimovitch and L. P. Sil'nikov, On attainable transitions from Morse-Smale systems to systems with many periodic motions, *Math. USSR Izvestija*, **8** (1974), 1235-1270.

[GW] J. Guckenheimer and R. Williams, Structural stability of Lorenz attractors, *Publ. Math. IHES*, **50** (1979), 59-72.

[LP] R. Labarca and M. J. Pacífico, Stability of singular horseshoes, *Topology*, **25** (1986), 337-352.

[MT] J. Milnor and W. Thurston, *On iterated maps on the interval*, I, II, Preprints, Princeton, 1977.

[NP1] S. Newhouse and J. Palis, Bifurcations of Morse-Smale Dynamical Systems, *Dynamical Systems*, edited by M. M. Peixoto, Academic Press, 303-366, 1973.

[NP2] S. Newhouse and J. Palis, Cycles and bifurcation theory, *Astérisque*, **31** (1976), 43-140.

[NPT] S. Newhouse, J. Palis and F. Takens, Bifurcations and stability of families of diffeomorphisms, *Publ. Math. IHES*, **57** (1983), 5-72.

[PT] J. Palis and F. Takens, Cycles and measure of bifurcation sets for two dimensional diffeomorphisms *Inventiones Math.*, **82** (1985), 397-422.

[PR] M. J. Pacífico and A. Rovella, Contracting singular cycles, *Ann. Ec. Norm. Sup.*, to appear.

[S] S. Sternberg, On the structure of local homeomorphisms of euclidean n-space II, *Am. J. Math.*, **80** (1958), 623-631.

R. B.
Departamento de Matematica
Universidad de Chile
Casilla 653, Santiago, Chile

R. L.
Departamento de Matematica y CC
Universidad de Santiago de Chile
Casilla 307, Correo 2, Santiago, Chile

R. M.
IMPA
Estrada Dona Castorina, 110
Cep 22460, Rio de Janeiro, RJ, Brasil

M. J. P.
Departamento de Matematica Aplicada
Instituto de Matematica — UFRJ
Caixa Postal 68530, Cep 21910
Rio de Janeiro, RJ, Brasil

Manuscrit reçu le 16 décembre 1991.

Nova Série
BOLETIM
DA SOCIEDADE BRASILEIRA DE MATEMÁTICA

Bol. Soc. Bras. Mat., Vol. 24, N. 1, 1-11
© 1993, Sociedade Brasileira de Matemática

On a Theorem of Fatou

Ricardo Mañé

Abstract. We prove a result on the backward dynamics of a rational function nearby a point not contained in the ω-limit set of a recurrent critical point. As a corollary we show that a compact invariant subset of the Julia set, not containing critical or parabolic points, and not intersecting the ω-limit set of *recurrent* critical points, is expanding, thus extending a classical criteria of Fatou. We also prove that the boundary of a Siegel disk is always contained in the ω-limit set of a *recurrent* critical point.

1. The Results

Let $\overline{\mathbf{C}}$ be the Riemann sphere. Given a rational map $f\colon \overline{\mathbf{C}} \hookleftarrow$, we say that $\Lambda \subset \overline{\mathbf{C}}$ is an *expanding set* of f if it is compact, invariant (i.e. $f(\Lambda) = \Lambda$) and there exists a positive integer $n > 0$ such that $|(f^n)'(z)| > 1$ for every $z \in \Lambda$. A not very difficult to prove, and yet very useful criteria to recognize this fundamental class of invariant sets was given by Fatou (P.Fatou, Sur les équations fonctionelles, Bull. Soc. Math. France 47, 1919, 161–271). He proved that if a compact invariant set Λ of a rational map $f\colon \overline{\mathbf{C}} \hookleftarrow$, is contained in its Julia set $J(f)$, and the ω-limit set $\omega(c)$ of every critical point of f satisfies $\omega(c) \cap \Lambda = \phi$, then Λ is expanding. Here we shall provide an improvement of this result, that roughly speaking states that one has to be concerned only about intersections with *recurrent* critical points. Recall that a periodic point p of a rational map f is said to be parabolic if $(f^n)'(p)$ is a root of unity when n is the period of p.

Theorem I. *Let $f\colon \overline{\mathbf{C}} \hookleftarrow$ be a rational map and $\Lambda \subset J(f)$ a compact invariant set not containing critical points or parabolic periodic points.*

Received 10 November 1992.

Originally published in: *Bol. Soc. Brasil. Mat. (N.S.) 24 (1993), no. 1, 1–11,*

Then either Λ *is expanding or* $\Lambda \cap \omega(c) \neq \phi$ *for some recurrent critical point* c *of* f.

It follows immediately from Theorem I that every periodic orbit of f, that is not parabolic or a source, is contained in the ω-limit set of a *recurrent* critical point.

This theorem will be a corollary of the following result:

Theorem II. *Let* $f: \overline{\mathbf{C}} \hookleftarrow$ *be a rational map. If a point* $x \in J(f)$ *is not a parabolic periodic point and is not contained in the* ω*-limit set of a recurrent critical point, then for all* $\varepsilon > 0$ *there exists a neighborhood* U *of* x *such that:*

(a) *For all* $n \geq 0$, *every connected component of* $f^{-n}(U)$ *has diameter* $\leq \varepsilon$;

(b) *There exists* $N > 0$ *such that for all* $n \geq 0$ *and every connected component* V *of* $f^{-n}(U)$ *the degree of* f^n/V *is* $\leq N$;

(c) *For all* $\varepsilon_1 > 0$ *there exists* $n_0 > 0$ *such that every connected component of* $f^{-n}(U)$, $n \geq n_0$, *has diameter* $\leq \varepsilon_1$.

The core of the theorem is (a), from which properties (b) and (c) will easily follow.

We begin by proving Theorem II. The deduction of Theorem I from Theorem II will be given afterwards, when we shall also prove the following corollary to Theorem II.

Corollary. *If* Γ *is the boundary of a Siegel disk or a connected component of the boundary of a Herman ring, there exists a recurrent critical point* c *such that* $\omega(c) \supset \Gamma$.

2. The Proofs

Given an open set $U \subset \overline{\mathbf{C}}$ denote $c(U, n)$ the set of connected components of $f^{-n}(U)$. Observe that $V \in c(U, n)$ implies $f^j(V) \in c(U, n-j)$ for all $0 \leq j \leq n$. If $V \in c(U, n)$ define $\Delta(V, n) = \#\{x \in V \mid (f^n)'(x) = 0\}$ counted with algebraic multiplicity. Given x as is the statement of Theorem II we can assume without loss of generality that $f(\infty) = \infty$ and $\infty \neq x$. Hence, if U is a neighborhood of x not containing ∞, it follows that $\infty \notin f^{-n}(U)$ for all $n \geq 0$. This means that from now on we shall have to deal only with subsets of the complex plane $\mathbf{C}$. A square is a set S of the form $S = \{z \in \mathbf{C} \mid |Re(z-p)| < \delta, |Im(z-p)| < \delta\}$. The point p is the center of S and δ its radius. Given a square S with center p and radius δ, then, given $k > 0$, denote by S^k the square with center p and radius $k\delta$.

Lemma 1. *Given $\varepsilon > 0$, $0 < k < 1$, $c > 0$ and $N > 0$, there exists $\delta = \delta(N, \varepsilon, k, c) > 0$ such that if S is s square of radius $< \delta$ such that $d(S,p) > c$ for every parabolic or attracting periodic point p, and, if for some $n > 0$, $V \in c(S,n)$ is such that $\Delta(V,n) \leq N$, then* $\operatorname{diam} W \leq \varepsilon$ *for all $W \in c(S^k, n)$ contained in V.*

Remark. Clearly, if follows from the lemma that $\operatorname{diam} f^i(W) \leq \varepsilon$ for all $0 \leq i \leq n$ because $f^i(V) \in c(S, n-i)$ and $f^i(W) \in c(S^k, n-i)$ and is contained in $f^i(V)$, which has at most N points x where $(f^{n-i})'(x) = 0$.

Proof. Suppose that for $\varepsilon > 0$, $0 < k < 1$, $c > 0$ and $N > 0$ the Lemma is false. Then, there exists a sequence S_n, $n = 1, 2, \ldots$ of squares with radius $\delta_n \to 0$ such that $d(S_n, p) > c$ for every parabolic or attracting periodic point p and sets $V_n \in c(S_n, m_n)$ satisfying $\Delta(V_n, m_n) \leq N$, but with $\operatorname{diam} W_n > \varepsilon$ for some $W_n \in c(S_n^k, m_n)$ contained in V_n. The condition $\delta_n \to 0$ implies that $m_n \to +\infty$. Take $0 < \gamma < 1$ such that $\gamma^{N+1} = k$. Then, there exists $0 \leq j \leq N+1$ such that, denoting by $\hat{S}_n = S_n^{\gamma^j}$, the set $\hat{S}_n - \hat{S}_n^\gamma$ doesn't contain critical values of f^{m_n}/V_n. Obviously, $\hat{S}_n^\gamma \supset S_n^k$. Take $\hat{W}_n \in c(\hat{S}_n^\gamma, m_n)$, such that $W_n \subset \hat{W}_n \subset V_n$. Then, $\operatorname{diam} \hat{W}_n > \varepsilon$.

Take $0 \leq j_n \leq m_n$ such that

$$\operatorname{diam} f^{j_n}(\hat{W}_n) > \varepsilon \tag{a}$$

$$\operatorname{diam} f^{j_n+1}(\hat{W}_n) \leq \varepsilon \tag{b}$$

observe that

$$f^{j_n}(\hat{W}_n) \in c(\hat{S}_n^\gamma, m_n - j_n).$$

Taking $\varepsilon > 0$ small enough, it follows from (a) and (b) that $f^{j_n}(\hat{W}_n)$ is a topological disk. Set $\tilde{W}_n = f^{j_n}(\hat{W}_n)$. Since $\hat{S}_n - \hat{S}_n^\gamma$ doesn't contain critical values of f^{m_n}/V_n, the element $\tilde{W}_n^+ \in c(\hat{S}_n, m_n - j_n)$ that contains

$\tilde{W}_n \in c(\hat{S}_n^k, m_n - j_n)$ is also a topological disk. Let $D = \{z | |z| < 1\}$ be the unit disk. Take conformal representations $\psi_n: D \to \hat{S}_n$, $\varphi_n: D \to \tilde{W}_n^+$ with $\psi_n(0) = x_n$, $f^{m_n - j_n}(\varphi_n(0)) = x_n$. Define:

$$F_n := \psi_n^{-1} f^{m_n - j_n} \varphi_n : D \hookleftarrow$$

Then $F_n : D \hookleftarrow$ is a Blaschke product of degree $\leq N$. Hence, $\{F_n\}$ is obviously a normal family and then there exists disks $D \supset D_1 \supset D_2$ centered at zero such that:

$$D_2 \supset F_n^{-1}(\psi_n^{-1}\hat{S}_n^k) \supset D_1.$$

Moreover,

$$\tilde{W}_n = \varphi_n F_n^{-1}(\psi_n^{-1}\hat{S}_n^k) \supset \varphi_n(D_1) \tag{1}$$

$$\varphi_n(D_2) \supset \varphi_n(F_n^{-1}(\psi_n^{-1}\hat{S}_n^k)) = \tilde{W}_n. \tag{2}$$

Now observe that $\{\varphi_n\}$ is a normal family because there exist three periodic orbits not intersecting S_n and then $\varphi_n(D)$ doesn't intersect them. Hence, we can assume that $\varphi_n \to \varphi$ uniformly on compact subsets of D and, by (2), φ is non constant because for all n:

$$\operatorname{diam} \varphi_n(D_2) \geq \operatorname{diam}(\tilde{W}_n) \geq \varepsilon.$$

From

$$0 = \lim_{n\to+\infty} \operatorname{diam}(\hat{S}_n) = \lim_{n\to+\infty} \operatorname{diam} f^{m_n - j_n}(\tilde{W}_n^+)$$

follows that

$$\tilde{W}_n^+ \subset J(f)^c \tag{4}$$

for all n. Moreover, (1) and the fact that φ is non constant imply that $\tilde{W}_{n_1} \cap \tilde{W}_{n_2} \neq \varnothing$ for some $n_1 < n_2$. Since $\tilde{W}_{n_1} = f^m(\tilde{W}_{n_2})$ for some m, this and (4) imply that there exists a connected component U of $J(f)^c$ such that

$$f^m(U) = U \tag{5}$$

and

$$U \supset \tilde{W}_n \qquad \forall n. \tag{6}$$

But (3) implies

$$\lim_{n\to+\infty} \operatorname{diam} f^{m_n - j_n}(\tilde{W}_n) = 0. \tag{7}$$

Moreover,

$$\inf_n \operatorname{diam}(\tilde{W}_n) > 0 \tag{8}$$

and

$$\lim_{n\to+\infty} d(x, f^{m_n - j_n}(\tilde{W}_n)) = 0. \tag{9}$$

From (5)–(9), it follows that x is either a sink or a parabolic point, thus contradicting the property $d(S_n, p) \geq c > 0$ for all n and every parabolic or attracting periodic point p.

Now let us prove Theorem II. If S is a square with radius δ, denote by $\mathcal{L}(S)$ the family of squares contained in $S^{3/2} - S$ and having radius $\delta/4$. Denote by $\mathcal{L}^*(S)$ the family of squares $S_0^{3/2}$ with $S_0 \in \mathcal{L}(S)$. Suppose that x is not a parabolic periodic point or is contained in the ω-limit set of a recurrent critical point. Then there exists $\delta_0 > 0$ such that

1) There is no critical point c of f such that there exist $0 < n_1 \leq n_2$ satisfying

$$d(f^{n_1}(c), c) < \delta_0$$
$$d(f^{n_2}(c), x) < \delta_0$$

2) $|x - p| > 10\delta_0$ for every parabolic or attracting periodic point p.

Given $\varepsilon > 0$ take $\varepsilon_1 > 0$ satisfying

3) $0 < \varepsilon_1 < \min\{\varepsilon/10, \delta_0/10\}$
4) If U is an open connected set with $\operatorname{diam} U \leq 2\varepsilon_1$ then $\operatorname{diam} W \leq \delta_0$ for all $W \in c(U, 1)$

Let N_0 be the number of critical points of f. Take $N_1 > 2$ such that

5) If S is a square and $V \in c(S, n)$ satisfies $\Delta(V, n) \leq N_0 + 1$ then the number of connected components of $f^{-n}(S^{2/3})$ contained in V is $\leq N_1$

Finally, take δ given by

6) $\delta = \min\{\delta_0/10, \varepsilon_1/10, \delta(2N_0, \frac{\varepsilon_1}{20N_1}, \frac{2}{3}, \delta_0)\}$ where $\delta(2N_0, \frac{\varepsilon_1}{20N_1}, \frac{2}{3}, \delta_0)$ is given by the lemma.

Let S_0 be the square of center x and radius δ. Suppose that Theorem II fails for $U = S_0$. Then there exist $n > 0$ and $V \in c(S_0, n)$ with $\operatorname{diam} V \geq \varepsilon \geq 10\varepsilon_1$. On the other hand, by (1), $\operatorname{diam} S_0 = 2\sqrt{2}\delta < 3\delta < \varepsilon_1$. Hence, there exists an integer $n_0 \geq 0$ such that there exists

$V_0 \in c(S_0^{3/2}, n_0)$ satisfying

7) $\operatorname{diam}(f^{-(n_0-i)}(S_0) \cap f^i(V_0)) \le \varepsilon_1$ for all $1 \le i \le n_0$, and

8) $\operatorname{diam}(f^{-n_0}(S_0) \cap V_0) > \varepsilon_1$

Since $\operatorname{diam} S_0 < \varepsilon_1$ it follows that $n_0 > 0$. Now, starting with S_0 we shall construct a sequence of squares $S_0, S_1, S_2, \ldots$ and strictly positive integers $n_0 \ge n_1 \ge n_2 \ge \ldots$ satisfying

9) $S_{j+1} \in \mathcal{L}^*(S_j)$

10) There exists $V_j \in c(S_j^{3/2}, n_j)$ such that

$$\operatorname{diam}(f^{-(n_j-i)}(S_j) \cap f^i(V_j)) \le \varepsilon_1$$

for all $1 \le i \le n_j$ and

$$\operatorname{diam}(f^{-n_j}(S_j) \cap V_j) > \varepsilon_1.$$

From (7) and (8), it follows that S_0 satisfies (10). If we construct such a sequence of squares and integers, then Theorem II will be proved by contradiction because the condition $n_0 \ge n_1 \ge \cdots \ge n_m \ge \cdots > 0$ implies that $n_j = n_i$ for all $j \ge i$ for a certain i. But (a) implies that the radius of S_j is $(3/8)^j$; in particular $\operatorname{diam} S_j \to 0$ when $j \to +\infty$. But by (10),

$$\varepsilon_1 < \operatorname{diam}(f^{-n_j}(S_j) \cap V_j) = \operatorname{diam}(f^{-n_i}(S_j) \cap V_j),$$

$$V_j \in c(S_j^{3/2}, n_j) = c(S_j^{3/2}, n_i).$$

Taking $j \to +\infty$, and recalling that i is constant and $\lim_{j \to +\infty} \operatorname{diam} S_j = 0$, we conclude that the inequality above cannot hold.

The sequences $\{S_j\}$ and $\{n_j\}$ will be constructed by induction starting with S_0. Suppose S_i and n_i constructed for $0 \le i \le j$. To find S_{j+1} and n_{j+1} we begin by observing that from (a) it follows that if $p \in S \in \mathcal{L}^*(S_j)$, then, by

$$d(p, x) \le \operatorname{diam} S + \sum_{i=0}^{j} \operatorname{diam} S_i$$

$$= \sum_{i=0}^{j+1} (\frac{3}{8})^i \operatorname{diam} S_0 = 2\sqrt{2} \sum_{i=0}^{j+1} (\frac{3}{8})^i \delta \le 4\sqrt{2}\delta.$$

Hence, if a point q satisfies $d(q,S) \le \delta_0$, we have

$$d(q,x) \le 4\sqrt{2}\delta + \delta_0 \le 2\delta_0.$$

By (2), this means that

11) $d(q,S) > \delta_0$ for all $S \in \mathcal{L}^*(S_j)$ and all parabolic or attracting periodic point q.

For the induction step (i.e. the construction of S_{j+1} and n_{j+1}), we shall use the following easy lemma.

Lemma 2. *If $U \subset \mathbf{C}$ is an open set and $V \in c(U,n)$ satisfies*

$$\operatorname{diam} f^i(V) \le \delta_0 \qquad 0 \le i \le n$$

then

$$\Delta(V,n) \le N_0.$$

Proof. If $\Delta(V,n) \ge N_0 + 1$, there exist N_0+1 different points x_i, $1 \le i \le N_0+1$, in V_j such that $(f^{n_j})'(x_j) = 0$. This means that for each $1 \le i \le N_0+1$ there exist $1 \le m_i < n$, such that $f^{m_i}(x_i)$ is a critical point. Recalling that N_0 is the number of critical points of f, it follows that there exist two different points in the set $\{x_i | 1 \le i \le N_0+1\}$, that we shall denote by x_1, x_2, and a critical point c such that $f^{m_1}(x_1) = f^{m_2}(x_2) = c$. Assume $0 \le m_1 \le m_2$. Then

$$d(f^{m_2-m_1}(c),c) = d(f^{m_2}(x_1), f^{m_2}(x_2)) \le \operatorname{diam} f^{m_2}(V_j) \le \delta_0$$

and

$$\begin{aligned} d(f^{n-m_1}(c),x) &= d(f^{n-m_1}(f^{m_1}(x_1)),x) \\ &= d(f^n(x_1),x) \le \delta_0 \end{aligned}$$

contradicting property (1) of δ_0. □

Now, to find S_{j+1} and n_{j+1} we first claim that there exists $S \in \mathcal{L}(S_j)$ that for some $0 < n \le n_j$ has $V \in c(S,n)$ with $\operatorname{diam} V \ge \varepsilon_1/10N_1$. Suppose that the claim is false. Then, for all $1 \le i \le n_j$,

$$\begin{aligned} \operatorname{diam} f^i(V_j) &\le \operatorname{diam}(f^{-(n_j-i)}(S_j) \cap f^i(V_j)) \\ &\quad + \sup\{\operatorname{diam} W | W \in c(S, n_j - i), S \in \mathcal{L}(S_j)\} \\ &\le \varepsilon_1 + \frac{\varepsilon_1}{10N_1} \le 2\varepsilon_1. \end{aligned}$$

From this inequality applied to $i = 1$ and property (4), we have

$$\operatorname{diam} V_j \leq \delta_0.$$

Moreover, since $2\varepsilon_1 \leq \delta_0$ (by (3)),

$$\operatorname{diam} f^i(V_j) \leq \delta_0$$

for all $1 \leq i \leq n_j$, hence for all $0 \leq i \leq n_j$. By Lemma 2 this proves $\Delta(V_j, n_j) \leq N_0$.

Then, since $V_j \in c(S_j^{3/2}, n_j)$, it follows from (5), (11) and the lemma that

$$W \in c(S_j, n_j), W \subset V_j \Rightarrow \operatorname{diam} W \leq \varepsilon_1/10N_1.$$

Moreover, by the way N_1 was chosen, we have

$$\#\{W \in c(S_j, n_j) | W \subset V_j\} \leq N_1$$

and we are assuming that

$$S \in \mathcal{L}(S_j), U \in c(S, n_j) \Rightarrow \operatorname{diam} U \leq \varepsilon_1/10N_1.$$

Now observe that V_j is the union of sets $U \in c(S, n_j)$, $U \subset V_j$, $S \in \mathcal{L}(S_j)$ and the sets $W \in c(S_j, n_j)$, $W \subset V_j$. Moreover, for any two sets W', W'' in this family there exist $W' = W_0, W_1, \ldots, W_k = W''$ in $c(S_j, n_j)$ and contained in V_j such that for all $0 \leq i < k$ there exist $S_i \in \mathcal{L}(S_j)$ and $U_i \in c(S_i, n_j)$, such that $\overline{U}_i \cap \overline{W}_i \neq \phi$, $\overline{U}_i \cap \overline{W}_{i+1} \neq \phi$. Then

$$\operatorname{diam} V_j \leq N_1 \left(\frac{\varepsilon_1}{10N_1} + \frac{\varepsilon_1}{10N_1} \right) = \frac{\varepsilon_1}{5}$$

contradicting the last inequality in condition (11). This completes the proof of the claim. Now we can take $S \in \mathcal{L}(S_j)$ such that $\operatorname{diam} V \geq \varepsilon_1/10N_1$ for some $V \in c(S, n)$, $0 \leq n \leq n_j$. Take $\hat{V} \in c(S^{3/2}, n)$ containing V. Suppose that $\Delta(\hat{V}, n) \leq N_0$. Then, by Lemma 1 and condition (5)

$$\operatorname{diam} V \leq \varepsilon_1/20N_1$$

because $V \in c((S^{3/2})^{2/3}, n)$ and is contained in $\hat{V}$. This contradicts $\operatorname{diam} V \geq \varepsilon_1/10N_1$ and proves $\Delta(\hat{V}, n) \geq N_0 + 1$. From Lemma 2, it follows that

$$\operatorname{diam} f^i(\hat{V}) > \delta_0$$

for some $0 \leq i \leq n$. Now we define $S_{j+1} = S^{3/2}$. Then $f^i(\hat{V}) \in c(S^{3/2}, n-i)$ and $\operatorname{diam} f^i(\hat{V}) > \delta_0 \geq 10\varepsilon_1$. Moreover $\operatorname{diam} S^{3/2}_{j+1} \leq 2\delta < \varepsilon_1$. Then there exists $0 \leq n_{j+1} \leq n-i \leq n_j - i$ and $V_{j+1} \in c(S^{3/2}_{j+1}, n_{j+1})$ such that

$$\operatorname{diam}(f^{-n_{j+1}}(S_{j+1}) \cap V_{j+1}) > \varepsilon_1$$

and

$$\operatorname{diam}(f^{-n_{j+1}+i}(S_{j+1}) \cap f^i(V_{j+1})) \leq \varepsilon_1.$$

Observe that $n_{j+1} > 0$ because $\operatorname{diam} S_{j+1} < 2\delta < \varepsilon_1$. This completes the construction of the sequences $\{S_j\}$ and $\{n_j\}$ and the proof of part (a) of Theorem II.

Property (b) of Theorem II follows from (a) and Lemma 2. To prove (c), note that (b) implies that there exists $n_1 > 0$ such that $V \in c(U, n)$, $n \geq n_1$ implies that V doesn't intersect the forward orbits of the critical points. Now (c) follows from this fact and the following classical property (Fatou, loc. cit.): If V is an open set such that $V \cap J(f) \neq \varnothing$ and V doesn't intersect the forward orbit of the critical points, then given an open set $V_0 \subset \overline{V}_0 \subset V$ and $\varepsilon_0 > 0$ there exists $n_0 \geq 0$ such that every $W \in c(V_0, n)$, $n \geq n_0$, satisfies $\operatorname{diam}(W) \leq \varepsilon_0$. The proof of theorem is now complete.

Proof of the Corollary. To prove the corollary of Theorem II, suppose that Γ is the boundary of a Siegel disk or a connected component of a Herman ring, say, B, that, without loss of generality, we shall assume it satisfies $f(B) = B$, $f(\Gamma) = \Gamma$. Let $c_1, \ldots, c_k$ be the recurrent critical points of f. Set $\Gamma_i = \omega(c) \cap \Gamma$. Obviously

$$f(\Gamma_i) \subset \Gamma_i \qquad 1 \leq i \leq k. \tag{13}$$

Let us prove that

$$B \subset \bigcup_i \Gamma_i. \tag{14}$$

Suppose by contradiction that there exists

$$x \in B - \bigcup_i \Gamma_i.$$

Let U be the neighborhood of x given by Theorem II. Take $y \in B \cup U$ and take $\delta > 0$ such that $d(z, \Gamma) > \delta$ for all $y \in f^{-n}(x) \cap B$ and $n \geq 0$. By Theorem II there exists $n_0 \geq 0$ such that

$$V \in c(U, n), \quad n \geq n_0 \Rightarrow \operatorname{diam}(V) \leq \delta. \tag{15}$$

Take $z \in B$ such that $f^{n_0}(z) = y$. Let $V \in c(n_0, U)$ be such that $z \in V$. Since $\operatorname{diam}(V) \leq \delta$ and $d(z, \Gamma) > \delta$ it follows that $V \subset B$. Take $w \in V \subset B$ such that $f^{n_0}(w) = x \in \Gamma$. Then $w \in B$ and $w \in J(f)$. This contradiction proves (14). Now observe that Γ supports an invariant ergodic probability, namely that induced by the boundary map of the conformal representation $\varphi: A \to B$, where φ is an annulus or a disk (according to whether B is a Siegel disk or Herman ring), normalized, if B is a Siegel disk, by mapping the center of the disk in the fixed point contained in B. Then (13) and (14) imply $\Lambda_i = \Gamma$ for some $1 \leq i \leq k$, thus concluding the proof of the Corollary. □

Proof of Theorem I. To prove Theorem I consider a compact invariant set $\Lambda \subset J(f)$ not containing critical points, parabolic periodic points and not intersecting the ω-limit set of recurrent critical points. We want to prove that Λ is expanding. Suppose it is not. Then there exists a sequence of points $\{x_n\} \subset \Lambda$ such that $|(f^n)'(x_n)| \leq 1$ for all n. Take a subsequence $\{x_{n_j}\}$ such that the sequence $\{f^{n_j}(x_{n_j})\}$ converges to a point $x \in \Lambda$. Take $\varepsilon > 0$ such that $d(\Lambda, c) > 2\varepsilon$ for every critical point c. Let U be the neighborhood of x associated to $\varepsilon > 0$ given by Theorem II. Without loss of generality we can assume that U is a disk centered at x. For j large, we have $f^{n_j}(x_{n_j}) \in U$. Let $V_j \in c(U, n_j)$ be such that $x_{n_j} \in V_j$. Then $f^i(V_j) \in c(U, n_j - i)$ for all $0 \leq i \leq n_j$. By Theorem II, $\operatorname{diam} f^i(V_j) \leq \varepsilon$ for all $0 \leq i \leq n_j$. Hence $f^i(V_j)$ doesn't contain critical points for $0 \leq i \leq n_j$ thus implying that f^{n_j} has no critical points on V_j. Since U is a disk and $V_j \in c(U, n_j)$, it follows that $f^{n_j}: V_j \to U$ is a bijection. Let $\varphi_j: U \to V_j$ be its holomorphic inverse. Then the family $\{\varphi_j\}$ is normal because $\operatorname{diam} V_j \leq \varepsilon$ for all j. Therefore we can assume that it converges to a holomorphic map $\varphi: U \to \overline{\mathbf{C}}$. But

$$\left|\varphi_j'(f^{n_j}(x_{n_j}))\right| = \left|(f^{n_j})'(x_{n_j})\right|^{-1} \geq 1$$

and $\lim_{j\to+\infty} f^{n_j}(x_{n_j}) = x$. Then $|\varphi'(x)| \geq 1$ and φ is non constant. Moreover

$$\varphi(x) = \lim_{j\to+\infty} \varphi_j(f^{n_j}(x_{n_j})) = \lim_{j\to+\infty} x_{n_j} \in \Lambda.$$

Then $\varphi(U)$ covers a compact neighborhood W of $\varphi(x) \in \Lambda$. Since $x \in \Lambda \subset J(f)$ there exists $N > 0$ such that $f^n(W) \supset J(f)$ for all $n \geq N$. Since the univalent maps $\varphi_j: U \to \overline{\mathbf{C}}$ converge to $\varphi: U \to \overline{\mathbf{C}}$, it follows that $\varphi_j(U) \supset W$ for j large. Hence

$$f^{n_j}(W) \subset f^{n_j}\varphi_j(U) = U.$$

For j large, we have $n_j \geq N$ and then $f^{n_j}(W) \supset J(f)$. Hence $U \supset J(f)$ contradicting the fact that U can be taken arbitrarily small in diameter. The proof of Theorem I is complete. □

Ricardo Mañé

Instituto de Matemática Pura e Aplicada, IMPA
Estrada Dona Castorina 110
22460-320, Rio de Janeiro-RJ
Brasil

Ergodic Variational Methods: New Techniques and New Problems

RICARDO MAÑÉ †

Former address: I.M.P.A.
E. Dona Castorina 110, J. Botânico
22460 Rio de Janeiro, Brazil

Introduction

Our subject will be *minimizing measures* of Lagrangian dynamical systems. This is a class of invariant probabilities μ of the flow generated by the Euler-Lagrange equation associated to a periodic Lagrangian on a closed manifold, selected by the property of minimizing the μ-average of the Lagrangian among all the invariant probabilities with a given asymptotic cycle (in the sense of Schwartzmann [S]). This concept was introduced by Mather [Ma 2] in a successful attempt to produce an analog of the Aubry-Mather theory for systems with more than one degree of freedom. His results on minimizing measures, when applied to periodic Lagrangians on the circle, recover the main results of the Aubry-Mather theory of twist maps. The main concepts and results of Mather's work will be recalled below.

Our objective is to show, through properties proved in [M 2], that the theory of minimizing measures becomes substantially sharper when restricted to *generic* Lagrangians; where genericity will be understood in a natural way to be defined in the next section. On the other hand, while producing new and stronger results, the generic viewpoint raises many interesting questions that seem beyond the range of the techniques employed in [M 2]. Some of them will be posed below; chosen to exhibit the most visible aspects of the incompleteness of our knowledge on which the ultimate sharpening possibilities of the generic approach really are.

I Minimizing Measures

Let M be a closed Riemannian manifold. By a Lagrangian on M we shall mean a C^∞ function $L: TM \times \mathbf{R} \to \mathbf{R}$, periodic (say of period 1) in the time variable, such that the second derivative of its restriction to the fibers T_xM is uniformly positive definite. In fact the weaker hypothesis required in [Ma 2] would suffice. Such a Lagrangian generates a flow $\varphi_t: TM \times S^1 \hookleftarrow$ (where $S^1 = \mathbf{R}/\mathbf{Z}$) defined by:

$$\varphi_t(x_o, v_o, t_o) = (x(t+t_o), \dot{x}(t+t_o), (t+t_o) \bmod \mathbf{Z})$$

where $x: \mathbf{R} \to M$ is the solution of the Euler-Lagrange equation associated to L, with $x(t_o) = x_o$, $\dot{x}(t_o) = v_o$.

Proceedings of the International Congress
of Mathematicians, Zürich, Switzerland 1994

Originally published in: *Proceedings of the International Congress of Mathematicians, Vol. 1, 2 (Zürich, 1994), 1216–1220,*

Let $\mathcal{M}^*(L)$ be the set of φ-invariant probabilities with compact support. To each $\mu \in \mathcal{M}^*(L)$ we associate its action $S_L(\mu)$ defined by:

$$S_L(\mu) = \int L\, d\mu$$

and its homology (or asymptotic cycle; [S], [Ma 2]) $\rho(\mu)$ defined by satisfying:

$$\langle [\omega], \rho(\mu) \rangle = \int \omega\, d\mu$$

for every closed 1-form ω on M. It always exists, is unique, and $\rho: \mathcal{M}^*(L) \to H_1(M, \mathbf{R})$ is surjective [Ma 2].

DEFINITION A *minimizing measure* of L is a $\mu \in \mathcal{M}^*(L)$ such that:

$$S_L(\mu) = \min\{S_L(\nu) \,/\, \nu \in \mathcal{M}^*(L), \rho(\nu) = \rho(\mu)\}.$$

For every $\gamma \in H_1(M, \mathbf{R})$, the set $\mathcal{M}_\gamma(L)$ of $\mu \in \mathcal{M}^*(L)$ with $\rho(\mu) = \gamma$ is *nonempty* [Ma 2].

The *action function* of L, $\beta_L: H_1(M, \mathbf{R}) \to \mathbf{R}$, is defined by:

$$\beta_L(\gamma) = \min\{S_L(\nu) \,/\, \nu \in \mathcal{M}^*(L), \rho(\nu) = \gamma\}.$$

This is a *convex* and *superlinear* function [Ma 2], and there exist interesting connections between the convex theoretical properties of β_L at a point γ and the dynamical properties of measures in $\mathcal{M}_\gamma(L)$. For instance, if γ is an *extremal* point of β_L (i.e. $(\gamma, \beta_L(\gamma))$ is an extremal point of the epigraph of β_L) then $\mathcal{M}_\gamma(L)$ contains ergodic elements, and if γ is a *strictly extremal* point of β_L (i.e. if there exists a hyperplane in $H_1(M, \mathbf{R}) \times \mathbf{R}$ touching the graph of β_L at $(\gamma, \beta_L(\gamma))$) then $\mathcal{M}_\gamma(L)$ contains measures supported in minimal sets [M 2], [Ma 2]. For properties concerning the vertex and smoothness of β_L, see [Ma 1] and [D].

Now we can introduce and exemplify the generic viewpoint. Let $C^\infty(M \times S^1)$ be the space of C^∞ functions on $M \times S^1$ endowed with the C^∞ topology. We say that a property holds for *generic* Lagrangians (or generically) if given any Lagrangian L there exists a residual subset $\mathcal{A} \subset C^\infty(M \times S^1)$ such that the property holds for every Lagrangian of the form $L_o + \psi$, with $\psi \in \mathcal{A}$.

THEOREM I.1. [M 2]. *Generically, the set* $\operatorname{Ext}(L)$ *of extremal points of* β_L *contains a residual subset* $\mathcal{A} \subset \operatorname{Ext}(L)$ *such that if* $\gamma \in \mathcal{A}$, $M_\gamma(L)$ *consists of a single measure that moreover is uniquely ergodic and its support is the limit in the Hausdorff metric of a sequence of periodic orbits.*

Observe that $\operatorname{Ext}(L)$ is always a Baire set, i.e. its residual subsets are dense.

PROBLEM I For a generic L, does the property of I.1 hold for every extremal point?

PROBLEM II For a generic L, does $\operatorname{Ext}(L)$ contain a dense subset $\mathcal{A}_o \subset \operatorname{Ext}(L)$ such that $\gamma \in \mathcal{A}_o$ implies that $\mathcal{M}_\gamma(L)$ contains a measure supported in a periodic orbit?

THEOREM I.2. *If L is a generic Lagrangian, for every ergodic minimizing measure μ there exists a sequence $\{\mu_n\}$ of uniquely ergodic minimizing measures such that $\mu_n \to \mu$ weakly and* $\mathrm{supp}(\mu_n) \to \mathrm{supp}(\mu)$ *in the Hausdorff metric.*

PROBLEM III For generic Lagrangians, is every ergodic minimizing measure uniquely ergodic?

II The Cohomological Approach and the Coboundary Property

The dual approach to minimizing measures is based on the fact that if ω is a closed 1-form on M then the Lagrangians L and $L - \omega$ generate the same flow. Then $\mathcal{M}^*(L) = \mathcal{M}^*(L - \omega)$ and:

$$S_{L-\omega}(\mu) = S_L(\mu) - \langle \omega, \rho(\mu) \rangle$$

for all $\mu \in \mathcal{M}^*(L)$. This also shows that $S_{L-\omega}(\mu)$ depends only on $[\omega] \in H^1(M, \mathbf{R})$. Define $\mathcal{M}^\omega(L)$ as the set of $\mu \in \mathcal{M}^*(L)$ such that:

$$S_{L-\omega}(\mu) = \min\{S_{L-\omega}(\nu) \,/\, \nu \in \mathcal{M}^*(L)\}.$$

Then $\mathcal{M}^\omega(L)$ depends only on $[\omega]$. To relate the sets $\mathcal{M}^\omega(L)$ with the sets $\mathcal{M}_\gamma(L)$ we define $c(\omega)$ as the maximum of the $c \in \mathbf{R}$ such that $\beta_L(x) \geq \langle [\omega], x \rangle + c$ for all $x \in H_1(M, \mathbf{R})$, and $K(\omega)$ by

$$K(\omega) = \{x \in H_1(M, \mathbf{R}) \,/\, \beta_L(x) = \langle \omega, x \rangle + c(\omega)\}.$$

Then it can be checked that:

$$\mathcal{M}^\omega(L) = \bigcup \{\mathcal{M}_\gamma(L) \,/\, \gamma \in K(\omega)\},$$
$$\bigcup_\omega \mathcal{M}^\omega(L) = \bigcup_\gamma \mathcal{M}_\gamma(L).$$

Therefore all the minimizing measures are found through the sets $\mathcal{M}^\omega(L)$. Denote by $\Lambda^\omega(L) \subset TM \times S^1$ the closure of the union of the supports of the measures in $\mathcal{M}^\omega(L)$. Among the main results of Mather in [Ma 2] are the *compacity* of $\Lambda^\omega(L)$ and the *graph property*, i.e. that if $\pi: TM \times S^1 \to M$ denotes the canonical projection then $\pi / \Lambda^\omega(L)$ is injective and its inverse is Lipschitz. The Aubry-Mather theory follows from this remarkable property and a theorem of Moser [Mo] describing twist maps as time one maps of flows of periodic Lagrangians on the circle.

The "converse" of the definition of $\Lambda^\omega(L)$ is true:

THEOREM II.1. *If the support of $\mu \in \mathcal{M}^*(L)$ is contained in $\Lambda^\omega(L)$, then $\mu \in \mathcal{M}^\omega(L)$.*

From this it follows easily that:

COROLLARY II.2. *Every $\mathcal{M}^\omega(L)$ contains a measure supported in a minimal set. If $\mathcal{M}^\omega(L)$ contains a single measure μ, then μ is uniquely ergodic.*

The theorems stated in Section I rely on the following property, which has its own intrinsic interest.

THEOREM II.3. (Coboundary Property [M 2]). *If $\mu \in \mathcal{M}^\omega(L)$ is ergodic, then $(L - \omega - c(\omega))/\operatorname{supp}(\mu)$ is a Lipschitz coboundary, i.e. there exists a Lipschitz function $V\colon \operatorname{supp}(\mu) \to \mathbf{R}$ such that:*

$$(L - \omega - c(\omega))/\operatorname{supp}(\mu) = \frac{dV}{d\varphi}$$

where

$$\frac{dV}{d\varphi}(\theta) := \lim_{t\to 0} \frac{1}{t}(V(\varphi_t(\theta)) - V(\theta)).$$

The following result shows that for generic Lagrangians, the set $\mathcal{M}^\omega(L)$ consists, for most $[\omega] \in H^1(M, \mathbf{R})$, of a single measure, which, by Corollary II.2, has to be uniquely ergodic.

THEOREM II.4. [M 2]. *(a) Given a closed 1-form ω on M, the set $\mathcal{M}^\omega(L)$ consists, for a generic L, of a single measure.*

(b) For a generic Lagrangian L, there exists a residual subset $\mathcal{A} \subset H^1(M, \mathbf{R})$ such that $[\omega] \in \mathcal{A}$ implies that $\mathcal{M}^\omega(L)$ consists of a single measure.

PROBLEM IV Is it true that for a generic Lagrangian L there exists a dense set $\mathcal{A} \subset H^1(M, \mathbf{R})$ such that $[\omega] \in \mathcal{A}$ implies that $\mathcal{M}^\omega(L)$ consists of a single measure supported in a hyperbolic periodic orbit?

It follows from the results in [D] that the set of classes $[\omega]$ with this property is open for generic Lagrangians.

When $M = S^1$ the answer is positive. This follows from Mather's result on the differentiability of the action function, when $M = S^1$, at irrational values of $H_1(M, S^1)$, plus the generic properties of the vertex of β_L (where also another proof of the differentiability is given). Observe that when $M = S^1$, the answer to Problems I, II, and III is affirmative for every *Lagrangian.*

The proof of Theorem II.3 uses a reformulation of the space on which the variational principle is applied. We introduce a space $\mathcal{M}$ of probabilities on $TM \times S^1$, so large that $\mathcal{M} \supset \mathcal{M}^*(L)$ for *every* L, and such that for every L and ω there exists $\mu \in \mathcal{M}$ satisfying:

$$(*) \qquad \int (L - \omega)\, d\mu = \min\{\int (L - \omega)\, d\nu \,/\, \nu \in \mathcal{M}\}.$$

Moreover, it is proved that this property *implies* the L-invariance of μ. Then $\mu \in \mathcal{M}^\omega(L)$. The existence of μ's satisfying (*) follows from the fact that $\mathcal{M}$ is naturally embedded as a convex subset of a locally convex space and sets of the form

$$\{\nu / \int (L - \omega)\, d\nu \le C\}$$

are compact for all L, ω, and C. This translates the analysis of $\mathcal{M}^\omega(L)$ into questions on strictly extremal points (also called exposed points) of compact convex sets in locally convex spaces. Some care has to be taken with the fact that compact convex subsets of locally convex spaces (even Fréchet spaces) may have no strictly extremal points (as opposed to the case of Banach spaces).

References

[D] J. Delgado, *Vertex and differentiability of the action function of Lagrangians*, Thesis, IMPA 1993.

[M 1] R. Mañé, *Minimizing measures of Lagrangian systems*, Nonlinearity **5** (1992), 623–638.

[M 2] R. Mañé, *Generic properties and problems of minimizing measures*, preprint I.M.P.A. 1993.

[Ma 1] J. Mather, *Differentiability of the minimum average action as a function of the rotation numbers*, Bol. Soc. Brasil. Mat. **21** (1990), 57–70.

[Ma 2] J. Mather, *Action Minimizing measures for positive definite Lagrangian systems*, Math. Z. **207** (1993), 169–207.

[Mo] J. Moser, *Monotone twist mappings and the calculus of variations*, Ergodic Theory and Dynamical Systems **6** (1986), 401–413.

[S] S. Schwartzmann, *Asymptotic cycles*, Ann. of Math. (2) **66** (1957), 27–284.

J. DIFFERENTIAL GEOMETRY
Vol. 45 (1997) 74-93

ON THE TOPOLOGICAL ENTROPY OF GEODESIC FLOWS

RICARDO MAÑÉ

1. Introduction

Let M^n be a closed connected C^∞ manifold and let SM be its unit tangent bundle, defined as usual as $SM = \{\theta = (x,v) : x \in M,\ v \in T_xM,\ \|v\| = 1\}$. The geodesic flow $\varphi_t : SM \to SM$ is defined by $\varphi_t(x,v) = (\gamma(t), \dot\gamma(t))$, where $\gamma : \mathbf{R} \to M$ is the geodesic with initial conditions $\gamma(0) = x$ and $\dot\gamma(0) = v$.

Given x and y in M, define $n_T(x,y)$ as the number of geodesics of length $\leq T$ (parametrized by arc length) joining x and y. A standard application of Sard's Theorem to the exponential maps of M shows that $n_T(x,y)$ is finite and locally constant on an open full measure subset of $M \times M$.

Our aim is to relate *the exponential growth rate of* $n_T(x,y)$, as a function of T, with the *topological entropy of the geodesic flow* $h_{top}(\varphi)$. In that direction, among other results, we shall prove that

$$h_{top}(\varphi) = \lim_{T\to+\infty} \frac{1}{T} \log \int_{M\times M} n_T(x,y)\, dxdy.$$

While proving this result, we shall also prove that *Przytycki's upper estimate* for the topological entropy of general C^2 flows [8], *is always an equality for C^∞ geodesic flows.* Since Przytycki's inequality will be a key tool in our proofs we begin by recalling its statement. Given a

Received November 21, 1994, and, in revised form July 11, 1995.
The author died while the paper was being refereed. The suggested revisions were made by Keith Burns, Gabriel Paternain and Miguel Paternain.

Originally published in: *J. Differential Geom. 45 (1997), no. 1, 74–93,*

linear map $L : E \to F$ between finite dimensional Hilbert spaces, we define its *expansion* $ex(L)$ by

$$ex(L) = \max_S |\det(L|_S)|,$$

where the maximum is taken over all subspaces $S \subset E$. Przytycki's inequality states that *for a C^2 flow $\psi_t : N \to N$ on a closed manifold N,*

$$h_{top}(\psi) \leq \liminf_{t\to\infty} \frac{1}{t} \log \int_N ex(d_x\psi_t)\, dx.$$

The main result of this paper is Theorem 1.1 below. Half of its proof relies on a combination of Yomdin's theorem [9] and a formula due to Berger and Bott [2] that, although not difficult to prove, provides the link between the numbers $n_T(x, y)$ and the dynamics of the geodesic flow. We took this combination from [6], but for our purposes an improved version of Yomdin's theorem will be needed. The other half relies on Przytycki's inequality, Berger and Bott's formula, and a careful change of variables, where the novelty of this work resides.

Theorem 1.1.

$$\begin{aligned} h_{top}(\varphi) &= \lim_{T\to+\infty} \frac{1}{T} \log \int_{M\times M} n_T(x,y)\, dxdy \\ &= \lim_{T\to+\infty} \frac{1}{T} \log \int_{SM} ex(d_\theta\varphi_T)\, d\theta. \end{aligned}$$

When the manifold has no conjugate points, it is easy to check, using the fact that the exponential map $exp_x : T_xM \to M$ is a covering map admitting a fundamental domain of diameter $\leq diam(M) \stackrel{\text{def}}{=} c$, that for any $x_1, y_1, x_2, y_2 \in M$,

$$n_T(x_1, y_1) \leq n_{T+2c}(x_2, y_2).$$

This property and Theorem 1.1 imply:

Corollary 1.2. *If M has no conjugate points, then*

$$\lim_{T\to+\infty} \frac{1}{T} \log n_T(x, y) = h_{top}(\psi),$$

for all $x, y \in M$.

From this corollary we can recover the results of Freire and Mañé [3] on the equality of the topological entropy of the geodesic flow and the volume growth rate of the manifold when there are no conjugate points. Recall that the volume growth rate $\lambda(M)$ of M is defined by

$$\lambda(M) = \lim_{r\to+\infty} \frac{1}{r} \log \operatorname{Vol}(B_r(x)),$$

where $B_r(x)$ denotes the ball of radius r and center x in the universal covering $\tilde{M}$ of M, and $\operatorname{Vol}(B_r(x))$ denotes its volume. Manning proved [5] that this limit exists and is independent of x. Moreover he proved that $\lambda(M) \le h_{top}(\varphi)$ for *every* closed Riemannian manifold and $\lambda(M) = h_{top}(\varphi)$ when M has sectional curvatures ≤ 0. In [3], Freire and Mañé extended this result to manifolds *without conjugate points* through a different technique (see Remark 1.5 below). Here we can obtain it from Corollary 1.2.

Corollary 1.3. *If M has no conjugate points, then*

$$\lambda(M) = h_{top}(\varphi).$$

To deduce Corollary 1.3 from Corollary 1.2, it suffices to show that $\lambda(M) \ge h_{top}(\varphi)$, because, as we explained above, the inequality $\lambda(M) \le h_{top}(\varphi)$ always holds. If $p : \tilde{M} \to M$ is the covering map and $x \in \tilde{M}$, then the number $n_T(p(x), p(x))$ is just the number of points in the set

$$G_T \stackrel{\text{def}}{=} \{z \in M : \ p(z) = p(x), \ d(z, x) \le T\}.$$

Observe that there exists $r_0 > 0$ such that any two distinct points z' and z'' with $p(z') = p(z'')$, satisfy $d(z', z'') \ge r_0$. Clearly

$$B_{T+r_0}(x) \supset \bigcup_{z\in G_T} B_{r_0}(z),$$

and the sets $B_{r_0/2}(z)$, $z \in G_T$ are disjoint. Hence if

$$k \stackrel{\text{def}}{=} \min\{\operatorname{Vol}(B_{r_0/2}(a)) : \ a \in \tilde{M}\},$$

we have

$$\operatorname{Vol}(B_{T+r_0}(x)) \ge k\#G_T = k\, n_T(p(x), p(x)).$$

Therefore

$$\begin{aligned}\lambda(M) &= \lim_{T\to+\infty} \frac{1}{T+r_0} \log \mathrm{Vol}(B_{T+r_0}(x)) \\ &\geq \lim_{T\to+\infty} \frac{1}{T+r_0} \log k\, n_T(p(x), p(x)) = h_{top}(\varphi),\end{aligned}$$

thus completing the proof of the corollary.

In the next statement, $V(\theta)$ will denote the vertical fibre at $\theta = (x, v) \in SM$, defined by

$$V(\theta) = d_\theta \pi^{-1}(\{0\}),$$

where π is the projection map $\pi : SM \to M$.

Theorem 1.4.

$$\begin{aligned}h_{top}(\varphi) &= \lim_{T\to+\infty} \frac{1}{T} \log \int_{SM} |\det(d_\theta \varphi_T|_{V(\theta)})|\, d\theta \\ &= \lim_{T\to+\infty} \frac{1}{T} \log \int_M \mathrm{Vol}(\varphi_T(S_x M))\, dx.\end{aligned}$$

The second equality is trivial because

$$\mathrm{Vol}(\varphi_T(S_x M)) = \int_{S_x M} |\det(d_{(x,v)} \varphi_T|_{V(x,v)})|\, dv.$$

Remark 1.5. Manning's proof of Corollary 1.3 for manifolds of non-positive curvature relies on the fact that for such manifolds, any two geodesics $\gamma_i : \mathbf{R} \to \tilde{M}$ in the universal covering satisfy:

$$d(\gamma_1(t), \gamma_2(t)) \leq d(\gamma_1(0), \gamma_2(0)) + d(\gamma_1(T), \gamma_2(T)),$$

for all $0 \leq t \leq T$. In [3] the authors observed that for manifolds where this property holds with the right term multiplied by a constant independent of the geodesics, Manning's proof can be applied with insignificant changes, thus providing a much simpler proof of their result *if* the existence of such a constant could be established for manifolds without conjugate points. However the example of Ballmann, Brin and Burns [1], that appeared four years later, proved that such a constant *does not exist*.

Remark 1.6. On the question of the exponential growth rate of $n_T(x, y)$, without assumptions like the absence of conjugate points we have little to say; only that (see Appendix) as a corollary of a result of G. Paternain [6], it follows that for every $x \in M$,

$$\limsup_{T \to +\infty} \frac{1}{T} \log n_T(x, y) \leq h_{top}(\varphi),$$

for a.e. $y \in M$. This naturally poses the following question:

Problem I. Is it true that for a.e. $(x, y) \in M \times M$,

$$\lim_{T \to +\infty} \frac{1}{T} \log n_T(x, y) = h_{top}(\varphi)? \tag{0.1}$$

Since an affirmative answer to this problem may sound too good to be true, a humbler, and more feasible question is the following one:

Problem II. Is it true that (0.1) holds for generic Riemannian metrics when $\dim M = 2$?

Remark 1.7. Theorem 1.1 above was announced, and its proof sketched, in a preprint of the author circulated in January 93. The proof that we shall use here is essentially shorter than that outlined there.

The proof of Theorem 1.1 relies on three inequalities. Two of them (inequalities (A) and (A′) below) are where the novelty of this paper resides. The other inequality (inequality (B) below) is a variation of an inequality due to G. Paternain [6], which in turn comes from Yomdin's theorem [9]. One way to prove this inequality is to combine G. Paternain's method with a certain uniformity in Yomdin's theorem, visible in Gromov's exposition of this celebrated result in the Bourbaki Seminaire [4]. Such was the method employed in our announcement. Afterwards, G. Paternain and M. Paternain [7] proved that inequality using directly Yomdin's theorem without having to appeal to the uniformity mentioned above.

2. Proof of Theorem 1.1

We begin by recalling the basic formalism of geodesic flows, stressing its symplectic properties, which will play a key role in our proofs.

Given $\theta = (x, v) \in SM$ define

$$E(\theta) = \{w \in T_x M : \langle w, v\rangle = 0\}.$$

Denote by $\pi : SM \to M$ the canonical projection and set

$$N(\theta) = d_\theta \pi^{-1}(E(\theta)),$$

$$V(\theta) = d_\theta \pi^{-1}(\{0\}).$$

If X is the geodesic vector field on SM (i.e., the vector field generated by the geodesic flow), then $T_\theta SM$ is the direct sum of $N(\theta)$ and the one-dimensional subspace spanned by $X(\theta)$. Moreover, N is φ_t-invariant, i.e.,

$$d_\theta \varphi_t(N(\theta)) = N(\varphi_t(\theta)),$$

for all $\theta \in SM$ and $t \in \mathbf{R}$. On each N_θ there exist an inner product $< , >_\theta$ and an isometry $J_\theta : N(\theta) \to N(\theta)$ of this inner product, such that for all θ:

a) $J_\theta^2 = -I$;

b) $V(\theta)$ and $J_\theta V(\theta)$ are orthogonal;

c) $\langle J_{\varphi_t(\theta)} d_\theta \varphi_t(\zeta), d_\theta \varphi_t(\eta)\rangle_{\varphi_t(\theta)} = \langle J_\theta \zeta, \eta\rangle_\theta$;

d) $\langle\ ,\ \rangle_\theta$ and J_θ are C^∞ functions of θ.

We shall say that a subspace $S \subset N(\theta)$ is Lagrangian if its orthogonal complement $S^\perp$ is $J_\theta S$.

These properties can be translated into symplectic terms by defining, for each $\theta \in SM$, a 2-form $\omega_\theta : N(\theta) \times N(\theta) \to \mathbf{R}$ by

$$\omega_\theta(\zeta, \eta) = \langle J_\theta \zeta, \eta\rangle_\theta.$$

Then ω_θ is non-degenerate (by (a)) and $d\varphi_t$-invariant (by (c)). A subspace S is now Lagrangian if and only if $\dim S = \dim N(\theta)/2$ and $\omega_\theta|_{S \times S} = 0$. By property (b), $V(\theta)$ is a Lagrangian subspace for all θ.

The above discussion presents the properties of $\langle\ ,\ \rangle_\theta$ and J_θ that will be used here. We now give their definitions. First we define the so-called connector map $K_\theta : T_\theta SM \to T_x M$. Given $\zeta \in TT_\theta SM$, let $\theta(s) = (x(s), v(s))$ be a curve in SM such that $\theta(0) = \theta$ and $\dot\theta(0) = \zeta$. Then $K_\theta(\zeta) = Dv/ds(0)$, i.e., $K_\theta(\zeta)$ is the covariant derivative at $s = 0$ of the vector field $v(s)$ along the curve $x(s)$. There is a linear isomorphism $i_\theta : N(\theta) \to E(\theta) \oplus E(\theta)$ defined by

$$i_\theta(\zeta) = (d_\theta \pi \zeta, K_\theta \zeta).$$

The map J_θ is the pullback by i_θ of the rotation $(v, v') \to (v', -v)$ on $E(\theta) \oplus E(\theta)$, and

$$\langle \zeta, \eta \rangle_\theta = \langle d_\theta \pi \zeta, d_\theta \pi \eta \rangle + \langle K_\theta \zeta, K_\theta \eta \rangle.$$

The proof of Theorem 1.1 consists of proving the following inequalities:

$$A) \quad \liminf_{T\to+\infty} \frac{1}{T} \log \int_{M\times M} n_T(x,y)\, dxdy \geq \liminf_{T\to+\infty} \frac{1}{T} \log \int_{SM} ex(d_\theta \varphi_T)\, d\theta.$$

$$A') \quad \limsup_{T\to+\infty} \frac{1}{T} \log \int_{M\times M} n_T(x,y)\, dxdy \geq \limsup_{T\to+\infty} \frac{1}{T} \log \int_{SM} ex(d_\theta \varphi_T)\, d\theta.$$

$$B) \quad \limsup_{T\to+\infty} \frac{1}{T} \log \int_{M\times M} n_T(x,y)\, dxdy \leq h_{top}(\varphi).$$

These inequalities, plus Przytycki's inequality imply Theorem 1.1.

First we shall prove (A) and (A′). For this purpose, and also for the proof of inequality (B), we shall introduce, following Berger and Bott [2], a number $A(\theta, t)$ associated to each $\theta = (x, v) \in SM$ and $t > 0$, defined by

$$A(\theta, t) = |\det(d_\theta(\pi \circ \varphi_t)|_{V(\theta)})|.$$

Berger and Bott proved ([2])

$$\int_0^T \int_{S_xM} A((x,v),t)\, dvdt = \int_M n_T(x,y)\, dy.$$

Integrating this equality over M we obtain

$$\int_0^T \int_{SM} A(\theta,t)\, d\theta dt = \int_{M\times M} n_T(x,y)\, dxdy.$$

Next observe that:

$$\liminf_{T\to+\infty} \frac{1}{T} \log \int_{SM} ex(d_\theta \varphi_T)\, d\theta = \liminf_{T\to+\infty} \frac{1}{T} \log \int_0^T \int_{SM} ex(d_\theta \varphi_t)\, d\theta dt,$$

and

$$\limsup_{T\to+\infty}\frac{1}{T}\log\int_{SM} ex(d_\theta\varphi_T)\,d\theta$$
$$=\limsup_{T\to+\infty}\frac{1}{T}\log\int_0^T\int_{SM} ex(d_\theta\varphi_t)\,d\theta dt.$$

Therefore the proof of (A) and (A′) is reduced to showing:

$$\liminf_{T\to+\infty}\frac{1}{T}\log\int_0^T\int_{SM} A(\theta,t)\,d\theta dt$$
$$\geq \liminf_{T\to+\infty}\frac{1}{T}\log\int_0^T\int_{SM} ex(d_\theta\varphi_t)\,d\theta dt, \tag{0.1}$$

and

$$\limsup_{T\to+\infty}\frac{1}{T}\log\int_0^T\int_{SM} A(\theta,t)\,d\theta dt \tag{0.2}$$
$$\geq \limsup_{T\to+\infty}\frac{1}{T}\log\int_0^T\int_{SM} ex(d_\theta\varphi_t)\,d\theta dt. \tag{0.3}$$

Before going into the formal proof of these inequalities, let us informally describe the strategy we shall follow. Obviously the proof will rely on the analysis of the maps $d_\theta\varphi_t : N(\theta) \to N(\varphi_t(\theta))$. The key idea is that the analysis becomes more accurate for our objective if we shift the initial point θ backwards to a certain point $\varphi_{-\tau_1}(\theta)$, $\tau_1 > 0$, and shift the final point $\varphi_t(\theta)$ forward to $\varphi_{t+\tau_2}(\theta)$, $\tau_2 > 0$. It will be important that these numbers τ_1 and τ_2 can be chosen so that they are measurable functions of the pair (θ, t), and their possible values lie in a finite set of the form $\{0, 1/m, 2/m, ..., 1\}$, where $m \geq 1$ is an integer. Clearly inequalities (0.1) and (0.3) follow from

Lemma 2.1. *There exists $C > 0$ such that for all $T > 0$,*

$$\int_0^{T+2}\int_{SM} A(\theta,t)\,d\theta dt \geq C\int_0^T\int_{SM} ex(d_\theta\varphi_t)\,d\theta dt.$$

To prove this lemma we first introduce a definition. For subspaces S_1, S_2 of $N(\theta)$ with $\dim S_1 = \dim S_2 = \dim N(\theta)/2$, we define

$$\alpha(S_1,S_2) = |\det(P|_{S_1})|,$$

where $P : N(\theta) \to S_2^{\perp}$ is the orthogonal projection. Clearly α depends continuously on the subspaces and $\alpha(S_1, S_2) = 0$ if and only if $S_1 \cap S_2 \neq \{0\}$.

The next two lemmas will be proved later.

Lemma 2.2. *For each $\theta \in SM$ and $t \in \mathbf{R}$, there is a Lagrangian subspace $R_t(\theta) \subset N(\theta)$, which depends measurably on t and θ, and satisfies:*

a) $|\det(d_\theta\varphi_t|_{R_t(\theta)})| = ex(d_\theta\varphi_t)$;

b) if S is a subspace of $N(\theta)$ with $\dim S = \dim N(\theta)/2$, then

$$|\det(d_\theta\varphi_t|_S)| \geq \alpha(S, R_t^{\perp}(\theta))ex(d_\theta\varphi_t).$$

Lemma 2.3. *There exist $\delta > 0$, an integer $m \geq 1$ and measurable functions $\tau_i : SM \times \mathbf{R} \to \{0, 1/m, 2/m, ..., 1\}$, $i = 1, 2$, such that, after abbreviating $\tau_i(\theta, t)$ to τ_i for $i = 1, 2$ and setting $\tau = \tau(\theta, t) = \tau_1 + \tau_2$,*

$$\theta_1 = \varphi_{-\tau_1}(\theta), \quad \theta_2 = \varphi_{t+\tau_2}(\theta), \quad \textit{and} \quad V_i = V(\theta_i) \textit{ for } i = 1, 2,$$

we have for all θ and t:

a) $\alpha((d_{\theta_1}\varphi_{\tau_1})V_1, R_t^{\perp}(\theta)) \geq \delta$

and

b) $\alpha((d_{\theta_1}\varphi_{t+\tau})V_1, V_2) \geq \delta$.

From these two lemmas we shall deduce the following:

Corollary 2.4. *There exists $C > 0$ such that for all $t > 0$ and $\theta \in SM$, the functions τ_1 and τ given by Lemma 2.3 satisfy*

$$A(\varphi_{-\tau_1(\theta,t)}(\theta), t + \tau(\theta, t)) \geq C\, ex(d_\theta\varphi_t).$$

Proof of Corollary 2.4. Set $S = (d_{\theta_1}\varphi_{\tau_1})V_1$. By Lemma 2.2 and property (a) of Lemma 2.3 we have

$$(0.4) \qquad |\det(d_\theta\varphi_t|_S)| \geq \alpha(S, R_t^{\perp}(\theta))\, ex(d_\theta\varphi_t) \geq \delta ex(d_\theta\varphi_t).$$

Take $C_1 > 0$ such that

$$(0.5) \qquad |\det(d_\zeta\varphi_s|_L)| \geq C_1,$$

for every $\zeta \in SM$, $s \in [0,1]$, and every non-trivial linear subspace $L \subset N(\zeta)$. Set $\widehat{\theta} = \varphi_t(\theta)$ and $\widehat{S} = d_\theta \varphi_t S$. Then equations (0.4) and (0.5) imply

$$\begin{aligned} &|\det(d_{\theta_1}\varphi_{t+\tau}|_{V_1})| \\ (0.6) \qquad &= |\det(d_{\widehat{\theta}}\varphi_{\tau_2}|_{\widehat{S}})| \cdot |\det(d_\theta \varphi_t|_S)| \cdot |\det(d_{\theta_1}\varphi_{\tau_1}|_{V_1})| \\ &\geq C_1^2 \delta ex(d_\theta \varphi_t). \end{aligned}$$

Now set $S_2 = (d_{\theta_1}\varphi_{t+\tau})V_1$. By property (b) of Lemma 2.3 and the definition of α we have respectively

$$\alpha(S_2, V_2) \geq \delta,$$

$$(0.7) \qquad |\det(d_{\theta_2}\pi|_{S_2})| > \delta,$$

which together with inequality (0.6) implies that

$$\begin{aligned} A(\varphi_{-\tau_1(\theta,t)}(\theta), t+\tau(\theta,t)) &= |\det(d_{\theta_1}(\pi \circ \varphi_{t+\tau})|_{V_1})| \\ &= |\det(d_{\theta_2}\pi|_{S_2})| \cdot |\det(d_{\theta_1}\varphi_{t+\tau}|_{V_1})| \\ &\geq \delta |\det(d_{\theta_1}\varphi_{t+\tau}|_{V_1}| \geq \delta^2 C_1^2 ex(d_\theta \varphi_t). \end{aligned}$$

Hence the proof of the corollary with $C = \delta^2 C_1^2$ is completed.

Before proving Lemmas 2.2 and 2.3, let us see how Lemma 2.1 follows from the corollary.

Define $F : SM \times [0,T] \to SM \times [0, T+2]$ by

$$F(\theta, t) = (\varphi_{-\tau_1(\theta,t)}(\theta), t + \tau(\theta, t))),$$

where $\tau_1(\theta,t)$ and $\tau(\theta,t)$ are defined as in Lemma 2.3. Given integers $0 \leq i \leq m$, $0 \leq j \leq m$, we define

$$A(i,j) = \{(\theta,t) \in SM \times [0,T] : \ \tau_1(\theta,t) = i/m, \ \ \tau_2(\theta,t) = j/m\}.$$

On each $A(i,j)$, F is injective, and if we set $d\mu = d\theta dt$, then $\mu(F(S)) = \mu(S)$ for every Borel set $S \subset A(i,j)$. Hence for any integrable function $\Phi : SM \times [0, T+2] \to \mathbf{R}$ we have

$$\int_{F(A(i,j))} \Phi \, d\mu = \int_{A(i,j)} (\Phi \circ F) \, d\mu.$$

Suppose now that $\Phi > 0$. Then

$$\begin{aligned}\int_{SM\times[0,T]}(\Phi\circ F)\,d\mu &= \int_{\bigcup_{i,j}A(i,j)}(\Phi\circ F)\,d\mu = \sum_{i,j}\int_{A(i,j)}(\Phi\circ F)\,d\mu\\ &= \sum_{i,j}\int_{F(A(i,j))}\Phi\,d\mu \le \sum_{i,j}\int_{SM\times[0,T+2]}\Phi\,d\mu\\ &= (m+1)^2\int_{SM\times[0,T+2]}\Phi\,d\mu.\end{aligned}$$

From this inequality and Corollary 2.4, it follows that

$$\begin{aligned}&\int_0^{T+2}\int_{SM}A(\theta,t)d\theta dt\\ &\qquad\ge \frac{1}{(m+1)^2}\int_{SM\times[0,T]}A(F(\theta,t))\,d\mu\\ &\qquad= \frac{1}{(m+1)^2}\int_0^T\int_{SM}A(\varphi_{-\tau_1(\theta,t)}(\theta),t+\tau(\theta,t))\,d\theta dt\\ &\qquad\ge \frac{C}{(m+1)^2}\int_0^T\int_{SM}ex(d_\theta\varphi_t)\,d\theta dt,\end{aligned}$$

thus completing the proof of Lemma 2.1.

Proof of Lemma 2.2. Consider the polar decomposition

$$d_\theta\varphi_t = L_t(\theta)O_t(\theta),$$

where $L_t(\theta) : N(\theta) \to N(\theta)$ is symmetric and positive, and $O_t(\theta) : N(\theta) \to N(\varphi_t(\theta))$ is an isometry, both being C^∞ functions of θ. Since $L_t(\theta) = ((d_\theta\varphi_t)^*(d_\theta\varphi_t))^{1/2}$ and $(d_\theta\varphi_t)^*$ is symplectic (because so is $d_\theta\varphi_t$), $L_t(\theta)$ is *symplectic and symmetric*. Thus, if ζ is an eigenvector of $L_t(\theta)$ associated to an eigenvalue λ, then $J_\theta\zeta$ is an eigenvector associated to the eigenvalue λ^{-1} because $L_t(\theta)J_\theta = J_\theta L_t(\theta)^{-1}$ (by the symmetry and the symplecticity of $L_t(\theta)$) and hence

$$L_t(\theta)J_\theta\zeta = J_\theta L_t(\theta)^{-1}\zeta = \lambda^{-1}J_\theta\zeta.$$

Using this property it is possible to construct for each t an orthonormal basis of $N(\theta)$ of the form $\{\zeta_1,\dots,\zeta_{n-1},J_\theta\zeta_1,\dots,J_\theta\zeta_{n-1}\}$, where ζ_i is an eigenvector of $L_t(\theta)$ associated to an eigenvalue $\lambda_i \ge 1$. Let $R_t(\theta)$ be the subspace spanned by $\{\zeta_1,\dots,\zeta_{n-1}\}$. Clearly $R_t(\theta)$ is Lagrangian and satifies property (a).

To prove property (b), observe first that

$$|\det(d_\theta\varphi_t|_S)| = |\det(L_t(\theta)|_S)|,$$

because $O_t(\theta)$ is an isometry. Notice also that $L_t(\theta)$ leaves $R_t(\theta)$ and $R_t^\perp(\theta)$ invariant, because both of these spaces are spanned by the eigenvectors of $L_t(\theta)$. Hence $L_t(\theta)$ commutes with the orthogonal projection $P : N(\theta) \to R_t(\theta)$, i.e., $L_t(\theta) \circ P = P \circ L_t(\theta)$. Let us suppose that $S \cap R_t^\perp(\theta) = \{0\}$, otherwise there is nothing to prove. Then $P(S) = R_t(\theta)$ and thus

$$\begin{aligned}|\det(L_t(\theta)|_{R_t(\theta)})||\det(P|_S)| &= |\det(P|_{L_t(\theta)(S)})||\det(L_t(\theta)|_S)|\\ &\leq |\det(L_t(\theta)|_S)|.\end{aligned}$$

Hence

$$ex(d_\theta\varphi_t)\alpha(S, R_t^\perp(\theta)) \leq |\det(d_\theta\varphi_t|_S)|.$$

Finally we show the measurability of $R_t(\theta)$ as a function of t and θ. Let $\mathcal{F}$ denote the vector bundle over SM consisting of pairs (θ, g), in which $\theta \in SM$ and $g : N(\theta) \to N(\theta)$ is a symmetric linear map. Given positive integers p and l_i, $1 \leq i \leq p$, let $\mathcal{F}(p, l_1, ...l_p)$ be the set of pairs $(\theta, g) \in \mathcal{F}$, where g has p eigenvalues $\lambda_1 < \cdots < \lambda_p$ with multiplicities $l_1, ..., l_p$. Then $\mathcal{F}(p, l_1, ...l_p)$ is a Borel set (check it) and so is the subset $\mathcal{P}(p, l_1, ..., l_p)$ of $SM \times \mathbf{R}$ defined by $\{(\theta, t) : \ L_t(\theta) \in \mathcal{F}(p, l_1, ..., l_p)\}$. Now observe that $R_t(\theta)$ can be chosen to be continuous on each set $\mathcal{P}(p, l_1, ..., l_p)$. Since these sets are Borel and there are finitely many of them, the measurability is proved.

Proof of Lemma 2.3. It suffices to prove that we can find $\delta_1, \delta_2 > 0$, integers $m_1, m_2 \geq 1$ and measurable functions $\tau_i : SM \times \mathbf{R} \to \{0, 1/m, 2/m, ..., 1\}$ such that properties (a) and (b) of Lemma 2.3 hold with δ changed to δ_1 in (a) and to δ_2 in (b). Then we can easily obtain Lemma 2.3 with $m = m_1 m_2$ and $\delta = \min(\delta_1, \delta_2)$.

We shall prove first the existence of τ_1 and τ_2. The measurability will be discussed after that. We shall use the following well known property of the vertical subbundle: if $\theta \in SM$ and $S \subset N(\theta)$ is a Lagrangian subspace, then the set of values $s \in \mathbf{R}$ such that

$$d_\theta\varphi_s(S) \cap V(\varphi_s(\theta)) \neq \{0\}$$

is discrete.

We begin by proving the existence of τ_1. Suppose by contradiction that it does not exist. Then for every integer $m \geq 1$, there exists $(\theta_m, t_m) \in SM \times \mathbf{R}$ such that for all $s \in \{0, 1/m, 2/m, ..., 1\}$ we have

$$\alpha(d_{\varphi_{-s}(\theta_m)}\varphi_s(V(\varphi_{-s}(\theta_m))),\ R_{t_m}^{\perp}(\theta_m)) \leq 1/m. \tag{0.8}$$

Since M is compact, the sequence $(\theta_m, R_{t_m}^{\perp}(\theta_m))$ has a subsequence $(\theta_{m_k}, R_{t_{m_k}}^{\perp}(\theta_{m_k}))$ that converges to (θ, S), where S is a Lagrangian subspace. From equation (0.8) and the continuity of α we deduce

$$\alpha(d_{\varphi_{-s}(\theta)}\varphi_s(V(\varphi_{-s}(\theta))), S) = 0,$$

for all $s \in [0,1]$. This is equivalent to

$$d_{\varphi_{-s}(\theta)}\varphi_s(V(\varphi_{-s}(\theta))) \cap S \neq \{0\},$$

for all $s \in [0,1]$. Hence

$$d_\theta\varphi_{-s}(S) \cap V(\varphi_{-s}(\theta)) \neq \{0\},$$

for all $s \in [0,1]$. This contradicts the property of Lagrangian subspaces mentioned above.

Now we shall prove the existence of τ_2. Suppose that it does not exist. Then there exists a sequence $(\theta_m, t_m) \in SM \times \mathbf{R}$, such that after setting $\tau_1(m) = \tau_1(\theta_m, t_m)$ and $\beta_m = \varphi_{-\tau_1(m)}(\theta_m)$, we have

$$\alpha(d_{\beta_m}\varphi_{\tau_1(m)+s+t_m}(V(\beta_m)),\ V(\varphi_{t_m+s}(\theta_m))) \leq 1/m, \tag{0.9}$$

for all $s \in \{0, 1/m, 2/m, \ldots, 1\}$. By compactness, there is a subsequence

$$\left(\varphi_{t_{m_k}}(\theta_{m_k}), d_{\beta_{m_k}}\varphi_{\tau_1(m_k)+t_{m_k}}(V(\beta_{m_k}))\right)$$

which converges to (θ, S), where S is a Lagrangian subspace. From equation (0.9) and the continuity of α, we deduce

$$\alpha(d_\theta\varphi_s(S), V(\varphi_s(\theta))) = 0,$$

for all $s \in [0,1]$. Therefore

$$d_\theta\varphi_s(S) \cap V(\varphi_s(\theta)) \neq \{0\},$$

for all $s \in [0,1]$. Since S is Lagrangian, this again contradicts the above property and completes the proof of the existence of the functions τ_1 and τ_2.

The measurability follows easily from observing that they can be taken locally constant on each of the subsets $\mathcal{P}(p, l_1, ..., l_p)$, which we defined while proving Lemma 2.2. This finishes the proof of Inequalities (A) and (A').

Now let us prove Inequality (B). We shall use a slightly strengthened version of Yomdin's theorem combined with the method employed by G. Paternain [6] in his proof that for every $x \in M$,

$$\limsup_{T\to+\infty} \frac{1}{T} \log \int_M n_T(x,y)\,dy \leq h_{top}(\varphi).$$

Let N be a closed manifold and $f : N \to N$ a C^r diffeomorphism. We shall consider N embedded in a Euclidean space $\mathbf{R}^m$ and f extended to a C^r map $f : U \to U$, where U is an open neighborhood of N. Given a C^r map $g : U \to \mathbf{R}^m$, we define

$$||d_r g|| = \sup\{||d_x^k g|| : \quad x \in U, \quad 1 \leq k \leq r\}.$$

Fix an integer $l \geq 1$. If $Y \subset N$, we can define the C^r-size of Y as an l-dimensional set to be the infimum of $s > 0$ such that there exists a C^r map $h : [0,1]^l \to \mathbf{R}^m$ satisfying

$$h([0,1]^l) \supset Y \qquad \text{and} \qquad ||d_r h|| \leq s.$$

If no such s exists, the C^r size of Y is ∞. If $Y \subset N$ is a submanifold, the C^r size of Y will mean the C^r size of Y as a $\dim Y$-dimensional set. These definitions are taken from Gromov [4]. We can assume that $dist(f(U), \partial U) \geq 1/\sqrt{l}$ (as required by Gromov in [4]) by rescaling the embedding and thus increasing $dist(f(U), \partial U)$.

The next result is the slight improvement of Yomdin's theorem that we shall need. For simplicity, we shall state it and prove it for the case of diffeomorphisms; however we shall use it for flows.

For the function f considered above, we define

$$K = K(f) = 1 + \limsup_{k\to+\infty} \frac{1}{k} \log ||df^k||. \tag{0.10}$$

Theorem 2.5. *Let $f : N \to N$ and $K = K(f)$ be as above. For any $S, \varepsilon > 0$ and any integers $r, l \geq 1$, there exist $C > 0$ and an integer $n_0 \geq 1$ such that, for every l-dimensional submanifold $Y \subset N$ with C^r size $\leq S$, we have*

$$\mathrm{Vol}(f^n(Y)) \leq C \exp\left\{\left(h_{top}(f|_N) + \varepsilon + \frac{lK}{r}\right) n\right\}$$

for all $n \geq n_0$.

Proof. As observed by Gromov [4, 3.2 p.231], every set of C^r-size $\leq S$ can be divided into j^l subsets of C^r-size $\leq S/j$ for all $j = 1, 2, \ldots$. Thus it suffices to prove the theorem in the case where Y has C^r-size ≤ 1. We shall need the following lemma proved by Gromov [4, 3.6 p.233].

Lemma 2.6. *There exists $C = C(l, m, r)$ independent of f, such that if $Y_0 \subset N$ is a C^r l-dimensional submanifold of C^r-size ≤ 1 and Q_β, $\beta = 1, \ldots, i$, are unit cubes contained in space $\mathbf{R}^m$ in which we embedded N, then*

$$\mathrm{Vol}(f^i(Y_0 \cap (\bigcap_{b=1}^{i} f^{-\beta}(Q_\beta)))) \leq (C\|d_r f\|^{l/r} + 1)^i.$$

Given a continuous map $g : X \to X$ of a compact metric space, denote by $n(\delta, i, g)$ the minimal cardinality of a (δ, i, g)-spanning set.

Assume that the points x_s, $s = 1, \ldots, n(1/2, i, f|_N)$, form a $(1/2, i, f|_N)$-spanning set. Then the manifold N can be covered by sets of the form

$$A_s \stackrel{\text{def}}{=} \bigcap_{\beta=1}^{i} f^{-\beta}(Q_\beta), \qquad 1 \leq s \leq n(1/2, i, f|_N),$$

where Q_β is the unit cube centered at $f^\beta(x_s)$. If Y_0 is as in Lemma 2.6, we have

$$\begin{aligned} \mathrm{Vol}(f^i(Y_0)) &\leq \sum_s \mathrm{Vol}(f^i(Y_0 \cap A_s)) \\ &\leq n(1/2, i, f|_N)(C\|d_r f\|^{l/r} + 1)^i. \end{aligned} \tag{0.11}$$

Now define $f_j : jU \to jU$, for $j \geq 1$, by $f_j(x) = jf(j^{-1}x)$. Then $f_j(jN) = jN$. Let Y_0 be a submanifold of C^r-size ≤ 1. Observe that jY_0 can be covered by j^l-sets, Y_0^k, $k = 1, \ldots, j^l$, with C^r-size ≤ 1 [4]. Therefore we can apply equation (0.11) to each Y_0^k, obtaining

$$\begin{aligned} \mathrm{Vol}(f_j^i(jY_0)) &\leq \sum_k \mathrm{Vol}(f_j^i(Y_0^k)) \\ &\leq j^l n(1/2, i, f_j|_{jN})(C\|d_r f_j\|^{l/r} + 1)^i. \end{aligned} \tag{0.12}$$

But

$$n(1/2, i, f_j|_{jN}) = n(\frac{1}{2j}, i, f|_N).$$

By the definition of topological entropy, given $\delta > 0$, there exist $i_0 \geq 1$ and $j_0 \geq 1$ such that

$$n(\frac{1}{2j}, i, f|_N) \leq \exp\{(h_{top}(f|_N) + \delta)i\},$$

for all $i \geq i_0$, $j \geq j_0$. Then for $i \geq i_0$, $j \geq j_0$, inequality (0.12) implies

$$\begin{aligned} \mathrm{Vol}(f^i(Y_0)) &= \mathrm{Vol}(j^{-1} f_j^i(jY_0)) = j^{-l}\mathrm{Vol}(f_j^i(jY_0)) \\ &\leq n(\frac{1}{2j}, i, f|_N)(C||d_r f_j||^{l/r} + 1)^i \\ &\leq (\exp\{(h_{top}(f|_N) + \delta)i\}) \cdot (C||d_r f_j||^{l/r} + 1)^i. \end{aligned} \tag{0.13}$$

The definition of K in equation (0.10) allows us to choose k so that

$$\frac{1}{k}\log(C(2||df^k||)^{l/r} + 1) \leq \frac{lK}{r}.$$

Observe that $||d^s f_j|| = j^{1-s}||d^s f||$ for $s = 1, 2, \ldots$. We see from this that we can choose $j \geq j_0$ such that

$$||d_r f_j^k|| \leq 2||df^k||.$$

If $i \geq i_0$, then inequality (0.13) applied to f^k with $\delta = k\varepsilon$ gives

$$\begin{aligned} \mathrm{Vol}(f^{ki}(Y_0)) &\leq (\exp\{(h_{top}(f^k|_N) + k\varepsilon)i\}) \cdot (C||d_r f_j^k||^{l/r} + 1)^i \\ &\leq \exp\left\{\left(kh_{top}(f|_N) + k\varepsilon + \log(C(2||df^k||)^{l/r} + 1)\right) i\right\} \\ &\leq \exp\left\{\left(h_{top}(f|_N) + \varepsilon + \frac{lK}{r}\right) ki\right\}. \end{aligned}$$

If $n \geq ki_0$, we can choose $i \geq i_0$ such that $0 \leq n - ki < k$. For $n \geq ki_0$, we have

$$\begin{aligned} \mathrm{Vol}((f^n(Y_0)) &= \mathrm{Vol}(f^{n-ki}(f^{ki}(Y_0))) \\ &\leq ||df||^{l(n-ki)} \exp\left\{\left(h_{top}(f|_N) + \varepsilon + \frac{lK}{r}\right) ki\right\} \\ &\leq C\exp\left\{\left(h_{top}(f|_N) + \varepsilon + \frac{lK}{r}\right) n\right\}, \end{aligned}$$

where $C = ||df||^{lk}$. This completes the proof of the theorem.

Now let us complete the proof of Inequality (B). Observe that

$$\int_{S_xM} A((x,v),t)\,dv \le \mathrm{Vol}(\varphi_t(S_xM)).$$

By the theorem above, given $\varepsilon > 0$ and an integer $r \ge 1$, there exist $t_0 \ge 0$ and $C_r > 0$ such that

$$\mathrm{Vol}(\varphi_t(S_xM)) \le C_r \exp\left\{\left(h_{top}(\varphi) + \varepsilon + \frac{(n-1)K}{r}\right)t\right\},$$

for all $t \ge t_0$ and all $x \in M$, since it is easy to see (using the compactness of M) that there exists $S > 0$ such that the $(n-1)$-dimensional manifold S_xM has C^r-size $\le S$ for all $x \in M$. Hence there is a constant C_r' such that, for all large enough T, we have

$$\begin{aligned}\int_{M\times M} n_T(x,y)\,dxdy &= \int_0^T \int_M \int_{S_xM} A(x,v,t)\,dvdxdt \\ &\le C_r' \exp\left\{\left(h_{top}(\varphi) + \varepsilon + \frac{(n-1)K}{r}\right)T\right\}.\end{aligned}$$

Therefore

$$\limsup_{T\to+\infty} \frac{1}{T}\log \int_{M\times M} n_T(x,y)\,dxdy \le h_{top}(\varphi) + \varepsilon + \frac{(n-1)K}{r},$$

concluding the proof of Inequality (B), since the above inequality holds for all $\varepsilon > 0$ and all $r \ge 1$.

3. Proof of Theorem 1.4

We shall use here the following lemma.

Lemma 3.1. *There exist a constant $K > 0$, an integer $m \ge 1$ and a measurable function $\tau : SM \times \mathbf{R}^+ \to \{0, 1/m, 2/m, \dots, 1\}$ such that, after setting*

$$G_T(\theta) = \varphi_{-\tau(\theta,T)}\theta,$$

we have

$$|\det(d_{G_T(\theta)}\varphi_T|_{V(G_T(\theta))})| \ge K\, ex(d_\theta\varphi_T)$$

for any $T > 0$.

We shall not prove this lemma because it is obtained by the same methods as the lemmas of Section 2.

Now, given $T > 0$, we prove as in Section 2, that there exists $K_1 > 0$ (independent of T) such that

$$\int_{SM} (f \circ G_T)\, d\theta \leq K_1 \int_{SM} f\, d\theta,$$

for every integrable function $f : SM \to (0, +\infty)$. Hence

$$\begin{aligned}\int_{SM} ex(d_\theta \varphi_T)\, d\theta &\leq K^{-1} \int_{SM} |\det(d_{G_T(\theta)} \varphi_T|_{V(G_T(\theta))})|\, d\theta \\ &\leq K_1 K^{-1} \int_{SM} |\det(d_\theta \varphi_T|_{V(\theta)})|\, d\theta,\end{aligned}$$

and in consequence of Theorem 1.1,

$$\begin{aligned}h_{top}(\varphi) &= \lim_{T\to+\infty} \frac{1}{T} \log \int_{SM} ex(d_\theta \varphi_T)\, d\theta \\ &\leq \liminf_{T\to+\infty} \frac{1}{T} \log \int_{SM} |\det(d_\theta \varphi_T|_{V(\theta)})|\, d\theta.\end{aligned}$$

Since $ex(d_\theta \varphi_T) \geq |\det(d_\theta \varphi_T|_{V(\theta)})|$, by definition of $ex(.)$, we also have:

$$\begin{aligned}\limsup_{T\to+\infty} \frac{1}{T} \log \int_{SM} |\det(d_\theta \varphi_T|_{V(\theta)})|\, d\theta &\leq \lim_{T\to+\infty} \frac{1}{T} \log \int_{SM} ex(d_\theta \varphi_T)\, d\theta \\ &= h_{top}(\varphi).\end{aligned}$$

4. Appendix: An upper bound for the growth rate of $n_T(x,y)$.

Here we shall prove the following property:

Proposition 4.1. *For every $x \in M$,*

$$\limsup_{T\to+\infty} \frac{1}{T} \log n_T(x,y) \leq h_{top}(\varphi),$$

for a.e. $x \in M$.

This is an immediate corollary of the following inequality due to G. Paternain [6]:

Proposition 4.2. *For every $x \in M$, we have*

$$\limsup_{T\to+\infty} \frac{1}{T} \log \int_M n_T(x,y)\, dy \leq h_{top}(\varphi),$$

and the following application of the Borel-Cantelli Lemma:

Lemma 4.3. *Let $(X, \mathcal{A}, \mu)$ be a probability space, and $f_n : X \to (0, +\infty)$ a sequence of integrable functions. Then*

$$\limsup_{n\to\infty} \frac{1}{n} \log f_n(x) \leq \limsup_{n\to\infty} \frac{1}{n} \log \int_X f_n\, d\mu,$$

for μ-a.e. $x \in X$.

Proof. Set

$$\sigma = \limsup_{n\to\infty} \frac{1}{n} \log \int_X f_n\, d\mu.$$

Define $S(n,\epsilon) = \{x : f_n(x) \geq \exp(\sigma+\epsilon)n\}$. Then

$$\begin{aligned}
\sigma = \limsup_{n\to\infty} \frac{1}{n} \log \int_X f_n\, d\mu &\geq \limsup_{n\to\infty} \frac{1}{n} \log \int_{S(n,\epsilon)} f_n\, d\mu \\
&\geq \limsup_{n\to\infty} \frac{1}{n} \log[\mu(S(n,\epsilon)) \exp(\sigma+\epsilon)n] \\
&= \sigma + \epsilon + \limsup_{n\to\infty} \frac{1}{n} \log \mu(S(n,\epsilon)).
\end{aligned}$$

Hence

$$\limsup_{n\to\infty} \frac{1}{n} \log \mu(S(n,\epsilon)) \leq -\epsilon,$$

which implies

$$\sum_n \mu(S(n,\epsilon)) < +\infty.$$

By the Borel-Cantelli Lemma, for a.e. x, there exists $m(x)$ such that $x \notin S(n,\epsilon)$ for all $n \geq m(x)$. This means that $f_n(x) \leq \exp(\sigma+\epsilon)n$ for all $n \geq m(x)$ and then

$$\limsup_{n\to\infty} \frac{1}{n} \log f_n(x) \leq \sigma + \epsilon.$$

Since this holds for every $\epsilon > 0$, the lemma is proved.

References

[1] W. Ballmann, M. Brin & K.Burns, *On surfaces without conjugate points*, J. Differential Geom. **25** (1987) 249-273.

[2] M. Berger & R. Bott, *Sur les variétés à courbure strictement positive*, Topology **1** (1962) 302-311.

[3] A. Freire & R. Mañé, *On the entropy of the geodesic flow in manifolds without conjugate points*, Invent. Math. **69** (1982) 375-392.

[4] M. Gromov, *Entropy, homology and semialgebraic Geometry*, Séminaire Bourbaki, 38éme année, 1985-86, Exp. 663, 225-240.

[5] A. Manning, *Topological entropy for geodesic flows*, Ann. of Math. **110** (1979) 567-573.

[6] G. P. Paternain, *On the topology of manifolds with completely integrable geodesic flows*, Ergodic Theory Dynamical Systems **12** (1992) 109-121.

[7] _____, M. Paternain, *Topological entropy versus geodesic entropy*, Internat. J. Math. **2** (1994) 213-218.

[8] F. Przytycki, *An upper estimation for topological entropy of diffeomorphisms*, Invent. Math. **59** (1980) 205-213.

[9] Y. Yomdin, *Volume growth and entropy*, Israel J. Math. **57** (1987) 287-300.

INST. MAT. PURA APL.
RIO DE JANEIRO, BRAZIL

Nova Série
BOLETIM
DA SOCIEDADE BRASILEIRA DE MATEMÁTICA

Bol. Soc. Bras. Mat., Vol. 28, N. 2, 155-196

Lagrangian flows: The dynamics of globally minimizing orbits - II

Gonzalo Contreras[1], Jorge Delgado and
Renato Iturriaga[2]

— *To the memory of Ricardo Mañé.*

Abstract. Define the critical level $c(L)$ of a convex superlinear Lagragian L as the infimum of the $k \in \mathbb{R}$ such that the Lagragian $L+k$ has minimizers with fixed endpoints and free time interval. We provide proofs for Mañé's statements [7] characterizing $c(L)$ in termos of minimizing measures of L, and also giving graph, recurrence covering and cohomology properties for minimizers of $L + c(L)$. It is also proven that $c(L)$ is the infimum of the energy levels k such that the following for of Tonelli's theorem holds: *There exists minimizers of the $L + k$-action joining any two points in the projection of $E = k$ among curves with energy k.*

Introduction

In this work we prove most of the theorems of Mañé's unfinished work "Lagrangian Flows the dynamics of Globally Minimizing Orbits", [**7**]. Exceptions are theorem III, whose proof is divided in [**6**] and [**3**] and theorem IV which was proved in [**7**]. Also, we provide proofs for slightly different statements of theorems VII, XI and XIV. We would like to emphasize that all the theorems in this paper are due to Mañé and all the responsibility of the proofs is ours.

We encourage the reader to use Mañé's original paper [**7**] as the introduction of this work. In section 1 we prove theorems I and II, in section 2 we prove theorem V, in section 3 we prove theorems VI, VII,

[1]partially supported by CNPq-Brazil.
[2]partially supported by Conacyt-Mexico, grant 3398E9307.
Received 24 October 1996.

Originally published in: *Bol. Soc. Brasil. Mat. (N.S.) 28 (1997), no. 2, 155–196,*

VIII and IX, in section 4 we prove theorems X and XI, and in section 5 we prove theorems XII, XIII and XIV.

The first and second authors want to thank the hospitality of CIMAT.

We want to use this space to say how much we admired Ricardo's clearness and brightness and how grateful we are to his enormous generosity. This paper is to his memory.

1. Basic properties of the critical value

Let M be a smooth closed manifold. We say that a smooth function $L : TM \to \mathbb{R}$ is a *Lagrangian* if it satisfies the following conditions:

(a) *Convexity:* For all $x \in M$, $v \in T_xM$, the Hessian matrix $\frac{\partial^2 L}{\partial v_i \partial v_j}(x, v)$ (calculated with respect to linear coordinates on T_xM) is positive definite.

(b) *Superlinearity:* $\lim\limits_{\|v\| \to +\infty} \frac{L(x,v)}{\|v\|} = +\infty$, uniformly on $(x, v) \in TM$.

Given an absolutely continuous curve $x : [0, T] \to M$ define its L-action by

$$S_L(x) := \int_a^b L\left(x(t), \dot{x}(t)\right)\, dt\,.$$

Fixing $p, q \in M$ and $T > 0$, the critical points of the action functional on the set

$$AC(p, q, T) := \left\{ x : [0, T] \to M \;\middle|\; \begin{array}{l} x(0) = p,\, x(T) = q, \\ x \text{ absolutely continuous} \end{array} \right\}.$$

are solutions of the Euler-Lagrange equation, which in local coordinates is given by

$$\frac{d}{dt} L_v = L_x\,. \tag{E-L}$$

Because of the convexity of the Lagrangian this equation can be thought as a first order differential equation on TM. The *Lagrangian flow* f_t on TM is defined by $f_t(x, v) = (\gamma(t), \dot{\gamma}(t))$, where γ is the solution of (E-L) with $\gamma(0) = x$ and $\dot{\gamma}(0) = v$. Define the *energy* function $E : TM \to \mathbb{R}$ as

$$E(x, v) := \frac{\partial L}{\partial v}(x, v) \cdot v - L\,.$$

It can be seen that the value of $E(x,v)$ is constant along the orbits of f_t. The superlinearity condition implies that the level sets of the energy function have bounded velocities and hence that are compact. This, in turn, implies that the solutions of (E-L) are defined for all values $t \in \mathbb{R}$, i.e., that the flow f_t is *complete.*

Finally, for $p, q \in M$, let

$$AC(p,q) := \bigcup_{T>0} AC(p,q,T),$$

and define the *action potential* as

$$\Phi_k(p_1,p_2) := \inf \{ S_{L+k}(x) \mid x \in AC(p_1,p_2) \}, \; k \in \mathbb{R}.$$

Theorem I. *There exists* $c(L) \in \mathbb{R}$ *such that*
(a) $k < c(L) \Rightarrow \Phi_k(p_1,p_2) = -\infty, \quad \forall p_1, p_2 \in M.$
(b) $k \geq c(L) \Rightarrow \Phi_k(p_1,p_2) > -\infty, \quad \forall p_1, p_2 \in M$ *and* Φ_k *is Lipschitz.*
(c) $k \geq c(L) \Rightarrow$

$$\Phi_k(p_1,p_3) \leq \Phi_k(p_1,p_2) + \Phi_k(p_2,p_3) \quad \forall p_1, p_2, p_3 \in M$$
$$\Phi_k(p_1,p_2) + \Phi_k(p_2,p_1) \geq 0 \quad \forall p_1, p_2 \in M$$

(d) $k > c(L) \Rightarrow \Phi_k(p_1,p_2) + \Phi_k(p_2,p_1) > 0 \quad \forall p_1 \neq p_2.$

1.1. Remark. This theorem, with the same proof, holds for coverings $\pi : \widehat{M} \to M$ of a compact manifold M, with the lifted Lagrangian $\widehat{L} = L \circ \pi$.

Proof. We first prove that if for some $p_1, p_2 \in M$, $\Phi_k(p_1,p_2) = -\infty$, then $\Phi_k(q_1,q_2) = -\infty$ for all $q_1, q_2 \in M$. Let γ, η: $[0,1] \to M$, $\gamma \in AC(q_1,p_1)$, $\eta \in AC(p_2,q_2)$. Let $x_n \in AC(p_1,p_2)$ be such that $\lim_{n\to\infty} S_{L+k}(x_n) = -\infty$. Then

$$\lim_{n\to\infty} S_{L+k}(\gamma * x_n * \eta) = S_{L+k}(\gamma) + S_{L+k}(\eta) + \lim_{n\to\infty} S_{L+k}(x_n) = -\infty.$$

Thus the number

$$c(L) := \inf \{ k \in \mathbb{R} \mid \Phi_k(p,q) > -\infty \}$$

does not depend on (p,q). We have to see that $-\infty < c(L) < +\infty$. Since the function $k \mapsto \Phi_k(p,q)$ is nondecreasing, it is enough to see that there exist $k_1, k_2 \in \mathbb{R}$ such that $\Phi_{k_1}(p,q) = -\infty$ and $\Phi_{k_2}(p,q) > -\infty$. We first prove the existence of k_1. Since L is bounded on

$$\{(x,v) \in TM \mid v \in T_xM\,,\ |v| \leq 2\}\,,$$

there exists $B > 0$ such that

$$|L(x,v)| < B \quad \text{if} \quad |v| < 2\,. \tag{1}$$

Let $x_n : [0,n] \to M$, $x_n \in AC(p,q)$ be such that $|\dot{x}_n| \leq 2$. Then $S_L(x_n) \leq B\,n$ and hence, for $k_1 = -B-1$, we have that

$$\begin{aligned}\Phi_{k_1}(p,q) &\leq \liminf_n S_{L-B-1}(x_n) \leq \liminf_n \left(\int_0^n L(x_n,\dot{x}_n)\,dt - (B+1)n\right)\\ &\leq \liminf_n \big(Bn - (B+1)n\big) = -\infty.\end{aligned}$$

Now we prove the existence of k_2. The superlinearity hypothesis implies that L is bounded below. Let A be a lower bound for L on TM. We claim that it is enough to take $k_2 > -A+1$. Indeed

$$S_{L+k_2}(x) \geq \int_0^T (A+k_2)\,dt \geq 0 \qquad \text{for all } x \in AC(p,q).$$

hence $\Phi_{k_2}(p,q) \geq 0$.

It remains to prove that $\Phi_c(p,q) > -\infty$ for all p, $q \in M$, where $c = c(L)$. Suppose not. Take $p \in M$, then $\Phi_c(p,p) = -\infty$. Let $\gamma \in AC(p,p)$ be such that $S_{L+c}(\gamma) < -a < 0$. Then there exists $\varepsilon > 0$ such that $S_{L+c+\varepsilon}(\gamma) < -\frac{1}{2}a < 0$. Let

$$\delta_N := \gamma * \overset{N}{\ldots} * \gamma,$$

then

$$\Phi_{c+\varepsilon}(p,p) \leq \lim_N S_{L+c+\varepsilon}(\delta_N) \leq \lim_N -\tfrac{1}{2}\,a\,N = -\infty\,.$$

This contradicts the definition of $c(L)$. In particular, we have also proven that $\Phi_c(p,p) \geq 0$ for all $p \in M$. By taking $\gamma \in AC(p,p)$ with bounded velocities and arbitrarily small parameter intervals, we have that

$$\Phi_c(p,p) = 0 \qquad \text{for all } p \in M\,. \tag{2}$$

Similarly

$$\Phi_k(p,p) = 0 \qquad \text{for all } p \in M \text{ for all } k \geq c(L). \tag{3}$$

We now prove (c). Let $k \geq c(L)$ and $p_1, p_2, p_3 \in M$. Let $x_n^{ij} : [0, T_n] \to M$, $x_n^{ij} \in AC(p_i, p_j)$, be such that

$$\lim_n S_{L+k}(x_n^{ij}) = \Phi_k(p_i, p_j) \quad i, j \in \{1, 2, 3\}.$$

Then $x_n^{12} * x_n^{23} \in AC(p_1, p_3)$ and

$$\Phi_k(p_1, p_3) \leq S_{L+k}(x_n^{12} * x_n^{23}) = S_{L+k}(x_n^{12}) + S_{L+k}(x_n^{23}).$$

Taking the limit when $n \to \infty$ we get that

$$\Phi_k(p_1, p_3) \leq \Phi_k(p_1, p_2) + \Phi_k(p_2, p_3). \tag{4}$$

Finally, using (3) and (4), we obtain that

$$0 = \Phi_k(p_1, p_1) \leq \Phi_k(p_1, p_2) + \Phi_k(p_2, p_1) \tag{5}$$

when $k \geq c(L)$.

We prove that Φ_k is Lipschitz when $k \geq c(L)$. Let $\gamma : [0, d(p, q)] \to M$ be a geodesic joining p and q, using (1) we obtain

$$\Phi_k(p, q) \leq \int_0^{d(p,q)} [L(\gamma(t), \dot\gamma(t)) + k] \, dt$$

$$\Phi_k(p, q) \leq (B + k) \; d(p, q) \qquad \text{for} \qquad k \geq c(L). \tag{6}$$

Therefore if $k \geq c(L)$,

$$\begin{aligned}
\Phi_k(p_1, p_2) - \Phi_k(q_1, q_2) &\leq \Phi_k(p_1, q_1) + \Phi_k(q_1, p_2) - \Phi_k(q_1, q_2) \\
&\leq \Phi_k(p_1, q_1) + \Phi_k(q_1, q_2) + \Phi_k(q_2, p_2) - \Phi_k(q_1, q_2) \\
&\leq \Phi_k(p_1, q_1) + \Phi_k(q_2, p_2) \\
&\leq (B + k)\big(d(p_1, q_1) + d(p_2, q_2)\big).
\end{aligned}$$

Changing the roles of p_i and q_i, $i = 1, 2$ we get that

$$|\Phi_k(p_1, p_2) - \Phi_k(q_1, q_2)| \leq (B + k) \left(d(p_1, q_1) + d(p_2, q_2)\right).$$

We now prove that if $k \geq c(L)$ and $p \neq q$, then the function $k \mapsto \Phi_k(p, q)$ is strictly increasing. By (5) this implies (d). Let $p \neq q$ and $\ell > k \geq c(L)$. Let $x_n : [0, T_n] \to M$, $x_n \in AC(p, q)$ be such that $\lim_n S_{L+\ell}(x_n) = \Phi_\ell(p, q)$. We have that

$$S_{L+\ell}(x_n) = S_{L+k}(x_n) + (\ell - k)T_n$$

$$\Phi_\ell(p, q) \geq \Phi_k(p, q) + (\ell - k) \liminf_n T_n$$

It is enough to prove that $\liminf_n T_n > 0$, because then $\Phi_\ell(p,q) > \Phi_k(p,q)$. Suppose that $\liminf_n T_n = 0$. By the superlinearity of L, for all $B > 0$ there exists $A > 0$ such that

$$|L(x,v)| \geq B\,|v| - A\,.$$

Then

$$\begin{aligned}\Phi_\ell(p,q) = \lim_n S_{L+\ell}(x_n) &\geq \liminf_n \left[\int_0^{T_n} B\,|\dot{x}_n|\,dt + (k-A)\,T_n \right] \\ &\geq B\,d(p,q) + 0\,,\end{aligned}$$

for all $B > 0$. Therefore $\Phi_{L+\ell}(p,q) = +\infty$, which contradicts (6). □

Through the rest of the paper we shall neeed the following results:

1.2. Theorem. (Mather [8].) *For all $C > 0$ there exists $A_1 = A_1(C)$ such that if $T > 0$, p, $q \in M$ and $x \in AC(p,q,T)$ satisfy*
(a) $S_L(x) = \min\{\, S_L(y) \mid y \in AC(p,q,T) \,\}$.
(b) $S_L(x) \leq C\,T$.
Then
(c) $\|\dot{x}(t)\| < A_1$ *for all* $t \in [0,T]$.
(d) $x|_{[0,T]}$ *is a solution of (E-L).*

1.3. Corollary. *There exists $A > 0$ such that if $T > 1$, $p,q \in M$ and $x \in AC(p,q,T)$ satisfy*

$$S_L(x) = \min\{\, S_L(y) \mid y \in AC(p,q,T) \,\}\,,$$

then $\|\dot{x}(t)\| < A$ for all $t \in [0,T]$.

Proof. Let

$$C := \sup\{\, |L(x,v)| \mid \|v\| < \mathrm{diam}\,(M) \,\}\,.$$

There exists a geodesic $\gamma \in AC(p,q,T)$ with

$$|\dot{\gamma}| \equiv \frac{d(p,q)}{T} < \mathrm{diam}(M) \quad \text{and} \quad S_L(\gamma) < C\,T\,.$$

Then the corollary follows from theorem 1.2. □

1.4. Corollary. *There exists $A > 0$ such that if $p,q \in M$ and $x \in AC(p,q,T)$ satisfy*
(a) $S_L(x) = \min\{\, S_L(y) \mid y \in AC(p,q,T) \,\}$.

(b) $S_L(x) < \Phi_c(p,q) + d_M(p,q)$.

Then

(c) $T > \frac{1}{A}\, d_M(p,q)$.

(d) $\|\dot{x}(t)\| < A$ *for all* $t \in [0,T]$.

Proof. Let B be from (1) and let $\gamma : [0, d_M(p,q)] \to M$ be a minimal geodesic with $|\dot\gamma| \equiv 1$ joining p and q. By the superlinearity of L, for $D = 2\,B + 1 > 0$ there exists $E > 0$ such that

$$|L(x,v)| > D\;|v| - E \quad \text{for all } (x,v) \in TM.$$

We have that

$$\begin{aligned} B\, d_M(p,q) > S_L(\gamma) \geq \Phi_c(p,q) &\geq S_L(x) - d_M(p,q) \\ &> \int_0^T (D\;|\dot{x}(t)| - E)\, dt - d_M(p,q) \\ &> D\, d_M(p,q) - E\,T - d_M(p,q)\,. \end{aligned}$$

Hence

$$T > \frac{(D-B-1)}{E}\, d_M(p,q) = \left(\frac{B}{E}\right)\, d_M(p,q)\,. \tag{7}$$

Now let

$$C := \max\{\, |L(x,v)| \mid |v| \leq \tfrac{E}{B}\,\}\,.$$

Let $\eta \in AC(p,q,T)$ be a minimal geodesic. Then, by (7),

$$|\dot\eta| \equiv \frac{d_M(p,q)}{T} < \frac{E}{B}$$

and hence $S_L(\eta) \leq C\,T$. Let $A_1 = A_1(C)$ be from theorem 1.2, then it is enough to use

$$A = \max\left\{\, A_1(C),\ \tfrac{E}{B}\,\right\}. \qquad \square$$

Let $\mathcal{M}(L)$ be the set of invariant Borel probability measures for the Lagrangian flow.

Theorem II.

$$c(L) = -\min\left\{ \int L\, d\mu \;\middle|\; \mu \in \mathcal{M}(L) \right\}.$$

Proof. Let $\mu \in \mathcal{M}(L)$ be ergodic. Let $(p,v) \in TM$ be such that

$$\lim_{T\to\infty} \frac{1}{T} \int_0^T L(f_t(p,v))\, dt = \int L\, d\mu\,.$$

Let $B > 0$ be such that

$$|L(x, v)| < B \quad \text{if } |v| \leq 2\,.$$

For $N > 0$ let $q_N := \pi(f_N(p, v))$ and let $\gamma_N : [0, d(p, q_N)] \to M$ be a geodesic joining q_N to p. Let $x_N : [0, N] \to M$ be defined by $x_N(t) = \pi(f_t(p, v))$. Then $f_t(p, v) = (x_N(t), \dot{x}_N(t))$. For $k \in \mathbb{R}$, we have that

$$S_{L+k}(\gamma_N) = \int_0^{d(p,q_N)} L(\gamma_N(t), \dot{\gamma}_N(t))\, dt \leq (B + k)\,\mathrm{diam}(M)\,.$$

$$\lim_N \frac{1}{N} S_{L+k}(x_N * \gamma_N) = \lim_N \frac{1}{N} S_{L+k}(x_N) + 0 = S_{L+k}(\mu) = S_L(\mu) + k\,.$$

If $k < -S_L(\mu)$, then

$$\Phi_k(p, p) \leq \lim_N S_{L+k}(x_N * \gamma_N) = -\infty\,.$$

Hence $k \leq c(L)$. Therefore

$$\begin{aligned} c(L) &\geq \sup\left\{-S_L(\mu) \mid \mu \in \mathcal{M}_{\mathrm{erg}}(L)\right\} \\ &\geq -\min\left\{S_L(\mu) \mid \mu \in \mathcal{M}(L)\right\}. \end{aligned}$$

Now let $k < c(L)$ and $p, q \in M$. Then $\Phi_k(p, q) = -\infty$ and there exists a sequence of paths $x_n : [0, T_n] \to M$, $x_n \in AC(p, q)$ such that

$$\lim_n S_{L+k}(x_n) = -\infty\,. \tag{8}$$

Since L is bounded below, we have that

$$\lim_n T_n = +\infty\,. \tag{9}$$

Let $y_n : [0, T_n] \to M$, $y_n \in AC(p, q)$ be a minimizer of the action among the curves in $AC(p, q, T_n)$. Then $(y_n(t), \dot{y}_n(t))$ is an orbit segment of f_t and by 1.3, $|\dot{y}|$ is bounded. Let ν_n be the probability measure defined by

$$\begin{aligned} \int h\, d\nu_n &= \frac{1}{T_n} \int_0^{T_n} h(y_n(t), \dot{y}_n(t))\, dt \\ &= \frac{1}{T_n} \int_0^{T_n} h\big(f_t(y_n(0), \dot{y}_n(0))\big)\, dt\,, \end{aligned}$$

for all $h : TM \to \mathbb{R}$ continuous. There exists a subsequence ν_{n_i} which converges weakly* to a probability measure μ. By (9), μ is invariant under the flow f_t. Since the velocities are bounded we have that

$$\lim_{n_i} \frac{1}{T_{n_i}} S_{L+k}(y_{n_i}) = S_{L+k}(\mu) = S_L(\mu) + k\,.$$

Since $\lim_n S_{L+k}(y_n) = \Phi_k(p,q) = -\infty$ and $T_n > 0$ for all n, then $S_L(\mu) + k \leq 0$. For any $k < c(L)$ we found an invariant measure μ such that $k \leq -S_L(\mu)$. Therefore

$$\begin{aligned} c(L) &\leq \sup\{-S_L(\mu) \mid \mu \in \mathcal{M}(L)\} \\ &\leq -\min\{S_L(\mu) \mid \mu \in \mathcal{M}(L)\}\,. \qquad \square \end{aligned}$$

We state now theorem III. The proof of theorem III is split in Mañé [**6**] and [**3**]. We say that a property holds for a *generic Lagrangian* if given any Lagrangian L, there exists a residual set $\mathcal{O} \subseteq C^\infty(M,\mathbb{R})$ such that the property holds for all the Lagrangians $L+\psi$ with $\psi \in \mathcal{O}$. We say that an invariant measure is *uniquely ergodic* if it is the only invariant measure on its support. A periodic orbit for the Lagrangian flow is said *hyperbolic* if it is a hyperbolic periodic orbit of the flow restricted to its energy level.

Definition. We say that $\mu \in \mathcal{M}(L)$ is a *minimizing measure* if

$$\int L\, d\mu = -c(L).$$

Denote by $\widehat{\mathcal{M}}(L)$ the set of minimizing measures in $\mathcal{M}(L)$.

Theorem III.

(a) *For generic L, $\widehat{\mathcal{M}}(L)$ contains a single measure and this measure is uniquely ergodic.*

(b) *When this measure is supported on a periodic orbit, this orbit is hyperbolic.*

Item (a) is proved in Mañé [**6**] and item (b) is proved in [**3**].

Conjecture. **(Mañé.)** *For a generic L, $\widehat{\mathcal{M}}(L)$ consists on a single measure supported in a periodic orbit.*

2. Recurrence properties

The prerequisite of the following definition is this remark: since $d_k \geq 0$

for $k \geq c(L)$, then for any absolutely continuous curve $x : [a,b] \to M$ and $k \geq c(L)$ we have that

$$S_{L+k}(x) \geq \Phi_k(x(a), x(b)) \geq -\Phi_k(x(b), x(a)) \,. \tag{10}$$

Definition. Set $c = c(L)$. We say that $x : [a,b] \to M$ is a *semistatic curve* if it is absolutely continuous and:

$$S_{L+c}\left(x|_{[t_0,t_1]}\right) = \Phi_c\left(x(t_0), x(t_1)\right) \,, \tag{11}$$

for all $a < t_0 \leq t_1 < b$; and that it is a *static* curve if

$$S_{L+c}\left(x|_{[t_0,t_1]}\right) = -\Phi_c(x(t_1), x(t_0)) \tag{12}$$

for all $a < t_0 \leq t_1 < b$.

By (10), equality (12) implies (11). Hence static curves are semistatic. Semistatic curves are solutions of (E-L) because they minimize the action of $L+c$ in $AC(x(t_0), x(t_1), t_1 - t_0)$. If p, $q \in M$ are on a *static* curve then $d_c(p,q) = 0$.

If $w \in TM$, denote by $x_w : \mathbb{R} \to M$ the solution of (E-L) with $\dot{x}_w(0) = w$.

Definition.

$$\Sigma(L) := \{\, w \in TM \mid x_w : \mathbb{R} \to M \text{ is semistatic } \},$$
$$\widehat{\Sigma}(L) := \{\, w \in TM \mid x_w : \mathbb{R} \to M \text{ is static } \},$$
$$\Sigma^+(L) := \{\, w \in M \mid x_w|_{[0,\infty[} \text{ is semistatic } \}.$$

Remark. Replacing c by any other real number in the definition of semistatic solution, the set $\Sigma^+(L)$ (and then $\widehat{\Sigma}(L) \subset \Sigma(L) \subset \Sigma^+(L)$) becomes empty.

For $k > c(L)$ this remark follows from the following estimates:

$$\begin{aligned} +\infty > \max\nolimits_{p,q\in M} \Phi_k(p,q) &\geq S_{L+k}(x|_{[0,T]}) = \Phi_k(x(0), x(T)) \\ &\geq \Phi_c(x(0), x(T)) + (k-c)\,T \\ &\geq \max\nolimits_{p,q\in M} \Phi_c(p,q) + (k-c)\,T. \end{aligned}$$

The following theorem is proven in Mañé [**7**].

Theorem IV. (Characterization of minimizing measures.) *A measure $\mu \in \mathcal{M}(L)$ is minimizing if and only if supp$(\mu) \subseteq \widehat{\Sigma}(L)$.*

We include a proof below, using theorem V(c).

Given an f_t invariant subset $\Lambda \subseteq TM$, and $\varepsilon > 0$, $T > 0$, an (ε, T)-*chain joining* $\xi_a \in \Lambda$ *and* $\xi_b \in \Lambda$ is a finite sequence $\{(\zeta_i, t_i)\}_{i=1}^{N} \subset \Lambda \times \mathbb{R}$ such that

$$\zeta_1 = \xi_a, \quad \zeta_{N+1} = \xi_b, \quad t_i > T \quad \text{and} \quad d(f_{t_i}(\zeta_i), \zeta_{i+1}) < \varepsilon,$$

for $i = 1, \ldots, N$. We say that a set $\Lambda \subset TM$ is *chain transitive* if for all ξ_a, $\xi_b \in \Lambda$, and all $\varepsilon > 0$, $T > 0$ there exists an (ε, T)-chain in Λ joining ξ_a and ξ_b. When this condition holds only for all $\xi_a = \xi_b \in \Lambda$ we say that Λ is *chain recurrent*.

Theorem V. (Recurrence Properties.)

a) *$\Sigma(L)$ is chain transitive.*

b) *$\widehat{\Sigma}(L)$ is chain recurrent.*

c) *The α and ω-limit sets of a semistatic orbit are contained in $\widehat{\Sigma}(L)$.*

Proof. Lets first prove (c). Let $w \in \Sigma$ and let $u \in \omega(w)$. We prove that $\omega(w) \subseteq \widehat{\Sigma}(L)$, the proof that $\alpha(w) \subseteq \widehat{\Sigma}(L)$ is similar. It is enough to prove that for all $T > 0$, $x_u|_{[0,T]}$ is static. Let $p := x_u(0)$, $q := x_u(T)$. Let $p_n = x_w(s_n)$, $q_n = x_w(t_n)$ be sequences of points in M with

$$s_n < s_n + T = t_n < s_{n+1}, \quad s_n,\, t_n \underset{n}{\longrightarrow} +\infty$$

and

$$\dot{x}_w(s_n) \underset{n}{\longrightarrow} \dot{x}_u(0), \quad \dot{x}_w(t_n) \underset{n}{\longrightarrow} \dot{x}_u(T).$$

We have that

$$\begin{aligned} S_{L+c}\left(x_w|_{[s_n,t_n]}\right) &= \Phi_c(p_n, q_n)\,, \\ x_w|_{[s_n,t_n]} &\overset{C^1}{\longrightarrow} x_u|_{[0,T]}\,, \\ S_{L+c}\left(x_u|_{[0,T]}\right) &= \lim_n S_{L+c}\left(x_w|_{[s_n,t_n]}\right) \\ &= \lim_n \Phi_c(p_n, q_n) \\ &= \Phi_c(p, q)\,. \end{aligned}$$

Moreover, by the continuity of Φ_c, we have that

$$\begin{aligned}
\Phi_c(p,q)+\Phi_c(q,p) &= \lim_n\{\Phi_c(p_n,q_n)+\Phi_c(q_n,p_{n+1})\}\\
&= \lim_n\left\{S_{L+c}\left(x_w|_{[s_n,t_n]}\right)+S_{L+c}\left(x_w|_{[t_n,s_{n+1}]}\right)\right\}\\
&= \lim_n\left\{S_{L+c}\left(x_w|_{[s_n,s_{n+1}]}\right)\right\}\\
&= \lim_n \Phi_c(p_n,p_{n+1})\\
&= \Phi_c(q,q)=0\,.
\end{aligned}$$

Proof of theorem IV. Suppose that $\mu\in\mathcal{M}(L)$ and $\operatorname{supp}(\mu)\in\hat{\Sigma}(L)$. Let θ be a generic point for μ and $L+c$. Then θ is static and if $x_\theta(t)=\pi\circ\varphi_t(\theta)$, we have

$$\begin{aligned}
\int(L+c)\,d\mu &= \lim_{T\to+\infty}\frac{1}{T}\int_0^T\left[L(x_\theta,\dot{x}_\theta)+c\right]dt\\
&= \lim_{T\to+\infty}\frac{1}{T}\,\Phi_c\big(x_\theta(0),x_\theta(T)\big)\\
&\leq \lim_{T\to+\infty}\frac{1}{T}\sup_{p,q\in M}\Phi_c(p,q)=0.
\end{aligned}$$

where the second equality is because θ is (semi)static. Hence μ is minimizing.

Now suppose that $\mu\in\mathcal{M}(L)$ is minimizing. Applying lemma 2.2 in Mañé [**6**], we get that for μ-almost every θ there is a sequence $T_j=T_j(\theta)\to+\infty$ such that

$$\lim_{j\to+\infty} d_{TM}\big(\theta,\varphi_{T_j}(\theta)\big)=0\,, \tag{13}$$

$$\lim_{j\to+\infty}\int_0^{T_j}\left[L(x_\theta,\dot{x}_\theta)+c\right]dt=0\,. \tag{14}$$

By theorem V(c), it is enough to prove that $\operatorname{supp}(\mu)\subseteq\Sigma(L)$. Since Φ_c is continuous, it is enough to prove that μ-almost every θ is semistatic. Now let θ satisfy (13), (14) and define

$$\delta_\theta(T):=S_{L+c}\left(x_\theta|_{[0,T]}\right)-\Phi_c\big(x_\theta(0),x_\theta(T)\big)\,.$$

Then $\delta_\theta(t)$ is non-derecasing and $\delta_\theta(t)\geq 0$.

By the continuity of Φ_c and (13) we have that

$$\lim_{j\to+\infty}\Phi_c\big(x_\theta(0),x_\theta(T_j)\big)=0\,. \tag{15}$$

From (15) and (14) we have that $\lim_{j\to+\infty}\delta_\theta(T_j)=0$. Since $\delta_\theta(t)$ is non-decerasing, then $\delta_\theta(t)\equiv 0$ for all $t\geq 0$. Hence θ is semistatic. □

Proof of theorem V(a). Given $\varepsilon>0$, $S>0$, u, $v\in\Sigma(L)$, we have to find an (ε,S)-chain in $\Sigma(L)$ joining u to v. It is easy to see that such (ε,S)-chain exists if $\omega(u)\cap\alpha(v)\neq\varnothing$. Let

$$A:=\alpha(v)\quad,\quad\Omega:=\omega(u) \tag{16}$$

and suppose that $A\cap\Omega=\varnothing$. Let $p\in\pi(\Omega)$, $q\in\pi(A)$. Let $\eta_n:=[0,T_n]\to M$ be such that

$$p:=\eta_n(0)\in\pi(\Omega)\quad,\quad q:=\eta_n(T_n)\in\pi(A)$$

$$S_{L+c}(\eta_n)\leq\Phi_c(p,q)+\frac{1}{n}. \tag{17}$$

By corollary 1.4, we can assume that η_n is a solution of (E-L) and satisfies

$$|\dot\eta_n(t)|<A\qquad\text{for all } 0\leq t\leq T_n. \tag{18}$$

Given $\frac{\varepsilon}{2}>\delta_0>0$ there exists $0<\delta<\delta_0$ such that if $|v|$, $|w|<A$ and $d_{TM}(v,w)<\delta$ then

$$d_{TM}(f_t(v),f_t(w))<\delta_0\ \text{for all}\ |t|\leq S. \tag{19}$$

Let $\mathcal{M}$ be the union of A, Ω and the set of accumulation points of the tangent vectors of the η_n's:

$$\mathcal{M}:=A\cup\Omega\cup\left\{v\in TM\;\middle|\;\begin{array}{ll}\exists\,\langle n_k\rangle\subseteq\mathbb{N} & 0\leq t_{n_k}\leq T_{n_k}\\ \exists\,\langle t_{n_k}\rangle\subseteq\mathbb{R} & v=\lim_k\dot\eta_{n_k}(t_{n_k})\end{array}\right\}.$$

Then

$$\mathcal{M}\subseteq\{\,v\in TM\mid \|v\|\leq A\,\}. \tag{20}$$

We shall need the following lemmas

2.1. Lemma.

(a) $\mathcal{M}\subseteq\Sigma(L)$.

(b) $\mathcal{M}$ *is invariant.*

Let $\mathbb{K}$ be the set of vectors which are on the ω-limit of vectors of $\mathcal{M}$.

$$\mathbb{K}:=\bigcup_{v\in\mathcal{M}}[\omega(v)]\,.$$

Since $\mathcal{M}$ is closed and foward invariant, then the closure $\overline{\mathbb{K}} \subseteq \mathcal{M} \subseteq \Sigma(L)$. Moreover, the vectors in $\mathbb{K}$ are chain recurrent. By (20) $\overline{\mathbb{K}}$ is compact. Given $\delta > 0$ let

$$\mathbb{K}_\delta := \{\, v \in TM \mid d_{TM}(v, \overline{\mathbb{K}}) < \delta \,\}.$$

Since $\overline{\mathbb{K}}$ is compact, the number of connected components of $\mathbb{K}_\delta$ is finite. Let $\Lambda_i = \Lambda_i(\delta)$, $i = 0, 1, \dots, N$, $\Omega \subseteq \Lambda_0$, be the connected components of $\mathbb{K}_\delta$.

2.2. Lemma. *Each component Λ_i is $(2\delta, T)$-chain transitive for any $T > 0$.*

If $A \subset \Lambda_0$ the proposition is proved. Suppose that $A \cap \Lambda_0 = \varnothing$. Consider the oriented graph $\boldsymbol{\Gamma}$ with vertices Λ_i, $i = 0, 1, \dots, N$ and an edge $\Lambda_i \to \Lambda_j$ if there exists $v \in \mathcal{M}$ and $t_i < t_j$ such that $f_{t_i}(v) \in \Lambda_i$ and $f_{t_j}(v) \in \Lambda_j$.

2.3. Lemma. *There exists a path in the graph joining $\Lambda_0 \supseteq \Omega$ to $\Lambda_j \supseteq A$.*

We need that the connecting orbits between the Λ_j's have time intervals greater than S. Let $\Lambda_i \to \Lambda_j$ be an edge of the graph. Take a connecting orbit $f_t(v)$, $t_i < t_j$, $v \in \mathcal{M}$, from $\Lambda_i(\delta)$ to $\Lambda_j(\delta)$. Since $f_{t_j}(v)$ is in a δ-neighbourhood of $\overline{\mathbb{K}} \cap \Lambda_j(\delta)$ and $\overline{\mathbb{K}}$ is invariant, (19) implies that $f_{t_j+S}(v)$ is in a δ_0-neighbourhood of $\overline{\mathbb{K}} \cap \Lambda_i(\delta)$.

Now we can construct an (ε, S)-chain joining $u \in A$ to $v \in \Omega$ by using the connecting orbits between the Λ_i's and joining them by (ε, S)-chains inside each Λ_i. This completes the proof of theorem V(a). □

Now we prove the lemmas

Proof of lemma 2.1(a). Equation (17) implies that

$$S_{L+c}(\eta_n|_{[\alpha,\beta]}) \le \Phi_c(\eta_n(\alpha), \eta_m(\beta)) + \frac{1}{n}$$

for all $0 < \alpha < \beta < T_n$. If $0 < a_{n_k} < T_{n_k}$, $\dot\eta_{n_k}(a_{n_k}) \xrightarrow{k} v$, $b_{n_k} - a_{n_k} \xrightarrow{k} \tau$, then

$$(\eta_{n_k}(t), \dot\eta_{n_k}) \to (x_v(t), \dot x_v(t)) = f_t(v) \text{ for } 0 \le t < b_n - a_n$$

and by the continuity of Φ_c we have that

$$S_{L+c}(x_v|_{[0,T]}) \le \Phi_c(x_v(0), x_v(T)).$$

Hence $x_v|_{[0,T]}$ is semistatic. $\square$

Proof of 2.1(b). Since the set $A \cup \Omega \subseteq \mathcal{M}$ is invariant, we only have to prove that $\mathcal{M} \setminus (A \cup \Omega)$ is invariant. Let $\omega \in \mathcal{M} \setminus (A \cup \Omega)$ and $\langle k \rangle \subseteq \mathbb{N}$, $\langle t_{n_k} \rangle \subseteq \mathbb{R}^+$ be such that $\dot\eta_{n_k}(t_{n_k}) \underset{k}{\to} w$. It is enough to show that

$$\liminf_k t_{n_k} = +\infty\,, \tag{21}$$

$$\liminf_k \left|T_{n_k} - t_{n_k}\right| = +\infty\,. \tag{22}$$

We need the folowing claim

Claim. *The limit points of* $\{\, \dot\eta_n(0) \mid n \in N \,\}$ *and* $\{\, \dot\eta(T_n) \mid n \in \mathbb{N} \,\}$ *are in* Ω *and* A *respectively.*

Proof. We prove it only for limit points $\dot\eta_n(0) \to u \in TM$. Since $\eta_n(0) \in \pi(\Omega)$, then $\pi(u) \in \pi(\Omega)$. Since $\Omega \subseteq \widehat{\Sigma}(L)$ and $\mathcal{M} \subseteq \Sigma^+(L)$, then the claim follows from part (b) of theorem VI. $\square$

We only prove (21). Suppose that

$$\liminf_k t_{n_k} = a < +\infty\,.$$

Then there exists a convergent subsequence $\langle \eta_{n_\ell}|_{[0,a]} \rangle$ in the C^1 topology, to a semistatic solution γ such that

$$\dot\gamma(a) = \lim_\ell \dot\eta_{n_\ell}(a) = w\,.$$

Then by the claim we have that

$$\dot\gamma(0) = f_{-a}(w) = \lim_\ell \dot\eta_{n_\ell}(0) \in \Omega\,.$$

and hence $w \in \Omega$. This contradicts our previous assumption that $w \in \mathcal{M} \setminus (A \cup \Omega)$. $\square$

Proof of lemma 2.2. Let $T > 0$. Since $\Lambda_i = \Lambda_i(\delta)$ is open and connected, it is pathwise connected. Let ξ, $\zeta \in \Lambda_i$ and let $\Gamma : [0, S] \to TM$ be a continuous path such that $\Gamma(0) = \xi$, $\Gamma(s) = \zeta$. Let $0 = s_0 < s_1 < \cdots < s_n = S$ be such that $d_{TM}(\Gamma(s_i), \Gamma(s_{i+1})) < \frac{\delta}{2}$. For each s_i let $v_i \in \Lambda_i(\delta) \cap \overline{\mathbb{K}}$ be such that $d(\Gamma(s_i), v_i) < \delta$. Since the orbits of $\overline{\mathbb{K}}$ are recurrent, there exist $\tau_i > T$ such that $d(f_{\tau_i}(v_i), v_i) < \frac{\delta}{2}$. We have that

$$\begin{aligned} d(f_{\tau_i}(v_i), v_{i+1}) &< d(f_{\tau_i}(v_i), \Gamma(s_i)) + d(\Gamma(s_i), \Gamma(s_{i+1})) + d(\Gamma(s_{i+1}), v_{i+1}) \\ &< 2\,\delta. \end{aligned}$$

The orbit segments $\{ f_t(v_i) \,|\, 0 \leq t \leq \tau_i,\, i = 0, \dots, n-1 \}$ give the required $(2\,\delta, T)$-chain. □

In order to prove lemma 2.3 we need first

2.4. Lemma. *For all $\delta > 0$ there exists $N = N(\delta) > 0$ and $S = S(\delta) > 0$ such that if $n > N$, then for all $c \in [0, T - n]$ we have that*

$$\mathbb{K}_\delta \cap \{(\eta_n(t), \dot\eta_n(t)) \,|\, t \in [c - S, c + S] \cap [0, T_n] \} \neq \varnothing .$$

Proof. Suppose it is not true. Then there exists $\delta > 0$ and sequences $n_k \to \infty$ and $c_k \in [0, T_{n_k}]$, such that $0 < c_k - k < c_k + k < T_{n_k}$ and

$$\mathbb{K}_\delta \cap \{ (\eta_{n_k}(t), \dot\eta_{n_k}(t)) \,|\, t \in [c_k - k, c_k + k] \} = \varnothing .$$

Let $\{ v \in TM \,|\, \|v\| \leq A,\, v \notin \mathbb{K}_\delta \} = B$, then $(\eta_{n_k}(t), \dot\eta_{n_k}(t)) \in B$ for all $k \in \mathbb{N}$ and $t \in [c_k - k, c_k + k]$. Consider the measures μ_k defined by

$$\int_B \phi \, d\mu_k := \frac{1}{2k} \int_{c_k - k}^{c_k + k} \phi(\eta_{n_k}(t), \dot\eta_{n_k}(t)) \, dt .$$

Since the η_{n_k}'s are solutions of (E-L), B is compact and $k \to +\infty$, then there exists a convergent subsequence $\nu_k \to \nu$ in the weak* topology to an invariant measure ν of the Lagrangian flow, with $\operatorname{supp}(\nu) \subset B$. Moreover, $\operatorname{supp}(\nu) \subset \mathcal{M}$. Let $v \in \operatorname{supp}(\nu)$. Since B is compact, we have that the ω-limit $\varnothing \neq \omega(v) \subset \mathbb{K}$ and

$$\varnothing \neq \omega(v) \subseteq \operatorname{supp}(\nu) \subset B \cap \mathbb{K} \subseteq B \cap \mathbb{K}_\delta = \varnothing .$$

This is a contradiction. □

Proof of lemma 2.3. Suppose that $A \cap \Lambda_0 = \varnothing$, otherwise there is nothing to prove. For each $n \in \mathbb{N}$ let

$$a_n := \sup\{ t \in [0, T_n] \,|\, (\eta_n(t), \dot\eta_n(t)) \in \Lambda_0 \} ,$$
$$a_n + s_n := \inf\{ t \in]a_n, T_n] \,|\, (\eta_n(t), \dot\eta_n(t)) \in \mathbb{K}_\delta \} .$$

By lemma 2.4, for $n > N(\delta)$ we have that $0 < s_n < S(\delta)$. Choose a sequence η_n^1 such that $(\eta_n^1(a_n), \dot\eta_n^1(a_n))$ converges. Renumbering the Λ_i's if necessary, we can assume that $s_n \to s^1 \in [0, S(\delta)]$ and $(\eta_n^1(a_n + s^1), \dot\eta_1^n(a_n + s^1)) \in \overline{\Lambda_1}$. By lemma 2.1(a), the sequence

$$(\eta_n^1, \dot\eta_n^1)|_{[a_n, a_n + s^1]}(s - a_n)$$

converges to a semistatic solution of (E-L) $(\gamma^1,\dot\gamma^1)|_{[0,s^1]}$, whose points are in $\mathcal{M}$ and such that $\gamma^1(0)\in\Lambda_0$, $\gamma^1(s^1)\in\Lambda_1\neq\Lambda_0$. Hence there is and edge $\Lambda_0\longrightarrow\Lambda_1$ in the graph $\mathbf{\Gamma}$.

Suppose that $\Lambda_1\cap A=\varnothing$, otherwise the lemma is proved. Let

$$b_n:=\sup\{\,t\in[0,T_n]\,|\,(\eta_n^1(t),\dot\eta_n^1(t))\in\Lambda_0\cap\Lambda_1\,\}\,.$$

The same arguments show that there exists $0<s^2\le S(\delta)$ and a subsequence $\eta_n^2|_{[b_n,b_n+s^2]}$ such that $\eta_n^2(b_n+s^2)\in\Lambda_2$, $\Lambda_2\neq\Lambda_i$, $i=0,1$, and an edge $\Lambda_1\longrightarrow\Lambda_2$ or $\Lambda_0\longrightarrow\Lambda_2$. We can repeat this argument each time that the final Λ_j does not contain A. Since the number of the Λ_j's is finite, we obtain a path in the graph $\mathbf{\Gamma}$ from Λ_0 to Λ_j with $A\subseteq\Lambda_j$. □

Proof of theorem V(b). Let $w\in\widehat\Sigma(L)$ and let $A=\alpha(w)$, $\Omega=\omega(w)$. It is easy to see that if $A\cap\Omega\neq\varnothing$ then w is chain recurrent in $\widehat\Sigma(L)$. Suppose that $A\cap\Omega=\varnothing$. For all $s<t$ and all $\varepsilon>0$ there exist $T=T(\gamma)>0$ and $\gamma=\gamma_{\varepsilon,s,t}:[0,T]\to M$ such that $\gamma(0)=x_w(t)$, $\gamma(T)=x_w(s)$ and

$$S_{L+c}(\gamma)\le-\Phi_c\big(x_w(t),x_w(s)\big)+\varepsilon$$

Let

$$\eta_n:=\gamma_{\frac{1}{n},-n,n},\qquad\qquad T_n:=T(\gamma_{\frac{1}{n},-n,n})\tag{23}$$

We can assume that $|\dot\eta(t)|<A$ for all $0\le t\le T_n$.

The rest of the proof is similar to item (a), but now the corresponding $\mathcal{M}\subseteq\widehat\Sigma(L)$. □

3. Graph, covering and coboundary properties

Theorem VI. (Graph Properties.)

(a) *If $\gamma(t)$, $t\ge0$ is an orbit in $\Sigma^+(L)$, then, denoting $\pi:TM\to M$ the canonical projection, the map $\pi|_{\{\dot\gamma|t>0\}}$ is injective with Lipschitz inverse.*

(b) *Denoting $\Sigma_0(L)\subset M$, the projection of $\widehat\Sigma(L)$, for every $p\in\Sigma_0(L)$ there exists a unique $\xi(p)\in T_pM$ such that*

$$(p,\xi(p))\in\Sigma^+(L)\,.$$

Moreover

$$(p,\xi(p))\in\widehat\Sigma(L)\,,$$

and the vector field ξ is Lipschitz. Obviously

$$\widehat{\Sigma}(L) = \text{graph}(\xi)\,.$$

For a proof of the following lemma see Mather [8] or Mañé [4].

3.1. Lemma. ([8].) *Given $A > 0$ there exists $K > 0$ $\varepsilon_1 > 0$ and $\delta > 0$ with the following property: if $|v_i| < A$, $(p_i, v_i) \in TM$, $i = 1, 2$ satisfy $d(p_1, p_2) < \delta$ and $d((p_1, v_1), (p_2, v_2)) \geq K^{-1} d(p_1, p_2)$ then, if $a \in \mathbb{R}$ and $x_i : \mathbb{R} \to M$, $i = 1, 2$, are the solutions of L with $x_i(a) = p_i$, $\dot{x}_i(p_i) = v_i$, there exist solutions $\gamma_i : [a - \varepsilon, a + \varepsilon] \to M$ of L with $0 < \varepsilon < \varepsilon_1$, satisfying*

$$\gamma_1(a-\varepsilon) = x_1(a-\varepsilon) \quad , \quad \gamma_1(a+\varepsilon) = x_2(a+\varepsilon)\,,$$
$$\gamma_2(a-\varepsilon) = x_2(a-\varepsilon) \quad , \quad \gamma_2(a+\varepsilon) = x_1(a+\varepsilon)\,,$$
$$S_L(x_1|_{[a-\varepsilon,a+\varepsilon]}) + S_L(x_2|_{[a-\varepsilon,a+\varepsilon]}) > S_L(\gamma_1) + S_L(\gamma_2)$$

Proof of theorem VI.

(a). Since the curve γ is semistatic, corollary 1.4 implies that there exists $A > 0$ such that $|\dot{\gamma}(t)| < A$. Let $K > 0$, $\varepsilon_1 = 1$, $\delta > 0$ be from lemma 3.1. We prove that if (p, v), $(q, w) \in \{\dot{\gamma}(t) \,|\, t > 0\}$ and $d_M(p, q) < \delta$, then

$$d_{TM}\big((p,v),(q,w)\big) < K\, d_M(p,q)\,.$$

This implies item(a). Suppose it is false. Then there exist $0 < t_1 < t_2$ such that

$$d_M\big(\gamma(t_1), \gamma(t_2)\big) < \delta$$
$$d_{TM}\big(\dot{\gamma}(t_1), \dot{\gamma}(t_2)\big) > K\, d_M\big(\gamma(t_1), \gamma(t_2)\big)$$

By lemma 3.1 there exist $0 < \varepsilon < t_1$, and solutions α, $\beta : [-\varepsilon, \varepsilon] \to M$, of (E-L) such that

$$\alpha(-\varepsilon) = \gamma(t_1 - \varepsilon) =: p \qquad\qquad \alpha(\varepsilon) = \gamma(t_2 + \varepsilon) =: q$$
$$\beta(-\varepsilon) = \gamma(t_2 - \varepsilon) =: s \qquad\qquad \beta(\varepsilon) = \gamma(t_1 + \varepsilon) =: r$$

$$S_L(\alpha) + S_L(\beta) < S_L(\gamma|_{[t_1-\varepsilon\,,\,t_1+\varepsilon]}) + S_L(\gamma|_{[t_2-\varepsilon\,,\,t_2+\varepsilon]})$$

Since $\gamma|_{]0,+\infty[}$ is semistatic, we have that

$$\begin{aligned}\Phi_c(p,q)+\Phi_c(s,r) &\leq S_{L+c}(\alpha)+S_{L+c}(\beta)\\ &< S_{L+c}(\gamma|_{[t_1-\varepsilon\,,\,t_1+\varepsilon]})+S_{L+c}(\gamma|_{[t_2-\varepsilon\,,\,t_2+\varepsilon]})\\ &< \Phi_c(p,r)+\Phi_c(s,q)\end{aligned}$$

Since $0\leq\Phi_c(s,r)+\Phi_c(r,s)$,

$$\begin{aligned}\Phi_c(p,q) &< \Phi_c(p,r)-\Phi_c(s,r)+\Phi_c(s,q)\\ \Phi_c(p,q) &< \Phi_c(p,r)+\Phi_c(r,s)+\Phi_c(s,q) \qquad (24)\end{aligned}$$

Since $\gamma|_{]0,+\infty[}$ is semistatic, we have that

$$\begin{aligned}\Phi_c(p,q) &= S_{L+c}(\gamma|_{[t_1-\varepsilon\,,\,t_1+\varepsilon]})+S_{L+c}(\gamma|_{[t_1+\varepsilon\,,\,t_2-\varepsilon]})+S_{L+c}(\gamma|_{[t_2-\varepsilon\,,\,t_2+\varepsilon]})\,,\\ \Phi_c(p,q) &= \Phi_c(p,r)+\Phi_c(r,s)+\Phi_c(s,q)\,.\end{aligned}$$

This contradicts (24).

(b). Now we prove item (b). We prove that if $p,\,q\in\Sigma_0(L)$, $(p,v)\in\widehat{\Sigma}(L)$, $(q,w)\in\Sigma^+(L)$, and $d(p,q)<\delta$, then

$$d_{TM}\big((p,v),(q,w)\big)<K\,d_M(p,q)\,.$$

Observe that this implies item (b). Suppose it is false. Then by lemma 3.1 there exist $\alpha,\,\beta:[-\varepsilon,\varepsilon]\to M$ such that

$$\begin{aligned}\alpha(-\varepsilon) &= f_{-\varepsilon}(q,w)=:q_{-\varepsilon} &\qquad \alpha(\varepsilon) &= f_\varepsilon(p,v)=:p_{-\varepsilon}\\ \beta(-\varepsilon) &= f_{-\varepsilon}(p,v)=:p_\varepsilon &\qquad \beta(\varepsilon) &= f_\varepsilon(q,w)=:q_\varepsilon\end{aligned}$$

and

$$S_L(\alpha)+S_L(\beta)<S_L(x_w|_{[-\varepsilon,\varepsilon]})+S_L(x_v|_{[-\varepsilon,\varepsilon]})$$

So

$$\begin{aligned}\Phi_c(q_{-\varepsilon},p_\varepsilon)+\Phi_c(p_{-\varepsilon},q_\varepsilon) &< \Phi_c(q_{-\varepsilon},q_\varepsilon)+\Phi_c(p_{-\varepsilon},p_\varepsilon)\\ &= \Phi_c(q_{-\varepsilon},q_\varepsilon)-\Phi_c(p_\varepsilon,p_{-\varepsilon})\end{aligned}$$

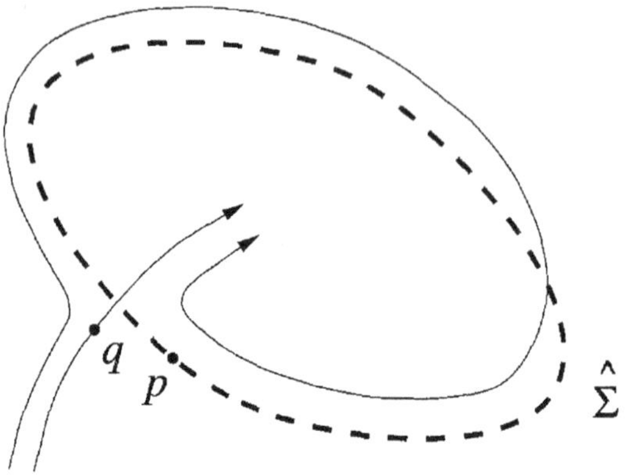

Theorem VI B.

Thus

$$\Phi_c(q_{-\varepsilon}, q_\varepsilon) \leq \Phi_c(q_{-\varepsilon}, p_\varepsilon) + \Phi_c(p_\varepsilon, p_{-\varepsilon}) + \Phi_c(p_{-\varepsilon}, q_\varepsilon) < \Phi_c(q_{-\varepsilon}, q_\varepsilon)$$

which is a contradiction. □

Denote by $\mathcal{L}$ the set of equivalence classes of the equivalence relation $d_c(a,b) = 0$ lifted to $\widehat{\Sigma}(L)$ by the vectorfield ξ of theorem VI. Denote by $\omega(v)$ the ω-limit of $v \in TM$ by the flow f_t. If $\Gamma \in \mathcal{L}$ set

$$\Gamma^+ := \{\, w \in \Sigma^+(L) \,|\, \omega(w) \in \Gamma \,\}.$$

Clearly Γ^+ is foward invariant. Let

$$\Gamma_0^+ := \bigcup_{t>0} \pi\, f_t\, \Gamma^+ \,.$$

Theorem VII. **(Covering Property.)**
(a) $\pi\Sigma^+(L) = M$.
(b) *For all* $p \in \Gamma_0^+$, *there exists a unique* $\xi_\Gamma(p) \in T_pM$ *such that*

$$(p, \xi_\Gamma(p)) \in \Gamma^+ \,.$$

Moreover, ξ_Γ *is Lipschitz.*

Observe that $\pi : \Sigma^+(L) \to M$ is not necessarily injective. We recall that Mañé stated item (a) in a stronger form: $\pi\Gamma = M$ for every equivalence class Γ. This may not be true as the following example shows. Let $M = S^1 = \mathbb{R}/\mathbb{Z}$ be the unit circle and $L = \frac{1}{2}v^2 - \cos 6\pi x$. Then the three maximums of the potential $A = 0 = 2\pi$, $B = \frac{2\pi}{3}$ and $C = \frac{4\pi}{3}$ are singular (hyperbolic saddle) points of the Lagrangian flow. For *mechanical Lagrangians* $L = \frac{1}{2}v^2 - \phi(x)$, we have that

$$c(L) = e_0 = -\min_{x \in M} L(x,0)\,,$$

and the static points are the critical points $(x,0)$ of the Lagrangian flow such that $L(x,0) = -e_0$. In this example $c(L) = 1$ and $\widehat{\Sigma}(L) = \{(A,0),(B,0),(C,0)\}$. We shall prove below that $d_1(A,B) = \Phi_1(A,B) + \Phi_1(B,A) > 0$. Hence $(A,0)$ and $(B,0)$ are not in the same equivalence class. By (30), $\Sigma^+(L) \subseteq E^{-1}\{c(L)\}$, hence if Γ is the equivalence class of $(A,0)$, then $\pi\Gamma^+ \subseteq [0, \frac{2\pi}{3}[\cup]\frac{2\pi}{3}, 2\pi] \neq S^1$. This example can easily generalized to higher dimensions.

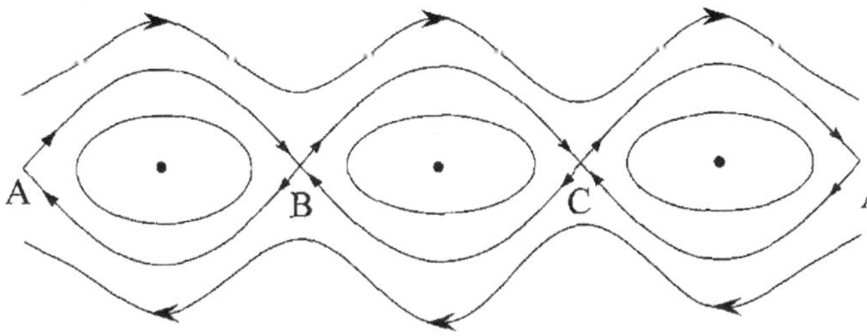

We show now that in the example above $d_1(A,B) > 0$. By theorem V, the set $\Sigma(L)$ is chain transitive and contains

$$\widehat{\Sigma}(L) = \{(A,0),(B,0),(C,0)\}.$$

Moreover, by (30), $\Sigma^+(L) \subseteq E^{-1}\{c(L)\}$. In our example this is only possible if $\Sigma(L)$ contains complete components of the saddle connections (so that $\pi\Sigma^+(L) = M$). From the symmetry of this Lagrangian we get that $\Sigma(L)$ contains all the saddle connections. Hence $\Sigma(L) = E^{-1}\{c(L)\}$. Let $(x,\dot{x})$ be the orbit on $E^{-1}\{c(L)\}$ with α-limit $(A,0)$ and ω-limit $(B,0)$. Then it is semistatic and hence

$$L(x,\dot{x}) + c(L) = \dot{x}\, L_v(x,\dot{x}) = |\dot{x}|^2\,,$$

$$\Phi_1(A,B) = \lim_{T\to+\infty} S_{L+c}(x|_{[-T,+T]}) = \int_{-\infty}^{+\infty} |\dot{x}(t)|^2\, dt > 0\,.$$

The same argument gives that $\Phi_1(B,A) > 0$ and thus $d_1(A,B) > 0$.

Proof of theorem VII. We prove (a). We may assume that $\pi\widehat{\Sigma}(L) \neq M$ otherwise the proof is trivial. Let $p \in M \setminus \pi\widehat{\Sigma}(L)$ and $q \in \pi\widehat{\Sigma}(L) \neq \varnothing$. For each $n \in \mathbb{N}$ let $\gamma_n : [0, T_n] \to M$ be such that

1. $\gamma_n(0) = p$, $\gamma_n(T_n) = q$.
2. $S_{L+c}(\gamma_n) \leq \Phi_c(p,q) + \frac{1}{n}$.
3. γ_n minimizes $L + c$ in $AC(p, q, T_n)$.

Then

(i) γ_n is a solution of (E-L).
(ii) $|\dot{\gamma}_n| < A$ for all $t \in [0, T_n]$.

Since $\widehat{\Sigma}(L)$ is invariant under the Lagrangian flow, item (b) of theorem VI implies that

(iii) $T_n \to +\infty$.

Let $\eta(t) := \pi f_t(p, v)$. By (i) and (iii), for any fixed $T > 0$ we have that $\gamma_n|_{[0,T]} \to \eta|_{[0,T]}$ in the C^1 topology, and hence

$$S_{L+c}(\eta|_{[0,T]}) = \lim S_{L+c}(\gamma_n|_{[0,T]})\,.$$

Clearly $\eta(0) = p$. It is enough to show that $\eta|_{[0,+\infty[}$ is semistatic. For, we have that

$$\begin{aligned} S_{L+c}(\gamma_n) &= S_{L+c}(\gamma_n|_{[0,T]}) + S_{L+c}(\gamma_n|_{[T,T_n]}) \\ &\leq \Phi_c(p,q) + \tfrac{1}{n} \\ &\leq \Phi_c(p, \gamma_n(T)) + \Phi_c(\gamma_n(T), q) + \tfrac{1}{n} \\ &\leq \Phi_c(p, \gamma_n(T)) + S_{L+c}(\gamma_n|_{[T,T_n]}) + \tfrac{1}{n}\,. \end{aligned}$$

Hence

$$S_{L+c}(\gamma_n|_{[0,T]}) \leq \Phi_c(p, \gamma_n(T)) + \tfrac{1}{n}\,.$$

Taking the limit when $n \to +\infty$ we obtain that η is semistatic.

We now prove item (b). Let $A > 0$ be from corollary 1.4 and $K = K(A) > 0$, $\varepsilon_1 = \varepsilon_1(A) > 0$, $\delta = \delta(A) > 0$ be from Mather's lemma 3.1. It is enough to prove that if $p_1, p_2 \in \Gamma_0^+$, $v_1, v_2 \in \Gamma^+$, $\pi(v_i) = p_i$, $i = 1, 2$ and $d_M(p_1, p_2) < \delta$, then $d_{TM}(v_1, v_2) \leq K\, d_M(p_1, p_2)$.

Suppose it is false. Then there exists $p_i \in \Gamma_0^+$, $v_i \in \Gamma^+$, $\pi(v_i) = p_i$ such that $d_M(p_1, p_2) < \delta$ and $d_{TM}(v_1, v_2) > K\, d_M(p_1, p_2)$. Since $p_i \in \Gamma_0^+$, there exists $0 < \varepsilon < \varepsilon_1$ such that $f_t(v_i) \in \Gamma^+$ for $t > -\varepsilon$.

Let $q_i := \pi f_{-\varepsilon}(v_i)$, $r_i := f_{+\varepsilon}(v_i)$. By Mather's lemma 3.1, there exist $\eta_i : [-\varepsilon, \varepsilon] \to M$ such that $\eta_i(-\varepsilon) = q_i$, $\eta_i(+\varepsilon) = r_i$ and

$$S_{L+c}(\eta_1) + S_{L+c}(\eta_2) < S_{L+c}(x_{v_1}|_{[-\varepsilon,+\varepsilon]}) + S_{L+c}(x_{v_2}|_{[-\varepsilon,+\varepsilon]}).$$

Thus

$$\Phi_c(q_1, r_2) + \Phi_c(q_2, r_1) < \Phi_c(q_1, r_1) + \Phi_c(q_2, r_2). \tag{25}$$

Let $u_i \in \omega(v_i)$ and $z_i := \pi(u_i)$. If $T^i_n \xrightarrow{n} +\infty$ is such that $f_{T^i_n}(v_i) \xrightarrow{n} u_i$, we have that

$$\begin{aligned}\Phi_c(q_i, z_i) &= \lim_n \Phi_c(q_i, x_{v_i}(T^i_n)) = \lim_n S_{L+c}(x_{v_i}|_{[-\varepsilon,T^i_n]}) \\ &= S_{L+c}(x_{v_i}|_{[-\varepsilon,+\varepsilon]}) + \lim_n S_{L+c}(x_{v_i}|_{[+\varepsilon,T^i_n]}) \\ &\geq \Phi_c(q_i, r_i) + \Phi_c(r_i, z_i).\end{aligned}$$

By the triangle inequality for Φ_c we get that

$$\Phi_c(q_i, z_i) = \Phi_c(q_i, r_i) + \Phi_c(r_i, z_i) \tag{26}$$

From (26) and (25) we have that

$$\begin{aligned}\Phi_c(q_1, z_1) + \Phi_c(q_2, z_2) &= \Phi_c(q_1, r_1) + \Phi_c(r_1, z_1) + \Phi_c(q_2, r_2) + \Phi_c(r_2, z_2) \\ &> \Phi_c(q_1, r_2) + \Phi_c(q_2, r_1) + \Phi_c(r_1, z_1) + \Phi_c(r_2, z_2) \\ &\geq \Phi_c(q_1, z_2) + \Phi_c(q_2, z_1).\end{aligned} \tag{27}$$

Since $u_i \in \Gamma$, then $d_c(z_1, z_2) = \Phi_c(z_1, z_2) + \Phi_c(z_2, z_1) = 0$. Adding $d_c(z_1, z_2) = 0$ to the right of inequality (27), we obtain that

$$\begin{aligned}\Phi_c(q_1, z_1) + \Phi_c(q_2, z_2) &> \Phi_c(q_1, z_2) + \Phi_c(z_2, z_1) + \Phi_c(z_1, z_2) + \Phi_c(q_2, z_1) \\ &\geq \Phi_c(q_1, z_1) + \Phi_c(q_2, z_2).\end{aligned}$$

This is a contradiction. □

Theorem VIII. **(Generic Structure of $\widehat{\Sigma}(L)$.)** *For a generic Lagrangian L, $\widehat{\Sigma}(L)$ is a uniquely ergodic set. If it is a periodic orbit then it is a hyperbolic periodic orbit.*

Proof. This should be thought as a corollary of theorems III and IV. Take the generic set given by theorem III of Lagrangians L that satisfy $\#\widehat{\mathcal{M}}(L) = 1$ and call this unique minimizing measure $\mu(L)$. Then if μ is an invariant measure of L and it is supported in $\widehat{\Sigma}(L)$ then by theorem IV it is minimizing. Thus $\mu = \mu(L)$. This proves the theorem. □

Theorem IX. **(Coboundary Property.)** *If $c = c(L)$, then $(L+c)|_{\widehat{\Sigma}(L)}$ is a Lipschitz coboundary. More precisely, taking any $p \in M$ and defining $G : \widehat{\Sigma}(L) \to \mathbb{R}$ by*

$$G(w) = \Phi_c(p, \pi(w))$$

Then

$$(L+c)|_{\widehat{\Sigma}(L)} = \frac{dG}{df},$$

where

$$\frac{dG}{df}(w) := \lim_{h\to 0} \frac{1}{h}\big[\, G(f_h(w)) - G(w)\,\big].$$

Proof. Let $w \in \widehat{\Sigma}(L)$ and define $F_w(v) := \Phi_c\big(\pi(w), \pi(v)\big)$. We have that

$$\begin{aligned}
\left.\frac{dF_w}{df}\right|_w &= \lim_{h\to 0} \frac{1}{h}\left[\, F_w(f_h\, w) - F_w(w)\,\right] \\
&= \lim_{h\to 0} \frac{1}{h}\left[\, \Phi_c(\pi w, \pi f_h\, w) - \Phi_c(\pi w, \pi w)\,\right] \\
&= \lim_{h\to 0} \frac{1}{h}\left[\, S_{L+c}\left(x_w|_{[0,h]}\right)\,\right] \\
&= \lim_{h\to 0} \frac{1}{h}\int_0^h \left[L\big(x_w(s), \dot{x}_w(s)\big) + c\right] ds \\
&= L(w) + c.
\end{aligned}$$

We claim that for any $p \in M$ and any $w \in \widehat{\Sigma}(L)$, $h \in \mathbb{R}$,

$$\begin{aligned}
G(f_h\, w) &= \Phi_c\big(p, \pi(f_h\, w)\big) = \Phi_c\big(p, \pi(w)\big) + \Phi_c\big(\pi(w), \pi(f_h\, w)\big) \\
G(f_h\, w) &= \Phi_c\big(p, \pi(w)\big) + F_w\big(f_h(w)\big)\,. \qquad (28)
\end{aligned}$$

This is enough to prove the theorem because then

$$\left.\frac{dG}{df}\right|_w = \left.\frac{d}{dh} F_h(f_h\, w)\right|_{h=0} = \left.\frac{F_w}{df}\right|_w = L(w) + c,$$

and G is Lipschitz by theorem I.

We now prove (28). Let $q := \pi(w)$, $x := \pi(f_h\, w)$. We have to prove that

$$\Phi_c(p,x) = \Phi_c(p,q) + \Phi_c(q,x)\,. \qquad (29)$$

Since the points q and x can be joined by the static curve $x_w|_{[0,h]}$, then

$$\Phi_c(x,q) = -\Phi_c(q,x)\,.$$

Using twice the triangle inequality for Φ_c we get that

$$\Phi_c(p,q) \le \Phi_c(p,x) + \Phi_c(x,q) = \Phi_c(p,x) - \Phi_c(q,x) \le \Phi_c(p,q)\,.$$

This implies (29). □

4. Connecting orbits inside fixed energy levels

We quote a paragraph from Mañé [**7**]:

"Exploiting that the energy, $E : TM \to \mathbb{R}$, defined as usual by $E(x,v) = \frac{\partial L}{\partial v}\, v - L$, is a first integral of the flow generated by L, leads to *information on the position of* $\Sigma^+(L)$. First observe that it is easy to check that a semistatic curve $x : [a,b] \to M$ satisfies:

$$E(x(t), \dot{x}(t)) = c(L)\,. \tag{30}$$

This follows from calculating the derivative at $\lambda = 1$ of the function $F : \mathbb{R} \to \mathbb{R}$ given by:

$$F(\lambda) = \int_a^{\frac{b}{\lambda}} (L+c)(x_\lambda(t), \dot{x}_\lambda(t))\, dt\,,$$

where $x_\lambda : [a, \frac{b}{\lambda}] \to M$ is given by $x_\lambda(t) = x(\lambda t)$. From (30) follows that:

$$\Sigma^+(L) \subset E^{-1}(c)\,,$$

that together with $\pi\,\Sigma^+(L) = M$ implies:

$$\pi\, E^{-1}(c) = M\,.$$

Hence,

$$c \ge \max_q E(q,0)\,.$$

Moreover $\widehat{\Sigma}(L) \subset E^{-1}(c)$ implies:"

4.1. Corollary. *$\mu \in \mathcal{M}(L)$ is minimizing if and only if*

$$\int \left(\frac{\partial L}{\partial v}\right) v\, d\mu = 0\,, \qquad supp(\mu) \subset E^{-1}(c(L))\,.$$

Theorem X. *If $k > c(L)$ then for all p, $q \in M$, $p \neq q$ there exists a solution $x(t)$ of (E-L) such that $x \in AC(p,q)$ and*

$$\Phi_k(p,q) = S_{L+k}(x)$$

Moreover the solution $(x(t), \dot{x}(t))$ is contained in the energy level $E^{-1}\{k\}$. ∎

4.2. Remark. If $p = q$ and $k > c(L)$, then the infimum

$$\Phi_k(p,p) = \inf\{S_{L+k}(\gamma) \mid \gamma \in AC(p,q)\} = 0$$

can not be realized by a path defined on an interval with nonzero length. For otherwise if $\gamma : [0,T] \to M$, $T > 0$ is a minimum, then

$$\Phi_c(p,p) \leq S_{L+c}(\gamma) = \Phi_k(p,p) - (k - c(L))\,T < \Phi_k(p,p) = 0\,.$$

Contradicting theorem I.

4.3. Remark. This theorem, with the same proof, holds for coverings $\pi : \widehat{M} \to M$ of a compact manifold M, with the lifted Lagrangian $\widehat{L} = L \circ \pi$.

Proof. Suppose that p, $q \in M$ and $p \neq q$. For each $T > 0$ there exists a minimizer x_T of S_L on $AC(p,q,T)$. Then

$$\Phi_k(p,q) = \inf_{T>0} S_L(x_T) + k\,T\,.$$

Observe that

$$\begin{aligned}\lim_{T\to+\infty} \left(S_L(x_T) + k\,T\right) &= \lim_{T\to+\infty} \left(S_{L+c}(x_T) + (k-c)\,T\right)\\ &\geq \lim_{T\to+\infty} \left(\Phi_c(p,q) + (k-c)\,T\right)\\ &= +\infty\,.\end{aligned}$$

Choose a sequence $\{T_i\}$ such that $\lim_i S_L(x_{T_i} + k\,T_i) = \Phi_k(p,q)$. Then $\{T_i\}$ must be bounded and we may assume that it has a limit T_0. Since $p \neq q$, by Corollary 1.4 we have that $\|\dot{x}_{T_i}\| < A$ and then $T_0 > 0$. By Theorem 1.2, x_{T_i} is a solution of (E-L). Write $w_i = \dot{x}_{T_i}(0)$, then $x_{T_i}(s) = \pi \circ f_s(w_i)$, where f_s is the Euler-Lagrange flow. Choose a convergent subsequence $w_i \to w$, then $x_{T_i} \to x_w$ in the C^1 topology, where $x_w(s) = \pi \circ f_s(w)$, $s \in [0, T_0]$. Hence

$$S_{L+k}(x_w) = \lim_i S_{L+k}(x_{T_i}) = \Phi_k(p,q).$$

We now prove that minimizers of S_{L+k} are in the energy level $E = k$. Suppose that $x \in AC(p,q,T)$ is such that

$$S_{L+k}(x) = \min_{y\in AC(p,q)} S_{L+k}(y)\,.$$

Define

$$F(\lambda) = \int_0^{\lambda T} (L+k)(x_\lambda, \dot{x}_\lambda)$$

where $x_\lambda(t) : [0, \lambda T] \to M$ is defined as $x_\lambda(t) = x(\frac{t}{\lambda})$. By the minimizing condition $F'(1) = 0$. On the other hand

$$F'(\lambda) = T\, L\big(x_\lambda(\lambda T), \dot{x}_\lambda(\lambda T)\big) + \int_0^{\lambda T} \frac{\partial L}{\partial \lambda}\, dt + kT$$

now

$$\begin{aligned}\frac{\partial L}{\partial \lambda} &= \frac{\partial L}{\partial x}\frac{\partial x_\lambda}{\partial \lambda} + \frac{\partial L}{\partial v}\frac{\partial \dot{x}_\lambda}{\partial \lambda}\\ &= -\frac{\partial L}{\partial x}\dot{x}_\lambda\left(\frac{t}{\lambda}\right)\frac{t}{\lambda^2} - \frac{\partial L}{\partial v}\left(\frac{1}{\lambda^3}\ddot{x}_\lambda\left(\frac{t}{\lambda}\right)t + \frac{\dot{x}_\lambda(\frac{t}{\lambda})}{\lambda^2}\right)\end{aligned}$$

So

$$\begin{aligned}0 &= T\, L\big(x(T), \dot{x}(T)\big) + T\, k - \int_0^T \left(\frac{\partial L}{\partial x}\dot{x} + \frac{\partial L}{\partial v}\ddot{x}\right) t\, dt - \int_0^T \frac{\partial L}{\partial v}\dot{x}\, dt\\ &= T\, L\big(x(T), \dot{x}(T)\big) + T\, k - \int_0^T \frac{\partial L}{\partial v}\dot{x}\, dt - \int_0^T \left(\frac{dL}{dt}\right) t\, dt\\ &= T\, L\big(x(T), \dot{x}(T)\big) + T\, k - \int_0^T \frac{\partial L}{\partial v}\dot{x}\, dt + \int_0^T L\, dt - L\, t\big|_0^T\\ &= Tk - \int_0^T E\, dt\\ &= T(k - E).\end{aligned}$$

This proves that the energy level of the solution x is k. □

Observe that

$$L + k = (\partial L/\partial v)v \quad \text{on } E^{-1}\{k\}\,. \tag{35}$$

4.5. Corollary.

(a) *If $k > c(L)$ and $a, b \in M$, there exists a solution $x(t)$ of (E-L) such that $x(0) = a$, $x(T) = b$ for some $T > 0$, $E(x(t), \dot{x}(t)) = k$ for all*

$t \in \mathbb{R}$, and:

$$\int_0^T \frac{\partial L}{\partial v}(x,\dot{x})\,\dot{x}\,dt = \min \int_0^{T_1} \frac{\partial L}{\partial v}(y,\dot{y})\,\dot{y}\,dt\,, \tag{36}$$

where the minimum is taken over all the absolutely continuous $y : [0,T_1] \to M$, $T_1 \geq 0$, with $y(0) = a$, $y(T_1) = b$ and $E(y(t),\dot{y}(t)) = k$ for a.e. $t \in [0,T_1]$.

(b) *Conversely, if given $k > c(L)$ and $a, b \in M$, there exists an absolutely continuous $x : [0,T] \to M$ with $x(0) = a$, $x(T) = b$, $E(x(t),\dot{x}(t)) = k$ for a.e. $t \in [0,T]$ and satisfying the minimization property (36), then $x(t)$ is a solution of (E-L).*

If $p, q \in \pi E^{-1}\{k\}$, define

$$ACE(p,q;k) := \{x \in AC(p,q) \,|\, E(x,\dot{x}) = k \text{ a.e.}\,\}\,.$$

Item (a) of corollary 4.5 follows from (35) and item (b) follows from the fact that since minimizers of S_{L+k} have energy k, then minimizing S_{L+k} on $AC(p,q)$ is equivalent to minimize it on $ACE(p,q;k)$.

Given $\mu \in \mathcal{M}(L)$ define its *homology* or *asymptotic cycle* in M (cf. Schwartzman [**10**]), by $\rho(\mu) \in H_1(M,\mathbb{R}) = H^1(M,\mathbb{R})^*$ such that

$$\rho(\mu)\cdot[\theta] = \int_{TM} \theta\,d\mu \quad , \quad \forall\,[\theta] \in H^1(M,\mathbb{R})\,,$$

where θ is a closed 1-form and $[\theta]$ its homology class. Define the *Mather's beta function* $\beta : H_1(M,\mathbb{R}) \to \mathbb{R}$, by

$$\beta(h) := \min\{\, S_L(\mu) \,|\, \mu \in \mathcal{M}(L),\ \rho(\mu) = h\,\}\,.$$

Since, for any $h \in H_1(M,\mathbb{R})$ the set $K(h) := \{\, \mu \in \mathcal{M}(L) \,|\, \rho(\mu) = h\,\}$ is convex, it follows that β is a convex function. Let β^* be its Legendre transform: $\beta^* : H^1(M,\mathbb{R}) \to \mathbb{R}$,

$$\beta^*(w) := \max_{h\in H_1(M,\mathbb{R})}\{\, w(h) - \beta(h)\,\}\,.$$

The reader can check that

$$\beta^*([\theta]) = -c(L-\theta), \quad \forall\,[\theta] \in H^1(M,\mathbb{R})\,.$$

Define the *strict critical value* $c_0(L)$ by

$$c_0(L) = \min_\theta c(L-\theta)\,.$$

Then

$$c_0(L) = -\beta^*(0) = -\min\{\, S_L(\mu) \mid \mu \in \mathcal{M}(L),\ \rho(\mu) = 0 \,\}\,.$$

Observe that

$$\frac{d}{dt} E(x, tv)|_{t=1} = v \cdot L_{vv}(x, v) \cdot v > 0\,, \tag{37}$$

Therefore if (x, v) is a critical point of the energy function E then $v = 0$ and $\frac{\partial L}{\partial x}(x, 0) = 0$. Let

$$e_0 := \max_{p \in M} E(p, 0).$$

By (37), we have that

$$E(p, 0) = \min_{v \in T_pM} E(p, v)\,. \tag{38}$$

In particular

$$e_0 = \min\{\, k \in \mathbb{R} \mid \pi(E^{-1}\{k\}) = M \,\}\,.$$

Let θ_0 be a closed 1-form such that $c_0(L) = c(L - \theta_0)$. Then the energy function and the Euler-Lagrange equations for $L - \theta_0$ and L are the same. Theorem X implies that $\pi(E^{-1}\{k\}) = M$ for all $k > c(L - \theta_0) = c_0(L)$. Hence

$$e_0 \le c_0(L)\,.$$

As observed in Mañé [**7**], for mechanical Lagrangians

$$L(x, v) = \frac{1}{2}\langle v, v\rangle_x - \phi(x),$$

with $\langle\ ,\ \rangle_x$ a riemannian metric, we have that $d_0 = e_0 = c_0(L) = c(L)$. There is an example in [7] of a Lagrangian L with $e_0 < c_0(L)$. It uses the following corollary to the proof of theorem X:

4.6. Corollary. *If $k > c_0(L)$, for every free homotopy class $H \neq 0$ of M, there exists a periodic orbit in $E^{-1}\{k\}$ such that its projection on M belongs to that free homotopy class.*

Proof. Fix $k > c_0(L)$. By adding a closed 1-form we can assume that $c(L) = c_0(L)$. Let $AC(H)$ be the set of absolutely continuous closed curves in M with free homotopy class H. Let $x_n \in AC(H)$ with x_n :

$[0,T_n] \to M$ and

$$\lim_{n\to\infty} S_{L+k}(x_n) = \inf_{x\in AC(H)} S_{L+k}(x).$$

Let $\widetilde{x}_n$ be a lift of x_n to the universal cover $\widetilde{M}$ of M. We can assume that $\widetilde{x}_n$ is a minimizer of $S_{\widetilde{L}+k}$ on $AC(\widetilde{x}_n(0), \widetilde{x}_n(T_n); T_n)$ and in particular, that it is a solution of (E-L). Then the same arguments as in the proof theorem X yield that $\{T_n\}$ must be bounded and $\|\dot{x}_n\| < A$. We can assume that $\lim_n T_n = T_0$. Moreover

$$T_0 > \frac{1}{A} \min\{\, \text{length}(\gamma) \,|\, \gamma \in AC(H)\,\} > 0.$$

The same arguments as in theorem X give a closed curve $\gamma : [0,T_0] \to M$ which is a uniform limit of a subsequence of x_n and hence $\gamma \in AC(H)$. Moreover

$$S_{L+k}(\gamma) = \min\nolimits_{x\in AC(H)} S_{L+k}(x)\,, \tag{39}$$

and $(\gamma,\dot\gamma)$ is in the energy level $E^{-1}\{k\}$.

It remains to prove that $\dot\gamma(0) = \dot\gamma(T_n)$. Suppose that $\dot\gamma(0) \neq \dot\gamma(T_n)$. Let $\widetilde{M}$ be the universal cover of M and $\widetilde{L}$ the lift of L. Let $\widetilde{\gamma}$ be a lift of γ. Consider the path $\eta|_{[-\varepsilon,\varepsilon]} = \widetilde{\gamma}|_{[T_0-\varepsilon,T_0]} * \widetilde{\gamma}|_{[0,\varepsilon]}$. We have that η is not C^1 and hence it is not a solution of (E-L). Since $k > c_0(L) \geq e_0$, then k is a regular value of the energy $\widetilde{E}$ of $\widetilde{L}$. By the Maupertius principle (see theorem 3.8.5 of Abraham & Marsden [1]), $\eta|_{[-\varepsilon,\varepsilon]}$ is not a minimizer of the $(\widetilde{L}+k)$-action on $\widetilde{E}^{-1}\{k\}$. Then there exists $\xi \in (\eta(-\varepsilon), \eta(\varepsilon))$, with energy $\widetilde{E}(\xi,\dot\xi) \equiv k$ and $S_{\widetilde{L}+k}(\xi) < S_{L+k}(\eta)$. Moreover, since $\widetilde{M}$ is simply connected, the paths ξ and η are homotopic by a homotopy which fixes their endpoints. Hence $\pi(\xi * \widetilde{\gamma}|_{[-\varepsilon,T_0-\varepsilon]})$ and $\dot\gamma|_{[0,T_0]}$ are in the same free homotopy class of M. We have that

$$\begin{aligned} S_{L+k}\left(\pi(\xi * \widetilde{\gamma}|_{[\varepsilon,T_0-\varepsilon]})\right) &= S_{\widetilde{L}+k}(\xi * \widetilde{\gamma}|_{[\varepsilon,T_0-\varepsilon]}) \\ &= S_{\widetilde{L}+k}(\xi) + S_{L+k}(\widetilde{\gamma}|_{[\varepsilon,T_0-\varepsilon]}) \\ &< S_{L+k}(\gamma)\,. \end{aligned}$$

This contradicts the minimizing property (39) of γ. □

"An interesting characterization of the critical value $c(L)$, in terms of an analogous to Tonelli's theorem (Mather [8]) in a prescribed energy level is given by the following result:" — *Mañé* [7]

Theorem XI. *Assume that k is a regular value of E and* $\dim M > 1$. *Suppose that k has the following property: for all a, b in $\pi E^{-1}(k)$ there exists an absolutely continuous curve $x : [0,T] \to M$ such that:*

(i) $x(0) = a$ *and* $x(T) = b$.

(ii) $E(x(t), \dot{x}(t)) = k$ *a.e. in* $[0,T]$.

(iii) $\displaystyle\int_0^T \frac{\partial L}{\partial v}(x(t),\dot{x}(t))\,\dot{x}(t)\,dt = \min \int_0^{T_1} \frac{\partial L}{\partial v}(y(t),\dot{y}(t))\,\dot{y}(t)\,dt,$ *where the minimum is taken over all absolutely continuous $y : [0,T_1] \to M$, with $y(0) = a, y(T_1) = b$ and $E(x(t),\dot{x}(t)) = k$ a.e. in $[0,T_1]$.*

Then $k > c(L)$ and $x(t)$ is a solution of the Euler Lagrange equation.

The hypothesis $\dim M > 1$ is necessary as the example of a simple pendulum shows. Indeed, for $L(x,v) = \frac{1}{2}|v|^2 - \cos x$ and any regular value $k < e_0 = -\min_{p\in S^1} L(p,0) = c(L) = 1$ there are such minimizers. This is because a non-empty energy level $E^{-1}(k)$ with $k < e_0$ is a topological circle and given $a, b \in \pi(E^{-1}(k))$ there are two injective paths on $E^{-1}(k)$ from $\pi^{-1}\{a\}$ to $\pi^{-1}\{b\}$. One of them must minimize the $(L+k)$-action because $L + k = v \cdot L_v = \frac{1}{2}|v|^2 \geq 0$ on $E^{-1}(k)$. In fact, the minimizer is the one whose projection on S^1 is injective.

Another example in $M = S^1$ is $\mathbb{L} = L + \theta$ where $\theta_x(v)$ is a (closed) 1-form such that $c(\mathbb{L}) = c(L+\theta) > c(L) = e_0$. The energy functions and the Lagrangian flows for $\mathbb{L}$ and L are the same. For $e_0 < k < c(\mathbb{L})$ the energy levels $E^{-1}(k)$ support two periodic orbits. Lemma 4.8 below shows that these orbits have negative $(L+k)$-action, and hence there are no minimizers on these levels. On the other hand, a regular energy level $k < e_0$ consists of one periodic orbit $(\gamma,\dot{\gamma})$. By the symmetry of $E^{-1}(k)$, we have that $\int_\gamma v\,\mathbb{L}_v = \int_\gamma \frac{1}{2}|v|^2 + 0 > 0$. Then the same arguments as for L show that there exists minimizers for $k < c(L)$.

We comment now the hypothesis of the regularity of the energy value k. Recall that all the critical points of E are in the zero section of TM. The following lemma shows that if $(p_0, 0)$ is a critical point of E, then it is a singularity of the Lagrangian flow and that the point p_0 can not be joined to other points by a path with energy k which is differentiable at p_0. Thus completing the picture of theorem XI: If we require the differentiability of the minimizers $x(t)$ at the endpoints, then the same statement holds for critical values k of E if $E^{-1}\{k\}$ consists of more than one point, and if $\#E^{-1}\{k\} = 1$ the (unique) minimizing curve is a (singular) solution of (E-L). Conversely, if $(q_0, 0) \in E^{-1}\{k\}$ is not a critical point, then it is necessarily reached on finite time by curves on $E^{-1}\{k\}$.

Define

$$d_0 := \min_{p\in M} E(p,0) = -\max_{p\in M} L(p,0)$$

If $k < d_0$ then $E^{-1}(k) = \varnothing$. Given $d_0 \leq k \leq e_0$ define

$$\theta(k) := \{\, p \in M \mid E(p,0) = k \,\},$$

then $\theta(k) \neq \varnothing$. The proof of the following lemma is delayed to the end of the section.

4.7. Lemma.

(a) *If $p_0 \in \theta(k)$ and $D_pE(p_0,0) = 0$, then the only curve $p : [0,\delta] \to M$ with $E(p(t),\dot p(t)) = k$ and $p(0) = p_0$ is the constant curve $p(t) = p_0$.*

(b) *If $p : [0,t_0[\to M$ is such that $E(p,\dot p) \equiv k$, $\lim_{t\to t_0^-} p(t) = q_0 \in \theta(k)$ and $D_pE(q_0,0) \neq 0$, then t_0 is finite.*

Proof of theorem XI. For $k \leq c(L)$ we have to show that there are points in $E^{-1}\{k\}$ which can not be joined by a curve which minimizes the action of S_{L+k} on $ACE(p,q;k)$.

Observe that d_0 is necessarily a critical value of E. Suppose first that $d_0 < k < e_0$. Since k is a regular value of E, by the Maupertius principle (see theorem 3.8.5 of Abraham & Marsden [1]), the critical points of the functional S_{L+k} on $ACE(p,q,k)$ are solutions of (E-L). Let $p \in \theta(k) \neq \varnothing$. Then $\{(p,0)\} = T_pM \cap E^{-1}\{k\}$. Hence there is only one solution x of (E-L) with $E(x,\dot x) \equiv k$ and $x(0) = p$. The set $\{\, x(t) \mid \dot x = 0 \,\} = \{\, x(t) \mid x(t) \in \theta(k) \,\}$ has at most countably many points. But since k is a regular value of E then $\theta(k)$ is a submanifold of M of codimension 1 and $\dim \theta(k) > 0$. Hence there are points in $\theta(k) = \partial\pi E^{-1}(k)$ which can not be joined to p by $(L+k)$-minimizers with $E(x,\dot x) \equiv k$.

Suppose now that $k = c(L) =: c$. The arguments of equation (30) show that minimizers of S_{L+c} in $AC(p,q)$ are in the energy level $E = c(L)$. Thus minimizers of S_{L+c} on $ACE(p,q;c(L))$ are also minimizers on $AC(p,q)$ and hence semistatic curves. Let $(q,\xi(q)) \in \widehat{\Sigma} \subseteq E^{-1}\{k\}$. Let $p \notin \pi\{f_t(q,\xi(q)) \,|\, t \in \mathbb{R}\}$. Suppose that there exists a semistatic curve $x \in AC(p,q,T)$. Then by theorem VI (b), we have that $\dot{x}(T) = \xi(q)$. This contradicts the choice of p.

Now suppose that $e_0 < k < c(L)$. We shall prove in lemma 4.8 that there exists a closed curve $\gamma(t)$ such that $E(\gamma,\dot{\gamma}) \equiv k$ and

$$\int_\gamma (L+k) < 0\,.$$

By making a large number of loops along γ one can produce a curve with arbitrarily negative action. By adding two connecting segments this implies that for any $p,q \in M$ there are not minimizers of S_{L+k} on $ACE(p,q;k)$.

This completes the proof of theorem XI. □

Given an absolutely continuous closed curve $\Gamma : [0,T] \to M$, define the probability measure μ_Γ on TM by

$$\int_{TM} \varphi \, d\mu_\Gamma := \frac{1}{T}\int_0^T \varphi(\Gamma(t),\dot{\Gamma}(t))\, dt\,.$$

Let $\mathcal{C}(k)$ be the set of measures μ_Γ supported on the energy level $E^{-1}(k)$ and let $\overline{\mathcal{C}(k)}$ be its closure in the weak* topology. Define

$$\gamma(k) := \min_{\mu\in\overline{\mathcal{C}}(k)} \int v \cdot L_v \, d\mu\,.$$

Measures realizing this minimum may not be invariant. For example if $L(x,v) = \frac{1}{2}|v|^2 - \phi(x)$ is a mechanical lagrangian and $k < e_0$ is a regular energy level, then any such measure will be supported on $\partial E^{-1}\{k\}$, where $v = 0$ and $vL_v = \frac{1}{2}|v|^2 = 0$. Nevertheless, $\partial E^{-1}\{k\}$ has no invariant subsets.

In the following lemma we use ideas of Dias Carneiro [**2**].

4.8. Lemma. *For all $e_0 < k < c(L)$ we have that $\gamma(k) < 0$.*

Proof. From (38) we get that given any $k > e_0$ and $(p,v) \in TM$, there exists a unique $\lambda > 0$ such that $E(p,\lambda v) = k$. Moreover, $\lambda = \lambda(p,v,k)$:

$TM \times]e_0, +\infty[\to]0, +\infty[$ is a smooth function.

Let μ be an invariant measure $\mu \in \mathcal{M}(L)$ such that

$$\int (L + c(L))\, d\mu = 0\,. \tag{40}$$

By theorem 4 and the fact that static curves have energy level $c(L)$ we have that $\mathrm{supp}(\mu) \subseteq E^{-1}(c(L))$, and by the Poincaré recurrence theorem we have that $\mu \in \overline{\mathcal{C}(c(L))}$.

For $k > e_0$ define the measure ν_k on $E^{-1}(k)$ by

$$\int_{E^{-1}(k)} f\, d\nu_k := A(k) \int_{E^{-1}(k)} \frac{f(p, \lambda(p,v,k)v)}{\lambda(p,v,k)}\, d\mu(p,v) \tag{41}$$

for any continuous function $f : E^{-1}(k) \to \mathbb{R}$, where

$$A(k) := \left(\int \frac{1}{\lambda(p,v,k)}\, d\mu(p,v) \right)^{-1}.$$

Then $\nu_k \in \overline{\mathcal{C}(k)}$ and $\nu_{c(L)} = \mu$. This measure ν_k is just the (probability) measure obtained by reparametrizing the solutions of (E-L) on $\pi(\mathrm{supp}(\mu))$ so as to have energy k. This process is reversible, i.e. we can recover μ by reparametrizing ν_k. Let

$$g(k) := \int v \cdot L_v\, d\nu_k = k + \int L\, d\nu_k\,.$$

Then g is a differentiable function with derivative

$$g'(k) = 1 + A(k) \int \frac{\partial}{\partial k}\left(\frac{L(p,\lambda v)}{\lambda} \right) d\mu + A'(k) \int \frac{L(p,\lambda v)}{\lambda}\, d\mu\,.$$

If we change the reference energy level $c(L)$ to $k_1 > e_0$, we can use ν_{k_1} instead of μ on formula (41). The function $g(k)$ does not change but now $\lambda(k_1) \equiv 1$, $A(k_1) = 1$ and

$$g'(k_1) = 1 + \int \frac{\partial}{\partial k}\left(\frac{\mathbb{L}}{\lambda} \right) d\nu_{k_1} + A'(k_1) \int L\, d\nu_{k_1}\,,$$

where $\mathbb{L}(p,v) := L(p, \lambda\, v)$. We compute this derivative:

$$\frac{\partial}{\partial k}\left(\frac{\mathbb{L}}{\lambda} \right)\Big|_{k=k_1} = \frac{\mathbb{L}_k\, \lambda - \mathbb{L}\, \lambda_k}{\lambda^2} = \mathbb{L}_k - \lambda_k\, \mathbb{L}|_{k=k_1}\,,$$
$$\mathbb{L}_k = v \cdot L_v\, \lambda_k\,.$$

Since $E(p, \lambda v) = k$, we have that

$$\frac{\partial E}{\partial k} = E_v \cdot (\lambda_k v) = (v \cdot L_{vv} \cdot v)\ \ \lambda_k = 1\,,$$
$$\lambda_k = \frac{1}{v \cdot L_{vv} \cdot v} \quad \text{and} \quad \mathbb{L}_k = \frac{v \cdot L_v}{v \cdot L_{vv} \cdot v}\,.$$

Moreover

$$\frac{\partial}{\partial k}\left(\frac{1}{\lambda}\right)\Big|_{k=k_1} = -\frac{\lambda_k}{\lambda^2}\Big|_{k=k_1} = -\frac{1}{v \cdot L_{vv} \cdot v}$$
$$A'(k_1) = -\frac{1}{A(k_1)} \int \frac{\partial}{\partial k}\left(\frac{1}{\lambda}\right)\, d\mu = \int \frac{1}{v \cdot L_{vv} \cdot v}\, d\nu_{k_1}\,.$$

Therefore

$$g'(k) = 1 + \int \frac{v \cdot L_v}{v \cdot L_{vv} \cdot v}\, d\nu_k - \int \frac{L}{v \cdot L_{vv} \cdot v}\, d\nu_k +$$
$$+ \left(\int \frac{1}{v \cdot L_{vv} \cdot v}\, d\nu_k\right)\left(\int L\, d\nu_k\right)$$
$$g'(k) = 1 + \int \frac{k}{v \cdot L_{vv} \cdot v}\, d\nu_k + \left(\int \frac{1}{v \cdot L_{vv} \cdot v}\, d\nu_k\right)\left(\int L\, d\nu_k\right)$$
$$g'(k) = \left(\int (L+k)\, d\nu_k\right)\left(\int \frac{1}{v \cdot L_{vv} \cdot v}\, d\nu_k\right) + 1$$
$$g'(k) = b(k)\, g(k) + 1\,, \tag{42}$$

where

$$b(k) := \int (v \cdot L_{vv} \cdot v)^{-1}\, d\nu_k > 0.$$

Let

$$B(k) := \int_{e_0}^{k} b(s)\, ds$$

and $h(k) := e^{-B(k)}\, g(k)$. From (42) we get that

$$h'(k) = e^{-B(k)} > 0\,.$$

By (40), $h(c(L)) = 0$, therefore $h(k) < 0$ for all $e_0 < k < c(L)$. And thus

$$\gamma(k) \le g(k) = e^{B(k)}\, h(k) < 0$$

for all $e_0 < k < c(L)$. □

Proof of lemma 4.7. We first prove (a). Using local coordinates we can assume that $M = \mathbb{R}^n$ and $TM = \mathbb{R}^n \times \mathbb{R}^n$. For v small and p near p_0

such that $E(p,v) = k$, we have that

$$\begin{aligned}
E &= v \cdot L_v - L \,, \\
k &= \Big(v \cdot L_v(p,0) + v \cdot L_{vv}(p,0) \cdot v + \mathcal{O}(v^3) \Big) \\
&\qquad - \left(L(p,0) + v \cdot L_v(p,0) + \frac{1}{2}\, v \cdot L_{vv}(p,0) \cdot v + \mathcal{O}(v^3) \right) \,, \\
\frac{1}{2}\, v &\cdot L_{vv}(p,0) \cdot v + \mathcal{O}(v^3) = k + L(p,0) \,.
\end{aligned}$$

Since $k = -L(p_0, 0)$, there exists a function $F(p,v)$ defined on a neighbourhood of $(p_0, 0)$ on the energy level $E(p,v) = k$ by

$$\begin{aligned}
F(p,v)\; |v|^2 &:= k + L(p,0) \\
&= L_p(p_0,0)\, \Delta p + \frac{1}{2}\, \Delta p\, L_{pp}(p_0,0)\, \Delta p + \mathcal{O}(\Delta p^3) \,,
\end{aligned} \tag{43}$$

where $\Delta p = p - p_0$, $F(p,v)$ is smooth and $F(p_0, v) > 0$. We have that

$$D_p E(p_0, 0) = -L_p(p_0, 0) = 0 \,. \tag{44}$$

Since the left hand side of (43) is positive, using (44) we have that $\Delta p\, L_{pp}(p_0, 0)\, \Delta p > 0$. Then there exists a function $G(p,v) > A > 0$ such that for v small and p near p_0 such that $E(p,v) = k$, we have

$$\begin{aligned}
|v|^2 &= G(p,v)^2\; |\Delta p|^2 \,, \\
|v| &\geq A\; |\Delta p| \,.
\end{aligned} \tag{45}$$

Suppose that there exists a differentiable curve $p : [0, \delta] \to M$ such that $p(0) = p_0$ and $E(p(t), \dot{p}(t)) = k$. For simplicity suppose that $\delta = 2$. Let $x(t) := |p(t) - p_0|$. Writing $v(t) = \dot{p}(t)$, we have that

$$\begin{aligned}
\frac{d}{dt} x(t)^2 &= 2\, x\, \dot{x} = 2\, \langle p(t) - p_0, v(t) \rangle \,, \\
\dot{x}(t) &= \left\langle \frac{p(t) - p_0}{|p(t) - p_0|}, v(t) \right\rangle \geq -B\; |v(t)| \,.
\end{aligned}$$

for some $B > 0$, because $p(t)$ is differentiable at $t = 0$. From (45), we have that

$$\begin{aligned}
\dot{x}(t) &\geq -B\; |v(t)| \geq -AB\, x(t) \,, \\
x(t) &\geq x(1)\, \exp\big(-AB\, (t-1)\big) \,.
\end{aligned}$$

In particular $x(0) \neq 0$. This contradicts the choice of $p(t)$.

Now we prove (b). Since $D_pE(q_0, 0) = L_p(q_0, 0) \neq 0$, then $\theta(k)$ is locally a codimension 1 submanifold near q_0. Let $q(t) \in \theta(k)$ be given by the condition

$$L_p(q(t), 0)\,(p(t) - q(t)) = 0 .$$

Using formula (43) with $\Delta p = p(t) - q(t)$, we have that

$$F(p, v)\;|v|^2 = L_p(q(t), 0)\,\Delta p + \mathcal{O}\left(|\Delta p|^2\right)$$

and then

$$|v| \geq A\,\sqrt{|\Delta p|}\,,$$

for some $A > 0$. Let $y(t) := |p(t) - q(t)|$. Computing $\frac{d}{dt}\,y(t)^2$, we obtain

$$\dot{y}(t) = \left\langle \frac{p(t) - q(t)}{|p(t) - q(t)|}\,,\; v(t) \right\rangle \leq -B\;|v(t)| \leq -A\,B\,\sqrt{y(t)}\,,$$

for some $B > 0$. Therefore

$$\frac{\dot{y}}{2\,y^{\frac{1}{2}}} \leq -\frac{A\,B}{2} \qquad \text{and} \qquad y(t) \leq \frac{A^2\,B^2}{4}\,(t - t_0)^2\,.$$

For future reference, we also note that

$$|\dot{y}(t)| \geq b\;|v(t)| \geq b_1\,\sqrt{y(t)}\,, \tag{46}$$

$$y(t) \geq b_2\,(t - t_0)^2\,, \tag{47}$$

for some b, b_1, $b_2 > 0$. □

5. Properties of weaker global minimizers

Definition. We say that a solution $x(t)$ of (E-L) is a *minimizer* (resp. *forward minimizer*) if

$$S_L(x|_{[t_0,t_1]}) \leq S_L(y)$$

for every $t_0 \leq t_1$ (resp. $0 < t_0 \leq t_1$) and every absolutely continuous $y : [t_0, t_1] \to M$, with $y(t_i) = x(t_i)$, $i = 1, 2$.

Denote by $\Lambda(L)$ (resp. $\Lambda^+(L)$) the set of $(p, v) \in TM$ such that the solution $x(t)$ of (E-L) with initial condition $(x(0), \dot{x}(0)) = (p, v)$ is a minimizer (resp. forward minimizer).

Let $w \in \Lambda^+(L)$ for $0 \le s \le t$ define $\delta_w(s,t)$ by

$$S_{L+c}(x_w|_{[s,t]}) = \Phi_c(x_w(s), x_w(t)) + \delta_w(s,t). \tag{48}$$

It is clear that

$$\delta_w(s,t) \ge 0. \tag{49}$$

The triangle inequality for Φ implies that

$$\delta_w(s,t) + \delta_w(t,r) \le \delta_w(s,r) \tag{50}$$

for any $w \in \Lambda^+(L)$ and $0 \le s \le t \le r$.

Claim. *$\delta_w(s,t)$ is uniformly bounded on $w \in \Lambda^+(L)$, and $0 \le s \le t < \infty$.*

We shall prove first theorems XIV, XII and XIII and then the claim.

Theorem XIV.

(a) *There exists $C > 0$ such that setting $c = c(L)$*

$$|S_{L+c}(x_w|_{[s,t]})| \le C$$

for every $w \in \Lambda^+(L)$ and all $0 \le s \le t$.

(b) *If $w \in \Lambda^+(L)$ and $p \in M$ is such that $p = \lim_{n\to\infty} x_w(t_n)$ for some sequence $t_n \to \infty$ then the limit*

$$\lim_{n\to\infty} S_{L+c}(x|_{[0,t_n]})$$

exists and does not depend in the sequence $\{t_n\}$.

Proof. Item (a) is straightforward consequence of the claim and the definition of δ_w. From equation (50) it follows that the function $t \to \delta_w(0,t)$ is increasing. Then the claim implies that $D := \lim_{t\to+\infty} \delta_w(0,t)$ exists. Thus

$$\begin{aligned}\lim_{n\to\infty} S_{L+c}(x_w|_{[0,t_n]}) &= \lim_{n\to\infty} \Phi_c(x_w(0), x_w(t_n)) + \lim_{n\to\infty} \delta_w(0,t_n)\\ &= \Phi_c(x_w(0), p) + D\,. \qquad \square\end{aligned}$$

Theorem XII.

(a) *The ω-limit set of an orbit in $\Lambda^+(L)$ is contained in $\widehat{\Sigma}(L)$.*

(b) *The α and ω-limit sets of an orbit in $\Lambda(L)$ are contained in $\widehat{\Sigma}(L)$.*

Proof. We only prove (a). Let $w \in \Lambda^+(L)$. Let $(p,v) \in \omega(w)$, $T > 0$ and let $\langle n_k \rangle$ be a sequence in $\mathbb{R}^+$ such that

$$n_{k+1} > n_k + T\,, \qquad (p,v) = \lim_k \big(x_w(n_k), \dot{x}_w(n_k)\big)\,.$$

Let

$$\begin{aligned} p_k &:= x_w(n_k), \qquad q_k := x_w(n_k + T)\,, \\ (q,u) &:= \big(x_v(T), \dot{x}_v(T)\big) = \lim_k \big(q_k\,,\ \dot{x}_w(n_k + T)\big)\,. \end{aligned}$$

we have that

$$\begin{aligned} S_{L+c}(x_w|_{[n_k,n_k+T]}) &= \Phi_c(p_k,q_k) + \delta_w(n_k, n_k + T)\,, \\ S_{L+c}(x_w|_{[n_k+T,n_{k+1}]}) &= \Phi_c(q_k,p_k) + \delta_w(n_k + T, n_{k+1})\,, \\ S_{L+c}(x_w|_{[n_k,n_{k+1}]}) &= \Phi_c(p_k,p_{k+1}) + \delta_w(n_k, n_{k+1})\,. \end{aligned}$$

By the claim and equation (50) there is a constant Q such that

$$\begin{aligned} \sum_{k=1}^{\infty} \big(\delta_w(n_k, n_k + T) + \delta_w(n_k + T, n_{k+1})\big) &\le \sum_{k=1}^{\infty} \delta_w(n_k, n_{k+1}) \\ &\le \lim_{t\to\infty} \delta_w(0,t) \le Q\,. \end{aligned}$$

Therefore

$$\begin{aligned} \lim_{k\to\infty} \delta_w(n_k, n_{k+1}) &= \lim_{k\to\infty} \delta_w(n_k, n_k + T) \\ &= \lim_{k\to\infty} \delta_w(n_k + T, n_{k+1}) = 0\,. \end{aligned} \tag{51}$$

Hence

$$\begin{aligned} S_{L+c}(x_w|_{[0,T]}) &= \lim_k S_{L+c}(x_w|_{[n_k,n_k+T]}) \\ &= \lim_k \Phi_c(p_k,q_k) = \Phi_c(p,q)\,. \end{aligned}$$

and from (51),

$$\begin{aligned} \Phi_c(p,q) + \Phi_c(q,p) &= \lim_k \Phi_c(p_k,q_k) + \lim_k \Phi_c(q_k,p_k) \\ &= \lim_k \big[\, S_{L+c}(x_w|_{[n_k,n_k+T]}) + S_{L+c}(x_w|_{[n_k+T,n_{k+1}]}) \big] \\ &= \lim_k \big[\, \Phi_c(p_k,p_k) + \delta_w(n_k, n_{k+1}) \big] \\ &= \Phi_c(p,p) \\ &= 0\,. \end{aligned}$$

This implies that $\omega(w) \subseteq \widehat{\Sigma}(L)$. □

Theorem XIII. *$f_t|_{\Lambda(L)}$ is chain transitive.*

Proof. Let $v, w \in \Lambda(L)$ and $\varepsilon > 0$. It is enough to prove that there exists an ε-chain in $\Lambda(L)$ joining $\omega(w)$ to $\alpha(v)$. By theorem XIII, $\alpha(v) \cup \omega(w) \subseteq \widehat{\Sigma}(L) \subseteq \Sigma(L)$. By theorem V there exists such ε-chain contained in $\Sigma(L) \subsetneq \Lambda(L)$. This completes the proof. □

Proof of the claim. Let

$$\begin{aligned} A &:= 2 \max\{\, L(p,v) + c(L) \mid \|v\| \leq \operatorname{diam} M + 1 \,\} \\ B &:= \max\{\, |\Phi_c(p,q)| \mid p,q \in M \,\} \\ Q &:= 3 \max\{\, A, B \,\} + 2\,. \end{aligned}$$

Suppose that there exist $\omega \in \Lambda^+(L)$, $0 \leq a \leq b$ such that $\delta_w(a,b) > Q$.

Let $\gamma : [a, T_b] \to M$ be in $AC(x_w(a), x_w(b))$ and such that

$$S_{L+c}(\gamma) < \Phi_c(x_w(a), x_w(b)) + 1\,.$$

We have that

$$S_{L+c}(\gamma) < S_{L+c}(x_w|_{[a,b]}) - \delta_w(a,b) + 1\,,$$

$$S_{L+c}(\gamma) < S_{L+c}(x_w|_{[a,b]}) - Q + 1\,, \tag{52}$$

$$S_{L+c}(\gamma) < S_{L+c}(x_w|_{[a,b]}) - (A+B)\,. \tag{53}$$

Suppose that $T_b \geq b - a$. Let $\eta : [0,1] \to M$ be a geodesic on M such that

$$\eta(0) = x_w(b) \quad , \quad \eta(1) = x_w(T_b + 1) \quad , \quad \|\dot\eta\| \leq \operatorname{diam}(M)\,.$$

Then

$$S_{L+c}(\eta) < A\,.$$

Since

$$-B \leq \Phi_c(x_w(b), x_w(T_b+1)) \leq S_{L+c}(x_w|_{[b,T_b+1]})\,,$$

using (53) (or (54)), we have that

$$\begin{aligned} S_{L+c}(\gamma * \eta) &< S_{L+c}(x_w|_{[a,b]}) - (A+B) + A \\ &< S_{L+c}(x_w|_{[a,b]}) + S_{L+c}(x_w|_{[b,T_b+1]}) \\ S_{L+c}(\gamma * \eta) &< S_{L+c}(x_w|_{[a,T_b+1]})\,. \end{aligned}$$

This contradicts the hypothesis $w \in \Lambda^+(L)$.

Now suppose that $T_b < b - a$. Let $u \in \widehat{\Sigma}(L)$ and let $\lambda : [0,1] \to M$ be a geodesic on M such that

$$\lambda(0) = x_w(b) = \gamma(T_b), \quad \lambda(1) = \pi\, u, \quad \|\dot{\lambda}\| \le \operatorname{diam} M\,,$$

and let $\overline{\lambda} : [0,1] \to M$ be $\overline{\lambda}(t) := \lambda(1-t)$. Then

$$S_{L+c}(\lambda) + S_{L+c}(\overline{\lambda}) \le \frac{A}{2} + \frac{A}{2} = A\,.$$

Let $\tau > (b-a) - T_b$ and let $\sigma : [0, T_\sigma] \to M$ be a curve in $AC(x_u(\tau), \pi\, u)$ such that

$$S_{L+c}(\sigma) < -\Phi_c(\pi\, u, x_u(\tau)) + 1\,.$$

Let $\overline{\gamma} := \gamma * \lambda * x_u|_{[0,\tau]} * \sigma * \overline{\lambda}$. This curve is in $AC(x_w(a), x_w(b))$, it is defined on a time interval of length

$$T_b + 1 + \tau + T_\sigma + 1 > T_b + \tau > b - a\,,$$

and has $(L+c)$-action

$$\begin{aligned}
S_{L+c}(\overline{\gamma}) &\le S_{L+c}(\gamma) + \big(S_{L+c}(\lambda) + S_{L+c}(\overline{\lambda})\big) + \big(S_{L+c}(x_u|_{[0,\tau]}) + S_{L+c}(\sigma)\big)\\
&\le S_{L+c}(\gamma) + A + \big(\Phi_c(\pi\, u, x_u(\tau)) - \Phi_c(\pi\, u, x_u(\tau)) + 1\big)\\
&< \big(S_{L+c}(x_w|_{[a,b]}) - Q + 1\big) + A + 1\,,\\
S_{L+c}(\overline{\gamma}) &< S_{L+c}(x_w|_{[a,b]}) - (A+B)\,. \qquad (54)
\end{aligned}$$

Now the same argument as in the case $T_b \ge b$, using $\overline{\gamma}$ instead of γ, gives a contradiction. □

References

[1] R. Abraham & J. E. Marsden. Foundations of Mechanics. Benjamin: London, 1978.

[2] M.J. Dias Carneiro. *On Minimizing Measures of the Actions of Autonomous Lagrangians.* Nonlinearity, **8** (1995) no. 6, 1077-1085.

[3] G. Contreras, R. Iturriaga. *Convex Hamiltonians without conjugate points.* Preprint. Available via internet at http://www.ma.utexas.edu.

[4] R. Mañé. Global Variational Methods in Conservative Dynamics. 18º Coloquio Bras. de Mat. IMPA. Rio de Janeiro, 1991.

[5] R. Mañé. *On the minimizing measures of Lagrangian dynamical systems.* Nonlinearity, **5**, (1992), no. **3**, 623-638.

[6] R. Mañé. *Generic properties and problems of minimizing measures of lagrangian systems.* Nonlinearity, **9**, (1996), no.2, 273-310.

[7] R. Mañé. *Lagrangian Flows: The Dynamics of Globally Minimizing Orbits.* In International Congress on Dynamical Systems in Montevideo (a tribute to Ricardo Mañé), F. Ledrappier, J. Lewowicz, S. Newhouse eds., Pitman Research Notes in Math. **362** (1996) 120-131. Reprinted in Bol. Soc. Bras. Mat. Vol **28**, N. 2, 141–153.

[8] J. Mather. *Action minimizing invariant measures for positive definite Lagrangian systems.* Math. Z. **207**, (1991), no. 2, 169-207.

[9] M. Morse. Calculus of variations in the large. Amer. Math. Soc. Colloquium Publications vol. XVIII, 1934.

[10] S. Schwartzman. *Asymptotic cycles.* Ann. of Math. (2) **66** (1957) 270, 284.

Gonzalo Contreras
Depto. de Matemática. PUC-Rio
R. Marquês de São Vicente, 225
22453-900 Rio de Janeiro
Brasil
E-mail: gonzalo@mat.puc-rio.br

Jorge Delgado
Depto. de Matemática. PUC-Rio
R. Marquês de São Vicente, 225
22453-900 Rio de Janeiro
Brasil
E-mail: jdelgado@mat.puc-rio.br

Renato Iturriaga
CIMAT
A.P. 402, 3600
Guanajuato. Gto
México
E-mail: renato@fractal.cimat.mx

Proceedings of the International Congress of Mathematicians
August 16–24, 1983, Warszawa

RICARDO MAÑÉ

Oseledec's Theorem from the Generic Viewpoint

Let f be a C^1 diffeomorphism of a compact boundaryless manifold M. A point $x \in M$ is said to be *regular* if the sequence of derivatives at x of the iterates of f admits the following description: there exists a splitting $T_xM = \bigoplus_{i=1}^{m} E_i(x)$ (the Lyapunov splitting at x) and numbers $\lambda_1(x) > \ldots > \lambda_m(x)$ (the Lyapunov exponents at x) such that $\lim_{n\to\pm\infty} n^{-1}\log\|(D_xf^n)v\| = \lambda_i(x)$ for every $1 \leqslant i \leqslant m$ and $0 \neq v \in E_i(x)$. The dimension of $E_i(x)$ is called the *multiplicity* of x. It is easy to check that the Lyapunov splitting and exponents are unique and that if x is regular so is $f(x)$ with Lyapunov. splitting $T_{f(x)}M = \bigoplus_{i=1}^{m} (D_xf)E_i(x)$ and exponents $\lambda_i(f(x)) = \lambda_i(x)$, $1 \leqslant i \leqslant m$ Denote by $\Lambda(f)$ the set of regular points. In the late sixties Oseledec proved that the set $\Lambda(f)$ has *total probability*, i.e. $\mu(\Lambda(f)) = 1$ for every f-invariant probability measure μ on the Borel σ-algebra of M [4]. Since then this theorem evolved into one of the central tools in smooth ergodic theory. Recently Oseledec's theorem was extended to compact maps of infinite dimensional Banach manifolds (Ruelle [6], Mañé [3]). This extension can be useful for the analysis of the dynamical systems generated by retarded functional differential equation or semilinear parabolic P.D.E's.

Oseledec's theorem is essentially a measure theoretical result and therefore the information it provides holds only in that category. For instance, the Lyapunov splitting is just a measurable function of the point and the limits defining the Lyapunov exponents are not uniform. It is clear that this is not a defficiency of the theorem but the natural counterweight to its remarkable generality. However, one can pose the problem, and this is the purpose of this exposition, of whether these aspects can be substantially improved by working under generic conditions. There are two approaches to this problem. One is to study Oseledec's theorem for

Originally published in: *Proceedings of the International Congress of Mathematicians, Vol. 1, 2 (Warsaw, 1983), 1269–1276, PWN, Warsaw, 1984,* generic invariant measures of generic diffeomorphisms. The other, to study Oseledec's theorem with respect to Lebesgue measure and generic volume preserving diffeomorphisms. Let us introduce some notation and terminology. Denote $\mathrm{Diff}^1(M)$ the space of C^1 diffeomorphisms of M endowed with the C^1-topology and let $\mathscr{M}$ be the space of probability measures on the Borel σ-algebra of M endowed with the weak topology. Let $d(\cdot, \cdot)$ be a metric in $\mathscr{M}$ associated to this topology. The support of a measure μ will be denoted by $s(\mu)$. Define $\mathscr{M}(f)$ to be the set of f-invariant elements of $\mathscr{M}$ and $\mathscr{M}_e(f)$ to be the set of ergodic measures in $\mathscr{M}(f)$. If $f \in \mathrm{Diff}^1(M)$ and $\gamma = \{x = f^m(x), f(x), \ldots, f^{m-1}(x)\}$ is a periodic orbit of f, denote by μ_γ the measure $\mu_\gamma = m^{-1} \sum_{j=0}^{m-1} \delta_{f^j(x)}$. Recall that a subset of a topological space is residual (or generic) if it contains the intersection of a countable family of open dense subsets. A Baire space is a topological space such that every residual subset is dense. $\mathscr{M}$ and $\mathscr{M}(f)$ are compact. $\mathscr{M}_e(f)$ is a Baire space (in fact, a residual subset of its closure).

Our main tool in developing the first approach described above is the following theorem; in the next section we shall show that it follows from the ergodic version of Pugh's Closing Lemma which we proved in [2]:

THEOREM A. *Given $f \in \mathrm{Diff}^1(M)$, $\varepsilon > 0$ and a neighborhood $\mathscr{U}$ of f, there exist $g \in \mathscr{U}$ and a g-periodic orbit γ such that $d(\mu, \mu_\gamma) < \varepsilon$.*

Applying standard methods involving upper semicontinuous set valued functions together with Theorem A, it is easy to prove the following corollary:

COROLLARY. *If $\mathscr{M}_p(f)$ denotes the set of measures μ_γ, where γ is a periodic orbit of $f \in \mathrm{Diff}^1(M)$, then $\mathscr{M}_p(f)$ is dense in $\mathscr{M}_e(f)$ for a residual set of diffeomorphisms f.*

If $f \in \mathrm{Diff}^1(M)$ and Λ is a compact invariant set we say that a splitting $TM/\Lambda = E_1 \oplus \ldots \oplus E_m$ is a *dominated splitting* if it is continuous, f-invariant (i.e. invariant under the derivative of f) and there exists $C > 0$, $0 < \lambda < 1$ such that, if $1 \leqslant i \leqslant m$, setting $E^+ = \bigoplus_{j \leqslant i} E_j$ and $E^- = \bigoplus_{j > i} E_j$ we have

$$\|(D_x f^n)/E_x^+\| \cdot \|(D_{f^n(x)} f^{-n})/E_{f^n(x)}^-\| \leqslant C\lambda^n$$

for all $x \in \Lambda$, $n \geqslant 0$.

In the next section we shall prove the following theorem:

THEOREM B. *For generic diffeomorphisms* $f \in \mathrm{Diff}^1(M)$, *there exists a residual set* $\mathscr{D}(f) \subset \mathscr{M}_e(f)$ *such that, if* $\mu \in \mathscr{D}(f)$, *there exists a dominated splitting* $TM/s(\mu) = \bigoplus_{i=1}^{m} E_i$ *which coincides with the Lyapunov splitting at* μ*-a.e. point of* $s(\mu)$.

Using the domination property it is possible to prove the existence of stable and unstable manifolds for a.e. point with respect to measures in $\mathscr{D}(f)$, even if f is only C^1 and not $C^{1+\varepsilon}$, as required in general. This follows from the results of Hirsch, Shub and Pugh [1].

Theorem B has a serious defficiency, namely, the fact that for a C^1 generic diffeomorphism, the entropy is zero with respect to generic invariant measures. This property follows easily from the methods used by Sigmund in [7] to prove the same statement for Axiom A, replacing the specification property by the corollary of Theorem A. This means that generic elements of $\mathscr{M}_e(f)$ fail to reflect the dynamic complexity of f. To avoid this problem one should work in the space $\mathscr{M}_e^c(f) = \{\mu \in \mathscr{M}_e(f) | h_\mu(f) > c\}$, proving that generic measures in $\mathscr{M}_e^c(f)$ satisfy a strong form of Oseledec's theorem. So far we have obtained no results in this direction.

Now let us consider the second approach we proposed: Oseledec's theory for generic volume preserving diffeomorphisms. For technical reasons which we discuss later we shall work with sympletic instead of volume preserving diffeomorphism. Given $f \in \mathrm{Diff}^1(M)$ and $x \in \Lambda(f)$, define $E^s(x) = \bigoplus \{E_i(x): \lambda_i(x) < 0\}$, $E^u(x) = \bigoplus \{E_i(x): \lambda_i(x) > 0\}$, $E^c(x) = E_j(x)$ if $\lambda_j(x) = 0$. When f is sympletic its Lyapunov exponents have the following symmetry property: If $\lambda_i(x)$ is a Lyapunov exponent then $-\lambda_i(x)$ is also a Lyapunov exponent with the same multiplicity. Thus $\dim E^s(x) = \dim E^u(x)$. Define the elliptic, hyperbolic and partially hyperbolic regions respectively by $\Lambda_E(f) = \{x \in \Lambda(f): E^c(x) = T_xM\}$, $\Lambda_H(f) = \{x \in \Lambda(f): T_xM = E^s(x) \oplus E^u(x)\}$, $\Lambda_P(f) = \{x \in \Lambda(f): 0 < \dim E^c(x) < \dim M\}$. Let $\mathrm{Symp}^1(M, \omega_0)$ be the space of symplectic diffeomorphisms of the symplectic compact manifold (M, ω_0) and let λ be the Lebesgue measure (associated to the volume form $\omega = \overbrace{\omega_0 \wedge \ldots \wedge \omega_0}^{k}$, $k = \dim M/2$).

THEOREM C. *There exists a residual set of* $\mathrm{Symp}^1(M, \omega_0)$ *whose elements satisfy one of the following properties:*

(a) $\lambda(\Lambda_E(f)) = 1$,

(b) f *if Anosov (and then* $\Lambda_E(f) = \Lambda_P(f) = \emptyset$),

(c) $\lambda(\Lambda_H(f)) = 0$, $\lambda(\Lambda_P(f)) > 0$ *and for all* $\varepsilon > 0$ there *exists a Borel f-invariant set* $\Lambda \subset \Lambda_P(f)$ *with* $\lambda(\Lambda_P(f) - \Lambda) < \varepsilon$ *and such that there exists* *a dominated splitting of $TM/\bar{\Lambda}$ which coincides with the Lyapunov splitting at a.e. point of Λ.*

Moreover, Λ is a uniformly partially hyperbolic set, i.e. there exist $C>0$ and $0<\lambda<1$ such that

(1) $\|(D_x f^n)/E^s(x)\| \leqslant C\lambda^n$,

(2) $\|(D_x f^{-n})/E^u(x)\| \leqslant C\lambda^n$

for all $x \in \Lambda$, $n \geqslant 0$.

The following corollary is interesting:

COROLLARY. *C^1-generically, area preserving diffeomorphisms of compact two-dimensional manifolds are either Anosov or satisfy* $\lim_{n\to\pm\infty} n^{-1}\log\|(D_x f^n)v\| = 0$ *for λ-a.e. x and every $0 \neq v \in T_x M$.*

We do not know (but this seems likely) if C holds for volume preserving diffeomorphisms. The symplectic property has two important features in its proof. First, it makes possible to deduce hyperbolicity conditions ((1) and (2) in (c)) from the domination property, that is what we can actually prove. Second, during the proof we need to approximate a C^1-symplectic diffeomorphism by a C^2-symplectic diffeomorphism. This is possible in the symplectic case (Zehnder [8]) but is unknown in the volume preserving case. The proof of Theorem C, much longer and more complicated than that of B, will appear elsewhere.

Proofs of Theorems A *and* B. If $f \in \mathrm{Diff}^1(M)$, denote by $\Sigma(f)$ the set of points $x \in M$ such that for every neighborhood $\mathscr{U}$ of x and all $\varepsilon > 0$ there exist $g \in \mathscr{U}$ and a g-periodic point y such that $d(f^j(x), g^j(y)) < \varepsilon$ for all $0 \leqslant j \leqslant m$ where m is the g-period of y. It is conjectured that $\Sigma(f)$ coincides with the set of recurrent points of f. For our purpose, it will be sufficient, combined with the result proved in [2], that $\Sigma(f)$ is a total probability set. Now suppose that $\mu \in \mathscr{M}_e(f)$ and that $\varphi\colon M \to R$ is a continuous function. Since $\mu(\Sigma(f)) = 1$, we can take $x \in \Sigma(f)$ such that

$$\lim_{n\to\pm\infty} \frac{1}{n}\sum_{j=0}^{n-1} \varphi(f^j(x)) = \int_M \varphi\, d\mu.$$

Given $\varepsilon > 0$, we can choose $N > 0$ such that

$$\left|\frac{1}{n}\sum_{j=0}^{n-1} \varphi(f^j(x)) - \int_M \varphi\, d\mu\right| < \varepsilon/2 \tag{1}$$

if $n \geqslant N$. If $\mathscr{U}$ is a neighborhood of f, there exists, by the definition of

$\Sigma(f)$, a diffeomorphism $g \in \mathcal{U}$ and a g-periodic point y such that $d\left(f^j(x), g^j(x)\right)$ is so small for all $0 \leqslant j \leqslant m$ (where m is the g-period of y) that

$$\left| \frac{1}{m} \sum_{j=0}^{m-1} \varphi\left(g^j(y)\right) - \frac{1}{m} \sum_{j=0}^{m-1} \varphi\left(f^j(x)\right) \right| < \varepsilon/2. \tag{2}$$

Moreover, observe that if x is f-periodic, there is nothing to prove, and that if it is not, then m is very large if g and y are close enough to f and x. Therefore we can suppose that $m \geqslant N$. Then (1) and (2) imply (denoting $\gamma = \{x, g(x), \dots, g^{m-1}(x)\}$)

$$\left| \int_M \varphi \, d\mu_\gamma - \int_M \varphi \, d\mu \right| < \frac{\varepsilon}{2} + \frac{\varepsilon}{2} = \varepsilon.$$

This concludes the proof of Theorem A. To prove B we shall use the following elementary lemma:

LEMMA I. *Let F be a Baire space, K a compact space and $S \subset F \times K$ a subset such that the function $F \ni x \to (\{x\} \times K) \cap S$ has a residual set of points of lower semicontinuity. Then, if $S_0 \subset S$ is a residual subset of S (in the relative topology of S), $S_0 \cap (\{x\} \times K)$ is a residual subset of $S \cap (\{x\} \times K)$ for a residual set of values of $x \in F$.*

We shall apply this lemma to $\mathrm{Diff}^1(M) \times \mathcal{M}$ and the subset $S = \{(f, \mu): f \in \mathrm{Diff}^1(M), \mu \in \mathcal{M}_e(f)\}$. Observe that at points $f \in \mathrm{Diff}^1(M)$ where every periodic point is hyperbolic and $\mathcal{M}_p(f)$ is dense in $\mathcal{M}_e(f)$, the function $\mathrm{Diff}^1(M) \ni f \to S \cap (\{f\} \times \mathcal{M}) = \mathcal{M}_e(f)$ is lower semicontinuous. This set of points is residual in $\mathrm{Diff}^1(M)$ by the corollary of Theorem A. Then we can apply the lemma and the proof of B is reduced to finding a residual subset $S_0 \subset S$ such that, if $(f, \mu) \in S_0$ then μ satisfies the properties in Theorem B. To find S_0, we define a function $\Phi: S \to \mathbf{R}^l$, where $l = \dim M$, which associates with each $(f, \mu) \in S$ the vector $\left(\lambda_1(f, \mu), \dots, \lambda_l(f, \mu)\right)$, where $\lambda_1(f, \mu) \geqslant \dots \geqslant \lambda_l(f, \mu)$ are the Lyapunov exponents of f with respect to μ, repeated according to their multiplicities. Denote by $D^k f: TM \wedge \overset{k}{\dots} \wedge TM$ the k-th exterior power of Df and define

$$\Phi_k(f, \mu) = \inf_{n \geqslant 1} \frac{1}{n} \int_M \log \|(D^k f)\| \, d\mu, \qquad k = 1, 2, \dots$$

It is known that

$$\Phi_k(f, \mu) = \sum_{j=1}^{k} \lambda_j(f, \mu).$$

Then

$$\Phi(f,\mu) = \big(\Phi_1(f,\mu), \Phi_2(f,\mu)-\Phi_1(f,\mu), \ldots, \Phi_l(f,\mu)-\Phi_{l-1}(f,\mu)\big). \quad (3)$$

Moreover, the functions Φ_k are upper-semicontinuous because they are lower bounds of continuous functions. Consequently there exists a residual set $S_k \subset S$ of points of continuity of Φ_k. Write $S_0 = \bigcap_{k\geq 1} S_k$. By (3) Φ is continuous at every point of S_0. Now we need the following lemma:

LEMMA II. *Let Λ be a compact invariant set of $f \in \mathrm{Diff}^1(M)$ such that there exists a sequence of diffeomorphisms $\{f_n: n \in \mathbf{Z}^+\}$ and periodic orbits γ_n of f_n, with period m_n, satisfying the following properties:*

(a) *For every $x \in \Lambda$ there exist $x_n \in \gamma_n$, $n = 1, 2, \ldots$ such that $\lim x_n = x$;*

(b) *There exist $\beta > 0$ and a neighborhood $\mathcal{U}$ of f such that if $g \in \mathcal{U}$ and, for some n, γ_n is a periodic orbit of g, then the eigenvalues of the Poincaré map of g at γ_n have moduli $\neq \beta^{m_n}$;*

(c) *There exists an integer $k(\beta)$ such that, for all n, the dimension of the subspace associated with the eigenvalues of modulus $< \beta^{m_n}$ is $k(\beta)$.*

Then there exists a dominated splitting $TM/\Lambda = E^+ \oplus E^-$ such that $\dim E_x^+ = k(\beta)$ for all $x \in \Lambda$.

The proof of this lemma is a direct application of the results of [2], Section 3. In fact, in [2] we worked with 1 playing the role of the constant β in Lemma II, but it is clear that the method of construction of dominated splittings developed in [2] works also in this case.

Now suppose that $(f, \mu) \in S_0$. Let $\lambda_1 > \ldots > \lambda_m$ be the Lyapunov exponents of f with respect to μ and let $k_1, \ldots, k_m$ be their multiplicities. Take numbers $\beta_1 > \ldots > \beta_{m-1}$ separating the Lyapunov exponents of f with respect to μ (i.e. $\lambda_1 > \beta_1 > \lambda_2 > \beta_2 > \ldots > \lambda_{m-1} > \beta_{m-1} > \lambda_m$). Let $f_n \to f$ be a sequence of diffeomorphisms with periodic orbits γ_n such that $\mu_{\gamma_n} \to \mu$. Such sequences exist by Theorem A. Then $\Phi(f_n, \mu_{\gamma_n})$ is near to $\Phi(f, \mu)$ for large values of n. But it is easy to see that $\Phi(f_n, \mu_{\gamma_n}) = (|a_1|^{1/m_n}, \ldots, |a_l|^{1/m_n})$ where m_n is the f_n-period of γ_n and $a_1, \ldots, a_l$ are the eigenvalues of the Poincaré map of f_n at γ_n repeated according to their multiplicities and ordered so that $|a_1| \geq \ldots \geq |a_l|$. This means that $\beta_1^{m_n} > \beta_2^{m_n} > \ldots > \beta_{m-1}^{m_n}$ are not moduli of eigenvalues of the Poincaré map of f_n at γ_n, and moreover, the dimension of the subspace associated with the eigenvalues of modulus $< \beta_i^{m_n}$ of the Poincaré map of f_n at γ_n is $\sum_{j>i} k_j$. Since this holds for any sequence of diffeomorphisms $f_n \to f$ and periodic orbits γ_n satisfying $\mu_{\gamma_n} \to \mu$, it follows that if we fix one such sequence then this sequence satisfies the hypothesis of Lemma II for

$\beta = \beta_i$ and $k(\beta) = \sum_{j>i} k_j$ for every $1 \leqslant i \leqslant m-1$. Thus fix i and apply Lemma II to $\beta = \beta_i$ and $k(\beta) = \sum_{j>i} k_j$. We obtain a dominated splitting $TM/s(\mu) = E^+ \oplus E^-$ with $\dim E_x^+ = k(\beta)$ for all $x \in s(\mu)$. We claim that $E_x^+ = E_{i+1}(x) \oplus \ldots \oplus E_m(x)$ for μ-a.e. $x \in s(\mu)$. To simplify the notation set $S(x) = E_{i+1}(x) \oplus \ldots \oplus E_m(x)$, $G(x) = E_1(x) \oplus \ldots \oplus E_i(x)$. Take a Borel set $K \subset \Lambda(f)$ such that the subspaces $S(x)$ and $G(x)$ are continuous functions of x when x varies in K and for all $x \in K$ there exists a sequence $n_j \to +\infty$ such that $f^{n_j}(x) \in K$ for all j and $\lim_{j\to+\infty} f^{n_j}(x) = x$. Since there exist sets with this property and μ-measure arbitrarily close to 1, it is sufficient, in order to prove the claim, to show that $S(x) = E_x^+$ and $G(x) = E_x^-$ when $x \in K$. Write $S(x) = S^+ \oplus S_0$, $G(x) = G^- \oplus G_0$ where $S^+ \subset E_x^+$, $G^- \subset E_x^-$, $S_0 \cap E_x^+ = \{0\}$, $G_0 \cap E_x^- = \{0\}$. We know that

$$\lim_{n\to+\infty} \|(D_x f^n)/E_x^+\| \cdot \|(D_{f^n(x)} f^{-n})/E_{f^n(x)}^-\| = 0 \tag{1}$$

$$\lim_{n\to-\infty} \|(D_x f^{-n})/E_x^-\| \cdot \|(D_{f^{-n}(x)} f^n)/E_{f^{-n}(x)}^+\| = 0. \tag{2}$$

Choose a sequence n_j such that $f^{n_j}(x) \to x$ if $j \to +\infty$ and $f^{n_j}(x) \in K$ for all j. From (1) it follows that $(D_x f^{n_j}) S_0$ converges to a subspace $S_\infty \subset E_x^-$. By the continuous dependence of $S(x)$ when x varies in K, this implies that

$$S(x) = S^+ \oplus S_\infty.$$

In a similar way we prove that

$$G(x) = G^- \oplus G_\infty$$

with $G_\infty \subset E_x^+$. But using the definition of $S(x)$, $G(x)$ and the Lyapunov exponents we obtain

$$\lim_{n\to+\infty} \|(D_x f^n)/S(x)\| \cdot \|(D_{f^n(x)} f^{-n})/G(x)\| = 0$$

or

$$\lim_{n\to+\infty} \frac{\|(D_x f^n) v\|}{\|(D_x f^n) w\|} = 0$$

if $0 \neq v \in S(x)$, $0 \neq w \in G(x)$. However, if we take $v \in S_\infty \subset E_x^-$ and $w \in G_\infty \subset E_x^+$, it follows from (1) that this limit is ∞. This contradiction shows that we must have $S_\infty = \{0\}$, $G_\infty = \{0\}$ and this will only happen if $S_0 = \{0\}$, $G_0 = \{0\}$. This means that $S(x) \subset E_x^+$, $G(x) \subset E_x^-$ and, because of the dimensions of these subspaces, we conclude $S(x) = E_x^+$, $G(x) = E_x^-$. This proves the claim. Applying this property to all the β_i's from $i = 1$ to $i = m-1$, we prove that μ satisfies the assertion of the theorem.

References

[1] Hirsch M., Shub M. and Pugh C., *Invariant Manifolds*, Lecture Notes in Mathematics 583, Springer-Verlag, 1977.
[2] Mañé R., An Ergodic Closing Lemma, *Ann. of Math.* **116** (1982), pp. 503–541.
[3] Mañé R., Lyapounov Exponents and Stable Manifolds for Compact Transformations, to appear in the *Proceedings of the Symposium on Dynamical Systems, Rio de Janeiro 1981*, Lecture Notes in Mathematics, Springer-Verlag.
[4] Oseledec V. I., A Multiplicative Ergodic Theorem, *Trans. Moscow Math. Soc.* **19** (1968), pp. 197–231.
[5] Pesin Ya., Invariant Manifolds Associated to Non Vanishing Exponents, *Math. USSR* **10** (1976).
[6] Ruelle D., Characteristic Exponent and Invariant Manifolds in Hilbert Spaces, *Ann. of Math.* **115** (1982).
[7] Sigmund K., Generic Properties for Axiom A Diffeomorphisms, *Inventiones Mathematicae* **11** (1970), pp. 99–109.
[8] Zehnder E., Note on Smoothing Symplectic and Volume Preserving Diffeomorphisms, *Geometry and Topology*, Lecture Notes in Mathematics 597, pp. 828–855. Springer-Verlag, 1977.

INSTITUTO DE MATEMÁTICA PURA E APLICADA (IMPA)
RIO DE JANEIRO,
RJ, BRAZIL

Topology, Vol. 17, pp. 383–396
© Pergamon Press Ltd., 1978. Printed in Great Britain

0040-9383/78/1201-0383/$02.00/0

CONTRIBUTIONS TO THE STABILITY CONJECTURE

RICARDO MAÑÉ

(*Received 25 November 1977*)

IN THE qualitative theory of the orbit structure of diffeomorphisms the central role is played by the concept of equivalence: two homeomorphisms f_i: $T_i \hookleftarrow$ of the topological spaces T_i, $i = 1, 2$ are equivalent if there exists a homeomorphism h: $T_1 \to T_2$ such that $f_2 h = h f_1$. When M is a smooth compact boundaryless manifold and $Diff^1(M)$ is the space of C^1 diffeomorphisms of M endowed with the C^1 topology, we say that f is structurally stable if every g nearby f (in the topology of $Diff^1(M)$) is equivalent to f. A closely related weaker concept is that of Ω-stability: f is Ω-stable if for every g nearby f, $g/\Omega(g)$ is equivalent to $f/\Omega(f)$ where $\Omega(f)$ denotes the set of nonwandering points of f (i.e. the points $x \in M$ such that for all neighborhood V of x there exists $n > 0$ satisfying $f^n(V) \cap V \neq \emptyset$). The main problem of the qualitative perturbation theory is the characterization of structurally and Ω-stable diffeomorphisms. Sufficient conditions are known: Smale proved that diffeomorphisms satisfying Axiom A and the no cycle condition[15] are Ω-stable, and Robbin and Robinson proved that replacing the no cycle condition by the strong transversality condition[14] the diffeomorphism is structurally stable. The converse results, known as the stability conjecture, are open (except for the case dim $M = 1$ where an affirmative answer is easily deduced from Peixoto's density theorem) the purpose of this paper is to prove this conjecture when dim $M = 2$ and $\Omega(f) = M$. This result follows from another one that shows that Ω-stability plus certain topological property of f is equivalent to Axiom A and the no cycles condition.

Before stating the theorems recall some results about the tability conjecture. In [8] Palis proved that Axiom A plus Ω-stability implies the no cycle condition and in [2] Franks proved that Ω-stable diffeomorphisms belong to $\mathcal{F}(M)$, where $\mathcal{F}(M)$ denotes the set of diffeomorphisms $f \in Diff^1(M)$ such that there exists a neighborhood $\mathfrak{U}$ of f such that every periodic point of every $g \in \mathfrak{U}$ is hyperbolic[15]. Then an affirmative answer to the following problem would imply the solution of the stability conjecture:

Problem. Does $f \in \mathcal{F}(M)$ imply that f satisfies Axiom A?

It is important to remark that the similar version for flows of this problem has a negative answer, even in S^3, as shown by an example due to Guckenheimer[3].

THEOREM A. *If* dim $M = 2$, *a diffeomorphism* $f \in \mathcal{F}(M)$ *such that* $\Omega(f)$ *is isolated* (*i.e. if* $\bigcap_n f^n(U) = \Omega(f)$ *for some neighborhood* U *of* $\Omega(f)$) *satisfies Axiom A.*

If dim $M = 2$, $f \in Diff^1(M)$ is Ω-stable and $\Omega(f) = M$ then by [2] $f \in \mathcal{F}(M)$ and then satisfies the hypothesis of A. Hence f satisfies Axiom A.

From Theorem A we shall deduce the following corollaries:

COROLLARY 1. *If* dim $M = 2$ *then* $f \in Diff^1(M)$ *satisfies Axiom A and the no cycles condition if and only if there exist neighborhoods* U *of* $\Omega(f)$ *and* $\mathfrak{U}$ *of* f *such that* $\bigcap_n g^n(U) = \Omega(g)$ *for all* $g \in \mathfrak{U}$.

The next corollary answers Problem 30 in [10]. Let $D^2 = \{x \in \mathbf{R}^2 \mid \|x\| \leq 1\}$ and let $Emb(D^2)$ be the space of embeddings f: $D^2 \to \mathbf{R}^2$ endowed with the C^1 topology.

Originally published in: *Topology 17 (1978), no. 4, 383–396,*

COROLLARY 2. *If $f \in Emb(D^2)$ satisfies $f(D^2) \subset \mathrm{Int}(D^2)$ and $\bigcap_n g^n(D^2) = \Omega(g)$ for every g near f then $\# \bigcap_n f^n(D^2) = 1$.*

The example in [3] shows that Theorem A and Corollary 1 are false for flows even when $\dim M = 3$. For diffeomorphisms Corollary 1 is false when $\dim M = 3$. This follows from the next example that also answers Problem 27 in [10].

THEOREM B. *There exists an open set $\mathfrak{U} \subset Diff^1(T^3)$ such that every $f \in \mathfrak{U}$ is topologically mixing and it is not Anosov.*

For the proof of A the main tool will be next theorem. In its statement we shall use the following definition: if Λ is is a compact invariant (i.e. $f(\Lambda) = \Lambda$) set of $f \in Diff^1(M)$ we say that the Λ-germ of f is expansive if there exists $\delta > 0$ such that if $x \in \Lambda$, $y \in M$ and $d(f^n(x), f^n(y)) < \delta$ for all n then $x = y$. In this case we say that δ is an expansivity constant for the Λ-germ of f.

THEOREM C. *If $f \in \mathcal{F}(M)$ and the $\Omega(f)$-germ of f is expansive then f satisfies Axiom A.*

A slightly less general version of C was stated in [5] with a sketch of the proof in the case $\Omega(f) = M$. A result closely related to theorem A was proved by Pliss[11]. He proved that a C^2 diffeomorphisms in $\mathcal{F}(M)$, with $\dim M = 2$ that has a measure zero non wandering set satisfies Axiom A.

Theorems A, B, C will be proved in §§4, 5, 3 respectively. Sections 1 and 2 are dedicated to proving preliminary results to be used in the proofs of A and C.

§1. HOMEOMORPHISMS WITH CANONICAL COORDINATES

Let K be a compact metric space and let $f: K \hookleftarrow$ be a homeomorphism. For $\epsilon > 0$ define the stable and unstable sets of $x \in K$ by:

$$W_\epsilon^s(x) = \{y \in K \mid d(f^n(y), f^n(x)) \le \epsilon \forall n \ge 0\}$$

$$W_\epsilon^u(x) = \{y \in K \mid d(f^n(y), f^n(x)) \le \epsilon \forall n \le 0\}.$$

As usual, we shall say that f is expansive if there exists $\epsilon_0 > 0$ (called an expansivity constant for f) such that $W_{\epsilon_0}^s(x) \cap W_{\epsilon_0}^u(x) = \{x\}$ for all $x \in K$. Following [1] we shall say that f has canonical coordinates if it is expansive and for all $\epsilon > 0$ there exists $\delta > 0$ such that $W_\epsilon^s(x) \cap W_\epsilon^u(y) \neq \phi$ whenever $d(x, y) \le \delta$. It is easy to see that if 2ϵ is an expansivity constant for f then $W_\epsilon^s(x) \cap W_\epsilon^u(y)$ contains at most one point.

The purpose of this section is to show that homeomorphisms with canonical coordinates have periodic orbits "shadowing" every large arc of recurrent orbit. More precisely, we have the following result:

PROPOSITION 1.1. *Let $f: K \hookleftarrow$ be a homeomorphism with canonical coordinates. Then if $x \in \omega(x)$ (where $\omega(x)$ denotes the ω-limit set of x) for all $\delta > 0$ there exists a periodic point y such that $d(f^n(x), f^n(y)) \le \delta$ for all $0 \le n \le m$ where m is the period of y.*

In [1] Bowen proved this result assuming the apparently stronger assumption that f has a hyperbolic metric. However it is unknown if every homeomorphism with canonical coordinates has a hyperbolic metric.

For the proof of Proposition 1.1 the following property will be useful:

PROPOSITION 1.2. *If $f: K \hookleftarrow$ is an expansive homeomorphism and $\delta > 0$ is an expansivity constant for f then for all $\epsilon > 0$ there exists $m > 0$ such that $f^n(W_\delta^s(x)) \subset W_\epsilon^s(f^n(x))$ for all $n \ge m$, $x \in K$.*

We shall first prove 1.2. Suppose that it is false. Then there exists an expansivity constant $\epsilon > 0$, a constant $\delta > 0$, a sequence $x_n \in K$, $n \ge 0$, and a sequence m_n, $n \ge 0$ such that $m_n \to +\infty$ when $n \to +\infty$, and $f^{m_n}(W_\delta^s(f^{m_n}(x_n)) - W_\epsilon^s(f^{m_n}(x_n)) \neq \emptyset$. Let $y_n \in f^{m_n}(W_\delta^s(x_n)) - W_\epsilon^s(f^{m_n}(x_n))$. Then $y_n \notin W_\epsilon^s(f^{m_n}(x_n))$ implies that for some $j_n \ge 0$ we have:

$$d(f^{j_n}(y_n), f^{j_n}(f^{m_n}(x_n))) > \epsilon \tag{1}$$

and $y_n \in f^{m_n}(W^s_\delta(x_n))$ implies that for all $j \geq -m_n$:

$$d(f^j(y_n), f^j(f^{m_n}(x_n))) \leq \delta. \tag{2}$$

Assume that $f^{i_n}(y_n) \to y$, $f^{i_n+m_n}(x_n) \to x$ when $n \to +\infty$. Then by (2):

$$d(f^j(y), f^j(x)) = \lim_{n\to+\infty} d(f^{j+i_n}(y), f^{j+i_n+m_n}(x_n)) \leq \delta \tag{3}$$

for all j. By (1) $d(x, y) \geq \epsilon$. Then (3) contradicts the fact that δ is an expansivity constant for f.

Now we shall prove proposition 1.1. If $x \in \omega(x)$ and $\epsilon > 0$ define:

$$U(\epsilon) = \{W^s_\epsilon(y_1) \cap W^u_\epsilon(y_2) | y_1 \in W^u_\epsilon(x), y_2 \in W^s_\epsilon(x)\}.$$

If ϵ is small enough the canonical coordinates property implies that $U(\epsilon)$ is a neighborhood of x. Moreover take ϵ so small that 4ϵ is an expansivity constant for f and $2\epsilon \leq \delta$. Take $n > 0$ such that $f^n(x) \in U(\epsilon/3)$. Moreover, by Proposition 1.2, assume that

$$f^n(W^s_\epsilon(p)) \subset W^s_{\epsilon/3}(f^n(p))$$

$$f^{-n}(W^u_\epsilon(p)) \subset W^u_{\epsilon/3}(f^{-n}(p)). \tag{4}$$

Define π: $W^s_\epsilon(x) \hookleftarrow$ by the property:

$$f^n(p) \in W^u_{\epsilon/3}(\pi(p)). \tag{5}$$

This map is well defined by (4). Define $\Lambda = \bigcap_{i\geq 0} \pi^i(W^s_\epsilon(x))$ and $S = \bigcup\{W^u_\epsilon(p) | p \in f^n(\Lambda)\}$. We claim that $f^{-n}(S) \subset S$. To prove this observe that $p \in f^n(\Lambda)$ implies $f^{-n}(p) = \pi(p)$ for some $q \in \Lambda$. By (5) that means:

$$f^n(q) \in W^u_{\epsilon/3}(\pi(q)) = W^u_{\epsilon/3}(f^{-n}(p)).$$

Hence:

$$W^u_{\epsilon/3}(f^{-n}(p)) \subset W^u_{2\epsilon/3}(f^n(q)).$$

Therefore by (4) $f^{-n}(W^u_\epsilon(p)) \subset W^u_\epsilon(f^n(q))$, with $q \in \Lambda$ and then $f^{-n}(S) \subset S$. Moreover observe that $a \in f^{-n}(S)$ implies:

$$d(f^j(a), f^j(x)) \leq 2\epsilon \tag{6}$$

for all $0 \leq j \leq n$. To see this take $q \in f^n(\Lambda)$ such that $a \in f^{-n}(W^u_\epsilon(q))$. Then:

$$d(f^j(a), f^j(x)) \leq d(f^j(a), f^{j-n}(q)) + d(f^{j-n}(q), f^j(x));$$

but $a \in f^{-n}(W^u_\epsilon(q))$ implies:

$$d(f^j(a), f^{j-n}(q)) \leq \epsilon$$

for all $0 \leq j \leq n$, and $q \in f^n(\Lambda)$ implies $f^{-n}(q) \in \Lambda \subset W^s_\epsilon(x)$. Hence:

$$d(f^{j-n}(q), f^j(x)) \leq \epsilon$$

for all $j \geq 0$. Then (6) is proved. Now define $S_0 = \bigcap_{k\geq 0} f^{-kn}(S)$. Clearly $f^n(S_0) = S_0$. Let $a \in S_0$, $b \in S_0$. By (6) we have

$$d(f^j(a), f^j(b)) \leq 2\epsilon \tag{7}$$

for all $0 \leq j \leq n$. But $f^{kn}(a) \in S_0$, $f^{kn}(b) \in S$ for all $k \geq 0$. Then (7) is satisfied for all $j \geq 0$. Finally, since $f^{-kn}(a) \in S_0$, $f^{-kn}(b) \in S_0$ for all $k \geq 0$ it follows that (7) is true for all j. Since 4ϵ is an expansivity constant for f this implies that S_0 contains only one point y. Then y is periodic and by (6) we have $d(f^j(y), f^j(x)) \leq 2\epsilon$ for all $0 \leq j \leq n$ and recalling that we took $\epsilon \leq \delta/2$ the Proposition is proved.

§2. LOCALLY INVARIANT DISCS

In this section M will denote a compact smooth boundaryless manifold (no restrictions on the dimension). First we shall state two propositions that were proved in [5, 11] (see also [7]). We shall use the following notation: if $f \in \mathit{Diff}^1(M)$ and $x \in M$ let $B_x^s(f)(B_x^u(f))$ the set of $v \in T_xM$ such that $\{\|(Tf)^n v\| n \geq 0\}$ $(\{\|(Tf)^{-n} v\| n \geq 0\})$ is bounded. Let $Per_j(f)$ the set of $x \in Per(f)$ such that $\dim B_x^s(f) = j$ and $\Lambda_j(f)$ its closure. Finally, if $x \in Per(f)$ let $\pi(f, x)$ be its prime period.

PROPOSITION 2.1. *If $f \in \mathcal{F}(M)$ there exist a neighborhood $\mathfrak{U}$ of f, and constants $C < 0$, $0 < \lambda < 1$ such that:*

$$\|(Tg)^{\pi(g,x)} / B_x^s(g)\| \leq C\lambda^{\pi(g,x)}$$

$$\|(Tg)^{-\pi(g,x)} / B_x^u(g)\| \leq C\lambda^{\pi(g,x)}$$

$$\|(Tg)^n / B_x^s(g)\| \cdot \|(Tf)^{-n} / B^u_{g^n(x)}(g)\| \leq C\lambda^n$$

for all $x \in Per(g)$, $n > 0$, $g \in \mathfrak{U}$.

PROPOSITION 2.2. *If $f \in \mathcal{F}(M)$ for all $0 < j < \dim M$ there exists a continuous Tf-invariant splitting $TM/\Lambda_j(f) = E^s \oplus E^u$ satisfying:*

(a)

$$E_x^s = B_x^s(f)$$

$$E_x^u = B_x^u(f)$$

if $x \in Per_j(f)$.

(b) *There exist constants $C > 0$, $0 < \lambda < 1$ such that:*

$$\|(Tf)^n / E_x^s\| \cdot \|(Tf)^{-n} / E^u_{f^n(x)}\| \leq C\lambda^n$$

for all $x \in \Lambda_j(f)$, $n > 0$.

Our goal in this section is to construct, with the help of Propositions 2.1 and 2.2 a family of discs $\tilde{W}_\epsilon^s(x)$, $\tilde{W}_\epsilon^u(x)$ tangent to E_x^s, E_x^u at x, with some invariance property, that will play the central role in the proof of Theorems A and C. These discs are constructed using only the domination condition (b) of 2.2 insteady of the usual hyperbolicity hypothesis. We don't know whether these discs are "true" stable and unstable discs in the sense that points in $\tilde{W}_\epsilon^s(x)$ are asymptotic with x. Most of the work in the proof of C will be to prove such properties, with the help of the assumption on the expansivity of the $\Omega(f)$-germ.

Let $f \in \mathcal{F}(M)$. Given $0 < j < \dim M$ let $\Lambda = \Lambda_j(f)$ and E^s, E^u the subbundles of TM/Λ given in Proposition 2.2. Denote $Emb_\Lambda(D^j, M)$ the space of C^1 embeddings β: $D^j \to M$, where $D^j = \{x \in \mathbf{R}^j \| \|x\| \leq 1\}$, such that $\beta(0) \in \Lambda$. Consider $Emb_\Lambda(D^j, M)$ endowed with the C^1 topology. Then π: $Emb_\Lambda(D^j, M) \to \Lambda$ defined by $\pi(\beta) = \beta(0)$ is a fibration.

PROPOSITION 2.3. *There exist sections φ^u: $\Lambda \to Emb_\Lambda(D^{\dim M - j}, M)$, φ^s: $\Lambda \to Emb_\Lambda(D^j, M)$ such that defining $\tilde{W}_\epsilon^s(x) = \varphi^s(x) D_\epsilon^j$, $\tilde{W}_\epsilon^u(x) = \varphi^u(x) D_\epsilon^{\dim M - j}$, we have the following properties:*

(a)

$$T_x\tilde{W}_1^s(x) = E_x^s$$

$$T_x\tilde{W}_1^u(x) = E_x^u.$$

(b) *There exists a power $g = f^N$ of f such that for all $0 < \epsilon_1 < 1$ there exists $0 < \epsilon_2 < 1$ satisfying*

$$g(\tilde{W}^s_{\epsilon_2}(x)) \subset \tilde{W}^s_{\epsilon_1}(g(x))$$

$$g^{-1}(\tilde{W}^u_{\epsilon_2}(x)) \subset \tilde{W}^u_{\epsilon_2}(g^{-1}(x))$$

for all $x \in \Lambda$.

Proof. Take $g = f^N$, $N > 0$ such that

$$\|(Tg)/E_x^s\| \cdot \|(Tg)^{-1} / E^u_{g(x)}\| < 1$$

for all $x \in \Lambda$. Such N exists by Proposition 2.2.

Then, using the notation of [4, p. 63], Λ is an immediate relative pseudo hyperbolic set for g. By Theorem 5.5 of [4] (see also the remark at pp. 62–63) there exist φ^s, φ^u satisfying all the required conditions.

As we mentioned before 2.3 doesn't give much information on the asymptotic properties of the discs $\tilde{W}_\epsilon^s(x)$, $\tilde{W}_\epsilon^u(x)$.

But when $x \in Per(f)$ then at least we have the following property:

PROPOSITION 2.4. *Let $C>0$, $0<\lambda<1$ be the constants given by Proposition 2.1. Let $G_j(M)$ be the Grassmann bundle of j-dimensional subspaces of TM and let $d_j(.,.)$ be a metric on $G_j(M)$ generating its usual topology. Then there exists $\bar{\epsilon}>0$ such that if $E\in G_j(M)$, $x\in Per_j(f)$ and $d_j((Tf)^nE, E^s_{f^n(x)})\le\bar{\epsilon}$ for all $0\le n\le\pi(f,x)$ then:*

$$\|(Tf)^{\pi(f,x)}/E\|\le C\lambda^{\pi(f,x)}.$$

Proof. Suppose that the proposition is false. Then for all $\epsilon>0$ we can find $x\in Per_j(f)$ and $E\in G_j(M)$ satisfying $d_j((Tf)^nE, E^s_{f^n(x)})\le\epsilon$ for all $0\le n\le\pi(f,x)$ and:

$$\|(Tf)^{\pi(f,x)}/E\|> C\lambda^{\pi(f,x)}. \tag{1}$$

If $\epsilon>0$ is small enough by Lemma 1.1 of [2] there exists $\hat{f}\in\mathfrak{U}$ (where $\mathfrak{U}$ is the neighborhood of f given by Proposition 2.1) satisfying

(a) $\hat{f}^n(x)=f^n(x)$for all $n\ge 0$.
(b) $(T\hat{f})^n/E_x^u(f)=(Tf)^n/E_x^u(f)$ for all $n\ge 0$.
(c) $T\hat{f}/E^s_{f^n(x)}=A^{-1}_{n+1}(Tf)A_n$.

for all $0\le n\le\pi(f,x)$, where A_n: $E^s_{f^n(x)}(f)\to(Tf)^nE$ is an isometry. Therefore $x\in Per(\hat{f})$ by (a). By (b):

$$E_x^u(\hat{f})\supset E_x^u(f) \tag{2}$$

and by (c)

$$(T\hat{f})^{\pi(f,x)}E_x^s(f)=E_x^s(f) \tag{3}$$

$$\|(T\hat{f})^{\pi(f,x)}/E_x^s(f)\|=\|(Tf)^{\pi(f,x)}/E\|>C\lambda^{\pi(f,x)}. \tag{4}$$

Inequalities (3) and (4) imply that $E_x^s(f)$ cannot be contained in $E_x^s(\hat{f})$, and the invariance (3) of $E_x^s(f)$ then proves that $E_x^s(f)$ contains vectors in $E_x^u(\hat{f})$. This together with (2) implies

$$\dim E_x^u(\hat{f})>\dim E_x^u(f). \tag{5}$$

Take a continuous arc f_λ in $\mathfrak{U}$ such that $f_0=f$, $f_1=\hat{f}$, and $f^n(x)=f(x)$ for all $\lambda\in[0,1]$, $n\ge 0$. Then (5) proves that for some $\lambda\in[0,1]x$ is a non hyperbolic periodic point of f_λ.

A straightforward combination of the mean value theorem and Proposition 2.4 gives the following corollary:

COROLLARY. *There exist $\bar{C}>0$, $0<\bar{\sigma}<1$, and $\epsilon_1>0$ such that if $x\in Per_j(f)$ and $\epsilon>0$ satisfy:*

$$g^n(\tilde{W}_\epsilon^s(x))\subset\tilde{W}^s_{\epsilon_1}(g^n(x))$$

for all $0\le n\le\pi(g,x)$, then

$$g^{\pi(g,x)}(\tilde{W}_\epsilon^s(x))\subset\tilde{W}_\mu^s(x)$$

where $\mu=\bar{C}\bar{\sigma}^{\pi(g,x)}\epsilon$.

§3. PROOF OF THEOREM C

In this section we shall prove Theorem C. We start proving some lemmas.

LEMMA 3.1. *$f\in\mathcal{F}(M)$ implies $\overline{Per(f)}=\Omega(f)$.*

Proof (after C. Pugh). Suppose $f \in \mathcal{F}(M)$, $x \in \Omega(f) - \overline{Per(f)}$. Let $\mathfrak{U}$ be a connected neighborhood of f contained in $\mathcal{F}(M)$. By the proof of the closing lemma[13] there exists $n > 0$ such that for all family U_j of neighborhoods of $f^j(x)$, $|j| \le n$, there exists $g \in \mathfrak{U}$ such that $g(y) = f(y)$ if $y \notin \bigcup_{-n}^{n} U_j$ and $Per(g) \cap U_0 \neq \emptyset$. Take the neighborhoods U_j satisfying $U_j \cap Per(f) = \emptyset$ for all $|j| \le n$ and let $q \in U_0 \cap Per(g)$. Then the number of fixed points of $f^{\pi(g,q)}$ is less than the number of fixed points of $g^{\pi(g,q)}$. On the other hand it is easy to see that for all $m > 0$ the number of fixed points of f^m is constant on each connected component of $\mathcal{F}(M)$.

LEMMA 3.2. *Let $f \in \mathcal{F}(M)$ with expansive $\Omega(f)$-germ. If $0 < j < \dim M$ and $\tilde{W}_\epsilon^s(x)$, $\tilde{W}_\epsilon^u(x)$ are the families of discs given in Proposition 2.3 then for all $\epsilon_1 > 0$ there exists $\epsilon_2 > 0$ such that:*

(a) *$d(f^n(x), f^n(y)) \le \epsilon_1$ for all $x \in \Lambda_j(f)$, $y \in \tilde{W}_{\epsilon_2}^s(x)$ and $n \ge 0$.*

(b) *If ϵ_1 is an expansivity constant for the $\Omega(f)$-germ of f (see definition in the Introduction) then $\lim_{n\to+\infty} d(f^n(x), f^n(y)) = 0$ for all $x \in \Lambda_j(f)$, $y \in \tilde{W}_{\epsilon_2}^s(x)$.*

Proof. To prove (a) it is sufficient to show that for all $\epsilon_1 > 0$ there exists $\epsilon_2 > 0$ such that

$$g^n(\tilde{W}_{\epsilon_2}^s(x)) \subset \tilde{W}_{\epsilon_1}^s(g^n(x)) \tag{1}$$

for all $x \in \Lambda_j(f)$, $n \ge 0$. To prove this property define, for $x \in Per_j(f)$, the number $\epsilon(x)$ by:

$$\epsilon(x) = \sup\{\epsilon > 0 | g^n(\tilde{W}_\epsilon^s(x)) \subset \tilde{W}_{\epsilon_1}^s(g^n(x)) \text{ for all } n \ge 0\}$$

It is easy to see that $\epsilon(x) > 0$ for all $x \in Per_j(f)$. If we prove that $\inf\{\epsilon(x) | x \in Per_j(f)\} > 0$ then $\epsilon_2 = \inf\{\epsilon(x) | x \in Per_j(f)\}$ satisfies (1). Suppose on the contrary that for some sequence $x_n \in Per_j(f)$ we have $\epsilon(x_n) \to 0$ when $n \to +\infty$. Let $m_n > 0$ and $y_n \in \tilde{W}_{\epsilon(x_n)}^s(x_n)$ such that

$$g^{m_n}(y_n) \in \partial \tilde{W}_{\epsilon_1}^s(g^{m_n}(x_n)) \tag{2}$$

By the Corollary to 2.4 we have:

$$g^{\pi(g,x_n)}(\tilde{W}_{\epsilon(x_n)}^s(x_n)) \subset \tilde{W}_\mu^s(x_n) \tag{3}$$

where $\mu = \bar{C}\bar{\sigma}^{\pi(g,x_n)}\epsilon(x_n)$. By (3) we can suppose without loss of generality that $0 < m_n < \pi(g, x_n)$.

The condition $\epsilon(x_n) \to 0$ and (2) imply $m_n \to +\infty$ and (3) implies $\pi(g, x_n) - m_n \to +\infty$. Then assume that $g^{m_n}(y_n) \to y$ and $g^{m_n}(x_n) \to x$ when $n \to +\infty$. If $r = \sup\{\operatorname{diam} \tilde{W}_{\epsilon_1}^s(x) | x \in \Lambda_j(f)\}$ we obtain:

$$d(g^j(g^{m_n}(y_n)), g^j(g^{m_n}(x_n))) \le r$$

for all $-m_n \le j \le \pi(g, x_n) - m_n$. Since $m_n \to +\infty$, $\pi(g, x_n) - m_n \to +\infty$, this implies:

$$d(g^j(y), g^j(x)) \le r \tag{4}$$

for all j. If we choose ϵ_1 so small that r is an expansivity constant for g, (4) implies $y = x$. But by (2) $y \in \partial \tilde{W}_{\epsilon_1}^s(x)$. To prove (b) assume that there exists $x \in \Lambda_j(f)$, $y \in \tilde{W}_{\epsilon_2}^s(x)$ such that $\limsup_{n\to+\infty} d(f^n(x), f^n(y)) = r > 0$. Then there exist a sequence $m_n > 0$ satisfying $m_n \to +\infty$ when $n \to +\infty$ and $d(f^{m_n}(x), f^{m_n}(y)) > r/2$ for all $n > 0$.

We can suppose that $f^{m_n}(x)$, $f^{m_n}(y)$ converge to $z \in \Lambda_j(f)$, $w \in M$ respectively when $n \to +\infty$. But by (a) we have $d(f^n(f^{m_n}(x)), f^{m_n}(y))) \le \epsilon_1$ for all $-m_n \le n$. Since $m_n \to +\infty$ this implies $d(f^n(z), f^n(w)) \le \epsilon_1$ for all n. But ϵ_1 is an expansivity constant for the $\Omega(f)$-germ of f. Hence $z = w$. On the other hand $d(z, w) = \lim d(f^{m_n}(x), f^{m_n}(y)) \ge r/2 > 0$.

LEMMA 3.3. *If $f \in \mathcal{F}(M)$ and the $\Omega(f)$-germ of f is expansive then $\Lambda_j(f) \cap Per_i(f) = \emptyset$ if $j \neq i$.*

Proof. Suppose $\Lambda_j(f)\cap Per_i(f)\neq\emptyset$ and $i<j$. Let $x\in\Lambda_j(f)\cap Per_i(f)$. If $W^s(x)$ is the stable manifold of x (i.e. the set of $y\in M$ satisfying $d(f^n(x),f^n(y))\to 0$ when $n\to+\infty$) by 3.2 (b) $W^s(x)\supset \tilde W^s_\epsilon(x)$ for small values of ϵ. Then $i=\dim W^s(x)\geq \dim \tilde W^s_\epsilon(x)=j$ contradicting $i<j$. If $\Lambda_i(f)\cap Per_j(f)\neq\emptyset$ a similar contradiction is obtained replacing in the argument $W^s(x)$, $\tilde W^s_\epsilon(x)$ by $W^u(x)$, $\tilde W^u_\epsilon(x)$.

LEMMA 3.4. *If $f\in\mathcal{F}(M)$ and the $\Omega(f)$-germ of f is expansive then $f/\Lambda_j(f)$ has canonical coordinates for all $0<j<\dim M$.*

Proof. Since the $\Omega(f)$-germ of f is expansive $f/\Omega(f)$ is expansive. Hence we have to verify that $f/\Lambda_j(f)$ satisfies condition (b) of Definition 2.1 for all $0<j<\dim M$ i.e. that given $\epsilon_1>0$ there exists $\delta>0$ such that

$$W^s_{\epsilon_1}(x)\cap W^u_{\epsilon_1}(y)\neq\emptyset \tag{5}$$

whenever $x,y\in\Lambda_j(f)$ and $d(x,y)\leq\delta$, where $W^s_{\epsilon_1}(x)$, $W^u_{\epsilon_1}(x)$ denote the local stable and unstable sets of x defined for f/Λ_j as in §1. We can assume without loss of generality that ϵ_1 is an expansivity constant for the $\Omega(f)$-germ of f. Then take $\epsilon_2>0$ given by Lemma 3.2 and $\delta>0$ such that $\tilde W^s_{\epsilon_2/2}(x)\cap\tilde W^u_{\epsilon_2/2}(y)$ contains one and only one point whenever $x,y\in\Lambda_j(f)$ and $d(x,y)<\delta$. We are going to prove that $x,y\in\Lambda_j(f)$ and $d(x,y)<\delta$ imply (5). Take sequences x_n, y_n in $Per_j(f)$ such that $x_n\to x$, $y_n\to y$ when $n\to+\infty$. Let z_n, w_n be sequences satisfying:

$$z_n\in\tilde W^s_{\epsilon_2}(x_n)\cap\tilde W^u_{\epsilon_2}(y_n)$$
$$w_n\in\tilde W^u_{\epsilon_2}(x_n)\cap\tilde W^s_{\epsilon_2}(y_n)$$
$$z=\lim z_n=\tilde W^s_{\epsilon_2/2}(x)\cap\tilde W^u_{\epsilon_2/2}(y)$$
$$\lim w_n=\tilde W^u_{\epsilon_2/2}(x)\cap\tilde W^s_{\epsilon_2/2}(y).$$

If $W^s(x_n)$, $W^u(x_n)$, $W^s(y_n)$, $W^u(y_n)$ denote the stable and unstable manifolds of the periodic points x_n, y_n we have by Lemma 3.2:

$$W^s(x_n)\supset\tilde W^s_{\epsilon_2}(x_n)$$
$$W^u(x_n)\supset\tilde W^u_{\epsilon_2}(x_n)$$
$$W^s(y_n)\supset\tilde W^s_{\epsilon_2}(y_n)$$
$$W^u(y_n)\supset\tilde W^u_{\epsilon_2}(y_n).$$

Hence $z_n\in W^s(x_n)\cap W^u(y_n)$, $w_n\in W^s(y_n)\cap W^u(x_n)$. Moreover these intersections are transversal because the dimension of $W^s(x_n)$ (resp. $W^u(x_n)$, $W^s(x_n)$, $W^u(y_n)$) is the same that the dimension of $\tilde W^s_{\epsilon_2}(x_n)$($\tilde W^u_{\epsilon_2}(x_n)$, $\tilde W^s_{\epsilon_2}(y_n)$, $\tilde W^u_{\epsilon_2}(y_n)$). Therefore, by well known results on homoclinic transversal points z_n can be approximated by points in $Per_j(f)$. Since $z_n\to z$ this implies $z\in\Lambda_j(f)$.

Now we are in condition to prove Theorem C. If $f\in\mathcal{F}(M)$ and the $\Omega(f)$-germ is expansive we know by 3.1 that $\overline{Per(f)}=\Omega(f)$. Hence it is sufficient to show that $\Lambda_j(f)$ is a hyperbolic set for all $0\leq j\leq\dim M$. But by [11] $\Lambda_0(f)$ and $\Lambda_{\dim M}(f)$ are finite sets of sources and sinks.

Hence $\Lambda_{\dim M}(f)=Per_{\dim M}(f)$, $\Lambda_0(f)=Per_0(f)$. Now let us consider the case $0<j<\dim M$. Let $TM/\Lambda_j(f)=E^s\oplus E^u$ be the splitting given by Proposition 2.2 We shall prove that there exists $K>0$, $0<\mu<1$ satisfying

$$\|(Tf)^n/E^s_x\|\leq K\mu^n \tag{6}$$

$$\|(Tf)^{-n}/E^u_x\|\leq K\mu^n \tag{7}$$

for all $x\in\Lambda_j(f)$, $n\geq 0$. Let $\mathcal{S}$ be the family of compact f-invariant subsets Σ of $\Lambda_j(f)$ such that there don't exist $K>0$, $0<\mu<1$ satisfying (6), (7) for all $x\in\Sigma$. It is easy to see that if $\{\Sigma_\alpha|\alpha\in\mathcal{A}\}\subset\mathcal{S}$ satisfies $\Sigma_{\alpha'}\subset\Sigma_{\alpha''}$ or $\Sigma_{\alpha''}\subset\Sigma_{\alpha'}$ for all $\alpha',\alpha''\in\mathcal{A}$ then $\bigcap_\alpha\Sigma_\alpha\in\mathcal{S}$. Hence by Zorn's lemma there exists $\Sigma\in\mathcal{S}$ such that $\Sigma'\in\mathcal{S}$ and $\Sigma'\subset\Sigma$

imply $\Sigma' = \Sigma$. Moreover it is proved in [6] that $\Sigma \in \mathcal{S}$ implies that $\{\|(Tf)^{-n}v\| | n \geq 0\}$ is bounded for some $0 \neq v \in E^s/\Sigma$ or $\{\|(Tf)^n v\| | n \geq 0\}$ is bounded for some $0 \neq v \in E^u/\Sigma$. Suppose that we have the first case. Take $x \in \Lambda_j(f)$ such that $v \in E_x^s$. Clearly $\alpha(x) \in \mathcal{S}$ (otherwise $\lim_{n\to+\infty} \|(Tf)^{-n}v\| = \infty$). Then $\alpha(x) = \Sigma$. In particular $x \in \alpha(x)$ and x cannot be periodic because then it would be a point of $Per_j(f)$ by Lemma 3.3 and then Proposition 2.3 $E_x^s = E_x^s(f)$ thus implying that $\{\|(Tf)^{-n}v\| | n \geq 0\}$ is unbounded. But by Proposition 1.1 (applied to $f^{-1}/\Lambda_j(f)$) and Lemma 3.4 for all $\epsilon \geq 0$ there exists $y \in Per(f/\Lambda_j(f))$ such that $d(f^{-n}(x), f^{-n}(y)) \leq \epsilon$ for all $0 \leq n \leq \pi(f, y)$. Put $m = \pi(f, y)$, $z = f^{-m}(x)$. By Lemma 3.3 $y \in Per_j(f)$ and since the spaces $(Tf)^n E_z^s = E_{f^n(z)}^s$ are near to $E_{f^n(y)}^s$ for all $0 \leq n \leq m$, by Proposition 2.4 we have:

$$\|(Tf)^{\pi(f,y)}/E_z^s\| \leq C\lambda^{\pi(f,y)}. \tag{8}$$

On the other hand the existence of v implies:

$$\|(Tf)^{\pi(f,y)}/E_z^s\| \geq (\sup\{\|(Tf)^{-n}v\|/\|v\| \mid 0 \leq n \leq m\})^{-1}. \tag{9}$$

Since x is not periodic, taking ϵ small we can make $m = \pi(f, y)$ arbitrarily large thus obtaining a contradiction between (8) and (9).

§4. PROOF OF THEOREM A

Let M be a 2-dimensional compact smooth boundaryless manifold and $f \in \mathcal{F}(M)$ such that $\Omega(f)$ is isolated. We want to prove that f satisfies Axiom A. By Theorem C it is sufficient to show that the $\Omega(f)$-germ of f is expansive. By Lemma 3.1 $\overline{Per(f)} = \Omega(f)$. Hence $\Omega(f) = \Lambda_0(f) \cup \Lambda_1(f) \cup \Lambda_2(f)$. But by [11] $\Lambda_0(f)$ and $\Lambda_2(f)$ are finite sets of sources and sinks. Therefore $\Lambda_0(f) \cap \Lambda_1(f) = \Lambda_2(f) \cup \Lambda_1(f) = \emptyset$. This implies that we have to show only that the $\Lambda_1(f)$-germ of f is expansive. By the isolation of $\Omega(f)$ it follows (because $\Lambda_1(f) \cap \Lambda_0(f) = \Lambda_1(f) \cap \Lambda_2(f) = \emptyset$) that $\Lambda_1(f)$ is isolated. Hence it is sufficient to show that $f/\Lambda_1(f)$ is expansive. To prove this property we need the next lemma. In its statement let $\bar{W}_\epsilon^s(x)$, $\bar{W}_\epsilon^u(x)$, g, E_x^s, E_x^u denote the discs, subbundles and the power of f given by Propositions 2.2 and 2.3. Fix $\epsilon_0 > 0$, $\delta > 0$ such that $x, y \in \Lambda_1(f)$ and $d(x, y) \leq \delta$ imply that $\bar{W}_{\epsilon_0}^s(x)$ intersects transversely $\bar{W}_{\epsilon_0}^u(y)$ at one and only one point. In these conditions put $[x, y] = \bar{W}_{\epsilon_0}^s(x) \cap \bar{W}_{\epsilon_0}^u(y)$. Moreover if $x \in \Lambda_1(f)$ and $y \in \bar{W}_{\epsilon_0}^s(x)$ let $d_s(x, y) = \inf\{\epsilon | y \in \bar{W}_\epsilon^s(x)\}$. For $x \in \Lambda_1(f)$, $y \in \bar{W}_{\epsilon_0}^u(x)$ define $d_u(x, y) = \inf\{\epsilon | y \in \bar{W}_\epsilon^u(x)\}$. We shall assume that $\delta > 0$ is so small that $d(x, y) < \delta$ implies:

$$g([x, y]) = [g(x), g(y)] \tag{1}$$

$$g^{-1}([x, y]) = [g^{-1}(x), g^{-1}(y)]. \tag{2}$$

Finally if $x \in \Lambda_1(f)$, $y \in \bar{W}_{\epsilon_0}^s(x)(\bar{W}_{\epsilon_0}^u(x))$ let $\lambda(x, y)$ be the compact arc in $\bar{W}_{\epsilon_0}^s(x)$ $(\bar{W}_{\epsilon_0}^u(x))$ with boundary $\{x\} \cup \{y\}$. For $w \in \lambda(x, y)$ define $\bar{E}_w^s = T_w W_{\epsilon_0}^s(x)$.

The following Lemma is a non linear version of the domination inequality in Proposition 2.2:

LEMMA 4.1. *There exist $K > 0$, $0 < \mu < 1$, $\epsilon_1 > 0$ such that if $x_i \in \Lambda_1(f)$ $i = 1, 2$, $m > 0$ and $y = [x_1, x_2]$ satisfy*

$$d_s(g^n(x_1), g^n(y)) \leq \epsilon_1 \tag{3}$$

$$d_u(g^n(x_2), g^n(y)) \leq \epsilon_1 \tag{4}$$

for all $0 \leq n \leq m$ then

$$\frac{d_s(g^n(x_1), g^n(y))}{d_s(x_1, y)} \leq K\mu^n \frac{d_u(g^n(x_2), g^n(y))}{d_u(x_2, y)} \tag{5}$$

for all $0 \leq n \leq m$.

Proof. By the mean value theorem there exist $K > 0$, $\epsilon_1 > 0$ such that $x \in \Lambda_1(f)$,

$y \in \tilde{W}^s_{\epsilon_0}(x)$ and $d_s(g^n(x), g^n(y)) \le \epsilon_1$ for all $0 \le n \le m$ imply that for all $0 \le n \le m$:

$$\frac{d_s(g^n(x), g^n(y))}{d_s(x, y)} \le K \sup \{\|(Tg)^n / \tilde{E}^s_z\| \, | z \in \lambda(x, y)\}.$$

But $\|(Tg)^n / \tilde{E}^s_z\| = \pi_{j=0}^{n-1} \|(Tg)/\tilde{E}^s_{g^j(z)}\|$. Given any $c > 0$ we can take ϵ_1 so small that:

$$\|(Tg)^n / \tilde{E}^s_z\| \le (1+c)^n \|(Tg)^n / E^s_x\|$$

for all $z \in \lambda(x, y)$, $0 \le n \le m$. Then for all $0 \le n < m$:

$$d_s(g^n(x), g^n(y)) \le K(1+c)^n \|(Tg)^n / E^s_x\| d_s(x, y).$$

By a similar argument if $\bar{x} \in \Lambda_1(f)$, $\bar{y} \in \tilde{W}^u_{\epsilon_0}(x)$ and $d_u(g^{-n}(\bar{x}), g^{-n}(\bar{y})) \le \epsilon_1$ for all $0 \le n \le m$, we have:

$$d_u(g^{-n}(\bar{x}), g^{-n}(\bar{y})) \le K(1+c)^n \|(Tg)^{-n} / E^u_{\bar{x}}\| d_u(\bar{x}, \bar{y})$$

for all $0 \le n \le m$. Then if $x \in \Lambda_1(f)$, $y \in \tilde{W}^u_{\epsilon_0}(x)$ and $d_u(g^n(x), g^n(y)) \le \epsilon_1$ for all $0 \le n \le m$, applying the last inequality to $\bar{x} = g^n(x)$, $\bar{y} = g^n(y)$ we obtain:

$$d_u(g^n(x), g^n(y)) \ge K^{-1}(1+c)^{-n} \|(Tg)^{-n} / E^u_{g^n(x)}\|^{-1} d_u(x, y)$$

for all $0 \le n \le m$. Hence if x_1, x_2, y satisfy (3), (4):

$$\frac{d_s(g^n(x_1), g^n(y))}{d_u(g^n(x_2), g^n(y))} \le K^2(1+c)^{2n} \|(Tg)^n / E^s_{x_1}\| \|(Tg)^{-n} / E^u_{g^n(x_2)}\| \cdot \frac{d_s(x_1, y)}{d_u(x_2, y)}.$$

But if ϵ_1 is small enough (3) and (4) mean that $d(g^n(x_1), g^n(x_2))$ is small for all $0 \le n \le m$. This implies (using again the fact that E^s, E^u are one dimensional) that we can assume

$$\|(Tg)^n / E^s_{x_1}\| \cdot \|(Tg)^{-n} / E^u_{g^n(x_2)}\| \le (1+c)^n \|(Tg)^n / E^s_{x_1}\| \cdot \|(Tg)^{-n} / E^u_{g^n(x_1)}\|.$$

Therefore, if $C > 0$, $0 < \lambda < 1$ are the constants given by Proposition 2.2 and $f^N = g$:

$$\frac{d_s(g^n(x_1), g^n(y))}{d_s(g^n(x_2), g^n(y))} \le K^2(1+c)^{3n} C\lambda^{Nn} \frac{d_s(x_1, y)}{d_u(x_2, y)}$$

Take then $K_1 = K^2 C$, $\mu = (1+c)^3 \lambda^N$ and Lemma 4.1 is proved.

As we mentioned before, to prove Theorem A it is sufficient to show that $f/\Lambda_1(f)$ is expansive. Clearly it is sufficient to show that $g/\Lambda_1(f)$ is expansive. We shall show that $g/\Lambda_1(f)$ admits as expansivity constant a number $0 < \delta_1 < \delta$ satisfying the following conditions:

(a) $d(x, y) \le \delta_1$ and $x, y \in \Lambda_1(f)$ imply

$$d_s(x, [x, y]) \le \epsilon_1, \; d_u(y, [x, y]) \le \epsilon_1.$$

(b) Let $\bar{\epsilon}$ be the constant given by Proposition 2.4 applied to $\Lambda_1(f)$. Then $d(x, y) < \delta_1$ and $x, y \in \Lambda_1(f)$ imply $d_1(\tilde{E}^u_{w_1}, \tilde{E}^u_{w_2}) < \bar{\epsilon}/2$ $(d_1(\tilde{E}^s_{w_1}, E^s_{w_2}) < \bar{\epsilon}/2)$ for all w_1, $w_2 \in \lambda(y, [x, y])(\lambda(x, [x, y]))$.

Suppose that $x, y \in \Lambda_1(f)$ and $d(g^n(x), g^n(y)) \le \delta_1$ for all n. If $x \ne y$ we must have $y \notin \tilde{W}^s_{\epsilon_0}(x)$ or $y \notin \tilde{W}^u_{\epsilon_0}(x)$. Suppose $y \notin \tilde{W}^s_{\epsilon_0}(x)$. Let $z = [x, y]$. By (1)

$$g^n(z) = [g^n(x), g^n(y)]$$

for all n. Since $d(g^n(x), g^n(y)) \le \delta_1$ (a) implies

$$d_u(g^n(y), g^n(z)) \le \epsilon_1 \tag{6}$$

for all n. Since $\Lambda_1(f)$ is isolated, inequality (6) implies that $\lambda(y, z) \subset \Lambda_1(f)$ if ϵ_1 was taken small enough. Hence we can find $p \in Per_1(f)$ such that $[p, y] \in \lambda(y, z)$. Let $q = [p, y]$. We claim that we can choose p such that:

$$d_s(g^n(p), g^n(q)) \le \epsilon_1 \tag{7}$$

$$d_u(g^n(y), g^n(q)) \leq \epsilon_1 \tag{8}$$

for all $n \geq 0$. The second inequality is an immediate consequence of (6). To prove (7) assume that it is true for $0 \leq n \leq m$ and we shall prove that it holds also for $n = m+1$. Take $\epsilon_2 > 0$ such that $d_s(a, b) \leq \epsilon_2$ implies $d_s(g(a), g(b)) \leq \epsilon_1$. By Lemma 4.1:

$$d_s(g^n(p), g^n(q)) \leq K\mu^n \frac{d_u(g^n(y), g^n(q))}{d_u(y, q)} d_s(p, q)$$

$$\leq K\mu^n \epsilon_1 \frac{ds(p, q)}{d_u(y, q)} \tag{9}$$

for all $0 \leq n \leq m$. But if p was chosen such that $K\mu^n\epsilon_1 d_s(p, q)/d_u(y, q) \leq \epsilon_2$ for all $n \geq 0$ then (9) proves that $d_s(g^m(p), g^m(q)) \leq \epsilon_2$ and this, by the definition of ϵ_2 implies $d_s(g^{m+1}(p), g^{m+1}(q)) \leq \epsilon_1$. So (7), (8) hold for all $n \geq 0$ and then (9) holds for all $n \geq 0$ thus implying $d_s(g^n(p), g^n(q)) \to 0$ when $n \to +\infty$. Moreover by condition (b) of the definition of δ_1 we have $d_1(\tilde{E}^u_{g^n(w)}, \tilde{E}^u_{g^n(q)}) < \bar{\epsilon}/2$ for all $w \in \lambda(y, z)$. Then $d_1(\tilde{E}^u_{g^n(w)}, E^u_{g^n(p)}) \leq \bar{\epsilon}$ for all $w \in \lambda(y, z)$ and n large. Then by Proposition 2.4

$$\lim_{n \to +\infty} \|(Tg)^n/\tilde{E}_w{}^u\|^{1/n} \geq \lambda^{-N} > 1$$

thus contradicting the fact that the length of $g^n(\lambda(y, z)) = \lambda(g^n(y), g^n(z))$ remains bounded when $n \to +\infty$.

Proof of Corollary 1. Let f, $\mathfrak{U}$ and U be as in the statement of Corollary 1. By Theorem A, if we prove $f \in \mathcal{F}(M)$, we are done. Suppose on the contrary that some $g \in \mathfrak{U}$ has a periodic point x_0 that is not hyperbolic. Then either the spectral radius of $(Tg)^m/T_{x_0}M$ (where m is the period of x_0) is ≤ 1 or the spectral radius of $(Tg)^{-m}/T_{x_0}M$ is ≤ 1. Suppose that we have the first situation. We claim that there exists $\hat{g}$ arbitrarily near to g (so that we can suppose $\hat{g} \in \mathfrak{U}$) such that

(a) $x_o \in Per(\hat{g})$ and the eigenvalues of $(T\hat{g})^m/T_{x_0}M$ have modulus <1.

(b) There exists $y_0 \in Per(\hat{g})$ such that $\hat{g}^n(y_0) \in U$ for all n, and z_0 such that $\hat{g}^n(z_0) \in U$ for all n, $\lim d(\hat{g}^n(z_0), \hat{g}^n(x_0)) = 0$ and $\lim_{n \to +\infty} d(\hat{g}^{-n}(y_0), \hat{g}^{-n}(z_0)) = 0$. We shall divide the proof of the claim in two cases.

(1) Both eigenvalues of $(Tg)^m/T_{x_0}M$ have modulus one. Then we can take $g_1 C^1$ near to g and polar coordinates (r, θ) with center at x_0 such that g_1 has the form $g_1{}^m(r, \theta) = (r, \theta + \alpha)$. Then we can approximate g_1 by $\hat{g}$ such that $g_1{}^n(x_0) = \hat{g}^n(x_0)$ for all n and in the same polar coordinates satisfies $\hat{g}^m(r, \theta) = (\psi(r), \theta + \bar{\alpha})$ where $\bar{\alpha} = 2\pi q$, q rational and for some small $r_0 \psi(r) < r$ for $r < r_0$ and $\psi(r) > r$ if $r > r_0$. The points y_0, z_0 with coordinate $(r_0, 0)$, $(r_1, 0)$ with $r_1 < r_0$ satisfy condition (b).

(2) $(Tg)^m/T_{x_0}M$ has eigenvalues $\lambda_0 = \pm 1$ and λ with $|\lambda| < 1$. Then we can find $g_1 C^1$ near to g and coordinates (t, u) in a neighborhood of x_0 such that $g_1{}^m(t, u) = (\lambda_0 t, \lambda u)$. Now take a C^1 perturbation $\hat{g}$ of g satisfying $\hat{g}^n(x_0) = g_1{}^n(x_0)$ for all n and such that in the coordinates system (t, u) has the form $\hat{g}^m(t, u) = (\varphi(t), \lambda u)$ where φ: $\mathbf{R} \hookleftarrow$ satisfies $\varphi(0) = 0$, $|\varphi'(\theta)| < 1$ and for some small t_0, $|\varphi(t)| < |t|$ if $|t| < |t_0|$, $|\varphi(t)| > |t|$ if $|t| > t_0$. Then the points y_0, z_0 with coordinates $(t_0, 0)$, $(t_1, 0)$ with $|t_1| < |t_0|$ satisfy the required condition.

If $\hat{g} \in \mathfrak{U}$ we should have $\Omega(\hat{g}) = \bigcap_n \hat{g}^n(U)$. Therefore by (b) $z_0 \in \Omega(\hat{g})$. But z_0 is asympotic with the sink x_0 thus implying $z_0 \notin \Omega(\hat{g})$.

Proof of Corollary 2. We can suppose that D^2 is embedded in S^2 (the two-dimensional sphere) and f_0 is extended to a diffeomorphism $f \in Diff^1(S^2)$ such that

(a)

$$f(D^2) \subset \mathrm{Int}\,(D^2)$$

$$f^{-1}(S^2 - D^2) \subset \mathrm{Int}\,(S^2 - D^2).$$

(b) $\bigcap_{n\geq 0} f^{-n}(S^2 - D^2)$ is a fixed point p and the eigenvalues of $(Tf)/T_pS^2$ have modulus >1.

(c) Every g near to f satisfies:

$$\bigcap_{n\geq 0} g^n(D^2) \subset \Omega(g).$$

By well known properties of hyperbolic fixed points it follows that $g(D^2) \subset \operatorname{Int}(D^2)$, $g^{-1}(S^2 - D^2) \subset S^2 - D^2$ and that $\bigcap_{n\geq 0} g^{-n}(S^2 - D^2)$ reduces to a fixed point p_g for all g nearby f. Since the eigenvalues of $(Tf)/T_pS^2$ have modulus >1 there exists a basis of compact neighborhoods V of p such that $f^{-1}(V) \subset \operatorname{Int}(V)$. Suppose V so small that $f(V) \cap D^2 = \emptyset$. We claim that if g is near to f then:

$$\bigcap_n g^n(V \cup D^2) = \{p_g\} \cup \left(\bigcap_{n\geq 0} g^n(D^2)\right).$$

Suppose $g^n(x) \in V \cup D^2$ for all n. If for some n_0, $g^{n_0}(x) \in D^2$ then $g^n(x) \in D^2$ for $n \geq n_0$ because $g(D^2) \subset D^2$, and $g^n(x) \in D^2$ also for $n < n_0$ because if for some $n < n_0$ $g^{-n}(x) \notin D^2$ then $g^{-n}(x) \in V$ and since we can assume that $g^{-n+1}(x) \notin V$, we obtain $g^{-n+1}(x) \in D^2$ and then $g(V) \cap D^2 \neq \emptyset$. In a similar way we can prove that if $g^{n_0}(x) \in V$ for some n_0 then $g^n(x) \in V$ for all n. Hence we have proved: $\bigcap_n g^n(V \cup D^2) = \left(\bigcap_n g^n(V)\right) \cup \left(\bigcap_n g^n(D^2)\right)$. Then

$$\bigcap_n g^n(V \cup D^2) = \left(\bigcap_{n\geq 0} g^{-n}(V)\right) \cup \left(\bigcap_{n\geq 0} g^n(D^2)\right) = \{p_g\} \cup \left(\bigcap_{n\geq 0} g^n(D^2)\right) = \Omega(g).$$

Applying Corollary 1 we obtain that f satisfies Axiom A. Then $\Lambda = \bigcap_{n\geq 0} f^n(D^2)$ is a hyperbolic attractor. But since D^2 is a disc the Lefschetz number $L(f^n/D^2)$ of f^n/D^2 is ± 1. On the other hand by [15] $L(f^n/D^2) = \Sigma(-1)^{u_i}\sigma_i$ where $p_1, \ldots, p_l$ are the fixed points of f^n, u_i the dimension of the unstable subspace $E^u_{p_i}$ and $\sigma_i = \pm 1$ if $(Tf)^n/E^u_{p_i}$ preserves or reverses orientation. Since Λ is a hyperbolic set $u_1 = \cdots = u_l$ and if $TM/\Lambda = E^s \oplus E^u$ is the hyperbolic splitting of TM/Λ observe that we can extend E^u to a neighborhood of Λ. Since the sets $g^n(D^2)$ are a basis of neighborhoods of Λ we then have a orientable splitting because $g^n(D^2)$ is simply connected. Then $\sigma_1 = \cdots = \sigma_l$. Hence f^n has a unique fixed point for all n. This implies that Λ contains a unique point.

§5. PROOF OF THEOREM B

A diffeomorphism f: $M \hookleftarrow$ is topologically mixing if for all open sets U, V there exists N such that $f^n(U) \cap V \neq \emptyset$ for all $n \geq N$. In this section we give a sufficient condition for a diffeomorphism to be topologically mixing and we shall prove that this condition is open. Finally, we shall construct a DA[16] diffeomorphism f of T^3 satisfying that condition and having two hyperbolic fixed points x_0, x_1 with stable manifolds of different dimension. Then every diffeomorphism $g \in \mathit{Diff}^1(T^3)$ nearby f is topologically mixing and it is not Anosov.

Let $f \in \mathit{Diff}^1(M)$ such that:

(a) There exists a continuous Tf-invariant splitting $TM = E^s \oplus E^u \oplus E^c$, with E^u, E^c one dimensional, and a constant $\lambda > 1$ satisfying:

$$\|(Tf)^{-1}/E^u_x\| < \lambda^{-1}$$

$$\|(Tf)/E^s_x\| < 1$$

$$\|(Tf)/E^s_x\| \cdot \|(Tf)^{-1}/E^c_{f(x)}\| < 1$$

$$\|(Tf)^{-1}/E^u_x\| \cdot \|(Tf)/E^c_{f^{-1}(x)}\| < 1$$

for all $x \in M$.

(b) There exists an f-invariant lamination[4] $\mathcal{F}^c$ of M such that every laminae[4] of f is dense, tangent to E^c and f is plaque expansive.

(c) By [4] there exists an f-invariant lamination $\mathcal{F}^u$ tangent to E^u and the laminaes of $\mathcal{F}^u$ are 1–1 immersions of $\mathbf{R}$. Let us say that λ is a u-interval if it is the bounded connected component of a laminae of $\mathcal{F}^u$ without two points. Denote $l(\lambda)$ the length of the u-interval λ, and suppose that there exist an open set $U_0 \subset M$ and constants $\delta_1 > 0$, $\delta_2 > 0$ such that every u-interval λ with $l(\lambda) \geq \delta_1$ contains a u-interval β with $l(\beta) > \delta_2$ such that $\beta \cap \bar{U}_0 = \emptyset$. Moreover $\lambda\delta_2 > \delta_1$.

(d) Let us say that $\lambda \subset M$ is a c-interval if λ is the bounded connected component of a laminae of $\mathcal{F}^c$ without two points. Let $l(\lambda)$ be its length. Then there exist an open set $U_1 \subset U_0$ and constants $\mu > 1$, $\delta_3 > 0$, $\delta_4 > 0$ such that $\|(Tf)^{-1}/E_x^c\| < \mu^{-1}$ for all $x \notin U_1$ and for every c-interval λ with $l(\lambda) \geq \delta_3$ there exists a c-interval $\beta \subset \lambda$ with $\lambda(\beta) > \delta_4$ such that $\beta \cap \bar{U}_1 = \emptyset$. Moreover $\mu\delta_4 > 2\delta_3$.

(e) There exists $\delta_5 > 0$ satisfying $\mu\delta_5 > \delta_3$ such that for every laminae $l \in \mathcal{F}^c$, every bounded connected component of $l \cap (U_0 - \bar{U}_1)$ has length $> \delta_5$.

We shall prove Theorem B in three steps:

Step I

A diffeomorphism satisfying conditions (a)–(c) is topologically mixing. For the proof of this property we shall need the following lemmas:

LEMMA 5.1. *For all open sets $U \subset M$ there exists $k > 0$ such that every c-interval with length $\geq k$ intersects U.*

Proof. Suppose that for all integer $n > 0$ there exists a c-interval λ_n with $l(\lambda_n) > n$ and $\lambda_n \cap U = \emptyset$. Take $x_n \in \lambda_n$. We can suppose that the sequence x_n converges to a point $x \in M$ when $n \to +\infty$. Since the laminae α of $\mathcal{F}^c$ containing x is dense, there exists a c-interval $\beta \subset \alpha$ such that $\beta \cap U \neq \emptyset$. Hence if x_n is near to x and n is so large that $l(\lambda_n) > 2l(\beta)$ it follows that $\lambda_n \cap \neq \emptyset$.

LEMMA 5.2. *For all u-interval λ there exists $x \in \lambda$ and $N > 0$ such that $f^n(x) \notin U_0$ for all $n \geq N$.*

Proof. By condition (a) we have $l(f^n(\lambda)) > \lambda^n l(\lambda)$. Take N such that $l(f^N(\lambda)) > \delta_1$. We claim that for all $n \geq N$ there exists a u-interval $\lambda_n \subset f^n(\lambda)$ such that $l(\lambda_n) > \delta_2$, $\lambda_{n+1} \subset f(\lambda_n)$ and $\lambda_n \cap U_0 = \emptyset$. For $n = N$ the property is true by condition (c). Suppose that we have found $\lambda_N, \lambda_{N+1}, \ldots, \lambda_m$. Then $l(f(\lambda_m)) > \lambda l(\lambda_m) > \lambda\delta_2 > \delta_1$. Then by (c) the u-interval $f(\lambda_m)$ contains a u-interval λ_{m+1} such that $l(\lambda_{m+1}) > \delta_2$ and $\lambda_{m+1} \cap U_0 = \emptyset$ and this proves our claim. Now take $x \in \bigcap_{n \geq N} f^{-n}(\lambda_n)$. Then for $n \geq N$ $f^n(x) \in \bar{\lambda}_n \subset M - U_0$.

Now let us prove Step I. Let U, V be open subsets of M. Take a u-interval $\lambda \subset U$ and $x \in \lambda$ such that for large values of n, say $n \geq n_0$, $f^n(x) \notin U_0$. Take a c-interval $\beta \subset U$ containing x. If we prove $\lim_{n \to +\infty} l(f^n(\beta)) = \infty$ by Lemma 5.1 we have $f^n(U) \cap V \supset f^n(\beta) \cap V \neq \emptyset$ for large values n and we are done. To prove $\lim l(f^n(\beta)) = \infty$ observe that $f^{n_1}(\beta) \cap U_1 \neq \emptyset$ for some $n_1 > 0$. Otherwise we should have $f^n(\beta) \cap U_1 = \emptyset$ for all $n \geq n_0$ and then by (d) $l(f^n(\beta)) > \mu^{n-n_0} l(f^{n_0}(\beta))$ and by condition (d) this implies $f^n(\beta) \cap U_1 \neq \emptyset$ for large values of n. But $f^{n_1}(\beta) \cap U_1 \neq \emptyset$ for some $n_1 \geq n_0$ implies by (c) that $f^{n_1}(\beta)$ contains a c-interval β_0 containing $f^{n_1}(x)$ and not intersecting U_1. Then $l(f^{n_1+1}(\beta)) > \mu l(f^{n_1}(\beta_0)) > \mu\delta_5 > \delta_3$. Now we claim $l(f^{n_1+j}(\beta)) > 2^{(j-1)}\delta_3$ for all $j \geq 1$. For $j = 1$ it is already proved. Suppose that it is true for $j = 1, \ldots, l-1$. Then $f^{n_1+l-1}(\beta)$ contains disjoint c-intervals $\beta_1, \ldots, \beta_2 l-2$ such that $l(\beta_i) > \delta_3$ for $i = 1, \ldots, 2^{l-2}$. Then, by (d), $\beta_i \supset \beta_i'$ such that $l(\beta_i') > \delta_4$, $\beta_i' \cap U_1 = \emptyset$. Therefore: $f^{n_1+l}(\beta) \supset \bigcup f(\beta_i')$ and since $\beta_i' \cap U_1 = \emptyset$: $l(f^{n_1+l}(\beta)) \geq \Sigma_i l(f(\beta_i')) > \Sigma_i \mu l(\beta_i') > \mu 2^{l-2}\delta_4 > 2^{l-1}\delta_3$.

Step II

The set of diffeomorphisms of M satisfying conditions (a)–(c) is open in $\mathit{Diff}^1(M)$. If $f \in \mathit{Diff}^1(M)$ satisfies (a)–(c) then every g nearby f has a continuous Tf-

invariant splitting $TM = \tilde{E}^s \oplus \tilde{E}^u \oplus \tilde{E}^c$ [4], and also by [4] there exist laminations $\tilde{\mathcal{F}}^c$, $\tilde{\mathcal{F}}^u$ tangent to E^c, E^u, invariant under f, and (by the plaque expansivity of f) a homeomorphism $h: M \hookleftarrow$ such that if λ is a laminae of $\mathcal{F}^c$ then $h(\lambda)$ is a laminae of $\tilde{\mathcal{F}}^c$ and $g(h(\lambda)) = h(f(\lambda))$ for every laminae λ. In particular g is again plaque expansive respect to $\tilde{\mathcal{F}}^c$. Define $\tilde{U}_0 = h(U_0)$, $\tilde{U}_1 = h(U_1)$. Then g satisfies conditions (a)–(c) with respect to the splitting $\tilde{E}^s \oplus \tilde{E}^u \oplus \tilde{E}^c$, the lamination $\tilde{\mathcal{F}}^c$, the open sets $\tilde{U}_0$, $\tilde{U}_1$ and the constants $\lambda, \mu, \delta_1, \ldots, \delta_5$ (the same as before).

Step III

There exists $f \in Diff^1(T^3)$ satisfying (a)–(c) and having fixed points x_0, x_1 with stable manifolds of different dimensions.

The idea of the construction of f is to take a linear Anosov diffeomorphism of T^3 with three different eigenvalues $\lambda_1 > \lambda_2 > 1 > \lambda_3$ and to make an isotopy of f_0 with support contained in a neighborhood of a fixed point x_0 (in order to change the dimension of the stable manifold of x_0). This deformation of f_0 can be constructed preserving the foliation associated to the central eigenvalue λ_2 and the foliation associated to λ_2 will be $\mathcal{F}^c$.

Let f_0: $T^3 \hookleftarrow$ be a linear Anosov diffeomorphism with eigenvalues λ_1, λ_2, λ_3 satisfying $\lambda_1 > \lambda_2 > 1 > \lambda_3$. Let $T(T^3) = E_1 \oplus E_2 \oplus E_3$ be the splitting associated to these eigenvalues. Take fixed points x_0, x_1, a neighborhood V of x_0 not containing x_1 and a diffeomorphism φ: $V \to D_1$ (where $D_r = \{x \in \mathbf{R}^3 | \|x\| < r\}$ satisfying:

$$(\varphi f_0 \varphi^{-1})(x_1, x_2, x_3) = (\lambda_1 x_1, \lambda_2 x_2, \lambda_3 x_3)$$

for all $x = (x_1, x_2, x_3) \in D_r$ where r is so small that $f_0 \varphi^{-1}(D_r) \subset D_1$. Define $U_0 = \varphi^{-1}(D_{r/2})$, $U_1 = \varphi^{-1}(D_{r/4})$. Take positive constants δ_1, δ_2, δ_3, δ_4 such that if λ is an interval contained in a leaf of the foliation associated to E_2 (resp. E_1) with length $> \delta_1(\delta_2)$ then λ contains an interval with length $> \delta_3(\delta_4)$ such that $\beta \cap U_0 = \emptyset$ $(\beta \cap U_1 = \emptyset)$. Take $\delta_5 > 0$ such that for every leaf l of the foliation associated to E_2, every connected component of $l \cap (U_1 - \bar{U}_0)$ has length $u > \delta_5$. Now take $N > 0$ such that:

$$\lambda_1^N \delta_2 > \delta_1$$

$$\lambda_2^N \delta_4 > 2\delta_3$$

$$\lambda_2^N \delta_5 > \delta_3.$$

Let $f_1 = f_0^N$, f_1 satisfies properties (a)–(c) respect to the neighborhoods U_0, U_1 and the constants $\delta_1, \ldots, \delta_5$, $\lambda = \lambda_1^N$, $\mu = \lambda_2^N$. Now define f by:

$$f(x) = f_1(x) \qquad x \notin U_0$$

$$\varphi f(x) = (\lambda_1 \varphi_1(x), \mu k(\varphi(x)), \lambda_3^N \varphi_3(x))$$

if $x \in U_0$, where $\varphi(x) = (\varphi_1(x), \varphi(x), \varphi_3(x))$ and k: $\mathbf{R}^3 \to \mathbf{R}$ satisfies $(D_2 k)(0) \leq 1$, $k(x) = \mu x_2$ if $\|x\| > 1/4$ and $(D_2 k)(x) > 0$ for all $x \in \mathbf{R}^3$. Then f is a diffeomorphism because if $L = \varphi f \varphi^{-1}$, then $\det L'(x) = \lambda \lambda_3^N (D_2 k)(x) > 0$. Moreover x_0, x_1 are fixed points of f and $\dim W^s(x_0) = 2$, $\dim W^s(x_1) = 1$. Finally let us show that f satisfies (a)–(c). Define the cones:

$$C_\epsilon^s(x) = \{v \in T_x(T^3) | \epsilon \|\pi_3 v\| \geq \|(\pi_1 + \pi_2)v\|\}$$

$$C_\epsilon^u(x) = \{v \in T_x(T^3) | \epsilon \|\pi_1 v\| \geq \|(\pi_2 + \pi_3)v\|\}$$

where π_i: $T(T^3) \to E_i$ are the projections associated to be splitting $T(T^3) = E_1 \oplus E_2 \oplus E_3$. Given $\epsilon > 0$, taking k such that $\sup\{(D_i k)(x) | x \in \mathbf{R}^3, i = 1, 3\}$ is small enough, it is easy to verify that:

$$(Tf)^{-1} C_\epsilon^s(x) \subset C_\epsilon^s(f^{-1}(x)), (Tf) C_\epsilon^u(x) \subset C_\epsilon^u(f(x)). \tag{1}$$

Using the methods in [16] it follows that there exist continuous Tf-invariant subbundles such that $\tilde{E}_x^s \subset C_\epsilon^s(x)$, $\tilde{E}_x^u \subset C_\epsilon^u(x)$ for all $x \in T^3$. Observing that $(Tf)E_2 = E_2$ and defining $\tilde{E}^c = E_2$ we have a continuous Tf-invariant splitting $T(T^3) = \tilde{E}^s \oplus \tilde{E}^c \oplus \tilde{E}^u$ and a foliation $\mathcal{F}^c$ tangent to $\tilde{E}^c$ with all leaves dense.

Since $\mathcal{F}^c$ is a normally hyperbolic foliation for f it follows[4] that f is plaque expansive. Finally since f_1 satisfies (a)–(e) with respect to the constants $\lambda, \mu, \delta_1, \ldots, \delta_5$ and the sets U_0, U_1 it follows from (1) that taking ϵ small enough f satisfy again conditions (a)–(e).

REFERENCES

1. R. Bowen: Topological entropy and axiom A, Global analysis, *Proc. Symp. Pure Math.* (1970) 14.
2. J. Franks: Necessary conditions for Ω-stability of diffeomorphisms, *Trans. Am. math. Soc.* **158**(1971), 301–308.
3. J. Guckenheimer: A strange, strange attractor, preprint Santa Cruz (1975).
4. M. Hirsch, C. Pugh and M. Shub: Invariant manifolds. *Lecture Notes*, Vol. 583. Springer Verlag.
5. R. Mañé: Expansive diffeomorphisms, *Dynamical Systems*, Warwick 1974. Lecture Notes, 468. Springer Verlag.
6. R. Mañé: Quasi-Anosov diffeomorphisms and hyperbolic manifolds, *Trans. Am. math. Soc.* **229** (1977), 351–370.
7. S. Newhouse: Quasi-elliptic periodic points in Conservative Dynamical Systems, to appear.
8. J. Palis: A note on Ω-stability, Global analysis, *Proc. Symp. Pure Math.* (1970) 14.
9. J. Palis and S. Smale: Global analysis. *Proc. Symp. Pure Math.* (1970) 14.
10. J. Palis and C. Pugh: Fifty problems in dynamical systems, Warwick, *Lecture Notes*, Vol. 468, pp. 345–365. Springer Verlag.
11. V.A. Pliss: Analysis of the necessity of the conditions of Smale and Robbin for structural stability for periodic systems of differential equations. *Diff. Uravnenija* **8**(1972).
12. V.A. Pliss: On a conjecture due to Smale, *Diff. Uravnenija* **8**(1972).
13. C. Pugh: An improved closing lemma and a general density theorem. *Am. J. Math.* **89**(1967) 1010–1021.
14. C. Robinson: Structural stability of C^1 diffeomorphisms. *J. diff. Eqns.* **22** (1976), 28–73.
15. S. Smale: Differentiable dynamical systems, *Bull. Am. math.* Soc. **73**(1967), 747–817.
16. R. Williams: The *DA* maps of Smale and structural stability, Global analysis, *Proc. Symp. Pure Math.* (1970) 14.

I.M.P.A. Rio de Janeiro

Ergod. Th. & Dynam. Sys. (1981), **1**, 95–102
Printed in Great Britain

A proof of Pesin's formula

RICARDO MAÑÉ†
Instituto de Matemática Pura e Aplicada, Rio de Janeiro, Brazil

(*Received 2 October* 1980)

Abstract. Let $f: M \to M$ be a diffeomorphism of a compact manifold M and let $\chi: M \to R$ be defined by putting $\chi(x)$ equal to the sum of the non-negative characteristic exponents of f at x, each being counted with its multiplicity. If μ is an f-invariant probability of M which is absolutely continuous relative to Lebesgue measure, then Pesin has proved the entropy, $h_\mu(f)$, is given by $h_\mu(f) = \int_M \chi \, d\mu$. We prove this formula without using the theory of stable manifolds.

The Lyapunov exponents of a diffeomorphism f of a compact boundaryless manifold M are defined by Oseledec's theorem which states that, for any f-invariant probability measure μ on M, for almost every point $x \in M$ there exists a unique family of numbers $\lambda_1(x) > \cdots > \lambda_m(x)$ (the Lyapunov exponents of f at x) and a unique splitting $T_xM = E_1(x) \oplus \cdots \oplus E_m(x)$ such that

$$\lim_{n \to \pm\infty} (1/n) \log \|(D_xf^n)v\| = \lambda_j(x)$$

for all $0 \neq v \in E_j(x)$, $1 \leq j \leq m$. The entropy $h_\mu(f)$ and the Lyapunov exponents are connected by the inequality [**2**]

$$h_\mu(f) \leq \int_M \chi \, d\mu, \tag{1}$$

where χ is defined by

$$\chi(x) = \sum_{\lambda_j(x) \geq 0} \lambda_j(x) \dim E_j(x).$$

Pesin's formula [**1**] states that, when μ is absolutely continuous with respect to the Lebesgue measure of M and the derivative of f is Hölder continuous, then (1) is an equality

$$h_\mu(f) = \int_M \chi \, d\mu. \tag{2}$$

The proof of this result given by Pesin relies heavily on the stable manifold theory, especially in the absolute continuity of the measurable foliation given by the stable

† Address for correspondence: Dr R. Mañé, Instituto de Matemática Pura e Aplicada, Rua Luiz de Camões 68, 20060 Rio de Janeiro, RJ, Brazil.

Originally published in: *Ergodic Theory Dynamical Systems 1 (1981), no. 1, 95–102,*

manifolds. The purpose of this note is to give a different, simpler proof of this result which does not make use of the stable manifolds. Clearly the stable manifold theory is of great intrinsic interest and its absolute continuity properties are a fundamental tool in the remarkable theory that Pesin developed for the case when (besides the hypothesis of (2)) the Lyapunov exponents are $\neq 0$ a.e. Nevertheless, it seems interesting to have a proof of (2) exhibiting the simple, elementary nature of this formula.

Our proof will be based on a general lower estimate for metric entropy that makes it possible to avoid the use of partitions. Let us suppose that $g: M \circlearrowleft$ is a map and $\rho: M \to (0, 1)$ a function. If $x \in M$ and $n \geq 0$, define

$$S_n(g, \rho, x) = \{y \mid d(g^j(x), g^j(y)) \leq \rho(g^j(x)), 0 \leq j \leq n\}.$$

If μ is a measure on M and g and ρ are measurable, define

$$h_\mu(g, \rho, x) = \limsup_{n \to +\infty} (1/n)(-\log \mu(S_n(g, \rho, x))).$$

PROPOSITION. *If $g: M \circlearrowleft$ is measurable, μ is a g-invariant probability measure on M and $\nu \gg \mu$ is another measure on M (not necessarily g-invariant), then*

$$h_\mu(g) \geq \int_M h_\nu(g, \rho, x)\, d\mu(x)$$

for every function $\rho: M \to (0, 1)$ such that $\log \rho$ is μ-integrable.

The proof of this proposition requires several lemmas.

LEMMA 1. *If $\sum_1^\infty x_n$ is a series with $0 < x_n < 1$ for all n such that $\sum_1^\infty n x_n < +\infty$, then*

$$\sum_1^\infty x_n \log(1/x_n) < +\infty.$$

Proof. Let S be the set of integers $n \geq 1$ such that $\log(1/x_n) < n$. Then $n \notin S$ implies $x_n \leq \exp(-n)$. Hence

$$\sum_1^\infty x_n \log(1/x_n) \leq \sum_{n \in S} n x_n + \sum_{n \notin S_n} (\sqrt{x_n})(\sqrt{x_n}) \log(1/x_n).$$

But $(\sqrt{t}) \log(1/t) \leq 2e^{-1}$ for all $t > 0$. Hence

$$\sum_1^\infty x_n \log(1/x_n) \leq \sum_1^\infty n x_n + 2e^{-1} \sum_{n \notin S_n} \sqrt{x_n}$$

$$\leq \sum_1^\infty n x_n + 2e^{-1} \sum_1^\infty \exp(-\tfrac{1}{2} n). \qquad \square$$

LEMMA 2. *If μ is a probability measure on M and $\rho: M \to (0, 1)$ is such that $\log \rho$ is μ-integrable, then there exists a countable partition $\mathcal{P}$ of M with entropy $H(\mathcal{P}) < +\infty$ such that, if $\mathcal{P}(x)$ denotes the atom of $\mathcal{P}$ containing x, then*

$$\operatorname{diam} \mathcal{P}(x) \leq \rho(x)$$

for a.e. $x \in M$.

Proof. There exists $C>0$ and $r_0>0$ such that, for all $0<r\le r_0$, there exists a partition $\mathcal{P}_r$ of M whose atoms have diameter less than or equal to r and such that the number of atoms in $\mathcal{P}_r$ (denoted by $|\mathcal{P}_r|$) satisfies

$$|\mathcal{P}_r|\le C(1/r)^{\dim M}.$$

Now define $U_n=\{x\mid \exp(-(n+1))<\rho(x)\le\exp(-n)\}$, $n\ge 0$. Since $\log\rho$ is integrable, we have

$$\sum_1^\infty n\mu(U_n)<\infty. \tag{3}$$

By lemma 1,

$$\sum_1^\infty \mu(U_n)(-\log\mu(U_n))<+\infty. \tag{4}$$

Finally, define $\mathcal{P}$ as the partition given by the sets $Q\cap U_n$, with $n\ge0$ and $Q\in\mathcal{P}_{r_n}$, $r_n=\exp(-(n+1))$, such that $\mu(Q\cap U_n)\ne 0$. Then

$$H(\mathcal{P})=\sum_{n=0}^\infty\Bigg(-\sum_{\substack{P\in\mathcal{P}\\ P\subset U_n}}\mu(P)\log\mu(P)\Bigg).$$

Recalling the well known inequality

$$-\sum_{i=1}^m x_i\log x_i\le\left(\sum_1^m x_i\right)\left(\log m-\log\sum_1^m x_i\right)$$

which holds for any set of real numbers $0<x_i\le1$, $i=1,\dots,m$, we obtain

$$H(\mathcal{P})\le\sum_0^\infty\mu(U_n)[\log|\mathcal{P}_{r_n}|-\log\mu(U_n)]$$

$$\le\sum_0^\infty\mu(U_n)[\log C+(\dim M)(n+1)-\log\mu(U_n)].$$

By (3), (4) and lemma 1, the last series converges and then $H(\mathcal{P})<+\infty$. Moreover, $x\in U_n$ implies that $\mathcal{P}(x)$ is contained in an atom of $\mathcal{P}_{r_n}$. Then $\operatorname{diam}\mathcal{P}(x)<r_n=\exp(-(n+1))<\rho(x)$. □

Proof of proposition. To prove the proposition take the partition $\mathcal{P}$ given by lemma 2. Denote by $\mathcal{P}_n(x)$ the atom of $\mathcal{P}\vee g^{-1}(\mathcal{P})\vee\cdots\vee g^{-n}(\mathcal{P})$ which contains x. By the Shannon–McMillan–Breiman theorem,

$$h_\mu(g)\ge h_\mu(\mathcal{P},g)=\int_M\lim_{n\to+\infty}\frac{1}{n}[-\log\mu(\mathcal{P}_n(x))]\,d\mu(x).$$

To complete the proof we have to show that

$$\lim_{n\to+\infty}\frac{1}{n}[-\log\mu(\mathcal{P}_n(x))]\ge h_\nu(g,\rho,x)\qquad \mu\text{-a.e.} \tag{5}$$

Let $\mathcal{P}_\infty$ be the σ-algebra generated by the partitions $\mathcal{P}_n$, $n\ge0$. Let $k:M\to\mathbb{R}$ be a ν-integrable function, measurable with respect to the σ-algebra $\mathcal{P}_\infty$, such that

$$\int_A k\,d\nu=\mu(A) \tag{6}$$

for all $A \in \mathcal{P}_\infty$. Such a function exists by the Radon–Nikodym theorem. It follows from (6) that

$$\lim_{n\to+\infty} \frac{\mu(\mathcal{P}_n(x))}{\nu(\mathcal{P}_n(x))} = k(x) \qquad \nu\text{-a.e.} \tag{7}$$

Hence

$$\frac{1}{n}\log \mu(\mathcal{P}_n(x)) = \frac{1}{n}\log \nu(\mathcal{P}_n(x)) + \frac{1}{n}\log \frac{\mu(\mathcal{P}_n(x))}{\nu(\mathcal{P}_n(x))} \tag{8}$$

and, using (7), we obtain

$$\lim_{n\to+\infty} \frac{1}{n}[-\log \mu(\mathcal{P}_n(x))] = \lim_{n\to+\infty} \frac{1}{n}[-\log \nu(\mathcal{P}_n(x))] \qquad \mu\text{-a.e.} \tag{9}$$

Moreover, it is easy to see that $\mathcal{P}_n(x) \subset S_n(g, \rho, x)$. Therefore, (5) follows from (9). □

Before going into the proof of Pesin's formula we shall prove a technical lemma. The reader familiar with the Hadamard graph transform method for constructing invariant manifolds will recognize in this lemma one of the steps of that method. In the statement of the lemma we shall use the following definitions.

Definition. If E is a normed space and $E = E_1 \oplus E_2$ is a splitting, we define $\gamma(E_1, E_2)$ as the supremum of the norms of the projections $\pi_i : E \to E_i$, $i = 1, 2$, associated with the splitting. Moreover, we shall say that a subset $G \subset E$ is a (E_1, E_2)-graph if there exists an open $U \subset E_2$ and a C^1 map $\Psi : U \to E_1$ satisfying $G = \{x + \Psi(x) \mid x \in U\}$. The number $\sup\{\|\Psi(x) - \Psi(y)\| / \|x - y\| \mid x, y \in U\}$ will be called the *dispersion* of G.

LEMMA 3. *Given $\lambda > \beta > 1$, $\alpha > 0$ and $c > 0$, there exists $\delta > 0$ with the following property. If E is a finite-dimensional normed space and $E = E_1 \oplus E_2$ a splitting with $\gamma(E_1, E_2) \le \alpha$, and F is a C^1 embedding of a ball $B_r(0) \subset E$ into another Banach space E' satisfying*

(i) *D_0F is an isomorphism and $\gamma((D_0F)E_1, (D_0F)E_2) \le \alpha$;*
(ii) *$\|(D_0F) - (D_xF)\| \le \delta$ for all $x \in B_r(0)$;*
(iii) *$\|(D_0F)v\| \ge \lambda \|v\|$ for all $v \in E_2$;*
(iv) *$\|(D_0F)/E_1\| \le \beta$;*

then for every (E_1, E_2)-graph G with dispersion $\le c$ contained in the ball $B_r(0)$, its image $F(G)$ is a $((D_0F)E_1, (D_0F)E_2)$-graph with dispersion $\le c$.

Proof. Identity E with $E_1 \times E_2$ and E' with $(D_0F)E_1 \times (D_0F)E_2$. Write the map F in the form

$$F(x, y) = (Lx + p(x, y), Ty + q(x, y)),$$

where $L = (D_0F)/E_1$, $T = (D_0F)/E_2$. It follows that the partial derivatives of p and q with respect to x and y have norm $\le \delta\alpha$. Let $U \subset E_2$ be an open set and $\Psi : U \to E_1$ a map whose graph $\{(\Psi(v), v) \mid v \in U\}$ is G. Then,

$$F(G) = \{(L\Psi(v) + p(\Psi(v), v), Tv + q(\Psi(v), v)) \mid v \in U\}.$$

To study this set define $\Phi : U \to (D_0F)E_2$ by

$$\Phi(v) = Tv + q(\Psi(v), v).$$

If $v, w \in U$,

$$\|\Phi(v)-\Phi(w)\| \geq \|T(v-w)\| - \|q(\Psi(v), v) - q(\Psi(w), w)\|.$$

Using the fact that the partial derivatives of q are $\leq \delta\alpha$ and hypothesis (iii) we obtain

$$\begin{aligned}\|\Phi(v)-\Phi(w)\| &\geq \lambda\|v-w\| - \delta\alpha(\|\Psi(v)-\Psi(w)\| + \|v-w\|)\\ &\geq (\lambda - \delta\alpha(1+c))\|v-w\|.\end{aligned}$$

Hence, if δ is so small that $\lambda - \delta\alpha(1+c) > 1$, Φ is a homeomorphism of U onto $\Phi(U)$ whose inverse has Lipschitz constant $\leq (\lambda - \delta\alpha(1+c))^{-1}$. In particular, $\Phi(U)$ is open. Now define $\hat{\Psi}: \Phi(U) \to (D_0F)E_1$ by

$$\hat{\Psi}(v) = (L\Psi\Phi^{-1})v + p(\Psi(\Phi^{-1}(v)), \Phi^{-1}(v)).$$

Clearly, $F(G) = \{(\hat{\Psi}(x), x) \mid x \in \Phi(U)\}$. To calculate the dispersion of $F(G)$, write $\hat{\Psi} = \tilde{\Psi} \circ \Phi^{-1}$ where $\tilde{\Psi}(w) = L\Psi(w) + p(\Psi(w), w)$. Then

$$\begin{aligned}\|\tilde{\Psi}(x) - \tilde{\Psi}(y)\| &\leq \beta\|\Psi(x)-\Psi(y)\| + \delta\alpha(\|\Psi(x)-\Psi(y)\| + \|x-y\|)\\ &\leq (c\beta + \delta\alpha(1+c))\|x-y\|.\end{aligned}$$

Then the dispersion of $F(G)$ is less than or equal to

$$c\frac{(\beta+\delta\alpha(1+c)/c)}{(\lambda-\delta\alpha(1+c))} = c\frac{\beta}{\lambda}\cdot\frac{(1+\delta\alpha(1+c)/(c\beta))}{(1-\delta\alpha(1+c)/\lambda)}.$$

Since $\beta/\lambda < 1$, taking δ small enough, the factor of c is <1 and the lemma is proved. □

Now we are ready to prove Pesin's formula. Let f and μ satisfy the hypothesis of (2). Write

$$E^u(x) = \oplus\{E_j(x) \mid \lambda_j(x) > 0\}, \qquad E^0(x) = \oplus\{E_j(x) \mid \lambda_j(x) \leq 0\}.$$

Put $\Sigma_j = \{x \mid \dim E^u(x) = j\}$ and let $S = \{j \geq 0 \mid \mu(\Sigma_j) > 0\}$. If $j \in S$, let μ_j be the measure on M given by $\mu_j(A) = \mu(A \cap \Sigma_j)/\mu(\Sigma_j)$ if A is a Borel subset of M. Then

$$h_\mu(f) = \sum_{j\in S} \mu(\Sigma_j) h_{\mu_j}(f).$$

Thus, all we have to show is that

$$h_{\mu_j}(f) \geq \int_M \chi\, d\mu_j$$

This inequality obviously holds for $j = 0$. Suppose $j > 0$. To simplify the notation put $\mu = \mu_j$, $\Sigma = \Sigma_j$. Fix any $\varepsilon > 0$. By a standard combination of Egorov's theorem and Birkhoff's ergodic theorem, we can take a compact set $K \subset M$ with $\mu(K) \geq 1 - \varepsilon$ such that the splitting $E^0(x) \oplus E^u(x) = T_xM$ is continuous when x varies in K and, for some $N > 0$, there exist constants $\lambda > \beta > 1$ such that, if $g = f^N$, the inequalities

$$\|(D_xg^n)v\| \geq \lambda^n\|v\| \tag{10}$$

$$\|(D_xg^n)/E^0(x)\| \leq \beta \tag{11}$$

$$\log|\det(D_xg^n)/E^u(x)| \geq N(\chi(x) - \varepsilon)n \tag{12}$$

hold for all $x \in K$, $n \geq 0$ and $v \in E^u(x)$. In what follows, in order to avoid a cumbersome and conceptually unnecessary use of coordinate charts, we shall treat M as if it were a Euclidean space. The reader will observe that all our arguments can be easily formalized by a completely straightforward use of local coordinates.

Lemma 4. *For all $c>0$ there exists $\xi>0$ such that, if $x\in K$ and $g^m(x)\in K$ for some $m>0$, then if a set $G\subset M$ is contained in the ball $B_{\xi^m}(x)$ and is a $(E^0(x), E^u(x))$-graph with dispersion $\le c$, then $g^m(G)$ is a $((D_xg^m)E^0(x), (D_xg^m)E^u(x))$-graph with dispersion $\le c$.*

Proof. Let $\alpha=\sup\{\gamma(E^0(x), E^u(x))\,|\,x\in K\}$. Let $\delta>0$ be given by lemma 3 as a function of α, c and the constants β, λ which satisfy (10) and (11). We claim that there exist $C>0$ and $0<t\le 1$ such that

$$\|(D_xg^n)-(D_yg^n)\|\le C^n\|x-y\|^t \tag{13}$$

for all $x, y\in M$. To prove this claim take $0<t\le 1$ and $C_0>0$ satisfying $\|(D_xg)-(D_yg)\|\le C_0\|x-y\|^t$ for all $x, y\in M$. Let A be a Lipschitz constant for g. In particular, $A\ge\|(D_xg)\|$ for all $x\in M$. Choose C satisfying for all $n\ge 0$

$$C\ge A+C_0\left(\frac{A^{t+1}}{C}\right)^n, \tag{14}$$

then (13) holds for $n=1$. Suppose it holds for $n=1,\dots,m$; then

$$\begin{aligned}\|(D_xg^{m+1})-(D_yg^{m+1})\| &\le\|(D_{g^m(x)}g)-(D_{g^m(y)}g)\|\,\|(D_yg^m)\|\\ &\quad+\|(D_{g^m(x)}g)\|\,\|(D_xg^m)-(D_yg^m)\|\\ &\le A^mC_0\|g^m(x)-g^m(y)\|^t+AC^m\|x-y\|^t\\ &\le A^mC_0A^{mt}\|x-y\|^t+AC^m\|x-y\|^t.\end{aligned}$$

By (14) the last expression is $\le C^{m+1}\|x-y\|^t$, completing the proof of the claim. Now take $\xi>0$ such that $(C\xi^t)^n<\delta$ for all $n\ge 1$. Hence

$$\begin{aligned}\|(D_yg^m)-(D_xg^m)\| &\le C^m\|x-y\|^t\\ &\le C^m\xi^{mt}<\delta\end{aligned}$$

if $y\in B_{\xi^m}(x)$. Therefore lemma 4 follows from lemma 3. □

Fix the constant c of the statement of lemma 4 so small that there exists $a>0$ with the following property. If $x\in K$, $y\in M$ and $d(x,y)<a$, then for every subspace $E\subset T_yM$ which is a $(E^0(x), E^u(x))$-graph with dispersion $\le c$ we have

$$\big|\log|\det(D_yg)/E|-\log|\det(D_xg)/E^u(x)|\big|\le\varepsilon. \tag{15}$$

If $x\in K$, write

$$D_r(x)=\{x+y_1+y_2\,|\,y_1\in E^0(x),\ y_2\in E^u(x),\ \|y_1\|\le r,\ \|y_2\|\le r\}$$

and let $k_2>k_1>0$, $r_1>0$ be such that

$$B_{k_1r}(x)\subset D_r(x)\subset B_{k_2r}(x) \tag{16}$$

for all $x\in K$, $0<r\le r_1$. Finally, if $x\in K$, define $N(x)$ as the minimum integer ≥ 1 such that $g^{N(x)}(x)\in K$.

This function is well defined for a.e. $x\in K$ and is integrable because, if $W_j=\{x\in K\,|\,N(x)=j\}$ it is easy to see that

$$\bigcup_{n\ge 0} g^n(K)=\bigcup_{j=1}^{\infty}\bigcup_{i=0}^{j-1} g^i(W_j)\qquad \text{mod } (0)$$

and that the sets in the union on the right are disjoint. Hence,

$$1>\mu\Big(\bigcup_{n\geq 0} g^n(K)\Big)=\sum_{j=1}^{\infty}\sum_{i=0}^{j-1}\mu(g^i(W_j))=\sum_{j=1}^{\infty} j\mu(W_j)=\int_M N\,d\mu.$$

Extend N to M, putting $N(x)=0$ when $x\notin K$, and define $\rho: M\to(0,1)$ as

$$\rho(x)=\min(a,(k_1/k_2)\xi^{N(x)}).$$

The integrability of N implies that of $\log\rho$. Let ν be the Lebesgue measure on M. We claim that there exists $K'\subset K$ with $\mu(M-K')\leq 2\sqrt{\varepsilon}$ such that

$$h_\nu(g,\rho,x)\geq N(\chi(x)-\varepsilon-4C\sqrt{\varepsilon})-\varepsilon$$

for μ-a.e. point $x\in K'$, where $C=\sup\{\log|\det(D_pf)/E|\,|p\in M, E\subset T_pM\}$. By the proposition, this property will imply

$$\begin{aligned} h_\mu(g)&\geq\int_M h_\nu(g,\rho,x)\,d\mu(x)\geq\int_{K'} h_\nu(x)\\ &\geq N\int_{K'}\chi\,d\mu-N(\varepsilon+4C\sqrt{\varepsilon})-\varepsilon\\ &\geq N\int_M\chi\,d\mu-N(\varepsilon+6C\sqrt{\varepsilon})-\varepsilon.\end{aligned}$$

Hence,

$$h_\mu(f)\geq(1/N)h_\mu(g)\geq\int_M\chi\,d\mu-(\varepsilon+6C\sqrt{\varepsilon})-\varepsilon/N.$$

Since ε is arbitrary this completes the proof of (2).

It remains to prove the claim. Observe that Birkhoff's theorem implies

$$\mu(\{x|\lim_{n\to+\infty}(1/n)\#\{0\leq j\leq n\,|\,g^j(x)\notin K\}\leq\sqrt{\varepsilon}\})\geq 1-\sqrt{\varepsilon}.$$

Then there exists a compact set $K_1\subset K$ with $\mu(K_1)\geq 1-2\sqrt{\varepsilon}$ and $N_0>0$ such that, if $n>N_0$,

$$\#\{0\leq j\leq n\,|\,g^j(x)\notin K\}\leq 2n\sqrt{\varepsilon} \tag{17}$$

for all $x\in K_1$. Define $K'=K\cap K_1$. Clearly, $\mu(K-K')\leq\mu(M-K_1)\leq 2\sqrt{\varepsilon}$. Fix any $x\in K'$. There exists $B>0$ satisfying

$$\nu(S_n(g,\rho,x))=B\int_{E^0(x)}\nu[(y+E^u(x))\cap S_n(g,\rho,x)]\,d\nu(y)$$

for all $n\geq 0$, where ν also denotes the Lebesgue measure in the subspaces $E^0(x)$, $y+E^u(x)$. Thus the claim is reduced to showing that

$$\limsup_{n\to+\infty}\inf_{y\in E^0(x)}\frac{1}{n}[-\log\nu((y+E^u(x))\cap S_n(g,\rho,x))]\geq N(\chi(x)-\varepsilon-4C\sqrt{\varepsilon})-\varepsilon.$$

If $y\in E^0(x)$, denote by $\Lambda_n(y)$ the set of points $w\in y+E^u(x)$ such that

$$g^j(w)\in D_{\rho(g^j(x))/k_1}(g^j(x))$$

for all $0\leq j\leq n$. By (16) we have

$$\Lambda_n(y)\supset(y+E^u(x))\cap S_n(g,\rho,x).$$

Therefore it is sufficient to prove that

$$\limsup_{n\to+\infty} \inf_{y\in E^0(x)} \frac{1}{n}[-\log \nu(\Lambda_n(y))] \geq N(\chi(x)-\varepsilon-4C\sqrt{\varepsilon})-\varepsilon. \tag{18}$$

LEMMA 5. *If $g^n(x)\in K$ and $y\in E^0(x)$ then, if $\Lambda_n(y)$ is not empty, $g^n(\Lambda_n(y))$ is a $(E^0(g^n(x)), E^u(g^n(x)))$-graph with dispersion $\leq c$.*

Proof. The property is trivial for $n=0$. Suppose that it holds for some $m>0$. Then, if $g^{\bar m}(x)\in K$ and $g^j(x)\notin K$ for $m<j<\bar m$ and $\Lambda_{\bar m}(y)\neq\emptyset$, we have

$$g^{\bar m}(\Lambda_{\bar m}(y)) \subset g^{\bar m-m}(g^m(\Lambda_m(y))) \cap D_{\rho(g^{\bar m}(x))/k_1}(g^{\bar m}(x)).$$

But $g^m(\Lambda_m(y))$ is a $(E^0(g^m(x)), E^u(g^m(x)))$-graph with dispersion $\leq c$, and

$$\begin{aligned} g^m(\Lambda_m(y)) &\subset D_{\rho(g^m(x))/k_1}(g^m(x)) \\ &\subset B_{k_2\rho(g^m(x))/k_1}(g^m(x)) \subset B_{\xi^{\bar m-m}}(g^m(x)). \end{aligned}$$

Then, using the fact that $N(g^m(x))=\bar m-m$ and lemma 4, it follows that $g^{\bar m}(\Lambda_m(y))= g^{\bar m-m}(g^m(\Lambda_m(y)))$ is a $(E^0(g^{\bar m}(x)), E^u(g^{\bar m}(x)))$-graph with dispersion $\leq c$. Therefore, so is its open subset $g^{\bar m}(\Lambda_{\bar m}(y))$.

To complete the proof of (18) take $D>0$ such that $D>\operatorname{vol}(G)$ (where $\operatorname{vol}(\cdot)$ denotes volume) for every $(E^0(w), E^u(w))$-graph G with dispersion $\leq c$ contained in $D_{\rho(w)}(w)$, $w\in K$. Then, if $g^n(x)\in K$ and $y\in E^0(x)$, we have

$$D>\operatorname{vol}(g^n(\Lambda_n(y))) = \int_{\Lambda_n(y)} |\det(D_zg^n)/T_z\Lambda_n(y)|\, d\nu(z). \tag{19}$$

Let $S_n=\{0\leq j<n \mid g^j(x)\in K\}$. If $n\geq N_0$ it follows from (15), (12), and (17) that

$$\begin{aligned} \log|\det(D_zg^n)/T_z\Lambda_n(y)| &= \sum_{j=0}^{n-1} \log|\det(D_{g^j(z)}g)/T_{g^j(z)}g^j(\Lambda_n(y))| \\ &\geq \sum_{j\in S_n} \log|\det(D_{g^j(z)}g)/T_{g^j(z)}g^j(\Lambda_n(y))| - NC(n-\#S_n) \\ &\geq \sum_{j\in S_n} \log|\det(D_{g^j(x)}g)/E^u(g^j(x))| - \varepsilon n - NC(n-\#S_n) \\ &\geq \sum_{j=0}^{n-1} \log|\det(D_{g^j(x)}g)/E^u(g^j(x)) - \varepsilon n - 2NC(n-\#S_n) \\ &\geq nN(\chi(x)-\varepsilon)-\varepsilon n-2NC(n-\#S_n) \\ &\geq nN(\chi(x)-\varepsilon)-\varepsilon n-4NCn\sqrt{\varepsilon}. \end{aligned}$$

Then

$$D>\nu(\Lambda_n(y))\cdot\exp(n(N(\chi(x)-\varepsilon-4C\sqrt{\varepsilon})))$$

for all $y\in E^0(x)$, $n\geq N_0$. This completes the proof of (18). □

REFERENCES

[1] Y. Pesin. Characteristic Lyapunov exponents and smooth ergodic theory. *Russian Math. Surveys* **32** (1977), 55–114.

[2] D. Ruelle. An inequality for the entropy of differentiable maps. *Bol. Soc. Bras. Mat.* **9** (1978), 83–87.

Ergod. Th. & Dynam. Sys. (1985), **5**, 71–88
Printed in Great Britain

On the Bernoulli property for rational maps

RICARDO MAÑÉ
Instituto de Matemática Pura e Aplicada, Rio de Janeiro, Brasil

(*Received* 19 *March* 1984)

Abstract. Every rational function f with degree $d \geq 2$ has a unique invariant probability μ_f that maximizes entropy. It has been conjectured that the system (f, μ_f) is equivalent to the one sided Bernoulli shift $\sigma: B^+(1/d, \ldots, 1/d) \rightleftarrows$. In this paper we prove that there exists $m > 0$ such that (f^m, μ_f) is equivalent to $\sigma: B^+(1/d^m, \ldots, 1/d^m) \rightleftarrows$.

0. *Introduction*

Let f be an analytic endomorphism of the Riemann sphere $\bar{\mathbb{C}} = \mathbb{C} \cup \{\infty\}$, of topological degree $d \geq 2$. Such an endomorphism is always given by a rational function $f(z) = P(z)/Q(z)$, where P and Q are polynomials with no common roots and max {degree (P), degree (Q)} = d. In [**5**] and independently in [**2**], [**6**] it was proved that there exists an f-invariant probability μ_f on the Borel σ-algebra of $\bar{\mathbb{C}}$ exhibiting several interesting properties. First, *it is the unique f-invariant probability on the Borel σ-algebra such that*

$$\mu_f(f(A)) = d\mu_f(A) \tag{1}$$

for every Borel set A such that $f|A$ is injective. Second, *it is the unique f-invariant probability that maximizes entropy* i.e. $h_\mu(f) = \log d$ (recall that by [7] and [3] the topological entropy of f is $\log d$). Third, with respect to μ_f, *f is exact* i.e. if $\mathscr{A}$ denotes the Borel σ-algebra of $\bar{\mathbb{C}}$ then every set $A \in \bigcap_{n\geq 1} f^{-n}(\mathscr{A})$ satisfies $\mu_f(A) = 0$ or 1. Finally, *μ_f describes the asymptotic distribution of the roots of the equation $f^n(z) = a$.* More precisely, if $z_i^{(n)}(a)$, $1 \leq i \leq d^n$ are the roots of the equation $f^n(z) = a$, define a probability

$$\mu^{(n)}(a) = d^{-n} \sum_{i=1}^{d^n} \delta_{z_i^{(n)}(a)}.$$

Then $\mu^{(n)}(a) \to \mu_f$ for every $a \in \bar{\mathbb{C}}$ with two possible exceptions that can be characterized explicitly. From this property it is not difficult to deduce that the support of μ_f is exactly the Julia set $J(f)$ of f. When f is a polynomial, Brolin [**1**] proved that μ_f is the equilibrium distribution of the Julia set (in the sense of potential theory, see [**1**] for the precise statement).

In both [2] and [**5**] it was conjectured that the system (f, μ_f) is equivalent to the *one sided* shift $\sigma: B^+(1/d, \ldots, 1/d) \rightleftarrows$ (or, what is the same, the transformation $z \to z^d$ of the circle $S^1 = \{z \in \mathbb{C}: |z| = 1\}$ endowed with Lebesgue measure). We shall give below some arguments to sustain this conjecture. The purpose of this paper is to prove that *for some $m \geq 1$, the system (f^m, μ_f) is equivalent to the one sided Bernoulli shift* $\sigma: B^+(1/d^m, \ldots, 1/d^m) \rightleftarrows$. In particular the natural extension of f^m is an invertible Bernoulli shift. Since roots of invertible Bernoulli shifts are also

Bernoulli [8], we conclude that the natural extension of (f, μ_f) is Bernoulli. Moreover we shall prove that *the equivalence between* $(f^n|J(f), \mu_f)$ *and* $\sigma: B^+(1/d^n, \ldots, 1/d^n) \circlearrowleft$ *can be realized by a topological equivalence between full measure invariant residual sets.*

Before going into the formal statement of our result and discussing some questions raised by the equivalence conjecture, we shall give a brief idea of why this conjecture is so natural. To prove the conjecture it suffices to find a Borel set X, with $\mu_f(X) = 1$ and $f^{-1}(X) = X$, and a partition $X = \bigcup_{i=1}^{d} X_i$ in disjoint Borel sets satisfying:

(a) for all $1 \le i \le d$, $f|X_i$ is injective and $f(X_i) = X$;

(b) if $n \ge 1$ and $1 \le i_j \le d$, $j = 1, \ldots, n$:

$$\mu_f\left(\bigcap_{j=1}^{n} f^{-j}(X_{i_j})\right) = d^{-n};$$

(c) if $\mathcal{P}$ denotes the partition $\{X_1, \ldots, X_d\}$ then $\bigvee_{n=0}^{\infty} f^{-n}(\mathcal{P})$ coincides mod (0) with the Borel σ-algebra of $\bar{\mathbb{C}}$.

Now observe that (b) is an easy consequence of (a) and (1). Then we have to worry only about (a) and (c). As we shall see, (a) is quite easy to satisfy and (c) becomes, at least in this approach, the obstruction to solving the conjecture. Take an arc γ containing all the critical values of f (i.e. the images under f of its critical points) satisfying $\mu_f(\gamma) = 0$. Standard measure-theoretical arguments show the existence of such arcs. Then $\hat{X} = \bar{\mathbb{C}} - \gamma$ is a topological disk (i.e. a set homeomorphic to the unit disk) and doesn't contain critical values of f. Then, there exist branches $\varphi_i : \hat{X} \to \bar{\mathbb{C}}$ $i = 1, \ldots, d$, of $f^{-1}|\hat{X}$. Set $X_i = \varphi_i(\hat{X})$. It is easy to see that $f|X_i$ is injective and $f(\hat{X}_i) = \hat{X}$ for every $1 \le i \le d$. Moreover

$$\mu_f(\hat{X}) = 1 - \mu_f(\gamma) = 0.$$

But it is not necessarily true that $f^{-1}(\hat{X}) = \hat{X}$. To correct this, define

$$X = \left(\bigcup_{n\ge0} f^{-n}\left(\bigcup_{m\ge0} f^m(\hat{X}^c)\right)\right)^c.$$

This set has full measure. This follows from the fact X^c has zero measure, the f-invariance of μ_f, and the property (not listed above but proved in [2] or [5]) that f transforms zero measure sets into zero measure sets. It also satisfies $f^{-1}(X) = X$. Now set

$$X_i = \hat{X}_i \cap X.$$

Property (a) still holds for $X, X_1, \ldots, X_d$. *The problem is to decide whether* (c) *is satisfied.* It is reasonable to think that the answer may depend on the choice of γ. We shall show that, for high powers of f, this method works for a suitable choice of the curve γ.

An alternative approach to solve the conjecture could be the following. Suppose that defining $\hat{X}, X, X_1, \ldots, X_d$ as above, condition (c) is not satisfied. Write $\mathcal{A}_0 = \bigvee_{n\ge0} f^{-n}(\mathcal{P})$. Then $\mathcal{A} \ne \mathcal{A}_0$ (where $\mathcal{A}$ denotes the Borel σ-algebra). If we can find an f-invariant probability measure $\mu : \mathcal{A} \to [0, 1]$ such that $\mu|\mathcal{A}_0 = \mu_f|\mathcal{A}_0$ but $\mu \ne \mu_f$, the conjecture will be proved because $h_\mu(f, \mathcal{P}) = h_{\mu_f}(f, \mathcal{P})$ and this last term is $\log d$ (by (a) and (b)). Then we shall have another f-invariant probability with maximal entropy, contradicting the uniqueness of the invariant probability with this property.

THEOREM. *There exists* $N \geq 1$, *Borel sets* $\Lambda_1 \subset J(f)$, $\Lambda_2 \subset B^+(1/d^N, \ldots, 1/d^N)$ *and a homeomorphism* $h: \Lambda_1 \to \Lambda_2$ *such that*:

(a) $f^{-N}(\Lambda_1) = \Lambda_1$, $\sigma^{-1}(\Lambda_2) = \Lambda_2$;

(b) $\mu_f(\Lambda_1) = 1$, $\mu(\Lambda_2) = 1$ (*where* μ *denotes the Bernoulli measure of* $B^+(1/d^m, \ldots, 1/d^m)$);

(c) Λ_1 *and* Λ_2 *are residual*;

(d) $\sigma h(x) = hf^N(x)$, *for all* $x \in \Lambda_1$.

The map h is constructed, as usual in this kind of result, using symbolic dynamics. In fact, given an analytic endomorphism $g: \bar{\mathbb{C}} \rightleftarrows$ of degree $l \geq 2$ and a topological disk U with $\mu_g(U) = 1$ not containing critical values of g we can define a measurable map $h: \Lambda \to B^+(1/l, \ldots, 1/l)$ where $\Lambda \subset J(g)$ is a residual set satisfying $g(\Lambda) \subset \Lambda$, such that $\sigma h(x) = hg(x)$ for all $x \in \Lambda$ and $h^*(\mu_g)$ is the Bernoulli measure of $B^+(1/l, \ldots, 1/l)$. This map is constructed as follows. Let $\varphi_i: U \to \bar{\mathbb{C}}$, $1 \leq i \leq l$, be the branches of $f^{-1}|U$. Then define $U_i = \varphi_i(U)$, $\Lambda = J(g) \cap (\bigcap_{n \geq 0} g^{-n}(\bigcup_i U_i))$ and $h: \Lambda \to B^+(1/l, \ldots, 1/l)$ by the condition:

$$g^n(x) \in U_{h(x)(n)}$$

for all $n \geq 0$. It is easy to check that h has the previous properties. The proof of the theorem consists of finding a power $N \geq 1$ and a topological disk $U \supset J(f^N)$ such that the map h constructed above satisfies the properties in the theorem. The boundary of the set U will be a finite union of continuous curves.

Our theorem naturally raises the question of whether a measure preserving map $T: B^+(1/d^N, \ldots, 1/d^N) \rightleftarrows$ such that $T^N = \sigma$ is equivalent to the Bernoulli shift of $B^+(1/d, \ldots, 1/d)$. Another question is whether an exact measure preserving map T of a probability space $(X, \mathcal{A}, \mu)$ which for some integer $d > 1$ satisfies $\mu(T(A)) = d(A)$ for every $A \in \mathcal{A}$ where $T|A$ is injective and such that X can be partitioned into sets $X_1, \ldots, X_d$ such that $T|X_i$ is a bijection between X_i and X for every i, is equivalent to the Bernoulli shift of $B^+(1/d, \ldots, 1/d)$. Obviously an affirmative answer to any of these questions implies an affirmative answer of the conjecture. But as interesting as knowing that $(f|J(f), \mu_f)$ is equivalent to $\sigma: B^+(1/d, \ldots, 1/d) \rightleftarrows$ is to understand geometrically the map that realizes the equivalence. More precisely, from the viewpoint of the understanding of the dynamics of f, the proper statement of the conjecture is *whether a full measure topological disk* U (preferably having a nice geometry) *can be found such that the map* $h: \Lambda \to B^+(1/d, \ldots, 1/d)$ *associated to* U *as above, is a measure preserving topological equivalence between full measure residual subsets.*

I. *Proof of the theorem*

In this section we shall deduce the theorem from the following lemma, which will be proved in the following four sections. Recall that if $g: \bar{\mathbb{C}} \rightleftarrows$ is an analytic map and $U \subset \bar{\mathbb{C}}$ is an open set, we say that a function $\varphi: U \to \bar{\mathbb{C}}$ is a branch of $g^{-1}|U$ if φ is continuous and $g(\varphi(x)) = x$ for all $x \in U$ (and then φ is analytic).

FUNDAMENTAL LEMMA. *There exists* $N > 1$, $\alpha > 0$ *and a topological disk* $\tilde{U}$ *with* $\mu_f(\tilde{U}) > 0$ *such that for all* $n \geq 1$ *there exist branches* $\varphi_i^{(n)}: \tilde{U} \to \bar{\mathbb{C}}$ *of* $f^{-nN}|\tilde{U}$, $i = 1, \ldots, k_n$ *satisfying*:

(i) $\lim_{n\to+\infty} \max\{\operatorname{diam}(\varphi_i^{(n)}(\tilde{U})) | 1 \leq i \leq k_n\} = 0$;

(ii) $\mu_f(\bigcup_i \varphi_i^{(n)}(\tilde{U})) \geq \alpha\mu_f(\tilde{U})$;

(iii) $\varphi_i^{(n)}(\tilde{U}) \subset \tilde{U}$;

(iv) $f^N(\bigcup_i \varphi_i^{(n+1)}(\tilde{U})) \subset \bigcup_i \varphi_i^{(n)}(\tilde{U})$;

for all $n \geq 1$, $1 \leq i \leq k_n$.

The first step in applying the fundamental lemma to prove the theorem, is to show that $\mu_f(\tilde{U}) = 1$. This will follow from the following lemma:

LEMMA I.1. *Let* T *be a mixing transformation of the probability space* $(X, \mathcal{A}, \mu)$ *and suppose that* $A_n \in \mathcal{A}$, $n \geq 1$, *is a sequence of sets satisfying*:

(a) $\mu(A_0) > 0$;

(b) $A_n \subset T^{-1}(A_{n-1})$ *for all* $n \geq 1$;

(c) *there exists* $\alpha > 0$ *such that for all* $n \geq 1$:

$$\mu(A_n) \geq \alpha\mu(A_0).$$

Then $\mu(\bigcup_{j\geq l} A_j) = 1$ *for all* $l \geq 0$.

Proof. Define $\alpha_n = \mu(A_n)/\mu(A_0)$. By (b) $\alpha_1 \geq \alpha_2 \geq \cdots \geq \alpha > 0$. Define $\alpha_\infty = \lim_{n\to\infty} \alpha_n$. If $n \geq m \geq 0$, (b) implies

$$T^{-(n-m)}(A_m) \supset A_n$$

and

$$\mu(T^{-(n-m)}(A_m) - A_n) = \mu(A_m) - \mu(A_n) = (\alpha_m - \alpha_n)\mu(A_0).$$

Then for all $S \in \mathcal{A}$ and $n \geq m \geq 0$:

$$\begin{aligned}\mu\left(\left(\bigcup_{j\geq m} A_j\right) \cap S\right) &\geq \mu(A_n \cap S)\\ &= \mu(T^{-(n-m)}(A_m) \cap S) - \mu((T^{-(n-m)}(A_m) - A_n) \cap S)\\ &\geq \mu(T^{-(n-m)}(A_m) \cap S) - (\alpha_m - \alpha_n)\mu(A_0).\end{aligned}$$

Since T is mixing, taking the limit as $n \to +\infty$, we obtain

$$\begin{aligned}&\lim_{n\to+\infty} (\mu(T^{-(n-m)}(A_m) \cap S) - (\alpha_m - \alpha_n)\mu(A_0))\\ &\quad= \lim_{n\to+\infty} \mu(T^{-n}(T^{-m}(A_m)) \cap S) - (\alpha_m - \alpha_\infty)\mu(A_0)\\ &\quad= \mu(T^{-m}(A_m))\mu(S) - (\alpha_m - \alpha_\infty)\mu(A_0)\\ &\quad= \mu(A_m)\mu(S) - (\alpha_m - \alpha_\infty)\mu(A_0)\\ &\quad= (\alpha_m\mu(S) - (\alpha_m - \alpha_\infty))\mu(A_0) > 0,\end{aligned}$$

if m is so large that $\alpha_m\mu(S) - (\alpha_m - \alpha_\infty) > 0$. Hence

$$\mu\left(\left(\bigcup_{j\geq m} A_j\right) \cap S\right) > 0$$

for every $m \geq 0$, $S \in \mathcal{A}$. Then $\mu(\bigcup_{j\geq m} A_j) = 1$ for every $m \geq 0$. □

Thus the set $\tilde{U}$ in the fundamental lemma has full measure. This follows by applying I.1 to $f^N = T$, $A_0 = \tilde{U}$, $A_n = \bigcup_{j=1}^{k_n} \varphi_i^{(n)}(\tilde{U})$. Then

$$\mu_f\left(\bigcup_{n\geq m}\bigcup_{i=1}^{k_n} \varphi_i^{(n)}(\tilde{U})\right) = 1 \qquad \text{for all } m,$$

and (iii) implies $\mu_f(\tilde{U}) = 1$.

To prove the theorem set $g = f^N$ and take a topological disk $U \subset \tilde{U}$ such that $\mu_f(\tilde{U} - U) = 0$ and U doesn't contain critical values of g. Such a topological disk is easily obtained by deleting some arcs of measure zero joining the critical values of g in $\tilde{U}$ with $\partial\tilde{U}$. Since U doesn't contain critical values of g, there exist branches $\varphi_i^{(1)}: U \to \bar{\mathbb{C}}$, $i = 1, \ldots, d^N$ of $g^{-1}|U$. Set $V_i = \varphi_i^{(1)}(U) \cap J(f)$, $V = \bigcup_i V_i$ and define

$$\tilde{\Lambda} = J(f) - \bigcup_{n\geq 0} g^{-n}\left(\bigcup_{m\geq 0} g^m(V^c)\right).$$

It is easy to check that $\mu_f(\tilde{\Lambda}) = 1$ and $g^{-1}(\tilde{\Lambda}) = \tilde{\Lambda}$. Define

$$\Lambda_0 = \bigcap_{m=1}^{\infty}\bigcup_{n\geq m}\bigcup_{i=1}^{k_n} \varphi_i^{(n)}(U \cap J(f)).$$

Then $\mu_f(\Lambda_0) = 1$. But it is not true that $g^{-1}(\Lambda_0) = \Lambda_0$. To obtain this property too we introduce the set

$$\hat{\Lambda} = J(f) - \bigcup_{n\geq 0} g^{-n}\left(\bigcup_{m\geq 0} g^m(J(f) - \Lambda_0)\right).$$

This set satisfies $\mu_f(\hat{\Lambda}) = 1$ and $g^{-1}(\hat{\Lambda}) = \hat{\Lambda}$. Hence, if we set

$$\Lambda_1 = \hat{\Lambda} \cap \tilde{\Lambda}$$

we get $\mu_f(\Lambda_1) = 1$ and $g^{-1}(\Lambda_1) = \Lambda_1$. Now we can define a map $h: \tilde{\Lambda} \to B^+(1/d^N, \ldots, 1/d^N)$ by the condition

$$g^n(x) \in V_{h(x)(n)}$$

for all $n \geq 0$. It is clearly a measurable map and

$$\sigma h(x) = hg(x) \tag{1}$$

for all $x \in \tilde{\Lambda}$. Now we shall show that $\mu_f(h^{-1}(A)) = \mu(A)$ for every Borel set $A \subset B^+(1/d^N, \ldots, 1/d^N)$. It is sufficient to prove this relation when A has the form $A = \bigcap_{j=0}^{n} \sigma^{-j}(S_{i_j})$, where $S_i = \{\theta \in B^+(1/d^N, \ldots, 1/d^N): \theta(0) = i\}$, because these sets generate the Borel σ-algebra. We leave to the reader the task of verifying by induction on n that

$$h^{-1}(A) = \tilde{\Lambda} \cap \left(\bigcap_{j=0}^{n} g^{-j}(V_{i_j})\right) \tag{2}$$

and that $g^n|(\bigcap_{j=0}^{n} g^{-j}(V_{i_j}))$ is a bijection onto $J(f) \cap U$. Then

$$\mu_f(h^{-1}(A)) = \mu_f\left(\bigcap_{j=0}^{n} g^{-j}(V_{i_j})\right)$$

because $\mu_f(\tilde{\Lambda}) = 1$, and

$$\mu_f\left(\bigcap_{j=0}^{n} g^{-j}(V_{i_j})\right) = 1/d^{nN}$$

because of property (1) in the introduction. On the other hand $\mu(A) = \prod_{j=0}^{n} \mu(S_{i_j}) = 1/d^{nN}$, completing the proof that $\mu_f(h^{-1}(A)) = \mu(A)$. Now let us prove that $\hat{\Lambda}$ and

$\tilde{\Lambda}$ are residual subsets of $J(f)$. This will prove that so is $\Lambda_1=\hat{\Lambda}\cap\tilde{\Lambda}$. The sets $\bigcup_{n\geq m}\bigcup_i \varphi_i^{(n)}(U\cap J(f))$ are open in $J(f)$, and also dense because they have full measure. Then Λ_0, that is their intersection, is residual in $J(f)$. Therefore $J(f)-\Lambda_0$ is meagre and then so is the set $\bigcup_{n\geq 0} g^{-n}(\bigcup_{m\geq 0} g^m(J(f)-\Lambda_0))$. Hence $\hat{\Lambda}$ is residual in $J(f)$. Moreover V is open in $J(f)$. But it has full measure because

$$\mu_f(V)=\mu_f(g^{-1}(U)\cap J(f))=\mu_f(U)=1.$$

Then it is open and dense. Therefore V^c is closed and its interior is empty. It follows that $\bigcup_{n\geq 0} g^{-n}(\bigcup_{m\geq 0} g^m(V^c))$ is meagre and $\tilde{\Lambda}$ is residual in $J(f)$. Hence Λ_1 is residual in $J(f)$.

Our next step is to prove that h is injective and a homeomorphism onto $h(\Lambda_1)$. For this we shall need the following lemma:

LEMMA I.2. *For every* $x\in\Lambda_1$

$$\lim_{n\to+\infty}\operatorname{diam}\left(\bigcap_{j=0}^{n} g^{-j}(\varphi_{h(x)(j)}(U)\right)=0.$$

Proof. We claim that if $n\geq 1$ and $x\in\varphi_{i_n}^{(n)}(U)$ for some $1\leq i_n\leq k_n$:

$$\bigcap_{j=0}^{n-1} g^{-j}(\varphi^{(1)}_{h(x)(j)}(U))\subset\varphi^{(n)}_{i_n}(U).$$

We shall prove this by induction. Assume it is true for $n=m$. Suppose that $x\in\varphi^{(m+1)}_{i_{m+1}}(U)$ for some $1\leq i\leq k_{m+1}$. By property (iv) of the fundamental lemma $g(x)\in\varphi^{(m)}_{i_m}(\tilde{U})$ for some $1\leq i_m\leq k_m$. But we know that $x\in\Lambda_1$, that means that $g^j(x)\in U$ for all $j\geq 0$. Then $g(x)\in\varphi^{(m)}_{i_m}(U)$ because if $g(x)\in\varphi^{(m)}_{i_m}(\tilde{U})-\varphi^{(m)}_{i_m}(U)$ then $g^m(g(x))\in\tilde{U}-U$, hence $g^{m+1}(x)\notin U$. Now we apply the induction hypothesis to the point $g(x)$ and we obtain:

$$\bigcap_{j=0}^{m-1} g^{-j}(\varphi^{(1)}_{h(g(x))(j)}(U))\subset\varphi^{(m)}_{i_m}(U).$$

Then

$$\begin{aligned}x\in\bigcap_{j=0}^{m} g^{-j}(\varphi^{(1)}_{h(x)(j)}(U))\\ &=\varphi^{(1)}_{h(x)(0)}(U)\cap g^{-1}\left(\bigcap_{j=0}^{m-1} g^{-j}(\varphi^{(1)}_{h(g(x))(j)}(U)\right)\\ &\subset\varphi^{(1)}_{h(x)(0)}(U)\cap g^{-1}(\varphi^{(m)}_{i_m}(U)).\end{aligned}\tag{3}$$

Now observe that $\varphi^{(1)}_{h(x)(0)}\varphi^{(m)}_{i_m}|U$ is a branch of $g^{-(m+1)}|U$ and

$$\varphi^{(1)}_{h(x)(0)}\varphi^{(m)}_{i_m}(U)=\varphi^{(1)}_{h(x)(0)}(U)\cap g^{-1}(\varphi^{(m)}_{i_m}(U)).\tag{4}$$

By (3)

$$x\in\varphi^{(1)}_{h(x)(0)}(U)\cap g^{-1}(\varphi^{(m)}_{i_m}(U)).$$

Then by (4) $\varphi^{(1)}_{h(x)(0)}\varphi^{(m)}_{i_m}|U$ is a branch of $g^{-(m+1)}|U$ whose image covers x. So is $\varphi^{(m+1)}_{i_{m+1}}|U$. Then $\varphi^{(1)}_{h(x)(0)}\varphi^{(m)}_{i_m}|U=\varphi^{(m+1)}_{i_{m+1}}$. Hence, using (3) and (4):

$$\begin{aligned}\bigcap_{j=0}^{m} g^{-j}(\varphi^{(1)}_{h(x)(j)}(U))&\subset\varphi^{(1)}_{h(x)(0)}(U)\cap g^{-1}(\varphi^{(m)}_{i_m}(U))\\ &=\varphi^{(m+1)}_{i_{m+1}}(U),\end{aligned}$$

as we wished to prove. To prove the lemma we use the fact that $x \in \bigcap_{m\geq 0}\bigcup_{j\geq m} g^{-j}(U)$ means that $x \in \varphi_{i_n}^{(n)}(U)$ for some $1 \leq i_n \leq k_n$, for infinitely many values of n. Then the claim implies

$$\limsup_{n\to+\infty} \sup_{x\in\Lambda_1} \operatorname{diam}\left(\bigcap_{j=0}^{n} g^{-j}(\varphi_{h(x)(j)}^{(1)}(U))\right) \leq \limsup_{n\to+\infty} (\max_i \operatorname{diam}(\varphi_i^{(n)}(U))).$$

The last limit is zero by property (i) of the fundamental lemma. □

To prove that h has the desired properties we introduce the set:

$$\Sigma = \bigcap_{j\geq 0} g^{-j}(\bigcup_i V_i)$$

and the map $h_0 : \Sigma \to B^+(1/d^N, \ldots, 1/d^N)$ defined by the condition:

$$g^j(x) \in V_{h_0(x)(j)} = \varphi_{(j)}^{(1)}(U) \cap J(f).$$

Clearly $\Sigma \supset \Lambda_1$ and $h_0|\Lambda_1 = h$.

LEMMA I.3. (a) *If $x \in \Lambda_1$, $y \subset \Sigma$ and $y \neq x$, then $h_0(x) \neq h_0(y)$.*

(b) *If $\Sigma_0 \subset \Sigma$ and $x \in \Lambda_1$ is in the interior of Σ_0, then $h(x)$ is in the interior of $h(\Sigma_0)$ in $h(\Sigma)$.*

(c) *$h(A)$ is dense in $B^+(1/d^N, \ldots, 1/d^N)$ if $A \subset \Sigma$ is dense.*

(d) *For all $x \in \Lambda_1$:*

$$\lim_{n\to+\infty} \operatorname{diam} \bigcap_{j=0}^{n} g^{-j}(V_{h(x)(j)}) = 0.$$

(e) *$h(\Lambda_1)$ is a residual subset of $B^+(1/d^N, \ldots, 1/d^N)$.*

Proof. (a) If $x \in \Lambda_1$ then $x \in \bigcap_{m\geq 0}\bigcap_{n\geq m} g^{-n}(U)$. Therefore by I.2

$$y \notin \bigcap_{j=0}^{n} g^{-j}(\varphi_{h_0(x)(j)}^{(1)}(U))$$

if n is large. This means $g^j(y) \notin \varphi_{h_0(x)(j)}^{(1)}(U)$ for some $0 \leq j \leq n$. Therefore $h_0(y)(j) \neq h_0(x)(j)$, or $h(x) \neq h(y)$.

(d) follows from intersecting the set in I.2 with $J(f)$.

(c) We have to prove that for any $l \geq 0$ and $\theta \in B^+(1/d^N, \ldots, 1/d^N)$ there exists $x \in A$ with $h(x)(j) = \theta(j)$ if $0 \leq j \leq l$. The set

$$\bigcap_{j=0}^{l} g^{-j}(V_{\theta(j)}) \cap \Sigma$$

is open and non-empty in Σ. Then it contains a point $x \in A$ because A is dense in Σ. Hence $g^j(x) \in V_{\theta(j)}$ if $0 \leq j \leq l$. Then $h(x)(j) = \theta(j)$ for $0 \leq j \leq l$.

(b) If $x \in \operatorname{Int}(\Sigma_0)$, by (d):

$$\bigcap_{j=0}^{n} g^{-1}(V_{h(x)(j)}) \cap \Sigma \subset \Sigma_0$$

for n large enough. Consider the open set in $h(\Sigma)$ given by:

$$h(\Sigma) \cap \{\theta : \theta(j) = h(x), 0 \leq j \leq n\}.$$

If θ belongs to this set it means that $\theta=h(y)$, $y\in\Sigma$ and $h(y)(j)=h(x)(j)$ for $0\le j\le n$. Then $g^{-j}(y)\in V_{h(x)(j)}$ for $0\le j\le n$. This means that:

$$y\in\bigcap_{j=0}^{n} g^{-j}(V_{h(x)(j)})\cap\Sigma.$$

Therefore:

$$h(\Sigma_0)\supset h\left(\bigcap_{j=0}^{n} g^{-j}(V_{h(x)(j)})\cap\Sigma\right)\supset h(\Sigma)\cap\{\theta\colon \theta(j)=h(x), 0\le j\le n\}$$

(e) To show that $h(\Lambda_1)$ is a residual set introduce the sets Σ_n of points $\theta\in B^+(1/d^N,\dots,1/d^N)$ such that for some $m\ge n$:

$$\overline{\bigcap_{j=0}^{m} g^{-j}(V_{\theta(j)})}\subset\bigcap_{j=0}^{n} g^{-j}(V_{\theta(j)}).$$

Now we shall prove that $\bigcup_{n\ge1}\Sigma_n\supset h(\Lambda_1)$. We have to prove that $h(x)\in\Sigma_n$ for every $x\in\Sigma$ and $n\ge1$. If $x\in\Lambda_1$, $x\in\bigcap_{j=0}^{n-1} g^{-j}(V_{h(x)(j)})$. This set is open. Hence, by (d):

$$\overline{\bigcap_{j=0}^{m} g^{-j}(V_{h(x)(j)})}\subset\bigcap_{j=0}^{n-1} g^{-j}(V_{h(x)(j)})$$

if $m>n$ is large enough. Then $h(x)\in\Sigma_n$ as we wished to show. Moreover each Σ_n is open. Since it contains the set $h(\Lambda_1)$, which is dense by (c), it is also dense. Therefore $\Sigma_\infty=\bigcap_{n\ge1}\Sigma_n$ is a residual subset. Then, to prove that $h(\Lambda_1)$ is a residual subset of $B^+(1/d^N,\dots,1/d^N)$ it suffices to prove that is residual in Σ_∞. Since Λ_1 is residual in Σ we can write $\Lambda_1=\bigcap_{n\ge0}A_n$, where each A_n is open and dense in Σ. By (a)

$$h(\Lambda_1)=\bigcap_{n\ge0} h(A_n).$$

But $h(\Lambda_1)\subset\Sigma_\infty$. Then

$$h(\Lambda_1)=\bigcap_{n\ge0} h(A_n)\cap\Sigma_\infty.$$

Moreover $\Lambda_1\subset A_n=\operatorname{Int}(A_n)$ for all n. By (b), $h(\Lambda_1)\subset\operatorname{Int}(A_n)$. Again, since $h(\Lambda_1)\subset\Sigma_\infty$, we obtain:

$$h(\Lambda_1)\subset\bigcap_{n\ge1}\operatorname{Int}(A_n)\cap\Sigma_\infty.$$

Then:

$$\bigcap_{n\ge0} h(A_n)\cap\Sigma_\infty=h(\Lambda_1)\subset\bigcap_{n\ge0}\operatorname{Int} h(A_n)\cap\Sigma_\infty\subset\bigcap_{n\ge0} h(A_n)\cap\Sigma_\infty.$$

This implies

$$h(\Lambda_1)=\bigcap_{n\ge0}\operatorname{Int} h(A_n)\cap\Sigma_\infty.$$

Then $h(\Lambda_1)$ is the intersection of the open subsets of Σ_∞ given by $\operatorname{Int} h(A_n)\cap\Sigma_\infty$, $n\ge1$. Since it is also dense (by (c)) it follows that is residual. □

To complete the proof of the theorem, it remains only to show that $h^{-1}\colon h(\Lambda_1)\to\Lambda_1$ is continuous. Suppose that $\lim_{k\to+\infty}h(y_k)=h(x)$. We want to show that $\lim_{k\to+\infty}y_k=x$. But $h(y_k)\to h(y)$ means that $h(y_k)(j)=h_k(y)(j)$ for $0\le j\le n_k$, where $n_k\to+\infty$.

Then:

$$y_k \in \bigcap_{j=0}^{n_k} g^{-j}(V_{h(y_k)(j)}) = \bigcap_{j=0}^{n_k} g^{-j}(V_{h(x)(j)})$$

$$x \in \bigcap_{j=0}^{n_k} g^{-1}(V_{h(x)(j)}).$$

Then

$$d(y_k, x) \leq \operatorname{diam} \bigcap_{j=0}^{n_k} g^{-j}(V_{h(x)(j)})$$

By I.3(d), it follows that $\lim_{k\to+\infty} y_k = x$. □

II. *Proof of the fundamental lemma*

Suppose that $a_1, \ldots, a_r$ are the periodic critical points of f. Without loss of generality we can suppose that they are fixed points of f. Take disjoint closed disks $D_1, \ldots, D_r$ centred at $a_1, \ldots, a_r$, not intersecting the Julia set of f, and satisfying $f(D_i) \subset D_i$. Set $D = \bigcup_i D_i$. Observe that $\mu_f(D^c) = 1$. To prove the fundamental lemma we shall need two preliminary lemmas.

If $g: \bar{\mathbb{C}} \circlearrowleft$ is an analytic map, we say that a topological disk $U \subset \bar{\mathbb{C}}$ is *g-adapted* if it doesn't contain critical values of g and its boundary is a finite union of C^1 arcs. The first step in the proof of the fundamental lemma, (lemma II.1 below), says that there exists $m \geq 1$ such that for any f^m-adapted topological disk $\tilde{U} \subset D^c$ there exists a branch $\varphi: \tilde{U} \to \bar{\mathbb{C}}$ of $f^{-m}|\tilde{U}$ such that $\varphi(\tilde{U})$ has a neighbourhood U_1 where for any $n \geq 1$ a reasonable number of branches of $f^{-nm}|U_1$ can be defined and they contract U_1 exponentially with respect to n. The topological disk $\tilde{U}$ is not going to be the set U of the fundamental lemma, neither is n going to be the N in that lemma. We shall still have to decrease U_0 (by deleting certain continuous arcs) and increase n.

LEMMA II.1. *There exists $m \geq 1$ such that if $\tilde{U} \subset D^c$ is an f^m-adapted topological disk, then there exists a branch $\hat{\varphi}: \tilde{U} \to \bar{\mathbb{C}}$ of $f^{-m}|\tilde{U}$, a topological disk $U_1 \supset \hat{\varphi}(\tilde{U})$, constants $\hat{\alpha} > 0$, $C > 1$ and $0 < \lambda < 1$, and for every $n \geq 1$, branches $\hat{\varphi}_i^{(n)}: U_1 \to \bar{\mathbb{C}}$ of $f^{-mn}|U_1$, with $1 \leq i \leq \hat{k}_n$, satisfying:*

(i) $f^m(\bigcup_i \hat{\varphi}^{(n+1)}(U_1)) \subset \bigcup_i \hat{\varphi}_i^{(n)}(U_1)$;

(ii) $\mu_f(\bigcup_i \hat{\varphi}_i^{(n)} \hat{\varphi}(\tilde{U})) \geq \hat{\alpha} \mu_f(\hat{\varphi}(\tilde{U}))$;

(iii) $|(\hat{\varphi}_i^{(n)})'(z)| \leq C\lambda^n$;

for all $n \geq 1$, $1 \leq i \leq \hat{k}_n$, $z \in U_1$.

The gap between what the fundamental lemma requires and II.1 produces is essentially that the property $\hat{\varphi}_i^{(n)}(U_0) \subset U_0$ is not granted by II.1. To fill this gap we shall need two lemmas that will tell us how to delete certain arcs of U_0 in order to, after increasing m, satisfy this property.

LEMMA II.2. *For all $m \geq 1$ and $\varepsilon > 0$ there exists an f^m-adapted topological disk $\tilde{U} \subset D^c$ and $0 < t < 1$ such that writing $B(\alpha, \tilde{U}^c) = \{z: d(z, \tilde{U}^c) \leq \alpha\}$ then:*

$$\sum_{n=1}^{\infty} \mu_f(B(t^n, \tilde{U}^c)) \leq \varepsilon.$$

To prove the fundamental lemma take n given by II.1 and for this m take $\tilde{U}$ and $0<t<1$ given by II.2. Choose $s>0$ so large that

$$\sum_{j=1}^{\infty} \mu_f(B(t^{j+s}, \tilde{U}^c)) \leq \frac{\tilde{\alpha}}{2d^m} \mu_f(\tilde{U}). \tag{1}$$

Now choose a positive integer m_1 such that

$$C\lambda_1^{m_1 n-1} \operatorname{diam} \hat{\varphi}(\tilde{U}) \leq t^{n+s}, \tag{2}$$

where C is the constant given by II.1 when applied to $\tilde{U}$. Set $N=m_1 m$. To prove the fundamental lemma we have to find the branches $\varphi_i^{(n)}: \tilde{U} \to \bar{\mathbb{C}}$, $1 \leq i \leq k_n$, satisfying the required properties. We shall define k_n and the branches by induction on n. Let $\hat{\varphi}$, $\hat{k}_n$, $\hat{\varphi}_i^{(n)}$, $\hat{\alpha}$, C and $0<\lambda<1$ be given by lemma II.1 when applied to the $\tilde{U}$ chosen above. Observe that $\hat{\varphi}_i^{(m_1 n-1)}\hat{\varphi}: \tilde{U} \to \bar{\mathbb{C}}$ is a branch of $f^{-Nn}|\tilde{U}$. Based on this, we define

$$S_1 = \{1 \leq i \leq \hat{k}_{m_1-1} : \hat{\varphi}_i^{(m_1-1)}\hat{\varphi}(\tilde{U}) \subset \tilde{U}\}.$$

Reorder the indices i in order to have

$$S_1 = \{1, \ldots, k_1\}.$$

This defines k_1. As branches $\varphi_i^{(1)}$, $1 \leq i \leq k_1$ of $f^{-N}|\tilde{U}$ we take:

$$\varphi_i^{(1)} = \hat{\varphi}_i^{(m_1-1)}\hat{\varphi}.$$

Now define S_n as the set of indexes $1 \leq i \leq \hat{k}_{m_1 n} - 1$ such that:

$$\hat{\varphi}_i^{(m_1 n-1)}\hat{\varphi}(\tilde{U}) \subset \tilde{U}$$

and

$$f^N \hat{\varphi}^{(m_1 n-1)}\hat{\varphi} = \varphi_j^{(n-1)}$$

for some $1 \leq j \leq k_{n-1}$. Now reorder the set S_n to have

$$S_n = \{1, \ldots, k_n\}.$$

Take this as the definition of k_n and define the branches by:

$$\varphi_i^{(n)} = \hat{\varphi}_i^{(m_1 n-1)}\hat{\varphi}.$$

Clearly these definitions imply properties (i), (iii) and (iv) of the fundamental lemma. Let us prove that (ii) is also satisfied. Define the sets

$$A_n = \bigcup_i \varphi_i^{(n)}(\tilde{U})$$

$$B_n = \bigcup_i \{\varphi_i^{(m_1 n-1)}\hat{\varphi}(\tilde{U}) : \hat{\varphi}_i^{(m_1 n-1)}\hat{\varphi}(\tilde{U}) \cap \tilde{U}^c \neq \emptyset\}.$$

and observe that

$$B_n - A_n = C_n \cup f^{-1}(B_{n-1} - A_{n-1})$$
$$B_1 - A_1 = C_1.$$

Hence

$$\mu_f(B_n - A_n) \leq \mu_f(C_n) + \mu_f(B_{n-1} - A_{n-1}).$$

Therefore:

$$\mu_f(B_n - A_n) \leq \sum_{j=1}^{n} \mu_f(C_j) \leq \sum_{j=1}^{\infty} \mu_f(C_j) \tag{3}$$

But by property (iii) of II.1 and (2)

$$\operatorname{diam} \hat{\varphi}^{(m_1 n-1)}\hat{\varphi}(\tilde{U}) \le C\lambda^{m_1 n-1} \operatorname{diam} \hat{\varphi}(\tilde{U}) \le t^{n+s}$$

for all $n \ge 1$. Then

$$C_j \subset B(t^{j+s}, \tilde{U}^c),$$

and by (i)

$$\sum_{j=1}^{\infty} \mu_f(C_j) \le \sum_{j=1}^{\infty} \mu_f(B(t^{j+s}, \tilde{U}^c)) \le \frac{\tilde{\alpha}}{2d^m} \mu_f(\tilde{U}).$$

This inequality together with (3) implies:

$$\mu_f(A_n) \ge \mu_f(B_n) - \frac{\tilde{\alpha}}{2d^m} \mu_f(\tilde{U}).$$

Moreover, property (ii) of lemma II.1 implies:

$$\mu_f(B_n) \ge \tilde{\alpha}\mu_f(\hat{\varphi}(\tilde{U})) = \frac{\tilde{\alpha}}{d^m} \mu_f(\tilde{U}).$$

Hence

$$\mu_f(A_n) \ge \frac{\tilde{\alpha}}{d^m} \mu_f(\tilde{U}) - \frac{\tilde{\alpha}}{2d^m} \mu_f(\tilde{U}) = \frac{\tilde{\alpha}}{2d^m} \mu_f(\tilde{U}).$$

Therefore, taking $\alpha = \tilde{\alpha}/2d^m$, property (ii) of the fundamental lemma is satisfied.

□

III. *Proof of lemma* II.1

There exists $r > 0$ such that for any power f^m of f, the order of every critical point z_0 of f^m that is not a periodic critical point, is $\le r$. Therefore for all $m \ge 1$, every critical point $z_0 \notin D$ has order $\le r$. Take m satisfying:

$$2dmr < d^{m/2}, \tag{1}$$

$$2dm < (1 - 4d^{-m/2})d^m, \tag{2}$$

$$1/2 < 1 - 4d^{-m/2}. \tag{3}$$

Now let $\tilde{U}$ be an f^m-adapted topological disk contained in D^c. Since $\tilde{U}$ doesn't contain critical values of f^m, there exist d^m branches ψ_i: $U_0 \to \bar{\mathbb{C}}$, $i = 1, \ldots, d^m$ of $f^{-m}|\tilde{U}$. From the definition of f^m-adapted topological disk and the fact that $\tilde{U}$ is contained in D^c it follows that a point of $\bar{\mathbb{C}}$ can belong to at most r sets $\psi_i(\bar{\tilde{U}})$ (in fact to r sets of type $\partial\psi_i(\tilde{U})$ since the sets $\psi_i(\tilde{U})$, $1 \le i \le d^m$ are disjoint). Then we can take topological disks $W_i \supset \psi_i(\bar{\tilde{U}})$ such that a point can belong to at most r sets W_i. We claim there exists $1 \le i_0 \le d^n$, a sequence

$$0 \le \varepsilon_n \le 2d^{-mn/2} \qquad n = 1, 2, \ldots, \tag{4}$$

and branches $\varphi_i^{(n)}$: $W_{i_0} \to \bar{\mathbb{C}}$, $1 \le i \le \hat{k}_n$ of $f^{-mn}|W_{i_0}$ satisfying:

$$\hat{k}_n \ge \left(1 - \sum_{j=1}^{n} \varepsilon_j\right) d^{mn}, \tag{5}$$

$$f^m\left(\bigcup_i \varphi_i^{(n+1)}(W_{i_0})\right) \subset \bigcup_i \varphi_i^{(n)}(W_{i_0}), \tag{6}$$

$$a(\varphi_i^{(n)}(W_{i_0})) \le d^{-mn/2}, \tag{7}$$

for all $n \ge 1$ where $a(\cdot)$ denotes the Lebesgue measure normalized to have $a(\bar{\mathbb{C}}) = 1$.

To determine i_0 observe that the number of critical values of f^m is bounded by $2dm$. Since every critical value of f^m belongs to at most r sets W_i, it follows that at most $2dmr$ of these sets can contain critical values of f^m. By (1) there exists i_0 such that W_{i_0} doesn't contain critical values of f^m. Then we can take branches $\varphi_i^{(1)}\colon W_{i_0}\to\bar{\mathbb{C}}$, $j=1,\dots,d^m$ of $f^{-m}|W_{i_0}$. Since the sets $\varphi_i^{(1)}(W_{i_0})$ are disjoint:

$$1=a(\bar{\mathbb{C}})\geq\sum_{i=1}^{d^n}a(\varphi_i^{(1)}(W_{i_0})).$$

If $S=\{1\leq i\leq d^n\colon a(\varphi_i^{(1)}(W_{i_0}))\leq d^{-m/2}\}$ we obtain:

$$1\geq\sum_{i\notin S}a(\varphi_i^{(1)}(W_{i_0}))\geq(d^m-\#S)d^{-m/2}.$$

This implies:

$$\#S\geq(1-d^{-m/2})d^m. \tag{8}$$

Arrange the indices to have $S=\{1,\dots,\hat{k}_1\}$ and define $\varepsilon_1=d^{-m/2}$. By (8), $\hat{k}_1\geq(1-\varepsilon_1)d^m$. Then $\hat{k}_1$ and ε_1 satisfy (4) and (5). Moreover $a(\varphi_i^{(1)}(W_{i_0}))\leq d^{-m/2}$ for $1\leq i\leq\hat{k}_1$. Hence (7) also is satisfied. Now suppose we have constructed ε_j, k_j, $\varphi_i^{(j)}$ for $j=1,2,\dots,n$ satisfying (4), (5), (6) and (7). Since the number of critical values of f^m is bounded by $2dm$ and the sets $\varphi_i^{(n)}(W_{i_0})$, $1\leq i\leq\hat{k}_n$, are disjoint, there exist k_n-2dm of these sets not containing critical values of f^m. Arrange the index i in such a way that these sets are $\varphi_i^{(n)}(W_{i_0})$, $1\leq i\leq\hat{k}_n-2dm$. Observe that by (5), (4), (3) and (2):

$$\begin{aligned}k_n-2dm&\geq\left(1-\sum_{j=1}^{n}\varepsilon_j\right)d^{mn}-2dm\\&\geq\left(1-\sum_{j=1}^{n}2d^{-mj/2}\right)d^{mn}-2dm\\&\geq(1-2d^{-m/2}(1-d^{-m/2})^{-1})d^{mn}-2dm\\&\geq(1-4d^{-m/2})d^{mn}-2dm>0.\end{aligned}$$

Define $S_0=\{1\leq i\leq\hat{k}_n\colon\varphi_i^{(n)}(W_{i_0})$ doesn't contain critical values of $f^m\}$. Then for every $j\in S_0$ there exist branches $\varphi_{j,i}\colon\varphi_j^{(n)}(W_{i_0})\to\bar{\mathbb{C}}$ of $f^{-m}|\varphi_j^{(n)}(W_{i_0})$ with $i=1,\dots,d^m$. The compositions $\varphi_{i,j}\varphi_j^{(n)}\colon W_{i_0}\to\bar{\mathbb{C}}$ give us $d^m(\#S_0)$ branches of $f^{-(n+1)m}|W_{i_0}$, that we shall denote $\varphi_i^{(n+1)}$, $i=1,\dots,t_n$, where $t_n=d^m(\#S_0)$. Observe that:

$$f^m\left(\bigcup_{i=1}^{t_n}\varphi_i^{(n+1)}(W_{i_0})\right)\subset\bigcup_{i=1}^{\hat{k}_n}\varphi_i^{(n)}(W_{i_0}). \tag{9}$$

Define $S=\{1\leq i\leq t_n\colon a(\varphi_i^{(n+1)}(W_{i_0}))\leq D^{-m(n+1)/2}\}$. Then:

$$\begin{aligned}1=a(\bar{\mathbb{C}})&\geq\sum_{i=1}^{t_n}a(\varphi_i^{(n+1)}(W_{i_0}))\geq\sum_{i\notin S}a(\varphi_i^{(n+1)}(W_{i_0}))\\&\geq(t_n-\#S)d^{-m(n+1)/2}.\end{aligned}$$

Then:

$$\begin{aligned}\#S &\geq t_n - d^{m(n+1)/2} \geq d^m(\#S_0) - d^{m(n+1)/2}\\ &\geq (\hat{k}_n - 2dm)d^m - d^{m(n+1)/2}\\ &\geq \left(1 - \sum_{j=1}^{n} \varepsilon_j\right) d^{m(n+1)} - 2d^{m+1}m - d^{m(n+1)/2}\\ &\geq \left(1 - \sum_{j=1}^{n} \varepsilon_j - (2md^{1-mn} + d^{-m(n+1)/2})\right) d^{m(n+1)}. \end{aligned} \tag{10}$$

Define

$$\varepsilon_{n+1} = 2md^{1-mn} + d^{-m(n+2)/2} \tag{11}$$

and $\hat{k}_{n+1} = \#S$. Arrange the indexes to have $S = \{1, \ldots, \hat{k}_{n+1}\}$. Using the fact that $n > 1$ it is easy to check that (11) and (1) imply (4). Clearly (10) and (11) imply (5). Property (6) follows from (9) and (7) from the definition of S. This completes the proof of the claim. To complete the proof of II.1 we shall use Koebe's theorem ([3]) which says that given a topological disk D and a compact subset $K \subset D$ there exists $A > 0$ such that any injective analytic function $f: D \to \bar{\mathbb{C}}$ satisfies $|f'(z_1)| \geq A|f'(z_2)|$ for every z_1 and z_2 in K. Take $D = W_{i_0}$ and as K a topological disk U_1 satisfying

$$W_{i_0} \supset \bar{U}_1 \supset U_1 \supset \overline{\psi_{i_0}(U_0)}.$$

Then, for all $n \geq 1$, $z \in \varphi_i^{(n)}(U)$, $1 \leq i \leq k_n$:

$$\begin{aligned} d^{-mn/2} &\geq a(\varphi_i^{(n)}(U_1)) \geq a(U_1) \inf_{w \in U_1} |(\varphi_i^{(n)})'(w)|^2\\ &\geq a(U_1)A^2|(\varphi^{(n)})'(z)|^2. \end{aligned}$$

Therefore:

$$|(\varphi_i^{(n)})'(z)| \leq A^{-1}a(U_1)^{-\frac{1}{2}}(d^{-m/4})^n \tag{12}$$

for all $n \geq 1$, $1 \leq i \leq k_n$, $z \in U_1$. Then (12) implies the existence of C and λ satisfying

$$|(\varphi_i^{(n)})'(z)| \leq C\lambda^n \tag{13}$$

for all $n \geq 1$, $1 \leq i \leq \hat{k}_n$, $z \in U_1$. Moreover, property (1) in the introduction implies:

$$\mu_f(\varphi_i^{(n)}(\varphi_{i_0}(\tilde{U}))) = d^{-mn}\mu_f(\varphi_{i_0}(\tilde{U}))$$

for all $n \geq 1$, $1 \leq i \leq \hat{k}_n$. Hence, by (5),

$$\begin{aligned} \mu_f\left(\bigcup_{i=1}^{\hat{k}_n} \varphi_i^{(n)}(\varphi_{i_0}(\tilde{U}))\right) &= \hat{k}_n d^{-mn}\mu_f(\varphi_{i_0}(\tilde{U}))\\ &\geq \left(1 - \sum_{j=1}^{n} c_j\right)\mu_f(\varphi_{i_0}(\tilde{U}))\\ &\geq \left(1 - \sum_{j=1}^{\infty} c_j\right)\mu_f(\varphi_{i_0}(\tilde{U}))\\ &= \left(1 - \sum_{j=1}^{\infty} \varepsilon_j\right) d^{-m}\mu_f(\tilde{U}). \end{aligned} \tag{14}$$

Define $\hat{\alpha}=(1-\sum_{j=1}^{\infty}\varepsilon_j)d^{-m}$. By, (2) and (3)

$$\hat{\alpha}=\left(1-\sum_{j=1}^{\infty}\varepsilon_j\right)d^{-m}\geq(1-2d^{-m/2}(1-d^{-m/2})^{-1})d^m$$

$$>(1-4d^{-m/2})d^m>\tfrac{1}{2}d^m>0. \tag{15}$$

Now to prove II.1 we just take $\hat{\varphi}=\varphi_{i_0}$, the topological disk U_1 and the constants $\hat{\alpha}>0$, $C>0$, $0<\lambda<1$. Then (15) implies (ii), (6) implies (i) and (13) implies (iii). By (15), $\hat{\alpha}>0$. □

IV. *Proof of lemma* II.2

The proof of II.2 is based on the following property of probability measures on $\bar{\mathbb{C}}$.

LEMMA IV.1. *Let μ be a probability measure on the Borel σ-algebra of $\mathbb{C}$. Write $B(r)=\{z: |z|<r\}$. If for all $0<t<1$*

$$\sum_{n=1}^{\infty}\mu(B(t^n))<+\infty,$$

then given any $0<t<1$ there exists $-1\leq a\leq 1$ such that the segment $\gamma_a=\{u(1+ai): 0\leq u\leq 1\}$ satisfies

$$\sum_{n=1}^{\infty}\mu(B(t^n,\gamma_a))<+\infty$$

where $B(t^n,\gamma_a)$ denotes the set $\{z: d(z,\gamma_a)\leq t^n\}$.

In its turn, the proof of this lemma will require a preliminary result about finite measures on an interval.

LEMMA IV.2. *Let ν be a finite measure on the Borel σ-algebra of an interval $[a,b]$. Then for every $0<t<1$ and $0<\alpha<1$ the following property holds for almost every $x\in[a,b]$ with respect to Lebesgue measure: there exists $n(x)$ such that*

$$\nu((x-t^n,x+t^n))\leq t^{\alpha n}$$

for all $n\geq n(x)$.

Proof. By identifying a with b we can translate the problem to a circle S (where it will be easier to handle). Suppose that $S=\{z: |z|=1\}$. Write $S_n=\{x: \nu((x-t^n, x+t^n))\leq t^{\alpha n}\}$. For every $n\geq 1$ let J_i, $1\leq i\leq m_n$ be disjoint open intervals in S of length t^n, $m_n=[2\pi/t^n]$. For each $1\leq i\leq m_n$ let $\hat{J}_i$ be an open interval with the same mid-point as J_i and diam $(\hat{J}_i)=3$ diam J_i. Then:

$$\sum_{i=1}^{m_n}\nu(\hat{J}_i)\leq 3\sum_{i=1}^{m_n}\nu(J_i),$$

because every J_i appears contained in at most three intervals of type $\hat{J}_i$. Then:

$$\sum_{i=1}^{m_n}\nu(\hat{J}_i)\leq 3\nu(S).$$

Defining $H_n=\{1\leq i\leq m_n: \nu(\hat{J}_i)\leq t^{\alpha n}\}$ we obtain

$$3\nu(S)\geq\sum_{i\notin H_n}\nu(\hat{J}_i)\geq(m_n-\#H_n)t^{\alpha n}.$$

Therefore

$$\# H_n \geq m_n - 3\nu(S)t^{-\alpha n} = [2\pi/t^n] - 3\nu(S)t^{-\alpha n}.$$

Writing λ for the Lebesgue measure on S:

$$\lambda\left(\bigcup_{i\in H_n} J_i\right) = t^n \,\# H_n \geq t^n[2\pi/t^n] - 3\nu(S)t^{(1-\alpha)n}$$
$$\geq 2\pi - t^n - 3\nu(S)t^{(1-\alpha)n}.$$

But

$$\bigcup_{i\in H_n} J_i \subset S_n$$

because if $x\in J_i$ and $i\in H_n$ we have $(x-t^n, x+t^n)\subset \hat{J}_i$, and then $\nu((x-t^n, x+t^n))\leq \nu(\hat{J}_i)\leq t^{\alpha n}$. Then

$$\lambda(S_n^c)\leq \lambda\left(\left(\bigcup_{i\in H_n} J_i\right)^c\right)\leq t^n + 3\nu(S)t^{(1-\alpha)n}.$$

By the Borel-Cantelli lemma it follows that for λ-a.e. $x\in S$ there exists $n(x)$ such that x belongs to every set S_n for $n\geq n(x)$. This completes the proof of the lemma. □

Now let us prove IV.1. Consider the interval $L=\{2+ti: |t|<1\}$. If A is a subset of $\{2+ti: t\in\mathbb{R}\}$, set $\hat{A}=\{sz: 0<s<1, z\in A\}$. Define a measure ν on L by $\nu(A)=\mu(\hat{A})$ if $A\subset L$ is a Borel set. Now we shall use the following elementary property whose verification we leave to the reader: There exist $\alpha_0>0$, $\beta_0>1$ and $C>0$ such that if $0<\alpha<\alpha_0$, $\beta_0<\beta<1/\alpha$ and $-\frac{1}{2}<a<\frac{1}{2}$, then

$$B(\alpha, \gamma_a)\subset B(\beta\alpha)\cup \hat{A}(a,\beta)$$

where:

$$A(a,\beta)=\{2+si: 2a-C\beta^{-1}<s<2a+C\beta^{-1}\}.$$

Apply this property to $\alpha=t^n$, $\beta=t^{-n/2}$. Then:

$$B(t^n, \gamma_a)\subset B(t^{n/2})\cup \hat{A}(a, t^{-n/2}).$$

Hence

$$\mu(B(t^n,\gamma_a))\leq \mu(B(t^{n/2}))+\mu(\hat{A}(a, t^{-n/2}))$$
$$= \mu(B(t^{n/2}))+\nu(\{2+si: 2a-Ct^{n/2}<s<2a+Ct^{n/2}\}).$$

By IV.2 we can choose $-\frac{1}{2}<a<\frac{1}{2}$ such that

$$\nu(\{2s+i: 2a-t^{\alpha n}<s<2a+t^{\alpha n}\}\leq t^{\alpha^2 n}$$

for large n, say $n\geq n_1$. Then

$$\mu(B(t^n,\gamma_a))\leq \mu(B(t^{n/2}))+t^{\alpha^2 n}$$

for $n\geq n_1$. Therefore

$$\sum_{n=n_1}^{\infty}\mu(B(t^n,\gamma_a))\leq \sum_{n=n_1}^{\infty}\mu(B((\sqrt{t})^n))+\sum_{n=n_1}^{\infty}(t^{\alpha^2})^n<+\infty.$$

In order to apply IV.1 to μ_f we shall first show that

$$\sum_{n=1}^{\infty} \mu_f(B(t^n, z)) < +\infty$$

for all $0<t<1$ and every critical value z of f (where $B(t^n, z)=\{w: |w-z|<t^n\}$). We shall prove the integrability of $\log|f'|$ with respect to ergodic measures of positive entropy.

LEMMA IV.3. *Let* $g: \bar{\mathbb{C}} \circlearrowleft$ *be an analytic endomorphism and* μ *a* g*-invariant ergodic probability defined on the Borel* σ*-algebra of* $\bar{\mathbb{C}}$ *with* $h_\mu(g)>0$. *Then*:

$$h_\mu(g) \le 2 \int_{\bar{\mathbb{C}}} \log|g'|\, d\mu.$$

Proof. Observe that the function $z \to \log|g'(z)|$ is measurable and upper bounded. Then, even if its integral with respect to μ is $-\infty$, we can apply Birkhoff's theorem and get:

$$\lim_{n\to+\infty} \frac{1}{n} \log|(g^n)'(z)| = \int_{\bar{\mathbb{C}}} \log|g'|\, d\mu$$

for μ-a.e.z. Write λ for the integral on the right. Then λ is the Lyapunov exponent of g and its multiplicity is 2. By Ruelle's inequality:

$$h_\mu(g) \le 2 \max\{\lambda, 0\}.$$

From $h_\mu(g)>0$ it follows that $\lambda>0$. Hence:

$$h_\mu(g) \le 2\lambda = 2 \int_{\bar{\mathbb{C}}} \log|g'|\, d\mu. \qquad \square$$

LEMMA IV.4

$$\int_{\bar{\mathbb{C}}} \log d(z, w)\, d\mu_g(z) > -\infty$$

for every critical value w *of* g.

Proof. First observe that if w_0 is a critical point (i.e. $g'(w_0)=0$):

$$\int_{\bar{\mathbb{C}}} \log d(z, w_0)\, d\mu_g(z) > -\infty,$$

because in a sufficiently small disk D centred at w_0 there exists $C>0$ such that $\log d(z, w_0) \ge C \log|g'(z)|$ for $z \in D$. Then

$$\int_{\bar{\mathbb{C}}} \log d(z, w_0)\, d\mu_g(z) \ge \int_{D^c} \log d(z, w_0)\, d\mu_g(z) + C \int_D \log|g'(z)|\, d\mu_g(z).$$

But $z \to \log d(z, w_0)$ is continuous in D^c, hence the first integral is $>-\infty$, and so is the second by IV.3. Now set $w=g(w_0)$. Let D_1 be a disk centred at w, so small that it doesn't contain other critical values of g. Let γ be a segment joining w to ∂D_1 with $\mu_g(\gamma)=0$. Let $\varphi: D_1-\gamma \to \bar{\mathbb{C}}$ be a branch of $g^{-1}|(D_1-\gamma)$ such that $\lim_{z\to w} \varphi(z) = w_0$. By property (1) in the introduction $\mu_g(g(A)) = d\mu_g(A)$, where d is the degree of g, for every Borel set $A\subset\varphi(D_1-\gamma)$. Then:

$$\int_{\varphi(D_1-\gamma)} d\log d(g(z),w)\,d\mu_g(z)=\int_{D_1-\gamma}\log(z,w)\,d\mu_g(z)$$
$$=\int_{D_1}\log(z,w)\,d\mu_g(z).$$

Take $0<k<1$ such that $d(g(z),w)\geq d(z,w)^k$ for all $z\in\varphi(D_1,\gamma)$. Then:

$$\log d(g(z),w)\geq k\log d(z,w).$$

Hence:

$$\int_{D_1}\log(z,w)\,d\mu_g(z)\geq dk\int_{\varphi(D_1-\gamma)}\log d(z,w)\,d\mu_g(z)>-\infty.\qquad\square$$

LEMMA IV.5. *If w is a critical value of g and $0<t<1$,*

$$\sum_{n=1}^{\infty}\mu_g(B(t^n,w))<+\infty.$$

Proof. Write $A_n=\{z\colon t^{n+1}<|w-z|<t^n\}$ and set

$$C=\int_{\bar{\mathbb{C}}}|\log d(z,w)|\,d\mu_g(z).$$

Then

$$C\geq\sum_{n=1}^{\infty}\int_{A_n}|\log d(z,w)|\,d\mu_g(z)\geq\sum_{n=1}^{\infty}(n+1)\log t\mu_g(A_n).$$

But $B(t^n,w)=\bigcup_{j=n}^{\infty}A_j$ and thus $\mu_g(B(t^n,w))=\sum_{j=n}^{\infty}\mu_g(A_j)$. Hence

$$\sum_{n=1}^{\infty}\mu_g(B(t^n,w))=\sum_{n=1}^{\infty}\sum_{j=n}^{\infty}\mu_g(A_j)=\sum_{n=1}^{\infty}n\mu_g(A_n)\leq\frac{1}{\log t}C.\qquad\square$$

Now we are ready to prove II.2. Suppose first that D contains only one disk. Applying a C^∞ diffeomorphism of $\bar{\mathbb{C}}$ we can suppose that the D has the shape, and the critical values $w_1,\ldots,w_k$ of f^m are arranged with respect to D, as in figure 1.

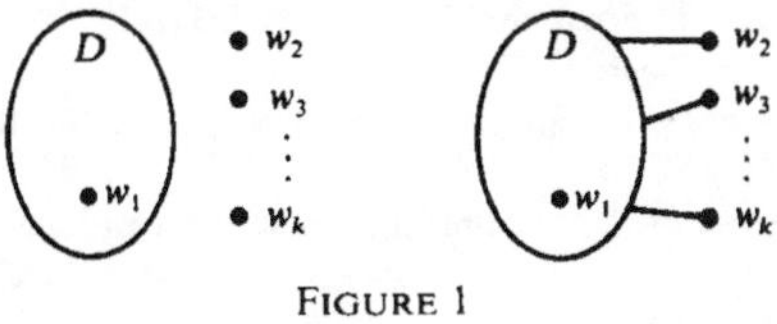

FIGURE 1

Using IV.5 and IV.1 we choose segments $\gamma_1,\ldots,\gamma_k$ as in figure 1 such that

$$\sum_{n=1}^{\infty}\mu_f(B(t^n,\gamma_i))<+\infty\qquad i=1,\ldots,k.\tag{1}$$

Then using (1) and the fact that D doesn't intersect the Julia set (and then has a neighbourhood with measure zero), $\tilde{U}=\bar{\mathbb{C}}-\bigcup_{i=1}^{k}\gamma_i$ satisfies II.2. If D is the union of two disks D_1, D_2, apply a diffeomorphism to place D_1, D_2 and the critical values as in figure 2. And choose $\gamma_1',\gamma_1,\ldots,\gamma_k$ with property (1) (again using IV.1). Then

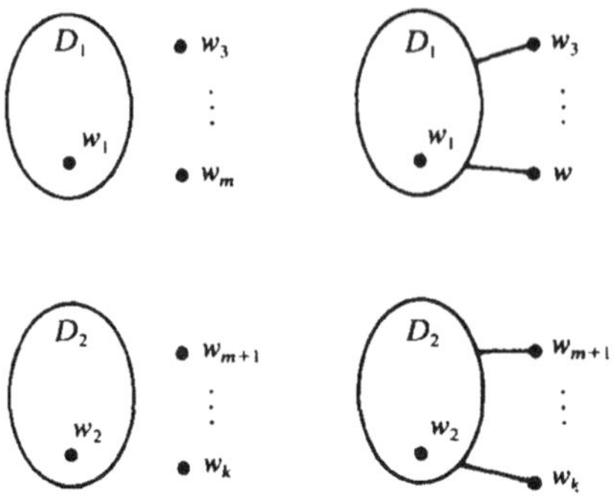

FIGURE 2

$\tilde{U} = \bar{\mathbb{C}} - \gamma_1' - \bigcup_{i=1}^{k} \gamma_i$ satisfies II.2. The case when D contains several disks is handled in a similar way. If $D = \varnothing$ take a segment γ situated with respect to $w_1, \ldots, w_k$ as in figure 3 (after applying a C^∞ diffeomorphism). Then choose $\gamma_1, \ldots, \gamma_k$ satisfying (1) as before and define $\tilde{U} = \bar{\mathbb{C}} - \bigcup_{i=1}^{k} \gamma_i$. □

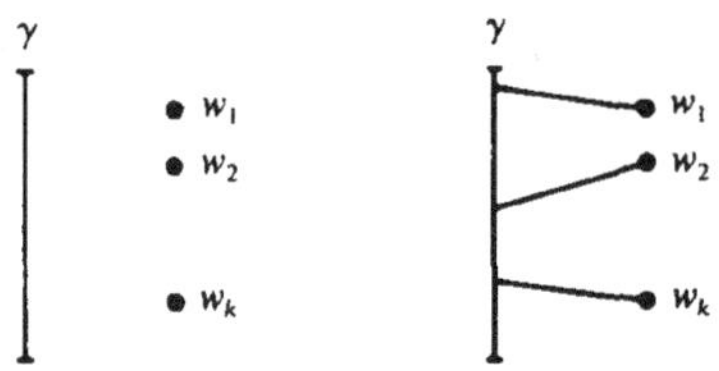

FIGURE 3

REFERENCES

[1] H. Brolin. Invariant sets under iteration of rational functions. *Arkiv für Matematik*, Band G, **6** (1966).

[2] A. Freire, A. Lopes & R. Mañé. An invariant measure for rational maps. *Bol. Soc. Bras. Mat.* **14** (1983) 45-62.

[3] M. Gromov. On the entropy of holomorphic maps. Preprint, IHES.

[4] E. Hille. *Analytic Functions Theory*.

[5] V. Ljubich. Entropy properties of rational endomorphisms of the Riemann sphere. *Ergod. Th. & Dynam. Sys.* **3** (1983), 351-386.

[6] R. Mañé. On the uniqueness of the maximizing measure for rational maps. *Bol. Soc. Bras. Mat.* **14** (1983) 27-43.

[7] M. Misiurewicz & F. Przyticki. Topological entropy and degree of smooth mappings. *Bull. Acad. Pol. Sci.* **25** (1977), 573-574.

[8] D. Ornstein. Factors of Bernoulli shifts are Bernoulli. *Adv. in Math.* **5** (1970), 349-364.

Nonlinearity **5** (1992) 623–638. Printed in the UK

On the minimizing measures of Lagrangian dynamical systems

Ricardo Mañé
IMPA, Estrada Donna Castorina, 110, Rio de Janeiro, RJ 22460, Brazil

Received 17 July 1991
Accepted by J Palis

Abstract. We consider dynamical systems generated by time-dependent periodic Lagrangians on a closed manifold M. An invariant probability μ of such a system has an homology $\rho(\mu) \in H_1(M, \mathbf{R})$ describing (roughly speaking) the average homological position of μ a.e. orbit, and an action $A(\mu)$ defined as the integral with respect to μ of the Lagrangian. The minimizing measures are defined as the invariant probabilities of the system that minimize the action among those that have a given homology γ. Their action $\beta(\gamma)$ define a convex function $\beta : H_1(M, \mathbf{R}) \to \mathbf{R}$. These concepts were introduced by Mather where he proved several theorems about them. In this paper we further develop Mather's theory, giving a characterization of minimizing measures and proving properties about minimizing measures whose homologies are strictly extremal points of the function β.

AMS classification scheme numbers: 58F05,58F11

1. Introduction

Let M be a compact boundaryless manifold and let $L : TM \times \mathbf{R} \to \mathbf{R}$ be a C^3 function (the Lagrangian) satisfying the following hypothesis.

(1) *Periodicity* $L(\theta, t+1) = L(\theta, t)$ for all $(\theta, t) \in TM \times \mathbf{R}$.

(2) *Positivity* For all $x \in M$ and $t \in \mathbf{R}$, if $v_1, \ldots, v_n$ are linear coordinates on T_xM, the Hessian matrix $((\partial^2 L/\partial v_i \partial v_j)(x, v, t))_{ij}$ is positive definite.

(3) *Completeness* All the maximal solutions of the Euler–Lagrange equation of L are defined on $\mathbf{R}$.

(4) *Superlinearity* $\lim_{v\to\infty} L(x, v, t)/\|v\| = +\infty$ uniformly on $x \in M, t \in \mathbf{R}$.

To study the dynamic of the periodic second-order differential equation on M defined by the Euler–Lagrange equation it is formally convenient to introduce a flow of diffeomorphisms $\varphi_t : TM \times S^1 \hookleftarrow$, $t \in \mathbf{R}$, where S^1 denotes the circle $S^1 = \mathbf{R}/\mathbf{Z}$, defined by

$$\varphi_t(x_0, v_0, t_0) = (x(t), \dot{x}(t), t \bmod \mathbf{Z})$$

where $x : \mathbf{R} \to M$ is the solution of the Euler–Lagrange equation with the initial condition $x(t_0) = x_0, \dot{x}(t_0) = v_0$. The completeness and periodicity hypotheses grant that this correctly defines a flow on $TM \times S^1$.

This setting, besides naturally covering many problems of analytical dynamics, contains some relevant conservative dynamical systems not directly presented in this form. For instance, every exact twist map of the cylinder $S^1 \times \mathbf{R}$ is the time one map of a periodic Hamiltonian $H(q, p, t)$, $q \in S^1$, $p \in \mathbf{R}$, to which the Legendre transform can be applied yielding a Lagrangian on S^1 satisfying the hypotheses (1)–(4) (see [4]). A similar result has been proved by Mather [2] for the dynamics of a symplectic diffeomorphism near a KAM torus.

For the study of the flow φ_t defined earlier, and in attempting an extension to higher dimension of the Aubry–Mather theory, Mather [3] introduced several concepts (homology of an invariant measure, minimizing measure, the β function) and through them proved several interesting properties that, when applied to twist maps (via the result of Moser [4]), cover the fundamental results of Aubry and Le Daeron [1].

The objective of this paper is to pursue this line of research, and we start by recalling the main concepts and results from Mather's paper [3].

Let $\mathcal{M}(L)$ be the set of probabilities on the Borel σ-algebra of $TM \times S^1$ that have compact support and are invariant under the flow φ_t.

Let $H_1(M, \mathbf{R})$ be the first real homology group of M. Given a closed 1-form ω on M and $\rho \in H_1(M, \mathbf{R})$ let $\langle \omega, \rho \rangle$ denote the integral of ω on any closed curve in the homology class ρ. If $\mu \in \mathcal{M}(L)$, its *homology* is defined as the unique $\rho(\mu) \in H_1(M, \mathbf{R})$ such that

$$\langle \omega, \rho(\mu) \rangle = \int \omega \, d\mu$$

for all closed 1-form ω on M. The integral on the right-hand side is with respect to μ with ω considered as a function $\omega : TM \times S^1 \to \mathbf{R}$ (i.e. independent of the second variable). For the proof of the existence and uniqueness of $\rho(\mu)$, see [3] where the surjectivity of $\rho : \mathcal{M}(L) \to H_1(M, \mathbf{R})$ is also proved.

From the definition of $\rho(\mu)$ and Birkhoff's ergodic theorem it follows that when $\mu \in \mathcal{M}(L)$ is ergodic then for the μ-a.e. point $(x_0, v_0, t_0) \in TM \times S^1$, the solution $x : \mathbf{R} \to M$ of the Euler–Lagrange equation of L with initial condition $x(t_0) = x_0, \dot{x}(t_0) = v_0$ satisfies

$$\langle \omega, \rho(\mu) \rangle = \lim_{T \to +\infty} \frac{1}{T} \int_{x/[t_0, t_0+T]} \omega \tag{1.1}$$

for every closed 1-form ω on M. A much stronger property will be proved later. We shall show that there exists a set $H \subset H_1(M, \mathbf{R})$, whose convex hull is all $H_1(M, \mathbf{R})$, such that if $\gamma \in H$ there exists $\mu \in \mathcal{M}(L)$ with $\rho(\mu) = \gamma$ and such that (1.1) *holds for all* (x_0, v_0, t_0) *in the support of* μ and moreover the limit in (1.1) is *uniform on* (x_0, v_0, t_0).

Since $L : TM \times \mathbf{R} \to \mathbf{R}$ is periodic with period 1 on the second variable, we can consider it to be a function $L : TM \times S^1 \to \mathbf{R}$. The *action* of $\mu \in \mathcal{M}(L)$ is defined by

$$A_L(\mu) = \int L \, d\mu.$$

Finally we define the function $\beta : H_1(M, \mathbf{R}) \to \mathbf{R}$ by

$$\beta(\gamma) = \inf\{A_L(\mu) | \rho(\mu) = \gamma\}.$$

The domain of β is $H_1(M, \mathbf{R})$ because $\rho : H_1(M, \mathbf{R}) \to \mathbf{R}$ is surjective. Moreover β is bounded below because our assumptions on L imply that $\inf L > -\infty$. In fact the infimum can be shown to be a minimum [3] and the measures at which the minimum is attained are called *minimizing measures*. In other words, $\mu \in \mathcal{M}(L)$ is a minimizing measure if

$$\beta(\rho(\mu)) = A_L(\mu).$$

Denote $\mathcal{M}_\gamma(L)$ the set of minimizing measures with homology γ. It is easy to see that β *is a convex function* (i.e. $\beta(t\gamma_1 + (1-t)\gamma_2) \leq t\beta(\gamma_1) + (1-t)\beta(\gamma_2)$ for all γ_1, γ_2 and $0 \leq t \leq 1$). Moreover β *is superlinear*, i.e. $\lim_{\gamma \to \infty} \beta(\gamma)/\|\gamma\| = +\infty$. As usual when dealing with convex functions, we say that $\gamma \in H_1(M, \mathbf{R})$ is an *extremal* point if $\beta(t\gamma_1 + (1-t)\gamma_2) < t\beta(\gamma_1) + (1-t)\beta(\gamma_2)$ for all $0 < t < 1$ and γ_1, γ_2 such that $\gamma = t\gamma_1 + (1-t)\gamma_2$ and that a convex set $S \subset H_1(M, \mathbf{R})$ is a *linear domain* of β if $\beta(t\gamma_1 + (1-t)\gamma_2) = t\beta(\gamma_1) + (1-t)\beta(\gamma_2)$ for all $\gamma_1, \gamma_2 \in S$. The superlinearity of β implies that every $\gamma \in H_1(M, \mathbf{R})$ *is contained in the convex hull of a set* S_0 *of extremal points with diameter* diam S_0 *bounded by a function of* $\|\gamma\|$. It is easy to see that if γ is extremal, $\mathcal{M}_\gamma(L)$ contains ergodic minimizing measures. In general there may exist γ's such that $\mathcal{M}_\gamma(L)$ does not contain any ergodic measures (see appendix).

The main result in Mather [3] is as follows:

If S is a linear domain and $\Lambda \subset TM \times S^1$ is the closure of the union of the supports of measures $\mu \in \mathcal{M}_\gamma(L)$, $\gamma \in S$, then Λ is bounded and, if $\pi : TM \times S^1 \to M \times S^1$ denotes the projection, π/Λ is injective and $(\pi/\Lambda)^{-1} : \pi(\Lambda) \to TM \times S^1$ is Lipschitz.

When $M = S^1$ this result, due to the particularly simple topological properties of flows on $S^1 \times S^1$, easily implies that every $\gamma \in H_1(M, \mathbf{R})$ is extremal and that the support of the ergodic measures $\mu \in \mathcal{M}(L)$ is minimal. Further applications of this theorem to the case of exact twist maps recover, as we stated earlier, the main results of the Aubry–Mather theory [3].

Finally, before stating the results, we recall the concept of minimizing orbit of the flow φ_t. Write $H_1(M, \mathbf{Z}) = E \oplus T$ where E is free and T is the torsion group. Take a smooth covering $p : \widetilde{M} \to M$ such that the group of deck transformations is E and satisfies

(1) If $\gamma \in E$, $x \in \widetilde{M}$ and α is an arc joining x and $\gamma \cdot x$, then $p \circ \alpha$ is a closed curve in the homology class γ.

(2) If α is a closed curve in $\widetilde{M}$, $p \circ \alpha$ defines the zero element in $H_1(M, \mathbf{R})$.

Clearly $\widetilde{M}$ is the universal covering if $\pi_1(M)$ is Abelian. In general $\widetilde{M}$ is obtained by taking the quotient of the universal covering under the action of the kernel of the Hurewicz homomorphism $h : \pi_1(M) \to H_1(M, \mathbf{Z})$.

Define $\tilde{L} = L \circ dp$. The *action* $A_{\tilde{L}}(u)$ of an absolutely continuous map $u : [a, b] \to \widetilde{M}$ is defined as

$$A_{\tilde{L}}(u) = \int_a^b \tilde{L}(u(t), \dot{u}(t), t)\, dt.$$

Recall that the classical calculus of variations proves that, under our assumptions on L, for every $a < b$ and $p_1, p_2 \in \widetilde{M}$ there exists a solution $x : \mathbf{R} \to \widetilde{M}$ of the Euler–Lagrange equations of $\tilde{L}$, such that $x(a) = p_1, x(b) = p_2$ and $x/[a, b]$ minimizes the action $A_{\tilde{L}}$, i.e.

$$A_{\tilde{L}}(x) \leq A_{\tilde{L}}(u)$$

for all absolutely continuous $u : [a,b] \to \widetilde{M}$ with $u(a) = p_1, u(b) = p_2$. For a clear and concise exposition of this result see [3]. The solution x is called a *minimizing* solution joining the pairs (a, p_1) and (b, p_2).

Every orbit $\gamma : \mathbf{R} \to TM \times S^1$ of the flow φ_t corresponds to a solution $x : \mathbf{R} \to M$ of the Euler–Lagrange equation of L sucht that $\gamma(t) = (x(t), \dot{x}(t), t \bmod \mathbf{Z})$. Its lifting $\tilde{x} : \mathbf{R} \to \widetilde{M}$ is a solution of the Euler–Lagrange equation of $\tilde{L}$. We say that x is *minimizing* if, for every interval $[a,b]$, $\tilde{x}/[a,b]$ is the minimizing solution joining the pairs $(a, \tilde{x}(a)), (b, \tilde{x}(b))$. Due to the similarity of this concept with the class A geodesics of Hedlund we shall call $A(L)$ the set of minimizing orbits of φ_t. Clearly it is invariant and closed. Mather proved in [3] that if $\mu \in \mathcal{M}(L)$ is minimizing then its support $\mathrm{Supp}(\mu)$ is contained in $A(L)$ and all its ergodic components are minimizing and the convex hull of the homologies $\rho(\nu)$, ν being an ergodic component of μ, is a linear domain of β. Our first result is the converse of that property.

Theorem 1. $\mu \in \mathcal{M}(L)$ is minimizing if and only if $\mathrm{Supp}(\mu) \subset A(L)$ and the convex hull of the set of homologies $\rho(\nu)$, ν being an ergodic component of μ, is a linear domain of β.

For the next theorem recall that an affine map $x \to Dx + c$, where $c \in \mathbf{R}$ and $D : H_1(M, \mathbf{R}) \to \mathbf{R}$ is linear, is called a *supporting affine map* of β if $\beta(x) \geq Dx + c$ for all $x \in H_1(M, \mathbf{R})$ and the equality holds at some x. In this case, the set $S = \{x | \beta(x) = Dx + c\}$ is called a *supporting domain of β*. Clearly every linear domain (in particular every point) is contained in a supporting domain. A point $\gamma \in H_1(M, \mathbf{R})$ such that $\{\gamma\}$ is a supporting domain is called *strictly extremal*. By the Straszewicz theorem [6], the strictly extremal points are dense in the set of extremal points and its convex hull is all $H_1(M, \mathbf{R})$.

Given a supporting domain S, define $\Lambda(S)$ as the closure of the union of the supports of the minimizing measures $\mu \in \mathcal{M}_\gamma(L)$ with $\gamma \in S$.

Theorem 2. If S is a supporting domain of β and $\nu \in \mathcal{M}(L)$ satisfies $\mathrm{Supp}(\nu) \subset \Lambda(S)$, then ν is minimizing and $\rho(\nu) \in S$.

Corollary. If $\gamma \in H_1(M, \mathbf{R})$ is strictly extremal the following properties hold:

(1) there exists an ergodic $\mu \in \mathcal{M}_\gamma(L)$ whose support is a minimal set;

(2) for every, closed 1-form ω on M, (1.1) holds uniformly for $(x_0, v_0, t_0) \in \Lambda(\{\gamma\})$; and

(3) if $(x_0, v_0, t_0) \in \Lambda(\{\gamma\})$ and $x : \mathbf{R} \to M$ is the solution of the Euler–Lagrange equation with initial condition $x(t_0) = x_0, \dot{x}(t_0) = v_0$ then

$$\beta(\gamma) = \lim_{t \to +\infty} \frac{1}{t} \int_{t_0}^{t_0+t} L(x(s), \dot{x}(s), s)\, ds$$

uniformly on $(x_0, v_0, t_0) \in \Lambda(\{\gamma\})$.

2. Proof of theorem 1

Let $\widetilde{M}$ be the covering of M described in the introduction. Write $H_1(M, \mathbf{Z}) = E \oplus T$, where T is the torsion group and E is free. Let $c_1, \ldots, c_k$ be a basis of E and let $\omega_1, \ldots, \omega_k$

be closed 1-forms satisfying $\langle \omega_i, c_i \rangle = \delta_{ij}$. Given $x, y \in \widetilde{M}$ define $\alpha(x,y) \in H_1(M, \mathbf{R})$ as

$$\alpha(x,y) = \sum_i \left(\int_\gamma \tilde{\omega}_i \right) c_i$$

where γ is a continuous arc joining x and y and $\tilde{\omega}_1, \ldots, \tilde{\omega}_k$ are the liftings to $\widetilde{M}$ of $\omega_1, \ldots, \omega_k$. Given $x, y \in \widetilde{M}$ and $a < b$ define $A(x, y, a, b)$ as the minimum of the actions

$$A_{\tilde{L}}(u) = \int_a^b \tilde{L}(u(t), \dot{u}(t), t)\, dt$$

where $u: [a, b] \to \widetilde{M}$ varies over all the absolutely continuous arcs with $u(a) = x, u(b) = y$. Under our assumptions on L this minimum is attained at certain solutions $\tilde{x} : [a, b] \to \widetilde{M}$ of the Euler–Lagrange equations of $\tilde{L}$.

For the proof of theorem 1 I shall use the following easy lemma. Its proof will be given after completing the proof of the theorem 1. Fix a Riemannian metric on M and denote $d(\cdot, \cdot)$ the distance induced by this metric on M(or on $\widetilde{M}$ by its lifting).

Lemma 1. Given $C > 0$ and $\delta > 0$, there exist $T_0 > 0$ and $\varepsilon > 0$ such that if $a, y, x \in \widetilde{M}$ and $T > T_0$ satisfy $A_{\tilde{L}}(a, y, 0, T) \le CT$ and $\|\alpha(y, a) - \alpha(x, a)\| \le \varepsilon T$ then

$$(1/T)(A_{\tilde{L}}(a, x, 0, T) - A_{\tilde{L}}(a, y, 0, T)) \le \delta$$

We shall also need the following important result from Mather [3].

Theorem 3. For all $C > 0$ there exists $C_1 = C_1(C)$ such that if $x: [a, b] \to \widetilde{M}$ is a minimizing solution of the Euler–Lagrange equation of L (in the sense defined in the introduction) with $A_{\tilde{L}}(x) \le C(b - a)$ then $\|\dot{x}(t)\| \le C_1$ for all $t \in [a, b]$.

To prove theorem 1 we shall first show that if $\mu \in \mathcal{M}(L)$ is ergodic and $\mathrm{Supp}(\mu) \subset A(L)$, then μ is minimizing. Set $\gamma = \rho(\mu)$. Let $S(\gamma)$ be a supporting domain containing γ. It is easy to see that the extremal points of the convex set $S(\gamma)$ are extremal points of β. Take a convex combination $\gamma = \sum_{i=1}^m \sigma_i \gamma_i$ of extremal points $\gamma_1, \ldots, \gamma_m$ of $S(\gamma)$. Since $\gamma_1, \ldots, \gamma_m$ are extremals, there exist ergodic measures $\mu_i \in \mathcal{M}_{\gamma_i}(L)$. Since $S(\gamma_i)$ is a linear domain,

$$\beta(\gamma) = \sum_i \sigma_i \beta(\gamma_i) = \sum \sigma_i A_L(\mu_i).$$

Due to the ergodicity of μ and Birkhoff's theorem there exists an orbit γ in the support of μ (in fact μ-almost-every-orbit) such that the corresponding solution $x : \mathbf{R} \to \widetilde{M}$ of the Euler–Lagrange equation of $\tilde{L}$ satisfies

$$\gamma = \rho(\mu) = \lim_{T \to \infty} (1/T) \alpha(x(0), x(t)) \tag{2.1}$$

$$A_L(\mu) = \lim_{T \to \infty} (1/T) \int_0^T \tilde{L}(x(t), \dot{x}(t), t)\, dt. \tag{2.2}$$

Moreover there exist solutions $x_i : \mathbf{R} \to \widetilde{M}$ of the Euler–Lagrange equations of $\tilde{L}$ corresponding to orbits contained in $\mathrm{Supp}(\mu_i)$, $1 \le i \le m$, such that

$$\gamma_i = \rho(\mu_i) = \lim_{T\to\infty} (1/T)\alpha(x_i(0), x_i(T)) \tag{2.3}$$

$$\beta(\gamma_i) = A_L(\mu_i) = \lim_{T\to\infty} (1/T) \int_0^T L(x_i(t), \dot{x}_i(t), t)\, dt. \tag{2.4}$$

Fix integers $T_i > 2$, $1 \le i \le m$. Define $\widehat{T}_i$, $1 \le i \le m$ by $\widehat{T}_0 = 0$ and

$$\widehat{T}_i = \sum_{j=1}^{i} T_j \qquad 1 \le i \le m.$$

Let K be the diameter of a fundamental domain of $\widetilde{M}$ as covering of M. Choose deck transformations $g_1 = \text{Identity}\ g_2, \ldots, g_m$ such that

$$d(g_i x_i(\widehat{T}_i), g_{i+1} x_{i+1}(0)) \le K$$

for $1 \le i \le m$. Set $T = \widehat{T}_m$ to simplify the notation. Define $y_0 : [0, \widehat{T}_m] \to \widetilde{M}$ by

$$y_0(t) = g_i x_i(t - \widehat{T}_{i-1})$$

for $t \in (\widehat{T}_{i-1}, \widehat{T}_i]$, $1 \le i \le m$, and $y_0(0) = x_1(0)$. This function is discontinuous. Define $y : [0, T] \to \widetilde{M}$ by $y(t) = y_0(t)$ when

$$t \in \sum_{i=1}^{m-1} [\widehat{T}_{i-1}, \widehat{T}_i - 1]$$

and $y(t) = u_i(t)$ if $t \in [\widehat{T}_i - 1, \widehat{T}_i]$, $1 \le i \le m-1$, where $u_i : [\widehat{T}_i - 1, \widehat{T}_i] \to \widetilde{M}$ is a minimizing solution of the Euler–Lagrange equation with boundary values $u_i(\widehat{T}_i - 1) = y_0(\widehat{T}_i - 1), u_i(\widehat{T}_i) = \lim_{t\to \widehat{T}_i^+} y_0(t)$. Now y is continuous. Let us see whether, given $\varepsilon > 0$ with a suitable choice of $T_1, \ldots, T_m$, we have, if T is large enough,

$$(1/T) A\tilde{L}(y) \le \beta(\gamma) + 4\varepsilon. \tag{2.5}$$

Take T and the T_i's such that

$$T/T_i = \sigma_i + \varepsilon_i \tag{2.6}$$

with

$$|\varepsilon_i| < \varepsilon \left| \sum_{i=1}^{m} \beta(\gamma_i) \right|^{-1} \qquad \text{and} \qquad 0 < \sum_i (\sigma_i + \varepsilon_i) < 2.$$

Then, from (2.4) follows that, if the T_i's are large enough,

$$
\begin{aligned}
\frac{1}{T}A(y) &\leq \frac{1}{T}\sum_{i=1}^{m} A(x_i) + \frac{1}{T}\sum_{i=1}^{m-1} A(u_i) \leq \sum_{i=1}^{m}(\beta(\gamma_i)+\varepsilon)\frac{T_i}{T} + \frac{1}{T}\sum_{i=1}^{m-1} A(u_i) \\
&= \sum_{i=1}^{m}(\beta(\gamma_i)+\varepsilon)(\sigma_i+\varepsilon_i) + \frac{1}{T}\sum_{i=1}^{m-1} A(u_i) \\
&= \sum_{i=1}^{m}\beta(\gamma_i)\sigma_1 + \sum_{i=1}^{m}\varepsilon_i\beta(\gamma_i) + 2\varepsilon + \frac{1}{T}\sum_{i=1}^{m-1} A(u_i) \\
&\leq \beta(\gamma) + 3\varepsilon + \frac{1}{T}\sum_{i=1}^{m-1} A(u_i).
\end{aligned}
$$

Let us estimate $A_L(u_i)$. We claim that there exists $C > 0$ such that $A_L(u_i) \leq C$ for all i independently of m and the selection of the T_i's. From theorem 3 and (2.4) it follows that there exists a $C_1 > 0$ such that $\|\dot{x}_i(t)\| \leq C_1$ for all t and i. Then

$$
\begin{aligned}
d(y_0(\widehat{T}_i - 1), \lim_{t\to\widehat{T}_i^+} y_0(t)) &= d(g_i x_i(T_i - 1), g_{i+1}x_{i+1}(0)) \\
&\leq d(x_i(T_i - 1), x_i(T_i)) + d(g_i x_i(T_i), g_{i+1}x_{i+1}(0)) \leq C_1 + K. \qquad (2.7)
\end{aligned}
$$

Then this proves that the points that the u_i's join are at a distance bounded independently of m and the choice of the T_i's, thus proving the claim.

Then

$$\frac{1}{T}A_L(y) \leq \beta(\gamma) + 3\varepsilon + \frac{(m-1)}{T}C_1.$$

Hence, if T is large, (2.5) is proved. Now we claim that, if T is large,

$$\left|\frac{1}{T}\alpha(y(0), y(T)) - \gamma\right| < \varepsilon.$$

Observe that

$$\alpha(y(0), y(T)) = \sum_{i=0}^{m-2}\alpha(y(\widehat{T}_i), y(\widehat{T}_{i+1} - 1)) + \alpha(y(\widehat{T}_{m-1}), y(\widehat{T}_m)) + \sum_{i=1}^{m-1}\alpha(u_i(\widehat{T}_i - 1), u(\widehat{T}_i)).$$

Taking the T_i's to be large enough, we have, by (2.3),

$$\alpha(y(\widehat{T}_i), y(\widehat{T}_{i+1} - 1)) = \alpha(x_{i+1}(0), x_{i+1}(T_{i+1} - 1)) = (\gamma_{i+1} + \widehat{\varepsilon}_{i+1})(T_{i+1} - 1).$$

with $|\widehat{\varepsilon}_i| < \varepsilon/2$.

Moreover, the norms $\|\alpha(u_i(\widehat{T}_i - 1), u(\widehat{T}_i))\|$ are bounded by a constant $C > 0$ independent of T and the choice of the T_i's. Then

$$
\begin{aligned}
\frac{1}{T}\alpha(y(0), y(T)) &= \sum_{i=10}^{m-1}\gamma_{i+1}\frac{T_{i+1}}{T} - \frac{1}{T}\sum_{i=10}^{m-1}\gamma_{i+1} + \sum_{i=0}^{m-1}\widehat{\varepsilon}_{i+1}\frac{T_{i+1}}{T} \\
&\quad - \frac{1}{T}\sum_{i=0}^{m-1}\widehat{\varepsilon}_{i+1} + \frac{1}{T}\sum_{i=0}^{m-1}\alpha(u_i(\widehat{T}_i - 1), u(\widehat{T}_i))
\end{aligned}
$$

and

$$\sum_{i=0}^{m-1} \gamma_{i+1} \frac{T_{i+1}}{T} = \sum_{i=0}^{m-1} \gamma_{i+1}\sigma_{i+1} + \sum_{i=0}^{m-1} \varepsilon_{i+1}\gamma_{i+1} = \overline{\gamma} + \sum_{i=10}^{m-1} \varepsilon_{i+1}\gamma_{i+1}.$$

Hence

$$\left| \frac{1}{T}\alpha(y(0), y(T)) - \overline{\gamma} \right| \le \frac{1}{T} \sum_i \|\gamma_i\| + \sup \|\widehat{\varepsilon}_i\| + \frac{m}{T} \sup \|\widehat{\varepsilon}_i\| + \frac{1}{T} mC \le \varepsilon$$

thus proving the claim.

Now observe that

$$\lim_{T\to\infty} \frac{1}{T}\alpha(y(0), x(T)) = \gamma \tag{2.8}$$

because

$$\frac{1}{T}\alpha(y(0), x(T)) = \frac{1}{T}\alpha(y(0), x(1)) + \frac{T-1}{T}\frac{1}{T-1}\alpha(x(1), x(T)).$$

Then, from (2.1) and (2.8), we have, for T large,

$$\left| \frac{1}{T}\alpha(y(0), y(T)) - \frac{1}{T}\alpha(y(0), x(T)) \right| \le \varepsilon. \tag{2.9}$$

Hence, by lemma 1, given $\delta > 0$, we can take $\varepsilon > 0$ such that (2.9) implies

$$(1/T)A_{\overline{L}}(y(0), x(T), 0, T) \le (1/T)A_{\overline{L}}(y(0), y(T), 0, T) + \delta.$$

Then

$$\frac{1}{T}A_{\overline{L}}(x(-1), x(T), -1, T) \le \frac{1}{T}A_{\overline{L}}(x(-1), y(0), -1, 0) + \frac{1}{T}A_{\overline{L}}(y(0), y(T), 0, T) + \delta.$$

Since x is minimizing

$$A_{\overline{L}}(x(-1), x(T), -1, T) = A_{\overline{L}}(x/[-1, T]).$$

Then, by (2.5)

$$A_L(\mu) = \lim_{T\to\infty} [1/(T+1)] A_{\overline{L}}(x/[-1, T]) \le \beta(\gamma) + \delta + 4\varepsilon.$$

Since δ and ε are arbitrarily small, $A_L(\mu) \le \beta(\gamma)$ and, since $\gamma = \rho(\mu)$, we have $A_L(\mu) \le \beta(\rho(\mu))$ thus proving that μ is minimal. Finally suppose that $\mu \in \mathcal{M}(L)$ has all its ergodic components with supports contained in $A(L)$ and the convex hull of its homologies is a linear domain of β. We can approximate (in the weak topology) μ by a convex combination $\nu = \sum_i t_i\mu_i$, where $\mu_1, \ldots, \mu_m$ are ergodic components of μ. Then $\mathrm{Supp}(\mu_i) \subset A(L)$ and this implies that μ_i is minimal for all i. Then, since the convex hull of $\{\rho(\mu_i) | 1 \le i \le m\}$ is a linear domain,

$$A_L(\nu) = \sum_i t_i A_L(\mu_i) = \sum_i t_i \beta(\rho(\mu_i)) = \beta\left(\sum_i t_i \rho(\mu_i)\right) = \beta\left(\rho\left(\sum_i t_i \mu_i\right)\right) = \beta(\rho(\nu)).$$

Then ν is minimizing. Since limits of minimizing measures are minimzing, μ is minimizing.

Now let us prove lemma 1. Endow $H_1(M, \mathbf{R})$ with the norm

$$\left\| \sum_{i=1}^{k} t_i c_i \right\| = \sum_{i+1}^{k} |t_i|$$

where $\{c_1 \ldots, c_k\}$ is the basis of $H_1(M, \mathbf{Z})$ chosen earlier.

Lemma 2. There exists $D > 0$ such that $\|\alpha(a,b)\| \leq 1$ implies $d(a,b) \leq D$.

Proof. Let K be the diameter of the orbit under the torsion group T of a fundamental domain of the covering $\widetilde{M}$ of M. Let $C > 0$ be such that $d(a,b) \leq K$ implies $\alpha(a,b) \leq C$. Set $K_1 = \max\{d(x, c_i \cdot x) | 1 \leq i \leq k, x \in \widetilde{M}\}$. Then

$$d(g \cdot x, x) \leq K_1 \|g\| \tag{2.10}$$

for all x and $g \in H_1(M, \mathbf{Z})$ because, by the definition of $\|g\|, g$ can be written as

$$g = \sum_{j=1}^{m} c_{i_j}$$

with $m = \|g\|$. Then

$$d(x, g \cdot x) \leq \sum_{s=1}^{m-1} d\left(\left(\sum_{j=1}^{s+1} c_{i_j}\right) \cdot x, \left(\sum_{j=1}^{s} c_{i_j}\right) \cdot x\right) + d(c_{i_1} \cdot x, x)$$

$$= \sum_{s=1}^{m-1} d\left(c_{i_{s+1}} \cdot \left(\left(\sum_{j=1}^{s} c_{i_j}\right) \cdot x\right), \left(\sum_{j=1}^{s} c_{i_j}\right) \cdot x\right)$$

$$+ d(c_{i_1} \cdot x, x) \leq mK_1 = K_1 \|g\|.$$

Hence, given $a, b \in \widetilde{M}$ we take $g \in H_1(M, \mathbf{Z})$ such that $d(g \cdot a, b) \leq K$, and then

$$\alpha(a,b) \leq \alpha(a, g \cdot a) + \alpha(g \cdot a, b)$$

and

$$\|\alpha(a,b)\| \geq \|\alpha(a, g \cdot a)\| - \|\alpha(g \cdot a, b)\| = \|g\| - \|\alpha(g \cdot a, b)\|$$

$$\geq \|g\| - C \geq \frac{1}{K_1} d(a, g \cdot a) - C \geq \frac{1}{K_1} d(a,b) - \frac{1}{K_1} d(b, g \cdot a) - C$$

$$\geq \frac{1}{K_1} d(a,b) - \frac{K}{K_1} - C.$$

Then, $\|\alpha(a,b)\| < 1$ implies

$$d(a,b) \leq K_1 \left(1 + \frac{K}{K_1} + C\right)$$

setting $D = K_1 + K + K_1 C$ and the lemma is proved. □

Lemma 3. There exists a continuous function $F: [0, +\infty) \to [0, +\infty)$ such that for all $x, y \in \widetilde{M}$ and $T > 0$

$$A(x, y, 0, T) \leq TF(d(x,y)/T).$$

Proof. Take the geodesic $\gamma: [0, T] \to \widetilde{M}$ with $\gamma(0) = x$, $\gamma(1) = y$, and $\|\dot{\gamma}(t)\| = T^{-1} d(x,y)$ for all t. Then

$$A\tilde{L}(x, y, 0, T) \leq \int_0^T \tilde{L}(\gamma, \dot{\gamma}, t)\, dt.$$

Take $c > 0$ and a continuous function $\Phi : [0, +\infty) \to [0, +\infty)$ such that

$$\tilde{L}(x, v, t) \leq c + \Phi(\|v\|)$$

for all x, v, t. Then

$$A_{\tilde{L}}(x, y, 0, T) \leq cT + \Phi(T^{-1} d(x,y))T.$$

Setting $F = c + \Phi$ the lemma is proved. □

Let $u: [0, T] \to \tilde{M}$ be a minimizing solution with $u(0) = a, u(T) = y$. Then $\|\dot{u}(t)\| \le C_1$ for all $t \in [0, T]$, where C_1 is a function of C given by theorem 2. Let $g \in H_1(M, \mathbf{Z})$ be the vector whose coordinates in the basis $\{c_1, \ldots, c_k\}$ are the integer part of $\alpha(a, x) - \alpha(a, y)$. The norm $\|g\|$ is an integer with

$$\|g\| \le \|\alpha(a, x) - \alpha(a, y)\|.$$

Take $0 < \delta < 1$ such that

$$\|g\| \le \delta T + 1 < T.$$

Moreover

$$d(u(0), u(\|g\|) \le C_1 \|g\|$$

and, if K_1 is chosen as in the proof of lemma 2, we have using (2.10)

$$d(g \cdot u(\|g\|), u(0)) \le d(g \cdot u(\|g\|), u(\|g\|)) + d(u(\|g\|), u(0)) \le K_1 \|g\| + C_1 \|g\|.$$

Take a minimizing solution $z : [0, \|g\|] \to \widetilde{M}$ with $z(0) = u(0), z(\|g\|) = g \cdot u(\|g\|)$. By lemma 3

$$A_{\tilde{L}}(z) = A_{\tilde{L}}(u(0), z(\|g\|), 0, \|g\|) \le F\left(\frac{d(z(\|g\|), u(0))}{\|g\|}\right) \|g\| \le F(K_1 + C_1) \|g\|.$$

Then

$$\begin{aligned} A_{\tilde{L}}(a, g \cdot y, 0, T) &\le A_{\tilde{L}}(a, g \cdot u(\|g\|), 0, \|g\|) + A_{\tilde{L}}(g \cdot u(\|g\|), g \cdot y, \|g\|, T) \\ &= A_{\tilde{L}}(u(0), z(\|g\|), 0, \|g\|) + A_{\tilde{L}}(u(\|g\|), u(T), \|g\|, T) \\ &\le F(K_1 + C_1) \|g\| + A_{\tilde{L}}(u(0), u(T), 0, T). \end{aligned}$$

Observe that

$$\alpha(x, g \cdot y) = \alpha(x, y) - g = \alpha(x, a) - \alpha(y, a) - g.$$

Hence

$$\|\alpha(x, g \cdot y)\| = \|\alpha(x, a) - \alpha(y, a) - g\| \le 1.$$

Then, by lemma 2,

$$d(x, g \cdot y) \le K_1.$$

Now define $w : [0, T] \to \widetilde{M}$ as

$$\begin{aligned} w(t) &= z(t) && 0 \le t \le \|g\| \\ w(t) &= g \cdot u(t) && \|g\| \le t \le T - 1 \\ w(t) &= s(t) && T - 1 \le t \le T \end{aligned}$$

where $s : [T-1, T] \to \widetilde{M}$ is a minimizing solution with $s(T-1) = g \cdot u(T-1)$, $s(T) = x$. But

$$\begin{aligned} d(s(T-1), s(T)) &= d(g \cdot u(T-1), x) \leq d(g \cdot u(T-1), g \cdot u(T)) + d(g \cdot u(T), x) \\ &= d(u(T-1), u(T)) + d(g \cdot y, x) \leq C_1 + K_1. \end{aligned}$$

Hence $A_{\tilde{L}}(s) \leq C_3$, where C_3 is a function of $C_1 + K_1$. Then

$$\begin{aligned} A_{\tilde{L}}(a, x, 0, T) &\leq A_{\tilde{L}}(w) \leq A_{\tilde{L}}(z) + A_{\tilde{L}}(g \cdot u) + A_{\tilde{L}}(s) = A_{\tilde{L}}(z) + A_{\tilde{L}}(u) + A_{\tilde{L}}(s) \\ &\leq F(K_1 + C_1)\|g\| + A_{\tilde{L}}(a, y, 0, T) + C_1 + K_1 \\ &\leq F(K_1 + C_1)\|\alpha(x, a) - \alpha(y, a)\| + A_{\tilde{L}}(a, y, 0, T) + C_1 + K_1 \end{aligned}$$

and

$$\frac{1}{T}(A_{\tilde{L}}(a, x, 0, T) - A_{\tilde{L}}(a, y, 0, T)) \leq \frac{1}{T} F(K_1 + C_1)\|\alpha(x, a) - \alpha(y, a)\| + \frac{1}{T}(C_1 + K_1)$$

proving lemma 1.

3. Proof of theorem 2

Let S be a supporting domain of β. Let us prove that, without loss of generality, we can assume that $\beta \geq 0$ and $S = \beta^{-1}(\{0\})$. In fact, from the definition of S there exist $c \in \mathbf{R}$ and a linear map $D : H_1(M, \mathbf{R}) \to \mathbf{R}$ satisfying

$$\beta(x) \geq Dx + c \qquad \text{for all } x$$

$$\beta(x) = Dx + c \Leftrightarrow x \in S.$$

Take a closed 1-form ω on M such that $Lx = \langle \omega, x \rangle$ for all $x \in H_1(M, \mathbf{R})$. Define $L_0 : TM \times \mathbf{R} \to \mathbf{R}$ by

$$L_0(x, v, t) = L(x, v, t) - \langle \omega, v \rangle - c.$$

The solutions of the Euler–Lagrange equations of L and L_0 coincide. Then $\mathcal{M}(L_0) = \mathcal{M}(L)$ and if $\mu \in \mathcal{M}(L_0) = \mathcal{M}(L)$,

$$A_{L_0}(\mu) = A_L(\mu) - \langle \omega, \rho(\mu) \rangle - c.$$

Hence, if β_0 denotes the β-function of L_0,

$$\beta_0(x) = \beta(x) - \langle \omega, x \rangle - c = \beta(x) - Dx - c.$$

Then S is a supporting domain of β_0 and $S = \beta_0^{-1}(\{0\})$. Therefore we shall assume $S = \beta^{-1}(\{0\})$. Suppose that $\nu \in \mathcal{M}(L)$ and $\mathrm{Supp}(\nu) \subset \Lambda(S)$. We shall prove $A_L(\nu) = 0$. Since $\beta \geq 0$ and $S = \beta^{-1}(\{0\})$, this implies that ν is minimizing and $\rho(\nu) \in S$, proving theorem 2. It suffices to prove $A_L(\nu_0) = 0$ for every ergodic component ν_0 of ν. Suppose by contradiction that $A_L(\nu_0) > 0$. From this property we shall show the existence of $b > 0$ and a sequence of curves $w_k : [a, T_k + a] \to M$, $k = 1, 2, \ldots$, such that $\lim_{k\to+\infty} T_k = +\infty$ and

$$(1/T_k)A_L(w_k) \le -b.$$

Once this is proved we can take minimizing solutions $x_k : [a, T_k + a] \to M$ of the Euler–Lagrange equations of L with $x_k(a) = w_k(a)$, $x_k(T_k + a) = w_k(T_k + a)$. Then

$$\frac{1}{T_k} A_L(x_k) \le \frac{1}{T_k} A_L(w_k) \le -b \tag{3.1}$$

and, if μ_k is the probability on $TM \times S^1$ given by

$$\int \psi \, d\mu_k = \frac{1}{T_k} \int_a^{T_k+a} \psi(x_k(s), \dot{x}_k(s), s) \, ds$$

for every continuous function $\psi : TM \times S^1 \to \mathbf{R}$, by theorem 3, (3.1) implies that $\sup_{t,k} \|\dot{x}_k(t)\| < +\infty$. Hence the supports of the μ_k's are contained in a compact set. Hence we can assume that $\mu_k \to \mu$ in the weak topology, where $\mu \in \mathcal{M}(L)$. Then, by (3.1)

$$A_L(\mu) = \lim_{k\to+\infty} A_L(\mu_k) = \lim_{k\to+\infty} \frac{1}{T_k} A_L(x_k) \le -b$$

implying $\beta(\rho(\mu)) < 0$ that contradicts $\beta \ge 0$ and proves theorem 2. To construct the curves w_k we take a solution $x : \mathbf{R} \to M$ of the Euler–Lagrange equation of L such that the orbit $(x(t), \dot{x}(t), t \bmod \mathbf{Z})$ is ν_0-generic. Given $\varepsilon > 0$ take an integer $N > 0$ so large that

$$d((x(0), \dot{x}(0)), (x(N), \dot{x}(N))) < \varepsilon$$
$$\frac{1}{N} A_L(x/[0, N]) > \frac{9}{10} A_L(\nu_0).$$

Take a neighbourhood B of $(x(0), \dot{x}(0))$ such that if a solution $y : \mathbf{R} \to M$ of the Euler–Lagrange equation of L satisfies $(y(0), \dot{y}(0)) \in B$, then

$$d((y(0), \dot{y}(0)), y(N), \dot{y}(N))) < \varepsilon$$
$$\frac{1}{N} A_L(y/[0, N]) > \frac{9}{10} A_L(\nu_0).$$

Since $\mathrm{Supp}(\nu_0) \subset \mathrm{Supp}(\nu) \subset \Lambda(S)$, there exists a minimizing ergodic μ such that

$$\mu(\{\varphi_t(p) | p \in B, 0 \le t \le 1\}) > 0. \tag{3.2}$$

Let $y : \mathbf{R} \to M$ be a solution of the Euler–Lagrange equation such that the orbit $(y(t), \dot{y}(t), t \bmod \mathbf{Z})$ is μ-generic. Then from (3.2) follows the existence of $c > 0$ and a sequence of integers $n_{k+1} > n_k + 2N$, for $k \ge 1$, and $n_0 = 0$, such that

$$(y(n_k), \dot{y}(n_k)) \in B \tag{3.3}$$

$$\lim_{k\to+\infty} \inf(k/n_k) > c. \tag{3.4}$$

From (3.3) it follows that

$$d((y(n_k),\dot y(n_k)),(y(n_k+N),\dot y(n_k+N)))\le \varepsilon.$$

Hence

$$d((y(n_k),\dot y(n_k)),(x(0),\dot x(0)))\le \operatorname{diam} B \tag{3.5}$$

$$d((y(n_k+N),\dot y(n_k+N)),(x(0),\dot x(0)))\le \varepsilon+\operatorname{diam} B. \tag{3.6}$$

Now take absolutely continuous curves $b_k : [n_k, +N, n_{k+1}] \to M$ such that

$$\begin{aligned} &b_k/[n_k+N+1, n_{k+1}-1] = y\\ &(b_k(n_k),\dot b_k(n_{k+1})) = (x(0),\dot x(0))\\ &(b_k(n_k+N),\dot b_k(n_{k+1}+N)) = (x(0),\dot x(0)).\end{aligned}$$

Observe that every μ-generic orbit $(z(t),\dot z(t), t \bmod \mathbf{Z})$ satisfies

$$\lim_{t\to+\infty}(1/t)A_L(z/[0,t]) = A_L(\mu)=0.$$

Hence, by theorem 3, there exists $C_0>0$ such that $\sup_t \|\dot z(t)\| < C_0$ for every μ-generic orbit $(z(t),\dot z(t), t \bmod \mathbf{Z})$. Clearly given $\varepsilon_1>0$ and $C>0$, there exists $\delta=\delta(\varepsilon_1,C)>0$ such that if $z:[0,1]\to M$ is a solution of the Euler–Lagrange equation of L with $\sup_t\|\dot z(t)\|\le C$ and $d(p,z(1))\le\delta$, then

$$A_L(0,1,z(0),p)\le A_L(z)+\varepsilon_1.$$

Pick ε and diam B such that

$$\varepsilon+\operatorname{diam} B<\delta\left(\frac{1}{20}A_L(\nu_0),C_0\right).$$

Hence, (3.4) and (3.5) imply that b_k can be taken such that

$$\begin{aligned} A_L(b_k)-A_L(y) &= A_L(b_k/[n_k+N,n_k+N+1]) - A_L(y/[n_k+N,n_k+N+1])\\ &\quad + A_L(b_k/[n_{k+1}-1,n_{k+1}]) - A_L(y/[n_{k+1}-1,n_{K+1}]) \le (1/10)A_L(\nu_0).\end{aligned}$$

Define $w_k:[0,n_k-kN]\to M$ by $w_k=b_0\cup\ldots\cup b_{k-1}$. Then

$$\begin{aligned} \frac{1}{n_k}A_L(w_k) &= \frac{1}{n_k}\sum_0^{k-1}A_L(b_i) = \frac{1}{n_k}\sum_0^{k-1}(A_L(b_i)+A_L(y/[n_i,n_i+N]))\\ &\quad - A_L(y/[n_i,n_i+N]) \le \frac{1}{n_k}\sum_0^{k-1}\Bigg(A_L(y/[n_i+N,n_{i+1}])+A_L(y/[n_i,n_i+N])\\ &\quad +\frac{1}{10}A_L(\nu_0)-A_L(y/[n_i,n_i+N])\Bigg)\le\frac{1}{n_k}A_L(y/[0,n_k])\\ &\quad +\frac{k}{n_k}\left(\frac{1}{10}A_L(\nu_0)-\frac{9}{10}A_L(\nu_0)\right).\end{aligned}$$

Set $T_k = n_k - kN$. Then

$$\frac{1}{T_k} A_L(w_k) = \frac{n_k}{T_k} \cdot \frac{1}{n_k} A_L(w_k)$$
$$\leq \frac{1}{1-(k/n_k)N} \cdot \left(\frac{1}{n_k} A_L(y/[0,n_k]) - \frac{k}{n_k} \cdot \frac{8}{10} A_L(\nu_0) \right).$$

But,

$$\lim_{k\to+\infty} \frac{1}{n_k} A_L(y/[0,n_k]) = A_L(\nu_0) = 0.$$

This and (3.3) imply

$$\limsup_{k\to+\infty} \frac{1}{T_k} A_L(w_k) \leq -\frac{8c}{1-cN} A_L(\nu_0) \tag{3.7}$$

But $n_{j+1} > n_j + 2N$. Hence $n_k \geq 2kN$ and then $cN < 1/2$. Therefore $1 - cN > 0$ and (3.6) proves that the sequence $\{w_k\}$ satisfies the desired condition for

$$b = \frac{c}{1-cN} A_L(\nu_0).$$

To prove part (1) of the corollary take $\mu \in \mathcal{M}_y(L)$ and let $\Gamma \subset \mathrm{Supp}(\mu)$ be a minimal set. Let $\nu \in \mathcal{M}(L)$ be ergodic with $\mathrm{Supp}(\nu) = \Gamma$. Then, by theorem 2, ν is minimizing with $\rho(\nu) = \gamma$.

To prove the second and third parts of the corollary, we use the following easy property whose proof is left to the reader.

Lemma 4. Let Λ be a compact metric space, $\varphi_t : \Lambda \hookleftarrow$, $t \in \mathbf{R}$, a flow of homeomorphism and $\psi : \Lambda \to \mathbf{R}$ a continuous function such that its integral with respect to every invariant probability of the flow has a value α independent of the invariant probability. Then

$$\lim_{t\to+\infty} \frac{1}{t} \int_0^t \psi(\varphi_s(x))\, ds = \alpha$$

uniformly on x.

Then part (2) of the corollary follows by applying this lemma to the function $\psi : \Lambda \to \mathbf{R}^k$ given by

$$\psi(x, v, t) = \omega(x)v$$

because if $\nu_0 \in \mathcal{M}(L)$ has $\mathrm{Supp}(\nu_0) \subset \Lambda(\{\gamma\})$, then $\rho(\nu_0) = \gamma$ by theorem 2. Hence

$$\int \psi\, d\nu_0 = \langle \omega, \gamma \rangle$$

for every invariant probability ν_0 of the flow φ_t/Λ. By lemma 4 this implies

$$\lim_{t\to+\infty} \frac{1}{t} \int_0^t \varphi_s(x_0, v_0, t_0)\, ds = \langle \omega, \gamma \rangle$$

uniformly on $(x_0, v_0, t_0) \in \Lambda$. But this proves the corollary because if $x : \mathbf{R} \to M$ is a solution of the Euler–Lagrange equation of L:

$$\lim_{t\to+\infty} \frac{1}{t} \int_{x/[a,t+a]} \omega = \lim_{t\to+\infty} \frac{1}{t} \int_0^t \omega(x(a+s))\dot{x}(a+s)\,\mathrm{d}s$$
$$= \lim_{t\to+\infty} \frac{1}{t} \int_0^t \psi(\varphi_s(x(a), \dot{x}(a), a))\,\mathrm{d}s.$$

Similarly (3) follows by applying lemma 4 to $\psi = L/\Lambda(\{\gamma\})$ because when ν_0 is an invariant probability of $\varphi_t/\Lambda(\{\gamma\})$ by theorem 2, ν_0 is minimizing and $\rho(\nu_0) = \gamma$. Hence $A_L(\nu_0) = \beta(\gamma)$ and

$$\int L\,\mathrm{d}\nu_0 = \beta(\gamma).$$

Acknowledgments

I wish to thank R Ures for several corrections to the proof of theorem 1 and P Duarte for substantially shortening the proof of theorem 2.

Appendix

There is an easy method of embedding any dynamical system on M in a Lagrangian satisfying our hypothesis. Through this method it is possible to construct some useful examples. Given a smooth time-dependent vector field X_t on M, with period 1 on t, consider the Lagrangian $L : TM \times \mathbf{R} \to \mathbf{R}$ given by $L(x, v, t) = \|v - X_t(x)\|^2$. It clearly satisfies our assumptions (the completeness hypothesis follows from the fact the Euler–Lagrange equation of L has linear growth on the fibers of TM). Every solution $x : \mathbf{R} \to M$ of the equation $\dot{x}(t) = X_t(x(t))$ is a minimizing solution of L because, for every interval $[a, b], A_L(x/[a, b]) = 0$ (since $L(x(t), \dot{x}(t), t) = \|\dot{x}(t) - X_t(x(t))\|^2 = 0$) and $A_L(y/[a, b]) > 0$ for every absolutely continuous $y : [a, b] \to M$ with $y(a) = x(a), y(b) = x(b), y \neq x$, because $y \neq x$ implies that $\dot{y}(t) \neq X_t(y(t))$ for a set of positive measure of values of $t \in [a, b]$ and then

$$A_L(y) = \int_a^b \|\dot{y}(t) - X_t(y(t))\|^2\,\mathrm{d}t > 0.$$

Define a submanifold $\widehat{M} \subset TM \times S^1$ by $\widehat{M} = \{(x, X_t(x), t \bmod \mathbf{Z}) | x \in M, t \in \mathbf{R}\}$. Then $\widehat{M}$ is invariant under the flow $\varphi_t : TM \times S^1 \hookleftarrow$ generated by L and the trajectories of φ_t contained in $\widehat{M}$ have the form $(x(t), \dot{x}(t), t \bmod \mathbf{Z})$, $t \in \mathbf{R}$, where $x : \mathbf{R} \to M$ is a solution of $\dot{x}(t) = X_t(x(t))$. Then all the trajectories of φ_t in $\widehat{M}$ are minimizing. Denote by $\widehat{\mathcal{M}}(L)$ the set of probabilities $\mu \in \mathcal{M}(L)$ with $\mathrm{Supp}(\mu) \subset \widehat{M}$. Moreover, the same argument used earlier shows that $A_L(\mu) = 0$ for all $\mu \in \widehat{\mathcal{M}}(L)$ and $A_L(\nu) > 0$ for all $\nu \in \mathcal{M}(L), \nu \notin \widehat{\mathcal{M}}(L)$. Hence every $\mu \in \widehat{\mathcal{M}}(L)$ is minimizing and the function β satisfies $\beta(\gamma) \geq 0$ for all γ and $\beta(\gamma) = 0$ if and only if $\gamma = \rho(\mu)$ for some $\mu \in \widehat{\mathcal{M}}(L)$. On the other hand, the vector field X_t generates a flow $\psi_t : M \times S^1 \hookleftarrow$ by $\psi_t(x_0, t_0) = (x(t), t_0 \bmod \mathbf{Z})$ where $x : \mathbf{R} \to M$ is the solution of $\dot{x}(t) = X_t(x(t))$, with $x(t_0) = x_0$. Then the canonical projection $\pi : TM \times S^1 \to M \times S^1$, conjugates $\varphi_t/\widehat{M}$ and ψ_t.

Using this method let us construct an example of a Lagrangian L on T^3 satisfying our hypothesis, whose function β satisfies $\beta(\gamma) \geq 0$ for all γ and $\beta^{-1}(\{0\})$ is a segment $[0, \gamma_0], \gamma_0 \neq 0$, such that $\mathcal{M}_{\gamma_0}(L)$ does not contain any measure whose support is minimal. Since $[0, \gamma_0] = \beta^{-1}(\{0\})$, the value γ_0 is extremal. Then, this example will show that the corollary of theorem 2 does not hold if the strict extremality of γ_0 is replaced by extremality. To construct L on $T^3 = \mathbf{R}^3/\mathbf{Z}^3$ take an irrational vector field X on T^3 (i.e. such that its lifting X_0 to T^3 is $\tilde{X}_0(p) = (1, \alpha)$, for all p, where α is irrational). Take $p \in T^3$ and a C^∞ function $\psi : T^3 \to [0, +\infty)$ such that $\psi^{-1}(\{0\}) = p$ and $1/\psi$ is integrable with respect to the Lebesgue measure. Then the vector field $X = \psi X_0$ has exactly two ergodic invariant probabilities: one, μ_0, concentrated as the fixed point p; and the other, μ_1, equivalent to the Lebesgue measure (if $1/\psi$, is *not* integrable with the Lebesgue measure, μ_0 is the unique invariant probability of X; for proofs of these easy properties see [5]). Then, $\mathcal{M}_{\gamma_0}(L)$ contains exactly two ergodic measures, namely $(\pi^{-1})^*\mu_0$ and $(\pi^{-1})^*\mu_1$, and, it is clear that $\rho((\pi^{-1})^*\mu_0) = 0$ and easy to prove that $\rho((\pi^{-1})^*\mu_1) \neq 0$. We proved previously that $\beta(\gamma) \geq 0$ for all γ and $\beta^{-1}(\{0\}) = \{\rho(\mu) | \mu \in \widehat{\mathcal{M}}(L)\}$. Hence, $\beta^{-1}(\{0\})$ is the segment $[0, \rho((\pi^{-1})^*\mu_1)]$ and the only minimizing measure with homology $\rho((\pi^{-1})^*\mu_1)$ is $(\pi^{-1})^*\mu_1$, whose support is $\widehat{M}$, which is not a minimal set because it contains the periodic orbit originated in the fixed point p.

References

[1] Aubry S and Le Daeron P Y 1983 *Phys. Rev.* D 381–422

[2] Mather J 1982 *Topology* **21** 457–67

[3] Mather J 1989 *Preprint* ETH

[4] Moser J 1986 *Ergod. Theor. Dynam. Sys.* **6** 401–13

[5] Nemitsky V V and Stepanov V V 1960 *Qualitative Theory of Dynamical Systems* (Princeton, NJ: Princeton University Press)

[6] Rockafellar R T 1970 *Convex Analysis* (Princeton, NJ: Princeton University Press)

RICARDO MAÑÉ

The Lyapunov exponents of generic area preserving diffeomorphisms

Editor's note. This article contains an outline of a proof of a result announced by the author around 1980. The theorem and its generalization to higher dimensions are reported in the author's address at the International Congress in 1983 [M]. The proofs have not been published. For the 2-dimensional case, the closest to a written draft that is available are some notes handed out by the author at a seminar in Paris. Except for some very minor editing, this article is a direct transcription of those notes. We stress that what is presented here is only the sketch of a proof, and that a substantial amount of work is needed to fill in the details.

Let M be a compact connected boundaryless smooth 2-manifold and let μ be a probability on its Borel σ-algebra generated by a smooth volume form. Let $D_\mu(M)$ be the space of μ-preserving C^1 diffeomorphisms endowed with the C^1 topology.

Theorem. *There exists a residual subset $\mathcal{A} \subset D_\mu(M)$ such that every $f \in \mathcal{A}$ is either Anosov or*

$$\lim_{n\to\pm\infty} \frac{1}{n} \log \|(D_x f^n)\| = 0$$

for a.e. x.

In these notes we shall sketch the proof of this result. This sketch will consist of the statements of all the lemmas conducing to the proof of the theorem. With few exceptions these lemmas will be stated without proof. The sequence of lemmas is divided in several steps. In the first step we introduce an "entropy" function $h : D_\mu(M) \to \mathbb{R}$ that is upper semicontinuous. Then its set $\mathcal{A}_0$ of points of continuity is residual. In the second step we shall prove the existence of a residual set $\mathcal{A}_1 \subset D_\mu(M)$ of diffeomorphisms such that all its hyperbolic sets have measure 0 or 1. The third, fourth and fifth steps are devoted to show that diffeomorphisms in $\mathcal{A}_0$ do not exhibit certain pathological behaviours, because these behaviours would produce discontinuities of h. The perturbations required to show this discontinuity

Originally published in: *International Conference on Dynamical Systems (Montevideo, 1995), 110–119, Pitman Res. Notes Math. Ser., 362,*

are constructed with the lemmas given in section III. Using this absence of pathological behaviour we shall show that a diffeomorphism $f \in \mathcal{A}_0 \cap \mathcal{A}_1$ satisfies one of the alternatives of the theorem.

§1 First step: the entropy function.

If $f \in D_\mu(M)$ define $\Sigma(f)$ as the set of points $x \in M$ such that the limits:

$$\lim_{n\to\pm\infty} \frac{1}{|n|} \log \|(D_x f^n)\|$$

exist and coincide. By Oseledec's theorem:

$$\mu(\Sigma(f)) = 1\,.$$

Define $\lambda^+ : \Sigma(f) \to M$ by

$$\lambda^+(x) = \lim_{n\to+\infty} \frac{1}{n} \log \|(D_x f^n)\|$$

and

$$\Sigma_H(f) = \{x \in \Sigma(f) \mid \lambda^+(x) \neq 0\}\,.$$

By Oseledec's theorem:

$$\lambda^+(x) \geq 0$$

for all $x \in \Sigma(f)$ and if $\lambda^+(x) \neq 0$ there exists a splitting $T_xM = E^s(x) \oplus E^u(x)$ such that:

$$\lim_{n\to\pm\infty} \frac{1}{n} \log \|(D_x f^n) \mid E^s(x)\| = -\lambda^+(x)$$

$$\lim_{n\to\pm\infty} \frac{1}{n} \log \|(D_x f^n) \mid E^u(x)\| = \lambda^+(x)\,.$$

Define the entropy function $h : D_\mu(M) \to \mathbb{R}$ by

$$h(f) = \int_{\Sigma(f)} \lambda^+ d\mu\,.$$

Lemma. *h is upper semicontinuous.*

Proof. Define

$$c_n = \int_{\Sigma(f)} \log \|(D_x f^n) \mid E^u(x)\| d\mu(x)$$

then:

$$c_{n+m} \leq c_n + c_m$$

for all $n > 0$, $m > 0$. Hence:

$$h(f) = \lim_{n\to+\infty} \frac{1}{n} \int_{\Sigma(f)} \log \|(D_x f^n) \mid E^u(x)\| d\mu(x) = \lim_{n\to+\infty} \frac{1}{n} c_n$$

$$= \inf_n c_n = \inf_n \frac{1}{n} \int_{\Sigma(f)} \log \|(D_x f^n) \mid E^u(x)\| d\mu(x)\,.$$

Definition. $\mathcal{A}_0$ = *set of points of continuity of* h.

Corollary. $\mathcal{A}_0$ *is residual.*

§2 Second step: avoiding "fat" hyperbolic sets

Proposition. *There exists a residual subset* $\mathcal{A}_1 \subset D_\mu(M)$ *such that if* $f \in \mathcal{A}_1$ *all its hyperbolic sets have measure 0 or 1.*

Proof. Let $\mathcal{B}$ be a basis of neighborhoods of M and let $\tilde{\mathcal{B}}$ be the set of all finite subsets of $\mathcal{B}$ such that $\overline{\cup_{B \in F} B} \neq M$. If $f \in D_\mu(M)$ and $F \in \tilde{\mathcal{B}}$ define:

$$S(F, f) = \cap_n f^n \left(\overline{\cup_{B \in F} B}\right) .$$

If $\varepsilon > 0$ and $F \in \tilde{\mathcal{B}}$ define $D(\varepsilon, F)$ as the set of diffeomorphisms $f \in D_\mu(M)$ such that one of the following properties is satisfied:

(1) There exists a neighborhood $\mathcal{U}$ of f such that $S(F, g)$ is not hyperbolic for all $g \in \mathcal{U}$.

(2) $\mu(S(F, f)) < \varepsilon$.

Obviously $D(\varepsilon, F)$ is open. Moreover it is dense because if $f \in D\mu(M)$ is C^2 (that is a dense property in $D_\mu(M)$ as proved by Zehnder) then every hyperbolic set of f has measure 0 or 1.

§3 The fundamental perturbation lemma

As we explained in the introduction in the next section we shall have to produce perturbations of diffeomorphisms exhibiting certain behaviours. The key tool in the construction of these perturbations will be the next lemma.

Definition. *If* $f : M \circlearrowleft$ *is a diffeomorphism and* $F \in \mathbb{Z}^+$ *we say that a sequence of linear maps* $L_j : T_{f^j(x)}M \to T_{f^{j+1}(x)}M$, $x \in M$, $j = 0, 1, \ldots m$, *is* F*-adapted if* $L_j = D_{f^j(x)} f$ *for all* j *except for at most* F *values of* j *or if*

$$(L_n L_{n-1} \cdots L_0) B_1(0) = (D_x f^n) B_1(0)$$

for all $0 \leq n \leq m$.

Fundamental Lemma. *If* $f \in D_\mu(M)$, $F \in \mathbb{Z}^+$ *and* $\mathcal{U}$ *is a neighborhood of* f *there exists* $\varepsilon > 0$ *such that if* $\Lambda_1 \subset M$ *is a Borel set,* $0 < \varepsilon_1 < \varepsilon$ *and* $N : \Lambda_1 \to \mathbb{Z}^+$ *is a measurable function with the property that for all* $x \in \Lambda_1$ *and* $m > N(x)$ *there exists an* F*-adapted sequence* $L_j : T_{f^j(x)}M \to T_{f^{j+1}(x)}M$, $0 \leq j \leq m$, *satisfying*

$$\|L_j - (D_{f^j(x)} f)\| \leq \varepsilon$$

for all $0 \leq j \leq m$ *and*

$$\|\prod_{j=0}^{m} L_j\| \leq \exp m\varepsilon_1$$

then there exists $g \in \mathcal{U}$ *such that*

$$h(g) \leq h(f) + 2\varepsilon_1 - \int_{\cup_{n\geq 0} f^n(\Lambda_1)} \lambda^+ d\mu\,.$$

This lemma is obtained from the following sequence of properties:

Lemma. *If* $f \in D_\mu(M)$, $0 < k < 1$ *and* $\mathcal{U}$ *is a neighborhood of* f, *there exist* $\varepsilon > 0$ *and* $r_0 > 0$ *such that if* $x \in M$, $0 < r < r_0$, *and* $L : T_xM \to T_{f(x)}M$ *is a linear map satisfying:*

$$\|L - (D_xf)\| < \varepsilon$$

then there exists $g \in \mathcal{U}$ *such that:*

$$g(y) = f(y)$$

when $y \notin B_r(x)$ *and*

$$D_yg = L$$

if $y \in B_{kr}(x)$.

Lemma. *Given* $f \in D_\mu(M)$, $F \in \mathbb{Z}^+$, $0 < k < 1$, $n > 0$ *and a neighborhood* $\mathcal{U}$ *of* f *there exist* $\varepsilon > 0$ *and* $r_0 > 0$ *such that for all* $0 < r < r_0$ *and every nonperiodic point* $x \in M$, *if* $L_j : T_{f^j(x)}M \to T_{f^{j+1}(x)}M$, $0 \leq j < n$, *is an* F*-adapted sequence satisfying*

$$\|L_j - (D_{f^j(x)}f)\| \leq \varepsilon$$

for all $0 \leq j < n$ *then for all* $0 < r < r_0$ *there exists* $g \in \mathcal{U}$ *and a compact set* $K \subset B_r(x)$ *with* $\mu(K) > k\mu(B_r(x))$ *such that*

$$f(y) = g(y)$$

if $y \notin \cup_{m=0}^{n} f^m(B_r(x))$ *and*

$$D_yg^m = L_{m-1}L_{m-2}\cdots L_0$$

if $y \in K$ *and* $0 \leq m < n$.

Definition. *Given* $f \in D_\mu(M)$ *we say that a family of Borel sets* $\{U_0, \ldots, U_m\}$ *is a tower if they are disjoint and* $U_j = f^j(U_0)$ *for all* $0 \leq j \leq m$.

Lemma. *Given $f \in D_\mu(M)$, $0 < k < 1$, $F \in \mathbb{Z}^+$ and a neighborhood $\mathcal{U}$ of f there exists $\varepsilon > 0$ such that if $\{U_0^{(n)}, \ldots, U_{j_n}^{(n)}\}$, $n = 1, \ldots, \ell$, is a family of disjoint towers and for all $x \in \cup_{n=1}^{\ell} U_0^{(n)}$ we have an F-adapted sequence $L_j(x)$: $T_{f^j(x)}M \to T_{f^{j+1}(x)}M$, $0 \leq j < j_n$, such that*

$$\|L_j(x) - (D_{f^j(x)}f)\| \leq \varepsilon$$

for all $0 \leq j < j_n$, then there exists a compact set $K \subset \cup_{n=1}^{\ell} U_0^{(n)}$ and $g \in \mathcal{U}$ satisfying:

(1) *$g(x) = f(x)$ for $x \in S$*

(2) *$\mu(K) \geq k\mu(\cup_{n=1}^{\ell} U_0^{(n)})$*

(3) *If $x \in K$ there exist $1 \leq n \leq \ell$ and $y \in U_0^{(n)}$ such that:*

$$D_x g^{j_n} = L_{j_n-1}(y) \cdots L_0(y) \,.$$

[Editor's remark: S apparently not defined.]

Lemma. *If $f \in D_\mu(M)$, $F \in \mathbb{Z}^+$ and $\mathcal{U}$ is a neighborhood of f there exists $\varepsilon > 0$ such that if $0 < \varepsilon_1 < \varepsilon$ and $\{U_0^{(n)}, \ldots, U_{j_n}^{(n)}\}$, $n = 1, \ldots,$ are disjoint towers whose union is f-invariant and for all $n \geq 1$ and $x \in U_0^{(n)}$ there exists an F-adapted sequence $L_j : T_{f^j(x)}M \to T_{f^{j+1}(x)}M$, $0 \leq j < j_n$, such that*

(1) *$\|L_j - (D_{f^j(x)}f)\| \leq \varepsilon$ for all $0 \leq j < j_n$*

(2) *$\|L_{j_n-1} \cdots L_0\| \leq \exp n\varepsilon_1$*

[Editor's remark: Author probably meant $\exp(j_n\varepsilon_1)$.]

then there exists $g \in \mathcal{U}$ satisfying

$$h(g) \leq h(f) + 2\varepsilon_1 - \int_{\cup_{n,j} U_j^{(n)}} \lambda^+ d\mu \,.$$

The fundamental lemma follows from the last one and the following well known result about decompositions of invariant sets in union of disjoint towers:

Lemma. *If $f \in D_\mu(M)$ and $\Lambda \subset M$ is a Borel set such that the measure of the set of periodic points of f contained in Λ is 0, then for every $N > 0$ Λ can be decomposed in a countable union of disjoint towers with its basis contained in Λ_1 and heights larger that N.*

[Editor's remark: The statement of this lemma is probably not intended as is.]

§4 third step: bounding the angles of the Oseledec splitting

If $f \in D_\mu(M)$ define $\delta_1 : \Sigma_H(f) \to \mathbb{R}$ by:

$$\delta_1(x) = \inf_{n \geq 0} \alpha(E^s(f^n(x)), E^u(f^n(x)))$$

where $\alpha(\cdot, \cdot)$ is the angle between the subspaces between brackets. Let $\Sigma_1(f) = \{x \in \Sigma_H(f) \mid \delta_1(x) = 0\}$. Then $f(\Sigma_1(f)) = \Sigma_1(f)$.

Proposition. *If $f \in D_\mu(M)$ and $\mu(\Sigma_1(f)) \neq 0$ then for every $\varepsilon_0 > 0$ and all neighborhood $\mathcal{U}$ of f there exists $g \in \mathcal{U}$ with*

$$h(g) \leq h(f) + \varepsilon_0 - \int_{\Sigma_1(f)} \lambda^+ d\mu .$$

Corollary. $f \in \mathcal{A}_0 \Rightarrow \mu(\Sigma_1(f)) = 0$.

Sketch of the proof. Given $f \in D_\mu(M)$ with $\mu(\Sigma_1(f)) > 0$ and a neighborhood $\mathcal{U}$ of f take $\varepsilon > 0$ given by the fundamental lemma. Fix any $c > 0$ and define:

$$\Lambda_c = \{x \in \Sigma_1(f) \mid \lambda^+(x) > c\} .$$

Then:

$$\lim_{n \to -\infty} \frac{1}{n} \log \frac{\|(D_x f^n) \mid E^s(x)\|}{\|(D_x f^n) \mid E^\mu(x)\|} \geq -2c$$

for all $x \in \Lambda_c$. [Editor's remark: Author probably meant "$\leq -2c$".] Then there exists $a > 0$ such that if $x \in \Lambda_c$ then $\alpha(E^s(f^m(x)),\ E^u(f^m(x))) \geq a$ for some $m > 0$. This means that if we define:

$$\Lambda_a = \{x \in \Lambda_c \mid \alpha(E^s(x),\ E^u(x)) > a\}$$

we have:

$$\Gamma_c = \cup_{n>0} f^n(\Gamma_a) .$$

Take $A > 0$ and $0 < \varepsilon_1 < \min(\varepsilon, a)$ and a measurable function $N : \Gamma_a \to \mathbb{Z}^+$ such that if $x \in \Gamma_a$ and $n > N(x)$ then in every interval $[j, j + n\varepsilon_1] \subset [0, n]$ there exists m such that:

$$\begin{aligned}
&\alpha(E^s(f^m(x)), E^u(f^m(x))) \leq \varepsilon_1 \\
&\|(D_{f^m(x)} f^{n-m}) \mid E^s(f^m(x))\| \leq \exp(n-m)(-\lambda^+(x) + A\varepsilon_1) \\
&\|(D_x f^m)\| \leq \exp m(\lambda^+(x) + A\varepsilon_1) \\
&\|(D_{f^m(x)} f^{n-m})\| \leq \exp(n-m)(\lambda^+(x) + A\varepsilon_1) \\
&\|(D_x f^m) \mid E^s(x)\| \leq \exp m(-\lambda^+(x) + A\varepsilon_1) .
\end{aligned}$$

Choose $m \in [(n - n\varepsilon_1)/2,\ (n + n\varepsilon_1)/2]$ with all these properties and define a sequence of linear mappings $L_j : T_{f^j(x)}M \to T_{f^{j+1}(x)}M$ putting $L_j = (D_{f^j(x)} f)$ when $0 \leq j \leq m-2$ or $m \leq j \leq n-1$ and

$$L_{m-1} = R(D_{f^{m-1}(x)} f)$$

where R is the rotation that sends $E^u(f^m(x))$ on $E^s(f^m(x))$. Then:

$$\|\prod_{j=0}^{n-1} L_j|E^s(x)\| \leq \exp 2(\lambda^+(x) + A\varepsilon_1)\varepsilon_1 n$$

$$\|\prod_{j=0}^{n-1} L_j|E^u(x)\| \leq \exp 2(\lambda^+(x) + A\varepsilon_1)\varepsilon_1 n$$

Since $\alpha(E^s(x), E^u(x)) \geq a$ we can choose $A = A(a)$ so small that these inequalities imply:

$$\|\prod_{j=0}^{n-1} L_j\| \leq \exp 3C\varepsilon_1 n$$

where:

$$C = \sup_{x \in \Sigma_\mu(f)} \lambda^+(x)\,.$$

If $\varepsilon_1 > 0$ is small enough, the rotation R is so near to the identity that:

$$\|L_j - (D_{f^j(x)} f)\| < \varepsilon$$

for all $0 \leq j \leq n-1$. Then the hypothesis of the fundamental lemma are satisfied and there exists $g \in \mathcal{U}$ with

$$\begin{aligned} h(g) &\leq h(f) + 6C\varepsilon_1 - \int_{\Lambda_c} \lambda^+ d\mu \\ &= h(f) + 6C\varepsilon_1 - \int_{\Sigma_1(f)} \lambda^+ d\mu + \int_{\Sigma_1(f)-\Lambda_c} \lambda^+ d\mu \\ &\leq h(f) + 6C\varepsilon_1 + 6\mu(\Sigma_1(f) - \Lambda_c) - \int_{\Sigma_1(f)} \lambda^+ d\mu\,. \end{aligned}$$

Since ε_1 is arbitrarily small the proposition is proved.

§5 Fourth step: the domination inequalities

Define $\delta_2 : \Sigma_H(f) \to \mathbb{R}^*$ (i.e. $\mathbb{R} \cup \{+\infty\}$) taking as $\delta_2(x)$ the supremum over all $n > 0$, $m > 0$ of

$$\|(D_{f^n(x)} f^m) \mid E^s(f^n(x))\| \cdot \|(D_{f^{n+m}(x)} f^{-m}) \mid E^u(f^{n+m}(x))\|\,.$$

Define $\Sigma_2(f)$ as the set of points $x \in \Sigma_H(f)$ where $\delta_2(x) = +\infty$ then $f(\Sigma_2(f)) = \Sigma_2(f)$.

Proposition. *If $f \in D_\mu(M)$ and $\mu(\Sigma_2(f) \cap \Sigma_1(f)^c) \neq 0$ then for every neighborhood $\mathcal{U}$ of f and every $\varepsilon > 0$ there exists $g \in \mathcal{U}$ such that:*

$$h(g) \leq h(f) + \varepsilon - \int_{\Sigma_2(f) \cap \Sigma_1(f)^c} \lambda^+ d\mu .$$

Corollary. $f \in \mathcal{A}_0 \Rightarrow \mu(\Sigma_2(f)) = 0$.

This proposition also follows from the fundamental lemma. To apply it take $K > 0$, $n_0 > 0$, $k > 0$ and define Λ as the set of points $x \in \Sigma_2(f) \cap \Sigma_1(f)^c$ such that

$$\|(D_x f^{n_0}) \mid E^s(x)\| \cdot \|(D_{f^{n_0}(x)} f^{-n_0}) \mid E^u(f^{n_0}(x))\| > K ,$$

$$\delta_1(x) > a .$$

[Editor's question: $k = a$?]

Let

$$\Lambda(K, n_0, k) = \cup_{n > 0} f^{-n}(\Lambda) .$$

If K and n_0 are large and k small, the measure of $\Sigma_2(f) \cap \Sigma_1(f)^c - \Lambda(K, n_0, k)$ is small. Now we take $\varepsilon_1 > 0$ and a measurable function $N : \Lambda(K, n_0, k) \to \mathbb{Z}^+$ such that if $x \in \Lambda(K, n_0, k)$ and $n > N(x)$ in every interval $[j, j + n\varepsilon_1]$ contains some m such that $f^m(x) \in \Lambda$. We define the sequence of linear maps L_j as follows: choose $m \in [(n - n\varepsilon_1)/2, (n + n\varepsilon_1)/2]$ with $f^m(x) \in \Lambda$. This property implies that there exists a subspace $S \in T_{f^m(x)}M$ with the properties described in the picture:

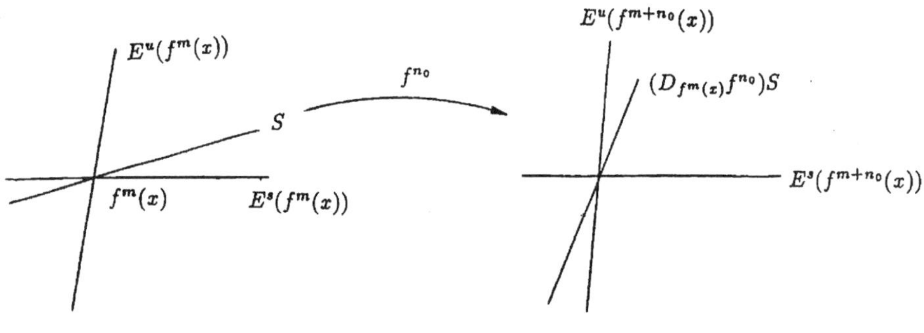

Define L_{m-1} as the composition of $(D_{f^{m-1}(x)} f)$ with a linear map near to I and sending $E^s(f^m(x))$ on S and leaving $E^u(f^m(x))$ invariant. Define L_{m+n_0-1} as $(D_{f^{m+n_0-1}(x)} f)$ composed with a rotation R mapping $(D_{f^m(x)} f^{n_0})(S)$ on

117

$E^u(f^{m+n_0}(x))$. Finally define L_{n-1} as the composition of $D_{f^{n-1}(x)}f$ with a linear map L near to the identity that maps $(D_{f^{m+n_0}(x)}f^{n-m})\mathrm{R}E^u(f^{m+n_0}(x))$ on $E^s(f^n(x))$ and leaves $E^u(f^n(x))$ invariant. The other linear maps L_j are defined as $(D_{f^j(x)}f)$. Then $\prod_{j=0}^{n-1} L_j$ maps $E^s(x)$ on $E^u(f^n(x))$ and $E^u(x)$ on $E^s(f^n(x))$ and the norm can be bounded by $\exp 3Cn\varepsilon_1$ as in the previous section. In fact this estimate as well as the possibility of taking L near to the identity require some more precautions in the choice of m that we didn't explain, hoping that this technical oversimplification may help to put in evidence the underlying idea of the proof.

§6 Fifth step: more domination inequalities

Define $\tilde{\delta}_3 : \Sigma_H(f) \to \mathbb{R}^*$ and $\delta_3 : \Sigma_H(f) \to \mathbb{R}^*$ by

$$\tilde{\delta}_3(x) = \sup\{n > 0 \mid \|(D_x f^m)|E^s(x)\| \cdot \|(D_{f^m(x)}f^{-m})|E^u(f^m(x))\| > 1/2 \quad \text{for all} \quad 0 \le m \le n\}$$

$$\delta_3(x) = \sup_{n>0} \tilde{\delta}_3(f^n(x)) .$$

Let $\Sigma_3(f) = \{x \mid \delta_3(x) = +\infty\}$.

Proposition. *If $f \in D_\mu(M)$ and $\mu(\Sigma_3(f) \cap \Sigma_2(f)^c \cap \Sigma_1(f)^c) > 0$* [then] *for all $\varepsilon > 0$ and every neighborhood $\mathcal{U}$ of f there exists $g \in \mathcal{U}$ satisfying.*

$$h(g) \le h(f) + \varepsilon - \int_{\Sigma_3(f)\cap\Sigma_2(f)^c\cap\Sigma_1(f)^c} \lambda^+ d\mu .$$

Corollary. $f \in \mathcal{A}_0 \Rightarrow \mu(\Sigma_3(f)) = 0$.

Lemma. *If $f \in D_\mu(M)$ then for all $c > 0$ the closure of the set*

$$\Lambda_c(f) = \{x \mid \delta_3(x) < c\}$$

is hyperbolic if $\Lambda_c(f) \ne \phi$.

Proof. The definition of $\Lambda_c(f)$ implies the existence of constants $K > 0$, $0 < \lambda < 1$ such that:

$$\|(D_x f^n)|E^s(x)\| \cdot \|(D_{f^n(x)}f^{-n})|E^u(f^n(x))\| \le K\lambda^n$$

for all $n \ge 0$, $x \in \Lambda_c(f)$. This condition implies

$$\inf_{x\in\Lambda_c(f)} \alpha(E^s(x), E^u(x)) > 0 .$$

Using the fact that f is area preserving and $\dim M = 2$ it follows that there exist $\tilde{K} > 0$, $0 < \tilde{\lambda} < 1$ such that:

$$\|(D_x f^n) \mid E^s(x)\| \le K\lambda^n$$
$$\|(D_{f^n(x)}f^{-n} \mid E^u(x)\| \le K\lambda^n$$

for all $x \in \Lambda_c(f)$, $n \geq 0$. From the uniformity of these estimates, the hyperbolicity of $\overline{\Lambda_c(f)}$ follows easily.

Proof of the Theorem. Suppose that $f \in \mathcal{A}_0 \cap \mathcal{A}_1$ and $\mu(\Sigma_H(f)) \neq 0$. We shall prove that f is Anosov. By the corollary in section 1, the hypothesis $f \in \mathcal{A}_0$ implies:

$$\mu(\Sigma_3(f)) = 0\,.$$

Hence $\mu(\Sigma_H(f) - \Sigma_3(f)) > 0$. But

$$\Sigma_H(f) - \Sigma_3(f) = \cup_{c>0}\Lambda_c(f)\,.$$

Therefore $\mu(\Lambda_c(f)) > 0$ for some $c > 0$. By the lemma the set $\overline{\Lambda_c(f)}$ is hyperbolic and since $f \in \mathcal{A}_1$ its measure must be 1. Then f is Anosov.

References

[M] R. Mañé, *Oseledec's theorem from the generic viewpoint*, Proceedings of the International Congress of Mathematicians, August 1983, Warszawa, Volume 2, p. 1259-1276 (Theorem C and its Corollary in particular).

Introduction and statement of results

Let D denote the open unit disc $\{z \in \mathbf{C} \mid |z| < 1\}$ in the complex plane, ∂D its boundary, the unit circle, and λ the Lebesgue probability on the Borel σ-algebra of ∂D. A classical result of Fatou states that a bounded holomorphic function $f: D \to \mathbf{C}$ possesses a radial limit $f^*(w) = \lim_{r \to 1} f(rw)$ at λ-almost every $w \in \partial D$. An *inner function* is a holomorphic function $f: D \hookleftarrow$ such that $f^*(w) \in \partial D$ for a.e. $w \in \partial D$. The map $f^*: \partial D \hookleftarrow$ thus induced is called the *boundary map* of f.

Our subject here is the dynamics of inner functions; more specifically, the ergodic theory of the boundary map of inner functions. Actually, the dynamics of the inner function itself is quite simple and fully explained by the following result due to Denjoy [**De**] and Wolff [**W**] (to be proved in Section 2): *Given a holomorphic map $f: D \hookleftarrow$ that is not a Möbius transformation, there exists a point $p \in \overline{D}$ (called the Denjoy-Wolff point of f) such that $\lim_{n \to \infty} f^n(z) = p$ uniformly on compact sets.* Obviously this implies that if a non-Möbius inner function $f: D \hookleftarrow$ has a fixed point, then it is unique and is the Denjoy-Wolff point of f, whereas if f has no fixed points, its Denjoy-Wolff point belongs to the boundary ∂D.

On the other hand, the boundary map $f^*: \partial D \hookleftarrow$ of an inner function $f: D \hookleftarrow$ very frequently exhibits interesting non-trivial

Originally published in: *Ensaios Matemáticos, Periodical Volume, 1–79,* forms of recurrence and ergodic phenomena. Since such boundary maps can easily be highly discontinuous, their ergodic theory falls, in general, far beyond the range of the theory of smooth maps of the circle. But, due to its extremely special origin, boundary maps of inner functions can be studied through the tools of classical complex analysis and a satisfactory ergodic theory can be obtained of them. Our first aim is to present an exposition of this theory, through proofs often different from those in the original papers, and afterwards to relate it to rational functions. Given a rational map $f: \overline{\mathbf{C}} \hookleftarrow$, where $\overline{\mathbf{C}} = \mathbf{C} \cup \{\infty\}$ is the Riemann sphere, an important aspect of its dynamics are the fixed components U of the complement of the Julia set, i.e., such that $f(U) = U$. By virtue of the uniformization theorem, $f|U: U \hookleftarrow$ lifts to an inner function whose properties yield relevant information on the ergodic theory (in the harmonic class) of $f|\partial U$.

In this introduction we present and comment on the results whose proof is the objective of the following sections. We begin with a brief review of the fundamental results of the classical theory of inner functions.

The simplest examples of inner functions are the finite Blaschke products

$$f(z) = \alpha \prod_{n=0}^{N} \frac{z - a_n}{1 - \overline{a}_n z} \quad ,$$

with $\alpha \in \partial D$ and $a_n \in D$ for $0 \le n \le N$. More generally, if $\{a_n\}$ is a sequence in D such that $\sum_{n \ge 0} (1 - |a_n|) < \infty$, it is easy to see that the (infinite) Blaschke product

$$B(z) = z^m \prod_{a_n \neq 0} \frac{|a_n|}{a_n} \frac{a_n - z}{1 - \bar{a}_n z} \tag{1}$$

converges (m is the number of a_n equal to zero). Moreover, $B\colon D \hookleftarrow$ is holomorphic, $B \not\equiv 0$ and (not trivially) B is an inner function. (For a proof, see [**Du**; Thm. 2.4] or [**G**; Thm. II.2.2].) Obviously not every inner function is a Blaschke product: simply take a Möbius map T of D onto the half plane $\{z \in \mathbf{C} \mid \operatorname{Re} z < 0\}$ and set $f(z) = \exp T(z)$. Then f is an inner function without zeros, hence not a Blaschke product. This example belongs to the class of holomorphic functions defined by

$$S_\mu(z) = \exp\left(-\int_{\partial D} \frac{w+z}{w-z} d\mu(w)\right), \tag{2}$$

where μ is a finite measure on ∂D. For the example above, take the unit mass μ concentrated at $w = 1$ and $f(z) = S_\mu(z) = \exp\left(\dfrac{z+1}{z-1}\right)$; observe that μ is singular with respect to λ, that is, there exists a set of full Lebesgue measure and zero μ measure.

Blaschke products and these functions S_μ with singular μ are the two conspicuous examples into which any inner function can be factored: the general analytical expression of inner functions is given by the following characterization theorem ([**Du**; Thm. 2.8], [**G**; Thm. II.5.5]): *If $f\colon D \hookleftarrow$ is holomorphic, then its sequence of zeros $\{a_n\}$, which may be empty, satisfies $\sum_{n \geq 0} (1 - |a_n|) < \infty$ and f can be written as*

$$f(z) = \alpha B(z) S_\mu(z), \tag{3}$$

where $|\alpha| \leq 1$, $B(z)$ *is the Blaschke product* (1) *and* $S_\mu(z)$ *is given by* (2), *with* μ *a finite measure on* ∂D. *Conversely, given a sequence* $\{a_n\}$ *in* D *satisfying* $\sum_{n\geq 0}(1-|a_n|) < \infty$, $|\alpha| \leq 1$ *and a finite measure* μ *on* ∂D, (3) *defines a holomorphic function* $f: D \hookleftarrow$. *This map* (3) *is an inner function if and only if* $|\alpha| = 1$ *and* μ *is singular with respect to Lebesgue measure.*

But *every inner function is conjugate to a Blaschke product.* More specifically, let $T_a: D \hookleftarrow$ denote the Möbius map $T_a(z) = (z-a)/(1-\bar{a}z)$, with $a \in D$; then, if f is an inner function, the composition $T_a \circ f$ is a Blaschke product for all $a \in D$, except possibly for a set of zero logarithmic capacity (Frostman; [**G**; Thm. II.6.4]). We shall not discuss the important concept of logarithmic capacity here, but we recall that sets of logarithmic capacity zero are extremely thin: they cannot contain connected sets with more than one point and their Hausdorff dimension is zero. Since $B \circ T_a$ is still a Blaschke product whenever B is one, it follows that if f is an inner function, then $T_a \circ f \circ T_a^{-1}$ is a Blaschke product except for a set of zero logarithmic capacity of values of $a \in D$.

The boundary map of an inner function $f: D \hookleftarrow$ can be highly discontinuous. In fact, the following radical dichotomy holds: *Given* $w \in D$, *either* f *has a holomorphic extension to* $D \cup V$ *for some open neighborhood* V *of* w *in* $\mathbf{C}$ *or else* f^* *maps every neighborhood* W *of* w *in* ∂D *onto almost all of* ∂D *(in the sense of Lebesgue measure)* ([**He**]). In the second case, we say that w is a *singular point* of f. Going back to the example $f(z) = \exp((z+1)/(z-1))$, at its unique singular point $w = 1$ we have $f^*(W) = \partial D$ for every neighborhood W of $w = 1$ in ∂D, a fact that can be read off the expression $f^*(e^{i\theta}) = \exp(-i \cot \frac{\theta}{2})$, $0 < \theta < 2\pi$, of the boundary map. In general, the singular points of an inner function given by (3) above, are the points in the closure of the support of the measure μ and the accumulation points of the zeros of f ([**G**; Thms. II.6.1,2]). Therefore it must be regarded as normal that every point of ∂D is a singular point of an inner function. Much more pathological behaviors may arise in special cases; for instance, there exist inner functions f such that f^* maps sets of zero measure onto sets of positive Lebesgue measure ([**He**]). These extremes of pathology can be avoided under supplementary hypotheses. We say that a holomorphic $F: D \to \mathbf{C}$ is a *Nevanlinna function* if

$$\sup_{0 \leq r < 1} \int_{\partial D} \log^+ |F(rw)| \, d\lambda(w) < \infty;$$

if F is a Nevanlinna function, the radial limit $F^*(w) = \lim_{r \to 1} F(rw)$ exists for λ-a.e. $w \in \partial D$ ([**Du**; Thm. 2.2]). When $f: D \hookleftarrow$ is an inner function and $f': D \to \mathbf{C}$ is Nevanlinna, f^* maps zero measure sets onto zero measure sets and ∂D can be covered, except for a set of measure zero, by a countable family $A_1, A_2, \ldots$ of disjoint Borel sets such that, for each n, $f^*|A_n$ is injective and maps Borel sets onto Borel sets ([**He**]). Moreover, because f' is Nevanlinna, $(f')^*$ exists and is a sort of derivative of f, in the following very weak sense: There exist Borel sets $S \subseteq \partial D$, with complement S^c in ∂D having arbitrarily small Lebesgue measure, such that

$$\lim_{z \to w} \frac{f^*(z) - f^*(w)}{z - w} = (f')^*(w)$$

for all $w \in S$, if $z \to w$ in the set S.

For the study of the ergodic properties of the boundary map, a key concept is that of harmonic measure. Given $p \in D$, the *harmonic measure* λ_p associated to p is the probability on the Borel σ-algebra of ∂D defined by

$$\lambda_p(A) = \int_A \frac{1-|p|^2}{|w-p|^2} d\lambda(w).$$

Clearly $\lambda_0 = \lambda$ is the Lebesgue probability. Moreover, if $\psi: \partial D \to \mathbf{R}$ is a continuous function and $\hat{\psi}: D \to \mathbf{R}$ denotes its harmonic extension (i.e., $\hat{\psi}$ is harmonic and for each $w \in \partial D$, $\lim_{z \to w} \hat{\psi}(z) = \psi(w)$), then Poisson's formula implies

$$\int \psi d\lambda_p = \hat{\psi}(p).$$

It is not difficult to prove, using this property, that λ_p is invariant under the boundary map of inner functions having $p \in D$ as fixed point (see Corollary 1.5 of Section 1 below). More than this, λ_p has good ergodic properties, as the next results will show. Before stating them, let us recall some definitions of abstract ergodic theory ([**Ho**]). Let $(X, \mathcal{A}, \mu)$ be a measure space and $T: X \hookleftarrow$ a measurable map, i.e., $A \in \mathcal{A} \Rightarrow T^{-1}(A) \in \mathcal{A}$. We say that T is *ergodic* if $A \in \mathcal{A}$ and $T^{-1}(A) = A$ imply $\mu(A) = 0$ or $\mu(A^c) = 0$, and that T is *exact* if every $A \in \bigcap_{n \geq 0} T^{-n}(\mathcal{A})$ satisfies $\mu(A) = 0$ or $\mu(A^c) = 0$; alternatively, we sometimes say that μ is ergodic (or exact) instead of T. Clearly exactness implies ergodicity, because $A \in \mathcal{A}$ and $T^{-1}(A) = A$ imply $T^{-n}(A) = A$ for all $n \geq 0$ and then $A \in \bigcap_{n \geq 0} T^{-n}(\mathcal{A})$. The map T is said to be *recurrent* if for every $A \in \mathcal{A}$ and μ-a.e. $x \in A$, there exist infinitely many values of $n \geq 0$ such that $T^n(x) \in A$. Finally, if $\mu(T^{-1}(A)) = \mu(A)$

for each $A \in \mathcal{A}$, we say that μ is *invariant* under T or that T preserves μ. When μ is finite, i.e., $\mu(X) < \infty$, T is recurent whenever T preserves μ (Poincaré), but if $\mu(X) = \infty$, not even from T being exact and measure preserving follows that T is recurrent. We shall find examples of this antagonistic coexistence in the boundary behavior of certain inner functions.

The harmonic measure is unique in the following sense.

THEOREM A. *If $f: D \hookleftarrow$ is an inner function, then $f^*: \partial D \hookleftarrow$ has an invariant probability μ on the Borel σ-algebra of ∂D which is absolutely continuous with respect to λ if and only if f has a fixed point $p \in D$; in this case, $\mu = \lambda_p$ and f^* is exact.*

This theorem follows from a stronger result to be stated below (Theorem G).

When the inner function $f: D \hookleftarrow$ has a fixed point $p \in D$ a very complete picture of the ergodic theory of $f^*: \partial D \hookleftarrow$ with respect to λ_p is given by the following results of Craizer.

THEOREM B. *If $f: D \hookleftarrow$ is an inner function with a fixed point $p \in D$, then*

*a) ([**Cr2**]) with respect to λ_p, $f^*: \partial D \hookleftarrow$ is Bernoulli, i.e., its natural extension is isomorphic, in the measure theoretical sense, to a Bernoulli shift;*

*b) ([**Cr1**]) the entropy $h_{\lambda_p}(f^*)$ of λ_p is finite if and only if f' is Nevanlinna, in which case,*

$$h_{\lambda_p}(f^*) = \int_{\partial D} \log |(f')^*| \, d\lambda_p.$$

We shall not give here the proof of these important (and difficult) results. Our presentation will be mostly oriented toward the other alternative, when the Denjoy-Wolff point belongs to ∂D.

Given $p \in \partial D$, we define a measure μ_p on the Borel σ-algebra of ∂D by

$$\mu_p(A) = \int_A \frac{1}{|w-p|^2} d\lambda(w).$$

Observe that μ_p is precisely the image of the Lebesgue measure on $\mathbf{R}$ (multiplied by a scalar) under any Möbius map transforming the upper half plane onto D and mapping ∞ to p. Clearly $\mu_p(\partial D) = \infty$.

The interesting property of this measure is that, if p is the Denjoy-Wolff point of an inner function f, then f^* transforms μ_p via multiplication by a fixed constant depending upon the behavior of f near p.

THEOREM C. *Let $f: D \hookleftarrow$ be an inner function with Denjoy-Wolff point $p \in \partial D$. There exists $0 < c \leq 1$ such that*

a) $\mu_p(f^{*-1}(A)) = c\mu_p(A)$ for every Borel set $A \subseteq \partial D$;

b) $\lim\limits_{n\to\infty} (1 - |f^n(z)|)^{1/n} = c$ for all $z \in D$;

c) $\lim \dfrac{1-|f(z)|}{1-|z|} = c$, when $z \to p$ nontangentially.

For a proof of this theorem, see Section 4.

It is sometimes convenient to consider inner functions of the upper half plane $\mathbf{R}^2_+ = \{z \in \mathbf{C} \mid \operatorname{Im} z > 0\}$ instead of inner functions of the disc. An inner function of $\mathbf{R}^2_+$ is a holomorphic function $f: \mathbf{R}^2_+ \hookleftarrow$ such that the boundary map $f^*(x) = \lim\limits_{t\to 0} f(x+ti)$, which is well defined for a.e. $x \in \mathbf{R}$, satisfies $f^*(x) \in \mathbf{R}$ for a.e. $x \in \mathbf{R}$ with respect to Lebesgue measure λ on $\mathbf{R}$. Obviously the class of inner functions of D coincides, up to conjugacy by a conformal representation of D onto $\mathbf{R}^2_+$, with the class of inner functions of $\mathbf{R}^2_+$. Thus, formally, no advantage can be expected from replacing the disc by the upper half plane. However, there are interesting examples of inner functions of the upper half plane that admit easy analytical expressions. For instance, the map $f(z) = z + \tan z$ (whose ergodic properties will presently be discussed) or the map $f(z) = z - z^{-1}$, that goes back to Boole who, in 1857, proved that its restriction to $\mathbf{R}$ preserves the Lebesgue measure ([**Bo**], [**Ad-W**]). The reader may check this by brute force, i.e., verifying that $1 = |f'(x_1)|^{-1} + |f'(x_2)|^{-1}$, where x_1 and x_2 are the two roots of $f(x) = c$, for all $c \in \mathbf{R}$. More subtlety is required to answer the problem, posed by Pólya in 1931, of characterizing all the rational maps f such that $f(\mathbf{R}) \subset \mathbf{R}$ and $f|\mathbf{R}$ preserves the Lebesgue measure. The answer (Pólya [**Pa**], Szegö [**Sz**]; see also Section 5) is that f has these properties if and only if f is of the form

$$f(z) = \varepsilon\left(z + \beta - \sum_{i=1}^{n} \frac{\gamma_i}{z - a_i}\right),$$

where $\varepsilon = 1$ or $= -1$, $\beta \in \mathbf{R}$ and $a_i \in \mathbf{R}$, $\gamma_i > 0$ for all $1 \leq i \leq n$. When $\varepsilon = 1$, this is an inner function of the upper half plane. Observe that $f'(x) > 1$ for all $x \in \mathbf{R}$ and, if to study f nearby ∞ we consider $g(z) = -(f(-z^{-1}))^{-1}$, the Taylor series of g at $z = 0$ is

$$g(z) = z + \beta z^2 - \left(\sum_{i=1}^{n} \gamma_i\right) z^3 + \dots .$$

Using these properties, the standard theory of rational functions shows that, for $\beta = 0$, $\mathbf{R} \cup \{\infty\}$ is the Julia set of f and $\lim_{n\to\infty} f^n(z) = \infty$ for all $z \notin \mathbf{R}$, whereas for $\beta \neq 0$, the Julia set of f is a Cantor set Λ of $\mathbf{R} \cup \{\infty\}$ (containing the parabolic fixed point ∞) and $\lim_{n\to\infty} f^n(z) = \infty$ for all $z \notin \Lambda$. The ergodic properties of $f|\mathbf{R}$ with

$\beta = 0$ will be discussed below.

The general formula for inner functions of $\mathbf{R}^2_+$ is the following: *Every holomorphic function* $f: \mathbf{R}^2_+ \hookleftarrow$ *can be written as*

$$f(z) = \alpha z + \beta + \int_{\mathbf{R}} \frac{1 + zw}{w - z} d\mu(w), \tag{4}$$

where $\alpha \geq 0$, $\beta \in \mathbf{R}$ *and* μ *is a finite measure on the Borel* σ*-algebra of* $\mathbf{R}$. *Moreover,* f *is an inner function if and only if* μ *is singular with respect to the Lebesgue measure* λ *on* $\mathbf{R}$ ([**T**]). Theorem D below is an upper half plane version of Theorem C, with $\alpha = c^{-1}$ (to be proved in Section 5).

THEOREM D. *If the inner function* f *of the upper half plane is given by* (4) *with* $\alpha > 0$, *then*

$$\lambda(f^{*-1}(A)) = \frac{1}{\alpha}\lambda(A)$$

for every Borel set $A \subseteq \mathbf{R}$. *Moreover,* ∞ *is the Denjoy-Wolff point of* f *if and only if* $\alpha \geq 1$.

Let us now address the question of the *recurrence* of the boundary map of an inner function. Recurrence is completely characterized by the following results of Aaronson [**Aa2**] (to be proved in Section 4).

THEOREM E. *If* $f: D \hookleftarrow$ *is an inner function with Denjoy-Wolff point* $p \in \partial D$, *then either* f^* *is recurrent or* $\lim_{n \to \infty} f^{*n}(w) = p$ *for a.e.* $w \in \partial D$.

THEOREM F. *If* $f: D \hookleftarrow$ *is an inner function with Denjoy-Wolff point* $p \in \partial D$, *then the following conditions are equivalent:*

a) f^* *is recurrent;*

b) $\sum_{n\geq 0}(1-|f^n(z)|)=\infty$ *for all* $z\in D$;

c) *There exists* $z\in D$ *with* $\sum_{n\geq 0}(1-|f^n(z)|)=\infty$.

However, with or without recurrence, if the Denjoy-Wolff point p is in ∂D, it acts as an "attractor in the mean"; more specifically,

$$\lim_{n\to\infty}\frac{1}{n}\#\{0\leq j<n|f^{*j}(w)\in W\}=1 \tag{5}$$

for every neighborhood W of p in ∂D and a.e. $w\in\partial D$. This is proved in Section 2, but also follows from a stronger result (Theorem H) to be stated below.

As an example of application, let us prove that, restricted to $\mathbf{R}$, the inner function $f(z)=z+\tan z$ of the upper half plane preserves the Lebesgue measure and is recurrent. The function f preserves the imaginary axis because

$$f(it)=i\left(t-\frac{e^{-t}-e^{t}}{e^{-t}+e^{t}}\right);$$

if $g(z)=-(f(-z^{-1}))^{-1}$, the Taylor series of $g(it)$, for small t, is

$$g(it)=i(t-t^2+\dots).$$

Hence

$$\lim_{t\to 0}\frac{|g(it)|}{t}=1,$$

thus proving that g preserves μ_0 and then that f preserves λ. Moreover, the above expression for g implies that $g^n(it)\cong\frac{1}{n}$; it follows that

$$\sum_{n\geq 0}|g^n(it)|=\infty$$

and therefore g and also f are recurrent.

Assume now that f is a Pólya-Szegö map with $\varepsilon = 1$ and $\beta = 0$ and let us see that $f|\mathbf{R}$ (which we already know is measure preserving) is recurrent. Since the Taylor series of $g(z) = -(f(-z^{-1}))^{-1}$ at $z = 0$ is

$$g(z) = z - \Big(\sum_{i=1}^{n} \gamma_i\Big) z^3 + \ldots,$$

standard methods concerning parabolic points imply that there exists a sector $S = \{z \in \mathbf{C} \,|\, |z| < \delta,\ |\mathrm{Arg}\,(-iz)| < \varepsilon\}$ such that $g(S) \subseteq S$ and, for all $z \in S$, $g^n(z) \to 0$ with $n \to \infty$ and more: $g^n(z) \cong \frac{1}{\sqrt{n}}$; thus $\sum_{n \geq 0} |g^n(z)| = \infty$.

Now let us consider the question of the *ergodicity* and *exactness* of the boundary map of an inner function. The next result was proved (with a somehow different statement) by Aaronson in [**Aa3**], relying on his previous result [**Aa2**] and on a theorem of Pommerenke [**Pe**]. For its statement, let $d_P(\cdot,\cdot)$ denote the Poincaré metric on D (or on $\mathbf{R}^2_+$).

THEOREM G. *Let $f\colon D \hookleftarrow$ be an inner function. The following conditions are equivalent:*

a) f^* is ergodic;

b) f^* is exact;

c) $\lim_{n\to\infty} d_P(f^n(x), f^n(y)) = 0$ for all $x, y \in D$;

d) There exist $a, b \in D$ such that $f^n(a) \neq f^n(b)$ for $n \geq 0$ and $\lim_{n\to\infty} d_P(f^n(a), f^n(b)) = 0$.

COROLLARY 1. *Recurrence implies ergodicity.*

COROLLARY 2. *If $f: D \hookleftarrow$ is an inner function with Denjoy-Wolff point $p \in \partial D$ and f^* ergodic, then f^* preserves μ_p.*

The proof of Theorem G is in Section 3; the proof of Corollaries 1 and 2 is straightforward (by Theorems F and C, respectively) and can be found in Section 4.

Let us show here how Theorem A follows from Theorems G and E. Suppose that the boundary map $f^*: \partial D \hookleftarrow$ of an inner function $f: D \hookleftarrow$ admits a λ-absolutely continuous and invariant probability μ. If f has a fixed point $p \in D$, then λ_p is f^*-invariant, f^* is recurrent and, by Corollary 1, λ_p is ergodic. Hence $\mu << \lambda_p$ implies $\mu = \lambda_p$. Let us assume that f has no fixed point (and seek for a contradiction). Then the Denjoy-Wolff point p belongs to ∂D and, moreover, μ-invariance implies that μ-a.e. $w \in \partial D$ is recurrent for f^*. Since $\mu << \lambda$, this implies that the set of recurrent points has positive Lebesgue measure. Theorem E now guarantees that f^* is recurrent and then μ_p is f^*-invariant and ergodic by the two corollaries above. Since $\mu << \mu_p$, it follows that $\mu = \mu_p$, contradicting that μ is a probability and showing that f has a fixed point in D.

The main virtue of ergodicity of finite invariant measures, however, thoroughly fails with μ_p, as the following theorem (to be proved in Section 4) shows: there is no equality between time and space averages.

THEOREM H. *Let $f: D \hookleftarrow$ be an inner function with Denjoy-Wolff point $p \in \partial D$ and such that μ_p is f^*-invariant. Then*

$$\lim_{n\to\infty} \frac{1}{n} \sum_{j=0}^{n-1} \psi(f^{*j}(w)) = 0$$

for every μ_p-integrable function $\psi: \partial D \to \mathbf{R}$ and a.e. $w \in \partial D$.

From this theorem, property (5) follows as a corollary. In fact, (5) is obvious when f^* is not recurrent, since then $\lim_{n\to\infty} f^{*n}(w) = p$ for a.e. $w \in \partial D$ by Theorem E. If f^* is recurrent, then the two corollaries above imply that μ_p is f^*-invariant; applying Theorem H to the characteristic function ψ of the complement of a neighborhood W of p in ∂D, we get (5).

The weak nature of the ergodicity of f^* when its Denjoy-Wolff point is in ∂D, is further enhanced by an example of Aaronson [**Aa3**] of an inner function $f: D \hookleftarrow$ with Denjoy-Wolff point $p \in \partial D$ and such that f^* *is ergodic but nevertheless* $\lim_{n\to\infty} f^{*n}(w) = p$ *for a.e.* $w \in \partial D$. Let us roughly describe this example. Let $\{a_n\}$ be an increasing sequence of positive real numbers with $\sum_{n\geq 0} a_n^{-2} < \infty$ and consider the function

$$f(z) = z - \sum_{n\geq 0} \frac{2z}{z^2 - a_n^2} = z + \sum_{n\geq 0} \left(-\frac{1}{z - a_n} - \frac{1}{z + a_n} \right).$$

The series converges uniformly on compact sets of $\mathbf{C} - \{a_n\}$, defines a meromorphic function and obviously maps $\mathbf{R} - \{a_n\}$ onto $\mathbf{R}$ and $\mathbf{R}^2_+$ into $\mathbf{R}^2_+$ (because every map $z \mapsto -(z-\alpha)^{-1}$, $\alpha \in \mathbf{R}$, maps $\mathbf{R}^2_+$ into $\mathbf{R}^2_+$). Then f is an inner function. Moreover, $f'(t) > 1$ for all $t \in \mathbf{R} - \{a_n\}$ and f maps every interval $(-a_{n+1}, -a_n)$, $(-a_0, a_0)$ and (a_n, a_{n+1}) onto $\mathbf{R}$. Hence, by standard techniques, it can be proved that f is *transitive*, i.e., for a residual subset of $\mathbf{R}$, its forward orbit is dense in $\mathbf{R}$. To show that ∞ is the Denjoy-Wolff point of f, consider $g(z) = -(f(-z^{-1}))^{-1}$. Then

$$g(it) = i\frac{t}{1+\sum\limits_{n\geq 0}\dfrac{2t^2}{1+a_n^2t^2}}$$

shows that g leaves the imaginary axis invariant and $\operatorname{Im} g(it) < t$, implying that $\lim\limits_{n\to\infty} g^n(it) = 0$ and proving that ∞ is the Denjoy-Wolff point of f. To study the ergodic properties of f, we have to specify the sequence $\{a_n\}$. Set $a_n = n^{\frac{1}{2}\gamma}$ with $\gamma > 1$. Then it is not hard to obtain

$$\sum_{n\geq 0}\frac{2t^2}{1+n^\gamma t^2} \cong t^{2-\frac{2}{\gamma}}$$

for small t; it follows that $\beta = 3 - \frac{2}{\gamma}$ satisfies

$$g(it) \cong i(t - t^\beta).$$

This implies $\lim\limits_{n\to\infty} d_P(g^n(it), g^{n+1}(it)) = 0$ (and then that f is ergodic), since $d_P(g^n(it), g^{n+1}(it))$ is proportional to

$$\frac{|g^{n+1}(it) - g^n(it)|}{|g^n(it)|} \cong \frac{|g^n(it)|^\beta}{|g^n(it)|} = |g^n(it)|^{\beta-1}.$$

But the asymptotic expression for g implies that

$$g^n(it) \cong \left(\frac{1}{n}\right)^{\frac{1}{\beta-1}}$$

and therefore

$$\sum_{n\geq 0}|g^n(it)| \cong \sum_n \left(\frac{1}{n}\right)^{\frac{1}{\beta-1}}$$

converges if $(\beta-1)^{-1} > 1$, i.e., if $\gamma < 2$ and diverges if $\gamma \geq 2$. Hence f is recurrent for $\gamma \geq 2$ but, when $\gamma < 2$, $\lim_{n\to\infty} f^n(w) = \infty$ for a.e. $w \in \mathbf{R}$.

Now we shall present some applications of these results to rational maps of the Riemann sphere $\overline{\mathbf{C}} = \mathbf{C} \cup \{\infty\}$. First we recall ([T]) the concepts of Dirichlet regularity and harmonic measures. An open connected set $U \subseteq \overline{\mathbf{C}}$ is said to be *regular* (in the sense of Dirichlet) if every continuous function $\varphi: \partial U \to \mathbf{R}$ has an harmonic extension $\hat{\varphi}: \overline{U} \to \mathbf{R}$, i.e., $\hat{\varphi}$ is continuous, harmonic on U and $\hat{\varphi}|\partial U = \varphi$. If $U \subseteq \overline{\mathbf{C}}$ is an open regular set and $p \in U$, we define the *harmonic measure* λ_p on the Borel σ-algebra of ∂U as the unique one such that

$$\int \varphi d\lambda_p = \hat{\varphi}(p)$$

for every continuous $\varphi: \partial U \to \mathbf{R}$ with harmonic extension $\hat{\varphi}$. Harmonic measures may be defined even when U is not regular via more subtle methods but, for our purposes, this definition will suffice. It is easy to see that λ_p is positive on open sets. Moreover, the harmonic measures λ_p and λ_q are equivalent if $p, q \in U$. Even more, there exists $k > 0$ such that

$$k^{-1} \leq \frac{d\lambda_p}{d\lambda_q}(w) \leq k$$

for every $w \in \partial U$. To check this, it is enough to show the existence of $k > 0$ such that

$$k^{-1} \leq \frac{V(p)}{V(q)} \leq k$$

for every strictly positive harmonic function $V: U \to \mathbf{R}$. But the existence of such a constant is an immediate corollary of Harnack's inequality. The *harmonic class* of ∂U is the class of measures on the Borel σ-algebra of ∂U which are equivalent to a harmonic measure λ_p for some $p \in U$.

Let $\psi: D \to U$ be a uniformization mapping, i.e., a holomorphic covering map of U; it always exists if U^c contains at least three points, as in our case, since U is regular. The regularity of U also implies that the logarithmic capacity of U^c is positive and then the radial limit $\psi^*(w) = \lim_{r\to 1} \psi(rw)$ exists for a.e. $w \in \partial D$ and, since ψ is a covering map, $\psi^*(w) \in \partial D$ for a.e. $w \in \partial D$ (in fact, whenever $\psi^*(w)$ exists). Moreover ψ^* transforms a harmonic measure λ_p on ∂U in the harmonic measure λ_q on ∂D, where $q \in \psi^{-1}(\{p\})$. More precisely,

$$\lambda_q((\psi^*)^{-1}(A)) = \lambda_p(A) \tag{6}$$

for every Borel set $A \subseteq \partial U$. To see this, take a continuous function $\varphi: \partial U \to \mathbf{R}$ and let $\hat{\varphi}: \overline{U} \to \mathbf{R}$ be its harmonic extension. Then $\lim_{r\to 1}(\hat{\varphi} \circ \psi)(rw) = \varphi(\psi^*(w))$ for a.e. $w \in \partial D$ and, by a theorem of Fatou,

$$\int (\varphi \circ \psi^*) d\lambda_q = (\hat{\varphi} \circ \psi)(q) = \hat{\varphi}(p) = \int \varphi d\lambda_p \ .$$

From this, (6) follows applying standard approximation methods.

Now let us consider a rational map $f: \overline{\mathbf{C}} \hookleftarrow$ and let $J(f)$ be its Julia set. It is easy to see that f maps connected components of $J(f)^c$ onto connected components of $J(f)^c$. A fundamental and deep result of Sullivan [**Su**] states that if U is a connected component of $J(f)^c$, then, for some $N > 0$, the connected component $f^N(U)$ is periodic, that is, for some $m > 0$, $f^m(f^N(U)) = f^N(U)$;

more than this, that there are finitely many periodic connected components of $J(f)^c$. Sullivan also proved that if U is a fixed connected component of $J(f)^c$, i.e., $f(U) = U$, then $f|U$ belongs to one of the following four classes:

a) U is a *Siegel disc*: U is a disc and $f|U: U \hookleftarrow$ is conjugate, via a conformal representation, to an irrational rotation of D, i.e., $z \mapsto \alpha z$ with $|\alpha| = 1$ and $\alpha^n \neq 1$ for all n.

b) U is a *Herman ring*: $f|U: U \hookleftarrow$ is conjugate, via conformal representation, to an irrational rotation of an annulus $\{z \in \mathbf{C} |\ r < |z| < R\}$.

c) U is the *immediate basin of a sink*: there exists a sink $p \in U$, i.e., $f(p) = p$, $|f'(p)| < 1$ and $\lim_{n\to\infty} f^n(z) = p$ for all $z \in U$.

d) U is a *parabolic basin*: there exists a parabolic fixed point $p \in \partial U$ of f, i.e., $f'(p)$ is a root of unity and $\lim_{n\to\infty} f^n(z) = p$ for all $z \in U$.

Inner functions and harmonic measures have interesting applications to the study of ergodic properties of $f|\partial U$ when U is in the last two classes. In the first two cases, they yield, so far, no relevant contributions.

Suppose first that U is the immediate basin of a sink p. We shall show that λ_p *is* f*-invariant and exact.* To prove the invariance of λ_p, take any continuous function $\varphi: \partial U \to \mathbf{R}$ and let $\hat{\varphi}: \overline{U} \to \mathbf{R}$ be its harmonic extension. Recall that every component of $J(f)^c$ is regular; even more, its linear density of logarithmic capacity is positive (Mañé, Rocha [**M-R**]). Then $\hat{\varphi} \circ f$ is the harmonic extension of $\varphi \circ f$ and

$$\int (\varphi \circ f) d\lambda_p = (\hat{\varphi} \circ f)(p) = \hat{\varphi}(f(p)) = \hat{\varphi}(p) = \int \varphi d\lambda_p \ ,$$

proving the f-invariance of λ_p. To prove its exactness, consider a uniformization $\psi: D \to U$ of U and a lifting $\hat{f}: D \hookleftarrow$ of f by ψ: $\psi \circ \hat{f} = f \circ \psi$. Then $\hat{f}$ is an inner function with a fixed point $q \in \psi^{-1}(\{p\})$. By Theorem A, λ_q is an exact probability for $\hat{f}^*: \partial D \hookleftarrow$. But since $\psi^* \circ \hat{f}^* = f \circ \psi^*$, it follows from (6) that λ_p on ∂U is also exact.

Many other interesting properties of λ_p are known; for instance, its relation with the entropy maximizing measure of $f|\partial U$. It is known (Freire, Lopes, Mañé [**F-L-M**], Ljubich [**Lj**], Lopes [**Lo**] and Mañé [**M**]) that the topological entropy of $f|\partial U$ is $\log m$, where m is the degree of $f|U: U \hookleftarrow$ and that there exists a unique invariant probability μ^+ on ∂U such that $h_{\mu^+}(f|\partial U) = \log m$ and even more: μ^+ is exact. It is also known that the invariant probabilities λ_p and μ^+ are singular unless $f^{-1}(\{p\}) \cap U = \{p\}$, in which case they coincide (Mañé, Rocha [**M-R**]). In [**Pi**], Przytycki presents a very detailed analysis of the harmonic class on the boundary of an immediate basin of a sink.

For parabolic basins, the ergodic theory of the harmonic class is quite different. The following result will be proved in Section 6.

THEOREM I. *Let U be a fixed parabolic basin of a rational map $f: \overline{\mathbf{C}} \hookleftarrow$ with parabolic fixed point $p \in \partial U$. Let $\psi: D \to U$ be a uniformization and $\hat{f}: D \hookleftarrow$ be a lifting (i.e., $\psi \circ \hat{f} = f \circ \psi$). Then, for all $z \in D$ and all $\alpha > \frac{1}{2}$, the inequality*

$$|1 - \hat{f}^n(z)| \geq \frac{1}{n^\alpha}$$

holds for sufficiently large n.

This implies that the boundary map $\hat{f}^*$ is recurrent and that $\hat{f}$ has no fixed points (otherwise f would have a fixed point in U).

Using the boundary map $\psi^*: \partial D \to \partial U$ and (6), we easily obtain the following corollary.

COROLLARY. *If U is a fixed parabolic basin of a rational map $f: \overline{\mathbf{C}} \hookleftarrow$, then $f|\partial U: \partial U \hookleftarrow$ is exact and recurrent with respect to the harmonic class of ∂U.*

The following result (to be proved in Section 6) further characterizes the ergodic theory of parabolic basins.

THEOREM J. *Let U be a fixed parabolic basin of a rational map $f: \overline{\mathbf{C}} \hookleftarrow$. If $q \in U$, $\varphi \in L^\infty(\lambda_q)$ is positive and $\not\equiv 0$, then*

$$\lim_{n\to\infty} \frac{\int \left(\sum_{i=0}^{n-1} (\varphi \circ f^i)\psi\right) d\lambda_q}{\sum_{i=0}^{n-1} \int (\varphi \circ f^i) d\lambda_q} = \int \psi d\lambda_q$$

for every $\psi \in L^1(\lambda_q)$.

In other words, with $n \to \infty$,

$$\sum_{i=0}^{n-1} \varphi \circ f^i \Big/ \sum_{i=0}^{n-1} \int (\varphi \circ f^i) d\lambda_q \to 1$$

weakly in the dual of $L^1(\lambda_q)$. To this property, a.e. convergence *cannot be added* because of the following result of Aaronson [**Aa1**]: If $T: X \hookleftarrow$ is an ergodic and recurrent measure preserving map of a σ-finite non-atomic infinite measure space $(X, \mathcal{A}, \mu)$, then for every positive $\varphi \in L^1(X, \mathcal{A}, \mu)$ and every sequence $\{a_n\}$ of positive real numbers, one (or both) of the following properties holds for a.e. $x \in X$:

$$\limsup_{n\to\infty} \frac{1}{a_n} \sum_{j=0}^{n-1} (\varphi \circ T^j)(x) = \infty,$$

$$\liminf_{n\to\infty}\frac{1}{a_n}\sum_{j=0}^{n-1}(\varphi\circ T^j)(x)=0.$$

Then, considering the map $f(z)=z-z^{-1}$, for which $U=\mathbf{R}^2_+$ is a parabolic basin, and applying this result to $f|\partial U$, which preserves Lebesgue measure and is ergodic and recurrent, it is clear that a.e. convergence is impossible in Theorem J.

Theorem J has an interesting consequence for Pólya-Szegö maps

$$f(z)=z-\sum_{i=1}^{k}\frac{\gamma_i}{z-a_i}\ ,$$

with $a_i\in\mathbf{R}$ and $\gamma_i>0$ for all $1\le i\le k$, namely, that

$$\lim_{n\to\infty}\frac{1}{\sqrt{n}}\sum_{j=0}^{n-1}\lambda(f^{-j}(A)\cap B)=\frac{2}{\pi\sqrt{2\gamma}}\lambda(A)\lambda(B),\tag{7}$$

where $\gamma=\sum_{i=1}^{k}\gamma_i$, for all bounded Borel sets $A,B\subseteq\mathbf{R}$.

To prove this, let ψ_A,ψ_B stand for the characteristic functions of $A,B\subseteq\mathbf{R}$; then $(\psi_A\circ f^j)\psi_B$ is the characteristic function of $f^{-j}(A)\cap B$ and therefore,

$$\lambda(f^{-j}(A)\cap B)=\int(\psi_A\circ f^j)\psi_B\frac{d\lambda}{d\lambda_z}d\lambda_z$$

for $z\in\mathbf{R}^2_+$. Since B is bounded, $\psi_B\dfrac{d\lambda}{d\lambda_z}\in L^1(\lambda_z)$ and Theorem J implies that

$$\lim_{n\to\infty}\frac{\sum_{j=0}^{n-1}\lambda(f^{-j}(A)\cap B)}{\sum_{j=0}^{n-1}\int(\psi_A\circ f^j)d\lambda_z}=\int\psi_B\frac{d\lambda}{d\lambda_z}d\lambda_z=\lambda(B).\tag{8}$$

On the other hand, the Taylor development of $g(z) = -(f(-z^{-1}))^{-1}$ at $z = 0$ is

$$g(z) = z - \gamma z^3 + \dots .$$

Therefore every $z \in \mathbf{R}^2_+$ approaches the Denjoy-Wolff point of g (which is 0) satisfying

$$\lim_{n\to\infty} \sqrt{n}\,|g^n(z)| = \frac{1}{\sqrt{2\gamma}},$$

which for f decodes to

$$\lim_{n\to\infty} \frac{f^n(z)}{i\sqrt{n}} = \sqrt{2\gamma}.$$

From this limit we obtain, trivially

$$\lim_{j\to\infty} \frac{\text{Im } f^j(z)}{\sqrt{2\gamma}\sqrt{j}} = 1$$

and also, less obviously (but only using the expression of the derivative $d\lambda_z/d\lambda$; see Section 5 for details),

$$\lambda(A) = \lim_{j\to\infty} \pi \ \text{Im} f^j(z) \ \lambda_{f^j(z)}(A),$$

and therefore,

$$\lambda(A) = \lim_{j\to\infty} \pi\sqrt{2\gamma}\sqrt{j} \ \lambda_{f^j(z)}(A).$$

But then

$$\lambda(A) = \lim_{n\to\infty} \frac{\sum_{j=0}^{n-1} \lambda_{f^j(z)}(A)}{\frac{1}{\pi\sqrt{2\gamma}} \sum_{j=1}^{n} \frac{1}{\sqrt{j}}} = \lim_{n\to\infty} \frac{\sum_{j=0}^{n-1} \lambda_{f^j(z)}(A)}{\frac{2}{\pi\sqrt{2\gamma}}\sqrt{n}},$$

which, together with (8), proves (7), since

$$\lambda_{f^j(z)}(A) = \lambda_z(f^{-j}(A)) = \int (\psi_A \circ f^j) d\lambda_z.$$

§1 Nontangential limits and harmonic extensions

In this section we gather some basic results on holomorphic and harmonic functions, defined in the open unit disc D, which will be used in the forthcoming sections and also prove that the harmonic measure associated to the fixed point of an inner function is invariant under the boundary map.

For $w \in \partial D$ and $0 < \alpha < \frac{\pi}{2}$, let

$$S_\alpha(w) = \{z \in D \,|\, |\mathrm{Arg}\left(\tfrac{w-z}{w}\right)| < \alpha\}$$

denote the inner sector with vertex w and inner angle 2α. Given any function $f: D \to X$, with $X = \mathbf{C}$ or $\mathbf{R}$, we say that it has a nontangential limit at $w \in \partial D$ if

$$\lim_{\substack{z \to w \\ z \in S_\alpha(w)}} f(z)$$

exists for every $0 < \alpha < \frac{\pi}{2}$; it is easy to see that the limit is independent of α. Its value is denoted by $f^*(w)$ and called the *nontangential limit* of f at w.

The classical theorem of Fatou states that if $f: D \to \mathbf{C}$ is holomorphic and bounded then the nontangential limit of f exists at a.e. point of ∂D, with respect to the Lebesgue probability λ of the unit circle. (For a proof, see [**T**; Thm. IV.7].) (The same is true if instead of a bounded we have a Nevanlinna function, i.e., one for which

$$\sup_{0\le r<1}\int_{\partial D}\log^{+}|f(rw)|\,d\lambda(w)<\infty,$$

and it also holds if $\mathbf{C}-f(D)$ has positive logarithmic capacity.) We shall later have use for the following uniqueness theorem, due to F. and M. Riesz ([**T**; Thm. IV.9]).

THEOREM 1.1. *If $f\colon D\to\mathbf{C}$ is a bounded holomorphic function and $f^*(w)=0$ for each w in a set of positive Lebesgue measure, then $f\equiv 0$.*

Let us take a quick look at the behavior of harmonic functions with respect to nontangential limits. Given a continuous function $\psi\colon\partial D\to\mathbf{R}$, its *harmonic extension* is a harmonic function $\hat{\psi}\colon D\to\mathbf{R}$ such that $\lim_{z\to w}\hat{\psi}(z)=w$ for all $w\in\partial D$. It is well known that every continuous function in D has one and only one harmonic extension. Moreover, introducing the Poisson kernel $P\colon D\times\partial D\to\mathbf{R}$, defined by

$$P(z,w)=\frac{1-|z|^2}{|w-z|^2}=\operatorname{Re}\left(\frac{w+z}{w-z}\right),$$

the unique harmonic extension of ψ is given by

$$\hat{\psi}(z)=\int_{\partial D}\psi(w)P(z,w)\,d\lambda(w) \tag{1}$$

for each $z\in D$. More generally, we *define* the harmonic extension $\hat{\psi}\colon D\to\mathbf{R}$ of any $\psi\in L^1(\lambda)$ by (1); one checks easily that $\hat{\psi}$ is indeed harmonic.

The following results, due to Fatou and Schwarz, respectively, contain the essential properties of harmonic extensions ([**T**; Thms. IV.1,2]).

THEOREM 1.2. *Given* $\psi \in L^1(\lambda)$,

a) *for a.e.* $w \in \partial D$ *the nontangential limit* $\hat{\psi}^*(w)$ *exists and* $\hat{\psi}^* = \psi$ *in* $L^1(\lambda)$;

b) *if* ψ *is continuous at some* $w_0 \in \partial D$, *then* $\lim_{z \to w_0} \hat{\psi}(z) = \psi(w_0)$; *if* $J \subseteq \partial D$ *is an open set and* ψ *is continuous on* J, *then* $\lim_{z \to w} \hat{\psi}(z) = \psi(w)$ *uniformly in* $w \in J$.

For bounded functions, we again have a uniqueness result, also due to Fatou ([T; Thm. IV.6]).

THEOREM 1.3. *If* $u: D \to \mathbf{R}$ *is a bounded harmonic function, then the nontangential limit* $u^*(w)$ *exists at a.e.* $w \in \partial D$ *and* u *is the harmonic extension of* u^*, *i.e.,* $\widehat{u^*} = u$. *In particular, for* $\psi \in L^\infty(\lambda)$, $\hat{\psi}$ *is the unique bounded harmonic function defined in* D *and such that*

$$\lim_{r \to 1} \hat{\psi}(rw) = \psi(w)$$

for a.e. $w \in \partial D$.

Observe that this theorem is false for unbounded functions: simply take the harmonic function $u(z) = \mathrm{Re}((z+1)/(z-1))$, $z \in D$, for which $\lim_{z \to w} u(z) = 0$ at every $w \neq 1$ in ∂D and therefore $u^* = 0$ in $L^\infty(\lambda)$.

Given $z \in D$, we define the *harmonic measure* λ_z as the unique probability on ∂D satisfying

$$\frac{d\lambda_z}{d\lambda}(w) = P(z, w)$$

for $w \in \partial D$. Thus, for every Borel set $A \subseteq \partial D$,

$$\lambda_z(A) = \int_A P(z, w) d\lambda(w)$$

and, for every $\psi \in L^1(\lambda)$,

$$\int \psi d\lambda_z = \hat{\psi}(z),$$

since $\int \psi d\lambda_z = \int \psi \frac{d\lambda_z}{d\lambda} d\lambda = \int \psi(w) P(z,w) d\lambda(w) = \hat{\psi}(z)$.

Let us now prove that λ_z is invariant under the boundary map of an inner function having z as fixed point. In this and the next sections we shall (drop the asterisk and) simply write $f\colon \partial D \hookleftarrow$ for the boundary map of an inner function $f\colon D \hookleftarrow$, since there is hardly any risk involved. The essential property relating nontangential limits and harmonc extensions is the following commutative property.

THEOREM 1.4. *If $f\colon D \hookleftarrow$ is an inner function and $\psi \in L^1(\lambda)$, then $\psi \circ f \in L^1(\lambda)$ and*

$$\hat{\psi} \circ f = \widehat{\psi \circ f}.$$

COROLLARY 1.5. *If $f\colon D \hookleftarrow$ is an inner function and $z \in D$, then the harmonic measure satisfies:*

a) $\int (\psi \circ f) d\lambda_z = \int \psi d\lambda_{f(z)}$ for every $\psi \in L^1(\lambda)$;

b) $\lambda_z(f^{-1}(A)) = \lambda_{f(z)}(A)$ *for every Borel set* $A \subseteq \partial D$;

c) λ_z *is f-invariant if and only if* $f(z) = z$.

The corollary is an immediate consequence of the theorem: $\int (\psi \circ f) d\lambda_z = \widehat{\psi \circ f}(z) = (\hat{\psi} \circ f)(z) = \hat{\psi}(f(z)) = \int \psi d\lambda_{f(z)}$ proves a); if ψ is the characteristic function of $A \subseteq \partial D$, then $\psi \circ f$ is the characteristic function of $f^{-1}(A)$ and a) implies

$$\lambda_z(f^{-1}(A)) = \int (\psi \circ f) d\lambda_z = \int \psi d\lambda_{f(z)} = \lambda_{f(z)}(A),$$

which proves b); finally, c) follows from a) and b).

To prove the theorem we need two preliminary results.

LEMMA 1. *For* $\psi \in L^1(\lambda)$ *and* $z \in D$, $|\hat{\psi}(z)| \leq \frac{1+|z|}{1-|z|} \|\psi\|_1$.

PROOF: Given $\psi \in L^1(\lambda)$ and $z \in D$,

$$|\hat{\psi}(z)| = \left| \int \psi(w) P(z,w) d\lambda(w) \right| \leq \int |\psi(w)| \, P(z,w) d\lambda(w)$$

$$\leq \|\psi\|_1 \max_{w \in \partial D} P(z,w) \leq \frac{1+|z|}{1-|z|} \|\psi\|_1 .$$

□

LEMMA 2. *Suppose* $f: D \hookleftarrow$ *is an inner function and let* $\psi: \partial D \to \mathbf{R}$ *be given.*

a) If ψ *is continuous then* $\psi \circ f \in L^\infty(\lambda)$ *and* $\hat{\psi} \circ f = \widehat{\psi \circ f}$.

b) If $\psi \in L^1(\lambda)$ *then* $\psi \circ f \in L^1(\lambda)$ *and*

$$\|\psi \circ f\|_1 \leq \frac{1+|f(0)|}{1-|f(0)|} \|\psi\|_1 .$$

PROOF: a) Given an inner function $f: D \hookleftarrow$, we have $\lim_{r \to 1} f(rw) = f(w)$ for a.e. $w \in \partial D$. Given a continuous $\psi: \partial D \to \mathbf{R}$, $\hat{\psi}$ is harmonic and $\lim_{z \to w} \hat{\psi}(z) = \psi(w)$ for each $w \in \partial D$ by Theorem 1.2.b). Therefore

$$\lim_{r \to 1} (\hat{\psi} \circ f)(rw) = \lim_{r \to 1} \hat{\psi}(f(rw)) = \psi(f(w)) = (\psi \circ f)(w)$$

for a.e. $w \in \partial D$. But $\hat{\psi} \circ f$ is harmonic and bounded. Thus the uniqueness property of Theorem 1.3 guarantees that $\hat{\psi} \circ f$ is the harmonic extension $\widehat{\psi \circ f}$ of $\psi \circ f$.

To prove b) it is sufficient to prove the inequality for continuous ψ; the result then follows by density. For continuous ψ,

$$\|\psi \circ f\|_1 = \int |\psi \circ f| \, d\lambda = \int (|\psi| \circ f)(w) P(0,w) d\lambda(w)$$

$$= \widehat{|\psi| \circ f}(0) = \widehat{|\psi|}(f(0)),$$

where the last equality is secured by part a). Applying Lemma 1 we get

$$\|\psi \circ f\|_1 = \widehat{|\psi|}(f(0)) \leq \frac{1 + |f(0)|}{1 - |f(0)|} \|\psi\|_1 .$$

□

Proof of Theorem 1.4: Given $\psi \in L^1(\lambda)$, take a sequence of continuous $\psi_n\colon \partial D \to \mathbf{R}$ converging to ψ in $L^1(\lambda)$. By Lemma 1,

$$\sup_{|z|<r} \left|(\hat{\psi}_n - \hat{\psi})(z)\right| \leq \frac{1+r}{1-r} \|\psi_n - \psi\|_1$$

for all $0 < r < 1$. Hence $\hat{\psi}_n \to \hat{\psi}$ uniformly on compact sets and then also $\hat{\psi}_n \circ f \to \hat{\psi} \circ f$ uniformly on compact sets. Now for every $n \geq 0$, Lemma 2.a) yields

$$\hat{\psi}_n \circ f = \widehat{\psi_n \circ f}. \tag{2}$$

Moreover, $(\psi_n - \psi) \circ f \in L^1(\lambda)$ and

$$\|(\psi_n - \psi) \circ f\|_1 \leq \frac{1 + |f(0)|}{1 - |f(0)|} \|\psi_n - \psi\|_1$$

by Lemma 2.b). Therefore, for all $0 < r < 1$, Lemma 1 gives

$$\begin{aligned} \sup_{|z|<r} \left|(\widehat{\psi_n \circ f} - \widehat{\psi \circ f})(z)\right| &\leq \frac{1+r}{1-r} \|(\psi_n - \psi) \circ f\|_1 \\ &\leq \frac{1+r}{1-r}\, \frac{1 + |f(0)|}{1 - |f(0)|} \|\psi_n - \psi\|_1 . \end{aligned}$$

Hence, $\widehat{\psi_n \circ f} \to \widehat{\psi \circ f}$ uniformly on compact sets. Taking limits in (2) we obtain $\hat{\psi} \circ f = \widehat{\psi \circ f}$. □

§2 The Denjoy-Wolff theorem

The dynamics of a holomorphic function of D into itself is described by the following result due to Denjoy [**De**] and Wolff [**W**].

THEOREM 2.1. *Let $f: D \hookleftarrow$ be a holomorphic function that is not a Möbius transformation. Then there exists a point $p \in \overline{D}$ such that $\lim_{n\to\infty} f^n(z) = p$ for all $z \in D$, uniformly on compact sets. Moreover, if $p \in D$ then $|f'(p)| < 1$.*

The point $p \in \overline{D}$ given by the theorem is obviously unique and we shall call it the *Denjoy-Wolff point* of f. Thus, given any holomorphic function $f: D \hookleftarrow$, the sequence of its iterates $f^n: D \hookleftarrow$ converges, uniformly on compact sets, to the constant p; the only exception (besides the identity) is presented by Möbius transformations with elliptic fixed point in D, since clearly the other two possibilities of Möbius transformations $T: D \hookleftarrow$ (parabolic or hyperbolic) also possess a unique Denjoy-Wolff point $p \in \partial D$.

In the proof of Theorem 2.1, as well as in subsequent sections, we shall use the conformally invariant *Poincaré metric* d_P on D; $d_P(x, y)$ is defined as the infimum of the hyperbolic lenghts of the arcs in D joining x to y ([**G**; pg. 5]) and we have

$$\lim_{n\to\infty} d_P(x_n, y_n) = 0 \Leftrightarrow \lim_{n\to\infty} \frac{|x_n - y_n|}{|1 - x_n\overline{y}_n|} = 0$$
$$\Leftrightarrow \lim_{n\to\infty} \frac{|x_n - y_n|}{1 - |x_n|^2} = 0,$$

for all sequences of points $x_n, y_n \in D$.

PROOF: Let $f: D \hookleftarrow$ be a holomorphic function and let us assume that f is not a Möbius transformation.

Suppose first that there exist $x \in D$ and $n_1 < n_2 < \ldots$ such that $f^{n_j}(x)$ converges with $j \to \infty$ to a point $p \in D$. By virtue of Pick's invariant formulation of Schwarz's lemma,

$$d_P(f(x), f(y)) < d_P(x, y)$$

for all $x \neq y$ in D. Therefore, for all $n \geq 0$,

$$\begin{aligned} d_P(f^{n+2}(x), f^{n+1}(x)) &= d_P(f(f^{n+1}(x)), f(f^n(x))) \\ &\leq d_P(f^{n+1}(x), f^n(x)) \end{aligned}$$

and it follows that the sequence $d_P(f^{n+1}(x), f^n(x))$ converges with $n \to \infty$. Then

$$\begin{aligned} \lim_{n\to\infty} d_P(f^{n+1}(x), f^n(x)) &= \lim_{n\to\infty} d_P(f^{n_j+1}(x), f^{n_j}(x)) \\ &= d_P(f(p), p) \end{aligned}$$

and

$$\begin{aligned} \lim_{n\to\infty} d_P(f^{n+1}(x), f^n(x)) &= \lim_{n\to\infty} d_P(f^{n_j+2}(x), f^{n_j+1}(x)) \\ &= d_P(f^2(p), f(p)) \end{aligned}$$

imply $d_P(f^2(p), f(p)) = d_P(f(p), p)$. This gives $f(p) = p$, because otherwise

$$d_P(f^2(p), f(p)) = d_P(f(f(p)), f(p)) < d_P(f(p), p).$$

To prove that $\lim_{n\to\infty} f^n(z) = p$ uniformly on compact subsets of D, we will assume that $p = 0$. (If not, we conjugate f with an appropriate

Möbius transformation.) Then $f(0) = 0$ and Schwarz's lemma gives $|f'(0)| < 1$ and $|f(z)| < |z|$ for all $0 < |z| < 1$, since f is not Möbius. As in the proof of Schwarz's lemma, consider $F: D \to \mathbf{C}$ defined by $F(z) = \frac{1}{z}f(z)$ for $z \neq 0$ and $F(0) = f'(0)$. Then $\lim_{|z|\to 1} |F(z)| \leq 1$ and, since F cannot be constant because f is not a Möbius map, the maximum principle gives $|F(z)| < 1$ for $z \in D$. Therefore

$$c_R = \max_{|z|<R} |F(z)| < 1$$

and $|f(z)| \leq c_R |z| \leq c_R R < R$ for all $|z| < R < 1$; then we also have

$$|f^2(z)| = |f((f(z))| \leq c_R |f(z)| \leq c_R^2 |z| < R$$

when $|z| < R$. We conclude that

$$|f^n(z)| \leq c_R^n |z|$$

for all $n \geq 0$ and $|z| < R$, proving that $\lim_{n\to\infty} f^n(z) = p$ uniformly on compact subsets of D.

Now suppose that for all $z \in D$, the sequence $\{f^n(z)\}$ has no accumulation points in D. Then either there exists a point $p \in \partial D$ such that $\lim_{n\to\infty} f^n(z) = p$ for all $z \in D$ or else there exist $x, y \in D$, $p_1 \neq p_2$ in ∂D and sequences $n_1 < n_2 < \ldots$ and $m_1 < m_2 < \ldots$ such that $\lim_{j\to\infty} f^{n_j}(x) = p_1$ and $\lim_{j\to\infty} f^{m_j}(y) = p_2$. Let us derive a contradiction from this second possibility.

We start with the following: since for all $z \in D$ the sequence $\{f^n(z)\}$ accumulates nowhere in D, we have $\lim_{n\to\infty} |f^n(z)| = 1$. We claim that this convergence is uniform on compact subsets of D. If it is not, we can choose a sequence $\{z_j\}$ in D, bounded away from

∂D, and integers $k_1 < k_2 < \ldots$ such that $|f^{k_j}(z_j)| \le c < 1$. Since the sequence of iterates of f is uniformly bounded, we may assume that the sequence $\{f^{k_j}\}$ converges uniformly on compact sets to a holomorphic function $F: D \to \mathbf{C}$. Without loss of generality, we may also assume that z_j converges with $j \to \infty$ to some point $z \in D$. Then

$$\lim_{j\to\infty} f^{k_j}(z) = F(z) = \lim_{j\to\infty} f^{k_j}(z_j) \in D$$

shows that the sequence $f^n(z)$ has accumulation points in D, contrary to our hypothesis.

Assume now that $x, y \in D$, $p_1 \neq p_2$ in ∂D and $n_1 < n_2 < \ldots$ and $m_1 < m_2 < \ldots$ satisfy $\lim_{j\to\infty} f^{n_j}(x) = p_1$ and $\lim_{j\to\infty} f^{m_j}(y) = p_2$. Take $0 < r < 1$ sufficiently large to guarantee that x, y belong to

$$B = \{z \in D | \, |z| < r\}$$

and $f(B) \cap B \neq \phi$; it follows that $f^{n+1}(B) \cap f^n(B) \neq \phi$ for each $n \ge 0$. Then, for every $N \ge 0$, $\bigcup_{n\ge N} f^n(B)$ is an open connected set that accumulates on p_1 and p_2. Moreover, given any $\varepsilon > 0$, the annulus $A_\varepsilon = \{z \in \mathbf{C} | \, 1-\varepsilon < |z| < 1\}$ contains $\bigcup_{n\ge N} f^n(B)$ for sufficiently large N, by the uniform convergence of $\lim_{n\to\infty} |f^n(z)| = 1$, $z \in B$. Hence, given any $q_1, q_2 \in \partial D$ which separate p_1 and p_2, $\bigcup_{n\ge N} f^n(B)$ intersects the interval $S_1 = \{tq_1 | 1-\varepsilon < t < 1\}$ or the interval $S_2 = \{tq_2 | 1-\varepsilon < t < 1\}$, for sufficiently large N. Since this holds for all $\varepsilon > 0$, it follows that $\bigcup_{n\ge N} f^n(B)$ accumulates radially on q_1 or q_2. But q_1 and q_2 are arbitrary points separating

p_1 and p_2 and $\bigcup_{n\geq N} f^n(B)$ is connected. Thus there exists an interval $J = [p_1, p_2]$ in ∂D all of whose points are radially accumulated by $\bigcup_{n\geq N} f^n(B)$. Given any $w \in J$ such that the nontangential limit $f(w)$ exists, we choose points $z_j \in B$ and integers $k_1 < k_2 < \dots$ such that $f^{k_j}(z_j)$ converges radially to w with $j \to \infty$. Without loss of generality we may assume that z_j converges to $\tilde{z} \in D$ with $j \to \infty$ and that f^{k_j} converges uniformly on compact sets to a holomorphic function $F: D \to \mathbf{C}$. Since $\lim_{n\to\infty} |f^n(z)| = 1$ for all $z \in D$, we obtain $F(D) \subseteq \partial D$ and then F must be constant: $F(z) = w$ for all $z \in D$. It follows that

$$\begin{aligned} f(w) &= \lim_{j\to\infty} f(f^{k_j}(z_j)) = \lim_{j\to\infty} f^{k_j+1}(z_j) \\ &= \lim_{j\to\infty} f^{k_j}(f(z_j)) = F(f(\tilde{z})) = w, \end{aligned}$$

proving that the boundary map of f satisfies $f(w) = w$ for a.e. $w \in J$. But J has positive Lebesgue measure; by Theorem 1.1, necessarily $f(z) = z$ for all $z \in D$. Since we precluded this possibility, we arrived at the contradiction we stroved for. Thus there exists $p \in \partial D$ such that $\lim_{n\to\infty} f^n(z) = p$ for all $z \in D$; since f^n is uniformly bounded, the convergence is uniform on compact subsets of D. □

COROLLARY 2.2. *Let $f: D \hookleftarrow$ be an inner function and let $p \in \overline{D}$ be its Denjoy-Wolff point.*

a) If $p \in D$, then the boundary map of f preserves the harmonic measure λ_p.

b) If $p \in \partial D$, then for every neighborhood W of p in ∂D,

$$\lim_{n\to\infty} \frac{1}{n} \#\{0 \leq j < n | f^j(w) \in W\} = 1$$

for a.e. $w \in \partial D$.

PROOF: a) is part c) of Corollary 1.5. To prove b), let ψ be the characteristic function of W; since $f^j(w) \in W$ if and only if $(\psi \circ f^j)(w) = 1$, we have

$$\#\{0 \leq j < n | f^j(w) \in W\} = \sum_{j=0}^{n-1} (\psi \circ f^j)(w)$$

for a.e. $w \in \partial D$. Denoting $z_n = f^n(0)$, Corollary 1.5 implies

$$\int (\psi \circ f^j) d\lambda = \int \psi d\lambda_{z_j} = \hat{\psi}(z_j).$$

But $z_n \to p$ and therefore $\hat{\psi}(z_n) \to \psi(p) = 1$, with $n \to \infty$, by Theorem 1.2.b). Dominated convergence thence implies

$$\int \Big(\lim_{n\to\infty} \frac{1}{n} \sum_{j=0}^{n-1} (\psi \circ f^j) \Big) d\lambda = \lim_{n\to\infty} \frac{1}{n} \sum_{j=0}^{n-1} \int (\psi \circ f^j) d\lambda$$

$$= \lim_{n\to\infty} \frac{1}{n} \sum_{j=0}^{n-1} \hat{\psi}(z_j) = \lim_{n\to\infty} \hat{\psi}(z_n) = \psi(p) = 1,$$

proving that $\lim_{n\to\infty} \frac{1}{n} \sum_{j=0}^{n-1} (\psi \circ f^j)(w) = 1$ for a.e. $w \in \partial D$. □

§3 Ergodicity of inner functions

Recall that a measurable map $T: X \hookleftarrow$ of a measure space $(X, \mathcal{A}, \mu)$ is *ergodic* if $\mu(A) = 0$ or $\mu(X - A) = 0$ for every $A \in \mathcal{A}$ such that $T^{-1}(A) = A$ and is *exact* if the same occurs for $A \in \bigcap_{n \geq 0} T^{-n}(\mathcal{A})$. Exact maps are ergodic because $T^{-1}(A) = A$ implies $T^{-1}(A) = A$ for all $n \geq 0$ and then $A \in \bigcap_{n \geq 0} T^{-n}(\mathcal{A})$. If μ is a probability and T preserves μ then T is μ-recurrent and exactness also implies mixing.

In this section we give necessary and sufficient conditions for the ergodicity of the boundary map of an inner function and prove that, for such maps, ergodicity and exactness are the same. These results are due to Aaronson [**Aa3**] and Pommerenke [**Pe**]. Observe that ergodicity does not imply that a.e. boundary point is recurrent; this is true if the Denjoy-Wolff point belongs to D, since then the harmonic measure associated to it is invariant by Corollary 2.2 above, hence recurrent and, being equivalent to Lebesgue measure, a.e. boundary point is recurrent. Necessary and sufficient conditions for recurrence will be given in the next section.

THEOREM 3.1. *The following conditions are equivalent for inner functions $f: D \hookleftarrow$.*

a) *The boundary map of f is exact.*

b) *The boundary map of f is ergodic.*

c) There exist $a, b \in D$ *such that* $f^n(a) \neq f^n(b)$ *for* $n \geq 0$ *and* $\lim_{n\to\infty} d_P(f^n(a), f^n(b)) = 0$.

d) $\lim_{n\to\infty} d_P(f^n(x), f^n(y)) = 0$ *for all* $x, y \in D$.

COROLLARY 3.2. *The boundary map is exact whenever the inner function has a fixed point in* D.

The corollary is obvious: recall that in the presence of a fixed point $p \in D$, p is the Denjoy-Wolff point and $\lim_{n\to\infty} |f^n(z) - f^n(p)| = 0$, hence $\lim_{n\to\infty} d_P(f^n(z), f^n(p)) = 0$ for each $z \in D$, implying condition d).

PROOF: c) $\Rightarrow$ d) We shall use the following simple fact: if a sequence of holomorphic functions $g_n D \to \mathbf{C}$ converges uniformly on compact sets to $g: D \to \mathbf{C}$ and satisfies $g_{n+1}^{-1}(\{0\}) \supseteq g_n^{-1}(\{0\})$ for all $n \geq 0$, then either $g \equiv 0$ or $g^{-1}(\{0\}) = \bigcup_{n\geq 0} g_n^{-1}(\{0\})$.

To prove c) $\Rightarrow$ d) we may, and will, assume that $a = 0$. For each $n \geq 0$ choose a Möbius map $T_n: D \hookleftarrow$ satisfying $T_n f^n(0) = 0$ and select $n_0 < n_1 < \ldots$ in such a way that the sequence $g_k = T_{n_k} \circ f^{n_k}: D \hookleftarrow$ converges uniformly on compact sets to a map $g: D \hookleftarrow$. Then

$$g_k^{-1}(\{0\}) = f^{-n_k}(T_{n_k}^{-1}(\{0\})) = f^{-n_k}(\{f^{n_k}(0)\})$$

and it is easy to check that $g_{k+1}^{-1}(\{0\}) \supseteq g_k^{-1}(\{0\})$ for $k \geq 0$. Now the hypothesis $f^n(0) \neq f^n(b)$ guarantees $g_k(b) \neq 0$ for each k, therefore $b \notin \bigcup_{k\geq 0} g_k^{-1}(\{0\})$. On the other hand, $g(b) = \lim_{k\to\infty} g_k(b) = 0$, since $d_P(g_k(b), 0) = d_P(T_{n_k} f^{n_k}(b), T_{n_k} f^{n_k}(0)) = d_P(f^{n_k}(b), f^{n_k}(0))$ converges to 0 with $k \to \infty$ by the hypothesis

$\lim_{n\to\infty} d_P(f^n(0), f^n(b)) = 0$. Thus the above stated fact implies $g \equiv 0$ and we obtain, for $z \in D$:

$$\begin{aligned}0 = d_P(g(z), 0) &= \lim_{k\to\infty} d_P(g_k(z), 0)\\ &= \lim_{k\to\infty} d_P(T_{n_k} f^{n_k}(z), T_{n_k} f^{n_k}(0)) = \lim_{k\to\infty} d_P(f^{n_k}(z), f^{n_k}(0)).\end{aligned}$$

Since the sequence $d_P(f^n(z), f^n(0))$ is decreasing, its limit exists and it follows that $\lim_{n\to\infty} d_P(f^n(z), f^n(0)) = 0$ for all $z \in D$, which clearly implies condition d).

d) $\Rightarrow$ a) Let $\mathcal{A}$ denote the Borel σ-algebra of ∂D and suppose $A \in \bigcap_{n\geq 0} f^{-n}(\mathcal{A})$; then we can write $A = f^{-n}(A_n)$ with $A_n \in \mathcal{A}$, $n \geq 0$. To prove that $\lambda(A) = 0$ or $\lambda(A) = 1$, we consider the characteristic functions ψ and ψ_n of A and A_n respectively. Since $f^{-n}(A_n) = A$, we have $\psi_n \circ f^n = \psi$ and therefore $\hat{\psi} \circ f^n = \widehat{\psi_n \circ f^n} = \hat{\psi}$ by the commutative property of Theorem 1.4. Let $x, y \in D$ be given; for every $n \geq 0$,

$$(1) \qquad \hat{\psi}(x) - \hat{\psi}(y) = \hat{\psi}_n(f^n(x)) - \hat{\psi}_n(f^n(y)).$$

For each $n \geq 0$ we now choose a Möbius map $T_n: D \hookleftarrow$ with $T_n(0) = f^n(x)$; since $\widehat{\psi_n \circ T_n} = \hat{\psi}_n \circ T_n$, it follows that

$$\begin{aligned}\hat{\psi}_n(f^n(x)) - \hat{\psi}_n(f^n(y)) &= \widehat{\psi_n \circ T_n}(0) - \widehat{\psi_n \circ T_n}(T_n^{-1} f^n(y))\\ &= \int \psi_n(T_n w)[1 - P(T_n^{-1} f^n(y), w)] d\lambda(w).\end{aligned}$$

But $\lim_{n\to\infty} T_n^{-1} f^n(y) = 0$, since

$$d_P(T_n^{-1} f^n(y), 0) = d_P(f^n(y), T_n(0)) = d_P(f^n(y), f^n(x))$$

converges to 0 with $n \to \infty$ by hypothesis. Therefore

$$\lim_{n\to\infty} [\sup_{|w|=1} |1 - P(T_n^{-1} f^n(y), w)|] = 0,$$

which implies

$$\lim_{n\to\infty} [\hat{\psi}(f^n(x)) - \hat{\psi}(f^n(y))] = 0$$

and we obtain $\hat{\psi}(x) = \hat{\psi}(y)$ from (1). This proves that $\hat{\psi}$ is constant, hence $\psi = \hat{\psi}^*$ is constant a.e. by Theorem 1.2.a). Since ψ is the characteristic function of A, this means that $\lambda(A) = 0$ or $\lambda(A) = 1$.

a) $\Rightarrow$ b) Immediate, as above.

b) $\Rightarrow$ c) We shall prove that f is not ergodic when condition c) does not hold. Assume, therefore, that $\lim_{n\to\infty} d_P(f^n(x), f^n(y)) > 0$ for all $x, y \in D$ satisfying $f^n(x) \neq f^n(y)$ for each $n \geq 0$. Clearly we may choose $z \in D$ with $f^{n+1}(z) \neq f^n(z)$, $n \geq 0$; to simplify, we take $z = 0$. For each $n \geq 0$ we choose a Möbius map $T_n: D \hookleftarrow$ such that

$$T_n f^n(0) = 0 \quad \text{and} \quad T_n f^{n+1}(0) > 0.$$

Then

$$\begin{aligned} d_P(T_n T_{n+1}^{-1}(0), 0) &= d_P(T_n T_{n+1}^{-1}(0), T_n f^n(0)) \\ &= d_P(T_{n+1}^{-1}(0), f^n(0)) = d_P(f^{n+1}(0), f^n(0)) \end{aligned}$$

implies that the moduli $|T_n T_{n+1}^{-1}(0)|$ are decreasing and, since

$$T_n T_{n+1}^{-1}(0) = T_n f^{n+1}(0) > 0,$$

the sequence $\{T_n T_n^{-1}(0)\}$ converges to some $c \geq 0$ with $n \to \infty$. But our hypothesis gives

$$d_P(c, 0) = \lim_{n\to\infty} d_P(T_n T_{n+1}^{-1}(0), 0) = \lim_{n\to\infty} d_P(f^{n+1}(0), f^n(0)) > 0$$

and we have $c > 0$. These facts guarantee that the sequence $T_n \circ T_{n+1}^{-1}: D \hookleftarrow$ of Möbius maps converges to a Möbius map $T: D \hookleftarrow$; notice that $T(0) = c > 0$.

Next we prove that the sequence $g_n = T_n \circ f^n$ converges uniformly on compact sets. We have $g_n(0) = 0$ and, for every $z \in D$,

$$|g_n(z)| \leq |g_{n-1}(z)| \tag{2}$$

because

$$\begin{aligned} d_P(g_n(z), 0) &= d_P(T_n f^n(z), T_n f^n(0)) \\ &= d_P(f^n(z), f^n(0)) \leq d_P(f^{n-1}(z), f^{n-1}(0)) \\ &= d_P(T_{n-1} f^{n-1}(z), T_{n-1} f^{n-1}(0)) = d_P(g_{n-1}(0), 0). \end{aligned}$$

Since the sequence $\{g_n\}$ is uniformly bounded, to prove its convergence it is sufficient to show that any two convergent subsequences $g_{n_\ell} \to g^{(1)}$ and $g_{m_k} \to g^{(2)}$ have the same limit, i.e., $g^{(1)} = g^{(2)}$. But (2) implies that the sequence $\{|g_n|\}$ is convergent; hence $\lim_{\ell \to \infty} |g_{n_\ell}| = |g^{(1)}|$ and $\lim_{k \to \infty} |g_{m_k}| = |g^{(2)}|$ coincide with $\lim_{n \to \infty} |g_n|$. Then $|g^{(1)}| = |g^{(2)}|$. This implies that $g^{(1)} = \alpha g^{(2)}$ with $|\alpha| = 1$; if we show that $\alpha = 1$, we are done. Above we proved $0 < c = \lim_{n \to +\infty} T_n f^{n+1}(0)$, therefore

$$0 < c = \lim_{n \to +\infty} T_n f^n(f(0)) = g^{(1)}(f(0)) = g^{(2)}(f(0))$$

implies $\alpha = 1$ and completes the proof of the convergence of $\{g_n\}$. We define $g = \lim_{n \to \infty} g_n$ and obtain

$$g \circ f = T \circ g,$$

because

$$g(f(z)) = \lim_{n\to\infty} T_n f^n(f(z)) = \lim_{n\to\infty} T_n T_{n+1}^{-1} T_{n+1} f^{n+1}(z)$$
$$= Tg(z).$$

Finally we claim that

$$T = \lim_{n\to\infty} T_n \circ f \circ T_n^{-1}$$

uniformly on compact sets. Since the sequence $T_n \circ f \circ T_n^{-1}$ is uniformly bounded, it is sufficient to show that $T = \lim_{k\to\infty} T_{n_k} \circ f \circ T_{n_k}^{-1}$ for every convergent subsequence $T_{n_k} \circ f \circ T_{n_k}^{-1}$. Suppose $S = \lim_{k\to\infty} T_{n_k} \circ f \circ T_{n_k}^{-1}$ for some choice $n_1 < n_2 < \dots$. Since $f^{n_k}(0) = T_{n_k}^{-1}(0)$ and $g \circ f = T \circ g$, we get

$$S(0) = \lim_{k\to\infty} T_{n_k} f(T_{n_k}^{-1}(0)) = \lim_{k\to\infty} T_{n_k} f^{n_k}(f(0))$$
$$= g(f(0)) = Tg(0) = T(0)$$

and

$$S(T(0)) = \lim_{k\to\infty} T_{n_k} \circ f \circ T_{n_k}^{-1}(T(0))$$
$$= \lim_{k\to\infty} T_{n_k} \circ f \circ T_{n_k}^{-1} \circ T_{n_k} \circ T_{n_k+1}^{-1}(0)$$
$$= \lim_{k\to\infty} T_{n_k} \circ f \circ T_{n_k+1}^{-1}(0) = \lim_{k\to\infty} T_{n_k} \circ f \circ f^{n_k+1}(0)$$
$$= \lim_{k\to\infty} T_{n_k} \circ T_{n_k+1}^{-1} \circ T_{n_k+1} \circ f^{n_k+2}(0)$$
$$= T(\lim_{k\to\infty} T_{n_k+1} \circ f^{n_k+2}(0))$$
$$= T(\lim_{k\to\infty} T_{n_k+1} \circ T_{n_k+2}^{-1}(0)) = T^2(0).$$

Hence $T^{-1} \circ S\colon D \hookleftarrow$ has two fixed points, 0 and $T(0) = c > 0$; Schwarz's lemma now implies $T = S$ and this concludes the proof of the claim. In particular, T has no fixed points, because otherwise $T_n \circ f \circ T_n^{-1}$ would have a fixed point (for large n) and then its image under T_n^{-1} would be a fixed point of f. But this is impossible, because if f has a fixed point then $f^n(0)$ convergs to it, by the Denjoy-Wolff theorem, and therefore $\lim_{n\to\infty} d_P(f^{n+1}(0), f^n(0) = 0$, contradicting our hypothesis.

Thus we have constructed a holomorphic function $g: D \hookleftarrow$ and a Möbius map T of D onto D without fixed points satisfying $g \circ f = T \circ g$. Let us see, to complete the proof of the theorem, that these properties imply that f is not ergodic. Since it has no fixed points, the Möbius map $T: D \hookleftarrow$ is either parabolic (one fixed point in ∂D) or hyperbolic (two fixed points in ∂D) and in both cases it is not hard to find a Borel set $A \subseteq \partial D$ such that $0 < \lambda(A) < 1$ and $T^{-1}(A) = A$. Then the characteristic function ψ of A satisfies $\psi \circ T = \psi$ and, writing $h = \hat{\psi} \circ g$, we get

$$h \circ f = (\hat{\psi} \circ g) \circ f = \hat{\psi} \circ T \circ g = \widehat{\psi \circ T} \circ g = \hat{\psi} \circ g = h$$

by the commutative property of Theorem 1.4. Since h is bounded, Theorem 1.3 gives $h = \widehat{h^*}$ and the same commutative property implies $\widehat{h^* \circ f} = \widehat{h^*} \circ f = h \circ f = h$; therefore, taking nontangential limits, we obtain

$$h^* \circ f = h^*.$$

Thus h^* is f-invariant; if we prove that it is not constant a.e., the non-ergodicity of f follows. Now T has no fixed points; since $g \circ f = T \circ g$, g is not constant and $g(D)$ is a non-empty open set. But $0 < \lambda(A) < 1$, therefore ψ is not constant a.e., which implies that $\hat{\psi}$ and therefore $\hat{\psi}|g(D)$ are not constant. Hence $h = \hat{\psi} \circ g$ is not constant; being bounded, it follows that h^* is not constant. □

§4 Recurrence of inner functions

A measurable map $T: X \hookleftarrow$ of a measure space $(X, \mathcal{A}, \mu)$ is *recurrent* if the set

$$\tilde{A} = \bigcap_{N\geq 0} \bigcup_{n\geq N} T^{-n}(A)$$

of all points $x \in X$ such that $T^n(x) \in A$ for infinitely many values of $n \geq 0$ satisfies $\mu(A - \tilde{A}) = 0$ for every $A \in \mathcal{A}$. Equivalentely, T is recurrent if $\mu(S - T^{-1}(S)) = 0$ for every $S \in \mathcal{A}$ such that $T^{-1}(S) \subseteq S$. To see that recurrence implies this property it suffices to observe that $\tilde{S} = \bigcap_{n\geq 0} T^{-n}(S)$ and therefore $\mu(S - T^{-1}(S)) \leq \mu(S - \tilde{S})$ whenever $T^{-1}(S) \subseteq S$; the converse is obtained following the usual proof of Poincaré's recurrence theorem for measure preserving maps of probability spaces.

Here we shall consider the question of the recurrence of the boundary map of an inner function $f: D \hookleftarrow$. We will show that the series

$$\sum_{n\geq 0}(1 - |f^n(z)|)$$

either converges for all $z \in D$ or else diverges for all $z \in D$. In the first case, the Denjoy-Wolff point p of f necessarily belongs to the boundary (obviously, because $\lim_{n\to\infty} |f^n(z) - p| = 0$ and then $p \in D$ would imply $\inf_{n\geq 0}(1 - |f^n(z)|) > 0$) and

$$\lim_{n\to\infty} f^n(w) = p$$

for a.e. $w \in \partial D$. In the second case, we shall prove that the boundary map is recurrent and exact. Moreover, whenever the Denjoy-Wolff point p of f belongs to ∂D, then the infinite measure μ_p, given by its derivative

$$\frac{d\mu_p}{d\lambda}(w) = \frac{1}{|w-p|^2}, \qquad w \in \partial D,$$

with respect to the Lebesgue probability λ on ∂D, is f-invariant and, for every μ_p-integrable function $\psi\colon \partial D \to \mathbf{R}$ we have

$$\lim_{n\to\infty} \frac{1}{n} \sum_{j=0}^{n-1} \psi(f^j(w)) = 0$$

for a.e. $w \in \partial D$.

THEOREM 4.1. *If $f\colon D \hookleftarrow$ is an inner function and there exists a point $z \in D$ such that*

$$\sum_{n\geq 0} (1 - |f^n(z)|) < \infty,$$

then the Denjoy-Wolff point p of f belongs to ∂D and $\lim_{n\to\infty} f^n(w) = p$ for a.e. $w \in \partial D$.

PROOF: Assume $z \in D$ is such that $\sum_{n\geq 0}(1-|f^n(z)|) < \infty$; obviously this implies that the Denjoy-Wolff point p of f belongs to ∂D. Given a Borel set $S \subseteq \partial D$, let

$$\hat{S} = \bigcup_{N\geq 0} \bigcap_{n\geq N} f^{-n}(S)$$

denote the set of all points $w \in \partial D$ for which there exists an $N \geq 0$ such that $f^n(w) \in S$ for all $n \geq N$. It is enough to prove that

$\lambda(\hat{S}) = 1$ for every open neighborhood S of p, because taking a decreasing sequence of open neighborhoods S_m of p such that $\bigcap_{m\geq 0} S_m = \{p\}$, we then obtain

$$\lambda(\bigcap_{m\geq 0} \hat{S}_m) = 1;$$

all that remains to do is to observe that $w \in \bigcap_{m\geq 0} \hat{S}_m$ implies that for each neighborhood S_m of p there exists $N_m \geq 0$ such that $f^n(w) \in S_m$ for all $n \geq N_m$, which means that $\lim_{n\to\infty} f^n(w) = p$. Since $\lambda(\bigcap_{m\geq 0} \hat{S}_m) = 1$, this will prove the theorem.

To estimate $\lambda(\hat{S})$, we recall that the harmonic measure λ_z is equivalent to the Lebesgue probability λ and consider the characteristic function ψ of S^c. Then $\psi \circ f^n$ is the characteristic function of $f^{-n}(S^c)$ and Corollary 1.5 gives

$$\lambda_z(f^{-n}(S^c)) = \int (\psi \circ f^n) d\lambda_z = \int \psi d\lambda_{f^n(z)} = \hat{\psi}(f^n(z)).$$

Since

$$\hat{S}^c = \bigcap_{N\geq 0} \bigcup_{n\geq N} f^{-n}(S^c),$$

we obtain

$$\lambda_z(\hat{S}^c) \leq \inf_{N\geq 0} \sum_{n\geq N} \hat{\psi}(f^n(z)). \tag{1}$$

We also have

$$\frac{\hat{\psi}(f^n(z))}{1-|f^n(z)|} = \frac{1}{1-|f^n(z)|} \int_{S^c} P(f^n(z), w) d\lambda(w)$$
$$= (1+|f^n(z)|) \int_{S^c} \frac{1}{|w - f^n(z)|^2} d\lambda(w).$$

Let us now assume that S is an open neighborhood of p. Then $\lim_{n\to\infty} f^n(z) = p \notin \overline{S^c}$ implies

$$\lim_{n\to\infty} \frac{\hat{\psi}(f^n(z))}{1-|f^n(z)|} = 2\int_{S^c} \frac{1}{|w-p|^2} d\lambda(w).$$

Hence there exists $C > 0$ such that $\hat{\psi}(f^n(z)) \leq C(1-|f^n(z)|)$ for large $n \geq 0$ and therefore

$$\sum_{n\geq N} \hat{\psi}(f^n(z)) \leq C \sum_{n\geq N} (1-|f^n(z)|)$$

for large $N \geq 0$. But our hypothesis is $\sum_{n\geq 0}(1-|f^n(z)|) < \infty$, from which it follows that $\inf_{N\geq 0} \sum_{n\geq N} \hat{\psi}(f^n(z)) = 0$. By (1) we get $\lambda_z(\hat{S}^c) = 0$, implying $\lambda(\hat{S}^c) = 0$ and proving $\lambda(\hat{S}) = 1$. □

THEOREM 4.2. *If $f\colon D \hookleftarrow$ is an inner function and there exists a point $z \in D$ such that*

$$\sum_{n\geq 0}(1-|f^n(z)|) = \infty,$$

then the boundary map of f is recurrent.

PROOF: Assume $z \in D$ is such that $\sum_{n\geq 0}(1-|f^n(z)|) = \infty$. We shall prove that $\lambda(S - f^{-1}(S)) = 0$ for every Borel set $S \subseteq \partial D$ such that $f^{-1}(S) \subseteq S$; by our introductory remark, this ensures recurrence. Given a Borel set $S \subseteq \partial D$ such that $f^{-1}(S) \subseteq S$, assume $A \subseteq S - f^{-1}(S)$ is a Borel subset and ψ its characteristic function; then, as in the above proof,

$$\lambda_z(f^{-n}(A)) = \hat{\psi}(f^n(z))$$

and

$$\frac{\hat{\psi}(f^n(z))}{1-|f^n(z)|} = (1+|f^n(z)|)\int_A \frac{1}{|w-f^n(z)|^2}d\lambda(w).$$

Let $p = \lim_{n\to\infty} f^n(z)$ be the Denjoy-Wolff point of f; again $p \notin \overline{A}$ implies

$$(2) \qquad \lim_{n\to\infty}\frac{\hat{\psi}(f^n(z))}{1-|f^n(z)|} = (1+|p|)\int_A \frac{1}{|w-p|^2}d\lambda(w).$$

Now $A \subseteq S - f^{-1}(S)$ and $f^{-1}(S) \subseteq S$; therefore

$$f^{-j}(A) \cap f^{-i}(A) = \phi$$

for all $0 \leq j < i$ and we have

$$(3) \quad 1 \geq \lambda_z(\bigcup_{n\geq 0} f^{-n}(A)) = \sum_{n\geq 0} \lambda_z(f^{-n}(A)) = \sum_{n\geq 0} \hat{\psi}(f^n(z)).$$

Suppose that $\lambda(A) > 0$. Then from (2) it follows that

$$\lim_{n\to\infty}\frac{\hat{\psi}(f^n(z))}{1-|f^n(z)|} > 0,$$

which guarantees the existence of $C > 0$ and $N \geq 0$ such that $\hat{\psi}(f^n(z)) \geq C(1-|f^n(z)|)$ for all $n \geq N$. But then, from (3) and our hypothesis, we get

$$1 \geq \sum_{n\geq N} \hat{\psi}(f^n(z)) \geq C\sum_{n\geq N}(1-|f^n(z)|) = \infty.$$

This contradiction proves that $\lambda(A) = 0$ for every Borel set $A \subseteq S - f^{-1}(S)$ satisfying $p \notin \overline{A}$. Since $S - f^{-1}(S)$ (as, in fact, every subset of ∂D) is a countable union of such sets A, plus (maybe) the point p itself, it follows that $\lambda(S - f^{-1}(S)) = 0$, thus proving the theorem. □

COROLLARY 4.3. *If the boundary map of an inner function is recurrent then it is exact.*

PROOF: Let $f: D \hookleftarrow$ be an inner function with recurrent boundary map and let us prove it is exact; by virtue of Theorem 3.1, it suffices to prove ergodicity. Suppose $A \subseteq \partial D$ is a Borel set such that $f^{-1}(A) = A$ and let us consider its characteristic function ψ. Then $\psi \circ f = \psi$ and the commutative property gives $\hat{\psi} \circ f = \widehat{\psi \circ f} = \hat{\psi}$. Therefore, if $g: D \to \mathbf{C}$ is any holomorphic function with $\operatorname{Re} g = \hat{\psi}$, then $\operatorname{Re}(g \circ f) = (\operatorname{Re} g) \circ f = \hat{\psi} \circ f = \hat{\psi} = \operatorname{Re} g$ and we may choose $b \in \mathbf{R}$ such that $g \circ f = g + ib$. Let us first assume that $b \neq 0$. Defining $G: D \to \mathbf{C}$ by $G = \exp(-2\pi\, |b|^{-1}\, g)$, we obtain

$$|G(z)| = \exp\left(-\frac{2\pi}{|b|} \operatorname{Re} g(z)\right) = \exp\left(-\frac{2\pi}{|b|} \hat{\psi}(z)\right) \leq 1$$

and

$$\begin{aligned}(G \circ f)(z) &= \exp\left(-\frac{2\pi}{|b|} g(f(z))\right) = \exp\left(-\frac{2\pi}{|b|} g(z) \pm 2\pi i\right) \\ &= \exp\left(-\frac{2\pi}{|b|} g(z)\right) = G(z),\end{aligned}$$

that is, G is bounded and f-invariant. It follows that $G \circ f^n = G$ for all $n \geq 0$, which means that $G - G(a)$ is zero at every $f^n(a)$, for all $n \geq 0$ and every $a \in D$. Given $a \in D$, Blaschke's theorem ([**Du**; Thm. 2.3]) ensures that either $G - G(a) \equiv 0$ or else its sequence of zeros $a_0, a_1, \ldots$ satisfies $\sum_{n \geq 0} (1 - |a_n|) < \infty$. If the latter were the case, we would obtain $\sum_{n \geq 0} (1 - |f^n(z)|) < \infty$ and then, by Theorem 4.1, the Denjoy-Wolff point p of f would belong to ∂D and $\lim_{n\to\infty} f^n(w) = p$ for a.e. $w \in \partial D$. Since our hypothesis is the recurrence of f, this cannot happen and we are left with $G - G(a) \equiv 0$. But this implies that G, and then g and $\operatorname{Re} g = \hat{\psi}$, are all constant. Hence also $\psi = \hat{\psi}^*$ is constant, proving that $\lambda(A) = 0$ or $\lambda(A) = 1$. The other case, $b = 0$, is proved applying the same argument to $G = \exp(-g)$, which again is bounded and f-invariant.

□

Recall that, given $p \in \partial D$, we let μ_p denote the measure on ∂D whose derivative with respect to the Lebesgue probability λ is

$$\frac{d\mu_p}{d\lambda}(\omega) = \frac{1}{|\omega - p|^2}, \qquad \omega \in \partial D.$$

Clearly $\mu_p(\partial D) = +\infty$.

THEOREM 4.4. *Let $f\colon D \hookleftarrow$ be an inner function with Denjoy-Wolff point $p \in \partial D$. There exists $0 < c \leq 1$ such that*

a) *$\mu_p(f^{-1}(A)) = c\mu_p(A)$ for every Borel set $A \subseteq \partial D$;*

b) *$\lim_{n\to\infty} (1 - |f^n(z)|)^{1/n} = c$, for all $z \in D$;*

c) *$(1 - |f(z)|)/(1 - |z|)$ converges to c when z converges non-tangentially to p and $f(z)$ converges to p.*

COROLLARY 4.5. *μ_p is f-invariant if f is ergodic.*

The corollary is obvious: by Theorem 3.1, ergodicity implies $\lim_{n\to\infty} d_P(f^{n+1}(z), f^n(z)) = 0$ for all $z \in D$, hence

$$\lim_{n\to\infty} \frac{|f^{n+1}(z) - f^n(z)|}{1 - |f^n(z)|^2} = 0$$

and then

$$c = \lim_{n\to\infty} \frac{1 - |f^{n+1}(z)|}{1 - |f^n(z)|} = 1.$$

PROOF: Let $f: D \hookleftarrow$ be an inner function with Denjoy-Wolff point $p \in \partial D$. We define

$$\delta = \liminf_{\substack{z \to p \\ f(z) \to p}} \frac{1 - |f(z)|}{1 - |z|}.$$

Observe that $0 \leq \delta \leq 1$, because for $z_n = f^n(0)$ we have $\lim_{n\to\infty} f(z_n) = \lim_{n\to\infty} z_n = \lim_{n\to\infty} f^n(0) = p$, implying

$$\delta \leq \liminf_{n\to\infty} \frac{1 - |f(z_n)|}{1 - |z_n|} = \liminf_{n\to\infty} \frac{1 - |z_{n+1}|}{1 - |z_n|} \leq 1.$$

We claim that

$$\mu_p(f^{-1}(A)) \leq \delta \mu_p(A) \tag{4}$$

for every Borel set $A \subseteq \partial D$ with $p \notin \overline{A}$. To prove this, let us take any sequence $\{z_n\}$ in D with both z_n and $f(z_n)$ converging to p. Then $\lim_{n\to\infty} |f(z_n)| = \lim_{n\to\infty} |z_n| = |p| = 1$ and we have

$$\liminf_{n\to\infty} \frac{1 - |f(z_n)|^2}{1 - |z_n|^2} = \liminf_{n\to\infty} \frac{1 - |f(z_n)|}{1 - |z_n|} \leq \delta.$$

Given any Borel set $A \subseteq \partial D$ we also have

$$\begin{aligned} &\int_{f^{-1}(A)} \frac{1}{|w - z_n|^2} d\lambda(w) \\ &= \frac{1 - |f(z_n)|^2}{1 - |z_n|^2} \int_A \frac{1}{|w - f(z_n)|^2} d\lambda(w) \ , \end{aligned} \tag{5}$$

because

$$(1 - |z_n|^2) \int_{f^{-1}(A)} \frac{1}{|w - z_n|^2} d\lambda(w) = \int_{f^{-1}(A)} P(z_n, w) d\lambda(w)$$
$$= \lambda_{z_n}(f^{-1}(A)) = \lambda_{f(z_n)}(A)$$
$$= \int_A P(f(z_n), w) d\lambda w = (1 - |f(z_n)|^2) \int_A \frac{1}{|w - f(z_n)|^2} d\lambda(w),$$

where we used Corollary 1.5 to ensure the middle step. Finally, if $p \notin \overline{A}$, $\lim_{n\to\infty} f(z_n) = p$ gives

$$(6) \quad \lim_{n\to\infty} \int_A \frac{1}{|w - f(z_n)|^2} d\lambda(w) = \int_A \frac{1}{|w-p|^2} d\lambda(w) = \mu_p(A).$$

Since

$$\begin{aligned} \mu_p(f^{-1}(A)) &= \int_{f^{-1}(A)} \frac{1}{|w-p|^2} d\lambda(w) \\ &= \int_{f^{-1}(A)} \lim_{n\to\infty} \frac{1}{|w-z_n|^2} d\lambda(w) \ , \end{aligned}$$

Fatou's lemma thence gives

$$\begin{aligned} \mu_p(f^{-1}(A)) &\le \liminf_{n\to\infty} \int_{f^{-1}(A)} \frac{1}{|w-z_n|^2} d\lambda(w) \\ &= \liminf_{n\to\infty} \frac{1-|f(z_n)|^2}{1-|z_n|^2} \int_A \frac{1}{|w-f(z_n)|^2} d\lambda(w) \\ &\le \delta \lim_{n\to\infty} \int_A \frac{1}{|w-f(z_n)|^2} d\lambda(w) = \delta \mu_p(A), \end{aligned}$$

proving claim (4).

Now assume that $\{z_n\}$ is a sequence in D which converges nontangentially to p and is such that $f(z_n)$ converges to p and

$$\lim_{n\to\infty} \frac{1-|f(z_n)|}{1-|z_n|} = c$$

exists; we claim that

$$(7) \qquad \mu_p(f^{-1}(A)) = c\mu_p(A)$$

for every Borel set $A \subseteq \partial D$ such that $\mu_p(A) < \infty$. This obviously implies that c is independent of the sequence $\{z_n\}$, thus proving items a) and c) of the theorem. To prove (7), let $A \subseteq \partial D$ be a Borel set such that $p \notin \overline{A}$; by claim (4),

$$\mu_p(f^{-1}(A)) = \int_{f^{-1}(A)} \frac{1}{|w-p|^2} d\lambda(w) < \infty.$$

Since z_n converges nontangentially to p, there exists $K > 0$ such that

$$\frac{1}{|w-z_n|^2} \leq K \frac{1}{|w-p|^2}$$

for all $w \in \partial D$ and sufficiently large n; applying dominated convergence, we get

$$\mu_p(f^{-1}(A)) = \lim_{n\to\infty} \int_{f^{-1}(A)} \frac{1}{|w-z_n|^2} d\lambda(w).$$

Hence, using (5) and (6),

$$\begin{aligned}\mu_p(f^{-1}(A)) &= \lim_{n\to\infty} \frac{1-|f(z_n)|^2}{1-|z_n|^2} \int_A \frac{1}{|w-f(z_n)|^2} d\lambda(w) \\ &= c\mu_p(A).\end{aligned}$$

The general case of claim (7) follows easily, because if $A \subseteq \partial D$ is a Borel set with $\mu_p(A) < \infty$, then we can write it as $A = \bigcup_{n\geq 0} A_n$ with $A_0 \subseteq A_1 \subseteq \ldots$ and each A_n bounded away from p. Thus

$$\begin{aligned}\mu_p(f^{-1}(A)) = \mu_p\Big(\bigcup_{n\geq 0} f^{-1}(A_n)\Big) &= \lim_{n\to\infty} \mu_p(f^{-1}(A_n)) \\ &= c \lim_{n\to\infty} \mu_p(A_n) = c\mu_p(A).\end{aligned}$$

In particular, $\mu_p(f^{-1}(A)) < \infty$; iterating, we obtain

(8) $$\mu_p(f^{-n}(A)) = c^n \mu_p(A)$$

for all $n \geq 0$ and every $A \subseteq \partial D$ with $\mu_p(A) < +\infty$.

To prove item b), we take $z \in D$ and set

$$\alpha^+ = \limsup_{n\to\infty}(1 - |f^n(z)|)^{1/n},$$

$$\alpha^- = \liminf_{n\to\infty}(1 - |f^n(z)|)^{1/n}.$$

Let $A \subseteq \partial D$ a Borel set such that $p \notin \overline{A}$ and $\mu_p(A) > 0$. For some constant $K = K(z) > 0$ we have, for all $n \geq 0$,

$$\begin{aligned} K\mu_p(f^{-n}(A)) &\geq \lambda_z(f^{-n}(A)) = \lambda_{f^n(z)}(A) \\ &= (1 - |f^n(z)|^2)\int_A \frac{1}{|w - f^n(z)|^2}\,d\lambda(w). \end{aligned}$$

But $\limsup_{n\to\infty}(1 - |f^n(z)|^2)^{1/n} \geq \alpha^+$; therefore, using (6) and (8) we obtain

$$c = \limsup_{n\to\infty}(\mu_p(f^{-n}(A)))^{1/n} \geq \alpha^+.$$

Now we take $\varepsilon > 0$ and choose $m_1 < m_2 < \dots$ such that

$$1 - \left|f^{m_n+1}(z)\right| \leq (\alpha^- + \varepsilon)(1 - |f^{m_n}(z)|).$$

Proceeding as in the proof of claim (4) with the sequence $z_n = f^n(z)$ we get

$$\begin{aligned} c\mu_p(A) = \mu_p(f^{-1}(A)) &\leq \left(\liminf_{n\to\infty}\frac{1 - |f(z_n)|^2}{1 - |z_n|^2}\right)\mu_p(A) \\ &\leq (\alpha^- + \varepsilon)\mu_p(A), \end{aligned}$$

proving $c \leq \alpha^- + \varepsilon$. Since $\varepsilon > 0$ is arbitrary, we have

$$\alpha^- \leq \alpha^+ \leq c \leq \alpha^-,$$

completing the proof of item b).

To close the proof of the theorem, we observe that $0 < c$; otherwise we would have $\mu_p(f^{-1}(A)) = 0$ for some Borel set $A \subseteq \partial D$ with $p \notin \overline{A}$ and $\mu_p(A) > 0$. Then

$$0 = \mu_p(f^{-1}(A)) \geq \lambda_z(f^{-1}(A)) = \lambda_{f(z)}(A)$$

for $z \in D$, implying $\lambda(A) = 0$ and then $\mu_p(A) = 0$. □

THEOREM 4.6. *Let $f: D \hookleftarrow$ be an inner function with Denjoy-Wolff point $p \in \partial D$ and such that μ_p is f-invariant. For every $\psi \in L^1(\mu_p)$,*

$$\lim_{n\to\infty} \frac{1}{n} \sum_{j=0}^{n-1} \psi(f^j(w)) = 0$$

for a.e. $w \in \partial D$.

PROOF: Lef $f: D \hookleftarrow$ be an inner function with Denjoy-Wolff point $p \in \partial D$ and μ_p invariant. It is enough to prove the theorem for positive $\psi \in L^1(\mu_p)$. If $\psi \geq 0$ is zero in a neighborhood of p then $\lim_{n\to\infty} \hat{\psi}(f^n(0)) = 0$ and, using Fatou's lemma, we get

$$0 \leq \int \left(\lim_{n\to\infty} \frac{1}{n} \sum_{j=0}^{n-1} \psi \circ f^j\right) d\lambda \leq \liminf_{n\to\infty} \frac{1}{n} \sum_{j=0}^{n-1} \int (\psi \circ f^j) d\lambda$$

$$= \liminf_{n\to\infty} \frac{1}{n} \sum_{j=0}^{n-1} \hat{\psi}(f^j(0)) = 0,$$

hence

$$\lim_{n\to\infty} \frac{1}{n} \sum_{j=0}^{n-1} \psi(f^j(w)) = 0 \tag{9}$$

for a.e. $w \in \partial D$. Given any positive $\psi \in L^1(\mu_p)$, we choose a neighborhood W of p in ∂D and write $\psi = \psi_0 + \psi_1$, with $\psi_0(w) = 0$ for $w \in W$ and $\psi_1(w) = 0$ for $w \in W^c$. Since μ_p is f-invariant,

$$\int (\psi_1 \circ f^n) d\mu_p = \int \psi_1 d\mu_p;$$

therefore, Fatou's lemma and (9) imply

$$\int \Big(\lim_{n\to\infty} \frac{1}{n} \sum_{j=0}^{n-1} \psi \circ f^j \Big) d\mu_p = \int \Big(\lim_{n\to\infty} \frac{1}{n} \sum_{j=0}^{n-1} \psi_1 \circ f^j \Big) d\mu_p$$

$$\leq \liminf_{n\to\infty} \frac{1}{n} \sum_{j=0}^{n-1} \int (\psi_1 \circ f^j) d\mu_p = \int \psi_1 d\mu_p = \int_W \psi d\mu_p.$$

Since ψ is μ_p-integrable, the last expression can be made arbitrarily small; thus

$$\int \Big(\lim_{n\to\infty} \frac{1}{n} \sum_{j=0}^{n-1} \psi \circ f^j \Big) d\mu_p = 0,$$

proving the theorem. □

§5 Inner functions of the upper half plane

In this section we consider inner functions of the open upper half plane $\mathbf{R}^2_+ = \{z \in \mathbf{C} \mid \operatorname{Im} z > 0\}$ and characterize the Pólya-Szegö maps.

All the results of the previous sections translate into results for inner functions of $\mathbf{R}^2_+$ via the Möbius map $S: D \to \mathbf{R}^2_+$ given by $S(z) = i(1+z)/(1-z)$; an additional feature of the boundary $\mathbf{R} \cup \{\infty\}$ of $\mathbf{R}^2_+$, however, is the distinguished point at infinity which, for inner functions of $\mathbf{R}^2_+$, is far more conspicuous than its counterpart $w = 1$ is for inner functions of the unit disc. Thus, if the inner function $f: \mathbf{R}^2_+ \hookleftarrow$ has no fixed point, we may, naturally, choose ∞ as the Denjoy-Wolff point of f (conjugating by a Möbius map if necessary). In fact, the canonical representation ([**T**]) of inner functions $f: \mathbf{R}^2_+ \hookleftarrow$,

$$f(z) = \alpha z + \beta + \int_{\mathbf{R}} \frac{1+zw}{w-z} d\mu(w), \tag{1}$$

with $\alpha \geq 0$, $\beta \in \mathbf{R}$ and μ a singular finite measure on the Borel σ-algebra of $\mathbf{R}$, already enhances this choice for the Denjoy-Wolff point of f: α is characterized by

$$\lim_{t\to\infty} \frac{f(it)}{it} = \alpha$$

and we shall see that ∞ is the Denjoy-Wolff point of f if and only if $\alpha \geq 1$. Before we prove this, let us comment on the upper half plane versions of the two measures which played a major role in the unit disc scenario. The harmonic measure λ_z, for $z = a + ib \in \mathbf{R}^2_+$, $a, b \in \mathbf{R}$, now has, with respect to Lebesgue measure λ on $\mathbf{R}$, derivative

$$\frac{d\lambda_z}{d\lambda}(x) = \frac{b}{\pi}\frac{1}{(x-a)^2 + b^2}$$

for $x \in \mathbf{R}$ ([G; pg. 12]). Denoting, as before, the boundary map of an inner function $f: \mathbf{R}^2_+ \hookleftarrow$ simply by $f: \mathbf{R} \hookleftarrow$ again, Corollary 1.5 gives

$$\lambda_z(f^{-1}(A)) = \lambda_{f(z)}(A)$$

for every Borel set $A \subseteq \mathbf{R}$; if $z \in \mathbf{R}^2_+$ is a fixed point of f then λ_z is f-invariant. Now if $f: \mathbf{R}^2_+ \hookleftarrow$ has no fixed point then, associated to its Denjoy-Wolff point $p \in \mathbf{R} \cup \{\infty\}$, we have an infinite measure μ_p which satisfies

$$\mu_p(f^{-1}(A)) = c\mu_p(A)$$

for some $0 < c \leq 1$ and every Borel set $A \subseteq \mathbf{R}$ (Theorem 4.4). If $p = \infty$, we (normalize and) have for μ_p the Lebesgue measure on $\mathbf{R}$: $\lambda = \mu_\infty$. If $z \to \infty$ nontangentially, we have the following (compare with the relation (2) of Section 4):

LEMMA 1. *Let $\{z_n\}$ be any sequence in $\mathbf{R}^2_+$ such that $\operatorname{Im} z_n \to \infty$ and $\frac{|\operatorname{Re}\, z_n|}{\operatorname{Im}\, z_n} \to 0$ with $n \to \infty$. Then*

$$\lambda(A) = \lim_{n\to\infty} \pi \ \operatorname{Im}\, z_n \ \lambda_{z_n}(A)$$

for every Borel set $A \subseteq \mathbf{R}$.

PROOF: Write $z_n = a_n + ib_n$, $b_n > 0$, $a_n \in \mathbf{R}$. Then $|\frac{x-a_n}{b_n}| \leq \frac{|x|}{b_n} + \frac{|a_n|}{b_n} \to 0$ with $n \to \infty$ for each $x \in \mathbf{R}$. Given a Borel set

$A \subseteq \mathbf{R}$ we have

$$\lambda_{z_n}(A) = \int_A d\lambda_{z_n} = \frac{1}{\pi}\int_A \frac{b_n}{(x-a_n)^2 + b_n^2} d\lambda(x)$$

and therefore

$$\pi b_n \lambda_{z_n}(A) = \int_A \frac{1}{(\frac{x-a_n}{b_n})^2 + 1} d\lambda(x).$$

Dominated convergence then gives

$$\lim_{n\to\infty} \pi b_n \lambda_{z_n}(A) = \int_A d\lambda(x) = \lambda(A).$$

□

LEMMA 2. *If $f{:}\,\mathbf{R}^2_+ \hookleftarrow$ is an inner function given by* (1), *then*

$$\alpha = \lim_{n\to\infty} \frac{f(it)}{it} \quad .$$

PROOF: For each $t > 0$ we have

$$\begin{aligned} f(it) &= \alpha it + \beta + \int_{\mathbf{R}} \frac{1+iwt}{w-it} d\mu(w) \\ &= \alpha it + \beta + (1-t^2)\int_{\mathbf{R}} \frac{w}{w^2+t^2} d\mu(w) + it\int_{\mathbf{R}} \frac{1+w^2}{w^2+t^2} d\mu(w). \end{aligned}$$

For $t \geq 1$ we certainly have $(1+w^2)/(w^2+t^2) \leq 1$ for all $w \in \mathbf{R}$, and therefore dominated convergence gives

$$\lim_{t\to\infty} \int_{\mathbf{R}} \frac{1+w^2}{w^2+t^2} d\mu(w) = 0.$$

From the above expression for $f(it)$, we can thus read off $\lim_{t\to\infty} \frac{f(it)}{it} = \alpha$, as soon as we prove the following

Claim: There exists $C > 0$ such that for all $t > 0$

$$\int_{\mathbf{R}} \frac{|w|}{w^2+t^2} d\mu(w) \leq \frac{C}{t^2} .$$

Observe that

$$\begin{aligned}\int_{\mathbf{R}} \frac{|w|}{w^2+t^2} d\mu(w) &= \int_{-1}^{+1} \frac{|w|}{w^2+t^2} d\mu(w) + 2\int_{1}^{+\infty} \frac{w}{w^2+t^2} d\mu(w) \\ &\leq \frac{1}{t^2}\mu([-1,1]) + 2\int_{1}^{+\infty} \frac{w}{w^2+t^2} d\mu(w) .\end{aligned}$$

The last term we integrate by parts, obtaining

$$\begin{aligned}\int_{1}^{+\infty} \frac{w}{w^2+t^2} d\mu(w) &= \frac{F(1)}{1+t^2} - \int_{1}^{+\infty} F(w)\frac{t^2-w^2}{(w^2+t^2)^2} dw \\ &\leq \frac{F(1)}{1+t^2} + F(1)\int_{1}^{+\infty} \frac{w^2-t^2}{(w^2+t^2)^2} dw \\ &\leq \frac{F(1)}{1+t^2} + \frac{F(1)}{t}\int_{1/t}^{+\infty} \frac{s^2-1}{(s^2+1)^2} ds = \frac{2F(1)}{1+t^2},\end{aligned}$$

where we wrote $F(w) = \mu([w, +\infty))$. □

Theorem 5.1. *If the inner function* $f\colon \mathbf{R}^2_+ \hookleftarrow$ *is given by (1) and* $\alpha > 0$, *then for every Borel set* $A \subseteq \mathbf{R}$,

$$\lambda(f^{-1}(A)) = \frac{1}{\alpha}\lambda(A).$$

Moreover, $\alpha \geq 1$ *if and only if* ∞ *is the Denjoy-Wolff point of* f.

Proof: Let $f\colon \mathbf{R}^2_+ \hookleftarrow$ be given by (1) with $\alpha > 0$. Then Lemma 2 gives $\lim\limits_{b\to\infty} \dfrac{f(ib)}{ib} = \alpha > 0$, implying $\dfrac{\operatorname{Im} f(ib)}{b} \to \alpha$, $\operatorname{Im} f(ib) \to \infty$

and $\frac{|\text{Re}\, f(ib)|}{\text{Im}\ f(ib)} \to 0$ with $b \to \infty$, $b > 0$. Let $A \subseteq \mathbf{R}$ be a Borel set. Since $\lambda_{f(ib)}(A) = \lambda_{ib}(f^{-1}(A))$ for $b > 0$, Lemma 1 implies

$$\begin{aligned}\lambda(A) &= \lim_{b\to\infty} \pi \ \text{Im}\ f(ib)\ \lambda_{f(ib)}(A)\\ &= \lim_{b\to\infty} \frac{\text{Im}\ f(ib)}{b} \pi b \lambda_{ib}(f^{-1}(A)) = \alpha\lambda(f^{-1}(A)),\end{aligned}$$

proving the first part of the theorem.

Now assume that $\alpha \geq 1$ and let us prove that ∞ is the Denjoy-Wolff point of f. Suppose that it isn't. Let $p \neq \infty$ be the Denjoy-Wolff point of f and let us write ν for either the harmonic measure λ_p, if $p \in \mathbf{R}^2_+$, or the infinite measure μ_p, if $p \in \mathbf{R}$. From the results of the previous sections, we can choose a constant $0 < c \leq 1$ such that

$$\nu(f^{-1}(A)) = c\nu(A)$$

for every Borel set $A \subseteq \mathbf{R}$. Hence

$$\begin{aligned}c\int_A \frac{d\nu}{d\lambda} d\lambda = c\nu(A) = \nu(f^{-1}(A)) &= \int_{f^{-1}(A)} \frac{d\nu}{d\lambda} d\lambda\\ &= \frac{1}{\alpha}\int_A \left(\frac{d\nu}{d\lambda}\circ f\right) d\lambda\end{aligned}$$

for every Borel set $A \subseteq \mathbf{R}$, where in the last equality we used the first part of the theorem, since $\alpha > 0$. Then

$$\frac{d\nu}{d\lambda}\circ f = \alpha c \frac{d\nu}{d\lambda} \qquad \lambda\text{-a.e.}\ . \tag{2}$$

If $\alpha c < 1$, (2) implies

$$\lim_{n\to\infty} \frac{d\nu}{d\lambda}(f^n(x)) = 0,$$

and then

$$\lim_{n\to\infty} f^n(x) = \infty$$

for a.e. $x \in \mathbf{R}$, which means that ∞ is the Denjoy-Wolff point of f. If $\alpha > 1$ we have

$$\lambda(f^{-n}(A)) = \frac{1}{\alpha^n}\lambda(A)$$

for every Borel set $A \subseteq \mathbf{R}$ with $\lambda(A) < \infty$. This implies that for a.e. $x \in A$ there exists $N > 0$ such that $f^n(x) \notin A$ if $n \geq N$ and then, since this holds for every $A = [-a, a]$, that $\lim_{n\to\infty} f^n(x) = \infty$ for a.e. $x \in \mathbf{R}$; again ∞ is the Denjoy-Wolff point of f. Thus our hypotheses $\alpha \geq 1$ and ∞ is not the Denjoy-Wolff point of f imply $\alpha c \geq 1$ and $\alpha \leq 1$. It follows that $c = 1 = \alpha$ and then (2) implies that f =identity, which is the contradiction we wanted.

Conversely, suppose that ∞ is the Denjoy-Wolff point of f and let us prove that $\alpha \geq 1$. But ∞ being the Denjoy-Wolff point of f implies that $\lambda = \mu_\infty$ and there exists $0 < c \leq 1$ such that $\lambda(f^{-1}(A)) = c\lambda(A)$ for every Borel set $A \subseteq \mathbf{R}$. Suppose we prove that $\alpha > 0$; then, by the first part of the theorem, we will get

$$\lambda(A) = \alpha\lambda(f^{-1}(A)) = \alpha c\lambda(A)$$

for every Borel set $A \subseteq \mathbf{R}$, thence $\alpha c = 1$ and $\alpha = \frac{1}{c} \geq 1$. Therefore all we need to prove is that $\alpha > 0$. Suppose that $\alpha = 0$. Let $\{b_n\}$ be any sequence with $b_n > 0$ and $\lim_{n\to\infty} b_n = \infty$. Then

$$f(ib_n) = \beta + (1 - b_n^2)\int_{\mathbf{R}} \frac{w}{w^2 + b_n^2}d\mu(w) + ib_n \int_{\mathbf{R}} \frac{1 + w^2}{w^2 + b_n^2}d\mu(w),$$

where

$$\sup_{n\geq 0}\left|(1 - b_n^2)\int_{\mathbf{R}} \frac{w}{w^2 + b_n^2}d\mu(w)\right| < \infty$$

by the claim in the proof of Lemma 2. Let us assume first that

$$\lim_{n\to\infty} b_n \int_{\mathbf{R}} \frac{1+w^2}{w^2+b_n^2} d\mu(w) = \infty.$$

Then $\operatorname{Re} f(ib_n)$ is bounded and $\operatorname{Im} f(ib_n) \to \infty$ with $n \to \infty$; since also $\dfrac{\operatorname{Im} f(ib_n)}{b_n} \to \alpha = 0$ with $n \to \infty$ (by Lemma 2), Lemma 1 gives

$$\begin{aligned}\lambda(A) &= \lim_{n\to\infty} \pi \operatorname{Im} f(ib_n)\ \lambda_{f(ib_n)}(A)\\ &= \lim_{n\to\infty} \frac{\operatorname{Im} f(ib_n)}{b_n} \pi b_n \lambda_{ib_n}(f^{-1}(A))\\ &= \lambda(f^{-1}(A)) \lim_{n\to\infty} \frac{\operatorname{Im} f(ib_n)}{b_n} = 0,\end{aligned}$$

for every Borel set $A \subseteq \mathbf{R}$, clearly an impossibility. Therefore we are left with

$$\sup_{n\geq 0} b_n \int_{\mathbf{R}} \frac{1+w^2}{w^2+b_n^2} d\mu(w) < \infty.$$

But then $\{f(ib_n)\}$ is bounded and we may suppose that it converges to some $z \in \mathbf{C}$ with $\operatorname{Im} z \geq 0$. Assume that $\operatorname{Im} z > 0$; since ∞ is the Denjoy-Wolff point of f, this implies, by Lemma 1,

$$\begin{aligned}c\lambda(A) = \lambda(f^{-1}(A)) &= \lim_{n\to\infty} \pi b_n \lambda_{ib_n}(f^{-1}(A))\\ &= \lim_{n\to\infty} \pi b_n \lambda_{f(ib_n)}(A) = \lambda_z(A) \lim_{n\to\infty} \pi b_n = \infty\end{aligned}$$

for every Borel set $A \subseteq \mathbf{R}$, again an impossibility. Finally assume that $\operatorname{Im} z = 0$. Then, as above,

$$\lambda(f^{-1}(A)) = \lim_{n\to\infty} \pi b_n \lambda_{f(ib_n)}(A)$$

for every Borel set $A \subseteq \mathbf{R}$, but since now

$$\lim_{n\to\infty} \frac{\operatorname{Im} f(ib_n)}{b_n} = 0 = \lim_{n\to\infty} \operatorname{Re} f(ib_n),$$

we get

$$\lim_{n\to\infty} \pi \operatorname{Im} f(ib_n)\ \lambda_{f(ib_n)}(A) = \mu_z(A)$$

whenever $z \notin \overline{A}$ and A is bounded. Thus for such Borel sets with $0 < \mu_z(A) < \infty$, we obtain

$$c\lambda(A) = \lambda(f^{-1}(A)) = \lim_{n\to\infty} \frac{b_n}{\operatorname{Im} f(ib_n)} \pi \operatorname{Im} f(ib_n)\ \lambda_{f(ib_n)}(A) = \infty,$$

which is again impossible. We have run out of possibilities, thence $\alpha > 0$. □

Let us now turn to the problem, posed by Pólya (and answered by Pólya [**Pa**] and Szegö [**Sz**]), of characterizing the rational functions of the Riemann sphere $\overline{\mathbf{C}} = \mathbf{C} \cup \{\infty\}$ which preserve the real line and its Lebesgue measure.

THEOREM 5.2. *Let $f: \overline{\mathbf{C}} \hookleftarrow$ be a rational map. Then $f(\mathbf{R}) \subseteq \mathbf{R}$ and $f|\mathbf{R}$ preserves the Lebesgue measure if and only if there exist $\varepsilon = \pm 1$, $\beta \in \mathbf{R}$, $n \geq 0$, $a_i \in \mathbf{R}$ and $\gamma_i > 0$ for $1 \leq i \leq n$ such that, for all $z \in \mathbf{C}$,*

$$f(z) = \varepsilon \left(z + \beta - \sum_{i=1}^{n} \frac{\gamma_i}{z - a_i} \right). \tag{3}$$

PROOF: If $f: \overline{\mathbf{C}} \hookleftarrow$ is given by (3) then clearly $f(\mathbf{R}) \subseteq \mathbf{R}$ and $\varepsilon f(z)$, $z \in \mathbf{R}^2_+$, defines an inner function of the upper half plane, with $\alpha = 1$ in (1). Then Theorem 5.1 above implies that λ is εf-invariant, hence also f-invariant.

Let now $f:\overline{\mathbf{C}} \hookleftarrow$ be a rational map such that $f(\mathbf{R}) \subseteq \mathbf{R}$ and $\lambda(f^{-1}(A)) = \lambda(A)$ for every Borel set $A \subseteq \mathbf{R}$. Since f preserves the Lebesgue measure, we have

$$\sum_{f(x)=c} \frac{1}{|f'(x)|} = 1$$

for all $c \in \mathbf{R}$. Hence $|f'(x)| > 1$ for all x. In particular $\lim_{x\to+\infty} f(x) = +\infty$ or $-\infty$. Multiplying f by -1 if necessary, we can assume $\lim_{x\to+\infty} f(x) = +\infty$. This means that for some $a > 0$ and $m \in \mathbf{N}$,

$$f(x) = ax^m + \frac{P(x)}{Q(x)} ,$$

where P, Q are polynomials with no common root (in $\mathbf{C}$) and degree $P \leq$ degree Q. Let us prove that $m = 1$ and $a = 1$. Since $\lim_{x\to+\infty} f(x) = +\infty$ we have $m \geq 1$ and $a \geq 0$. Then, for $c > 0$ large, the real roots of $f(x) = c$ can be written as $\varphi_1(c) > \varphi_2(c) > \cdots > \varphi_k(c)$, k independent of c and $\varphi_i(c)$ being a C^∞ function of c. Clearly, when $c \to +\infty$, $\lim_{c\to+\infty} \varphi_1(c) = +\infty$, and $\lim_{c\to+\infty} \varphi_i(c)$ is, for $1 < i < k$, a real root of $Q(x)$. The same property holds for $i = k$ when m is odd and $\lim_{c\to+\infty} \varphi_k(c) = -\infty$ if m is even. Then

$$(4) \quad 1 = |f'(\varphi_1(c))|^{-1} + |f'(\varphi_k(c))|^{-1} + \sum_{1<i<k} |f'(\varphi_i(c))|^{-1} .$$

When $c \to +\infty$, clearly

$$\lim_{c\to+\infty} |f'(\varphi_i(c))|^{-1} = 0$$

and also for $i = k$ if m is odd. Moreover,

$$\lim_{c\to+\infty} \frac{|f'(\varphi_1(c))|}{ma\,|\varphi_1(c)|^{m-1}} = 1$$

and

$$\lim_{c\to+\infty} \frac{|f'(\varphi_k(c))|}{ma\,|\varphi_k(c)|^{m-1}} = 1$$

when m is even. Hence, if $m > 1$,

$$\lim_{c\to+\infty} |f'(\varphi_1(c))|^{-1} = \lim_{c\to+\infty} |f'(\varphi_k(c))|^{-1} = 0,$$

contradicting (4). Then $m = 1$ and

$$\lim_{c\to+\infty} |f'(\varphi_1(c))|^{-1} = \frac{1}{a}\ ;$$

taking the limit of (4) when $c \to +\infty$, we obtain $a = 1$. From the fact that $a = 1$ and $m = 1$, it follows that there exists a connected component V of $f^{-1}(\mathbf{R}^2_+)$ such that V contains a set of the form $\{z \in \mathbf{C} \mid |z| > \rho, \operatorname{Re} z > 0\}$. We claim that $V = \mathbf{R}^2_+$. Take any bounded interval $J_0 \subset \mathbf{R}$ and set $J_1 = f^{-1}(J_0) \cap \mathbf{R}$. Let ψ_{J_i} be the characteristic function of J_i, $i = 0, 1$. Let $\hat{\psi}_{J_i}\colon \mathbf{R}^2_+ \to \mathbf{R}$ be the harmonic extension of ψ_{J_i}. Then

$$\lambda(J_1) = \lim_{t\to+\infty} \frac{1}{t}\hat{\psi}_{J_1}(ti).$$

Now consider the harmonic function $\hat{\psi}_{J_0} \circ f\colon V \to \mathbf{R}$. Observe that $V \cap \mathbf{R} = \mathbf{R} \cap \partial V$ and $(\hat{\psi}_{J_0} \circ f)|(\mathbf{R} \cap \partial V) = \psi_{J_1}$. Moreover, if $V \neq \mathbf{R}^2_+$, ∂V contains arcs contained in $\mathbf{R}^2_+$, and these arcs contain arcs that are connected components of $f^{-1}(J_0)$. Hence $\hat{\psi}_{J_0} \circ f$ is equal to one on these arcs. On the other hand $0 < \hat{\psi}_{J_1}(z) < 1$ for all $z \in \mathbf{R}^2_+$, in particular on those arcs in ∂V where $\hat{\psi}_{J_0} \circ f$ is equal to one. Therefore $\hat{\psi}_{J_1}|\partial V \leq (\hat{\psi}_{J_0} \circ f)|\partial V$ and the inequality is strict on those arcs. Hence $\hat{\psi}_{J_1}|V < \hat{\psi}_{J_0} \circ f$ and even more:

$$\lim_{t\to+\infty} \frac{1}{t}\hat{\psi}_{J_1}(it) < \lim_{t\to+\infty} \frac{1}{t}(\hat{\psi}_{J_0} \circ f)(it).$$

$$= \lim_{t\to+\infty} \frac{|f(it)|}{t} \cdot \frac{1}{|f(it)|}(\hat{\psi}_{J_0} \circ f)(it).$$

But, since $m = 1$ and $a = 1$, $\lim_{t\to\infty} f(it)/t = i$. Then

$$\lim_{t\to+\infty} \frac{1}{|f(it)|}(\hat{\psi}_{J_0} \circ f)(it) = \lambda(J_0)$$

and

$$\lambda(J_1) < \lim_{t\to+\infty} \frac{|f(it)|}{t} \cdot \frac{1}{|f(it)|}(\hat{\psi}_{J_0} \circ f)(it) = \lambda(J_0).$$

This contradicts the measure preserving property of $f|\mathbf{R}$ and proves the claim. But $V = \mathbf{R}^2_+$ obviously implies that f is an inner function. Hence all its poles are real and simple, and, recalling that $m = 1$ and $a = 1$, we can write

$$f(z) = z + \beta + \sum_{i=1}^{n} \frac{\lambda_i}{z - a_i}$$

with $\beta \in \mathbf{R}$ and $a_i \in \mathbf{R}$, $1 \leq i \leq n$. Moreover $\lambda_i < 0$ for all i, because if some λ_i doesn't satisfy this condition, f maps the half disk $\{z \in \mathbf{C}|\, |z - a_i| < 1, \mathrm{Re}\, z > 0\}$ onto a set intersecting the *lower* half plane. □

§6 Inner functions and parabolic basins

Let $f: \overline{\mathbf{C}} \hookleftarrow$ be a rational map of the Riemann sphere $\overline{\mathbf{C}} = \mathbf{C} \cup \{\infty\}$ and let $U \subseteq \overline{\mathbf{C}}$ be a fixed connected component of $\overline{\mathbf{C}} - J(f)$, where $J(f)$ denotes the Julia set of f. In this section we suppose that U is a parabolic basin. (See [**Su**] for the classification of the components of $\overline{\mathbf{C}} - J(f)$.) Let $p \in \partial U$ be the fixed parabolic point of f, that is, $f'(p)$ is a root of unity and $\lim_{n\to\infty} f^n(z) = p$ for all $z \in U$. If we choose a uniformization $\psi: D \to U$ and lift $f: U \hookleftarrow$ to D, we obtain an inner function $\hat{f}: D \hookleftarrow$ of the unit disc such that $\psi \circ \hat{f} = f \circ \psi$, with Denjoy-Wolff point $\hat{p} \in \partial D$. Moreover, since the boundary map $\psi^*: \partial D \to \partial U$ of ψ satisfies

$$\lambda_{\hat{z}}((\psi^*)^{-1}(A)) = \lambda_z(A)$$

for each Borel set $A \subseteq \partial U$ and every $z \in U$, $\psi(\hat{z}) = z$, it is clear that the ergodic properties in the harmonic class of $f|\partial U$ are given by the ergodic properties of the boundary map of $\hat{f}$. Thus exactness and recurrence of $f|\partial U$ with respect to the harmonic class of ∂U are consequences of

$$\sum_{n\geq 0}(1 - |\hat{f}^n(z)|) = \infty$$

being valid for some $z \in D$, by Theorem 4.2 and Corollary 4.3. More than this is granted by the following two theorems.

THEOREM 6.1. *Let U be a fixed parabolic basin of a rational map $f: \overline{\mathbf{C}} \hookleftarrow$ and let $\hat{f}: D \hookleftarrow$ be a lifting of $f: U \hookleftarrow$ via a uniformization $\psi: D \to U : \psi \circ \hat{f} = f \circ \psi$. Then, for every $z \in D$ and all $\alpha > \frac{1}{2}$, the inequality*

$$1 - |\hat{f}^n(z)| \geq \frac{1}{n^\alpha}$$

holds for all sufficiently large n.

PROOF: We shall only prove that there exists a point $z_0 \in D$ such that, for all $\alpha > \frac{1}{2}$,

$$(1) \qquad 1 - |\hat{f}^n(z_0)| \geq \frac{1}{n^\alpha}$$

holds for n sufficiently large. From this, the same relation holds for all $z \in D$ because (1) implies the recurrence of $\hat{f}^*$ (by Theorem 4.2) and then its ergodicity (by Corollary 4.3); now Theorem 3.1 guarantees that $\lim_{n\to\infty} d_P(\hat{f}^n(z), \hat{f}^n(z_0)) = 0$ and therefore

$$\lim_{n\to\infty} \frac{|\hat{f}^n(z) - \hat{f}^n(z_0)|}{1 - |\hat{f}^n(z_0)|} = 0$$

for all $z \in D$. Hence, given $z \in D$, we have that

$$\begin{aligned} 1 - |\hat{f}^n(z)| &\geq 1 - |\hat{f}^n(z_0)| - |\hat{f}^n(z) - \hat{f}^n(z_0)| \\ &\geq (1-\varepsilon)(1 - |\hat{f}^n(z_0)|) \end{aligned}$$

holds for all $\varepsilon > 0$ and sufficiently large n. This implies that z also satisfies (1), because (1) holds for all $\alpha > \frac{1}{2}$.

To prove (1), let p be the parabolic fixed point of f; then $p \in \partial U \subseteq \overline{\mathbf{C}}$, $f'(p)$ is a root of unity and $\lim_{n\to\infty} f^n(z) = p$ for all

$z \in U$. Without loss of generality we can assume $p = 0$, $f'(p) = 1$ and that the Taylor series of f at $z = 0$ is

$$f(z) = z - z^k + \sum_{n>k} a_n z^n \tag{2}$$

for some $k \geq 2$. The basic theory of parabolic fixed points (see, for instance, Camacho [**Ca**]) implies that there exists an even $0 \leq m < 2(k-1)$ with the following property: for each $\varepsilon > 0$ there exists a $\delta > 0$ such that U contains the sector

$$S = \left\{ z \in \mathbf{C} \middle| |z| < \delta \ , \ \left| \operatorname{Arg}(z) - \frac{\pi m}{k-1} \right| < \frac{\pi(1-\varepsilon)}{k-1} \right\}$$

and, moreover, $f^n(S) \subseteq S$ for all large n, say $n \geq N$ and $\lim_{n\to\infty} f^n(z) = 0$ and $\lim_{n\to\infty} \left| \operatorname{Arg}(f^n(z)) - \frac{\pi m}{k-1} \right| = 0$ for all $z \in S$. Without loss of generality, we may assume that $m = 0$. From (2) it easily follows that for each $z \in S$ there exists $C > 0$ such that

$$|f^n(z)| \geq C \left(\frac{1}{n} \right)^{\frac{1}{k-1}} \tag{3}$$

for all n. Let $\psi: D \to U$ be the uniformization of U; write $\hat{S}$ for the connected component of $\psi^{-1}(S)$ such that $\hat{f}^n(\hat{S}) \subseteq \hat{S}$ for $n \geq N$. Set $\delta = \frac{1-\varepsilon}{k-1}$ and choose a branch $g: S \to \mathbf{C}$ of $z^{1/2\delta}$ which leaves the positive half line $\{t \in \mathbf{R} \mid t \geq 0\}$ invariant. Then $u(z) = \operatorname{Re} g(z)$ is a harmonic function on S that vanishes on the sides of S. Consider the harmonic function $u \circ \psi: \hat{S} \to \mathbf{R}$ and observe that there exists $K > 0$ such that

$$-K \log |z| \geq (u \circ \psi)(z)$$

for all $z \in \hat{S}$. In fact, to check this inequality, it sufficies to verify it at the boundary of $\hat{S}$. It holds in the part of $\partial\hat{S}$ mapped by ψ onto the sides of the sector S, since there $u \circ \psi$ vanishes. Taking K large enough we can make it true also on the part of $\partial \hat{S}$ that ψ maps onto $\left\{z \in \mathbf{C} \middle| |z| = \delta \ , \ |\mathrm{Arg}(z)| < \frac{\pi(1-\varepsilon)}{k-1}\right\}$, since $u \circ \psi$ is bounded there. Now, for $z_0 \in \hat{S}$ we have $\hat{f}^n(z_0) \in \hat{S}$ for $n \geq N$ and

$$\begin{aligned} -K \log |\hat{f}^n(z_0)| &\geq (u \circ \psi)(\hat{f}^n(z_0)) = u(f^n(\psi(z_0))) \\ &= \mathrm{Re}\, g(f^n(\psi(z_0))). \end{aligned}$$

Since $\lim_{n\to\infty} |\mathrm{Arg}(f^n(\psi(z_0)))| = 0$, it follows that also

$$\lim_{n\to\infty} |\mathrm{Arg}(g(f^n(\psi(z_0))))| = 0$$

and therefore

$$\begin{aligned} \mathrm{Re}\, g(f^n(\psi(z_0))) &\geq \frac{1}{2} |g(f^n(\psi(z_0)))| \\ &= \frac{1}{2} |f^n(\psi(z_0))|^{\frac{1}{2\delta}} \end{aligned}$$

for large n. Together with (3) we obtain, for large n,

$$\begin{aligned} -K \log \left|\hat{f}^n(z_0)\right| &\geq \frac{1}{2} C^{\frac{1}{2\delta}} \left(\frac{1}{n}\right)^{\frac{1}{2\delta(k-1)}} \\ &= \frac{1}{2} C^{\frac{1}{2\delta}} \left(\frac{1}{n}\right)^{\frac{1}{2\delta(1-\varepsilon)}} . \end{aligned}$$

Since $\varepsilon > 0$ is arbitrary, this proves (1). □

THEOREM 6.2. *Let U be a fixed parabolic basin of a rational map $f: \overline{\mathbf{C}} \hookleftarrow$. Given $x \in U$ and a positive $\varphi \in L^\infty(\lambda_x)$, $\varphi \not\equiv 0$,*

$$\lim_{n\to\infty} \frac{\int \left(\sum_{i=0}^{n-1} (\varphi \circ f^i)\psi\right) d\lambda_x}{\sum_{i=0}^{n-1} \int (\varphi \circ f^i) d\lambda_x} = \int \psi d\lambda_x$$

holds for every $\psi \in L^1(\lambda_x)$.

PROOF: Since $\varphi > 0$, clearly, for each $\psi \in L^1(\lambda_x)$,

$$\left| \frac{\int \left(\sum_{i=0}^{n-1} (\varphi \circ f^i)\psi \right) d\lambda_x}{\sum_{i=0}^{n-1} (\varphi \circ f^i) d\lambda_x} \right| \leq \|\psi\|_1 .$$

Hence it suffices to prove the theorem for a set of functions $\psi \in L^1(\lambda_x)$ that span a dense subspace of $L^1(\lambda_x)$. Let us see that the Radon-Nikodym derivatives $d\lambda_y/d\lambda_x$, $y \in U$, span such a dense subspace of $L^1(\lambda_x)$. To check its density, it is enough, by the Hahn-Banach theorem, to take any $\eta \in L^\infty(\lambda_x)$ such that

$$\int \eta d\lambda_y = \int \eta \left(\frac{d\lambda_y}{d\lambda_x} \right) d\lambda_x = 0$$

for every $y \in U$ and show that $\eta = 0$ a.e. . Let us first assume $U = D$. Then the harmonic extension $\hat{\eta}: D \to \mathbf{R}$ of η satisfies

$$\hat{\eta}(y) = \int \eta d\lambda_y = 0$$

for every $y \in D$, hence $\eta(w) = \lim_{r \to 1} \hat{\eta}(rw) = 0$ for a.e. $w \in \partial D$, by Theorem 1.3. For the general U the same follows using the uniformization theorem and this case $U = D$. Thus we proved the density in $L^1(\lambda_x)$ of the space spanned by $d\lambda_y/d\lambda_x$, with $y \in U$. To complete the proof of the theorem we have to check it for $\psi = d\lambda_y/d\lambda_x$. For these we have

$$\begin{aligned} \sum_{i=0}^{n-1} \int (\varphi \circ f^i) \frac{d\lambda_y}{d\lambda_x} d\lambda_x &= \sum_{i=0}^{n-1} \int (\varphi \circ f^i) d\lambda_y \\ &= \sum_{i=0}^{n-1} \int \varphi d\lambda_{f^i(y)} \end{aligned}$$

for each $y \in U$. Instead of a general $\varphi \in L^\infty(\lambda_x)$, let us first suppose that φ is continuous and let $\hat{\varphi}: U \to \mathbf{R}$ be its harmonic extension. Then

$$(4) \qquad \frac{\displaystyle\int\Big(\sum_{i=0}^{n-1}(\varphi\circ f^i)\frac{d\lambda_y}{d\lambda_x}\Big)d\lambda_x}{\displaystyle\sum_{i=0}^{n-1}\int(\varphi\circ f^i)d\lambda_x} = \frac{\displaystyle\sum_{i=0}^{n-1}\hat{\varphi}(f^i(y))}{\displaystyle\sum_{i=0}^{n-1}\hat{\varphi}(f^i(x))}.$$

We want to prove that this quotient converges to $\int \frac{d\lambda_y}{d\lambda_x} d\lambda_x = \int d\lambda_x = 1$ with $n \to \infty$. But the Poincaré distance $d_P(f^i(x), f^i(y))$ converges to 0 with $n \to \infty$, since the boundary map of $f: U \hookleftarrow$ is ergodic (Theorem 3.1). Hence, by Harnack's inequality,

$$(5) \qquad \lim_{n\to\infty}\frac{\hat{\varphi}(f^i(y))}{\hat{\varphi}(f^i(x))} = 1.$$

We claim that

$$(6) \qquad \sum_{i\geq 0}\hat{\varphi}(f^i(x)) = \infty.$$

This follows by lifting the problem to the disc D, via a uniformization. In fact, let us denote by $\hat{\varphi}_1$ the lifting of $\hat{\varphi}$ and by $\hat{f}$ the lifting of f. Now $\hat{\varphi}_1$ is harmonic and positive, therefore for some $C > 0$ we get $\hat{\varphi}_1(z) > C(1-|z|)$ for all $z \in D$. Since U is a fixed parabolic basin of f, Theorem 6.1 above gives $1 - |\hat{f}^i(z)| > i^{-\alpha}$ for all $\alpha > \frac{1}{2}$, $z \in D$ and large i, and therefore

$$\sum_{i\geq 0}\hat{\varphi}_1(\hat{f}^i(z)) = \infty$$

for all $z \in D$. This proves (6); since (6) together with (5) implies that (4) converges to 1 with $n \to \infty$, we proved the desired property for continuous φ. For the general case of $\varphi \in L^\infty(\lambda_x)$, we observe that the continuity of φ was used only to grant the existence of a harmonic function $\hat{\varphi}: U \to \mathbf{R}$ which satisfies

$$(7) \qquad \hat{\varphi}(p) = \int \varphi d\lambda_p$$

for every $p \in U$. Such a function, that extends the concept of harmonic extension, is easily obtained for every positive (hence for all) $\varphi \in L^1(\lambda_p)$, $p \in U$, by taking a sequence of continuous functions $\varphi_n: \partial U \to \mathbf{R}$ which are a Cauchy sequence in $L^1(\lambda_{p_0})$, for some $p_0 \in U$, and $\varphi_n \to \varphi$ a.e. (in the harmonic class). Then

$$\begin{aligned} |\hat{\varphi}_n(p_0) - \hat{\varphi}_m(p_0)| &= \left| \int \varphi_n d\lambda_{p_0} - \int \varphi_m d\lambda_{p_0} \right| \\ &\leq \int |\varphi_n - \varphi_m| \, d\lambda_{p_0}. \end{aligned}$$

Hence $\{\hat{\varphi}_n(p_0)\}$ is a convergent sequence. By Harnack's inequality, $\{\hat{\varphi}_n\}$ converges uniformly on compact sets to a harmonic function $\hat{\varphi}: U \to \mathbf{R}$ and, for all $p \in U$, we get

$$\hat{\varphi}(p) = \lim_{n \to \infty} \hat{\varphi}_n(p) = \lim_{n \to \infty} \int \varphi_n d\lambda_p = \int \varphi d\lambda_p,$$

thus providing the general $\varphi \in L^\infty(\lambda_p)$ with a harmonic function $\hat{\varphi}: U \to \mathbf{R}$ satisfying (7). □

REFERENCES

[Aa1] J. Aaronson, On the ergodic theory of non-integrable functions and infinite measure spaces. Israel Jour. of Math., 27 (1977), 163–173.

[Aa2] J. Aaronson, Ergodic theory for inner functions of the upper half plane. Ann. Inst. Henri Poincaré, Section B, vol. XIV (1978), 233–253.

[Aa3] J. Aaronson, A remark on the exactness of inner functions. J. London Math. Soc., 23 (1981), 469–474.

[Ad-W] R. Adler, B. Weiss, The ergodic, infinite measure preserving transformation of Boole. Israel Jour. of Math., 16 (1973), 263–278.

[B] G. Boole, On the comparison of transcendents with certain applications to the theory of definite integrals. Philos. Trans. Roy. Soc. London, 147 (1857), 745–803.

[Ca] C. Camacho, On the local structure of conformal mappings and holomorphic vector fields in $\mathbf{C}^2$. Asterisque, 59–60 (1978), 83–94.

[Cr1] M. Craizer, The entropy of inner functions. Preprint, IMPA (1989).

[Cr2] M. Craizer, Inner functions are Bernoulli. Preprint, PUC (1990).

[De] A. Denjoy, Sur l'itération des fonctions analytiques. C.R. Acad. Sci. Paris, 182 (1926), 255–257.

[Du] P. Duren, Theory of H^p spaces. Academic Press, New York, 1970.

[F-L-M] A. Freire, A. Lopes, R. Mañé, An invariant measure for rational maps. Bol. Soc. Bras. Mat., 14 (1983), 45–62.

[G] J. Garnett, Bounded analytical functions. Academic Press, New York, 1981.

[He] M. Heins, On the finite angular derivatives of an analytic function mapping the disk into itself. J. London Math. Soc., 15 (1977), 239–254.

[Ho] E. Hopf, Ergodentheorie. Springer Verlag, Berlin, 1937.

[M] R. Mañé, On the uniqueness of the maximizing measure for rational maps. Bol. Soc. Bras. Mat., 14 (1983), 27–43.

[M-R] R. Mañé, L.F. da Rocha, Julia sets are uniformly perfect. Preprint, IMPA (1990).

[Lj] M.J. Ljubich, Entropy properties of rational endomorphisms of the Riemann sphere. Ergod. Th. & Dynam. Sys., 3 (1983), 351–385.

[Lo] A. Lopes, Equilibrium measures for rational maps. Ergod. Th. & Dynam. Sys., 6 (1986), 393–399.

[Pa] G. Pólya, Problem 107. Jber. Deutsch. Math. Verein. 40, 2 Abt. (1931), p. 81.

[Pe] Ch. Pommerenke, On the iteration of analytic functions in a half plane, I. J. London Math. Soc., 19 (1979), 439–447.

[Pi] F. Przytycki, Hausdorff dimension of harmonic measure on the boundary of an attractive basin for a holomorphic map. Invent. Math., 80 (1985), 161–179.

[Su] D. Sullivan, Quasiconformal homeomorphisms and dynamics I. Solution of the Fatou-Julia problem on wandering domains. Ann. Math., 122 (1985), 401–418.

[Sz] G. Szegö, Problem. Jber. Deutsch. Math. Verein. 43, 2 Abt. (1934), 17–20.

[T] M. Tsuji, Potential theory in modern function theory. Maruzen, Tokyo, 1959.

[W] J. Wolff, Sur l'itération des fonctions holomorphes dans une region. C.R. Acad. Sci. Paris, 182 (1926), 42–43; 200-201.

List of Publications of Ricardo Mañé

Lagrangian flows: the dynamics of globally minimizing orbits. *Bol. Soc. Brasil. Mat. (N.S.)* 28 (1997), no. 2, 141–153.

On Ricardo Mañé. Edited by Jacob Palis. *Bol. Soc. Brasil. Mat. (N.S.)* 28 (1997), no. 2, i–iii.

On the topological entropy of geodesic flows. *J. Differential Geom.* 45 (1997), no. 1, 74–93.

Lagrangian flows: the dynamics of globally minimizing orbits. *International Conference on Dynamical Systems (Montevideo, 1995),* 120–131, Pitman Res. Notes Math. Ser., 362, *Longman, Harlow,* 1996.

The Lyapunov exponents of generic area preserving diffeomorphisms. *International Conference on Dynamical Systems (Montevideo, 1995),* 110–119, Pitman Res. Notes Math. Ser., 362, *Longman, Harlow,* 1996.

Generic properties and problems of minimizing measures of Lagrangian systems. *Nonlinearity* 9 (1996), no. 2, 273–310.

Ergodic variational methods: new techniques and new problems. *Proceedings of the International Congress of Mathematicians, Vol. 1, 2 (Zürich, 1994),* 1216–1220, *Birkhäuser, Basel,* 1995.

The explosion of singular cycles (with R. Bamón, R. Labarca and M.J. Pacífico). *Inst. Hautes Études Sci. Publ. Math.* No. 78 (1993), 207–232 (1994).

On a theorem of Fatou. *Bol. Soc. Brasil. Mat. (N.S.)* 24 (1993), no. 1, 1–11.

On the minimizing measures of Lagrangian dynamical systems. *Nonlinearity 5 (1992), no. 3,* 623–638.

Julia sets are uniformly perfect (with L.F. da Rocha) *Proc. Amer. Math. Soc.* 116 (1992), no. 1, 251–257.

The Hausdorff dimension of horseshoes of diffeomorphisms of surfaces. *Bol. Soc. Brasil. Mat. (N.S.)* 20 (1990), no. 2, 1–24.

M.J. Pacifico (ed.), *Ricardo Mañé – Selected Papers*

The Hausdorff dimension of invariant probabilities of rational maps. *Dynamical systems, Valparaiso 1986,* 86–117, *Lecture Notes in Math. 1331, Springer, Berlin,* 1988.

A proof of the C^1 stability conjecture. *Inst. Hautes Études Sci. Publ. Math.* No. 66 (1988), 161–210.

On the creation of homoclinic points. *Inst. Hautes Études Sci. Publ. Math.* No. 66 (1988), 139–159.

Erratum: "Hyperbolicity, sinks and measure in one-dimensional dynamics" [Comm. Math. Phys. 100 (1985), no. 4, 495–524; *Comm. Math. Phys.* 112 (1987), no. 4, 721–724.

On a theorem of Klingenberg. *Dynamical systems and bifurcation theory (Rio de Janeiro, 1985),* 319–345, Pitman Res. Notes Math. Ser., 160, *Longman Sci. Tech., Harlow,* 1987.

Ergodic theory and differentiable dynamics. Ergebnisse der Mathematik und ihrer Grenzgebiete (3) [Results in Mathematics and Related Areas (3)], 8. *Springer-Verlag, Berlin,* 1987.

On the instability of Herman rings. *Invent. Math.* 81 (1985), no. 3, 459–471.

Hyperbolicity, sinks and measure in one-dimensional dynamics. *Comm. Math. Phys.* 100 (1985), no. 4, 495–524.

On the Bernoulli property for rational maps. *Ergodic Theory Dynam. Systems* 5 (1985), no. 1, 71–88.

Oseledec's theorem from the generic viewpoint. *Proceedings of the International Congress of Mathematicians, Vol. 1, 2 (Warsaw, 1983),* 1269–1276,*PWN, Warsaw,* 1984.

Introdução à teoria ergódica. (Portuguese) [Introduction to ergodic theory] Projeto Euclides [Euclid Project], 14. *Instituto de Matemática Pura e Aplicada (IMPA), Rio de Janeiro,* 1983.

Errata to: "A proof of Pesin's formula" [Ergodic Theory Dynamical Systems 1 (1981), no. 1, 95–102]. *Ergodic Theory Dynam. Systems* 3 (1983), no. 1, 159–160.

An invariant measure for rational maps (with A. Freire, A. Lopes). *Bol. Soc. Brasil. Mat.* 14 (1983), no. 1, 45–62.

On the uniqueness of the maximizing measure for rational maps. *Bol. Soc. Brasil. Mat.* 14 (1983), no. 1, 27–43.

On the dynamics of rational maps (with P. Sad, D. Sullivan). *Ann. Sci. École Norm. Sup. (4)* 16 (1983), no. 2, 193–217.

Lyapounov exponents and stable manifolds for compact transformations. *Geometric dynamics (Rio de Janeiro, 1981),* 522–577, Lecture Notes in Math. 1007,*Springer, Berlin,* 1983.

An extension of Peixoto's structural stability theorem to open surfaces with finite genus (with C. Camacho, M. Krych, Z. Nitecki). *Geometric dynamics (Rio de Janeiro, 1981),* 60–87, Lecture Notes in Math. 1007, *Springer, Berlin,* 1983.

An ergodic closing lemma. *Ann. of Math. (2)* 116 (1982), no. 3, 503–540.

On the entropy of the geodesic flow in manifolds without conjugate points (with A. Freire). *Invent. Math.* 69 (1982), no. 3, 375–392.

On the dimension of the compact invariant sets of certain nonlinear maps. *Dynamical systems and turbulence, Warwick 1980 (Coventry, 1979/1980),* pp. 230–242, Lecture Notes in Math. 898, *Springer, Berlin-New York,* 1981.

A proof of Pesin's formula. *Ergodic Theory Dynamical Systems* 1 (1981), no. 1, 95–102.

Absolute stability and bifurcation theory. *Invent. Math.* 53 (1979), no. 1, 59–75.

Expansive homeomorphisms and topological dimension. *Trans. Amer. Math. Soc.* 252 (1979), 313–319.

Orbits of paths under hyperbolic toral automorphisms. *Proc. Amer. Math. Soc.* 73 (1979), no. 1, 121–125.

Contributions to the stability conjecture. *Topology* 17 (1978), no. 4, 383–396.

Persistent manifolds are normally hyperbolic. *Trans. Amer. Math. Soc.* 246 (1978), 261–283.

Invariant sets of Anosov's diffeomorphisms. *Invent. Math.* 46 (1978), no. 2, 147–152.

Quasi-Anosov diffeomorphisms and hyperbolic manifolds. *Trans. Amer. Math. Soc.* 229 (1977), 351–370.

Axiom A for endomorphisms. *Geometry and topology (Proc. III Latin Amer. School of Math., Inst. Mat. Pura Aplicada CNPq, Rio de Janeiro, 1976),* pp. 379–388. Lecture Notes in Math. 597, *Springer, Berlin,* 1977.

Characterizations of AS diffeomorphisms. *Geometry and topology (Proc. III Latin Amer. School of Math., Inst. Mat. Pura Aplicada CNPq, Rio de Janeiro, 1976),* pp. 389–394. Lecture Notes in Math. 597, *Springer, Berlin,* 1977.

Reduction of semilinear parabolic equations to finite dimensional *C*1 flows. *Geometry and topology (Proc. III Latin Amer. School of Math., Inst. Mat. Pura Aplicada CNPq, Rio de Janeiro, 1976),* pp. 361–378. Lecture Notes in Math. 597, *Springer, Berlin,* 1977.

Hyberbolic sets for semilinear parabolic equations. *Bol. Soc. Brasil. Mat.* 6 (1975), no. 2, 145–153.

Stability of endomorphisms (with C. Pugh). *Dynamical systems—Warwick 1974 (Proc. Sympos. Appl. Topology and Dynamical Systems, Univ. Warwick, Coventry, 1973/1974; presented to E. C. Zeeman on his fiftieth birthday),* pp. 175–184. Lecture Notes in Math. 468, *Springer, Berlin,* 1975.

Expansive diffeomorphisms. *Dynamical systems—Warwick 1974 (Proc. Sympos. Appl. Topology and Dynamical Systems, Univ. Warwick, Coventry, 1973/1974; presented to E. C. Zeeman on his fiftieth birthday),* pp. 162–174. Lecture Notes in Math. 468, *Springer, Berlin,* 1975.

On infinitesimal and absolute stability of diffeomorphisms. *Dynamical systems—Warwick 1974 (Proc. Sympos. Appl. Topology and Dynamical Systems, Univ. Warwick, Coventry, 1973/1974; presented to E. C. Zeeman on his fiftieth birthday),* pp. 151–161. Lecture Notes in Math. 468, *Springer, Berlin,* 1975.

Persitent manifolds are normally hyperbolic. *Bull. Amer. Math. Soc.* 80 (1974), 90–91.

A note on first integrals of vector fields and endomorphisms. *Bol. Soc. Brasil. Mat.* 3 (1972), no. 2, 127–132.

Acknowledgements

The Brazilian Mathematical Society (SBM), the Institute for Pure and Applied Mathematics (IMPA) and Springer-Verlag express their gratitude to the following copyright holders for granting permission to reprint these contributions.

American Mathematical Society

R. Mané: Persistent Manifolds are Normally Hyperbolic. *Trans. Amer. Math. Soc. 246 (1978), 261-28*

Annals of Mathematics, Princeton University

An ergodic closing lemma. *Ann. of Math. (2) 116 (1982), no. 3, 503–540*

Birkhäuser Verlag

Ergodic variational methods: new techniques and new problems. *Proceedings of the International Congress of Mathematicians, Vol. 1, 2 (Zürich, 1994),* 1216–1220, 1995

Brazilian Mathematical Society

An invariant measure for rational maps. *Bol. Soc. Brasil. Mat. 14 (1983), no. 1, 45–62*

On a theorem of Fatou. *Bol. Soc. Brasil. Mat. (N.S.) 24 (1993), no. 1, 1-11*

Lagrangian flows: the dynamics of globally minimizing orbits. II. *Bol. Soc. Brasil. Mat. (N.S.) 28 (1997), no. 2, 155–196*

The Dynamics of Inner Functions. *Ensaios Matemáticos, Periodical Volume, 1-79,* 1991

M.J. Pacifico (ed.), *Ricardo Mañé – Selected Papers*

Cambridge University Press

A proof of Pesin's formula. *Ergodic Theory Dynamical Systems* ***1*** *(1981), no. 1, 95–102*

On the Bernoulli property for rational maps. *Ergodic Theory Dynamical Systems 5 (1985), no. 1, 71–88*

Elsevier

Contributions to the stability conjecture. *Topology 17 (1978), no. 4, 383–396*

Gauthier-Villars

On the dynamics of rational maps. *Ann. Sci. Ecole Norm. Sup. (4) 16 (1983), no. 2, 193–21*

Institute des Hautes Études Scientifiques

On the creation of homoclinic points. *Publ. Math. Inst. Hautes Etudes Sci. No. 66 (1988), 139–159*

The explosion of singular cycles. *Publ. Math. Inst. Hautes Etudes Sci. No. 78 (1993), 207–232 (1994)*

International Press of Boston, Inc.

On the topological entropy of geodesic flows. *J. Differential Geom. 45 (1997), no. 1, 74–93*

IOP Publishing

On the minimizing measures of Lagrangian dynamical systems. *Nonlinearity 5(1992), no. 3, 623–638*

Longman

The Lyapunov exponents of generic area preserving diffeomorphisms. *International Conference on Dynamical Systems (Montevideo, 1995), 110–119, Pitman Res. Notes Math. Ser., 362,* 1996

Pearson Education

On a theorem of Klingenberg. *Dynamical systems and bifurcation theory (Rio de Janeiro, 1985), 319–345, Pitman Res. Notes Math. Ser., 160, Longman Sci. Tech., Harlow, 1987*

Polish Scientific Publishers PWN

Oseledec's theorem from the generic viewpoint. *Proceedings of the International Congress of Mathematicians, Vol. 1, 2 (Warsaw, 1983), 1269–1276, PWN, Warsaw, 1984*